ADVANCES IN FUZZY SET THEORY AND APPLICATIONS

ADVANCES IN FUZZY SET THEORY AND APPLICATIONS

edited by:

Madan M. GUPTA
University of Saskatchewan

associate editors:

Rammohan K. RAGADE
University of Louisville

Ronald R. YAGER
Iona College

[redacted]

1979

NORTH-HOLLAND PUBLISHING COMPANY – AMSTERDAM • NEW YORK • OXFORD

ISBN: 0 444 85372 3

Publishers:

NORTH-HOLLAND PUBLISHING COMPANY
AMSTERDAM • NEW YORK • OXFORD

Sole Distributors for the U.S.A. and Canada:

ELSEVIER NORTH-HOLLAND INC.
52 VANDERBILT AVENUE, NEW YORK, N.Y. 10017

Library of Congress Cataloging in Publication Data

Main entry under title:

Advances in fuzzy set theory and applications.

Bibliography: p.
1. Set theory--Addresses, essays, lectures.
2. System analysis--Addresses, essays, lectures.
3. Mathematical models--Addresses, essays, lectures. I. Gupta, Madan M. II. Ragade, Rammohan K., 1942- III. Yager, Ronald R., 1941-
QA248.A38 511'.3 79-17151
ISBN 0-444-85372-3

PRINTED IN THE NETHERLANDS

This volume is dedicated to

THE FUTURE GENERATION

for the preservation of essentials
of humanistic processes
in this advancing technological world.

Ā nō bhadrā ṛtvo yāntu viśvatāḥ

Let noble thoughts universally flow to us

Rg Veda

FOREWORD

Much progress has been made in the development of fuzzy set theory and its application since the first volume in this field was published by North-Holland in 1977. Considerable work has been done, for example, on the development of a well-defined body of knowledge on the properties and characteristics of fuzzy sets, fuzzy switching mechanisms and decision making in fuzzy situations. As well, a wide variety of applications has taken place in many areas of engineering and in medicine and the social sciences. A number of the more significant of these developments are reported in this Volume. They show a maturing of the field - the "fuzziness" of the fuzzy sets is being replaced by a much more rigorous mathematical basis and a much better appreciation as to how it can be applied. They also show that a fundamental relationship exists between many problems in these various fields - they have, in other words, much in common.

This Volume contains 35 papers written by 49 authors from 11 countries; thus, it is ample proof of the very rapid expansion of this field. It includes papers dealing with the basic theory and papers on application. As well, an up-to-date bibliography, which contains 1,800 references, is included. Hopefully, you will enjoy reading the works of the authors as much as they enjoyed doing the work and reporting it for the benefits of all of us.

I also wish to congratulate the editors of this book for a job well done. In particular, I wish to acknowledge Dr. M.M. Gupta who was largely the driving force behind it.

Peter N. Nikiforuk, Ph.D., D.Sc.
Dean of Engineering
University of Saskatchewan
Saskatoon, Saskatchewan
Canada, S7N OWO

PREFACE

This is a second volume on Fuzzy Systems published by North-Holland.

In many applied fields such as engineering, the social sciences and medical diagnostics, the sources of *vague* or *fuzzy data* are numerous and diverse both in origin and in magnitude. Conventionally, in the construction of mathematical models for most of these fuzzy systems, this imprecision is standardly portrayed as random processes. The introduction of Fuzzy Set Theory has changed this conventional approach. Fuzzy Set Theory has initiated a growing interest not only on the part of mathematicians, but as well among engineers and other applications-oriented workers. A major portion of this current interest and research is oriented toward the applications.

The mathematical sophistication required to understand the properties of *fuzzy processes* and the role they play in the corresponding applications is considerable; this tends to create a gap between mathematically oriented academics, with limited interest in design and development, and applications-oriented engineers, who tend to dismiss some of the available mathematical tools as being too complicated or of questionable usefulness.

This Volume on "Advances in Fuzzy Set Theory and Applications" has been edited to present a coherent view of this growing field. One purpose of the present collection of papers which has been drawn from mathematics, engineering, the social sciences and medical diagnostics is to contribute to the narrowing of the gap between theory and practice.

The literature on the subject of the present Volume is rich in contributions both in theory and applications of the fuzzy field. For this reason the selection of papers has been a difficult task. The principal criteria that have been used are the significance and quality of the contributions, their relevance to the central thesis of this Volume, and in the context of the historical evolution of the theory and its applications. Many outstanding contributions could not be included here because of limitations of space. An effort was made throughout to present a coherent sequence of the papers that enhances the aspect of interaction, between engineering considerations and the underlying mathematical ideas.

The first part of the Volume gives a broad perspective view of fuzzy set theory. The second part is concerned for the greater part, into recent theoretical developments, while in Part III the papers deal with applications of fuzzy sets. A synthesis of the Volume is presented in Chapter I. To further aid the reader in his/her studies, an extensive bibliography containing 1,799 references is included in this Volume.

The editors believe that this Volume of outstanding contributions should provide reference material which will be useful both to the theoretician and to the applications-oriented researchers. It is hoped that it will also be equally valuable to students, and teachers.

Madan M. Gupta
Rammohan K. Ragade
Ronald R. Yager

ACKNOWLEDGEMENTS

The task of organizing and editing a collection of papers on a subject as new and as diverse as the theory of fuzzy sets has not been an easy one to accomplish. I am deeply appreciative of the spirit of cooperation and understanding manifested by all of the contributors to this Volume. The editorial assistance of Dr. Rammohan Ragade and Dr. Ronald Yager is gratefully acknowledged. The Editor records his appreciations to Mrs. Janice Friesen who was very helpful during all the editorial phases of this Volume. I am indebted to my wife, Suman, who not only helped me since the inception of this Volume but also provided a moral support by not demanding much of my time.

Madan M. Gupta
Editor

TABLE OF CONTENTS

FOREWORD vii

PREFACE xi

ACKNOWLEDGMENTS xi

PART ONE : INTRODUCTION

Fuzzy Sets and Information Granularity
L.A. ZADEH 3

Fuzzy Set Theory and Applications: A Synthesis
R.K. RAGADE and M.M. GUPTA 19

Outline of Fuzzy Set Theory: An Introduction
D. DUBOIS and H. PRADE 27

Fuzzy Sets and the Social Nature of Truth
J. GOGUEN 49

On Fuzzy Systems
C.V. NEGOITA 69

A Survey of the Representation of Fuzzy Concepts and Its Applications
D. RALESCU 77

Fuzzy Logic and Its Application to Fuzzy Reasoning
J.F. BALDWIN 93

Some Methods of Fuzzy Reasoning
M. MIZUMOTO, S. FUKAMI and K. TANAKA 117

An Approach to Fuzzy Reasoning Method
Y. TSUKAMOTO 137

PART TWO : THEORY

Some Properties of Fuzzy Numbers
M. MIZUMOTO and K. TANAKA 153

Fuzzy Variables in a Random Environment
S. NAHMIAS 165

On Fuzzy Statistics
A. KANDEL 181

Extended Fuzzy Expression of Probabilistic Sets
K. HIROTA 201

Possibilistically Dependent Variables and a General Theory of Fuzzy Sets
E. HISDAL 215

Toward a Calculus of the Mathematical Notion of Possibility
H.T. NGUYEN 235

On the Theory of Fuzzy Switching Mechanisms (FSM's)
A. KANDEL 247

Fuzzy Sets and Languages
W. WECHLER 263

Decision-Making Under Fuzziness
D. DUBOIS and H. PRADE 279

Fuzzy Information and Decision in Statistical Model
H. TANAKA, T. OKUDA and K. ASAI 303

Entropy and Energy Measures of a Fuzzy Set
A. DE LUCA and S. TERMINI 321

Solution Concepts for n-Persons Fuzzy Games
D. BUTNARIU 339

Fuzzy Concepts: Their Structure and Problems of Measurement
M. NOWAKOWSKA 361

Effects of Context on Fuzzy Membership Functions
H.M. HERSH, A. CARAMAZZA and H.H. BROWNELL 389

Fuzzy Propositional Approach to Psycholinguistic Problems: An Application of Fuzzy Set Theory in Cognitive Science
G.C. ODEN 409

Compositions of Fuzzy Relations
E. SANCHEZ 421

PART THREE : APPLICATIONS

Medical Diagnosis and Composite Fuzzy Relations
E. SANCHEZ 437

The Application of Fuzzy Set Theory to Medical Diagnosis
W.A. FORDON and J.C. BEZDEK 445

Fuzziness and Catastrophe in Estimation and Decision Processes
M. KOKAWA, K. NAKAMURA and M. ODA 463

Contribution of the Fuzzy Sets Theory to Man-Machine System
D. WILLAEYS and N. MALVACHE 481

Exploring Linguistic Consequences of Assertions in Social Sciences
F. WENSTØP 501

An Example of Linguistic Modelling: The Case of Mulder's Theory of Power
W.J.M. KICKERT 519

Time-of-use Pricing of Electricity: A Policy Assessment Methodology
J. FIKSEL, J. DIFFENBACH and A. RENDA 541

The Construction and Evaluation of Fuzzy Models
R.M. TONG 559

Application of Fuzzy Sets for the Analysis of Complex Scenes
R. JAIN 577

A General Purpose Policy Capturing Device using Fuzzy Production Rules
J.J. WEISS and M.L. DONNELL 589

Fuzzy Clustering with a Fuzzy Convariance Matrix
D.E. GUSTAFSON and W.C. KESSEL 605

A 1979 BIBLIOGRAPHY ON FUZZY SETS, THEIR APPLICATIONS, AND RELATED TOPICS
A. Kandel and R.R. Yager 621

BIOGRAPHICAL INFORMATION ABOUT THE EDITORS AND CONTRIBUTING AUTHORS 745

****

PART ONE

INTRODUCTION

ADVANCES IN FUZZY SET THEORY AND APPLICATIONS
M.M. Gupta, R.K. Ragade, R.R. Yager (editors)

FUZZY SETS AND INFORMATION GRANULARITY*

L.A. Zadeh

Computer Science Division
Department of Electrical Engineering and Computer Sciences
and the Electronics Research Laboratory
University of California at Berkeley

1. Introduction

Much of the universality, elegance and power of classical mathematics derives from the assumption that real numbers can be characterized and manipulated with infinite precision. Indeed, without this assumption, it would be much less simple to define what is meant by the zero of a function, the rank of a matrix, the linearity of a transformation or the stationarity of a stochastic process.

It is well-understood, of course, that in most real-world applications the effectiveness of mathematical concepts rests on their robustness, which in turn is dependent on the underlying continuity of functional dependencies [1]. Thus, although no physical system is linear in the idealized sense of the term, it may be regarded as such as an approximation. Similarly, the concept of a normal distribution has an operational meaning only in an approximate and, for that matter, not very well-defined sense.

There are many situations, however, in which the finiteness of the resolving power of measuring or information gathering devices cannot be dealt with through an appeal to continuity. In such cases, the information may be said to be granular in the sense that the data points within a granule have to be dealt with as a whole rather than individually.

Taken in its broad sense, the concept of information granularity occurs under various guises in a wide variety of fields. In particular, it bears a close relation to the concept of aggregation in economics; to decomposition and partition--in the theory of automata and system theory; to bounded uncertainties--in optimal control [2], [3]; to locking granularity--in the analysis of concurrencies in data base management systems [4]; and to the manipulation of numbers as intervals--as in interval analysis [5]. In the present paper, however, the concept of information granularity is employed in a stricter and somewhat narrower sense which is defined in greater detail in Sec. 2. In effect, the main motivation for our approach is to define the concept of information granularity in a way that relates it to the theories of evidence of Shafer [6], Dempster [7], Smets [8], Cohen [9], Shackle [10] and others, and provides a basis for the construction of more general theories in which the evidence is allowed to be fuzzy in nature.

More specifically, we shall concern ourselves with a type of information granularity in which the data granules are characterized by propositions of the general form

$$g \triangleq X \text{ is } G \text{ is } \lambda \tag{1.1}$$

---

*To Professor J. Kampe de Feriet.

Research supported by Naval Electronic Systems Command Contract N00039-78-G0013 and National Science Foundation Grant ENG-78-23143.

in which X is a variable taking values in a universe of discourse U, G is a fuzzy subset of U which is characterized by its membership function μ_G, and the qualifier λ denotes a fuzzy probability (or likelihood). Typically, but not universally, we shall assume that U is the real line (or R^n), G is a convex fuzzy subset of U and λ is a fuzzy subset of the unit interval. For example:

$$g \triangleq X \text{ is small is likely}$$

$$g \triangleq X \text{ is not very large is very unlikely}$$

$$g \triangleq X \text{ is much larger than Y is unlikely}$$

We shall not consider data granules which are characterized by propositions in which the qualifier λ is a fuzzy possibility or fuzzy truth-value.

In a general sense, a <u>body of evidence</u> or, simply, <u>evidence</u>, may be regarded as a collection of propositions. In particular, the evidence is <u>granular</u> if it consists of a collection of propositions,

$$E = \{g_1,\ldots,g_N\} , \tag{1.2}$$

each of which is of the form (1.1). Viewed in this perspective, Shafer's theory relates to the case where the constituent granules in (1.2) are crisp (nonfuzzy) in the sense that, in each g_i, G_i is a nonfuzzy set and λ_i is a numerical probability, implying that g_i may be expressed as

$$g_i \triangleq \text{"Prob}\{X \in G_i\} = p_i\text{"} \tag{1.3}$$

where p_i, $i = 1,\ldots,N$, is the probability that the value of X is contained in G. In the theories of Cohen and Shackle, a further restriction is introduced through the assumption that the G_i are nested, i.e., $G_1 \subset G_2 \subset \cdots \subset G_N$. As was demonstrated by Suppes and Zanotti [11] and Nguyen [12], in the analysis of evidence of the form (1.3) it is advantageous to treat E as a random relation.

Given a collection of granular bodies of evidence $E = \{E_1,\ldots,E_K\}$, one may ask a variety of questions the answers to which depend on the data resident in E. The most basic of these questions--which will be the main focus of our attention in the sequel--is the following:

Given a body of evidence $E = \{g_1,\ldots,g_N\}$ and an arbitrary fuzzy subset Q of U, what is the probability--which may be fuzzy or nonfuzzy--that X is Q? In other words, from the propositions

$$\begin{array}{c} g_1 \triangleq X \text{ is } G_1 \text{ is } \lambda_1 \\ - - - - - - - - \\ g_N \triangleq X \text{ is } G_N \text{ is } \lambda_N \end{array} \tag{1.4}$$

we wish to deduce the value of $?\lambda$ in the question

$$q \triangleq X \text{ is Q is } ?\lambda \tag{1.5}$$

As a concrete illustration, suppose that we have the following granular information concerning the age of Judy ($X \triangleq \text{Age(Judy)}$)

$$\begin{array}{l} g_1 \triangleq \text{Judy is very young is unlikely} \\ g_2 \triangleq \text{Judy is young is likely} \\ g_3 \triangleq \text{Judy is old is very unlikely} \end{array} \tag{1.6}$$

The question is: What is the probability that Judy is not very young; or, equivalently: What is the value of $?\lambda$ in

$$q \triangleq \text{Judy is not very young is } ?\lambda$$

In cases where E consists of two or more distinct bodies of evidence, an important issue relates to the manner in which the answer to (1.5)--based on the information resident in E --may be composed from the answers based on the information resident in each of the constituent bodies of evidence $E_1,\ldots,E_K$. We shall consider this issue very briefly in Sec. 3.

In the theories of Dempster and Shafer, both the evidence and the set Q in (1.5) are assumed to be crisp, and the question that is asked is: What are the bounds on the probability λ that $X \in Q$? The lower bound, λ_*, is referred to as the <u>lower probability</u> and is defined by Shafer to be the <u>degree of belief</u> that $X \in Q$, while the upper bound, λ^*, is equated to the <u>degree of plausibility</u> of the proposition $X \in Q$. An extension of the concepts of lower and upper probabilities to the more general case of fuzzy granules will be described in Sec. 3.

As will be seen in the sequel, the theory of fuzzy sets and, in particular, the theory of possibility, provides a convenient conceptual framework for dealing with information granularity in a general setting. Viewed in such a setting, the concept of information granularity assumes an important role in the analysis of imprecise evidence and thus may aid in contributing to a better understanding of the complex issues arising in credibility analysis, model validation and, more generally, those problem areas in which the information needed for a decision or system performance evaluation is incomplete or unreliable.

2. Information Granularity and Possibility Distributions

Since the concept of information granularity bears a close relation to that of a possibility distribution, we shall begin our exposition with a brief review of those properties of possibility distributions which are of direct relevance to the concepts introduced in the following sections.

Let X be a variable taking values in U, with a generic value of X denoted by u. Informally, a possibility distribution, Π_X, is a fuzzy relation in U which acts as an elastic constraint on the values that may be assumed by X. Thus, if π_X is the membership function of Π_X, we have

$$\text{Poss}\{X = u\} = \pi_X(u) \;, \quad u \in U \tag{2.1}$$

where the left-hand member denotes the possibility that X may take the value u and $\pi_X(u)$ is the grade of membership of u in Π_X. When used to characterize Π_X, the function π_X: $U \to [0,1]$ is referred to as a <u>possibility distribution function</u>.

A possibility distribution, Π_X, may be induced by physical constraints or, alternatively, it may be epistemic in nature, in which case Π_X is induced by a collection of propositions--as described at a later point in this section.

A simple example of a possibility distribution which is induced by a physical constraint is the number of tennis balls that can be placed in a metal box. In this case, X is the number in question and $\pi_X(u)$ is a measure of the degree of ease (by some specified mechanical criterion) with which u balls can be squeezed into the box.

As a simple illustration of an epistemic possibility distribution, let X be a real-valued variable and let p be the proposition

$$p \triangleq a \le X \le b$$

where [a,b] is an interval in R^1. In this case, the possibility distribution

[1] A more detailed discussion of possibility theory may be found in [13]-[15].

induced by p is the uniform distribution defined by

$$\pi_X(u) = 1 \quad \text{for } a \le u \le b$$
$$= 0 \quad \text{elsewhere.}$$

Thus, given p we can assert that

$$\text{Poss}\{X = u\} = 1 \quad \text{for } u \text{ in } [a,b]$$
$$= 0 \quad \text{elsewhere.}$$

More generally, as shown in [16], a proposition of the form

$$p \triangleq N \text{ is } F \tag{2.2}$$

where F is a fuzzy subset of the cartesian product $U = U_1 \times \cdots \times U_n$ and N is the name of a variable, a proposition or an object, induces a possibility distribution defined by the <u>possibility assignment equation</u>

$$N \text{ is } F \rightarrow \Pi_{(X_1,\ldots,X_n)} = F \tag{2.3}$$

where the symbol $\rightarrow$ stands for "translates into," and $X \triangleq (X_1,\ldots,X_n)$ is an n-ary variable which is implicit or explicit in p. For example,

(a) $$X \text{ is small} \rightarrow \Pi_X = \text{SMALL} \tag{2.4}$$

where SMALL, the denotation of <u>small</u>, is a specified fuzzy subset of $[0,\infty)$. Thus, if the membership function of SMALL is expressed as μ_{SMALL}, then (2.4) implies that

$$\text{Poss}\{X = u\} = \mu_{SMALL}(u) \ , \quad u \in [0,\infty) \ . \tag{2.5}$$

More particularly, if--in the usual notation--

$$\text{SMALL} = 1/0 + 1/1 + 0.8/2 + 0.6/3 + 0.5/4 + 0.3/5 + 0.1/6 \tag{2.6}$$

then

$$\text{Poss}\{X = 3\} = 0.6$$

and likewise for other values of u.

Similarly,

(b) $$\text{Dan is tall} \rightarrow \Pi_{\text{Height(Dan)}} = \text{TALL} \tag{2.7}$$

where the variable Height(Dan) is implicit in the proposition "Dan is tall" and TALL is a fuzzy subset of the interval [0,220] (with the height assumed to be expressed in centimeters).

(c) $$\text{John is big} \rightarrow \Pi_{(\text{Height(John)},\text{Weight(John)})} = \text{BIG} \tag{2.8}$$

where BIG is a fuzzy binary relation in the product space $[0,220] \times [0,150]$ (with height and weight expressed in centimeters and kilograms, respectively) and the variables $X_1 \triangleq$ Height(John), $X_2 \triangleq$ Weight(John) are implicit in the proposition "John is big."

In a more general way, the translation rules associated with the meaning representation language PRUF [16] provide a system for computing the possibility distributions induced by various types of propositions. For example

$$X \text{ is not very small} \rightarrow \Pi_X = (\text{SMALL}^2)' \tag{2.9}$$

where $SMALL^2$ is defined by

$$\mu_{SMALL^2} = (\mu_{SMALL})^2 \qquad (2.10)$$

and ' denotes the complement. Thus, (2.10) implies that the possibility distribution function of X is given by

$$\pi_X(u) = 1 - \mu^2_{SMALL}(u) \; . \qquad (2.11)$$

In the case of conditional propositions of the form $p \triangleq$ If X is F then Y is G, the possibility distribution that is induced by p is a <u>conditional possibility distribution</u> which is defined by[2]

$$\text{If } X \text{ is } F \text{ then } Y \text{ is } G \rightarrow \Pi_{(Y|X)} = \bar{F}' \cup \bar{G} \qquad (2.12)$$

where $\Pi_{(Y|X)}$ denotes the conditional possibility distribution of Y given X, F and G are fuzzy subsets of U and V, respectively, $\bar{F}$ and $\bar{G}$ are the cylindrical extensions of F and G in $U \times V$, $\cup$ is the union, and the conditional possibility distribution function of Y given X is expressed by

$$\pi_{(Y|X)}(v|u) = \big(1 - \mu_F(u)\big) \vee \mu_G(v) \; , \qquad u \in U, \; v \in V \qquad (2.13)$$

where μ_F and μ_G are the membership functions of F and G, and $\vee \triangleq$ max. In connection with (2.12), it should be noted that

$$\pi_{(Y|X)}(v|u) = \text{Poss}\{Y = v | X = u\} \qquad (2.14)$$

whereas

$$\pi_{(X,Y)}(u,v) = \text{Poss}\{X = u, \; Y = v\} \; . \qquad (2.15)$$

A concept which is related to that of a conditional possibility distribution is the concept of a <u>conditional possibility measure</u> [13]. Specifically, let Π_X be the possibility distribution induced by the proposition

$$p \triangleq X \text{ is } G \; ,$$

and let F be a fuzzy subset of U. Then, the <u>conditional possibility measure</u> of F with respect to the possibility distribution Π_X is defined by

$$\text{Poss}\{X \text{ is } F | X \text{ is } G\} = \sup_u \big(\mu_F(u) \wedge \mu_G(u)\big) \; . \qquad (2.16)$$

It should be noted that the left-hand member of (2.16) is a set function whereas $\Pi_{(Y|X)}$ is a fuzzy relation defined by (2.12).

The foregoing discussion provides us with the necessary background for defining some of the basic concepts relating to information granularity. We begin with the concept of a <u>fuzzy granule</u>.

<u>Definition</u>. Let X be a variable taking values in U and let G be a fuzzy subset of U. (Usually, but not universally, $U = R^n$ and G is a convex fuzzy subset of U.) A <u>fuzzy granule</u>, g, in U is induced (or characterized) by a proposition of the form

$$g \triangleq X \text{ is } G \text{ is } \lambda \qquad (2.17)$$

---

[2] There are a number of alternative ways in which $\Pi_{(Y|X)}$ may be defined in terms of F and G [17], [18], [19]. Here we use a definition which is consistent with the relation between the extended concepts of upper and lower probabilities as described in Sec. 3.

where λ is a fuzzy probability which is characterized by a possibility distribution over the unit interval. For example, if $U = R^1$, we may have

$$g \triangleq X \text{ is small is not very likely} \tag{2.18}$$

where the denotation of small is a fuzzy subset SMALL of R^1 which is characterized by its membership function μ_{SMALL}, and the fuzzy probability not very likely is characterized by the possibility distribution function

$$\pi(v) = 1 - \mu^2_{LIKELY}(v) , \quad v \in [0,1] \tag{2.19}$$

in which μ_{LIKELY} is the membership function of the denotation of likely and v is a numerical probability in the interval [0,1].

If the proposition $p \triangleq X$ is G is interpreted as a fuzzy event [20], then (2.17) may be interpreted as the proposition

$$\text{Prob}\{X \text{ is } G\} \text{ is } \lambda$$

which by (2.3) translates into

$$\Pi_{\text{Prob}\{X \text{ is } G\}} = \lambda . \tag{2.20}$$

Now, the probability of the fuzzy event $p \triangleq X$ is G is given by [20]

$$\text{Prob}\{X \text{ is } G\} = \int_U p_X(u)\mu_G(u)du \tag{2.21}$$

where $p_X(u)$ is the probability density associated with X. Thus, the translation of (2.17) may be expressed as

$$g \triangleq X \text{ is } G \text{ is } \lambda \longrightarrow \pi(p_X) = \mu_\lambda\left(\int_U p_X(u)\mu_G(u)du\right) \tag{2.22}$$

which signifies that g induces a possibility distribution of the probability distribution of X, with the possibility of the probability density p_X given by the right-hand member of (2.22). For example, in the case of (2.18), we have

$$X \text{ is small is not very likely} \longrightarrow \pi(p_X) = 1 - \mu^2_{LIKELY}\left(\int_U p_X(u)\mu_{SMALL}(u)du\right). \tag{2.23}$$

As a special case of (2.17), a fuzzy granule may be characterized by a proposition of the form

$$g \triangleq X \text{ is } G \tag{2.24}$$

which is not probability-qualified. To differentiate between the general case (2.17) and the special case (2.24), fuzzy granules which are characterized by propositions of the form (2.17) will be referred to as πp-granules (signifying that they correspond to possibility distributions of probability distributions), while those corresponding to (2.24) will be described more simply as π-granules.

A concept which we shall need in our analysis of bodies of evidence is that of a conditioned π-granule. More specifically, if X and Y are variables taking values in U and V, respectively, then a conditioned π-granule in V is characterized by a conditional proposition of the form

$$g \triangleq \text{If } X = u \text{ then } Y \text{ is } G \tag{2.25}$$

where G is a fuzzy subset of V which is dependent on u. From this definition it follows at once that the possibility distribution induced by g is defined by the possibility distribution function

$$\pi_{(Y|X)}(v|u) \triangleq \text{Poss}\{Y=v|X=u\} = \mu_G(v) \ . \tag{2.26}$$

An important point which arises in the characterization of fuzzy granules is that the same fuzzy granule may be induced by distinct propositions, in which case the propositions in question are said to be semantically equivalent [16]. A particular and yet useful case of semantic equivalence relates to the effect of negation in (2.17) and may be expressed as ($\leftrightarrow$ denotes semantic equivalence)

$$X \text{ is } G \text{ is } \lambda \leftrightarrow X \text{ is not } G \text{ is ant } \lambda \tag{2.27}$$

where ant λ denotes the antonym of λ which is defined by

$$\mu_{\text{ant } \lambda}(v) = \mu_\lambda(1-v) \ , \quad v \in [0,1] \ . \tag{2.28}$$

Thus, the membership function of ant λ is the mirror image of that of λ with respect to the midpoint of the interval [0,1].

To verify (2.27) it is sufficient to demonstrate that the propositions in question induce the same fuzzy granule. To this end, we note that

$$\begin{aligned} X \text{ is not } G \text{ is ant } \lambda \rightarrow \pi(p_X) &= \mu_{\text{ant } \lambda}\left(\int_U p_X(u)(1-\mu_G(u))du\right) \\ &= \mu_{\text{ant } \lambda}\left(1 - \int_U p_X(u)\mu_G(u)du\right) \\ &= \mu_\lambda\left(\int_U p_X(u)\mu_G(u)du\right) \end{aligned} \tag{2.29}$$

which upon comparison with (2.22) establishes the semantic equivalence expressed by (2.27).

In effect, (2.27) indicates that replacing G with its negation may be compensated by replacing λ with its antonym. A simple example of an application of this rule is provided by the semantic equivalence

$$X \text{ is small is likely} \leftrightarrow X \text{ is not small is unlikely} \tag{2.30}$$

in which unlikely is interpreted as the antonym of likely.

A concept that is related to and is somewhat weaker than that of semantic equivalence is the concept of semantic entailment [16]. More specifically, if g_1 and g_2 are two propositions such that the fuzzy granule induced by g_1 is contained in the fuzzy granule induced by g_2, then g_2 is semantically entailed by g_1 or, equivalently, g_1 semantically entails g_2. To establish the relation of containment it is sufficient to show that

$$\pi_1(p_X) \leq \pi_2(p_X) \ , \quad \text{for all } \ p_X \tag{2.31}$$

where π_1 and π_2 are the possibilities corresponding to g_1 and g_2, respectively.

As an illustration, it can readily be established that ($\mapsto$ denotes semantic entailment)

$$X \text{ is } G \text{ is } \lambda \mapsto X \text{ is very } G \text{ is } {}^2\lambda \tag{2.32}$$

or, more concretely,

$$X \text{ is small is likely} \mapsto X \text{ is very small is } {}^2\text{likely} \tag{2.33}$$

where the left-square of λ is defined by

$$\mu_{2_\lambda}(v) = \mu_\lambda(\sqrt{v}) , \quad v \in [0,1]$$

and μ_λ is assumed to be monotone nondecreasing. Intuitively, (2.32) signifies that an intensification of G through the use of the modifier very may be compensated by a dilation (broadening) of the fuzzy probability λ.

To establish (2.32), we note that

$$X \text{ is } G \text{ is } \lambda \longrightarrow \pi_1(p_X) = \mu_\lambda\left(\int_U p_X(u)\mu_G(u)du\right) \quad (2.34)$$

$$X \text{ is very } G \text{ is } {}^2\lambda \longrightarrow \pi_2(p_X) = \mu_{2_\lambda}\left(\int_U p_X(u)\mu_G^2(u)du\right) \quad (2.35)$$

$$= \mu_\lambda\left(\sqrt{\int_U p_X(u)\mu_G^2(u)du}\right) .$$

Now, by Schwarz's inequality

$$\sqrt{\int_U p_X(u)\mu_G^2(u)du} \geq \int_U p_X(u)\mu_G(u)du \quad (2.36)$$

and since μ_λ is monotone nondecreasing, we have

$$\pi_1(p_X) \leq \pi_2(p_X)$$

which is what we wanted to demonstrate.

3. Analysis of Granular Evidence

As was stated in the introduction, a body of evidence or, simply, evidence, E, may be regarded as a collection of propositions

$$E = \{g_1,\dots,g_N\} . \quad (3.1)$$

In particular, evidence is granular if its constituent propositions are characterizations of fuzzy granules.

For the purpose of our analysis it is necessary to differentiate between two types of evidence which will be referred to as evidence of the first kind and evidence of the second kind.

Evidence of the first kind is a collection of fuzzy πp-granules of the form

$$g_i \triangleq Y \text{ is } G_i \text{ is } \lambda_i , \quad i = 1,\dots,N \quad (3.2)$$

where Y is a variable taking values in V, $G_1,\dots,G_N$ are fuzzy subsets of V and $\lambda_1,\dots,\lambda_N$ are fuzzy probabilities.

Evidence of the second kind is a probability distribution of conditioned π-granules of the form

$$g_i \triangleq Y \text{ is } G_i . \quad (3.3)$$

Thus, if X is taken to be a variable which ranges over the index set $\{1,\dots,N\}$, then we assume to know (a) the probability distribution $P_X = \{p_1,\dots,p_N\}$, where

$$p_i \triangleq \mathrm{Prob}\{X = i\} , \quad i = 1,\dots,N \quad (3.4)$$

and (b) the conditional possibility distribution $\Pi_{(Y|X)}$, where

$$\Pi_{(Y|X=i)} = G_i , \quad i = 1,\dots,N . \quad (3.5)$$

In short, we may express evidence of the second kind in a symbolic form as

$$E = \{P_X, \Pi_{(Y|X)}\}$$

which signifies that the evidence consists of P_X and $\Pi_{(Y|X)}$, rather than P_X and $P_{(Y|X)}$ (conditional probability distribution of Y given X), which is what is usually assumed to be known in the traditional probabilistic approaches to the analysis of evidence. Viewed in this perspective, the type of evidence considered in the theories of Dempster and Shafer is evidence of the second kind in which the G_i are crisp sets and the probabilities $p_1,\ldots,p_n$ are known numerically.

In the case of evidence of the first kind, our main concern is with obtaining an answer to the following question. Given E, find the probability, λ, or, more specifically, the possibility distribution of the probability λ, that Y is Q, where Q is an arbitrary fuzzy subset of V.

In principle, the answer to this question may be obtained as follows.

First, in conformity with (2.20), we interpret each of the constituent propositions in E,

$$g_i \triangleq Y \text{ is } G_i \text{ is } \lambda_i\ , \quad i = 1,\ldots,N \tag{3.6}$$

as the assignment of the fuzzy probability λ_i to the fuzzy event $q_i \triangleq Y$ is G_i. Thus, if $p(\cdot)$ is the probability density associated with Y, then in virtue of (2.22) we have

$$\pi_i(p) = \mu_{\lambda_i}\left(\int_V p(v)\mu_{G_i}(v)dv\right) \tag{3.7}$$

where $\pi_i(p)$ is the possibility of p given g_i, and μ_{λ_i} and μ_{G_i} are the membership functions of λ_i and G_i, respectively.

Since the evidence $E = \{g_1,\ldots,g_N\}$ may be regarded as the conjunction of the propositions $g_1,\ldots,g_N$, the possibility of $p(\cdot)$ given E may be expressed as

$$\pi(p) = \pi_1(p) \wedge \cdots \wedge \pi_N(p) \tag{3.8}$$

where $\wedge \triangleq \min$. Now, for a p whose possibility is expressed by (3.8), the probability of the fuzzy event $q \triangleq X$ is Q is given by

$$\rho(p) = \int_V p(v)\mu_Q(v)dv\ . \tag{3.9}$$

Consequently, the desired possibility distribution of $\rho(p)$ may be expressed in a symbolic form as the fuzzy set [21]

$$\lambda = \int_{[0,1]} \pi(p)/\rho(p) \tag{3.10}$$

in which the integral sign denotes the union of singletons $\pi(p)/\rho(p)$.

In more explicit terms, (3.10) implies that if ρ is a point in the interval [0,1], then $\mu_\lambda(\rho)$, the grade of membership of ρ in λ or, equivalently, the possibility of ρ given λ, is the solution of the variational problem

$$\mu_\lambda(\rho) = \mathrm{Max}_p(\pi_1(p) \wedge \cdots \wedge \pi_N(p)) \tag{3.11}$$

subject to the constraint

$$\rho = \int_V p(v)\mu_Q(v)dv\ . \tag{3.12}$$

In practice, the solution of problems of this type would, in general, require both discretization and approximation, with the aim of reducing (3.11) to a computationally feasible problem in nonlinear programming. In the longer run, however, a more effective solution would be a "fuzzy hardware" implementation which would yield directly a linguistic approximation to λ from the specification of q and E.

It should be noted that if we were concerned with a special case of evidence of the first kind in which the probabilities λ_i are numerical rather than fuzzy, then we could use as an alternative to the technique described above the maximum entropy principle of Jaynes [22] or its more recent extensions [23]-[26]. In application to the problem in question, this method would first yield a probability density $p(\cdot)$ which is a maximum entropy fit to the evidence E, and then, through the use of (3.12), would produce a numerical value for λ.

A serious objection that can be raised against the use of the maximum entropy principle is that, by constructing a unique $p(\cdot)$ from the incomplete information in E, it leads to artificially precise results which do not reflect the intrinsic imprecision of the evidence and hence cannot be treated with the same degree of confidence as the factual data which form a part of the database. By contrast, the method based on the use of possibility distributions leads to conclusions whose imprecision reflects the imprecision of the evidence from which they are derived and hence are just as credible as the evidence itself.

Turning to the analysis of evidence of the second kind, it should be noted that, although there is a superficial resemblance between the first and second kinds of evidence, there is also a basic difference which stems from the fact that the fuzzy granules in the latter are π-granules which are conditioned on a random variable. In effect, what this implies is that evidence of the first kind is conjunctive in nature, as is manifested by (3.8). By contrast, evidence of the second kind is disjunctive, in the sense that the collection of propositions in E should be interpreted as the disjunctive statement: g_1 with probability λ_1 or g_2 with probability λ_2 or ... or g_N with probability λ_N.

As was stated earlier, evidence of the second kind may be expressed in the equivalent form

$$E = \{P_X, \Pi_{(Y|X)}\}$$

where X is a random variable which ranges over the index set $U = \{1,\ldots,N\}$ and is associated with a probability distribution $P_X = \{p_1,\ldots,p_N\}$; and $\Pi_{(Y|X)}$ is the conditional possibility distribution of Y given X, where Y is a variable ranging over V and the distribution function of $\Pi_{(Y|X)}$ is defined by

$$\pi_{(Y|X)}(v|i) \triangleq \mathrm{Poss}\{Y=v|X=i\}\ , \quad i \in U,\ v \in V\ . \tag{3.13}$$

For a given value of X, $X=i$, the conditional possibility distribution $\Pi_{(Y|X)}$ defines a fuzzy subset of V which for consistency with (3.2) is denoted by G_i. Thus,

$$\Pi_{(Y|X=i)} = G_i\ , \quad i = 1,\ldots,N \tag{3.14}$$

and more generally

$$\Pi_{(Y|X)} = G_X\ . \tag{3.15}$$

As was pointed out earlier, the theories of Dempster and Shafer deal with a special case of evidence of the second kind in which the G_i and Q are crisp sets and the probabilities $p_1,\ldots,p_N$ are numerical. In this special case, the event $q \triangleq Y \in Q$ may be associated with two probabilities: the lower probability λ_* which is defined--in our notation--as

$$\lambda_* \triangleq \mathrm{Prob}\{\Pi_{(Y|X)} \subset Q\} \tag{3.16}$$

and the upper probability λ^* which is defined as[3]

$$\lambda^* \triangleq \text{Prob}\{\Pi_{(Y|X)} \cap Q \neq \theta\} \quad (\theta \triangleq \text{empty set}) . \tag{3.17}$$

The concepts of upper and lower probabilities do not apply to the case where the G_i and Q are fuzzy sets. For this case, we shall define two more general concepts which are related to the modal concepts of necessity and possibility and which reduce to λ_* and λ^* when the G_i and Q are crisp.

For our purposes, it will be convenient to use the expressions sup F and inf F as abbreviations defined by[4]

$$\sup F \triangleq \sup_v \mu_F(v) \ , \quad v \in V \tag{3.18}$$

$$\inf F \triangleq \inf_v \mu_F(v) \ , \quad v \in V \tag{3.19}$$

where F is a fuzzy subset of V. Thus, using this notation, the expression for the conditional possibility measure of Q given X may be written as (see (2.16))

$$\begin{aligned} \text{Poss}\{Y \text{ is } Q|X\} &= \text{Poss}\{Y \text{ is } Q|Y \text{ is } G_X\} \\ &= \sup(Q \cap G_X) \end{aligned} \tag{3.20}$$

Since X is a random variable, we can define the expectation of Poss{Y is Q|X} with respect to X. On denoting this expectation by $E\Pi(Q)$, we have

$$\begin{aligned} E\Pi(Q) &\triangleq E_X \, \text{Poss}\{Y \text{ is } Q|X\} \\ &= \sum_i p_i \sup(Q \cap G_i) \end{aligned} \tag{3.21}$$

We shall adopt the expected possibility, $E\Pi(Q)$, as a generalization of the concept of upper probability. Dually, the concept of lower probability may be generalized as follows.

First, we define the conditional certainty (or necessity) of the proposition $q \triangleq Y$ is Q given X by

$$\text{Cert}\{Y \text{ is } Q|X\} \triangleq 1 - \text{Poss}\{Y \text{ is not } Q|X\} . \tag{3.22}$$

Next, in view of the identities

$$\begin{aligned} 1 - \sup(F \cap G) &= \inf\big((F \cap G)'\big) \\ &= \inf(F' \cup G') \\ &= \inf(G \Rightarrow F') \end{aligned} \tag{3.23}$$

where the implication $\Rightarrow$ is defined by (see (2.13))

$$G \Rightarrow F' \triangleq G' \cup F' \tag{3.24}$$

we can rewrite the right-hand member of (3.22) as

---

[3] It should be noted that we are not normalizing the definitions of λ_* and λ^*--as is done in the papers by Dempster and Shafer--by dividing the right-hand members of (3.16) and (3.17) by the probability that $\Pi_{(Y|X)}$ is not an empty set. As is pointed out in [27], the normalization in question leads to counterintuitive results.

[4] The definitions in question bear a close relation to the definitions of universal and existential quantifiers in L_{Aleph_1} logic [28].

$$\mathrm{Cert}\{Y \text{ is } Q|X\} = \inf(G_X \Rightarrow Q) \ . \tag{3.25}$$

Finally, on taking the expectation of both sides of (3.22) and (3.25), we have

$$\begin{aligned} EC(Q) &\triangleq E_X \, \mathrm{Cert}\{Y \text{ is } Q|X\} \\ &= \sum_i p_i \inf(G_i \Rightarrow Q) \\ &= 1 - E\Pi(Q') \end{aligned} \tag{3.26}$$

As defined by (3.26), the expression EC(Q), which represents the expected certainty of the conditional event (Y is Q|X), may be regarded as a generalization of the concept of lower probability.

The set functions EΠ(Q) and EC(Q) may be interpreted as fuzzy measures. However, in general, these measures are neither normed nor additive. Instead, EΠ(Q) and EC(Q) are, respectively, superadditive and subadditive in the sense that, for any fuzzy subsets Q_1 and Q_2 of V, we have

$$EC(Q_1 \cup Q_2) \geq EC(Q_1) + EC(Q_2) - EC(Q_1 \cap Q_2) \tag{3.27}$$

and

$$E\Pi(Q_1 \cup Q_2) \leq E\Pi(Q_1) + E\Pi(Q_2) - E\Pi(Q_1 \cap Q_2) \ . \tag{3.28}$$

It should be noted that these inequalities generalize the superadditive and subadditive properties of the measures of belief and plausibility in Shafer's theory.

The inequalities in question are easy to establish. Taking (3.28), for example, we have

$$\begin{aligned} E\Pi(Q_1 \cup Q_2) &= \textstyle\sum_i p_i \sup_v \left[\left(\mu_{Q_1}(v) \vee \mu_{Q_2}(v) \right) \wedge \mu_{G_i}(v) \right] \\ &= \sum_i p_i \sup_v \left(\mu_{Q_1}(v) \wedge \mu_{G_i}(v) \vee \mu_{Q_2}(v) \wedge \mu_{G_i}(v) \right) \\ &= \sum_i p_i \left[\sup_v \left(\mu_{Q_1}(v) \wedge \mu_{G_i}(v) \right) \vee \sup_v \left(\mu_{Q_2}(v) \wedge \mu_{G_i}(v) \right) \right] \end{aligned} \tag{3.29}$$

Now, using the identity (a,b ≜ real numbers)

$$a \vee b = a + b - a \wedge b \tag{3.30}$$

the right-hand member of (3.29) may be rewritten as

$$\begin{aligned} E\Pi(Q_1 \cup Q_2) = \sum_i p_i \Big[&\sup_v \left(\mu_{Q_1}(v) \wedge \mu_{G_i}(v) \right) + \sup_v \left(\mu_{Q_2}(v) \wedge \mu_{G_i}(v) \right) \\ &- \left(\sup_v \left(\mu_{Q_1}(v) \wedge \mu_{G_i}(v) \right) \wedge \sup_v \left(\mu_{Q_2}(v) \wedge \mu_{G_i}(v) \right) \right) \Big] \end{aligned} \tag{3.31}$$

Furthermore, from the min-max inequality

$$\sup_v f(v) \wedge \sup_v g(v) \geq \sup_v \left(f(v) \wedge g(v) \right) \tag{3.32}$$

it follows that

$$\begin{aligned} &\sup_v \left(\mu_{Q_1}(v) \wedge \mu_{G_i}(v) \right) \wedge \sup_v \left(\mu_{Q_2}(v) \wedge \mu_{G_i}(v) \right) \\ &\geq \sup_v \left(\mu_{Q_1}(v) \wedge \mu_{Q_2}(v) \wedge \mu_{G_i}(v) \right) \end{aligned} \tag{3.33}$$

and hence that

$$E\Pi(Q_1 \cup Q_2) \leq \sum_i p_i \sup_v(\mu_{Q_1}(v) \wedge \mu_{G_i}(v)) + \sum_i p_i \sup_v(\mu_{Q_2}(v) \wedge \mu_{G_i}(v)) \tag{3.34}$$
$$- \sum_i p_i \sup_v(\mu_{Q_1}(v) \wedge \mu_{Q_2}(v) \wedge \mu_{G_i}(v)) \quad .$$

Finally, on making use of (3.21) and the definition of $Q_1 \cap Q_2$, we obtain the inequality

$$E\Pi(Q_1 \cup Q_2) \leq E\Pi(Q_1) + E\Pi(Q_2) - E\Pi(Q_1 \cap Q_2) \tag{3.35}$$

which is what we set out to establish.

The superadditive property of $EC(Q)$ has a simple intuitive explanation. Specifically, because of data granularity, if Q_1 and Q_2 are roughly of the same size as the granules $G_1,\ldots,G_N$, then $EC(Q_1)$ and $EC(Q_2)$ are likely to be small, while $E(Q_1 \cup Q_2)$ may be larger because the size of $Q_1 \cup Q_2$ is likely to be larger than that of $G_1,\ldots,G_N$. For the same reason, with the increase in the relative size of Q_1 and Q_2, the effect of granularity is likely to diminish, with $EC(Q)$ tending to become additive in the limit.

In the foregoing analysis, the probabilities $p_1,\ldots,p_N$ were assumed to be numerical. This, however, is not an essential restriction, and through the use of the extension principle [21], the concepts of expected possibility and expected certainty can readily be generalized, at least in principle, to the case where the probabilities in question are fuzzy or linguistic. Taking the expression for $E\Pi(Q)$, for example,

$$E\Pi(Q) = \sum_i p_i \sup(Q \cap G_i) \tag{3.36}$$

and assuming that the p_i are characterized by their respective possibility distribution functions $\pi_1,\ldots,\pi_N$, the determination of the possibility distribution function of $E\Pi(Q)$ may be reduced to the solution of the following variational problem

$$\pi(z) \triangleq \text{Max}_{p_1,\ldots,p_N} \; \pi_1(p_1) \wedge \cdots \wedge \pi_N(p_N) \tag{3.37}$$

subject to

$$z = p_1 \sup(Q \cap G_1) + \cdots + p_N \sup(Q \cap G_N)$$

$$p_1 + \cdots + p_N = 1$$

which upon solution yields the possibility, $\pi(z)$, of a numerical value, z, of $E\Pi(Q)$. Then, a linguistic approximation to the possibility distribution would yield an approximate value for $\Pi_{E\Pi(Q)}$ expressed as, say, <u>not very high</u>.

As was alluded to already, a basic issue in the analysis of evidence relates to the manner in which two or more distinct bodies of evidence may be combined. In the case of evidence of the second kind, for example, let us assume for simplicity that we have two bodies of evidence of the form

$$E = \{E_1, E_2\} \tag{3.38}$$

in which

$$E_1 = \{P_{X_1}, \Pi_{(Y|X_1)}\} \tag{3.39}$$

$$E_2 = \{P_{X_2}, \Pi_{(Y|X_2)}\} \tag{3.40}$$

where Y takes values in V; while X_1 and X_2 range over the index sets $U_1 = \{1,\ldots,N_1\}$ and $U_2 = \{1,\ldots,N_2\}$, and are associated with the joint probability distribution $P_{(X_1,X_2)}$ which is characterized by

$$p_{ij} \triangleq \text{Prob}\{X_1 = i,\ X_2 = j\} \ . \tag{3.41}$$

For the case under consideration, the expression for the expected possibility of the fuzzy event $q \triangleq Y$ is Q given E_1 and E_2 becomes

$$E\Pi(Q) = E_{(X_1,X_2)} \text{Poss}\{Y \text{ is } Q|(X_1,X_2)\} \tag{3.42}$$
$$= \sum_{i,j} p_{ij} \sup(Q \cap G_i \cap H_j)$$

where

$$\Pi_{(Y|X_1=i)} \triangleq G_i \tag{3.43}$$

and

$$\Pi_{(Y|X_2=j)} \triangleq H_j \; . \tag{3.44}$$

The rule of combination of evidence developed by Dempster [7] applies to the special case of (3.42) in which the sets G_i and H_j are crisp and X_1 and X_2 are independent. In this case, from the knowledge of $E\Pi(Q)$ (or $EC(Q)$) for each of the constituent bodies of evidence and $Q \subset V$, we can determine the probability distributions of X_1 and X_2 and then use (3.42) to obtain $E\Pi(Q)$ for the combined evidence. Although simple in principle, the computations involved in this process tend to be rather cumbersome. Furthermore, as is pointed out in [27], there are some questions regarding the validity of the normalization employed by Dempster when

$$G_i \cap H_j = \theta \tag{3.45}$$

for some i, j, and the probability of the event "Y is θ" is positive.

4. Concluding Remarks

Because of its substantial relevance to decision analysis and model validation, analysis of evidence is likely to become an important area of research in the years ahead.

It is a fact of life that much of the evidence on which human decisions are based is both fuzzy and granular. The concepts and techniques outlined in this paper are aimed at providing a basis for a better understanding of how such evidence may be analyzed in systematic terms.

Clearly, the mathematical problems arising from the granularity and fuzziness of evidence are far from simple. It may well be the case that their full solution must await the development of new types of computing devices which are capable of performing fuzzy computations in a way that takes advantage of the relatively low standards of precision which the results of such computations are expected to meet.

References and Related Papers

1. A.N. Tikhonov and V. Ya. Arsenin, Methods of Solution of Ill-Posed Problems, Nauka, Moscow, 1974.

2. S. Gutman, "Uncertain dynamical systems--a Lyapounov min-max approach," IEEE Trans. on Automatic Control, AC-24, 437-443, 1979.

3. F. Schlaepfer and F. Schweppe, "Continuous-time state estimation under distrubances bounded by convex sets," IEEE Trans. on Automatic Control, AC-17, 197-205, 1972.

4. D.R. Ries and M.R. Stonebraker, "Locking granularity revisited," ERL Memorandum M78/71, Electronics Research Laboratory, University of California, Berkeley, 1978.

5. R.E. Moore, Interval Analysis, Prentice-Hall, Englewood Cliffs, N.J., 1966.

6. G. Shafer, A Mathematical Theory of Evidence, Princeton University Press, 1976.

7. A.P. Dempster, "Upper and lower probabilities induced by a multivalued mapping," Ann. Math. Statist. 38, 325-329, 1967.

8. P. Smets, "Un modele mathematico-statistique simulant le processus du diagnostic medicale," Free University of Brussels, 1978.

9. L.J. Cohen, The Implications of Induction, Methuen, London, 1970.

10. G.L.S. Shackle, Decision, Order and Time in Human Affairs, Cambridge University Press, Cambridge, 1961.

11. P. Suppes and M. Zanotti, "On using random relations to generate upper and lower probabilities," Synthese 36, 427-440, 1977.

12. H.T. Nguyen, "On random sets and belief functions," J. Math. Anal. Appl. 65, 531-542, 1978.

13. L.A. Zadeh, "Fuzzy sets as a basis for a theory of possibility," Fuzzy Sets and Systems 1, 3-28, 1978.

14. H.T. Nguyen, "On conditional possibility distributions," Fuzzy Sets and Systems 1, 299-309, 1978.

15. E. Hisdal, "Conditional possibilities: independence and noninteraction," Fuzzy Sets and Systems 1, 283-297, 1978.

16. L.A. Zadeh, "PRUF--a meaning representation language for natural languages," Int. J. Man-Machine Studies 10, 395-460, 1978.

17. M. Mizumoto, S. Fukame and K. Tanaka, "Fuzzy reasoning methods by Zadeh and Mamdani, and improved methods," Proc. Third Workshop on Fuzzy Reasoning, Queen Mary College, London, 1978.

18. B.S. Sembi and E.H. Mamdani, "On the nature of implication in fuzzy logic," Proc. 9th International Symposium on Multiple-Valued Logic, Bath, England, 143-151, 1979.

19. W. Bandler and L. Kohout, "Application of fuzzy logics to computer protection structures," Proc. 9th International Symposium on Multiple-Valued Logic, Bath, England, 200-207, 1979.

20. L.A. Zadeh, "Probability measures of fuzzy events," J. Math. Anal. Appl. 23, 421-427, 1968.

21. L.A. Zadeh, "The concept of a linguistic variable and its application to approximate reasoning, Part I," Information Sciences 8, 199-249, 1975; Part II, Information Sciences 8, 301-357, 1975; Part III, Information Sciences 9, 43-80, 1975.

22. E.T. Jaynes, "Information theory and statistical mechanics," Parts I and II, Physical Review 106, 620-630; 108, 171-190, 1957.

23. S. Kullback, Information Theory and Statistics, John Wiley, New York, 1959.

24. M. Tribus, Rational Descriptions, Decisions and Designs, Pergamon Press, New York, 1969.

25. J.E. Shore and R.W. Johnson, "Axiomatic derivation of the principle of maximum entropy and the principle of minimum cross-entropy," NRL Memorandum Report 3-898, Naval Research Laboratory, Washington, D.C., 1978.

26. P.M. Williams, "Bayesian conditionalization and the principle of minimum information," School of Mathematical and Physical Sciences, The University of Sussex, England, 1978.

27. L.A. Zadeh, "On the validity of Dempster's rule of combination of evidence," Memorandum M79/24, Electronics Research Laboratory, University of California, Berkeley, 1979.

28. N. Rescher, Many-Valued Logic, McGraw-Hill, New York, 1969.

29. J. Kampé de Feriet and B. Forte, "Information et probabilité," Comptes Rendus Acad. Sci. A-265, 110-114, 142-146, 350-353, 1967.

30. M. Sugeno, "Theory of fuzzy integrals and its applications," Tokyo Institute of Technology, 1974.

31. T. Terano and M. Sugeno, "Conditional fuzzy measures and their applications," in Fuzzy Sets and Their Application to Cognitive and Decision Processes (L.A. Zadeh, K.S. Fu, K. Tanaka and M. Shimura, eds.), 151-170, 1975.

32. E. Sanchez, "On possibility-qualification in natural languages," Memorandum M77/28, Electronics Research Laboratory, University of California, Berkeley, 1977.

33. P.M. Williams, "On a new theory of epistemic probability (review of G. Shafer: A Mathematical Theory of Evidence)," Brit. J. for the Philosophy of Science 29, 375-387, 1978.

34. P. Suppes, "The measurement of belief," J. Roy. Statist. Soc. B 36, 160-175, 1974.

35. K.M. Colby, "Simulations of belief systems," in Computer Models of Thought and Language (R.C. Schank and K.M. Colby, eds.), W. Freeman, San Francisco, 1973.

36. B.C. Bruce, "Belief systems and language understanding," N.I.H. Report CBM-TR-41, Rutgers University, 1975.

37. R.P. Abelson, "The structure of belief systems," in Computer Models of Thought and Language (R.C. Schank and K.M. Colby, eds.), W. Freeman, San Francisco, 1973.

38. R.D. Rosenkrantz, Inference, Method and Decision, D. Reidel, Dordrecht, 1977.

39. N. Rescher, Plausible Reasoning, Van Gorcum, Amsterdam, 1976.

40. A. Tversky and D. Kahneman, "Judgment under uncertainty: heuristics and biases," Science 185, 1124-1131, 1974.

41. G. Banon, "Distinctions between several types of fuzzy measures," Proc. Int. Colloquium on the Theory and Applications of Fuzzy Sets, University of Marseille, Marseille, 1978.

42. I.J. Good, "Subjective probability as the measure of a non-measurable set," in Logic, Methodology and Philosophy of Science: Proceedings of the 1960 International Congress (E. Nagel, P. Suppes and A. Tarski, eds.), Stanford University Press, 1962.

ADVANCES IN FUZZY SET THEORY AND APPLICATIONS
M.M. Gupta, R.K. Ragade, R.R. Yager (editors)

FUZZY SET THEORY AND APPLICATIONS: A SYNTHESIS

Rammohan K. Ragade* and Madan M. Gupta**

*University of Louisville, Systems Science Institute
Belknap Campus, Louisville, Kentucky 40208, U. S. A.
**Cybernetics Research Laboratory, College of Engineering
University of Saskatchewan, Saskatoon, Sask., Canada S7N OWO

1. INTRODUCTION

Advances in Science and technology have made our modern society very complex, and with this decision processes have become increasingly vague and hard to analyze. The human brain possesses some special characteristics that enable it to learn and reason in a vague and fuzzy environment. It has the *ability* to arrive at decisions based on imprecise, qualitative data in contrast to formal mathematics and formal logic which demands precise and quantitative data. Modern computers possess capacity but lack the human-like ability. Undoubtedly, in many areas of cognition,human intelligence far excells the computer "intelligence" of today, and the development of fuzzy concepts is a step forward toward the development of tools capable of handling humanistic type of problems.

We do have sufficient mathematical tools and computer-based technology for analyzing and solving the problems embodied in deterministic and uncertain (probabilistic) environment. Here uncertainty may arise from the probabilistic behavior of certain physical phenomena in mechanistic systems. We know the important role that vagueness and inexactitude play in human decision making, but we did not know until 1965 how the vagueness arising from subjectivity which is inherent in human thought processes, can be modeled and analyzed.

In 1965, Professor Lotfi A. Zadeh laid the foundation of fuzzy set theory. In effect, fuzzy set theory is a body of concepts and techniques that laid a form of mathematical precision to human thought processes that in many ways are imprecise and ambiguous by the standards of classical mathematics. Today, these concepts are gaining a growing acceptability among engineers, scientists, mathematicians, linguists and philosophers. Since its inception, the research in fuzzy field has faced an increasing exponential growth as evidenced by an extensive bibliography appeared in this volume. This fuzzy field has blossomed into a many-faceted field of inquiry, drawing on and contributing to a wide spectrum of areas ranging from pure mathematics to human perception and judgment. Its influence in science, engineering and social sciences has been felt already, and is certain to grow in the decade to come.

Perhaps some people would realize that this theory has faced, like any new theory, an antagonistic attitude, which is definitely a reflection of prior prejudice, professed by those who did not want or did not have the time to study the theory or its ramifications.

Undoubtedly, the theory is still in its developing phases, but many more interesting and new theoretical developments and their applications are emerging to appear. A small sample of these new developments and applications appear in this Volume.

Dubois and Prade [2]* have an excellent introduction to the current state of knowledge in fuzzy set theory. They rightly remark, "... owing to Zadeh's fruitful efforts a theory of vagueness, what Zadeh calls "*fuzziness*" is sufficiently developed ..." Although "fuzzicists" converse about *fuzzy matters*, their discourses and debates have precise meanings!

It is not a paradox that a science of vagueness or fuzziness must be precise! Thus "fuzzy analysis" does considerably overlap diverse areas: interval analysis, probability theory, lattice theory, boolean algebraic analysis, statistical multivariate analysis, linguistic analysis, cluster analysis, pattern recognition, analysis of evidence etc. No wonder, those who work in each of these areas at first doubt the legitimacy of a theory of fuzzy analysis. Their first reaction is to ask for a concrete positive proof which ably demonstrates the uniqueness of fuzzy analysis.

Our current state of knowledge demonstrates, certainly to all fuzzicists and perhaps to quite a few others, that the faith and conviction in fuzzy analysis being a new approach, has been worth while.

The considerable debate between reductionists and antireductionists has resulted in the growth of General Systems Theory as understanding whole systems rather than the need to identify disjoint indivisible elementary units and to explain system behavior by a relation of their elementary unit behavior.

Fuzzy Analysis may be approached as a method of preserving the wholeness, while indicating the degrees of compatibility of components to understanding the whole system. Thus, whole systems are understood through partial truths provided by an understanding of components. This aspect is particularly made explicit in the application by Jain [33]to understand complex scenes.

More specifically, Part I of the volume contains introductory papers providing a broad perspective view of recent trends in the theoretical research. The papers in Part II, are addressed, for the most part, to the basic aspects of the theory and recent developments in the theory. Part III deals with the papers giving applications of the fuzzy set theory in medical diagnostics to complex scenes analysis.

In the following, we present a synthesis of the papers appearing in this Volume.

Dubois and Prade [2] scan the mathematics of fuzzy set theory which shows the comparatively greater richness afforded by fuzzy set theory compared to "ordinary fuzzy set theory." Measurement of fuzziness is of much concern to the empiricist and the applicationists. As yet, there isn't much sound satisfying foundational work. Credit must go to Nowakowska [21] and Sugeno who have given considerable thought to these issues. A probability measure is a fuzzy measure if the

* Numbers in brackets refer to the chapters in this Volume.

underlying measure space is the same, see Nahmia [10) and Kandel [11] but not vice versa. Other examples are similarity measures, relevance measures, degrees of match, Shafer's belief function, the λ-fuzzy measure of Sugeno, etc.

The theory of possibility is distinct from the theory of probability and is much richer. A consistency principle exists, see Dubois and Prade [2]. Ralescu [5] succintly introduces the utility of category theory for the representation problem. Thus, one is able to enhance through fuzzification most mathematical objects such as groups, rings, fields, topologies, etc.

A fuzzy number on the real line R is a fuzzy set characterized by a membership function $\mu_A: R \to (0,1)$. Both Mizumoto and Tanaka [9], and Dubois and Prade [2,17] use the extension principle of Zadeh to define operations, '+','-','x','÷','$\leq$'. It is inviting to think of μ_A as a distribution. However, it may be quickly observed that distributions in the usual sense of probability theory do not add, multiply, divide, etc., easily. Mizumoto and Tanaka [9] show algebraic properties of fuzzy numbers. For instance, a theorem is proved showing "the family of positive convex fuzzy numbers" forms a commutative semi-ring with zero '0' and unity '1' under '+' and 'x'. However, there are two types of orderings of fuzzy numbers written respectively $A \lneq B$, $A \dot{\leq} B$. Thus, triangular fuzzy numbers add to yield triangular fuzzy numbers.

Fuzzy sets have been suggested for handling the imprecision of real world situations by using truth values between the usual "true" and "false." Goguen [3] sketches a new approach to the foundations of fuzzy set theory. The main issue that he addresses is the determination and justification of fuzzy truth values arising in real social interaction.

There is an evidence of widespread concern over the usefulness of fuzzy systems as models of natural processes. Negoita [4] presents new conclusions on fuzzy systems theory that was obscured in the old system theory.

In parallel to this development is that of fuzzy logic and fuzzy reasoning methods as discussed in the papers by Baldwin [6]; Tsukamoto[8]; Mizumoto, Fukami, and Tanaka [7]. Fuzzy and vague knowledge are the basics for fuzzy logic. These arise through the interaction of an observer with the system under observation. Greater interaction concentrates only on certain parts or subsystems, hence clarity and increased precision in one subsystem is obtained at the expense of only vague and fuzzy information elsewhere. Thus, one is able to design adaptive man-machine systems, with a match between the competence of man to fuzzy environments and the machine to precise environments.

Problems of influence and reasoning with nebulous concepts are shown by Baldwin [6] and Mizumoto, et.al. [7], to be amenable to a fuzzy logic. Such are issues in many branches of human and social sciences. Baldwin [2] shows how a 'mixed input argument' can be formally tracked based on fuzzy logic. Thus, partial or restricted truth values are no longer based on a truth-functional modification and an inverse truth-functional modification.

Mizumoto, et.al. [7], consider various situations in which the methods of Zadeh and Mamdani are inadequate, e.g., statements such as x is *very A* implies Y is *very B*. By considering base relations between the universes of A and B, one is able to extend fuzzy conditional inferences. For example, their base universes are the same as in the case of first ten integers. Hence we can make statements if x is *small* Y is *middle*.

Nahmias [10] views fuzzy numbers as a special case of a more general concept of fuzzy variables. This he arrives by considering what a fuzzy set really is. He approaches this by constructing a mathematical structure analagous to the sample space of probability and names it the *pattern* space. Next he introduces the concept of a *scale*, which is a real valued set function from the discrete topology (defined on a set of base patterns) to the interval (0,1). Kandel [11], however, considers fuzzy variables as representing a more general case of boolean variables, and thus is able to extend boolean methods to fuzzy situations. Yet, when considering the problems presented by a data-set, such as ascertaining the central tendency, both Nahmias [10] and Kandel [11] base their approach on the triple (X,Γ,μ) where X is the set of base patterns, Γ is a discrete topology (Nahmias) or a Borel field (Kandel), μ is the scale function (Nahmias) or the fuzzy measure (Kandel). The function μ has identical requirements in both cases.

In this book Nahmias [10] is concerned with a problem of simultaneous occurrence of fuzziness and randomness, a problem first noted by Zadeh in 1968. This theme occurs in many papers in this book. Hirota [12]; Ralescu [5]; De Luca and Termini [19]; Dubois and Prade [17]; Tanaka, Okuda, and Asai [18]; and Kokawa, Nakamura and Oda [27].

In another paper Kandel [15] is concerned with the study of imprecise functions, their properties and possible applications, and introduces the theory of fuzzy switching mechanisms. Wechler [16] presents the development of fuzzy set theory based on many-valued logic. He found this approach very useful for investigating generalized languages.

Another analogous development, arising from this oft repeated criticism that fuzzy set theory is just probability theory in a new garb, is that of possibility theory. It was again Zadeh who saw the richness of exploring this concept further. It turns out, as was observed earlier, that possibility theory is more general than probability theory. Possibilities are like alternatives in decision theory. Probabilities relate to choices which reduce the number of alternatives to be considered.

Papers by Hisdal [13] and Nguyen [14] discuss new aspects of the theory of possibility, while the paper by Fiskel, et.al. [31] show an interesting application. Yet another concept with ties to probability theory is that of entropy. De Luca and Termini [27], Kokawa, et.al. [17], and Tanaka, et.al. [18] have applications in decision processes.

The basic decision theory model for fuzzy situations is considered in Tanaka, et.al. [18] as well as in Dubois and Prade [17].

Dubois and Prade [17] show the use of fuzzy numbers in extending the decision making model of decision theory. They consider both fuzzy utilities which are fuzzy numbers and linguistic probabilities. Thus, definitive methods are evolved in decision making when fuzziness blurs alternatives. Some do not require much computation and hence are quite appealing. Much work remains to be done in situations where fuzzy variables (in the sense of Nahmias) arise or where decisions are made based on fuzzy statistics instead of on the regular expected value methods.

Butnariu [20] presents a mathematical approach to the problems of rational behavior in n-persons fuzzy games encountered in political, economical and other environments. He gives a heuristical discussion and discusses a comparative analysis between fuzzy and non-fuzzy games.

In applications of fuzzy set concepts, one is often concerned with the problem of measurement of membership functions. Nowakowska [21] introduces a formalism to describe the structure of such concepts. She also shows that well-known psychological test methods are applicable to membership measurement.

Hersh, Caramazza and Brownell [22] investigate empirically the effects of context variation in describing terms such as *large* and *small*. To compare the results, they have successfully adapted a technique from multiple regression and probit analysis.

Oden [23] surveys his work in developing and empirically testing fuzzy propositional models of human semantic information processing. In this a new cognitive construct is proposed: that of *human fuzziness competency*.

Sanchez [24] presents some results related to sup-min (or max-min) compositions of fuzzy relations. The results of this paper are being used in a companion application paper [25] on medical diagnosis. Fordon and Bezdek [26] deal with a similar problem of organizing and analyzing large amounts of data in medical diagnosis. They demonstrate the inadequacy of the conventional parametric approach and give a successful application of fuzzy set theory in medical diagnosis.

Willaeys and Malvache [28] address a central problem in man-machine system design. How does one match a machine to human needs, when the latter are imprecisely known ahead of time? Such a study contributes to better systems development and allows for adaptivity to the human operator.

Most theories in social sciences are formulated verbally, making possible general and approximate statements. Yet, there is a barrier to wide ranging formal deduction due to the large number of assumptions and contexts. Numerical exemplification leads to unmeaningful constructs. Thus, numerical type simulations are not always helpful.

Wenstop [29] proposes a promising linguistic simulation methodology in a language called ZL - 2 based on APL. A parallel approach termed linguistic modeling is developed by Kickert [30].

To a systems modeler, the construction of systems models is a process, and as such this process calls for stages at which it can be evaluated. If one has a precise model, then why does one require a fuzzy model? The answer is in those contexts where operators in on-line control situations (on the shop-floor perhaps) need general descriptive qualitative guidelines, that they may adapt to unpredictable system uncertainties, in a simple yet clear verbal description. Thus, one requires verbal precision while not requiring numberical precision. Tong [32] approaches this problem in terms of modeling through finite discrete relations in a three stage process: verbalization, fuzzification and identification. Two well-known models: that of a gas furnace model of Box and Jenkins via time series analysis, and that of Beck's model for river quality modeling are tested for the proposed methodology. Although the test shows that fuzzy models are not better performers, they demonstrate how close to numerical models one might get by a set of simple rules and guidelines.

Jain [33] demonstrates a successful application of methods of approximate reasoning and fuzzy logic to integrate the imprecise knowledge concerning different pieces of complex scenes. Fuzzy properties such as monotonicities or fillness of a sequence are used to track motion of objects in complex real world scenes.

The work of Dubois and Prade [17] on fuzzy numbers is of particular relevance in systems modeling. This provides a convenient way of describing quasi-numeric and qualitative control approaches in industrial settings, which are natural to most "on the shop floor" operators. Thus, policies may be set to be soft rather than hard, allowing local adaptibility. Fiskel, Diffenbach and Renda [31] present an interesting application of possibility theory to explain policy issues. The methodology concerns tracing higher order impacts in potential future scenarios. Policy assessment is based on fuzzy impact chains affecting future scenarios.

Weiss and Donnell [34] consider another application to policy issues. It is the generation of a general purpose (as opposed to task-specific) "guide lines" development in management and planning. They use linguistic production rules.

Finally, Gustafson and Kessel [35] develop a class of fuzzy ISODATA clustering algorithm. They generalize this algorithm to include fuzzy convariances, and present an application to clustering problem.

In summary, we remark that a "take-off" stage has been reached for accelerated growth in numbers of applications, which range from problems in inference via natural languages, to those in operations research, systems analysis, cognitive science, decision analysis, classification and cluster analysis, control statistical and fuzzy inference, medical diagnosis, learning *systems*, and instructional systems, policy applications, medical geography and thus, in the biological, medical social and urban sciences. Readers may find the extensive bibliography appended with this Volume to be very helpful in exploring further theoretical developments and applications in this growing fuzzy field; this Volume is just a small sample of this.

LIST OF BOOKS ON FUZZY SYSTEMS

1. D. Dubois and H. Prade (1979): "Fuzzy Sets and Systems: Theory and Applications," Academic Press.

2. M.M. Gupta (Editor), G.N. Saridis and B.R. Gaines (Assoc. Editors) (1977): "Fuzzy Automata and Decision Processes," North-Holland.

3. A. Kaufmann (1972): "Theory of Fuzzy Sets," Merson, Paris.

4. A. Kaufmann (1975): "Theory of Fuzzy Subsets," Academic Press, N.Y.

5. A. Kaufmann (1973): "Introduction a la Theorie des Sous-Ensembles Flous, 1: Elements Theoretiques de Base," Masson et cie, Paris, France.

6. A. Kaufmann (1975A): "Introduction a la Theorie des Sous-Ensembles Flous, 2: Applications a la Linguistique et a la Semantique," Masson et cie, Paris, France.

7. A. Kaufmann (1975B): "Introduction a la Theorie des Sous-Ensembles Flous, 3: Applications a la Classification et la Reconnaisance des Formes, Aux Automates et Aux Systems, Aux Choix des Critares," Masson et cie, Paris, France.

8. A. Kaufmann (1975): "Introduction to the Theory of Fuzzy Subsets Vol. 1," Academic Press, New York.

9. A. Kaufmann (1976): "Introduction a la Theorie des Sous-Ensembles Flous," Tomes IV, Ed. Masson, Paris.

10. W.J.M. Kickert (1978): "Fuzzy Theories on Decision Making," Nijhoff, Leiden, Boston.

11. C.V. Negoita and D.A. Ralescu (1975): "Applications of Fuzzy Sets to Systems Analysis," Birkhauser Verlag.

12. L. A. Zadeh , K.S. Fu, K. Tanaka, and M. Shimura (eds) (1975): "Fuzzy Sets and Their Applications to Cognitive and Decision Processes," Academic Press.

13. International Journal of Fuzzy Sets and Systems, North-Holland.

ADVANCES IN FUZZY SET THEORY AND APPLICATIONS
M.M. Gupta, R.K. Ragade, R.R. Yager (editors)

OUTLINE OF FUZZY SET THEORY : AN INTRODUCTION

Didier DUBOIS
Laboratoire d'Informatique
Institut de Mathématiques
Appliquées de Grenoble
B.P. 53, 38041 Grenoble Cedex
FRANCE

Henri PRADE
Languages et Systèmes Informatiques
Université Paul Sabatier
118 route de Narbonne
31077 Toulouse Cedex
FRANCE

I. INTRODUCTION

Since its emergence in 1965, with an outstanding paper by L.A. Zadeh [Z1], fuzzy sets theory has gained ground. By now, more than one thousand works dealing with this topic have been published, and hundreds of researchers all over the world are still working on the theory itself or its applications. Besides, Zadeh's own contributions to the field have kept on putting forward new key-concepts.

The elaboration of the theory has led to question many admitted opinions and thinking habits, and to infringe upon the domains investigated by probability theory. Besides, the ultimate aim of fuzzy sets theory would be, according to Zadeh, to represent how the human mind perceives and manipulates information: "Indeed the pervasiveness of fuzziness in human thought processes suggests that much of the logic behind human reasoning is not the traditional two-valued or even multivalued logic, but a logic with fuzzy truths, fuzzy connectives, fuzzy rules of inference" [Z2].

The present paper is an extensive overview of existing results and future prospects of fuzzy sets theory and its applications, although may opened problems remain unsolved. This paper does not assume any specific prerequisite to be read. No proof explicitly appears in the text, for the sake of brevity. The reader is referred to the appended bibliography. However, it is hoped that clarity and precision are preserved, and his intuition does not wander too much.

As a matter of fact, the first section is intuitive, and discusses what fuzziness may mean. Section two is devoted to the mathematics of fuzzy sets theory in its various facets. The third part surveys the main fields of existing and potential applications.

II. INTUITIVE SETTING

Many philosophers and scientists have long since tried to investigate problems pertaining to uncertainty, ambiguity, vagueness. Actually, until recently, no mathematical formalism dealing with such questions existed - except probability theory. Owing to Zadeh's fruitful attempts, a theory of vagueness - what Zadeh calls "fuzziness" - is now sufficiently developed to stimulate researchers from a wide range of areas, although a lot of work remains to be done in this framework.

Fuzziness does bear a precise meaning - without joking! It must be distinguished from generality, ambiguity and error. A symbol is called "general" when it applies to a multiplicity of objects and retains only a common essential feature, it is ambiguous when it may denote several unrelated objects. But the fuzziness of a symbol lies in the lack of precise boundaries of the set of objects to which it applies. Besides, in tolerance analysis, the possibility of some error on the value of a parameter is expressed by an interval in which this value may lie. Fuzziness may occur only when this interval has no definite bounds.

For instance, consider the set X of people living in a given town, and the subset $\tilde{A}$ of X including only tall people. Obviously some people in X will be definitely tall, others definitely "not tall", but there will be borderline cases. Traditionally, members of the first category are assigned a membership grade 1, those of the second one, a membership grade 0. Very naturally, borderline cases are assigned numbers $\mu_{\tilde{A}}(x)$ lying between 0 and 1. The more an element x of X belongs to $\tilde{A}$, the closer to 1 is its membership grade. $\tilde{A}$ is called a fuzzy set. Obviously the choice of [0,1] as a numerical scale of membership is a matter of suitability. Precise membership values do not convey any absolute significance, they are context dependent and can be subjectively assessed. They only reflect an ordering of the universe (the reference set X) with respect to a vague predicate (the fuzzy set $\tilde{A}$). This ordering is more important than membership values by themselves.

It is easy to realize that man frequently uses fuzzy concept when he perceives the outside world and when he thinks. It seems that the human brain processes names of fuzzy sets ("tall", "large", "long", "similar to") more easily than numbers. Besides, quoting Zadeh [Z2], "one of the most important facets of human thinking is its ability to summarize information into labels of fuzzy sets which bear an approximate relation to the primary data". Linguistic descriptions of complex situations or strategies generally include fuzzy denotations. To be able to represent and manipulate them yield tools for modeling humanistic or man-machine systems in which information is often vague and/or whose complexity is a challenge for usual techniques.

Sugeno [S1] used the word "fuzziness" in a quite different context : considering an unlocated object x in a universe X, any classical subset of X is assigned a value $g_x(A) \in [0,1]$, called "grade of fuzziness" of the assertion "x belongs to A". Actually this number evaluates the certainty of the location of x in or out of A. According to the axioms defining the measure g_x, $g_x(A)$ can be the probability, belief degree, possibility for x to be in A. It is implicitly assumed that once x is located it will belong to A or its complement exclusively. A natural assumption for g_x is the monotonicity with respect to set inclusion, that is

$$\forall A, B \subseteq X, A \subseteq B \Longrightarrow g_x(A) \leq g_x(B) \quad (1)$$

this assumption is less strong than that of additivity which restricts g_x to a probability measure. It has been very often noticed that additivity is not consistent with human guessing (see Shafer [S2] for instance). By convention $g_x(\emptyset) = 0$ and $g_x(X) = 1$, express that x cannot belong to the empty set and must belong to the reference set. g_x, with the monocity assumption and this above convention, is called a fuzzy measure.

Fuzzy measures of fuzzy sets can be defined by means of a so-called "fuzzy integral" (Sugeno [S1]), thus clearly indicating both notions are distinct. They are also dual concepts since a fuzzy set A is defined by the set of numbers $\{\mu_A(x) | x \in X\}$ and a fuzzy measure g_x by the set $\{g_x(A) | A \subseteq X\}$. Actually, "fuzzy sets are closely related to a special class of fuzzy measures called "possibility measures" (Zadeh [Z3]).

As an illustration, X is supposed to be a set of pieces of furniture. Both points of view correspond to the following situations :

* <u>Fuzzy set</u> : the age of each item is precisely known. $\mu_{\tilde{A}}(x)$ is an assessment of the answer to the question : "you know the age of x ; do you consider it is old ?" $\tilde{A}$ is the fuzzy set of old pieces of furniture.

* <u>Fuzzy measure</u> : the age of each item is unknown. $g_x(A)$ is an assessment of the answer to the question : "By looking at x, do you consider it is more than 200 years old?" A is the non fuzzy set of pieces of furniture being exactly more than 200 years old.

The combined situation would yield $g_x(\tilde{A})$, an assessment of the answer to the question "by looking at x, do you consider it is old" ?

III. THE MATHEMATICS OF FUZZY SETS THEORY

We successively deal with sets, measures, relations and functions, and lastly the extension principle.

A) FUZZY SETS

a) Basic definitions

Let X be a set, from now on called universe. The characteristic function μ_A of a classical subset A of X takes its values in the two-element set {0,1}, and is such that $\mu_A(x) = 1$ if $x \in A$ and 0 otherwise. A fuzzy set A has a characteristic function taking its values in the interval [0,1]. μ_A is also called a membership function. $\mu_A(x)$ is the grade of membership of $x \in X$ in A.
A is symbolically denoted $\{(x, \mu_A(x)) | x \in X\}$. Other convenient notations are :

$$A = \int_X \mu_A(x)/x \quad \text{when X is a continuum}$$

$$A = \mu_A(x_1)/x_1 + \ldots + \mu_A(x_n)/x_n = \sum_{i=1}^{n} \mu_A(x_i)/x_i \quad \text{when X has n elements}$$

$(|X| = n)$.

In a fuzzy set, the transition between membership and non-membership is gradual rather than abrupt. Besides, note that X is not fuzzy.

$\{x | \mu_A(x) > 0\} = \text{Supp}(A)$ is called the <u>support</u> of A. The height of a fuzzy set A is $\text{hgt}(A) = \sup_{x \in X} \mu_A(x)$. hgt(A) evaluates the possibility of finding in X at least one element which fits the predicate A exactly. A is said normalized when such an element can be exhibited, that is $\mu_A(a) = 1$ for some a ; in that case, hgt(A) = 1.

<u>NB</u>. an often debated problem is the one of estimating membership values.
A qualitative estimation reflecting a given ordering is sufficient. However, one may use a similarity index to a prototype, or even some more sophisticated methods such as the one by Saaty [S3].

b) Set-theoretic operators

Several noticeable structures can be defined on the interval [0,1], each of which induces fuzzy set union and intersection operator, which coincide with the classical ones. Zadeh first used the lattice {[0,1], max, min} which remains the most frequently employed, i.e.

union : $\mu_{A \cup B}(x) = \max(\mu_A(x), \mu_B(x))$ (2)

intersection : $\mu_{A \cap B}(x) = \min(\mu_A(x), \mu_B(x))$ (3)

Theoretical justifications within specific contexts can be found in Bellman and Giertz [B1], Fung and Fu [F1].

Another possible choice, which sounds probabilistic, is (Zadeh [Z1])

: $\mu_{A.B}(x) = \mu_A(x) \cdot \mu_B(x)$ (product) (4)

: $\mu_{A \hat{+} B}(x) = \mu_A(x) + \mu_B(x) - \mu_A(x) \cdot \mu_B(x)$ (5)

These operators are said to be interactive, any change in A or B induces a change in A.B and $A \hat{+} B$.

Lastly we may also choose (Dubois Prade [D0], [D1]) :

union : $\mu_{A \Cup B}(x) = \min(1, \mu_A(x) + \mu_B(x))$ (Zadeh's bounded sum [Z4])

intersection : $\mu_{A \Cap B}(x) = \max(0, \mu_A(x) + \mu_B(x) - 1)$ (7)

The choice of a pair of operators is a matter of context.

$\cap$, $\cup$, $\hat{+}$, ., $\Cup$, $\Cap$ are commutative, associative, and satisfy

$A \perp \emptyset = A \qquad A \top \emptyset = \emptyset$

$A \perp X = X \qquad A \top X = A$

where ($\perp$, $\top$) denotes any of the pairs of union and intersection introduced above. Defining the complement of A, as $\bar{A}$ such that $\mu_{\bar{A}}(x) = 1 - \mu_A(x)$, it is obvious that the complementation is involutive, and that the pairs of operators satisfy De Morgan's Laws, that is :

$\overline{A \perp B} = \bar{A} \top \bar{B}$

$\overline{A \top B} = \bar{A} \perp \bar{B}$

Denoting $\tilde{\mathcal{P}}(X)$ the set of fuzzy sets in X, ($\tilde{\mathcal{P}}(X)$, $\cup$, $\cap$, $^{-}$) is a pseudo-complemented distributive lattice : $\cap$ and $\cup$ are idempotent and mutually distributive, but $A \cup \bar{A} \neq X$ and $A \cap \bar{A} \neq \emptyset$. The excluded middle laws no longer hold. An interpretation of this property can be : due to the lack of precise boundaries, complementary subsets are overlapping and cannot cover X perfectly. ($\tilde{\mathcal{P}}(X)$, $\hat{+}$, $\cdot$, $^{-}$) is a non complemented non distributive structure, $\hat{+}$ and $\cdot$ are not idempotent. ($\tilde{\mathcal{P}}(X)$, $\Cup$, $\Cap$, $^{-}$) is a complemented non distributive structure. $\Cup$ and $\Cap$ satisfy the excluded middle laws but are neither mutually distributive nor idempotent. $\hat{+}$, $\cdot$, $\Cup$, $\Cap$ do not satisfy absorption laws. As a matter of fact, it can be proved that an incompatibility exists between the excluded middle laws on one hand, and both distributivity and idempotency on the other hand, as soon as membership is gradual rather than abrupt.

<u>NB</u>. Each of the above introduced structures is closely related to a particular multivalent logic. Translate the union, intersection and complementation operators into the disjunction ($\vee$), conjunction ($\wedge$), and negation ($\neg$) connectives, and define the implication by the formal identity $P \to Q = \neg P \vee Q$. P and Q are fuzzy predicates, that is $v(P), v(Q) \in [0,1]$. We get $v(\neg P) = 1 - v(P)$, and :

- corresponding to ($\cup$,$\cap$) disjunction : $v(P \vee Q) = \max(v(P), v(Q))$ (8)

  conjunction : $v(P \wedge Q) = \min(v(P), v(Q))$ (9)

  implication : $v(P \to Q) = \max(1 - v(P), v(Q))$ (10)

- corresponding to ($\hat{+}$,$\cdot$) disjunction : $v(P \curlyvee Q) = v(P) + v(Q) - v(P).v(Q)$ (11)

  conjunction : $v(P \,\&\, Q) = v(P).v(Q)$ (12)

  implication : $v(P \twoheadrightarrow Q) = 1 - v(P) + v(P) \cdot v(Q)$ (13)

- corresponding to ($\Cup$, $\Cap$) (Giles [G1])

  disjunction : $v(P \veebar Q) = \min(1, v(P) + v(Q))$ (14)

  conjunction : $v(P \barwedge Q) = \max(0, v(P) + v(P) - 1)$ (15)

  implication : $v(P \Rightarrow Q) = \min(1, 1 - v(P) + v(Q))$ (16)

A more extensive study of these logics can be found in Dubois Prade [D0].

c) Inclusion-Equality

Usually, fuzzy set inclusion is defined by

$$\forall A, B \in \tilde{\mathcal{P}}(X) \quad A \subseteq B \text{ iff } \forall x \in X, \mu_A(x) \leq \mu_B(x) \qquad (17)$$

this inclusion is closely related to implication $\Longrightarrow$
the corresponding set equality is

$$\forall A, B \in \tilde{\mathcal{P}}(X),\ A = B \text{ iff } \forall x \in X, \mu_A(x) = \mu_B(x) \qquad (18)$$

Hence, Zadeh's fuzzy set theory [Z1] is closely related to an hybrid logical system using $\vee$, $\wedge$, $\neg$ as disjunction, conjunction and negation connectives together with implication $\Longrightarrow$. This logical system was studied in the thirties by Lukasiewicz under the name $L_{\text{aleph } 1}$.

Note that $A \cap\!\!\!\!\cdot\ B \subseteq A.B \subseteq A \cap B$ and $A \cup B \subseteq A \hat{+} B \subseteq A \uplus B$

Other inclusions and equalities can be introduced, for instance the ones related to implication$\rightarrow$. Besides (17) and (18) can be weakened into ε-inclusions and ε-equalities, using scalar inclusion or equality indices. For instance

$$A \subset_{\varepsilon} B \text{ iff } I(A,B) \geq \varepsilon \ (\varepsilon \in \,]0,1]\,). \qquad (19)$$

where I is an inclusion index. More details can be found in Dubois-Prade [DO].

d) Other important notions

* α-cut the α-cut A_α of a fuzzy set A is the classical set $\{x \in X, \mu_A(x) \geqslant \alpha\}$ for $\alpha \in]0,1]$. A can be expanded in terms of its α-cuts :

$$\forall x \in X,\ \mu_A(x) = \sup_{\alpha \in]0,1]} \min(\alpha, \mu_{A_\alpha}(x)) \qquad (20)$$

Moreover $(A \cup B)_\alpha = A_\alpha \cup B_\alpha$, $(A \cap B)_\alpha = A_\alpha \cap B_\alpha$, but $(\bar{A})_\alpha \neq \overline{(A_\alpha)}$

* Cardinal of a fuzzy set

The cardinality of the fuzzy set A is defined by $|A| = \sum_{x \in X} \mu_A(x)$ when A has a finite support. This definition can be extended to infinite supports: $\frac{|A|}{|X|}$ evaluates the proportion of elements of X having the property A when X is finite. Other definitions exist. For instance it may seem natural that the cardinal of a fuzzy set be fuzzy (see part D).

* Entropy of a fuzzy set

In order to evaluate how much fuzzy a fuzzy set A is, several authors have introduced measures of fuzziness. The corresponding indices "e" are such that e(A) = 0 iff A is not fuzzy. An exemple of such indices is the entropy

$$e(A) = -\sum_{x \in X} (\mu_A(x) \operatorname{Log}\mu_A(x) + (1-\mu_A(x)) \operatorname{Log}(1 - \mu_A(x))) \quad (X \text{ finite})$$

this non probabilistic entropy has numerous properties (De Luca, Termini [D2])
A more general formulation on a measurable set X can be found in Knopfmacher [K1].

* Fuzzy structured sets : subsets of X having algebraic structures can be turned into fuzzy sets which are closed in the sense of the structure. For instance a fuzzy monoïd M (Rosenfeld [R1]) can be defined on a monoïd X by the property :

$$\forall x, x' \in X,\ \mu_M(x * x') \geq \min(\mu_M(x), \mu_M(x'))$$

where * is the monoïd operation.

Similarly, if X is a euclidean space, a convex fuzzy set C is defined by (Zadeh [Z1])

$$\forall x, x' \in X,\ \forall \lambda \in [0,1],\ \mu_C(\lambda x + (1-\lambda)x') \geq \min(\mu_C(x), \mu_C(x'))$$

e) Other kinds of fuzzy sets

* when [0,1] is considered as a non suitable membership scale, it can be replaced by another set L equipped with a poorer structure, for instance a lattice.

The corresponding fuzzy sets are called L- fuzzy sets (Goguen [G2]).

* The use of precise grades of membership can be argued. Membership values $\mu_A(x)$ are themselves fuzzy, and can be represented as fuzzy sets of [0,1]. A is then said to be a type two fuzzy set, i.e. a L- fuzzy set with L = $\tilde{\mathcal{P}}([0,1])$(Mizumoto, Tanaka [M1]) when membership values are only closed intervals in [0,1], A is called a Φ-fuzzy set.

* A level two fuzzy set is a fuzzy set of fuzzy sets of X, i.e. a fuzzy set of $\tilde{\mathcal{P}}(X)$.

* A very simple way of relaxing the notion of crisp membership is to consider a set as containing a subset of central elements, while other are only peripheral. A is then a pair (E_1, E_2) where E_1 is the set of central elements, $E_1 \subseteq E_2$, and $E_2 - E_1$ is the set of peripheral elements. We get a special kind of L-fuzzy sets studied by Gentilhomme [G3].

B) FUZZY MEASURES - PROBABILITIES - POSSIBILITIES

We consider now the second point of view on fuzziness : the one of certainty degrees for assertions such as "$u \in A$" where A is a non fuzzy set and u is a variable whose value is not perfectly located in X, with respect to A.

a) Fuzzy measures

Let $\mathcal{P}(X)$ be the set of classical subsets of X and g a mapping from $\mathcal{P}(X)$ to [0,1]. g is called a fuzzy measure iff (Sugeno [S1]).

$$g(\emptyset) = 0 \; ; \; g(X) = 1 \tag{21}$$

$$\forall A, B \in \mathcal{P}(X) \text{ if } A \subseteq B \text{ then } g(A) \leq g(B) \text{ (monotonicity)} \tag{22}$$

Given a convergent sequence of sets $(A_i, i \in \mathbb{N})$ such that (23)
$A_1 \subseteq A_2 \ldots \subseteq A_n \subseteq \ldots$ or $A_1 \supseteq A_2 \supseteq \ldots \supseteq A_n \supseteq \ldots$ then $\lim_{i \to \infty} g(A_i) = g(\lim_{i \to \infty} A_i)$ (continuity).

NB. G is not a measure in the usual sense. It could also be defined on a Borel field. It is easy to see that $g(A \cup B) \geq \max(g(A), g(B))$;$g(A \cap B) \leq \min(g(A), g(B))$. The class of fuzzy measures is very large. Moreover, the specification of a fuzzy mea-sure requires the knowledge of g(A) for all subsets A in X. In order to reduce the quantity of primary data, an extra axiom can be added to 1)-2)-3) which allows the calculation of g(A) from $\{g(\{x\}) \mid x \in A\}$. Such an axiom was proposed by Sugeno [S1].

$$g_\lambda(A \cup B) = g_\lambda(A) + g_\lambda(B) + \lambda g_\lambda(A).g_\lambda(B) \qquad \lambda > -1 \tag{24}$$

when $A \cap B = \emptyset$.

g_λ is called a λ-fuzzy measure
it is easy to see that when $\lambda = 0$, g_0 is a probability measure, so that fuzzy measures are more general than probabilities.
Another possible axiom is :

$$g(A \cup B) = \max(g(A), g(B)), \forall A, B \subseteq X \tag{25}$$

g is then denoted Π and is called a possibility measure (Zadeh [Z3]). Other examples of fuzzy measures are Shafer [S2]'s belief functions (which include λ-fuzzy measures with $\lambda \geq 0$), and Shackle [S4]'s consonant belief functions (see also Shafer [S2]).

More details concerning these special fuzzy measures can be found in Banon [B2] and Dubois, Prade [DO].

The fuzzy measure of a fuzzy set (which is a scalar !) is defined by means of the so-called "fuzzy integral". Let h be a mapping from X to [0,1]. The fuzzy integral, in the sense of the fuzzy measure g, of h over a subset A of X is (Sugeno [S1])

$$\oint_A h(x) \circ g(\cdot) = \sup_{\alpha\in[0,1]} \min(\alpha, g(A \cap H_\alpha)) \tag{26}$$

where $H_\alpha = \{x \in X \mid h(x) \geq \alpha\}$

the fuzzy integral is a kind of weighted median (Kandel [K2]), and is similar to Lebesgue integral (Sugeno [S1]).

The fuzzy measure of the fuzzy set $\tilde{A} \in \tilde{\mathcal{S}}(X)$ is thus

$$g(\tilde{A}) = \oint_X \mu_{\tilde{A}}(x) \circ g(\cdot) = \sup_{\alpha\in[0,1]} \min(\alpha, g(\tilde{A}_\alpha)) \tag{27}$$

$g(\tilde{A})$ is the grade of certainty of the fuzzy event "u belongs to $\tilde{A}$".

N.B. (27) is very similar to (20). Hence $g(\tilde{A})$ may also be interpreted as the grade of membership of u in the (fuzzy) set of elements which "more or less surely belong" to $\tilde{A}$!

More properties of fuzzy measures can be found in Sugeno [S1], [S5], [S6].

b) Probabilities of fuzzy event (Zadeh [Z5])

Zadeh [Z5] defined the probability of a fuzzy event as the mathematical expectation of its membership function, i.e.

$$P(A) = \int_X \mu_A(x) p(x) dx \tag{28}$$

where A is a fuzzy set of X and p a probability distribution. When A is not fuzzy (28) gives back the usual P(A). Note that (28) is not a fuzzy integral in the sense of (27).

The following properties hold :

$$\begin{cases} P(A \cup B) = P(A)+P(B)-P(A \cap B), \forall A,B \in \tilde{\mathcal{S}}(X) \\ P(A \hat{+} B) = P(A)+P(B)-P(A.B), \forall A,B \in \tilde{\mathcal{S}}(X) \end{cases}$$

Two fuzzy events are independent when $P(A.B) = P(A).P(B)$ (and not $P(A \cap B) = \min(P(A), P(B))$, so that when A and B are independent, :

$$P(A|B) = \frac{P(A.B)}{P(B)} = P(A))$$

c) Theory of possibility (Zadeh [Z3])

To say that a variable u on X ranges over a non fuzzy subset A of X is equivalent, in terms of possibility, to

$$\Pi(u = x) = \pi_u(x) = 1 \text{ iff } x \in A \text{ and } 0 \text{ otherwise.}$$

The extension of this representation to a fuzzy set A yields naturally :

$$\Pi(u = x) = \pi_u(x) = \mu_A(x)$$

In some situations it may be interesting to interpret the membership function μ_A of a fuzzy set as a possibility distribution of a variable u. A is viewed as the set of more or less possible values for u.

Given a possibility distribution π_u, the possibility for x to belong to a non fuzzy set B is defined as : $\Pi(u \in B) = \sup_{x \in B} \pi_u(x)$. Which is consistent with our intuition of the possibility of any of several events ($\exists x \in B$, $u = x$) as the possibility of the most possible one.

Obviously $\Pi(u \in B \cup C) = \max(\Pi(B), \Pi(C)) = \Pi(B \cup C)$. Hence provided that $\sup_{x \in X} \pi_u(x) = 1$, Π is a possibility measure. A possibility distribution specifying a possibility measure is formally equivalent to a normalized fuzzy set.

NB. If A is a non normalized fuzzy set it is possible to define the corresponding possibility distribution as : $\pi(x) = \mu_A(x)/\text{hgt}(A)$; keeping $\pi(x) = \mu_A(x)$ implies $\Pi(X) < 1$ which implicitly locates the ideal element a such that $\mu_A(a) = \pi(u=a) = 1$ outside X. Using the fuzzy integral to calculate the possibility of a fuzzy event B yields (Dubois-Prade [DO]) : $\Pi(B) = \sup_{x \in X} \min(\mu_A(x), \pi(x))$ (29)

Also called consistency degree of μ_A and π.

Note that $\sup_{x \in X} \pi_u(x) = 1$ is similar to the probabilistic $\sum_{x \in X} p(x) = 1$ and (29) is similar to (28).

d) Probability - possibility consistency principle

It is generally admitted that what is possible may or not be probable, but what is probable must be possible, that is "probable implies possible". This general principle can be mathematically stated as :

$$\forall A \in \mathcal{P}(X), \ \Pi(A) \geq P(A) \tag{30}$$

which is consistent with implication $\Longrightarrow$ in A).

As a consequence, let h be a histogram function characteristic of the frequency of occurrence of the events $x_i \in X = \{x_1, \ldots, x_n\}$. (30) translates into :

$$\forall I \subseteq \{1,\ldots,n\}, \quad \frac{\sup_{i \in I} h(x_i)}{\sup_{i=1,n} h(x_i)} \geq \frac{\frac{\sum_{i \in I} h(x_i)}{n}}{\sum_{i=1} h(x_i)} \tag{31}$$

(31) clearly does not hold for any h. It represents an homogeneity condition on the population whose histogram is h. (31) forbids the existence of "large" $h(x_i)$ together with other small but not negligible ones.

Remark : conditional possibilities have recently received attention from Nguyen [N1] and Hisdal [H1]. Conditional fuzzy measures were previously investigated by Sugeno [S5], [S6].

C) FUZZY RELATIONS - FUZZY FUNCTIONS

In A) and B) we dealt with sets and measures. Now we consider how fuzziness modifies the concept of a correspondence.

a) Fuzzy relations

For the sake of simplicity we restrict ourselves to binary relations. A fuzzy relation R is a fuzzy set in a Cartesian product $X \times Y$ of universes X and Y. $\mu_R(x,y)$, the membership value of (x,y) in R estimates the strength of the link between x and y. Two major references on the topic are Zadeh [Z6], [Z7]. Fuzzy relations generalize ordinary relations. As such, they can be composed : Let R and S be two fuzzy relations on $X \times Y$ and $Y \times Z$ respectively, the membership function of the fuzzy relation R o S, on X × Z is defined by :

$$\mu_{RoS}(x,z) = \sup_{y \in Y} \min (\mu_R(x,y), \mu_S(y,z)) \quad (32)$$

(32) is interpreted as follows : a link between x and z is established through y, the strength of this link is that of the weakest of links x-y and y-z. But since any y may be used, the global link between x and z is that of the best chain x-y-z.

<u>NB</u>. Note that in (31) min could be replaced by a product or the algebraic operation underlying $\pitchfork$.

The <u>projection</u> of a fuzzy relation R on X is a fuzzy set $P_X(R)$ whose membership function is :

$$\mu_{P_X(R)}(x) = \sup_{y \in Y} \mu_R(x,y)$$

R is said <u>separable</u> iff :

$$\forall x, \forall y,\ \mu_R(x,y) = \min (\mu_{P_X(R)}(x), \mu_{P_Y(R)}(y)) \quad (33)$$

R can be interpreted as a fuzzy restriction on the value of a variable (u,v) ranging over (X × Y), i.e. R acts as an elastic constraint. When R is separable, u and v are said to be non-interactive in the sense that the choice of a value for u does not depend upon the choice of a value for v and conversely. Otherwise u and v are interactive.

A fuzzy binary relation R on X × Y is said to be :

reflexive	iff	$\forall x \in X,\ \mu_R(x,x) = 1$
symmetrical	iff	$\forall x,x' \in X, \mu_R(x,x') = \mu_R(x',x)$
anti-symmetrical	iff	$\forall x,x' \in X,\ x \neq x'$ and $\mu_R(x,x') > 0$, $\mu_R(x',x) = 0$
Δ-transitive	iff	$\forall x,x',x'' \in X, \mu_R(x,x'') \geq \mu_R(x,x') \Delta \mu_R(x',x'')$

reflexive and symmetrical fuzzy relations are called proximity relations. When R is transitive, its transitive closure is defined by

$$\hat{R} = R \cup R^2 \cup \dots \cup R^m \cup \dots \text{ where } R^m = R \text{ o } R^{m-1}$$

in the sense of sup-min composition. Due to the reflexivity of R, $\hat{R}$ exists and is min-transitive.

Reflexive, symmetrical and min-transitive fuzzy relations are called <u>similarities</u>. They can be obtained as transitive closures of proximities. Moreover, if R is a similarity, $\bar{R}$ is an ultrametric (see Zadeh [Z6]).

Reflexive, symmetrical, product-transitive fuzzy relations are weaker than similarities (see Zadeh [Z6]).

Reflexive, symmetrical, $\pitchfork$-transitives fuzzy relations have been recently considered by Bezdek and Harris [B3] and Ruspini [R2] under the name "likeness", they are weaker than product-transitive proximities, and closely related to pseudo-metrics.

A fuzzy partial ordering is a reflexive, antisymmetrical and min-transitive fuzzy relation. If antisymmetry does not hold, we only have a fuzzy preordering. Orlovsky [O1] showed that for any fuzzy preordering R on X × X,

$$\exists x \in X,\ \forall x' \in X,\ \mu_R(x,x') \geq \mu_R(x',x)$$

i.e. there always exist undominated elements in X. A fuzzy preordering is often interpreted as a fuzzy preference relation. Hence, fuzziness graduates preferences without blurring the choice of preferred elements.

Lastly, some works must be mentioned dealing with solution methods for fuzzy relation equations i.e. problems such as :

find R, knowing Q and S such that Q o R = S where o is the sup-min composition. Such problems appear in the context of diagnosis modeling (see Sanchez [S7], [S8], Tsukamoto, Terano [T1])

b) Fuzzy functions

There are several ways of viewing a fuzzy function, according to what is assumed to be fuzzy. We can distinguish 3 categories.

1) Non fuzzy functions with fuzzy domain and/or range (Negoita, Ralescu [NO]). Let f be a classical function from X to Y. The domain of f is the subset D of X containing x's such that f(x) exists. The range of f is $\{y \in Y | \exists x \in X, y = f(x)\}$ commonly denoted f(D) = I. When D and I are fuzzy, a consistency condition is

$$\forall x \in X, \mu_I(f(x)) \geq \mu_D(x)$$

Fuzzy functions (f,D,I) can model sentences such as : "Big trucks must go slowly" they are easily composed : (f,A,B) from X to Y composed with (g,B,C) from Y to Z yields (g o f, A,C) from X to Z. More details in [NO], [DO].

2) Fuzzy function of a non fuzzy variable or blurring function ([DO])

A blurring function $\tilde{f}$ from X to Y is a usual function from X to $\tilde{\mathcal{P}}(Y)$; the fuzzy image of x, $\tilde{f}(x)$, can be interpreted either as a fuzzy value or as the possibility distribution induced by a partially unknown or too-complex-to-be-precisely-defined function from X to Y.

$\tilde{f}$ can be viewed as a fuzzy relation R, such that $\tilde{f}(x) = \int_Y \mu_R(x,y)/y$ or as a fuzzy set of functions $\tilde{f} = \int_{Y^X} \mu_{\tilde{f}}(g)/g$ from X to Y.

When $\tilde{f}$ derives from a fuzzy relation R on X × Y and $\tilde{g}$ from a fuzzy relation Q on Y × Z, $\tilde{g}$ o $\tilde{f}$, the result of composing $\tilde{f}$ by $\tilde{g}$ is defined as deriving from R o Q (formula (32))

When $\tilde{f}$ is a fuzzy set of functions, it is equivalent to a multimodal fuzzy relation since it may exist some $x \in X$, g, $g' \in Y^X$ such that g(x) = g'(x) and $\mu_{\tilde{f}}(g) \neq \mu_{\tilde{f}}(g')$, i.e. $\mu_{\tilde{f}(x)}(g(x))$ can be either of $\mu_{\tilde{f}}(g)$ and $\mu_{\tilde{f}}(g')$. Hence a fuzzy set of functions is more general than a fuzzy relation; fuzzy sets of functions can also be composed owing to an extension principle (see D).

3) Function having fuzzy arguments and mapping on fuzzy sets

We way consider functions from $\tilde{\mathcal{P}}(X)$ to $\tilde{\mathcal{P}}(Y)$: $A \longrightarrow f(A)$ where A is a fuzzy set of X and f(A) a fuzzy set of Y. For instance X = Y and $f(A) = \bar{A}$, $\forall A \in \tilde{\mathcal{P}}(X)$. These functions can of course be composed as usual. Noticeable subclasses of such functions comprise extended ordinary functions from X to Y in the sense of D, and fuzzy relations R on X × Y, such that $A \longrightarrow A$ o R in the sense of sup-min composition. But other kinds do exist (the complementation function does not belong to any of these subclasses), more details may be found in [DO].

Remark : the above typology of fuzzy functions can be imbedded into category theory (see [NO]). Also worth noticing is a great amount of papers dealing with fuzzy topology (for a bibliography see [L1], or [DO])

D) THE EXTENSION PRINCIPLE

The extension principle, introduced by Zadeh [Z8], is one of the most important tools of fuzzy sets theory. Owing to this principle, any mathematical relationship between non-fuzzy elements can be fitted to deal with fuzzy entities. In this paragraph, the extension principle is stated, and the main applications existing in the literature are listed and sometimes briefly discussed.

a) Statement of the principle

Let $A_1,\ldots,A_n$ be fuzzy sets over $X_1,\ldots,X_n$ respectively. Their cartesian product is defined by : $A_1 \times \ldots \times A_n = \int_{X_1\times\cdots\times X_n} \min_{i=1,n} \mu_{A_i}(x_i)/(x_1,\ldots,x_n)$

Let f a function from $X_1\times\ldots\times X_n$ to Y.

The fuzzy image B of $A_1,\ldots,A_n$ through f has a membership function :

$$\forall y \in Y,\quad \mu_B(y) = \sup_{\substack{(x_1,\ldots,x_n)\\ \in X_1\times\ldots\times X_n}} \min_{i=1,n} \mu_{A_i}(x_i) \qquad (34)$$

under the constraint $y = f(x_1,\ldots,x_n)$
and the additional condition $\mu_B(y) = 0$ when $f^{-1}(y) = \{x_1,\ldots,x_n \mid y = f(x_1\ldots x_n)\} = \emptyset$

Remarks

* (34) is similar to (29) ; indeed (34) may be interpreted in terms of possibilities.
* (34) is a particular case of the composition of fuzzy relations ; we have :

$B = (A_1 \times A_2 \times \ldots \times A_n)\mathrm{o}F$ where F is the crisp relation

$\mu_F(x_1,\ldots,x_n,y) = 1$ iff $y = f(x_1,\ldots,x_n)$ and 0 otherwise.

* Other constraints may be added to "$y = f(x_1,\ldots,x_n)$" in (34) when there are other relationship linking the x_i's. For instance $\sum_{i=1}^{n} x_i = 1$ when the x_i's are probabilities. The practical computation of (34) will then be made more difficult.
* In (34) "min" could be replaced by "$\cdot$" or $\max(0, . + . -1)$
* Of course when the A_i's are just elements of the X_i's, $B = f(A_1,\ldots,A_n) \in Y$.

b) Applications

* The composition of fuzzy sets of functions may be defined using (34). Let $\tilde{f}$ be a fuzzy set of Y^X and $\tilde{g}$ a fuzzy set of Z^Y the composed "function" $\tilde{g} \circledcirc \tilde{f}$ of $\tilde{f}$ and $\tilde{g}$ is a fuzzy set of Z^X such that

$$\mu_{\tilde{g}\circledcirc\tilde{f}}(h) = \sup_{u,v:h=v\,\mathrm{o}\,u} \min(\mu_{\tilde{f}}(u), \mu_{\tilde{g}}(v)) \qquad (35)$$

this formula generalizes (32) to multimodal fuzzy relations.

* A very important application area of the extension principle is algebra and especially real algebra. For, let "$*$" be a composition law in the set of real numbers $\mathbb{R}$. "$*$" can be extended according to (34) to a composition law $\circledast$ on $\tilde{\mathcal{P}}(\mathbb{R})$ more specifically, let A and B two fuzzy sets in $\mathbb{R}$ - two fuzzy numbers - A $\circledast$ B is defined by

$$\mu_{A \circledast B}(z) = \sup_{x,y:z=x*y} \min(\mu_A(x), \mu_B(y))$$

Fuzzy real algebra has been extensively studied by Dubois and Prade ([D3],[D4],[D0]). It can be showed that the calculus on fuzzy numbers according to the extension principle (34) is a generalization of interval calculus in tolerance analysis. Moreover fast calculation formulae for extended addition, subtraction,product, division, max, min of fuzzy numbers have been found, which make possible the use of fuzzy algebra in pratical problems with fuzzy data (see also [D5] in this book).

* the consistency degree of two fuzzy sets A and B can be expressed as

$$\sup_{x=y} \min(\mu_A(x), \mu_B(y)) = \text{hgt}(A \cap B)$$

It is the application of the extension principle to the equality. Applying the same approach to the inequality $x \leq y$ on $\mathbb{R}$ yields a method for comparing fuzzy numbers (see [D0], [D8], or [D5] in this book).

* Real equations can be canonically extended to allow fuzzy coefficients. Their solutions are obtained by applying (34). Consider for instance the equation $a * x = b$ where $*$ is a group operation. If a and b are only approximately known and represented by fuzzy sets A and B, then the solution is a fuzzy set $\tilde{x}$ such that

$$\mu_{\tilde{x}}(x) = \sup_{a,b\ :\ a*x = b} \min(\mu_A(a), \mu_B(b)) \tag{36}$$

And since $a * x = b \Longleftrightarrow x = a^{-1} * b$, we get $\tilde{x} = A^{-1} \circledast B$

with $\mu_{A^{-1}}(u) = \sup_{a,\, a=u^{-1}} \mu_A(a) = \mu_A(u^{-1})$

<u>NB</u>. Extended equations must not be confused with fuzzy equations with a fuzzy unkown $\tilde{x}$, such as $\tilde{a} \circledast \tilde{x} = \tilde{b}$ (in the sense of set-equality), $\tilde{a} \circledast \tilde{x} \subseteq \tilde{b}$ (in the sense of set-inclusion) and $\tilde{a} \circledast \tilde{x} =_{\varepsilon} \tilde{b}$ (in the sense of a scalar similarity index S : $S(\tilde{a} \circledast \tilde{x}, \tilde{b}) \geq \varepsilon$). These kinds of equations are considered in [D6].

* The application of (34) to the connectives of propositionnal calculus, extended to fuzzy truth values which are normalized fuzzy sets of {0,1} has been carried out by Gaines [G4]. $\mu_P(0)$ and $\mu_P(1)$ are respectively the degree of falsity and truth of a proposition P. Assigning to P, the average truth value $v(P) = \frac{1-\mu_P(0)+\mu_P(1)}{2}$, the multivalent logic $(\vee, \wedge, \rightarrow)$ already mentioned in II)A)b) NB is obtained. Such multivalent logics can be extended to fuzzy truth values which are fuzzy sets over [0,1], and linguistically interpreted (see [D7]). Besides Zadeh [Z12] has introduced the idea of a relative (or local) truth value τ of a fuzzy predicate P with respect to another one Q taken as a reference. τ is a fuzzy compatibility value of P with Q, and is viewed as an extended membership value in P, that is

$$\mu_\tau(u) = \sup_{\substack{x \\ u=\mu_P(x)}} \mu_Q(x) \quad , \quad \forall u \in [0,1]$$

* The extension principle can also be applied to define the fuzzy maximum value of a function over a fuzzy domain (Orlovski [O2]), the fuzzy cardinality of a fuzzy set (Zadeh [Z12]),the fuzzy probability (resp : possibility) of a fuzzy event, the integral of a function over a fuzzy domain (Dubois Prade [D0]). In each case the fuzzy domain A is viewed as a fuzzy set of sets whose support is :

$$\text{supp}(A) = \{A_\alpha \mid \alpha \in]0,1]\}.$$

For instance the fuzzy cardinality of A is :

$$c(A) = \int_{\alpha \in]0,1]} \alpha / |A_\alpha|$$

The fuzzy probability of a fuzzy event A is $\tilde{P}(A) = \int_{\alpha \in]0,1]} \alpha / P(A_\alpha)$, etc...

* Many others mathematical concepts can be extended using (34) ; for instance

- the concept of distance between points becomes a fuzzy distance between fuzzy sets (Dubois Prade [DO]).
- the integral in the sense of Riemann of fuzzy real functions $\mathbb{R} \ni x \longrightarrow \tilde{f}(x) \in \tilde{\mathcal{P}}(\mathbb{R})$ over fuzzy or non fuzzy intervals has been studied by Dubois and Prade [D8], and their differentiation too.

IV. APPLICATIONS AND PROSPECTS

This part only intends to describe the main trends in past and present research concerning the applications of fuzzy sets theory. A more extensive bibliography can be found in [GO], [DO].

A) FORMAL SYSTEMS

The introduction of fuzziness in order to extend such well-known formal constructs as automata, formal grammars and algorithms is not new, since pioneering papers on these topics are respectively Wee and Fu [W1], Lee and Zadeh [L2], and Zadeh [Z10].

a) Fuzzy systems

The idea of a fuzzy system is due to Zadeh (for instance [Z9]). In short, fuzzy systems generalize non deterministic systems in the sense of a quantification of possible simultaneous behaviors. State transition and ouptut decision are governed in terms of sup-min compositions of those relations with fuzzy inputs (resp:states) which are fuzzy sets of inputs (resp : states). Such systems have been studied mainly in the framework of classical system theory and on a very abstract level. The reader is referred to Chang and Zadeh [C1], Negoita and Ralescu [NO], Arbib and Manes [A1].

An abundant literature deals with fuzzy automata, i.e. finite state fuzzy systems (Wee and Fu [W1], Santos [S9]) and their applications.

b) Fuzzy grammars and languages

A fuzzy language is a fuzzy set of the set of all finite strings obtained from a given alphabet (induding the null string), the membership value of a string expresses to what extent it is proper. Fuzzy languages can be generated from fuzzy grammars which are formal grammars whose production rules are valuated according to their properness. The grade of properness of a derivation chain yielding a given string x is the least grade of properness of the production rules involved, the membership value $\mu(x)$ is the grade of properness of the best derivation. Regular fuzzy languages (i.e. generated by regular fuzzy grammars) are accepted by fuzzy automata. This topic has been discussed at length in the literature, and many alternative models have been proposed (Santos [S10], Mizumoto et al. [M2]).

c) Fuzzy algorithms

Fuzzy Turing machines and Markov algorithms have been extensively studied by Santos [S11] on a rather abstract level. On the contrary, viewing a fuzzy algorithm as a sequence of fuzzy statements (instruction involving names of fuzzy sets) two kinds of fuzzy algorithms exist :

- those which manipulate fuzzy data in a deterministic fashion (owing to extended operations in the sense of (34)). The outputs are possibly fuzzy, such algorithms can be found in operations research for instance.

- those which manipulate non fuzzy data according to fuzzy rules. The results are generally precise, from this latter point of view, fuzzy algorithms are similar to unprecise rules of thumb employed by people in the performance of precise tasks (Zadeh [Z10]). Such algorithms could be useful in Artificial Intelligence and Robotics.

The above dichotomy does not forbid algorithms of a hybrid kind. Moreover, it reminds the well known opposition between procedural and declarative representations of knowledge in A.I..

B) NATURAL LANGAGE AND APPROXIMATE REASONING MODELS

Many sentences in natural language involve fuzzy denotations. Each fuzzy word can be associated with a possibility distribution over the objects to which it applies. For instance, the word "tall" in the sentence "Jack is tall" yields the possibility distribution $\pi = \mu_{tall}$ over Jack's possible sizes. In PRUF (Zadeh [Z13]), a meaning representation language for natural languages, sentences translate into possibility distributions over Cartesian products of universes of discourse which allow approximate inference from these sentences. Besides, fuzzy sets theory offers a conceptual framework for the representation of linguistic hedges such as "very" (Zadeh [Z11]).. If the membership function of A is μ_A, that of "very" A can be $(\mu_A)^2$; such assumptions, together with those underlying fuzzy set-theoretic operators have been experimentally verified (for instance Hersh and Caramazza [H2]). Much work is still to be done to identify suitable mathematical operators acting on possibility distributions (modification or aggregation), and expressing real life situations. Non fuzzy approximate inference models already exist in the literature : Bayesian, heuristic, multivalent logics-based models, etc... A new theory of approximate reasoning has been recently proposed by Zadeh and Bellman [B4], [Z12]. The basic tool is called generalized modus ponens, and works as follows, from proposition such as :

"X is A'" and "if X is A then Y is B"

where A, A', B are fuzzy sets A' has a meaning not very far from that of A(e.g : A = tall, A' = very tall) it is possible to calculate a fuzzy set B' such that "Y is B'" may be inferred from the two premisses.

The possibility distribution associated with the conditional proposition is found by means of an implication connective (see A.b, NB). For instance $\pi(x,y) = \min(1, 1 - \mu_A(x) + \mu_B(y))$ which is equivalent to a fuzzy relation R. B' is nothing but R o A'.

NB. When $A' = \bar{A}$ and $\bar{A}$ is normalized, B' is undeterminate ($\forall y, \mu_B(y) = 1$), i.e. any conclusion can be inferred. When A' = A, B' ≠ B due to fuzziness. However when A is crisp, A' = A implies B = B', as for the classical modus ponens.

C) APPLICATIONS TO SYSTEMS AND ARTIFICIAL INTELLIGENCE

a) Operations research

In many operation research problems issued from case studies, some of the data are often only approximately known. Such a lack of precision may greatly affect our knowledge of a solution ; moreover, the method of determination of this solution may itself depend upon the values of the data. Arbitrarily choosing those values when they are practically out of reach will yield unrealistic solutions. On the contrary, if such data are modelled by fuzzy numbers manipulated by extended operations (such as $\oplus$, $\odot$, $\widetilde{\max}$, etc...), the possibility distribution on the value of the solution can be obtained. An exemple of O.R. algorithm fitted to fuzzy data is given in Dubois and Prade [D9] (Ford's Algorithm for shortest paths, in a PERT problem) Although all O.R. algorithms are not so easy to fit to fuzzy data, one may think of considering such problems as the travelling salesman problem, Hitchcock problem, etc... from this point of view. Besides, Zimmermann [Z14] has dealt with soft (fuzzy) constraints (in the sense of (36) with A non fuzzy). Lastly, fuzzily-stated problems involving or not fuzzy data, could be solved by fuzzy algorithms (i.e. fuzzy solving rules).

b) Decision analysis :

A lot of works have been published dealing with decision analysis using fuzzy sets theory. A bibliography can be found in Dubois, Prade [DO]. Modelling real decision-making situations may involve fuzzy preference relations (e.g. Saaty [S12], Orlovsky [O1], fuzzy objective functions, fuzzy weightings and utilities (Jain [J1], Yager [Y1], fuzzy events (Okuda, Tanaka, Asai [O3]). Besides, fuzzy sets theory may be a helpful theoretic framework for the discussion of aggregation operators in multicriteria decision making or group opinion modelling (Bellman, Zadeh [B5], Fung, Fu [F1], Ragade [R3]. Lastly, the attractiveness of fuzzy evaluations of alternatives does not lie only on the realism of the resulting model but on the possibility of knowing when the final choice is blurred due to fuzziness.

c) Classification and clustering

The idea of a fuzzy set appeared in the framework of pattern classification problems (Bellman, Kalaba, Zadeh [B6]). The main point is that a pattern class is very often fuzzy by essence, and a pattern recognition (or classification) algorithm yields the membership value of an object under consideration in the pattern class, the feature values of the object can be linguistically assigned. The classification process can be described in terms of fuzzy rules subsumed in a fuzzy relation, the composition of the feature values with this relation yields the membership value of the object in the pattern class. The object is assigned to the pattern class to which it mostly belongs (Zadeh [Z15]). A lot of papers have been published on fuzzy pattern recognition : the semantic approach (Kotoh, Hiramatsu, [K3] Lee [L3]) and the syntactic approach (Thomason [T2], ET. Lee [L4], Kickert and Kopelaar [K4]). However, it seems that due to the rareness of convincing case studies, a great amount of work remains to be done in fuzzy pattern recognition. On the contrary the concept of a fuzzy partition (Ruspini [R4]) has somewhat renewed clustering analysis (Bezdek [B7], [B8], Dunn [D10], Ruspini [R5]) and has provided good methods in practice. A different but quite interesting approach based on the decomposition of multimodal fuzzy sets is proposed in Gitman, Levine [G5].

d) Control

Two kinds of applications of fuzzy sets theory to control theory exist :

- Bellman and Zadeh [B.5]'s approach in decision theory can be used in optimal control problems where the choice of optimality criteria is a matter of subjectivity and experience (Gluss [G6], Fung, Fu [F2]).

- The synthesis of linguistic control devices has been initiated by Mamdani and Assilian [M3] :the experimental control strategies are those of trained human operators faced to complex situations. They can be modelled by conditional rules such as "if X is A then Y is B" where A and B are fuzzy sets. The composition of the observed output of the process under control and the relation R built on the fuzzy rules yields a fuzzy control B' = A' o R to be applied, provided a prior "defuzzification". Surveys of this research area can be found in Mamdani [M4], Tong [T3], Dubois, Prade [DO].

e) Diagnosis

Owing to a causal interpretation of the equation A o R = B where R is a fuzzy relation, A is a fuzzy set of symptoms and B a fuzzy disease, automatic diagnosis models have been proposed by Sanchez ([S8]). R represents the medical knowledge and is estimated from a set of patients on whom symptoms are observed and who are diagnosed by reliable physicians. A similar approach, used in failure diagnosis of mechanical devices can be found in Tsukamoto and Terano [T1].

f) Learning

Fuzzy automata with a variable structure can be used alternatively to probabilistic automata to imbed reinforcement algorithms (Wee and Fu [W1]). Conditional fuzzy measures can be used as well (Sugeno and Terano [S13]). This approach can be useful in adaptive control (Saridis, Stephanou [S14]), pattern recognition (Wee and Fu [W1]), optimization of multi-modal functions (Asaï, Kitajima [A2])

Another kind of learning, closer to Artificial Intelligence, seems possible owing to fuzzy concepts : a fuzzy algorithm is an uncompletely or unperfectly specified procedure, hence more robust, i.e. capable of adapting itself to a larger class of situations. Memorizing fuzzy procedures and possibly unfuzzify them as information occurs is rather similar to human learning behavior.

g) Computers sciences and artificial intelligence

In order to make the implementation of fuzzy algorithms easier, specific programming languages involving fuzzy set representation and/or manipulation have been designed. Let us quote Le Faivre's FUZZY (issued from LISP [L5]) and Adamo's LPL [A3], a fuzzy version of PL1.

Fuzzy set theory may be useful in Artificial Intelligence for several purposes :

- Representation and manipulation of unprecise or ill-defined knowledge or informations.
- Natural language understanding and man-machine communication (Goguen [G6], Shaket [S15].
- Scene analysis (Jain and Nagel [J2])
- Production systems with fuzzy rules triggered by fuzzy pattern-matching (Cayrol, Farreny, Prade [C2]).

h) Other investigation domains

Fuzzy set theory can be interesting in other fields such as information retrieval (Negoita, Flondor [N2]), game theory (Aubin [A4], Dubois Prade [DO], structural identification (Tazaki Amagasa [T4] , questionnaire theory (Zadeh [Z16], Bouchon [B9]), catastrophe theory (Zwick et al. [Z17], Dubois Prade [DO]), damage assessment of structures (Blockley [B10]), economics (Ponsard [P1]).

CONCLUSION

Fuzzy set theory has become important first because it provides a tool for vagueness modelling, and also because it has allowed a questionning of admitted opinions on probability theory. Although already well developped, this theory of vagueness is may be not completely matured yet. It still needs a deeper understanding in order to be accurately applied. However, from economics, and social sciences to Artificial Intelligence, from control of industrial processes to computerized medical diagnosis, the fields of applications seem very large.

REFERENCES

Books :

K.O Kaufmann A. : "Introduction à la théorie des sous-ensembles flous" 4 volumes . Masson. 1973 - 1975 - 1975 - 1977.

N.O Negoita C.V., Ralescu D.A. : "Applications of fuzzy sets to systems analysis" Birkhaüser Verlag. 1975.

Z.O Zadeh L.A., Fu K.S., Tanaka K., Shimura M. (eds.) "Fuzzy sets and their applications to cognitive and decision processes" Academic Press. 1975.

G.O. Gupta M.M., Saridis G.N., Gaines B.R. : (eds) "Fuzzy automata and decision processes" North Holland. 1977.

D.O. Dubois D., Prade H. : "Fuzzy sets and systems : theory and applications". Academic Press. Forthcoming. 1979.

Introduction and intuitive setting :

Z.1 Zadeh L.A. : "Fuzzy sets" Inf. and Cont. Vol. 8, pp. 338-353. 1965.

Z.2 Zadeh L.A. : "Outline of a new approach to the analysis of complex systems and decision processes" I.E.E.E. Trans. S.M.C. Vol. 3, pp. 28-44. 1973.

S.1 Sugeno M. : "Fuzzy measures and fuzzy integrals : a survey" in [G.0] pp. 89-102

S.2 Shafer G. : "A mathematical theory of evidence" Princeton University Press. 1976

Z.3 Zadeh L.A. : "Fuzzy sets as a basis for a theory of possibility". Int. J. for Fuzzy Sets and Systems. Vol. 1 n° 1, pp. 3-28. 1978.

Fuzzy sets. Operators :

S.3 Saaty T.L. : "Measuring the fuzziness of sets" J. of Cybernetics Vol. 4 N° 4 pp. 53-61. 1974.

B.1 Bellman R.E., Giertz M. : "On the analytic formalism of the theory of fuzzy sets "Inf. Sci. Vol. 5, pp. 149-157. 1973.

F.1 Fung L.W., Fu K.S. : "An axiomatic approach to rational decision-making in a fuzzy environment" in [Z.0] pp. 227-256

D.1 Dubois D., Prade H. : "An alternative fuzzy logic" in "Fuzzy algebra, analysis, logics". Tech. Rep. TR-EE 78-13. Purdue University, Indiana 1978.

Z.4 Zadeh L. A. : "Theory of fuzzy sets". Memo UCB/ERL M77/1 Berkeley. 1977.

G.1 Giles R. : "Lukasiewicz logic and fuzzy theory" Int J. on Man-Machine Studies Vol. 8 pp. 313-327 (1976)

D.2 De Luca A., Termini S. : "a definition of a non-probabilistic entropy in the setting of fuzzy sets theory" Inf. and Cont. Vol. 20 pp. 301-312. 1972

K.1 Knopfmacher J. : "On measures of fuzziness". J. of Math. Anal. and Appl. Vol. 49, pp. 529-534. 1975.

G.2 Goguen J. A. : "L-fuzzy sets" J. of Math. Anal. and Appl. Vol. 18 pp. 145-174 (1967)

M.1 Mizumoto M., Tanaka K. : "Some properties of fuzzy sets of type 2" Inf. and Cont. Vol. 31 pp. 312-340 (1976)

G.3 Gentilhomme Y. : "Les ensembles flous en linguistique" Cahiers de linguistique Théorique et Appliquée, V. 47, Bucarest (Roumanie) 1968.

R.1 Rosenfeld A. "Fuzzy groups". J. of Math. Anal. and. Appl. Vol. 35. pp. 512-517 1971.

Fuzzy measures. Possibility. Probability :

S.4 Shackle G.L.S. : "Uncertainty in economics and other reflections". Cambridge U.P. U.K. 1955.

S.5 Sugeno M. "Theory of fuzzy integral and its applications" Ph. D. Thesis Tokyo Institute of Technology. 1974. Japan.

B.2 Banon G. : "Distinction entre plusieurs sous-ensembles de mesures floues" Colloque Int. sur la théorie et les Applications des Sous-Ensembles flous. Marseille, sept. 1978.

K.2 Kandel A. : "Fuzzy statistics and forecast evaluation" I.E.E.E. Trans S.M.C. Vol. 8 n° 5 pp. 396-401. 1978.

S.6 Sugeno M., Terano T. : "Conditional fuzzy measures and their applications" in [Z.0] pp. 151-170, 1975.

Z.5 Zadeh L.A. : "Probability measures of fuzzy events" J. of Math. Anal. and Appl. Vol. 23, pp. 421-427. 1968

N.1 Nguyen H.T. : "On conditional possibility distributions" Memo UCB/ERL M77/52 Berkeley. 1977. Also in Int. J. for Fuzzy Sets and Systems Vol. 1 N° 4 pp. 299-310. 1978

H.1 Hisdal E. : "Conditional possibility. Independence and non-interaction" Int. J. for Fuzzy Sets & Systems Vol. 1, pp. 283-297. 1978.

Fuzzy relations - Fuzzy functions :

Z.6 Zadeh L.A. : "Similarity relations and fuzzy orderings". Inf. Sci. Vol. 3, pp. 177-200. 1971

Z.7 Zadeh L.A. : "Calculus of fuzzy restrictions" in [Z.0] pp. 1-39.

B.3 Bezdek J.C., Harris J.D. : "Fuzzy partitions and relations : an axiomatic basis for clustering". Int. J. for Fuzzy Sets and Systems Vol. 1 n° 2, pp. 111-127. 1978.

R.2 Ruspini E.H. : "A theory of fuzzy clustering". Proc. I.E.E.E. Conf. on Decision and Control pp. 1378-1383. 1977.

O.1 Orlowsky S.A. : "Decision-making with a fuzzy preference relation". Int. J. for Fuzzy Sets and Systems Vol. 1. 1978.

S.7 Sanchez E. : "Resolution of composite fuzzy relation equations "Inf. and. Cont. Vol. 30, pp. 38-48. 1976.

S.8 Sanchez E. : "Solutions in composite fuzzy relation equations : application to medical diagnosis in Brouwerian logic" in [G.0] pp. 221-234.

T.1 Tsukamoto Y., Terano T. : "Failure diagnosis by using fuzzy logic". Proc. of the I.E.E.E. Conf. on Decision and Control, pp. 1390-1395. New Orleans. 1977.

L.1 Lowen R. : "A comparison of different compactness notions in fuzzy topological spaces". J. of Math. Anal. and Appl. Vol. 64, pp. 446-454. 1978

Extension principle :

Z.8 Zadeh L. A. : "The concept of a linguistic variable and its application to approximate reasoning" Inf. Sci, Part 1, Vol. 8 pp. 199-249 Part 2, Vol. 8 pp. 301-357, Part 3, Vol. 9, pp. 43-80 (1975)

D.3 Dubois D., Prade H. : "Operations on fuzzy numbers" Int. J. of Systems Science Vol. 9 n° 6, pp. 613-626. 1978.

D.4 Dubois D., Prade H. : "Fuzzy real algebra : some results". Int. J. for Fuzzy Sets & Systems Vol. 2. Forthcoming. 1979.

D.5 Dubois D., Prade H. : "Decision-making under fuzziness" See in this book.

D.6 Dubois D., Prade H. : "Systems of linear fuzzy constraints" Int. J. for Fuzzy Sets & Systems Vol. 2. Forthcoming. 1979.

G.4 Gaines B.R. : "Foundations of fuzzy reasoning" Int J. for Man - Machine Studies Vol. 8, pp. 623-668. 1976. Also in [G.0] pp. 19-75. 1977.

D.7 Dubois D., Prade H. : "Operations in a fuzzy-valued logic". in memo. TR-EE 78/13 Purdue. Indiana. 1978

O.2 Orlovsky S.A. : "On programming with fuzzy constraint sets". Kybernetes. Vol. 6. pp. 197-201. 1977.

D.8 Dubois D., Prade H. : "Towards fuzzy analysis : Integration and derivation of fuzzy functions". Communi.to Inf. Sci., memo. TR-EE 78/13 Purdue. Indiana 1978.

Formal systems :

Z.9 Zadeh L.A. : "Toward a theory of fuzzy systems" in "Aspects of Network and System Theory" (Kalman, De Claris eds.) pp. 469-490. 1971.

C.1 Chang S.S.L., Zadeh L.A. : "On fuzzy mapping and control". I.E.E.E. Trans S.M.C. Vol. 2 n° 1, pp. 30-34. 1972.

A.1 Arbib M.A ; Manes E.G. : "A category-theoretic approach to systems in a fuzzy world" Synthese Vol. 30, pp. 381-406. 1975.

W.1 Wee W.G., Fu K.S. : "A formulation of fuzzy automata and its application as a model of learning systems" I.E.E.E. Trans. on Sys. Sci. Cyber. Vol. 5, pp. 215-223. 1969

S.9 Santos E.S. : "Maximin automata" Inf. and Cont. 13, pp. 363-377. 1968

Fuzzy languages :

S.10 Santos E.S. : "Realization of fuzzy languages by probabilistic, max-product and maximin automata" Inf. Sci. Vol. 8, pp. 39-53. 1975

L.2 Lee E.T., Zadeh L.A. : "Note on fuzzy languages" Inf. Sci. Vol. 1, pp. 421-434 1969.

M.2 Mizumoto M., Toyoda J., Tanaka K. : "N-fold fuzzy grammars". Inf. Sci. Vol. 5 pp. 25-43. 1973.

Fuzzy algorithms :

S.11 Santos E.S. : "Fuzzy and probabilistic programs". Inf. Sci. Vol. 10, pp. 331-345. 1976. Aussi in [G.0] pp. 133-147. 1977.

Z.10 Zadeh L.A. : "Fuzzy algorithms" Inf. et Cont. Vol. 12, pp. 94-102. 1968.

Natural languages and approximate reasoning :

Z.11 Zadeh L.A. : "A fuzzy set - theoretic interpretation of linguistic hedges". J. of Cybernetics. Vol. 2 n° 3. pp. 4-34. 1972.

B.4 Bellman R.E., Zadeh L.A. : "Local and fuzzy logics" in "Modern uses of multiple-valued logic" (Dunn, Epstein eds.) R. Reidel (Hollande). pp. 103-165 1977.

Z.12 Zadeh L.A. : "A theory of approximate reasoning (AR)". Machine Intelligence Vol. 19. 1978.

Z.13 Zadeh L.A. : "PRUF - A meaning representation language for natural languages" Int J. for Man-Machine Studies Vol. 10 n° 4, pp. 395-460. 1978.

H.2 Hersh H.M., Caramazza A. : "A fuzzy-set approach to modifiers and vagueness in natural languages" J. Exp. Psych. General Vol. 105. pp. 254-276. 1976.

Operations research :

D.9 Dubois D., Prade H. : "Algorithmes de plus courts chemins pour traiter des données floues". RAIRO Série R.O. Vol. 12, pp. 213-227. 1978.

Z.14 Zimmermann H. -J. "Fuzzy programming and LP with several objective functions" Int. J. for Fuzzy Sets and Systems. Vol. 1, n° 1, p. 45 et seq. 1978

Decision theory :

B.5 Bellman R.E., Zadeh L.A. : "Decision-making in a fuzzy environment" Management Science Vol. 17, pp. B. 141 - B-164. 1970.

J.1 Jain R. : "A procedure for multiple aspect decision making" Int. J. Systems Sci. Vol. 8 n° 1, pp. 1-7. 1977

O.3 Okuda T., Tanaka H., Asai K. : "A formulation of fuzzy decision problems with fuzzy information, using probability measures of fuzzy events" Inf. and Cont. Vol. 38 n° 2, pp. 135-147. 1978.

S.12 Saaty T. : "Exploring the interface between hierarchies, multiple objectives and fuzzy sets". Int. J. for Fuzzy Sets and Systems Vol. 1, n° 1. 1978.

Y.1 Yager R.R. : "Fuzzy decision-making including unequal objectives" Int. J. for Fuzzy Sets and Systems Vol. 1, n° 2, pp. 87-95, 1978.

R.3 Ragade R.K. : "Profile transformation algebra and group consensus formation through fuzzy sets" in [G.0] pp. 331-356. 1977.

Classification and pattern recognition

B.6 Bellman R.E., Kalaba R., Zadeh L.A. : "Abstraction and pattern classification" J. Math. Anal. & Appl. Vol. 13, pp. 1-7. 1966.

Z.15 Zadeh L.A. : "Fuzzy sets and their application to pattern classification and cluster analysis" Memo UCB/ERL M-607 Berkeley. 1976

K.3 Kotoh K., Hiramatsu K. : "A representation of pattern classes using the fuzzy sets". Systems, Computers, Controls, pp. 1-8, 1973.

L.3 Lee E.T. : "Proximity measure for the classification of geometric figures." J. Cybernetics, Vol. 2, n° 4, pp. 43-59. 1972.

T.2 Thomason M.G. : "Finite fuzzy automata, regular fuzzy languages and pattern recognition" Pattern Recognition Vol. 5, pp. 383-390. 1973.

L.4 Lee E.T. : "Application of fuzzy languages to pattern recognition" Kybernetes Vol. 6, pp. 167-173, 1973.

K.4 Kickert W.J.M., Koppelaar H. : "Application of fuzzy set theory to syntactic pattern recognition of handwritten capitals" I.E.E.E. Trans. S.M.C. Vol. 6, n° 2, pp. 148-151. 1976.

R.4 Ruspini E.H. : "A new approach to clustering" Inf. Con. Vol. 15, pp. 22-32, 1969.

R.5 Ruspini E.H. : "Numerical methods for fuzzy clustering" Inf. Sci. Vol. 2, pp. 319-350, 1970.

B.7 Bezdek J.C. : "Numerical taxonomy with fuzzy sets". J. Math. Biology. Vol. 1 pp. 57-71, 1974.

B.8 Bezdek J.C. : "Cluster validity with fuzzy sets" J. of Cybernetics Vol. 3 n° 3, pp. 58-73. 1974.

D.10 Dunn J.C. : "A fuzzy relative of the ISODATA process and its use in detecting compact well-separated clusters" J. Of. Cybernetics Vol. 3, n° 3, pp. 32-57. 1974.

G.5 Gitman I., Levine M.D. : "An algorithm for detecting unimodal fuzzy sets and its application as a clustering technique" I.E.E.E. Trans. Comput. Vol. 19, pp. 583-593, 1970.

Control :

F.2 Fung L.W., Fu K.S. : "Characterization of a class of fuzzy optimal control problems" Proc. 8th Princeton Conf. on Information Sciences and Systems, 1974. Also in [.] pp. 209-219, 1977.

G.6 Gluss B. : "Fuzzy multistage decision-making" Int. J. on Control Vol. 17 pp. 177-192. 1973

M.3 Mamdani E.H., Assilian S. : "An experiment in linguistic synthesis with a fuzzy logic controller" Int. J. Man-Machine Studies Vol. 7, pp. 1-13, 1975.

M.4 Mamdani E.H. : "Application of fuzzy set theory to control systems : a survey". in [G.O] pp. 77-88

T.3 Tong R.M. : "A control engineering review of fuzzy systems" Automatica 13, 559-569, 1977.

Learning :

S.11 Sugeno M., Terano T. : "A model of learning based on fuzzy information". Kybernetes Vol. 6, pp. 157-166. 1977.

A.2 Asai K., Kitajima S. : "A method for optimizing control of multimodal systems using fuzzy automata" Inf. Sci. Vol. 3. pp. 343-353. 1971.

S.14 Saridis G.N., Stephanou H.E. : "Fuzzy decision-making of a prosthetic arm" I.E.E.E. Trans-S.M.C. Vol. 7, n° 6, pp. 407-420. 1977.

Computers sciences and Artificial Intelligence :

L.5 Le Faivre R.A. : "The representation of fuzzy knowledge" J. of Cybernetics Vol. 4 n° 2, pp. 57-66. 1974.

A.3 Adamo J.M. : "Semantics for a fuzzy programming language. Basic notions and logical expressions. Elementary statements and control structures. Sub. to Int. J. Syst. Science. 1978.

G.6 Goguen J.A. : "On fuzzy robot planning" in [Z.O] pp. 429-447.

S.15 Shaket E. : "Fuzzy semantics for a natural-like language defined over a world of blocks "Memo. n° 4 Art. Int. Comp. Science Dept. U.C.L.A. 1976.

J.2 Jain R., Nagel H.H. : "Analysing a real-world scene sequence using fuzziness" Proc I.E.E.E. Conf. on Decision and Control New Orleans, pp. 1367-1372. 1977

C.2 Cayrol M., Farreny H., Prade H. : "Fuzzy production rules directed by fuzzy pattern matching" LSI - Univ. Paul Sabatier Toulouse. 1979.

Other domains :

A.4 Aubin J.P. : "Fuzzy core and equilibria of games defined in strategic form". in "Directions in large-scale systems" Y.C. Ho, S.K. Mitter (eds). Plenum Press. New York. pp. 371-388. 1976.

T.4 Tazaki E., Amagasa M. : "Heuristic structure synthesis in a class of systems using a fuzzy automata" I.E.E.E. Trans. on S.M.C. Vol. 9 n° 2, pp. 73-79. 1979.

Z.16 Zadeh L.A. : "A fuzzy-algorithmic approach to the definition of complex or imprecise concepts". Int. J. Man-Machine Studies Vol. 8, pp. 249-291. 1976

B.9 Bouchon B. : "Longueur de questionnaires flous" Colloque Int. sur la Théorie et les Applications des Sous-Ensembles Flous Marseille. sept. 1978.

Z.17 Zwick M., Schwartz D.G., Lendaris G.G. : "Fuzziness and catastrophe" Proc. Int. Conf. on Cyber. & Society. Tokyo. pp. 1237-1241. 1978.

B.10 Blockley D.I. : "Analysis of subjective assessments of structural failures" Int. J. Man-Machine Studies. Vol. 10, pp. 185-195. 1978.

P.1 Ponsard C. : "Hiérarchie des places centrales et graphes Φ - flous" Environment and Planning A, 9, 1233-1252, 1977.

ADVANCES IN FUZZY SET THEORY AND APPLICATIONS
M.M. Gupta, R.K. Ragade, R.R. Yager (editors)

FUZZY SETS AND THE SOCIAL NATURE OF TRUTH

Joseph GOGUEN

Computer Science Department
University of California, Los Angeles

Fuzzy sets have been suggested for handling the imprecision of real world situations by using truth values between the usual "true" and "false." An extensive theory has grown up, exploring the properties of such a set theory, developing various logics for it, and "fuzzifying" various branches of pure and applied mathematics, including topology, graph theory, automata, and formal languages. This theory has been criticized for its remoteness from actual applications, but there is also a body of work which applies fuzzy sets to problems in control, medical diagnosis, pattern recognition, etc. However, this raises even more forcefully some fundamental questions about the basis of fuzzy set theory: How can one obtain and justify estimates of the truth values which are needed? How can one handle the variations of truth value with context? Is it a paradox that the "degree of membership" used to indicate a degree of uncertainty is itself very precisely given as a real number? More generally, is it a paradox that this theory of imprecision is very precise, and even based on ordinary "crisp" mathematics? The purpose of this paper is to give a new foundation for fuzzy set theory, based on the idea that "truth" is a socially determined judgement. Estimates of truth values can then be obtained from the analysis of conversational processes. This foundation resolves the problems listed above. It does not undermine the existing work in fuzzy set theory, but rather puts it in a new light. It also suggests new connections to the social sciences, particularly sociolinguistics, and it suggests some new approaches to applying fuzzy set theory. The paper includes new and succinct expositions of some basic mathematical developments in fuzzy set theory in order to show in detail that they do not solve the foundational difficulties. Perhaps these mathematical developments will be of major interest to some readers, but this paper is written so that those interested in our main points can skip the mathematics.

1. Introduction

This paper sketches a new approach to the foundations of fuzzy set theory. The main issue addressed is the determination and justification of fuzzy truth values. The proposed framework involves taking seriously the idea that assertions about truth arise in real social interaction, rather than in a vacuum, or in some constructed "ideal" or "objective" situation. Along the way, we explore some issues in the basic mathematics of fuzzy sets, including some implications of assuming the continuity of fuzzy sets; a general result relating the structure of the set of all fuzzy sets to the structure of the truth set; and an explication of "linguistic" truth values in terms of truth sets more general than the unit interval.

1.1 The Classical Theory of Truth

We shall not discuss the historical roots of the notion of truth in the classical civilizations of Greece and Rome, but rather, we shall list some simple properties of the notion of truth currently popular in our culture. These properties are, of course, not held by all contemporary philosophers; on the contrary, there seems to be a rising tide of doubt among philosophers about the nature of truth. Here are the properties:

(1) there are only two truth values, "true" and "false;"

(2) the truth value of a proposition does not change with time;

(3) the truth value of a proposition is independent of where it is said, and of who says it.

Some qualifications are necessary for (2) and (3): The propositions, "It is 2:30 p.m." and "I am John" have truth value depending on time and speaker, respectively. But "2+2=4" does not, nor does either "It was raining in Los Angeles on January 23, 1979" nor "John said 'I am John.'" Philosophical logic has carefully investigated the required qualifications, but the issues are more complex than the uninitiated might expect, and they are far from settled. Modal logic and tense logic are two of the more important theories in this area.

Property (1) above is the famous "Law of the Excluded Middle," and is of course directly challenged by fuzzy set theory.

Properties (2) and (3) point to the popular distinction between that which may be held, possibly inconsistently, by an individual - mere opinion or judgment - and that which transcends this - namely truth. There seems to be an attitude that the ordinary affairs of people, involving opinion, compromise, partial knowledge, emotional biases, subjectivity, etc., are in some way inferior to the realm of unchanging objective truth.

1.2 The Reality of Truth

The simple fact is, that there are no disembodied truths (or if there are, we can never know them). In fact, what we actually experience are not truths, but claims of truth, made by certain particular individuals in certain particular situations. I take this as including particular claims to oneself on particular occasions, e.g., experiences of the kinds ordinarily called "remembering," "imagining," "having an insight," "thinking," "being angry," "dreaming," etc. The trouble with this class of phenomena is that its particular instances are not replicable, or examinable, or verifiable by others, or even by oneself on another occasion; there is no way to be sure that one is having "the same" thought or experience again. Consequently, we shall restrict attention to statements (i.e., claims of truth) made in actual social contexts.

For these, one can in principle make a tape recording which permits repetition and also examination by others who can then debate our conclusions.

Given a particular statement made on a particular occasion, it makes sense to ask who made the statement, to whom, when, where, what happened before and after, and perhaps, why the statement was made. In an intellectual, academic context, we are used to thinking that statements occur as part of the processes of continual debate and discussion, the so-called "market place of ideas." But actually, this accounts for only a small part of the asserations made and heard by most people, even intellectuals. Statements, that is, claims of truth, are made in advertisements ("satisfaction guaranteed"), on soap boxes, and indeed, on an incredible variety of physical objects (e.g., "made in U.S.A." and "patent applied for"). These still resemble statements like "Socrates is a man" in appearing relatively impersonal. But statements also abound in our personal life. We say "I love you," or we try to explain why we are late; we say "I'm fine, thank you" or "That dress looks just lovely on you" or "I've always liked roast beef." These last few are members of a particularly interesting class of assertations which we all know may very well <u>not</u> be true in many common particular social situations.

To summarize, our ordinary lives contain many claims of truth of a highly mundane character, and often of such a nature as to render any attempt to assign a really accurate "truth value" rather problematical. We do not meet absolutely true statements in everyday life, or anywhere else, either.* Even the statement "2+2 = 4" is made in particular contexts to make particular points (e.g., as in this sentence).

1.3 Acknowledgements

I have been very fortunate in obtaining help and encouragement in trying to understand the issues addressed in this paper, and I would like to thank those without whom my present views would have been even more confused. First of all, I must thank Lotfi Zadeh, who aroused my interest in fuzzy set theory while I was a student at Berkely in the 1960's. Secondly, I would like to thank John Tait, Garret Birkoff, Saunders Mac Lane and William Lawvere, from whom I learned the elegance and power of algebraic, and particularly category theoretic, methods. Thirdly, I would like to thank George Lakoff, Charlotte Linde, Michael Moerman, Pete Becker, and Harold Garfinkel for their demonstrations of clear and relevant thinking in the social sciences; particular thanks to Dr. Linde for her efforts to educate me in sociolinguists, ethnomethodolgy, and discourse analysis. Fourthly, I would like to thank Francisco Varela and Fernando Flores for encouraging me to read hermeneutics. Last but surely not least, I would like to thank the administration, staff, and students of Naropa Institute, and particularly its founder, the Venerable Chögyam Trungpa, Rinpoche, for providing a spiritual and intellectual atmosphere within which it was possible for me to undertake such a radical reexamination of my own thought. I must also acknowledge receiving numerous specific valuable suggestions

* I do not wish to deny that there may be such a thing as an <u>experience</u> of absolute truth. I wish only to assert that there can be no <u>statement</u> of absolute truth. The preceeding statement may be as close to absolute truth as a statement can get.

from each of the persons mentioned above, although I have not always followed them.

2. Fuzzy Sets

This section briefly gives basic definitions and properties relevant to our discussion of the nature of truth. It also offers some new results and a succinct integration of many older results in the field.

2.1 Brief History and Exposition

It has been realized for some time that the phenomenon of vagueness presents a serious problem to the theory of truth (see Russell, 1923). This was part of the motiviation for the development of multi-valued logic, both finite (particulary three) valued logic, and infinite valued logic. It may have been Black (1937) who first tried to apply these ideas to particular semantic issues of vagueness in the real world. But it was Zadeh (1963) who took the decisive step of providing both genuine practical motivations (such as pattern recognition), and a calculus including union, intersection, etc. The fuzzy set literature has grown explosively in the decade and a half since then, and the field has its own journal, as well as several text books.

Now here is the basic definition. Let [0,1] denote the set of all real numbers between zero and one (inclusively). Then a fuzzy set is a function A: X -> [0,1] from a set X, called the domain, or universe of discourse, of X, to [0,1]. For $x \in X$, A(x) is the truth value of the statement "$x \in A$." A(x)=1 means that "$x \in A$" is true, and A(x) = 0 means that "$x \in A$" is false. The essential point is that A(x) can have intermediate values, such as 1/2, 8/9, 1/12, $\sqrt{1/2}$, or even $\pi/5$, in forthright contradiction to the Law of the Excluded Middle.

Here is a very general argument in favor of such a notion: All concepts as used in the real world are vague. For example, "square." There is no definite boundary between "square" and "rectangular." A figure which is 12" x 12" is square, and a figure which is 24" x 12" is definitely not square; but what about figures 12.01" x 12", or 12.1" x 12", or 12.5" x 12"? Thus, we need to be able to handle cases with intermediate truth values. (Notice that the concept "square" has been defined by humans and is used by humans; it is not a self-existing part of some non-human world, although we are free to pretend or to claim that it is. Our consistent use of "square" is social convention. That this convention can be embodied to some extent in machines or in logic, does not make its origin or essential nature any less social.)

To cover the same ground in a more abstract way, given a real-world set A, we can usually find points a_0 and a_1 in its domain X such that "$a_0 \in A$" is false and "$a_1 \in A$" is true. Moreover, X is often a connected topological space, so that we can find a path p in X from a_0 to a_1, that is, a continuous function p: [0,1] -> X such that $p(0) = a_0$ and $p(1) = a_1$. Our intuition suggests that a small variation in x should have at most a small effect on the truth of "$x \in A$", i.e., that A: X -> [0,1] is a continuous function. But now consider A(p(t)) for $t \in [0,1]$. It is a continuous [0,1]-valued function on [0,1] with A(p(0)) = 0 and A(p(1)) = 1. It follows that it must actually also take on all intermediate values $0 \leq r \leq 1$. Since real-world concepts are actually like this, fuzzy set theory should have some real-world relevance.

2.2 Generalized Truth Values

One difficulty with the definition of fuzzy sets given in the previous section is its inability to handle ambivalence, that is, conflict among differing measures of membership. For example, someone might say that a "good car" should be fast, reliable, inexpensive to run, easy to repair, cheap to buy, and beautiful to look at. Obviously these criteria conflict. Ambivalence is also important in dealing with bargaining, strategic planning, conflict resolution, etc.

Considerations of this kind motivate generalizing from using the elements of the unit interval [0,1] as truth values, to using elements of a partially ordered set or lattice L of truth values, as developed in Goguen (1967). It is actually convenient to go to a more abstract level, in order to avoid unnecessary detail and repetition. We shall assume of L that it has a kind of structure* S which is closed under arbitrary products. For example, S might be lattices, because we know that an arbitrary product of lattices is a lattice. Some other suitable structures are partially ordered sets, completely distributive lattices, Boolean algebras, complete lattice ordered monoids (or semigroups), Heyting algebras, topological spaces, semirings, and semilattices. An L-fuzzy set (with structure S) is a function A: X -> L (where L has structure S). As compensation for this abstraction, we get the following

Proposition. Let L have structure S, where S is closed under products, and let X be a set. Then the set of all L-fuzzy sets on X, here denoted L^X also has structure S, with all constants, operations, predicates and relations defined "pointwise."

---

*Roughly speaking, a structure includes some constants, operations, predicates, and/or relations (such as greatest lower bound, addition, less-than-or-equal, and zero), subject to some laws (such as the idempotent, communicative, or associative laws). More precisely, we can let S be a concrete category with products which are concretely Cartesian.

For example, if L has a constant $\perp$, which is a least element, then L^X also has a least element, also denoted $\perp$ defined by $\perp(x) = \perp$ for all $x \in X$. If L has a binary relation $\leq$, then so does L^X, defined by $A \leq B$ in L^X iff $A(x) \leq B(x)$ in L for all $x \in X$; moreover, $\leq$ in L^X is reflexive, or transitive, or symmetric, or antisymmetric if $\leq$ is in L. If L has a binary operation +, then so does L^X, defined by $(A + B)(x) = A(x) + B(x)$; and + is associative, commutative or idempotent in L^X if it is in L. Thus, if L is a Boolean algebra, or complete lattice ordered monoid, so is L^X.

This proposition generalizes the first few pages of many papers on fuzzy sets in a simple and elegant way. It is not actually necessary to define unions, intersections, and complements for each new kind of truth set, and then prove the usual rules again. (It is also worth noting that counter-examples, as well as equational laws, extend pointwise from the truth set L to the set L^X of L-fuzzy sets. For example, if L fails to satisfy the Law of the Excluded Middle, then so does L^X.)

2.3 Linguistic Truth Values

It is often objected that the classical definition of fuzzy sets in Subsection 2.1 is too precise, in that it demands the assignment of an infinite precision real number to each element of the domain, and surely this cannot be obtained empirically. This seems to contradict the original motivation behind fuzzy set theory, which was to admit imprecision. One effort to overcome this difficulty is the "linguistic" truth values of Zadah (1975). For example, the variable "size" can take on values <u>small</u>, <u>large</u>, <u>very small</u>, <u>very large</u>, <u>very very small</u>, and so on. It is not widely realized that this issue is quite closely related to the generalized truth values of Subsection 2.2.

The first step is to give a context free grammar G for the set L = L(G) of values of a variable S. For example, if G has the following productions

S -> <u>small</u>

S -> <u>large</u>

S -> <u>very</u> S

then L(G) is the language suggested above. (More precisely, it contains $(\underline{\text{very}})^n$ <u>small</u> and $(\underline{\text{very}})^n$ <u>large</u> for all integers $n \geq 0$). In general, we will want to assume that G contains the productions (called the Boolean productions)

S -> not S

S -> S and S

S -> S or S

S -> neither S nor S

plus (for purposes of disambiguation)

S -> **(**S**)**

Then L(G) will also contain expressions like very large or very small, and neither small nor very very large.

The next step is to give an interpretation for each element of L(G). This is accomplished by providing an interpretation for each production in G. For example, with the above grammar, let us interpret S as a set X (its elements might be boxes, for example) which will be the universe of discourse; also, let us use the classical unit interval [0,1] for truth values; finally, let us denote the interpretation by ϕ. Then ϕ(small): X -> [0,1] a fuzzy set which tells for each element x of X, the truth value of "x is small," and ϕ(large): x -> [0,1] tells the truth value of "x is large." Moreover, let us interpret

$$\phi(\underline{\text{very}}\ S) = \phi(S)^2$$

(following Zadeh (1972) even though Lakoff (1973) points out a number of difficulties with this), and for the Boolean productions,

$\phi(\underline{\text{not}}\ S) = 1 - \phi(S)$

$\phi(S1\ \underline{\text{and}}\ S2) = \min\{\phi(S1), \phi(S2)\}$

$\phi(S1\ \underline{\text{or}}\ S2) = \max\{\phi(S1), \phi(S2)\}$

$\phi(\underline{\text{neither}}\ S1\ \underline{\text{nor}}\ S2) = 1 - \phi(S1\ \underline{\text{and}}\ S2)$

(With for completeness

$\phi(\mathbf{(}S\mathbf{)}) = \phi(S)$

Hereafter, we will not use bold face parentheses.)

Mathematically speaking, the interesting point is that the above equations define a unique fuzzy set $\phi(t)$: X -> [0,1] for each unambiguous $t \in L(G)$; i.e., there is one and only one function $\phi: L_u(G) \to [0,1]^X$, where $L_u(G)$ is the set of unambiguous (uniquely parsable) expressions of sort S in L(G), satisfying the above conditions. The most elegant way to say this is, that there is a unique G-homomorphism from the initial G-algebra to the particular

G-algebra $[0,1]^X$. (This algebraic approach originates in Goguen (1974a), and is applied to fuzzy linguistic truth values in Goguen (1975)).

The last step is to notice that the structure of $[0,1]^X$ can be used to induce a structure on L = L(G). For example, we can define, for t, t' $\in L_u(G)$, t $\leq$ t' iff (t) $\leq$ (t') in $[0,1]^X$ (meaning $\phi(t)(x) \leq \phi(t')(x)$ for all $x \in X$). Moreover, if this ordering induces a lattice structure on $L_u(G)$, then we have natural interpretations for the Boolean operations <u>and</u> and <u>or</u>.

The difficulty with this approach is very much the same as with the simpler approaches discussed previously: how can we obtain justifiable interpretations for the basic constants and operations mentioned in G? Each $\phi(t)$ is still a [0,1]-valued fuzzy set. Evidentally, we could considerably change ϕ, and still induce the same ordering on $L_u(G)$, but also, sometimes arbitrarily small changes in ϕ will produce drastic changes in the ordering of $L_u(G)$. (Technically, the function from the space of interpretations to the space of orderings is not continuous.) These are (or should be) disturbing observations.

We can try to escape by getting still more abstract. It is not necessary to use [0,1] for the truth set with which linguistic values are interpreted; any truth set V with any structure $\underline{S}$ can be used, and we would then get : $L_u(G) \to V^X$. The "Boolean productions" would then be replaced by a context-free grammar corresponding to the operations in the structure $\underline{S}$ of V. For example, if we take V to be the non-negative intergers ω viewed as a semi-ring, then G should include productions

S -> S + S

S -> S * S

S -> 0

S -> 1

(As pointed out in Goguen (1974), this choice of V and $\underline{S}$ gives rise to a theory of "sets" in which objects can occur 0,1,2,3,... times; such sets are sometimes called "multisets," but in computer science, the term "bags" is becoming common.)

It does not appear that in itself this provides a satisfying solution to the problem of justifying choices of degree of membership, although it does open the door to some interesting applications, such as bags, where degree of membership is not a problem.

Zadeh (1975) suggests a clever way to avoid some of these difficulties, with what he calls "linguistic approximation." The basic idea can be stated simply as follows: Rather than interpret L(G) in the very large set $[0,1]^X$, we can interpret it in a smaller set V^X, where V is a set of "approximate truth values," such as _very true_, _more or less true_, _rather false_, etc. If V is finite, then the operations ϕ(_and_), etc., can be defined by finite tables. If V is infinite but described by a grammar G_V, then simple algorithms based on G_V can be used to define interpretations of the operations. (If V is finite, it can be defined by a very simple grammar G_V.)

Zadeh (1975) suggests the following way to obtain interpretations for the operations. First choose a set B of "basic" linguistic truth values, certainly including _true_ and _false_; then choose an interpretation for each of these in $[0,1]^{[0,1]}$; if there are operations in G_V, choose interpretations for these over [0,1] as well - probably the usual ones of fuzzy set theory for _and_, _or_, _not_, etc.; finally, for each non-basic linguistic truth value t, choose the basic linguistic truth value b with interpretation ϕ(b) which is closest to (best approximates) ϕ(t). This will give, for example, a truth table for ϕ(_and_) by finding the best basic approximation ϕ(b) to each ϕ(b1 _and_ b2) for b1 and b2 both basic. Zadeh (1975) gives a simple example in which B = {_true_, _false_, _true or false_}.

2.4 Fuzzy Fuzzy Sets

The purpose of this Subsection is to consider another move, first introduced in Goguen (1967), to handle the problem of the over-precision of membership functions. The basic idea is to use fuzzy sets as truth values. Thus, rather than assume that the truth value of the assertion "$x \in A$" is an element A(x) of L, i.e., that A is a function X -> L, we shall assume that we get a truth value for each such function; that is, we shall assume we have a function L^X -> L. Call such a function a _fuzzy fuzzy set_ on X.

Notice that this construction is very different from assigning a "fuzzy truth value," as a function L -> L, to each $x \in X$. Then A is a function X -> L^L, and can be viewed as a function X L -> L, so that the logical type is not raised as radically as with fuzzy fuzzy sets. The "type 2" fuzzy sets of Zadeh (1975) are L^L-valued sets, with L = [0,1]; They were, in effect, used to interpret the linguistic truth values discussed earlier.

Obviously, we can iterate both these constructions, to obtain fuzzy fuzzy fuzzy sets, type 3 fuzzy sets, $(\text{fuzzy})^n$ sets (as in Goguen (1967)) and type n fuzzy sets (as in Zadeh (1975)). Moreover, these two concepts could be combined, if desired.

What may not be so obvious, is that one can take the limit as $n \to \infty$ of $(\text{fuzzy})^n$ sets, provided that one restricts consideration to fuzzy sets (at each level) which are continuous, by using very sophisticated methods first introduced by Scott (1976). This construction is sketched in Goguen (1974) (see also Goguen (1979)).

However, the basic point here is that one cannot escape the difficulties of obtaining and justifying truth values simply by running up the logical type hierarchy (even if one takes it to the limit and goes over the top). In fact, since one actually gets more and more values which have to be dealt with, the problem becomes intensified.

2.5 Some Other Approaches

This subsection briefly discusses some approaches to fuzzy sets which involve abstracting in such a way that the actual truth values assigned are not important, but only some of their properties are.

The first approach looks only at properties of the class of all fuzzy sets, and never at particular truth values at all. Some examples of such properties are that the class is closed under appropriate kinds of products, unions, and intersections. It turns out that one can give an intuitively plausible set of axioms such that any class satisfying them is equivalent to the class of all L-fuzzy sets, for some completely distributive lattice L. (See Goguen 1968, 1974.) It turns out that _any_ completely distributive lattice can arise; the axioms do not impose any constraints on particular values at all.

The utility of this approach is in providing axiomatic foundations for fuzzy set theory; it does not help with practical problems which arise in using fuzzy sets. But it certainly does avoid problems of estimating truth values.

A second approach is, for a fixed truth set, let us say [0,1], to consider not individual fuzzy sets, but rather equivalence classes of them. An elegant way to do this is to let the classes be determined as invariants (or "orbits") under the action of a group. For example, one might use the group of all monotone bijections of [0,1] with itself. Then one is not concerned with the particular values of a fuzzy set, but only with the order relationships which hold among those values. This approach, sketched at the beginning of (Goguen 1969), is applied in resolving the classical paradox "falakros" (also known as "sorites," or "the bald man"). The method seems to be quite reasonable where it is applicable. Unfortunately, it does not seem to be applicable to most practical problems because one needs actual values to do calculations.

Zadeh (1978) suggests an approach quite different from those described above, which moreover employs a somewhat different terminology from that of this paper. Zadeh presents his system, called PRUF, as a "meaning representation language for natural languages." A (natural language) proposition p determines a procedure P for computing a "possibility distribution" π^p, which represents the "meaning conveyed by p." A "possibility distribution" in this sense is exactly what we would call a fuzzy set with linguistic truth values interpreted as fuzzy subsets of the unit interval. Zadeh reserves the phrase "truth value of a proposition p" for the "compatability" of p with a "reference proposition" r, which is conveniently given by the state of a fuzzy relational data base; compatability here turns out to be another possibility distribution. Thus, Zadeh uses "truth value" for a relatively high level concept in a rather complex particular system; whereas in this paper, "truth value" refers to the very low level concept of the value of a fuzzy set on one of its domain's elements. Zadeh's approach is intriguing and suggestive, particularly in that it permits "truth value" (in his sense) to vary with context. But there is still the vexing problem of obtaining and justifying all the actual truth value (in our sense) assignments involved, both numerical and linguistic.

2.6 Conclusions

Any approach to the theory of fuzzy sets which is based on ordinary "crisp" mathematics, as the various preceeding discussions have been, will not in itself be inherently fuzzy. However, I do not think that this is a difficulty. Using ordinary mathematics in this context, is merely a convenient way of trying to avoid unclarity, errors, and contradictions in what is being talked about.

Any approach to fuzzy sets which makes use of definite truth values is going to be in difficulty in regard to applications, concerning the estimation of justification of the values. It does not seem that any purely theoretical solution can help with this; we have argued that generalized "abstract" truth values, fuzzy truth values, and linguistic truth values, are still in need of justification. On the other hand, there is nothing wrong with these theoretical developments. It is just that the solution to the problem of justifying truth values is going to have to come from somewhere else.

It should be noted that the problem of what to use for truth values is also by no means closed; it is not obvious that the unit interval is the best possible choice. Indeed, if one wants to handle ambivalence, one must abandon it.

The problem of what to use for the structure of the truth set is also somewhat open. It will not do simply to say that every operation defined on [0,1] is potentially available, since some of them are very bizarre.

3. Some Applications of Fuzzy Sets

In this section, I would like to very briefly sketch some applications of fuzzy set theory, in answer to the criticism sometimes heard, that fuzzy set theory has not fulfilled its early promises in regard to practicality. The discussion will be limited to a small number of cases with which I happen to have some familiarity. There is no intention to be comprehensive, or even respresentative. Moreover, the descriptions are intentionally brief, and slanted toward particular points made in this paper.

We begin with the work of Mandami (1974) on the fuzzy control of a steam engine. So-called "optimal" control theory tends to yield control strategies which are extreme, for example, in trying to accomplish a desired change of state as quickly as possible. Often, this is "bang-bang" control, which switches the value of a control parameter from one extreme to its opposite. In the case of a steam engine, this could easily produce another kind of "bang," as the very complex physical limitations of the device are exceeded in unpredicted ways, so that the device might actually explode. In any case, it can be very difficult to compute an optimal control in real time, and the actual response time of a controller might not be quite what is needed to prevent disaster.

It is easily observed that a skilled human controller has no such difficulties. He does not attempt optimal control, but leaves adequate margins for error, and relies on simple and easily executed control strategies. What Mandami did, was to interview such a human operator and elicit rules not unlike "If P is large and Q is increasing not too quickly, then turn down V fairly slowly until A decreases to approximately R," where P,Q,R are meter readings and V is a valve. These constitute a fuzzy algorithm with all rules of the particular form, "If P then do A," where P is a fuzzy predicate and A is a fuzzy action. This algorithm might be converted into a simple tabular form, and then executed by a computer attached to the steam engine, with results much like those with the human operator, and with far smaller computational cost than with optimal control.

Zadeh (1972) proposes a "principle of maximum meaningfulness" which we can paraphrase as follows: Suppose that a fuzzy predicate P is asserted in a given context C. Let X be the universe of discourse of P, so that P is a function P: X -> L, and let the context be represented by a "fuzzy restriction" on P, more specifically as a function C: X -> L. Then the <u>meaning</u> of P in context C is the set (possibly fuzzy) of $x \in X$ which <u>maximize</u> the compatablity of P with C.

Shaket (1976) has implemented this principle in a computer program (written in APL) to determine the references of noun phrases in a world of blocks. Thus, X is a set of blocks, and C is given by a table (implemented as an APL array whose rows correspond to blocks, and whose columns correspond to attributes such as location, size, color and shape.) In Shaket's program, the context C is not fuzzy, but a wide variety of fuzzy precidates can be used in P, as for example, "The large rather green block which is very near a small blue cube on top of a very large red-orange slab." The meaning of such a noun phrase is the block to which it refers, and this is computed as the $x \in X$ for which P(x) is largest. In case of more than one block x for which P(x) is maximal, or close to maximal, the noun phrase is <u>ambiguous</u>; and in case there are no $x \in X$ for which P(x) is <u>much larger</u> than most others, the noun phrase is meaningless. The computation of the meaning makes use of fuzzy sets for the attributes occurring in the table, and for relations such as "near."

The approach is similar to that of Zadeh (1978) in that truth values (for expressions involving only constants, e.g., P(b), where b is a particular block) are computed relative to a "base" of given fuzzy assertations.

Gershman, Zamfir and Haake have implemented the system described in Goguen (1975). A "maze" is given in the form of: an initial node; a function $\Gamma(n)$, which gives the set of successor nodes of any already known node n; and a function GOAL(n) which tells whether or not node n is a goal node. The successors of a node may be to the North, South, East, or West of it, but cannot be in the opposite direction of the edge from its parent node (if any). Also, a "hint" is given, in the form of a sequence of "fuzzy vectors," where a fuzzy vector consists of a fuzzy length and a fuzzy direction, both in linguistic form. For example, "go a fairly short way almost south, then a long way approximately north-north-east, and then a very short way due west," is a typical hint. The implemented system uses the hint to decide what move to make next (this means, for which node n to compute $\Gamma(n)$ next), and we found that it took about as many moves to find the goal as human beings did, given the same information (Haake 1978).

The point of implementing and testing these two systems was to demonstrate the feasibility of <u>robust</u> programming languages, using fuzzy set methods. In a robust <u>system</u>, the result of executing command C in world W is the same as executing C' in W', for all C' in some neighborhood of C and W' in some neighborhood of W. This means that the user's command need not be tremendously precise, and similarly his idea of the state of the world: anything close will produce the same results. It is furthermore easier for humans to generate, remember, and understand fuzzy commands in natural-like languages (as in the above two systems) than rigid precise commands in the usual kind of programming languages; in many cases the fuzzy commands are also shorter, and sometimes enormously shorter.

In artificial intelligence, one finds a variety of well-known systems which, because they are dealing with the so-called "real world," must take account of the imprecision of our expectations about what will happen. Prominent among such systems today are "knowledge-based" systems, which provide "expert" level cosultation in specialized fields. These include: MYCIN, which does medical diagnosis, in particular, blood infections and meningitis; DENDRAL, which infers chemical structure from spectrometer data; and PROSPECTOR, which evaluates sites for geological exploration. These are all rule-based, or so-called "production" systems, in that knowledge is represented by rules, such as (in a rather free translation)

IF

1) CULTURE SITE IS BLOOD
2) ORGANISM GRAM STAIN IS GRAMNEG
3) ORGANISM MORPHOLOGY IS ROD
4) PATIENT IS COMPROMISED HOST

THEN

THERE IS SUGGESTIVE EVIDENCE (.6) THAT THE IDENTITY OF ORGANISM IS PSEUDOMONAS-AERUGINOSA.

Here .6 is a "certainty factor," that is, a fuzzy truth value. MYCIN, DENDRAL, PROSPECTOR and similar systems include algorithms for propagating such certainty factors. These algorithms are, however, _ad hoc_, and it seems clear that there could be much benefit from _explicitly_ recognizing their involvement with fuzzy logic, and investigating their inexact deduction methods in a more systematic way; for it is certainly not widely recognized either that fuzziness is a crucial feature of such systems, or that a great deal of progress has already been made in the study of fuzziness. Perhaps this situation will change when researchers in artificial intelligence feel compelled to give some justification for their _ad hoc_ techniques.

Fuzziness also appears implicitly in a number of artificial intelligence natural language processing systems. Among the least implicit of these are Wilks' "preference semantics" and Carbonell and Collins' SCHOLAR system. Even Evans' classic analogy system uses "degrees of matching" and (_ad hoc_) "scoring" techniques.

4. The Social Nature of Truth

Bertrand Russell (1923) has written, "All traditional logic habitually assumes that precise symbols are being employed. It is therefore not applicable to this terrestial life but only to an imagined celestial existence," and this has been quoted with approval several times in the fuzzy set literature. I would argue that we must go much further. Not only all traditional logic, but all of mathematics and science, assume that the symbols employed have precise meanings. But this meaning cannot be inherent in the symbols; rather, it must arise in the users, the readers, writers, speakers or listeners of the symbols, that is, in members of a community of human beings who share the ability to use the symbols "correctly." Actually, it is more accurate to say that the meaning is inherent in the community, rather than in the members, for it is the community which judges correctness of symbol use. Such judgements will, of course, vary as a function of the persons judging and being judged; but, nevertheless, it is generally the case that professional groups operate so as to achieve a group consensus on such issues. Indeed, professional societies often have institutionalized ways of recognizing various levels of professional competence (e.g., criteria for membership, grades of membership, various prizes). The standard processes of submitting a paper for publication in a professional journal, or a talk for presentation at a professional meeting, include a variety of occasions on which judgements of competence are made; refereeing is prominent among these. Returning to the main point now, whether or not "Fermat's last theorem is true," is determined by a social process, carried out in the community of professional mathematicians, primarily through judgements of the correctness of proferred proofs or counter examples. Notice that there are many cases in which the truth value assigned to an assertation by the mathematical community has changed with time.

The situation for a small group engaged in conversation is similar. Individuals may assert propositions during the course of the conversation; assertations may or may not be challenged by others in the group; there may be general agreement, or further discussion, or introduction of an alternative assertion; also, an assertion could simply be ignored. The point is, that the future development of the conversation can be seen as assigning a status to each assertation which occurs in it. This "socially determined truth value" might, of course, be <u>undetermined</u>, <u>controversial</u>, or <u>very controversial</u>, as well as more conventionally <u>true</u>, <u>very true</u>, <u>false</u>, and so on. Notice it is not the individuals who determine the truth value, but the conversation itself. This seems to be characteristic of truth value assessment. Moreover, we shall see that this provides a realistic foundation for fuzzy set theory, with a well-motivated methodology for determining the truth values of assertations.

4.1 How to Determine Truth Values

This is not the place for a detailed exposition of methodology, and we merely present some relevent background in linguistics and sociology. However, I would like to emphasize as strongly as possible my belief that this area should not be approached naively, on a purely intuitive basis, without any formal discipline or background. Unfortunately, there is not very much work concerned with the linguistic analysis of conversation, which maintains respect for the social character of the conversational process. Most contemporary linguistics is concerned only with isolated sentences, and moreover, is concerned with idealized linguistic "competence" rather than with actual linguistic performance. Also, traditional sociology has not been concerned with the details of actual conversations.

However, the branches of linguistics known as discourse analysis and sociolinguistics, and of sociology known as ethnomethodology and conversational analysis, do contain much that is relevant for us. From sociolinguistics, perhaps the most relavant references are Labov (1972), especially the article, "The Transformation of Experience in Narrative Syntax," and Labov and Fanshel (1977). In discourse analysis, an excellent survey and summary is Linde (1979).

The book edited by Turner (1974) gives a good survey of ethnomethodology, and conversational analysis is well represented by Sacks, Jefferson and Schegloff (1974).

There are a variety of things which the above mentioned literature suggest it would be valuable to look at in assessing truth values. Labov (1972) introduces the idea of an "evaluative clause," and Linde (1979) points out that material having evaluative impact is scattered throughout a text (i.e., a story or a conversation). By "evaluation" is meant some linguistic assessment of the importance or other value to be associated with more directly narrative material. Sometimes, this can be used to infer an assessment of truth value. "Hedges" are words or phrases which are used to modify truth values. Familiar examples include "very," "rather," "somewhat," and "fairly." These are often used in connection with assertions, as "I'm fairly sure that ...," to indicate a truth value assessment. Syntactic placement of an assertion in a sentence, or even in a discourse unit, can be used to infer its importance, with the general principles that items which are "fronted" are more important; the size and emphaticness of the syntactic unit in which an assertion occurs also has indicative value.

At a more interactional level, one can look at "agreement markers," the scattering of occurrences of "uh-huh," "ok," "yeah," etc. that occurs (or fails to occur) during any speaker's presentation in a group situation. One might also look at the hesitations, stutters, and corrections used by a speaker, as these tend to correlate with uncertainty or stress.

These, and a variety of similar techniques, permit an analyst to assess how an assertion is received by a group. For some examples, see Linde and Goguen (1978).

4.2 The Role of the Analyst

The above discussion has omitted any explicit recognition of the analyst's role. This is not the place for a full discussion; we wish merely to indicate the existence of a literature which refutes objections to our methodology, that it is "merely subjective," "not really scientific," "not falsifiable," etc. This literature constitutes an emerging and important "philosophy of understanding," called hermeneutics. It is surveyed in Palmer (1969). The fundamental idea is that there is a "hermeneutic circle" into which an analyst must enter, in order to understand a text. This subjective act necessarily precedes analytic or "objective" understanding. A similar issue is discussed in enthomethodology under the name "member's competence," referring to the need for the analyst's ability to function, with recognized social competence, in a group like that which he is analyzing. For if he lacks an understanding of the conventions and values by which the group operates (including at the most trivial level, the syntax, semantics, and phonology of the language they speak), he will not be able to understand what they are doing.

It is also helpful if the analyst is a member of a group of analysts, an "inquiring community," to whom he feels responsible to justify his judgements. As argued previously, this is the way that science proceeds in any case, not by there being an actual "objective truth," nor by completely unrestrained subjectivity, but rather by a group of interested and trained persons, reaching an "intersubjective agreement."

5. Conclusions

We have argued that fuzzy set theory does in fact have significant potential for soft science applications, but that its foundations need recasting, and the vexing question of how to estimate truth values points the way. For truth values are themselves vague, time varying, and enormously context dependent. For example, there is not some one single fixed eternal fuzzy set of "tall men." The degrees of membership assigned by this set must vary with time (medieval suits of armor show us clearly that men were shorter in medieval Europe than they are now); with place (there is actually a whole heirarchy of places to which the word "here" might possibly refer, for example, continent, nation, city, building); and more generally, with context (who is speaking to whom, what was said previously, etc.).

This paper suggests we must abandon classical presuppositions about truth, and view assertations in their social context. This does not undermine existing work in fuzzy set theory, but merely provides it with a new foundation which is both broader and more precise than the existing, largely unarticulated foundation in the classical philosophy of science. On the other hand, some doubt is cast upon applied studies in fuzzy set theory which estimate truth values with no account of social context, or lack rigor in their methodology for examining social interaction.

While the arguments of this paper have little effect on the hard sciences, including the theoretical development of fuzzy sets, they do suggest a view of the nature of the scientific enterprise completely different from that of traditional philosophies of science which take ideal entities, such as "scientific law" and "truth value of a scientific law," as actually existing. We claim they should rather be viewed as items on the language of the scientific community. To go a little further, we might suggest that the "logic of science" is not mathematical logic (or any variant thereof) but rather is given by the structure of the discourse of science.

As an application of this view, we would not say arithmetic works in practical situations because it is true, but rather that it is "true," in some particular social context, if it is accepted as working. This puts scientific truth on a basis of socially situated efficacy, and avoids the pseudo-problem of the nature of the relationship between an ideal eternal world of truth, and a so-called real world. The general point here is our intention to investigate science primarily as a social process, rather than to uncritically accept the terms of scientific discourse as *a priori* true.

I would like to conclude this essay on what may well seem a radically pessimistic note. There seems to be little room for belief that fuzzy set theory, even when placed on a proper social foundation, will provide an ultimate answer to mankind's problems and uncertainties about the future of large systems, either in general, or in particular important instances. Indeed, it might be argued that our present presuppositions, about truth, about progress, about control and about the future, have actually created insoluble, perhaps meaningless, problems for system theory, and that the most for which we can hope is to be able to see this clearly. It is not inconceivable that fuzzy set theory, particularly conceived as founded in linguistic theories of social process, may help in clearing away the tangles of unnecessary conceptualizations which characterize the psychological worlds in which we actually live.

References

Black, M. (1937) "Vagueness," Phil. of Science 4, 427-455.

Goguen, J. (1967) "L-Fuzzy Sets," J. Math. Analysis and Applications 18, 145-174.

Goguen, J. (1968) "Categories of Fuzzy Sets," Ph.D. Thesis, Dept. Math., Univ. of Calif., Berkeley.

Goguen, J. (1969) "The Logic of Inexact Concepts," Synthese 19, 325-373.

Goguen, J. (1974) "Concept Representation in Natural and Artificial Languages: Axioms, Extensions and Applications for Fuzzy Sets," Int. J. of Man-Machine Studies 6, 513-561.

Goguen, J. (1974a) "Semantics of Computation," Proc. First Int. Symp. on Category Theory Applied to Computation and Control, Univ. of Mass. at Amherst 234-249; also, Lecture Notes in Computer Science 25 (Springer-Verlag) 151-163.

Goguen, J. (1975) "On Fuzzy Robot Planning," in Fuzzy Sets and Their Applications to Cognitive and Decision Processes, ed. by L. A. Zadeh, K. -S. Fu, K. Tanaka and M. Shimura (Academic Press) 429-448.

Goguen, J. (1979) "reveiw of C. V. Negoita and D. A. Raliscu, Applications of Fuzzy Sets to Systems Analysis," to appear in J. of Symbolic Logic.

Haake, J. S. (1978) "Maze Running Using Fuzzy Logic," M.Sc. Thesis, Computer Science Dept., Univ. of Calif., Los Angeles.

Labov, W. (1972) "The Transformation of Experience in Narrative Syntax," in Language in the Inner City (Univ. of Penn. Press) 354-396.

Labov, W. and D. Fanshel (1977) Therapeutic Discourse (Academic Press).

Lakoff, G. (1973) "Hedges: A Study in Meaning Criteria and the Logic of Fuzzy Concepts" J. of Phil. Logic 2 458-508.

Linde, C (1979) "The Organization of Discourse" to appear in The English Language: English and its Social and Historical Context, ed. by T. Shopen, A. Zwicky and D. Griffen.

Linde, C. and J. Goguen (1978) "Structure of Planning Discourse," J. of Social and Biological Structures 1, 219-251.

Mandami, E. H. (1974) "Applications of Algorithms for Control of Simple Dynamic Plant," Proc. of IEEE 121. 1585-1588.

Palmer, R. E. (1969) Hermeneutics: Interpretation Theory, in Schleiermacher, Dilthey, Heidegger and Gadamer (Northwestern Univ Press).

Russell, B. (1923) "Vagueness," Australian J. Philosophy 1, 84-92.

Sacks, H., E. A. Schegloff and G. Jefferson (1974) "A Simplest Systematics for the Organization of Turn-Taking for Conversation," Language 50, 696-735.

Scott, D. (1976) "Data Types as Lattices," SIAM J. Computing 5, 522-587.

Shaket, E (1976) "Fuzzy Semantics for a Natural-like Language Defined over a Set of Blocks," M.Sc. Thesis, Computer Science Dept., Univ. of Calif., Los Angeles; also Artificial Intelligence Memo No. 4.

Turner, R. (ed.) (1974) Ethnomethodology (Penguin).

Zadeh, L.A. (1965) "Fuzzy Sets," Information and Control 8 338-353.

Zadeh, L. A. (1972) "Fuzzy Languages and their Relation to Human Intelligence," Proc. Int. Conf. Man and Computer, Berdeaux, France (S. Karger) 130-165.

Zadeh, L. A. (1972a) "A Fuzzy-Set-Theoretical Interpretation of Linguistic Hedges," J. Cybernetics 2, 4-34.

Zadeh, J. A. (1975) "The Concept of a Linguistic Variable and its Application to Approximate Reasoning," Information Sciences 8, 199-249, 301-357, and 9, 43-80.

Zadeh, L. A. (1978) "PRUF - A Meaning Representation Language for Natural Languages," Int. J. Man-Machine Studies 10, 395-460.

ADVANCES IN FUZZY SET THEORY AND APPLICATIONS
M.M. Gupta, R.K. Ragade, R.R. Yager (editors)

ON FUZZY SYSTEMS

Constantin Virgil Negoita

Faculty of Economic Cybernetics
Str. Caragea Voda 9-15, R-71149 Bucuresti 22, Romania

Abstract

There is an evidence of widespread concern over the usefulness of fuzzy systems as models of natural processes. In this paper fuzzy system theory is viewed as bringing about new conclusions that the old theory of systems could not yield. Fuzzy sets are subjective evaluations associated with undecidability generated by conflict. The lattice structure of the subjective evaluation induces a lattice structure of all possible evaluations. Structural stability means structure preserving in the continual transformation of partial evaluations by synthesis. A synthesis process is viewed as a dynamic system whose state is a fuzzy relation, or, more generally, a pullback, i.e. a regresses in the structure. Decision making is presented as a model of the theory.

Introduction

Fuzzy set theory has been with us now long enough to quiet critics who once prophesied its uselessness and short enough to warrant a serious philosophical concern. The challenge posed to traditional modelling theories by the deliberate employment of fuzzy sets is acknowledged and discussed.

Zeleny |1| points out the increasing complexity of fuzzy set theory, the trend toward mechanistic fuzzification of existing mathematical systems. He says that descriptive theory of fuzzy systems should not be confused with fuzzy versions of mathematical descriptions of systems. One must develop new descriptions of fuzzy events. Watanabe |2| says that a new theory is appreciated if it brings about new empirically verifiable conclusions that the old theory could not yield. In absence of such achievements, a new theory can only be evaluated by its formal beauty, and Zadeh's theory of fuzzy sets is beautiful.

Kickert |3| finds himself reacting with somewhat less alarm. He says that, although the fuzzy theories seem to have originated from the wish to apply fuzzy sets rather than to have been invented or developed to solve specific practical problems, they have indeed extended traditional mathematical theories so that they can cope with vagueness i.e. with a broader part of empiry. He states that fuzzy theories are just as previous theories but formulated in a new language. Therefore, the great barrier which seems to prevent rapid growth of fuzzy system research in soft sciences is the gap between mathematics and soft sciences.

However, alongside the movement toward fuzzification of classical models there has been another attempt to discover the fundamental principles that underlie some problems in soft sciences |4|. This paper questions the relevance of such a movement, to indicate why one might care to study fuzzy systems. The basic assumptions are that an evaluation is not there to be discovered but must be created, that possibility is an efficient uncertainty variable, and that a global approach is preferable to a local one.

Although I shall be primarily concerned with fuzzy systems as they arise in decision making, much of what emerges could, without excessive changes, be extended to cover analogous possibilities in other fields where mental images are also handled |5|.

Multiple evaluation

I shall try to show that, because of the lattice structure of the set of all subjective evaluations, naturally induced by the lattice structure underlying any subjective evaluation described by a fuzzy set, the subjective fuzziness is objectivised. That is, no matter how subjective one is looking at the real world, by aggregating many subjective evaluations, a final objective evaluation is reached.

Multiple evaluation is associated with undecidability generated by conflict, which is a clash of feelings. The principle of excluded middle is inapplicable in these situations. There is no contradiction. The same element can be evaluated in many ways. Such a clash as a source of action determines a process, a sequence of states, and a dynamic system. The states are evaluations, and dynamics means incorporation of alternate evaluations into a new one which refines them all. To avoid undecidability generated by conflict humans pull back on higher levels of synthesis. To get structural stability they concentrate partial evaluations in global ones. The lattice structure explains the movement from one level of synthesis to the next and how structural stability is achieved. One has a collection of partial evaluations, and, according to simple inreaction rules, given by the structure of the evaluation process, a new evaluation emerges spontaneously. The structure is preserved in a new, unique form.

The lattice structure of the evaluation process implies a purposeful system controlled by the difference between the actual state (evaluation) and a final one, interpreted as ideal. Each next state is a goal, and dynamics is goal seeking. The motivation of this movement is the difference between the desired and the actual goals. Purposefulness means here the will to eliminate ambiguity introduced by the partial evaluation, the multiple evaluation. Caught in ambiguity, there is only one possible outcome: to resolve the ambiguity |6|.

This is indeed a powerful model with lots of applications. Take, for instance, the case of planning based on multiple evaluation. Management scientists which distinguish between constraints and goals use extensively, in planning, linear programming. In the linear programming, the linear programming formulation is usually considered as a general model because of its wide acceptance, goals are viewed as functions and constraints as sets. In fuzzy programming |7| constraints are transformed in functions.

Constraints and goals have the same nature, there is no more a distinction between means and ends; we are speaking about confluence. Both constraints and goals are now evaluations (fuzzy sets). To solve a fuzzy programming problem means to move the decision on a higher level, to think broadly, i.e. across and between means and ends. The movement on a higher level is the transition in the next stage described by the dynamics on fuzzy systems. The new empirically verifiable conclusion, that the old theory of linear programming cannot yield is structural stability on this higher level. Synthesis cope with situations which cannot be foreseen in detail, with possible evaluations. This is robustness. Fuzziness is introduced by humans to cope with complexity, and by synthesis one gets structural stability.

A fuzzy system is an objective aggregation rule which puts together numerical subjective judgments. Treating auto-poietic processes as fuzzy system many applications arise so naturally as to seem to suggest themselves.

Pullback and synthesis

We shall denote by $F(X) = \{f:X \to [0,1]\}$ the set of all fuzzy sets defined as the carrier X. It is well known |8| that F(X) is a lattice such that all two subsets of F(X) have both a greatest lower bound and a least upper bound. The existence of a unique glb (lub) for each pair of fuzzy sets implies the existence of a mapping

$$F(X) \times F(X) \to F(X).$$

In this framework, three fundamental elements which characterize a dynamic process can be discovered. These elements are stage, state and transition. It is quite natural to think of lattice F(X) as describing a change and as a succession of states. Here, the states are evaluations given by axiologic option, dynamics is given by conflict resolution (incorporation of alternate descriptions into a new description which refines them all) and homeostasis means structural stability.

We use these ideas to motivate a general categorical construction. The order relation in the lattice F(X) will be considered as a morphism. Clearly, for every pair $f_1 \xrightarrow{A} g \xleftarrow{B} f_2$ of morphisms A and B with common domain g there is a fuzzy set p, called the pullback, such that the square on p commutes (AP = BQ as below) with the following universal property: given any commutative square (BK=AH) on the edges A and B with new corner z there is a unique morphism $S : z \to p$ such that the whole diagram commutes (PS = H, QS = K)

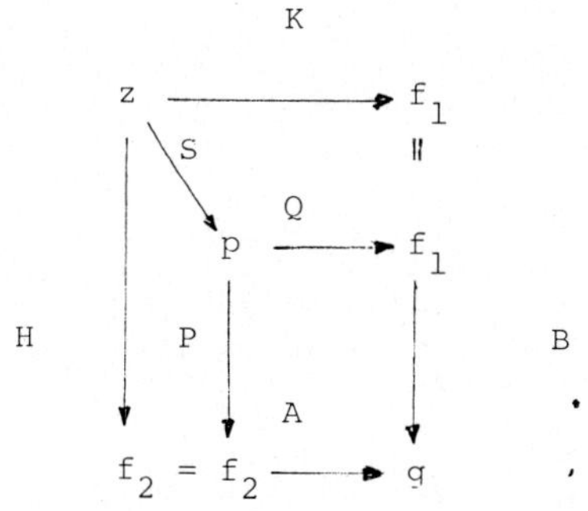

Immense clarity may be offered by this framework. As an example consider the case of flexible planning. |7|
It is well known that in any linear programming problem (find min $f_o(x)$ subject to $Ax \leq b$, $x \in R^n$) the inequalities specify a feasible

region and c is as close as possible to b. The meaning of the statement "as close as possible" is given by function $h : R \to [0,1]$ having the support $[b,B]$ such that $c \in [b,B]$ implies $h(c) > 0$. Bearing in mind that A is a matrix and c a vector each constraint may be written as a composition

$$R^n \xrightarrow{A_i} R \xrightarrow{h_i} [0,1]$$

In other words the diagram

$$\begin{array}{ccc} R^n & \longrightarrow & R \\ \uparrow & & \uparrow \\ A_i^{1}(h_i) & \longrightarrow & h_i \end{array}$$

commutes. Here $A_i^{-1}(h_i) \to R^n$ is the pullback of $h_i \to R$ along A_i. This procedure takes maximum advantage of the compromise and seems to correspond closely to the thinking procedure implicitly employed by humans.

Decision unfolding

Divide and conquer seems to be the first principle of scientific attack. Breaking a complex system into tractable pieces is often the only way to understand it. Complex systems are understood by first explaining the parts and then understanding the way they fit together. In addition to an ability to encode specific facts in simple concepts, a semantic system needs some facility for grouping sets of similar facts into units, allowing these facts to be conceptualized as an integrated whole.

The scheme is hierarchical, in that some concepts may be subconcepts of more general concepts. The lower a concept is in the concept hierarchy, the more alike its members must be. The conceptual system serves the important purpose of spotlighting similarities among objects and compressing redundant information by recording the common characterizing property.

The simplest way to describe evaluations is in terms of functions $f_i : X \to L$ where X is the set of objects and L is the unit interval. A general decision problem can be formulated as: given X find $\bar{x} \in X$ such that for all $x \in X$

$$f_i(\bar{x}) \leq f(x)$$

An element $\bar{x}$ of X satisfying the above inequality for all $x \in X$ is a solution of the optimization problem specified by the pair (f_i, X). When the function f_i is specified by two functions $M : X \to Y$ and $G_i : XxX \to L$, then

$$f_i(x) = G_i(x, M_i(x))$$

and M_i is referred to as the model of a control process and the function G_i is referred to as a performance function. Since the subsystems operate so as to achieve their own individual goals, a conflict generally develops among them. Absence of conflict means that a decrease in each of the partial functions does not cause an increase in an overall function p. Therefore, p(x) = min p(x) whenever the object $x \varepsilon X$ is such that $f_i(x) = \min f_i$, i = 1,...n. In other words the equation

$$p(x) = J\ (f_1(x),\ldots,f_n(x))$$

holds for each object x in X when J is monotone, i.e. order preserving. |9|

If p and f_i are fuzzy sets, $p,f_i : R \to [0,1]$ and we denote by $\wedge$ the operation of taking the minimum, then we face a particular case of pullback.

So far we have had a static image of what is called a decision. The decision was viewed as a final judgment, a settlement, or more precisely an attainment.

We shall consider now the decision making as an act. The term "act" convers both processes and attainments. It is the mark of a process that is has duration, but an attainment, although datable, is not time consuming. Decision making could be defined as a struggle to resolve a dilemma of conflicting objectives |10|. This distinction between campaigning and winning is signified by the terms "process" and attainment." The set of decision criteria is evolving and changing during the process of decision making. We speak about unfolding a decision. Partial decisioning includes a directional adjustment of the decision situation. It consists of adding criteria. The new criteria are melted in order to have finally only one criterion. The activity is analogous to building a bridge from the mainland to an island. The mainland is the body of evaluation which a person already processes, and the island is some knowledge that is as yet unassimilated. The aim of thought is to bring the new into intelligible relations with the old. Unfolding a decision is a criteria melting process. First there is a sense of conflict. Conflict provides the decision motivating tension, period of dissatisfaction with the status quo of a current situation. The underlying source of predicision conflict is structural instability. I think the source of predecision conflict which triggers the decision process is the infeasibility of structural stability. Because the experienced conflict of different criteria, one starts searching for a new representation, structuraly stable.
Let us consider a criterion $f_1 : X \to [0,1]$. We say that this criterion is structurally stable if, roughly speaking, the shape or the ordering is not affected by the presence of other criteria, f_2, by synthesis. If not, the new criterion $f_2 : X \to [0,1]$ is taken into account, i.e. the same carrier is viewed significantly from a second point of view, and one seeks an integration of the two descriptions f_1 and f_2. A new criterion $p = J\ (f_1,f_2)$ is constructed. It is the characteristic of a whole to bring diverse elements together into a higher unity. Initial properties are reconciled in a higher synthesis. One has a collection of descriptions, and, according to simple interaction rules between them, a new description emerges spontaneously. As a consequence of those interaction rules the structure is preserved. A dynamic series of descriptions is formed which continually permeate. As a result of this movement a compact descriptions is achieved.

Evaluations are abstractions, they protray what is common to objects. What we are doing is to try to capture an object by putting criteria together. A criterion can only circumscribe around an object; a frame, which is too large and does not fit exactly. Realizing this in the case of any one criterion, we add another one, which is also too large but which partially overlaps the previous frame, and thus cuts down the area within which object is to be found. We are moving around the set X. Because of spatial perspective an object appears different from various points of observation. To identify the object with any one of these appearances would be a mistake. All such knowledge is relative and partial. There could be an infinity of such criteria. Perhaps the most noteworthy feature of the decision unfolding based on synthesis is the stability of this movement. It is quite natural to think of this change as a succession of states in which each stage holds the stage for a while before it is followed by its successor. This change is a process in which states are fuzzy sets. With the notation $f_1 = x_k$, $f_2 = u_k$ and $p = s_{k+1}$ we can write the standard state equation

$$x_{k+1} = x_k \wedge u_k$$

Clearly, in this framework we can handle control problems. When small changes in control inputs u_k have negligible effects upon a state, we say that the state is structurally stable.

It is proposed here that the principle of homeostasis, the law that organisms seek to maintain inner equilibrium, can be expanded into a major generalization in the field of decision theory. This principle seems to have value in illuminating decision making considered both as a natural and as a unique phenomenon. The concept of structural stability gives a clear picture of the role played by homeostasis (homoios, of the same king + stasis, condition) in decision making. In attempting to reduce the variability of external stimuli the decision maker must deal with structrual stable logical objects. Criteria and objectives are concepts. They eliminate the need for treating each situation as new.

The innovative or learning step is necessary for the decision maker to remain stable under unusual or new environmental conditions. Control, then, includes not only regulation of routine events but also innovation. Therefore, the concept of structural stability will aid in distinguishing between systems that achieve stability under specific constant conditions and those that can learn or evolve new behaviours so as to remain stable under changing conditions. The latter is the focus here. Another way of expressing the above is to say that the open systems face the fluctuating environment through processes of learning and innovation. The emergence of stable, enduring patterns of decisioning is in part a process of conflict resolution. It is the supreme tendency of the decision maker to get rid of tensions and come to rest in a state of equilibrium.

Hence it can appear that unfolding a decision is attempting or seeking this point or as sometimes expressed, the system is seeking a goal. Every criterion is a model in the broad sense of the word, that is, a conceptual image intended to reflect certain aspects of reality. Similar climax formations may develop from different initial starting points and repeated disequilibration results in continuous modification of the global representation.

References

/1/ Zeleny M., Membership and Their Assesment, in : Rose J.(ed), Current Topics in Cybernetics and Systems (Springer Verlag, Berlin, 1978)

/2/ Watanabe S., A Generalized Fuzzy Set Theory, IEEE Trans. on Systhems, Man, and Cybernetics, SMC-8 (1978) 756-760

/3/ Kickert W., Fuzzy Theories on Decision Making (Martin Nijhoff, Amsterdam, 1978)

/4/ Negoita C.V., Fuzzy Systems (Abacus Press, Tunbridge Wells, Kent, 1979)

/5/ Negoita C.V., Pullback versus Feedback. An Essay in Human System Management (February 1979), to appear

/6/ Negoita C.V., Editorial Introduction, Kybernetes, 6 (1977) 145-146

/7/ Negoita C.V., Management Applications of System Theory, (Birkhäuser Verlag, Basel, 1979)

/8/ Negoita C.V. and Ralescu D.A., Applications of Fuzzy Sets to System Analysis (Birkhäuser Verlag, Basel and Halsted Press, New York, 1975)

/9/ Mesarovic M.D., Macko D. and Takahara Y., Theory of Hierarchical, Multilevel Systems (Academic Press, New York, 1970)

/10/ Zeleny M., The Theory of Displaced Ideal, in : Zeleny M.(ed), Multiple Criteria Decision Making (Springer Verlag, Berlin, 1976)

ADVANCES IN FUZZY SET THEORY AND APPLICATIONS
M.M. Gupta, R.K. Ragade, R.R. Yager (editors)

A SURVEY OF THE REPRESENTATION OF FUZZY CONCEPTS AND ITS APPLICATIONS

DAN RALESCU
Department of Mathematics, Indiana University
Bloomington, Indiana 47405, USA .

Abstract

Since the concept of a fuzzy set was introduced in 1965 by Zadeh, a great number of papers were devoted to extending classical concepts to the new field and to solving the new, "fuzzy problems". The concept of a level set was recognized to be important from the very beginning, but the representation theorem for fuzzy sets was only proved in the 74's (Negoita and Ralescu).

We shall present in this paper the representation theorem; roughly speaking, it says that a fuzzy set may be identified with a family of classical sets (levels). The same representation is true for a large class of fuzzy concepts, including fuzzy algebraic structures (as fuzzy groups), fuzzy relations, fuzzy events, and others.

One may think at this point that representation, as beautiful as it might be, is only of a theoretical interest. To prove the contrary, we shall see how representation of fuzzy concepts may be applied to solving new problems, especially in the fields of system analysis and decision-making.

We first show how to represent fuzzy systems as families of deterministic systems. Then we prove that such problems as fuzzy mathematical programming, and fuzzy linear programming, can be reduced to the corresponding classical problems.

Finally, we show that fuzzy optimization problems and equations involving fuzzy relations may be attacked, and solved, by using the representation theory.

1. INTRODUCTION

The concepts of fuzzy set and membership degree were introduced in 1965 by Zadeh [33] in order to provide a possible model for inexact concepts and subjective judgements. This model was intended to be used in situations when deterministic and/or probabilistic models do not provide a realistic description of the phenomena under study. Indeed, in a large area of situations, such as pattern recognition, decision-making, large-systems control, management problems, production scheduling, and others, human judgement is often imprecise and the decision-maker no more manipulates numbers, but vague concepts. Most often, on such premises, the decisions themselves are vague, imprecise. However, by manipulating vague concepts and inferring vague conclusions, it is not only possible to solve real problems, but this is most often the way decisions are taken. A great number of papers were devoted to such approaches; most of them are quoted in the bibliography by Gaines and Kohout [8].

In spite of the great amount of research in the field, two main difficulties are present when trying to use fuzzy sets theory:

(A) The assessment of the membership functions. Most often the degrees of membership are not given, they may not be assessed arbitrarily, and it is sometimes difficult to find them correctly.

(B) There is no general methodology for solving "fuzzy problems".

The main aim of this paper is to give an answer to the problem (B). What we call the representation theorem for fuzzy sets will assert that any fuzzy set can be identified with a family of classical sets. This result, due to Negoita and Ralescu [14], may be extended to represent fuzzy concepts, such as fuzzy algebraic structures, fuzzy topological spaces, fuzzy measurable sets (fuzzy events), similarity relations, fuzzy orderings, and others. Roughly speaking, any fuzzy concept is a family or ordinary concepts of the same kind. For example: a fuzzy group is a family of groups; a similarity relation is a family of equivalence relations, and so on.

It is in this sense that we speak about representation of fuzzy concepts as about a methodology to solving fuzzy problems. Roughly speaking, a fuzzy problem can be viewed as a family of classical problems of the same kind.

The idea of representing fuzzy sets is also related to another concept: that of a vague set. This concept was defined by Gentilhomme [9] under the name "ensemble flou", in French. Vague sets were found to be useful in linguistics and their underlying logic is three-valued.

By suitable generalization, we are led to the general concept of a vague set and the point is that this concept is proved to be equivalent to that of a fuzzy set. This equivalence holds at the more abstract level of the L-fuzzy sets, as defined by Goguen [10]. In the more concrete case of the fuzzy sets (i.e. [0,1]-valued functions), the representation theorem has a more convenient form, due to the presence of a nice topolgoy on $\mathbb{R}$ (the real numbers).

It may happen, sometimes, that the conditions of the theorem are difficult to be verified. Ralescu [24] showed that it is still possible to generate a fuzzy set by starting with a decreasing family of sets.

A great number of "fuzzy concepts" were defined so far in the literature. Some examples are: fuzzy algebraic structures (like fuzzy groups [27]), fuzzy topological spaces [3], fuzzy convex sets, similarity relations [35], fuzzy orderings [35], fuzzy measurable sets [34], and others. Apparently, the definitions of these concepts are different in nature; more precisely, there is not a unitary way in defining such fuzzy concepts. As we shall see later, the representation theorem unifies all these definitions and may be used to define new concepts. A fuzzy subobject of an object in an arbitrary category may be defined as in Ralescu [21].

By using a categorical framework and by defining the category of fuzzy sets like in Goguen [11], it is possible to prove a categorical representation theorem. Moreover, by using the concept of a C-set, due to Flondor [7] it is possible to generalize the representation, by showing that the category of C-sets is equivalent to a category of functors (see Ralescu [21]).

Recently, the language of elementary topoi was found to be useful for generalizing fuzzy sets. Negoita and Stefanescu [19] proved a representation theorem in this context.

The above discussion shows that mathematically, the representation of fuzzy sets is a rich subject, still growing.

From the practical point of view, the first application was to show how fuzzy systems can be represented by using families of deterministic systems (Negoita and Ralescu [16]). This representation is useful in trying to define what a control problem means for a fuzzy system.

Decision-making in a fuzzy environment is an important, practical problem. We show how representation is useful in the fuzzy mathematical programming ([32],[15]) and in solving the fuzzy linear programming problem ([18]).

The level sets were used by Ralescu [22], [23], [24] to define the "fuzzy optimal set" in an optimization problem with fuzzy constraints.

Another application of the representation is in solving equations involving fuzzy relations. Such a theory was found to be useful in medical diagnosis [28] and economics [20].

Both the theory and the applications of the representation of fuzzy concepts will be reviewed in the next sections.

It is to be expected that new applications are to come both in which definitions of new fuzzy concepts is concerned, and in solving new "fuzzy problems".

2. THE REPRESENTATION THEOREM

Throughout this section, X will be a fixed set, without any structure. By L we denote a completely distributive lattice.

A <u>fuzzy subset</u> of X (or an L-<u>fuzzy set</u>, or an L-<u>set</u>) is a function $u : X \to L$. We denote by

$$F_L(X) = \{u/u : X \to L\} \tag{2.1}$$

It is clear that $F_L(X)$ becomes a completely distributive lattice, with the operations:

$$(u \vee v)(x) = u(x) \vee v(x) \tag{2.2}$$

$$(u \wedge v)(x) = u(x) \wedge v(x) \tag{2.3}$$

These operations generalize the classical union and intersection of sets.

A particular, important case is $L = [0,1]$; we denote then by $F(X) = F_{[0,1]}(X)$ - these are the fuzzy sets defined by Zadeh [33]. .

A <u>vague subset</u> of X is a function $\phi : L \to P(X)$, with the properties:

(i) $\phi(0) = X$

(ii) $\phi(\sup_{i \in I} \alpha_i) = \bigcap_{i \in I} \phi(\alpha_i)$ for any $(\alpha_i)_{i \in I} \subset L$.

By $P(X)$ we have denoted the set of all subsets of X ; a vague subset may be thought as a family $(X_\alpha)_{\alpha \in L}$ of ordinary subsets, with:

(i) $X_o = X$

(ii) $X_{\sup_{i \in I} \alpha_i} = \bigcap_{i \in I} X_{\alpha_i}$

The set of all vague subsets of X will be denoted by $Fl_L(X)$; it is also a completely distributive lattice.

It is easy to see that to any fuzzy set $u \in F_L(X)$ we can associate a vague set $\phi_u \in Fl_L(X)$:

$$\phi_u(\alpha) = L_\alpha(u) = \{x \in X \,/\, u(x) \geq \alpha\},\ \alpha \in L \tag{2.4}$$

The set $L_\alpha(u)$ is called the α-<u>level set</u> of u ; so we can assign to any fuzzy set a family of ordinary sets. The converse problem is very important: given a family of set $(X_\alpha)_{\alpha \in L}$ is it possible to find a fuzzy set whose levels are exactly the X_α's ?

The answer is affirmative, if $(X_\alpha)_\alpha$ is a vague set. More precisely, we can prove the following:

<u>REPRESENTATION THEOREM 1</u>: The complete lattices $F_L(X)$ and $Fl_L(X)$ are isomorphic.

The proof is given in Negoita and Ralescu [14] .

In the particular case $L = [0,1]$, by a vague set $\phi \in Fl(X)$ we mean a function $\phi : L \to P(X)$ with properties

(i') $\phi(0) = X$

(ii') $\alpha \leq \beta \Rightarrow \phi(\alpha) \supseteq \phi(\beta)$

(iii') if $\{\alpha_n\}_n$ is an increasing sequence, $\alpha_n \to \alpha$, then $\phi(\alpha) = \bigcap_{n=1}^{\infty} \phi(\alpha_n)$.

Instead of property (ii) we now have (iii') which is easier to be checked. The representation theorem becomes in this case:

THEOREM 2: The complete lattices $F(X)$ and $Fl(X)$ are isomorphic.

We conclude this section with the following remark: if we take $L = \{0,1/2,1\}$, then a vague set can be identified with a pair (A,B) where A,B are ordinary subsets of X and $A \subseteq B$.

3. THE CATEGORICAL REPRESENTATION

We shall define the category of fuzzy sets $Set(L)$. Its objects are pairs (X,u), where X is a set and $u \in F_L(X)$. A morphism $f : (X,u) \to (Y,v)$ is a function $f : X \to Y$ such that $v \circ f \geq u$. This category was introduced by Goguen [11]; he also gave a characterization theorem which, roughly speaking, says the following: if a category $\mathcal{C}$ satisfies certain axioms, then $\mathcal{C}$ is equivalent to the category $Set(L)$, for some L. An interesting discussion, related to the "category of concepts", follows from this result (see Goguen [12]).

We shall be mainly concerned here with giving a representation theorem for $Set(L)$.

The category of vague sets, denoted by $Flou(L)$, has objects pairs (X,ϕ) where X is a set and $\phi \in Fl_L(X)$. A morphism $m : (X,\phi) \to (Y,\psi)$ is a function $m : X \to Y$, such that $m^*(\phi(\alpha)) \subseteq \psi(\alpha)$ for all $\alpha \in L$. By m^* we denote the extension of m to the power set $\mathcal{P}(X)$.

The representation is then described by the following

THEOREM 3: The categories $Set(L)$ and $Flou(L)$ are isomorphic.

For a proof, the reader may consult Ralescu [21].

This theorem says that, practically, the categories $Set(L)$ and $Flou(L)$ are the same.

4. REPRESENTATION OF FUZZY CONCEPTS

A great number of fuzzy concepts were defined in the literature, as generalizations of classical concepts. Examples are: fuzzy convex sets, fuzzy groups [27], similarity relations [35], fuzzy events [34]. It is possible, by using the representation theorem, to show a result of the following type: any "fuzzy concept" may be identified with a family of "classical concepts" of the same kind.

To be more specific, let us recall some definitions. If X is a linear space (over $\mathbb{R}$, for example), a fuzzy set $u \in F_L(X)$ is called convex if

$$u(\lambda x + (1-\lambda)y) \geq u(x) \wedge u(y) \tag{4.1}$$

for any $x,y \in X$ and $\lambda \in [0,1]$.

If $(X,.)$ is a group, a fuzzy set $u \in F_L(X)$ is called a fuzzy subgroup if

$$u(x \cdot y^{-1}) \geq u(x) \wedge u(y) \tag{4.2}$$

for any $x,y \in X$.

A fuzzy relation is a fuzzy subset of $X \times Y$, i.e. $R \in F_L(X \times Y)$. A similarity relation in X is $S \in F_L(X \times X)$ with

(1) $S(x,x) = 1$, $(\forall)\, x \in X$ (reflexivity)

(2) $S(x,y) = S(y,x)$, $(\forall)\, x,y \in X$ (symmetry)

(3) $\bigvee_{y \in X} [S(x,y) \wedge S(y,z)] \leq S(x,z)$, $(\forall) x,z \in X$ (transitivity)

If $\mathcal{A}$ is a σ-algebra of subsets of X, a fuzzy set $u \in F(X)$ is called measurable if $L_\alpha(u) = \{x \in X \,/\, u(x) \geq \alpha\} \in \mathcal{A}$ for all $\alpha \in [0,1]$.

Let us denote by $FC(X)$ the fuzzy convex subsets of X, and by $F\ell C(X)$ families $(C_\alpha)_{\alpha \in L}$ of convex subsets of X such that

(i) $C_o = X$

(ii) $C_{\sup_{i \in I} \alpha_i} = \bigcap_{i \in I} C_{\alpha_i}$, $(\forall)\{\alpha_i\}_{i \in I} \subset L$

Let us denote by $FG(X)$ the fuzzy subgroups of X, and by $F\ell G(X)$ families $(G_\alpha)_{\alpha \in L}$ of subgroups of X such that (i) and (ii) above are satisfied (with appropriate notations).

Let us denote by $FS(X)$ similarity relations in X, and by $F\ell S(X)$ families $(R_\alpha)_{\alpha \in L}$ of equivalence relations in X with the above properties (i), (ii).

The same notations may be introduced for fuzzy measurable sets and families of measurable sets, respectively (note the necessary changes in (i), (ii) due to the fact that $L = [0,1]$ in this case).

The following theorem is a sample of what representation of fuzzy concepts means:

Theorem 4: (a) $FC(X)$ and $F\ell C(X)$ are isomorphic.

(b) $FG(X)$ and $F\ell G(X)$ are isomorphic.

(c) $FS(X)$ and $F\ell S(X)$ are isomorphic.

Proofs may be found in Negoita and Ralescu [15].

An interesting remark is in order. The inequality (4.1) defining a fuzzy convex set has been found useful in mathematics, before. As a function, u is called quasi-concave (De Finetti [4]), if it satisfies (4.1).

An interesting problem is: under what conditions on a family of sets $(A_\alpha)_\alpha$, $A_\alpha = L_\alpha(u)$ for some convex function u ? (see Fenchel [5]).

By looking at the above definitions of the "fuzzy concepts", we see that the representation theorem can be embedded in a more general framework. The concept of a "closure system" will play an important role.

Let X be a set; we say that a family of subsets $\mathcal{T} \subset P(X)$ is a closure system, if:

(CS1) $X \in \mathcal{T}$

(CS2) $\bigcap_{i \in I} F_i \in \mathcal{T}$ for any $(F_i)_{i \in I}$, $F_i \in \mathcal{T}$

For example: the subgroups of a group, or the convex subsets of a linear space, are both closure systems.

We shall denote by $\mathcal{T}_L(X)$ the collection of those fuzzy subsets $u \in F_L(X)$ such that $L_\alpha(u) \in \mathcal{T}$ for all $\alpha \in L$. As analoguous to vague sets, we consider the collection of all functions $\phi : L \to \mathcal{T}$ such that $\phi(0) = X$ and $\phi(\sup_{i \in I} \alpha_i) = \bigcap_{i \in I} \phi(\alpha_i)$. This will be denoted by $\mathcal{F}$. We can state the following:

THEOREM 5: $\mathcal{T}_L(X)$ and $\mathcal{F}$ are isomorphic.

For a proof, see Negoita and Ralescu [15].

At the end of this section we mention that other fuzzy concepts can be represented by using the above theory. Examples are: fuzzy topological spaces, fuzzy orderings, fuzzy algebraic structures.

5. REPRESENTATION OF $\mathcal{C}$-SETS

The concept of a $\mathcal{C}$-set is a significant generalization of the fuzzy set (or L-fuzzy set) concept.

This section will be more technical and the reader is assumed to have some knowledge of categorical algebra. A good reference for this is Mac Lane [13].

Let $\mathcal{C}$ be a category; $|\mathcal{C}|$ will denote the class of objects of $\mathcal{C}$; for two objects $A, B \in |\mathcal{C}|$, $\mathcal{C}(A,B)$ will denote the set of morphisms from A into B.

If X is a set, then a $\mathcal{C}$-set (or $\mathcal{C}$-subset of X) is simply a function $u : X \to |\mathcal{C}|$.

This concept generalizes fuzzy sets in the following sense: any lattice L may be thought as a category, whose objects are elements $\alpha \in L$ and a morphism and only one exists from $\alpha \in L$ into $\beta \in L$, if and only if $\alpha \leq \beta$. Then an L-set becomes a fuzzy set.

In this new context, the membership degree of an element $x \in X$ is an object $u(x)$ of the category $\mathcal{C}$.

It is possible to define the category of $\mathcal{C}$-sets; it will be denoted by $F_{\mathcal{C}}(X)$ (see Ralescu [21]).

The vague sets have a nice counterpart: a category denoted by $Fl_{\mathcal{C}}(X)$. Its objects are (contravariant) functors $\phi : \mathcal{C} \to \mathcal{P}(X)$, (where $\mathcal{P}(X)$ being a lattice, is a category in the sense described above), with the properties:

(1) $\phi(0) = X$

(2) $\phi(\coprod_{i \in I} A_i) = \bigcap_{i \in I} \phi(A_i)$, $(\forall)\ (A_i)_{i \in I} \subset |\mathcal{C}|$

By 0 we have denoted the initial object of $\mathcal{C}$, and $\coprod$ stands for the coproduct. Of course, we assume that $\mathcal{C}$ has an initial object and arbitrary coproducts.

Property (2) above shows that what is essential (from an algebraic point of view) in the definition of a vague set, is the property of commuting with coproducts.

For $\mathcal{C}$-sets, it is no longer possible to prove a "strong" version of the representation, as in Theorems 1, 2, 3. The categories $F_{\mathcal{C}}(X)$ and $Fl_{\mathcal{C}}(X)$ are neither isomorphic, nor equivalent. We may state the following "soft" representation

THEOREM 6: There is a pair of functors $F_{\mathcal{C}}(X) \underset{V}{\overset{U}{\rightleftarrows}} Fl_{\mathcal{C}}(X)$ such that V is a right-inverse, left-adjoint to U.

The proof, as well as more details, are to be found in Ralescu [21].

To conclude this section, let us note that in defining a $\mathcal{C}$-set, different points of X are no more compared via an ordering (between their membership degrees), but via morphisms in a category $\mathcal{C}$. The membership degree, itself, becomes an object of $\mathcal{C}$. It is to be expected that different examples of $\mathcal{C}$'s will give more insight in the process of modeling inexactness by fuzzy sets.

6. REPRESENTATION OF FUZZY SYSTEMS

One of the earliest applications of the representation theory was in representing fuzzy systems. In fact, we shall speak about fuzzy subsystems of a given, deterministic system. The reason for doing so is that a global process may be modeled by a deterministic system, while we might be interested in its fuzzy subsystems.

The material presented in this section is described in greater detail in Negoita and Ralescu [16]. For related work on fuzzy systems, see Sugeno [31].

A (deterministic) system is a 6-uple

$$S = (X, U, Y, \delta, \beta, x_o)$$

where X is the state space, U is the input space, Y is the output space, $\delta : X \times U \to X$ is the dynamics, $\beta : X \to Y$ is the output map, and $x_o \in X$ is the initial state.

A subsystem of S is $S_o = (X_o, U_o, Y_o, \delta_o, \beta_o, x_o)$ such that $X_o \subseteq X$, $U_o \subseteq U$, $Y_o \subseteq Y$, $\delta|_{X_o \times U_o} = \delta_o$, and $\beta|_{X_o} = \beta_o$. The intersection of a family of subsystems is defined in an obvious way. Let us denote by $Fl(S)$ the set of all families $(S_\alpha)_{\alpha \in L}$ of subsystems of S, such that

(1) $S_o = S$

(2) $S_{\sup_i \alpha_i} = \bigcap_i S_{\alpha_i}$ for all $(\alpha_i)_i \subset L$

As in the previous sections, L is a fixed, completely distributive lattice.

A fuzzy subsystem of S is

$$S_f = (f, g, h, \delta, \beta)$$

where $f \in F_L(X)$, $g \in F_L(U)$, $h \in F_L(Y)$, $\delta : X \times U \to X$, $\beta : X \to Y$ such that:

$$f(\delta(x,u)) \geq f(x) \wedge g(u) \quad , (\forall)x \in X \ , u \in U \qquad (6.1)$$

$$h(\beta(x)) \geq f(x) \quad , \quad (\forall)x \in X \qquad (6.2)$$

Note that the above inequalities mean that δ , β are morphisms in the category Set(L) as defined in section 3 .

Denoting by F(S) the set of all fuzzy subsystems of S , we can prove the following:

THEOREM 7: There is a bijection between F(S) and Fl(S) .

The proof is in [16].

Roughly speaking, a fuzzy system is a family of (deterministic) systems. Such problems as controlling fuzzy systems may be defined by using this representation theorem.

7. OPTIMIZATION WITH FUZZY CONSTRAINTS

Decision-making in a fuzzy environment was first defined by Bellman and Zadeh [2]. They used the dynamic programming approach to solve the problem.

A restatement appears in Tanaka, Okuda and Asai [32] under the name fuzzy mathematical programming. This different point of view is essential to see the fundamental role of the level sets.

The result which will be presented is valid in a more general setting (see Ralescu [25], [26] for more details).

The main problem is to maximize a cost function $f : \mathbb{R}^n \to [0,1]$ over a fuzzy set $u \in F(\mathbb{R}^n)$: $\sup_u f$. This is defined as:

$$\sup_u f = \sup_{x \in \mathbb{R}^n} (\min (f(x),u(x))) \qquad (7.1)$$

If we think of f as a goal and of u as a (fuzzy) constraint, the above means the "confluence" of the goal and constraint.

For technical reasons, u is supposed to be <u>strictly fuzzy convex</u>

$$u(\lambda x + (1-\lambda)y) > u(x) \wedge u(y) \qquad (7.2)$$

for all $x,y \in \mathbb{R}^n$, $x \neq y$, and $\lambda \in (0,1)$.

The following result then holds:

<u>THEOREM 8</u>: If u is strictly fuzzy convex, there exists an $\bar{\alpha} \in (0,1]$ such that $\sup_u f = \sup_{L_{\bar{\alpha}}(u)} f$.

Different proofs may be found in [32] or [1], [15], respectively. New results are recently reported in [6].

Since it is difficult in practice to use $\bar{\alpha}$, it may be shown that $\sup_u f = \sup_A f$, where $A = \{x \in \mathbb{R}^n / u(x) - f(x) \geq 0\}$.

The meaning of the above theorem is the following: under some hypothesis about the fuzzy constraint u, we may reject part of the information contained in u and only concentrate ourselves on some level set $L_{\bar{\alpha}}(u)$. The function u models the inexact constraint; however only part of the information is needed to solve the above optimization problem.

8. FUZZY LINEAR PROGRAMMING

A classical linear programming problem can be written under the form $\sup_{x\in M} f(x)$ where $f(x) = (x\ ,\ c) = \sum_{i=1}^{n} x_i c_i$ is a linear cost function and M is given by the linear constraints:

$$M = \{x \in \mathbb{R}^n \,/\, x_1 a_1 + x_2 a_2 + \ldots + x_n a_n \leq b\} \tag{8.1}$$

By $a_1, a_2, \ldots, a_n$ we have denoted the columns of the matrix A such that the constraints can be written as $Ax \leq b$. Thus $a_1, \ldots, a_n, b \in \mathbb{R}^m$, $c \in \mathbb{R}^n$ are given vectors.

Related to the inexact linear programming, Soyster [30] defined a linear programming problem with set-inclusive constraints. The idea is to replace $a_1, \ldots, a_n, b$ by sets $A_1, \ldots, A_n, B \subset \mathbb{R}^m$. So M takes the form:

$$M' = \{x \in \mathbb{R}^n \,/\, x_1 A_1 + \ldots + x_n A_n \subseteq B\} \tag{8.2}$$

and the problem becomes: $\sup_{x\in M'} f(x) = \sup_{x\in M'} (x\ ,\ c)$.

The extension of these ideas to the fuzzy case was done by Negoita, Flondor and Sularia [18]. They defined the <u>fuzzy linear programming</u> in the following way: $\sup_{x\in M''} f(x) = \sup_{x\in M''} (x\ ,\ c)$, where

$$M'' = \{x \in \mathbb{R}^n \,/\, x_1 u_1 + \ldots + x_n u_n \leq v\} \tag{8.3}$$

Here $u_1, \ldots, u_n, v$ are fuzzy subsets of $\mathbb{R}^m$.

This point of view can be applied in many practical problems, where the coefficients $a_1, \ldots, a_n, b$ are only known within some inexact tolerances.

The addition of fuzzy sets and multiplication with a constant, are defined in the following way:

$$(u + v)(x) = \sup_{y+z=x} (\min (u(y), v(z))) \tag{8.4}$$

$$(\lambda \cdot u)(x) = \begin{cases} u(\lambda^{-1}x) , & \text{if } \lambda \neq 0 \\ \sup_y u(y) , & \text{if } \lambda = 0 , x = 0 \\ 0 , & \text{if } \lambda = 0 , x \neq 0 \end{cases} \tag{8.5}$$

By using the representation theorem, the following result may be proved easily:

THEOREM 9: The following programming problems are equivalent:

(a) $\sup_{x \in M''} f(x) \quad , \quad M'' = \{x \in \mathbb{R}^n / x_1 u_1 + \ldots + x_n u_n \leq v\}$

(b) $\sup_{x \in N} f(x) \quad , \quad N = \bigcap_{\alpha \in [0,1]} \{x \in \mathbb{R}^n / x_1 L_\alpha(u_1) + \ldots + x_n L_\alpha(u_n) \subseteq L_\alpha(v)\}$

If v is a "bad" membership function, problem (b) is complicated, since it involves infinitely many constraints. If v has a finite number of values, the number of constraints becomes finite. The fuzzy linear programming was applied in planning problems (see [18]) .

9. FUZZY OPTIMIZATION

In sections 7 and 8, the solution of a decision-making problem in a fuzzy environment is still a classical set, of optimal points.

A new approach is to consider a fuzzy optimal set; thus "degrees of optimality". This point of view may be found in Ralescu [22], [23], [24].

To sketch the ideas, let us consider $f : \mathbb{R}^n \to \mathbb{R}$ to be a cost function, and $u \in F(\mathbb{R}^n)$ a fuzzy constraint such that $L_\alpha(u)$ are compact (i.e. closed and bounded) for all $\alpha \neq 0$. The problem is to define $\sup_u f$ as a fuzzy set.

DEFINITION: The fuzzy optimal set for the problem $\sup_u f$ is $v \in F(\mathbb{R}^n)$ defined by:

$$v(\bar{x}) = \begin{cases} \sup\{\alpha \neq 0 \ / \ f(\bar{x}) = \sup_{x \in L_\alpha(u)} f(x)\} & \text{, if such an } \alpha \text{ exists} \\ 0 & \text{, otherwise} \end{cases}$$

It may be shown that v is a subset of u . An example concerning $\sup(x^2 + 3)$ subject to the constraint

u = " x is approximately between -1 and 1 "

is given in [22]. The fuzzy optimal set corresponds with our intuition, in the following sense: while $\sup_{-1 \leq x \leq 1} (x^2 + 3)$ has as optimal set $\{-1,1\}$, the fuzzy optimal set for $\sup_u (x^2 + 3)$ is:

" x is close to -1 or to 1 "

An important problem for further study is one of "stability" : if a set A approximates the fuzzy constraint u , does the (classical) optimal set $\{\bar{x} \ / \ f(\bar{x}) = \sup_{x \in A} f(x)\}$ approximate the fuzzy optimal set v ? See [17] for a definition of the term "approximate".

10. FUZZY EQUATIONS

By fuzzy equations we mean here equations involving fuzzy relations; the unknown is a fuzzy set. This subject was studied by Sanchez [29] and found its applicability in the medical diagnosis [28] and economics [20].

In this section we shall consider the so called "weak" level sets of a fuzzy set. For X a set and $u \in F_L(X)$, these are:

$$W_\alpha(u) = \{x \in X \,/\, u(x) > \alpha\} \tag{10.1}$$

for all $\alpha \in L$.

Let us consider $R : X \times Y \to L$ a fuzzy relation, and $f \in F_L(Y)$. The problem is to solve for $u \in F_L(X)$, the following equation:

$$R(u) = f$$

By $R(u)$ we mean the fuzzy subset of Y, defined by:

$$R(u)(y) = \bigvee_{x \in X} [u(x) \wedge R(x,y)] \quad , \quad y \in Y$$

At this level of generality, it is not to be expected to find "explicitely" the solutions u. However, by using the representation theory, we may prove that this problem can be reduced to the classical case. By this, we mean to solve equations of the form: $E(M) = A$, where $E \subset X \times Y$ (a relation), $M \subset X$, $A \subset Y$, and $E(M)$ is defined by:

$$E(M) = \{y \in Y \,/\, (\exists) x \in M \,,\, (x,y) \in E\} \tag{10.2}$$

We can prove the following

THEOREM 10: (a) If $u \in F_L(X)$ is a solution of $R(u) = f$ then, for each $\alpha \in [0,1]$, the level set $W_\alpha(u)$ is a solution of

$$W_\alpha(R)(M) = W_\alpha(f) \tag{10.3}$$

(b) If M_α is a solution of $W_\alpha(R)(M) = W_\alpha(f)$ for each $\alpha \in [0,1]$, and the family $(M_\alpha)_\alpha$ satisfies (i) $M_1 = \phi$, (ii) $\alpha \leq \beta \Rightarrow M_\alpha \supseteq M_\beta$, (iii) $x \in M_\alpha \Rightarrow (\exists)\beta > \alpha \,,\, x \in M_\beta$, then $u(x) = \sup\{\alpha/x \in M_\alpha\}$ is a solution of $R(u) = f$.

Proof: (a) We have to show that $W_\alpha(R)(W_\alpha(u)) = W_\alpha(f)$. This immediately follows from the equality $W_\alpha(R(u)) = W_\alpha(R)(W_\alpha(u))$.

(b) let us consider a family $(M_\alpha)_\alpha$ with properties (i)-(iii). To show that $R(u) = f$, where $u(x) = \sup\{\alpha/x \in M_\alpha\}$, observe that

$$R(u)(y) = \sup\{\alpha/y \in W_\alpha(R(u))\} = \sup\{\alpha/y \in W_\alpha(R)(W_\alpha(u))\}$$

But it may be shown that $W_\alpha(u) = M_\alpha$, $(\forall)\alpha$, therefore

$$R(u)(y) = \sup\{\alpha/y \in W_\alpha(R)(M_\alpha)\} = \sup\{\alpha/y \in W_\alpha(f)\} = f(y) \quad .$$

It follows, by this theorem, that the problem of solving $R(u) = f$ was reduced to that of finding a special family of solutions for the family of equations $W_\alpha(R)(M) = W_\alpha(f)$ involving sets and classical relations.

11. CONCLUDING REMARKS

Our aim in this paper was to survey the theory and application of the representation theorem for fuzzy sets. It is to be expected that more applications are to come.

The main advantage of this approach is in the possibility of reducing a problem in fuzzy sets theory to a family of classical problems.

Some complications can arise, however, in applying the representation theory. For example, the property $X_{\sup_i \alpha_i} = \bigcap_i X_{\alpha_i}$ for a family of sets $(X_\alpha)_\alpha$ is difficult to be checked, in some problems. It is easier to work with a decreasing family (X_α), i.e. $\alpha \leq \beta \Rightarrow X_\alpha \supseteq X_\beta$. Hints of how to use such families are to be found in [24].

The representation may also be used in assessing the membership functions (problem (A) as stated in the Introduction). Instead of trying to define a function $u : X \to [0,1]$, we may try to find the level sets $(X_\alpha)_\alpha$ and then to construct the function. The levels $(X_\alpha)_\alpha$ are to be thought as "nuances" or "interpretations" of the inexact concept that is to be modeled.

The representation theory looks therefore like a good methodology in coping with "inexact problems". However, this should not be viewed as a definitive tool; work is still to be done in this subject.

ACKNOWLEDGEMENTS

In developing the subject concerning the applications of the representation theory, the author received helpful comments and support during the past years. Special thanks are due to Claude Ponsard and to the hospitality of the Institut de Mathématiques Economiques, Université de Dijon, France.

The importance of studying the fuzzy optimal set was pointed out to me by Lotfi Zadeh.

Many of the topics presented here are based on my joint work with Dinu Negoita.

Finally, I thank my wife, Anca, for continuous encouragements throughout the period of writing this paper.

REFERENCES

[1]. K. Asai, H. Tanaka, C.V. Negoita and D. Ralescu (1978), "Introduction to Fuzzy Systems Theory", Ohm-Sha Co. Ltd., Tokoyo, (in Japanese).

[2]. R. Bellman and L.A. Zadeh (1970), "Decision-making in a fuzzy environment", Management Sciences, vol. 17 B (4), pp. 141-164.

[3]. C.L. Chang (1968), "Fuzzy topological spaces", J. of Math. An. and Appl., vol. 24, pp. 182-190.

[4]. B.De Finetti (1949), "Sulle stratificazioni convesse", Ann. Mat. Pura Appl. [4] vol. 30, pp. 173-183.

[5]. W. Fenchel (1953), "Convex Cones, Sets and Functions", (mimeographed lecture notes), Princeton Univ. Press, Princeton, N.J.

[6]. J. Flachs and M.A. Pollatschek (1978), "Further results on fuzzy-mathematical programming", Inf. and Control, vol. 38, pp. 241-257.

[7]. P. Flondor (1975), "On C-sets", Working Group on Fuzzy Systems, Bucharest, (unpublished).

[8]. B.R. Gaines and L.J. Kohout (1977), "The fuzzy decade: a bibliography of fuzzy systems and closely related topics", Int. J. Man-Machine Studies, vol. 9, pp. 1-68.

[9]. Y. Gentilhomme (1968), "Les ensembles flous en linguistique", Cahiers de Linguistique Théorique et Appliquée, vol. 5, #47.

[10]. J.A. Goguen (1967), "L-fuzzy sets", J. of Math. An. and Appl., vol. 18, pp. 145-174.

[11]. J.A. Goguen (1969), "Categories of V-sets", Bull. of the Amer. Math. Soc., vol. 75, pp. 622-624.

[12]. J.A. Goguen (1974), "Concept representation in natural and artificial languages: axioms, extensions and applications for fuzzy sets", Int. J. of Man-Machine Studies, vol. 6, pp. 513-561.

[13]. S.Mac Lane (1971), "Categories for the Working Mathematician", Springer-Verlag, Berlin.

[14]. C.V. Negoita and D.A. Ralescu (1975), "Representation theorems for fuzzy concepts", Kybernetes, vol. 4, pp. 169-174.

[15]. C.V. Negoita and D.A. Ralescu (1975), "Applications of Fuzzy Sets to Systems Analysis", Birkhäuser Verlag, Basel.

[16]. C.V. Negoita and D.A. Ralescu (1977), "Some results in fuzzy systems theory", in J. Rose and C. Bilciu (eds.): "Modern Trends in Cybernetics and Systems", Springer-Verlag, Berlin.

[17]. C.V. Negoita and D.A. Ralescu (1977), "On fuzzy optimization", Kybernetes, vol. 6, pp. 193-195.

[18]. C.V. Negoita, P. Flondor and M. Sularia (1977), "On fuzzy environment in optimization problems", in J. Rose and C. Bilciu (eds.): "Modern Trends in Cybernetics and Systems", Springer-Verlag, Berlin.

[19]. C.V. Negoita and Al. Stefanescu, "Fuzzy objects in topoi: a generalization of fuzzy sets" (to appear).

[20]. Cl. Ponsard (1978), "Jalons pour une théorie spatiale du comportement du consommateur", Document de Travail, IME, Dijon.

[21]. D. Ralescu (1978), "Fuzzy subobjects in a category and the theory of C-sets", Int. J. of Fuzzy Sets, vol. 1, pp. 193-202.

[22]. D. Ralescu (1978), "The interface between orderings and fuzzy optimization", paper presented at the ORSA/TIMS National Meeting, Los Angeles, California.

[23]. D. Ralescu (1978), "Orderings, preferences, and fuzzy optimization", Proc. of the 4th Int. Congress of Cybern. and Syst., Amsterdam, Aug. 21-24 (to appear).

[24]. D. Ralescu (1977), "Inexact solutions for large-scale control problems", Proc. of the First World Conf. on Math. at the Service of Man, Barcelona, Spain, July.

[25]. D. Ralescu, "Integration on fuzzy sets" (to appear).

[26]. D. Ralescu, "Toward a general theory of fuzzy variables" (forthcoming).

[27]. A. Rosenfeld (1971), "Fuzzy groups", J. of Math. An. and Appl., vol. 35, pp. 512-517.

[28]. E. Sanchez and R. Sambuc (1976), "Relations floues, fonctions Φ-floues. Application à l'aide au diagnostic en pathologie thyroïdienne", in Medical Data Processing Symposium, IRIA, Toulouse, 1976(Taylor and Francis Ltd., London).

[29]. E. Sanchez (1976), "Resolution of composite fuzzy relation equation", Inf. and Control, vol. 30, pp. 38-48.

[30]. A.L. Soyster (1973), "Convex programming with set-inclusive constraints; application to inexact linear programming", Oper. Research, vol. 21, pp. 1154-1157.

[31]. M. Sugeno (1977), "Fuzzy systems with underlying deterministic systems", Report #3: Summary of papers on general fuzzy problems, Tokyo, Japan.

[32]. H. Tanaka, T. Okuda and K. Asai (1974), "On fuzzy mathematical programming", J. of Cybernetics, vol. 3, pp. 37-46.

[33]. L.A. Zadeh (1965), "Fuzzy sets", Inf. and Control, vol. 8, pp. 338-353.

[34]. L.A. Zadeh (1968), "Probability measures of fuzzy events", J. of Math. An. and Appl., vol. 23, pp. 421-427.

[35]. L.A. Zadeh (1971), "Similarity relations and fuzzy orderings", Inf. Sci., vol. 3, pp. 177-200.

ADVANCES IN FUZZY SET THEORY AND APPLICATIONS
M.M. Gupta, R.K. Ragade, R.R. Yager (editors)

FUZZY LOGIC AND ITS APPLICATION TO FUZZY REASONING

J.F. BALDWIN

Department of Engineering Mathematics,
University of Bristol,
U.K.

Summary

A fuzzy logic based upon truth value restrictions, which are fuzzy sets on truth space is proposed, and a method of analysis for this logic, which is independent of any particular multi-valued base logic, is given. It is then shown how this fuzzy logic and method of analysis can be used for theorem proving and reasoning with vague statements and loose concepts.

Applications of such an approach of fuzzy reasoning are discussed with special reference to data processing, pattern recognition, fuzzy concept definition and hypothesis testing, control of complex man/machine systems, decision theory for decision making with multiple and conflicting goals and hierarchical modelling of complex systems.

Introduction

The precision required by modern scientific analytical methods, although in one sense an admirable quality, in that, it motivates careful rather than rash reasoning, inevitably is too restricting, since it only allows application to that which can be precisely defined and therefore, in the limit, to nothing at all.

Heisenberg's uncertainty principle gives an upper bound to the precision of knowledge concerning even the state of an elementary particle and this limitation is a result of interaction of the system and observer. This interaction is necessary for observation and without it no knowledge would be obtained. The principle of uncertainty is applicable to all Man/Machine systems in which to know something about the system is to necessarily interact with that system. For example, if the system is a child and the observer his parent, then interaction can take the form of a conversation between child and parent or simply the presence of the two in the same room. If the system is observed from afar with only weak interaction, then only very vague and fuzzy knowledge will be obtained. As the observer strengthens his interaction with the system, this interaction becomes more concentrated about a smaller and smaller part of the system, so learning more and more about less and less.

Thus, when observing and analysing complex systems, whether in the technological, scientific, medical or economic fields, one must balance the need for precision with the need for general perception. Over precision will lead to an over specialised perception while a too vague analysis may give such a general perception as not to be useful. A photograph of too large a landscape gives only a very general impression, while a snapshot of a face tells much about the face but nothing of the surroundings. To make a summary of a large mass of data, the use of fuzzy concepts will generally be necessary to give an overall interpretation and this, expressed in the form of fuzzy linguistic statements, can provide an economical representation of the data. This summary can then be used with similar summaries from other sources for modelling and fuzzy reasoning.

Reasoning with loose concepts is therefore not to be deplored but controlled so that adequate precision is obtained but not at the expense of tunnel vision. Zadeh introduced fuzzy set theory [1], [2] and initiated much of the research in fuzzy systems [3], [4]. Fuzzy set theory assumes that not all objects can be given precise classification. For example it is possible to be neither short nor not short. The law of excluded middle does not always apply. It has been stated by Russell [5] that "All traditional logic habitually assumes that precise symbols are being employed. It is therefore not applicable to this terrestrial life, but only to an imagined celestial existence". Consider for example the following educationalist's remark: "Children from underprivileged homes and difficult home environments require nursery education to compensate for their lack of educational stimulation during early years of childhood". One cannot associate with this proposition a truth value of true or false without giving much greater precision to many of the terms used. Such concepts as underprivileged, difficult home environments, nursery education, lack .... stimulation, early years, are all fuzzy concepts and although a further nesting of definition can be given, this nesting would never lead to precise definition. The statement is extremely economical and is probably a good summary of many other statements of less generality. High level decision makers have a need for general statements like this which summarises numerous more detailed statements. See Bellman [6] for comment on ambiguity.

Man/Machine systems will play an ever important role in years to come. Man will input to the computer impressions, general observations, etc. and will wish to receive outputs from the computer in a suitable form for human reasoning. Imagine a man/machine complex playing chess against another man/machine complex. An optimal division of labour would be to allow the machine to do what it can do best and man do what he can do best. Communication between the two would be of a form suitable to both. The conversation will be extremely limited if man is restricted to too rigid a language in which each term must be precisely defined, each category crisp and every statement true or false with the law of excluded middle always to hold.

Imprecision because of vagueness of definition is different to the uncertainty associated with an event because of randomness or the uncertainty of the truth of a statement, which is known to be true or false, because of the lack of information. For example, the written 'a' cannot be precisely defined since if it were then most of us would not be able to write. Its definition has a flexibility which allows many variations without defining what these variations are to be. I can write a letter 'a' in a way which has never been constructed previously and yet the reader will have no difficulty in identifying it. This is because a letter 'a' has a few robust characteristics which distinguish it from other letters. These characteristics are not given to, nor confused by, subtleties. Furthermore for identifying a written letter 'a', after a certain point of precision, greater precision of the defining standard will lead to more errors of identification. Of course semantic, in addition to structural, information is used for identification purposes. The possibility of it being a letter 'a' is also concerned with the possibility of an 'a' making up a word which makes sense in the sentence the word is in, and in the general context of the piece of writing as a whole. This again can only be decided using a fuzzy analysis except for simple tutorial type examples. Of course a probability analysis is also possible for this example since it is true that the written letter is either an 'a' or not an 'a' assuming the script is written English. Even then fuzzy probabilities would be required and the best model would probably use fuzzy set theory, fuzzy logic and fuzzy probability theory.

A suitable model for defining the letter 'a' might be as follows

letter 'a' $\supset$ an approximate circular part is very true

a joining vertical line on the right
$\supset$ which is approximately the same size is fairly true
as the diameter of the circular part

approximate circular part $\supset$ roughly speaking a closed curve is true
$\supset$ no major sharp corners is very true

etc.

when $\supset$ stands for material implication.

This hierarchically structured model is the type of model which the method of fuzzy reasoning, described later, will process and the result will be a certain falsification of the fact that the written letter is an 'a'. The concepts at the bottom of the hierarchy will be directly measurable and these measurements will be inputs to the model. The fuzzy truth values in this model can be thought of as a measure of importance of the corresponding proposition. A similar approach for the letter 'd' could be given and all other letters. Suppose for a given symbol the results showed letter 'a' to be fairly false, letter 'd' to be very false and other letters to be absolutely false then from structural considerations letter 'a' would be chosen. The meaning of these fuzzy truth values will be given later but here they can be interpreted as we use them in every day conversation. Of course from semantic considerations a different conclusion might be drawn and a conflict situation arises which can be solved using a fuzzy compromise model where semantic and structural considerations are related.

We are thus accepting the fact that a proposition can have truth values other than true and false and that these truth values can themselves be fuzzy in nature. This fuzziness is necessary when dealing with models involving propositions containing fuzzy terms and also fuzzy data to be used with the model, which will be understood more clearly later. For example we may wish to know the truth of proposition P where

P : John is tall

when we are given the data D where

D : John is not more than about 6 ft but much greater than 5 ft.

A conclusion might be

John is tall is fairly true

and the process for determining this will be fully defined below. It should be pointed out, however, that a multi-valued logic without a fuzzification is not enough. Fuzzy logic as given in this paper is a fuzzification of a multi-valued logic and can be viewed as dealing with truth value distributions. With fuzzy logic and fuzzy set theory fuzzy concepts can be modelled, fuzzy algorithms and programs produced and fuzzy controllers constructed. A completely different outlook can be given to such areas as data processing, hypothesis testing, decision and control theory, pattern recognition, all of which are of the utmost importance in modelling real life problems.

The approach of falsification indicated briefly in the letter 'a' example above can be interpreted as a generalisation of Popper's approach to scientific method [7] where fuzzy concepts are allowed both in the basic theory and in the observation data of a test experiment. A fuzzy falsification will be given. This gives rise to the notion of a fuzzy demarcation of science and such a fuzzy classification of science-v-non-science would seem very reasonable for areas such as the social sciences and engineering.

Wiener's cybernetic theory which is basically a theory of messages for the control of and communication between man/machine systems can be extended to incorporate fuzzy uncertainties in addition to statistical uncertainties. The concept of fuzzy entropy has been discussed by Kauffmann [8], Zadeh [9], De Luca and Termini [10] and Baldwin and Pilsworth [11], and Ragude [29].

The fact that uncertainty is many faceted has been discussed by Bellman and Zadeh [12] and Gaines [13]. Human reasoning is more sophisticated than classical logic reasoning as exemplified by the ability to reason with imprecisely defined and robust concepts and also imprecise data. Furthermore this reasoning with uncertainties is not easily explained in terms of probabilistic models. Fuzzy logic provides another way of treating uncertainty and we think more applicable to those situations which lack preciseness of definition.

Two concepts of importance to man are

(i) approximate reasoning

(ii) feedback.

Both are required for a control theory which is to be applicable to complex man/machine systems. Desired system performance, concepts of a stable system, a reliable system, an adaptive system are all fuzzy terms.

Consider for example the possible definition of a humane industrial society:

> "A humane industrial society is one for which everyone has at least a reasonable standard of living with a fair distribuiton of income where all people have the opportunity of satisfying and interesting work and pastimes. It should be flexible but stable both industrially and socially and be adaptable to outside influences. Equal educational opportunities should be available of high standard, good medical facilities provided and the dignity of man respected."

Almost every term here is fuzzy in nature, lacking precision in definition and this will be so when attempting to define the required performance of any system. Furthermore many of the criteria are in conflict with each other and the system performance cannot be described in such elementary terms as satisfied or not satisfied.

This, then, is the background and motivation for establishing a fuzzy logic which in conjunction with fuzzy set theory can be used for approximate or fuzzy reasoning, modelling complex systems, fuzzy data processing, designing fuzzy controllers and perhaps to give a logic nearer to human reasoning than either classical or probability logics.

In section 2 we introduce the idea of a truth value restriction, truth functional modification and inverse truth functional modification. These are necessary to the development of fuzzy logic given in section 3. Examples of fuzzy theorem proving using the method of analysis described in section 3 are given and other applications indicated. Section 4 deals with fuzzy reasoning based upon fuzzy logic and fuzzy set theory and many examples are given to illustrate the method. Computational details are not stressed but these can be found in references [14], [15] as can details of a general fuzzy reasoning program called FUZLOG which we have recently written. The method is computationally efficient because it works in a simple truth space and many of the calculations can be done with pencil and paper. Section 5 briefly discusses applications of the method of fuzzy reasoning described here. We have so far applied it to several problems including fuzzy pattern recognition, military strategy design and an ergonomic design of a ship control room.

2. Truth Value Restrictions

The concept of truth value restriction is fundamental to the approach of fuzzy logic reasoning given below and has been described in Baldwin [16], [17], [18].

Let U denote the set of possible truth values, so that for classical logic U = {0,1} with 0 denoting false and 1 denoting true while for a non-denumerably infinite valued logic U = [0,1].

Definition 2.1

A truth value restriciton τ is a subset of U, denoted by $\tau \subset U$, and can be defined by its membership function, χ_τ, which is a mapping

$$\chi_\tau \; : \; U \to \{0,1\}$$

Definition 2.2

A fuzzy truth value restriction $\underset{\sim}{\tau}$ is a fuzzy subset of U, denoted by $\underset{\sim}{\tau} \underset{\sim}{\subseteq} U$ and can be defined by its membership function, $\chi_{\underset{\sim}{\tau}}$, which is a mapping

$$\chi_{\underset{\sim}{\tau}} \; : \; U \to [0,1]$$

Definition 2.1 is thus relevant to binary logic with U = {0,1} and a non-denumerably infinite valued logic with U = [0,1] while definition 2.2 is relevant to fuzzy logic with U = [0,1]. If we add the further restriction that $\chi_{\underset{\sim}{\tau}}(0) + \chi_{\underset{\sim}{\tau}}(1) = 1$ then definition 2.2 with U = {0,1} is relevant to probability logic.

Definition 2.3

Let X be a universe of discourse given by the interval [a,b]; a, b ε R' when R' is the real line. Let $\underset{\sim}{x}$ be a label associated with the function $\chi_{\underset{\sim}{x}}$ defined by

$$\chi_{\underset{\sim}{x}}(\eta) = \frac{\eta - a}{b - a} \; ; \forall \eta \in [a,b]$$

then $\underset{\sim}{x}$ is said to be a fuzzy subset of X, denoted by $\underset{\sim}{x} \underset{\sim}{\subseteq} X$ and is called the reference ramp of X.

Definition 2.4

For the universe of discourse X defined above, if $\chi_{\underset{\sim}{y}}$ is a mapping

$$\chi_{\underset{\sim}{y}} \; : \; X \to U$$

then $\underset{\sim}{y}$ is said to be a fuzzy subset of X, written as $\underset{\sim}{y} \underset{\sim}{\subseteq} X$.

Definition 2.5

Let $\underset{\sim}{y} \underset{\sim}{\subseteq} X$ and $\underset{\sim}{\tau} \underset{\sim}{\subseteq} U$ then

$$\underset{\sim}{y}(\underset{\sim}{\tau}) \underset{\sim}{\subseteq} X \text{ where } \quad \chi_{\underset{\sim}{y}(\underset{\sim}{\tau})}(\eta) = \chi_{\underset{\sim}{\tau}}(\chi_{\underset{\sim}{y}}(\eta))$$

and this process of modifying $\underset{\sim}{y}$ to $\underset{\sim}{y}(\underset{\sim}{\tau})$ by means of $\underset{\sim}{\tau}$ is called truth functional modification (T.F.M.). We will sometimes write $\underset{\sim}{y}(\underset{\sim}{\tau})$ = T.F.M. ($\underset{\sim}{y}/\underset{\sim}{\tau}$). Truth functional modification, first introduced by Zadeh [3] and its inverse, introduced by Baldwin [16], which will be discussed later, play an important role in the theory to follow.

We might ask what is the fuzzy truth value restriction $\underset{\sim}{\tau}_s$ which when applied to $\underset{\sim}{y} \underset{\sim}{\subseteq} X$ in the sense just described gives $\underset{\sim}{y} \underset{\sim}{\subseteq} X$

$$\text{i.e.} \quad \underset{\sim}{y}(\underset{\sim}{\tau}_s) = \underset{\sim}{y}$$

Obviously $\underset{\sim}{\tau}_s$ is defined by

$$\chi_{\underset{\sim}{\tau}_s}(\eta) = \eta \; ; \; \forall \eta \in U$$

and we give $\underset{\sim}{\tau}_s$ the special label true.
Thus for $\underset{\sim}{y} \subseteq X$, $\underset{\sim}{y}(\text{true}) = \underset{\sim}{y}$.
Other special truth value restrictions can be defined.

Definition 2.6

The fuzzy truth value restriction true, false, unrestricted, impossible, absolutely true and absolutely false are defined as

$$\left.\begin{array}{ll} \chi_{\text{true}}(\eta) & = \eta \\ \chi_{\text{false}}(\eta) & = 1 - \eta \\ \chi_{\text{unrestricted}}(\eta) & = 1 \\ \chi_{\text{impossible}}(\eta) & = 0 \\ \chi_{\text{absolutely true}}(\eta) & = \begin{cases} 1 & \text{if } \eta = 1 \\ 0 & \text{otherwise} \end{cases} \\ \chi_{\text{absolutely false}}(\eta) & = \begin{cases} 1 & \text{if } \eta = 0 \\ 0 & \text{otherwise} \end{cases} \end{array}\right\} \quad \forall\ \eta \in U$$

See Figure 1 for pictorial representation.

Theorem 2.1

For a given fuzzy subset $\underset{\sim}{y} \subseteq X$ there exists a fuzzy truth value restriction $\underset{\sim}{\tau} \subseteq U$ such that $\underset{\sim}{x}(\underset{\sim}{\tau}) = \underset{\sim}{y}$ when $\underset{\sim}{x} \subseteq X$ is the reference ramp of X.

A fuzzy subset $\underset{\sim}{y} \subseteq X$ can thus be represented by the two-tuple $(X, \underset{\sim}{\tau})$.

This theorem is obvious and the representation of fuzzy subsets as two-tuples has considerable computational advantages. In Figure 2 examples for tall and very tall are given and these definitions are purely subjective.

Definition 2.7

If $\underset{\sim}{y} \subseteq X$ then NOT $\underset{\sim}{y} \subseteq X$ is defined as NOT $\underset{\sim}{y} = \underset{\sim}{y}$ (false).

Theorem 2.2

If $\underset{\sim}{y} \subseteq X$ then NOT $\underset{\sim}{y}$ has the membership function $\chi_{\text{NOT } \underset{\sim}{y}}(\eta) = 1 - \chi_{\underset{\sim}{y}}(\eta)$; $\forall\ \eta \in U$.

Proof:

$$\chi_{\underset{\sim}{y}(\text{false})}(\eta) = \chi_{\text{false}}\left(\chi_{\underset{\sim}{y}}(\eta)\right) = 1 - \chi_{\underset{\sim}{y}}(\eta)$$

In what follows we will be concerned with propositions containing fuzzy terms such as John is tall, short, heavy, etc. or John and Bill have approximately the same height which requires the concept of a fuzzy relation. We can generalise the definition of fuzzy subset given above to product spaces.

Definition 2.8

A fuzzy binary relation $\underset{\sim}{R}$ from a universe of discourse L to a universe of discourse M is a fuzzy subset of $M \times L$ characterised by a membership function

$$\chi_{\underset{\sim}{R}} : L \times M \rightarrow U$$

and denoted by $\underset{\sim}{R} \subseteq L \times M$.

If $L = M$ then $\underset{\sim}{R}$ is a fuzzy relation on L.
Similarly a n-ary fuzzy relation $\underset{\sim}{R}$ can be defined on the produce space $X_1 \times X_2 \times \ldots \times X_n$ by means of the mapping

$$\chi_{\underset{\sim}{R}} \; : \; X_1 \times X_2 \times \ldots \times X_n \;\rightarrow\; U$$

It will be possible in most cases to define a fuzzy relation in terms of a reference ramp fuzzy subset and a fuzzy truth value restriction. For example, consider the fuzzy relation approximately equal, denoted by $\underset{\sim}{R}$. Then $\underset{\sim}{R} \subseteq X \times Y$. Let $L \doteq [-a, a]$; $a, -a \in R'$; and let $\underset{\sim}{\ell} \subseteq L$ be the reference ramp of L. Then we can chose a fuzzy truth value restriction $\underset{\sim}{\tau} \subseteq U$ such that $\underset{\sim}{R}$ is defined by means of the two-tuple $(L, \underset{\sim}{\tau})$ where

$$\chi_{\underset{\sim}{R}}(x,y) \;=\; \chi_{\underset{\sim}{\tau}}(\chi_{\underset{\sim}{\ell}}(x-y)) \;;\; \forall\, x \in X \text{ and } \forall\, y \in Y$$

where 'a' is chosen such that

$$x - y \in L \;; \quad \forall\, x \in X, \quad \forall\, y \in Y.$$

This example is shown in Figure 3.

In what follows we will be concerned with allocating fuzzy truth value restrictions to propositions. These propositions may be atomic or compound and will involve fuzzy concepts in the form of fuzzy subsets and fuzzy relations. Compound propositions will be dealt with fully in the next section and we restrict the discussion in this section to atomic proposition.

Definition 2.9

The meaning of a proposition $P \doteq u$ is p, where u is the name of an object (or a construct) and $\underset{\sim}{p} \subseteq X$, is expressed by the relational assignment equation

$$R\,(A(u)) \;=\; \underset{\sim}{p}$$

Where A is an implied attribute of u, i.e. an attribute which is implied by u and p; and R denotes a fuzzy restriction on A(u) to which the value p is assigned.

This definition is that given by Zadeh [3] and has been fully discussed by him. The implied attribute u may have a nested structure, i.e. may be of the general form

$$A_k \left(A_{k-1} \,(\ldots(A_2(A_1(u))\,\ldots\right) .$$

For example

1. Anita's piano playing is fairly good is written as

   R (Quality (piano playing (Anita))) = fairly good

2. John and Bill have approximately the same height

   R (Relation (Height (John, Bill))) = approximately equal.

We now introduce the concept of inverse truth functional modification first introduced by Baldwin [16].

Definition 2.10

If P : u is $\underset{\sim}{p}$; $\underset{\sim}{p} \subseteq X$

and further it is known that u is $\underset{\sim}{p}'$; $\underset{\sim}{p}' \subseteq X$ then $v(P|u \text{ is } \underset{\sim}{p}')$, read as the truth value restriction of proposition P given that u is $\underset{\sim}{p}'$, is defined by the membership function

$$\chi_{\underset{\sim}{\tau}}(\eta) \;=\; \bigvee_{\substack{x \in X \\ \chi_{\underset{\sim}{p}}(x)=\eta}} \left[\chi_{\underset{\sim}{p}'}(x)\right] \;;\quad \forall\, \eta \in U$$

This is termed inverse truth functional modification (I.T.F.M.) and the proposition P, given the data u is $\underset{\sim}{p}'$, can be replaced by the proposition P' where

$$P' \; : \; (u \text{ is } \underset{\sim}{p}) \text{ is } \underset{\sim}{\tau} \quad \text{or} \quad v(P) = \underset{\sim}{\tau}.$$

We can thus write $v(P) = \text{I.T.F.M.} (\underset{\sim}{p} \,/\, \underset{\sim}{p}')$.

Theorem 2.3

If $\underset{\sim}{p} \doteq (\tau_{\underset{\sim}{p}}, X)$ and $\underset{\sim}{p}' \doteq (\tau_{\underset{\sim}{p}'}, X)$ for the last definition then

$$\chi_{v(P|\, u \text{ is } \underset{\sim}{p}')}(\eta) = \bigvee_{\substack{x \in U \\ \chi_{\tau_{\underset{\sim}{p}}}(x)=\eta}} \left[\chi_{\tau_{\underset{\sim}{p}'}}(x)\right]$$

The proof is obvious.

Figure 4 illustrates an example for the truth of the proposition 'John is tall' given the data 'John is very tall'. Other examples and discussion can be found in Baldwin [16], [19]. Here we refrain from defining very true, fairly true etc. since these definitions will vary from one situation to another. Nevertheless, the approximate description of terms like this are shown in Figure 5.

Finally in this section we consider compound linguistic statements such as tall but NOT very tall, tall OR short. These take the form of

if P : u is $\underset{\sim}{p}$ where $\underset{\sim}{p} \doteq (\tau_{\underset{\sim}{p}}, X)$

and P' : u is $\underset{\sim}{p}'$ where $\underset{\sim}{p}' \doteq (\tau_{\underset{\sim}{p}'}, X)$

then (i) Q : P AND P' is $\underset{\sim}{\tau}$

or (ii) Q : P or P' is $\underset{\sim}{\tau}$

where $\underset{\sim}{\tau}$ is the fuzzy truth value restriction associated with the compound statement. These are special cases of compound propositions since $\underset{\sim}{p}$ and $\underset{\sim}{p}'$ are defined on the same universe of discourse. They are equivalent to

(i) Q : u is ($\underset{\sim}{p}$ AND $\underset{\sim}{p}'$ is $\underset{\sim}{\tau}$)

(ii) Q : u is ($\underset{\sim}{p}$ or $\underset{\sim}{p}'$ is $\underset{\sim}{\tau}$)

The following definition, which defines the conjunction and disjunction of linguistic statements, is a special case of a more general definition given in the next section for the disjunction and conjunction of propositions associated with different universes of discourse.

Definition 2.11

If $\underset{\sim}{p} \doteq (\tau_{\underset{\sim}{p}}, X)$ and $\underset{\sim}{q} \doteq (\tau_{\underset{\sim}{q}}, X)$

then $\underset{\sim}{p}$ AND $\underset{\sim}{q}$ is $\underset{\sim}{\tau} \doteq (\tau_{\underset{\sim}{p}} . \cap_{(\underset{\sim}{\tau})} . \tau_{\underset{\sim}{q}})$ and

$\underset{\sim}{p}$ OR $\underset{\sim}{q}$ is $\underset{\sim}{\tau} \doteq (\tau_{\underset{\sim}{p}} . \cup_{(\underset{\sim}{\tau})} . \tau_{\underset{\sim}{q}})$

where $\chi_{\tau_{\underset{\sim}{p}} \cap_{(\underset{\sim}{\tau})} \tau_{\underset{\sim}{q}}}(\eta) = \chi_{\underset{\sim}{\tau}}(\chi_{\tau_{\underset{\sim}{p}}}(\eta) \wedge \chi_{\tau_{\underset{\sim}{q}}}(\eta))$ and

$\chi_{\tau_{\underset{\sim}{p}} \cup_{(\underset{\sim}{\tau})} \tau_{\underset{\sim}{q}}}(\eta) = \chi_{\underset{\sim}{\tau}}(\chi_{\tau_{\underset{\sim}{p}}}(\eta) \vee \chi_{\tau_{\underset{\sim}{q}}}(\eta))$

where $(a \wedge b) \doteq \text{MIN}(a,b)$ and $a \vee b \doteq \text{MAX}(a,b)$.

This is a generalised intersection and union of fuzzy subsets. In fact

$$\cap_{(\text{true})} = \cap \text{ and } \cup_{(\text{true})} = \cup$$

Figure 6 gives an illustration of tall AND NOT very tall is $\underset{\sim}{\tau}$ for $\underset{\sim}{\tau}$ = true and a $\underset{\sim}{\tau}$ which normalises the resultant fuzzy subset. This normalisation can be thought of as resulting from a compromise in the truth of the compound statement from accepting it as true to accepting it as $\underset{\sim}{\tau}$. If the compound statement arises from different sources then a truth value restriction can be allocated to each statement separately and a normalised subset obtained by separate compromising of the truth of each source. For example suppose a little old lady makes statement P and a policeman P'

P : The criminal was very tall ($\underset{\sim}{\tau}$)
P' : The criminal was tall but NOT very tall ($\underset{\sim}{\tau}'$)

then we can give greater credibility to P' than P when choosing $\underset{\sim}{\tau}$ and $\underset{\sim}{\tau}'$ to give a normalised fuzzy subset. In this example we take AND as true and give an illustration in Figure 7. This concept of compromise can be used to model bargaining and other aspects of conflict between several parties.

3. Fuzzy Logic

A multi-valued logic is used as the base logic for the development of a fuzzy logic. No particular base logic is assumed, although the development below is in terms of the non-denumerably infinite multi-valued logic system of Lukasiewicz as a base logic [28]. The general method of analysis must be independent of any choice of base logic so that it is generally applicable. The method below is suitable with any particular base logic and can also be used for classical logic theorem proving to avoid truth table analysis in the case of the propositional calculus.

It is assumed that a compound proposition is composed of atoms connected through the logic connectives of conjunction (AND), disjunction (OR), implication ($\supset$), equivalence ($\equiv$) and negation ($\neg$). We also associate with any compound proposition a truth value restriction. For example

P AND Q is $\underset{\sim}{\tau}$.

This is not required in classical logic since P AND Q when used as an axiom means P AND Q is true.

We will associate with any given fuzzy logic statement a fuzzy logic relation defined on the appropriate product space. For example if a statement $\sum$ is in terms of atoms $P_1, P_2, \ldots P_n$ then a fuzzy relation $\underset{\sim}{R}$, $\underset{\sim}{R} \subseteq U_{P_1} \times U_{P_2} \ldots \times U_{P_n}$ will be defined where U_{P_i} is the truth space associated with the atomic proposition P_i. How to determine this fuzzy relation $\underset{\sim}{R}$ will be discussed below.

Definition 3.1

We assume that using the base logic being considered, a fuzzy relation can be defined for each of the logic operators.

Let $\underset{\sim}{C}$, $\underset{\sim}{D}$, $\underset{\sim}{I}$, $\underset{\sim}{E}$ be fuzzy binary relations corresponding to conjunction, disjunction, implication and equivalence. Let $\underset{\sim}{N}$ be a fuzzy unary relation corresponding to negation.

Definition 3.2

If the non-denumerably infinite multi-valued logic system of Lukasiewicz is used as base logic then

$$\begin{aligned}
\underset{\sim}{C} &\subseteq U_1 \times U_2\ ; & \chi_{\underset{\sim}{C}}(x,y) &= x \wedge y \\
\underset{\sim}{D} &\subseteq U_1 \times U_2\ ; & \chi_{\underset{\sim}{D}}(x,y) &= x \vee y \\
\underset{\sim}{I} &\subseteq U_1 \times U_2\ ; & \chi_{\underset{\sim}{I}}(x,y) &= (1 - x + y) \wedge 1 \\
\underset{\sim}{E} &\subseteq U_1 \times U_2\ ; & \chi_{\underset{\sim}{E}}(x,y) &= (1 - x + y) \wedge (1 - y + x)
\end{aligned} \quad ; \forall x \in U\ ,\ \forall y \in U$$

when $a \wedge b = \mathrm{MIN}(a,b)$ and $a \vee b = \mathrm{MAX}(a,b)$; $U_1 = U_2 = [0,1]$.

Furthermore $\underset{\sim}{N} \subseteq U$; $\chi_{\underset{\sim}{N}}(x) = 1 - x$; $\forall x \in U$

Definition 3.3

The fuzzy relations $\underset{\sim}{R}$ associated with P AND Q is *true*, P OR Q is *true*, $P \supset Q$ is *true*, $P \equiv Q$ is *true* are $\underset{\sim}{C}$, $\underset{\sim}{D}$, $\underset{\sim}{I}$, $\underset{\sim}{E}$ respectively. Furthermore the fuzzy relation associated with ($\neg P$ is *true*) is $\underset{\sim}{N}$.

Definition 3.4

If $\underset{\sim}{R}_1 \subseteq U_1 \times U_2 \times \ldots \times U_n$ is the fuzzy relation associated with a compound proposition $\sum$ involving atoms $P_1, P_2, \ldots P_n$ then the fuzzy relation associated with the proposition

$$\textstyle\sum \text{ is } \underset{\sim}{\tau}$$

is given by $\underset{\sim}{R} = \underset{\sim}{R}_1(\underset{\sim}{\tau})$ where

$$\chi_{\underset{\sim}{R}}(u_1, u_2, \ldots u_n) = \chi_{\underset{\sim}{\tau}}\left[\chi_{\underset{\sim}{R}_1}(u_1, \ldots u_n)\right].$$

This generalises the concept of T.F.M.

We can thus associate with any compound statement a fuzzy relation. For example

$$\{[(P \text{ and } Q) \text{ is } \underset{\sim}{\tau}_1] \supset S\} \text{ is } \underset{\sim}{\tau}_2$$

will have the associated fuzzy logic relation $\underset{\sim}{R} \subseteq U_P \times U_Q \times U_S$ where

$$\chi_{\underset{\sim}{R}}(u_1, u_2, u_3) = \chi_{\underset{\sim}{\tau}_2}\left[(1 - (\chi_{\underset{\sim}{\tau}_1}(u_1 \wedge u_2)) + u_3) \wedge 1\right]$$

where $u_1 \in U_P$, $u_2 \in U_Q$, $u_3 \in U_S$; $U_P = U_Q = U_S = [0,1]$.

Definition 3.5

The relation associated with the atomic proposition

$$P \text{ is } \underset{\sim}{\tau}$$

is the unary relation $\underset{\sim}{R} \subseteq U$ where $\underset{\sim}{R} = \underset{\sim}{\tau}$.

The following definitions will be used to establish a method of analysing compound statements to establish a truth value restriction to be associated with them and also theorem proving.

Definition 3.6

The global projection of $\underset{\sim}{R} \subseteq U_1 \times U_2 \times \ldots \times U_n$, denoted by $h(\underset{\sim}{R})$ is defined by

$$h(\underset{\sim}{R}) = \bigvee_{u_1 \in U_1} \bigvee_{u_2 \in U_2} \cdots \bigvee_{u_n \in U_n} \chi_{\underset{\sim}{R}}(u_1, u_2, \ldots u_n)$$

Definition 3.7

The projection of $\underset{\sim}{R} \subseteq U_1 \times U_2 \times \ldots \times U_n$ on U_i is a fuzzy subset of U_i denoted by $\text{Proj}_{U_i}(\underset{\sim}{R})$ and defined by its membership function

$$\chi_{\text{Proj}_{U_i}(\underset{\sim}{R})}(u_i) = \bigvee_{\left\{\begin{smallmatrix} u_j \in U_j \\ j \neq i \end{smallmatrix}\right\}} \left[\chi_{\underset{\sim}{R}}(u_1, u_2, \ldots u_n)\right] \quad ; \forall u_i \in U_i$$

Definition 3.8

If $\underset{\sim}{R} \subseteq U_1 \times U_2 \times \ldots \times U_n$ is the fuzzy relation associated with a compound fuzzy logic proposition $\sum$ involving atoms $P_1, P_2, \ldots P_n$, then the least restrictive fuzzy relation $\underset{\sim}{R}_1 \subseteq U_1 \times \ldots \times U_r$, for $r < n$, associated with atoms $P_1, P_2, \ldots P_r$ is given by

$$\underset{\sim}{R}_1 = \text{Proj}_{U_1 \times U_2 \times \ldots \times U_r}(\underset{\sim}{R})$$

and more specifically the least restrictive truth value for P_i is

$$v(P_i) = \text{Proj}_{U_i}(\underset{\sim}{R})$$

Furthermore, if a compound fuzzy logic proposition $\sum_1$ involving atoms $P_1, \ldots P_r$ has an associated fuzzy relation $\underset{\sim}{R}_2 \subseteq U_1 \times \ldots \times U_r$ then the least restrictive truth value for $\sum_1$ denoted by $v(\sum_1|\sum)$ is defined by means of the membership function

$$\chi_{v(\sum_1|\sum)}(\eta) = h\left[\underset{\sim}{R}_2(e_\eta) \cap \mathrm{Proj}_{U_1 \times U_2 \times \ldots \times U_r}(\underset{\sim}{R})\right] \; ; \; \forall \; \eta \in U$$

where e_η is a singleton truth value restriction defined by

$$\chi_{e_\eta}(u) = \begin{cases} 1 & \text{if } u = \eta \\ 0 & \text{otherwise.} \end{cases}$$

This is a generalisation of I.T.F.M. and we can write

$$v(\textstyle\sum_1|\sum) = \text{I.T.F.M.}\left[\underset{\sim}{R}_2 | \mathrm{Proj}_{U_1 \times \ldots \times U_r}(\underset{\sim}{R})\right]$$

Theorem 3.1

Let $v(P * Q \text{ is } \underset{\sim}{\tau} \mid v(P) \text{ and } v(Q)) \subseteq U$ represent the fuzzy truth value restriction for the proposition $P * Q$ is $\underset{\sim}{\tau}$, where $*$ is any logic connective, when given $v(P)$ and $v(Q)$. Then

$$\chi_{v(P*Q \text{ is } \underset{\sim}{\tau}|v(P),v(Q))}(\eta) = h\left[\underset{\sim}{R}(e_\eta) \cap (v(P) \times v(Q))\right] \; ; \; \forall \; \eta \in U$$

where $\underset{\sim}{R} = \underset{\sim}{R}_1(\underset{\sim}{\tau})$; $\underset{\sim}{R} \subseteq U_P \times U_Q$

and $\underset{\sim}{R}_1$ is the fuzzy logic relation corresponding to $*$; $\underset{\sim}{R}_1 \subseteq U_P \times U_Q$.

Proof

We interpret the given information as

$(v(P) \text{ AND } v(Q))$ is $\underset{\sim}{\text{true}}$

so that the corresponding fuzzy relation is

$$(v(P) \times U_Q) \cap (U_P \times v(Q)) = v(P) \times v(Q)$$

and hence theorem follows from definition 3.8.

Example 3.1

$$\frac{\sum \; : \; P \supset Q \text{ is } \underset{\sim}{\tau} \qquad v(P), \; v(Q)}{\chi_{v(P \supset Q \text{ is } \underset{\sim}{\tau}|v(P),v(Q))}(\eta) = h\left[\underset{\sim}{R}(e_\eta) \cap (v(P) \times v(Q))\right] \; ; \; \forall \; \eta \in U}$$

where $\underset{\sim}{R} = \underset{\sim}{I}(\underset{\sim}{\tau})$.

Theorem 3.2

(a) The conjunction of two propositions P and Q, written as P AND Q, with fuzzy truth value restrictions $v(P) \subseteq U_P$ and $v(Q) \subseteq U_Q$, has a truth value restriction $v(P \text{ AND } Q) \subseteq U$ with membership function

$$\chi_{v(P \text{ AND } Q)}(\eta) = \bigvee_{\substack{y \wedge z = \eta \\ y \in U_P; \; z \in U_Q}} \left[\chi_{v(P)}(y) \wedge \chi_{v(Q)}(z)\right] \; ; \; \forall \; \eta \in U$$

(b) The disjunction of two propositions P and Q, denoted by P OR Q, with fuzzy truth value restrictions $v(P) \subseteq U_P$ and $v(Q) \subseteq U_Q$ has a truth value restriction $v(P \text{ OR } Q) \subseteq U$ with membership function

$$\chi_{v(P \text{ OR } Q)}(\eta) = \bigvee_{\substack{y \vee z = \eta \\ y \in U_P; \; z \in U_Q}} \left[\chi_{v(P)}(y) \wedge \chi_{v(Q)}(z)\right] \; ; \; \forall \; \eta \in U$$

Proof:

Using definition 3.8

$$\chi_{v(P \text{ AND } Q)}(\eta) = h\left[\underset{\sim}{C}(e_\eta) \cap (v(P) \times v(Q))\right]$$
$$= h(\underset{\sim}{R}) \text{ say}$$
$$= \bigvee_{y \in U_P;} \bigvee_{z \in U_Q} \left[\chi_{\underset{\sim}{R}}(y,z)\right]$$

But

$$\chi_{\underset{\sim}{C}(e_\eta)}(y,z) = \chi_{e_\eta}(\chi_{\underset{\sim}{C}}(y,z)) = \chi_{e_\eta}(y \wedge z)$$
$$= \begin{cases} 1 & \text{if } y \wedge z = \eta \\ 0 & \text{otherwise} \end{cases}$$

so that $\chi_{\underset{\sim}{R}}(y,z) = \begin{cases} \chi_{v(P)}(y) \wedge \chi_{v(Q)}(z) & \text{if } y \wedge z = \eta \\ 0 & \text{otherwise} \end{cases}$

and thus $\chi_{v(P \text{ AND } Q)}(\eta) = \bigvee_{\substack{y \wedge z = \eta \\ y \in U_P; \; z \in U_Q}} \left[\chi_{v(P)}(y) \wedge \chi_{v(Q)}(z)\right]$

and hence theorem (a).

Theorem (b) follows in a similar manner, since

$$\chi_{v(P \text{ OR } Q)}(\eta) = h\left[\underset{\sim}{D}(e_\eta) \cap (v(P) \times v(Q))\right] ; \forall \eta \in U$$

and

$$\chi_{\underset{\sim}{D}(e_\eta)}(y,z) = \chi_{e_\eta}(\chi_{\underset{\sim}{D}}(y,z)) = \chi_{e_\eta}(y \vee z)$$
$$= \begin{cases} 1 & \text{if } y \vee z = \eta \\ 0 & \text{otherwise} \end{cases}$$

Hence theorem Q.E.D.

Theorem 3.3

Using a similar notation to above

$v(P * Q \text{ is } \underset{\sim}{\tau} \mid v(P), v(Q)) \subseteq U$ is defined by the membership function

$$\chi_{v(P*Q \text{ is } \underset{\sim}{\tau} | v(P), v(Q))}(\eta) = \bigvee_{\substack{\chi_{\underset{\sim}{R}}(x,y) = \eta \\ x \in U_P, \; y \in U_Q}} \left[\chi_{v(P)}(x) \wedge \chi_{v(Q)}(y)\right]; \forall \eta \in U$$

where $\underset{\sim}{R} = \underset{\sim}{R}_1(\underset{\sim}{\tau})$ and $\underset{\sim}{R}_1$ corresponds to the logic connective *.

Proof

For definition 3.8

$$\chi_{v(P*Q \text{ is } \underset{\sim}{\tau} | v(P), v(Q))}(\eta) = h\left[\underset{\sim}{R}(e_\eta) \cap (v(P) \times v(Q))\right] = h(\underset{\sim}{R}') \text{ say}$$
$$= \bigvee_{x \in U_P} \bigvee_{y \in U_Q} \left[\chi_{\underset{\sim}{R}'}(x,y)\right]$$

But

$$\chi_{\underset{\sim}{R}(e_\eta)}(x,y) = \chi_{e_\eta}(\chi_{\underset{\sim}{R}}(x,y)) = \begin{cases} 1 & \text{if } \chi_{\underset{\sim}{R}}(x,y) = \eta \\ 0 & \text{otherwise} \end{cases}$$

so that

$$\chi_{\underset{\sim}{R}'}(x,y) = \begin{cases} \chi_{v(P)}(x) \wedge \chi_{v(Q)}(y) & \text{if } \chi_{\underset{\sim}{R}}(x,y) = \eta \\ 0 & \text{otherwise} \end{cases}$$

Hence theorem Q.E.D.

Example 3.2

$v(P \supset Q$ is true | P is absolutely true, Q is true) = true

Definition 3.9

The MAX-MIN composition of $\underset{\sim}{R}_1 \subseteq U_1 \times U_2$ and $\underset{\sim}{R}_2 \subseteq U_2 \times U_3$, denoted by $\underset{\sim}{R}_1 \circ \underset{\sim}{R}_2$, is given by

$$\underset{\sim}{R} = \underset{\sim}{R}_1 \circ \underset{\sim}{R}_2 \; ; \quad \underset{\sim}{R} \subseteq U_1 \times U_3$$

where
$$\chi_{\underset{\sim}{R}}(x,y) = \bigvee_{\eta} \left\{ \chi_{\underset{\sim}{R}_1}(x,\eta) \wedge \chi_{\underset{\sim}{R}_2}(\eta,y) \right\}$$

Theorem 3.4

$$\chi_{v(P*Q \text{ is } \underset{\sim}{\tau} | v(Q), v(P))}(\eta) = v(P) \circ \underset{\sim}{R}(e_\eta) \circ v(Q) \; ; \quad \forall \; \eta \in U$$

where $\underset{\sim}{R} = \underset{\sim}{R}_1(\underset{\sim}{\tau})$ and $\underset{\sim}{R}_1$ corresponds to *.

Proof

Using theorem 3.3

$$\chi_{v(P*Q \text{ is } \underset{\sim}{\tau} | v(P), v(Q))}(\eta) = \bigvee_{\substack{x \quad y \\ \chi_{\underset{\sim}{R}}(x,y) = \eta}} \bigvee \left[\chi_{v(P)}(x) \wedge \chi_{v(Q)}(y) \right]$$

Also

$$v(P) \circ \underset{\sim}{R}(e_\eta) \circ v(Q) = v(P) \circ \underset{\sim}{\tau}_1$$

where
$$\chi_{\underset{\sim}{\tau}_1}(\eta) = \bigvee_y \left[\chi_{\underset{\sim}{R}(e_\eta)}(\eta,y) \wedge \chi_{v(Q)}(y) \right]$$

so that

$$v(P) \circ R(e_\eta) \circ v(Q) = \bigvee_x \bigvee_y \left[\chi_{v(P)}(x) \wedge \chi_{\underset{\sim}{R}(e_\eta)}(x,y) \wedge \chi_{v(Q)}(y) \right]$$

$$= \bigvee_{\substack{x \quad y \\ \chi_{\underset{\sim}{R}}(x,y) = \eta}} \bigvee \left\{ \chi_{v(P)}(x) \wedge \chi_{v(Q)}(y) \right\}$$

since
$$\chi_{\underset{\sim}{R}(e_\eta)}(x,y) = \begin{cases} 1 \text{ if } \chi_{\underset{\sim}{R}}(x,y) = \eta \\ 0 \text{ otherwise} \end{cases}$$

Hence theorem Q.E.D.

Theorem 3.5

$$\chi_{v(\neg P | v(P))}(\eta) = \chi_{v(P)}(1 - \eta) \; ; \quad \forall \; \eta \in U$$

Proof

$$\chi_{v(\neg P | v(P))}(\eta) = h\left(\underset{\sim}{N}(e_\eta) \cap v(P)\right) \; ; \quad \forall \; \eta \in u$$

$$= \bigvee_x \left\{ \chi_{\underset{\sim}{N}(e_\eta)}(x) \wedge \chi_{v(P)}(x) \right\}$$

But
$$\chi_{\underset{\sim}{N}(e_\eta)}(x) = \chi_{e_\eta}\left(\chi_{\underset{\sim}{N}}(x)\right) = \begin{cases} 1 \text{ if } \chi_{\underset{\sim}{N}}(x) = \eta \\ 0 \text{ otherwise} \end{cases} = \begin{cases} 1 \text{ if } 1 - x = \eta \\ 0 \text{ otherwise} \end{cases}$$

so that
$$\chi_{\underset{\sim}{N}(e_\eta)}(x) \wedge \chi_{v(P)}(x) = \begin{cases} \chi_{v(P)}(1 - \eta) \text{ if } x = 1 - \eta \\ 0 \text{ otherwise} \end{cases}$$

Hence
$$\chi_{v(\neg P | v(P))}(\eta) = \chi_{v(P)}(1 - \eta)$$
Q.E.D.

The following definitions and theorems are fundamental to the proposed method of fuzzy reasoning. They provide a means of proving theorems and reasoning with fuzzy statements. First the definitions will be presented and examples then given to show their use.

Definition 3.10

If $\underset{\sim}{R}_1 \subseteq U_1 \times U_2 \times \ldots \times U_n$ is the fuzzy relation corresponding to the compound proposition Σ_1 involving atoms $P_1, P_2, \ldots, P_n$ and $\underset{\sim}{R}_2 \subseteq U_1 \times U_2 \times \ldots \times U_n$ is the fuzzy relation corresponding to the compound proposition Σ_2 involving atoms $P_1, P_2, \ldots, P_n$ and $\underset{\sim}{R}(\underset{\sim}{\tau})$ is the fuzzy relation corresponding to

$$\Sigma_1 \,\square\, \Sigma_2 \quad \text{is} \quad \underset{\sim}{\tau}$$

where $\square$ is a logic connective and $\underset{\sim}{R}_3$ is the fuzzy relation corresponding to $\square$ then

$$\underset{\sim}{R}(\underset{\sim}{\tau}) = \underset{\sim}{R}_1 \cap_{\square} (\underset{\sim}{\tau})\, \underset{\sim}{R}_2$$

where

$\cap_{\square}(\underset{\sim}{\tau})$ is defined as a generalised intersection with

$$\chi_{\underset{\sim}{R}(\underset{\sim}{\tau})}(u_1, u_2, \ldots, u_n) = \chi_{\underset{\sim}{R}_3(\underset{\sim}{\tau})}\left(\chi_{\underset{\sim}{R}_1}(u_1, u_2, \ldots u_n), \chi_{\underset{\sim}{R}_2}(u_1, u_2, \ldots u_n)\right)$$

It can be noted that

$$\cap_{AND}(\underset{\sim}{true}) = \cap \; ; \quad \cap_{OR}(\underset{\sim}{true}) = \cup$$

Definition 3.11

The MAX-$\square$ composition of $\underset{\sim}{R}_1 \subseteq U_1 \times U_2$ and $\underset{\sim}{R}_2 \subseteq U_2 \times U_3$, denoted by $\underset{\sim}{R}_1 \circ_{\square} \underset{\sim}{R}_2$, is given by

$$\underset{\sim}{R} = \underset{\sim}{R}_1 \circ_{\square} \underset{\sim}{R}_2; \quad \underset{\sim}{R} \subseteq U_1 \times U_3$$

where

$$\chi_{\underset{\sim}{R}}(u_1, u_3) = \bigvee_{n}\left[\chi_{\underset{\sim}{R}_3}\{\chi_{\underset{\sim}{R}_1}(u_1,n), \chi_{\underset{\sim}{R}_2}(n,u_3)\}\right] \; ; \; \forall\, u_1 \in U_1 \; ; \; \forall\, u_3 \in U_3$$

where

$\underset{\sim}{R}_3$ is the fuzzy relation associated with the logic connective $\square$. To obtain the MAX-$\square$ $(\underset{\sim}{\tau})$ composition we replace $\underset{\sim}{R}_3$ with $\underset{\sim}{R}_3(\underset{\sim}{\tau})$.

In particular

$$\circ_{AND}(\underset{\sim}{true}) = \circ \qquad \text{the MAX-MIN composition.}$$

Theorem 3.6

If $[(P \bullet Q \text{ is } \underset{\sim}{\tau}_1) * (Q \,\square\, S \text{ is } \underset{\sim}{\tau}_2)]$ is $\underset{\sim}{\tau}$ is represented by the fuzzy relation

$$\underset{\sim}{R} \subseteq U_P \times U_Q \times U_S \quad \text{and}$$

$$\underset{\sim}{R}_1 = \text{Proj}_{U_P \times U_S}(\underset{\sim}{R})$$

then

$$\underset{\sim}{R}_1 = \underset{\sim}{R}_{\bullet}(\underset{\sim}{\tau}_1) \circ_{*}(\underset{\sim}{\tau})\, \underset{\sim}{R}_{\square}(\underset{\sim}{\tau}_2)$$

where

$\underset{\sim}{R}_{\bullet} \subseteq U_P \times U_Q$ is the fuzzy relation corresponding to $\bullet$

and

$\underset{\sim}{R}_{\square} \subseteq U_Q \times U_S$ is the fuzzy relation corresponding to $\square$

and $\bullet$, $\square$ are any logic connectives.

Proof

$$\underset{\sim}{R} = \underset{\sim}{R}_{\bullet}(\underset{\sim}{\tau}_1) \cap_{*}(\underset{\sim}{\tau})\, \underset{\sim}{R}_{\square}(\underset{\sim}{\tau}_2) \quad \text{using definition 3.10}$$

so that

$$\chi_{\underset{\sim}{R}}(x,y,z) = \chi_{\underset{\sim}{R}_3(\underset{\sim}{\tau})}\left(\chi_{\underset{\sim}{R}_{\bullet}(\underset{\sim}{\tau}_1)}(x,y), \chi_{\underset{\sim}{R}_{\square}(\underset{\sim}{\tau}_2)}(y,z)\right) \; ; \forall x \in U_P ; \forall y \in U_Q ; \forall z \in U_S$$

where

$\underset{\sim}{R}_3$ is the fuzzy relation corresponding to $*$

and further

$$\chi_{\mathrm{Proj}_{U_P \times U_S}(\tilde{R})}(x,z) = \bigvee_y \left[\chi_{\tilde{R}}(x,y,z)\right] \; ; \; \forall\, x \in U_P; \; \forall\, z \in U_S$$

$$= \bigvee_y \left[\chi_{\tilde{R}_2(\tilde{\tau})}\left(\chi_{\tilde{R}_\bullet(\tilde{\tau}_1)}(x,y),\; \chi_{\tilde{R}_\square(\tilde{\tau}_2)}(y,z)\right)\right]$$

$$= \chi_{\tilde{R}_\bullet(\tilde{\tau}_1)\; o_*(\tilde{\tau})\; \tilde{R}_\square(\tilde{\tau}_2)}$$

Hence theorem · Q.E.D.

In the examples that follow, theorem 3.6 is used as often as possible to reduce the dimensionality of the computation. Care must be taken to produce the most restrictive projection in the sense that if a proposition is to be eliminated all compound statements involving that proposition must be considered together.

Examples 3.3

1. $(P \supset Q \text{ is } \tilde{\tau}_1)$ AND $(P \text{ OR } Q \text{ is } \tilde{\tau}_2)$ is $\tilde{\tau}$

If the associated fuzzy relation is $\tilde{R}$ then

$$\tilde{R} = \tilde{I}(\tilde{\tau}_1).\, \cap_{\mathrm{AND}}(\tilde{\tau}).\; \tilde{D}(\tilde{\tau}_2); \quad \tilde{R} \subseteq U_P \times U_Q$$

so that

$$v(P) = \mathrm{Proj}_{U_P}(R) = \left(\tilde{I}(\tilde{\tau}_1).\, \cap_{\mathrm{AND}}(\tilde{\tau}).\; \tilde{D}(\tilde{\tau}_2)\right) \; o \; \underline{\text{unrestricted}}$$

and

$$v(Q) = \mathrm{Proj}_{U_Q}(R) = \underline{\text{unrestricted}} \; o \; \left(\tilde{I}(\tilde{\tau}_1).\, \cap_{\mathrm{AND}}.(\tau).\tilde{D}(\tilde{\tau}_2)\right).$$

This and related examples have been discussed by Baldwin [18]. In the classical logic case where $\tilde{\tau}_1 = \tilde{\tau}_2 = \tilde{\tau} =$ absolutely true then

v(P) = unrestricted and v(Q) = absolutely true.

2. In this example we show how the method can be used for classical logic theorem proving

$P \supset Q$
Q AND R
$P \supset R$

Using the above theorems and definitions

$$v(P \supset R) = \text{I.T.F.M.}\ (I \mid I \circ C)$$

where

$$I \circ C = U_P \begin{matrix} & U_Q \\ & \begin{matrix} 0 & 1 \end{matrix} \\ \begin{matrix} 0 \\ 1 \end{matrix} & \begin{bmatrix} 1 & 1 \\ 0 & 1 \end{bmatrix} \end{matrix} \circ U_Q \begin{matrix} & U_{R_1} \\ & \begin{matrix} 0 & 1 \end{matrix} \\ \begin{matrix} 0 \\ 1 \end{matrix} & \begin{bmatrix} 0 & 0 \\ 0 & 1 \end{bmatrix} \end{matrix} = U_P \begin{matrix} U_R \\ \begin{bmatrix} 0 & 1 \\ 0 & 1 \end{bmatrix} \end{matrix}$$

$$\text{so that } v(P \supset R) = \text{I.T.F.M.} \left\{ \begin{bmatrix} 1 & 1 \\ 0 & 1 \end{bmatrix} \middle| \begin{bmatrix} 0 & 1 \\ 0 & 1 \end{bmatrix} \right\} = U \begin{matrix} 0 \\ 1 \end{matrix} \begin{bmatrix} 0 \\ 1 \end{bmatrix} = \text{true}$$

The fuzzy logic equivalent to this is

If $(P \supset Q \text{ is } \tilde{\tau}_1)$ AND $(Q \text{ AND } R \text{ is } \tilde{\tau}_2)$ is $\tilde{\tau}$

then $v(P \supset R \text{ is } \tilde{\tau}_3) = \text{I.T.F.M.}\ \{\tilde{I}(\tilde{\tau}_3) \mid \tilde{I}(\tilde{\tau}_1).\ o_{\mathrm{AND}}(\tilde{\tau}).\ \tilde{C}(\tilde{\tau}_2)\}$

The point about this example is that a conclusion not involving Q is required and this is eliminated using theorem 3.6

3. P OR Q is $\underset{\sim}{\tau}_1$ .... (i)
$\neg P \supset R$ is $\underset{\sim}{\tau}_2$ .... (ii)
$R \supset S$ is $\underset{\sim}{\tau}_3$ .... (iii)
$Q \supset \neg S$ is $\underset{\sim}{\tau}_4$ .... (iv)

We interpret the conjunction of the statements (i), (ii), (iii) and (iv) as (((i) AND (ii) is true) AND (iii)) is true AND (iv)) is true.

For determining v(P) we first use (ii) and (iii) to eliminate R, then the result of this with (iv) to eliminate S and the result of this with (i) to produce the relation corresponding to P * Q. We then project onto U_P to determine v(P). So that

$$v(P) = \text{Proj}_{U_P} \left[\underset{\sim}{D}(\underset{\sim}{\tau}_1) \cap \{(\underset{\sim}{I}^{T_r}(\underset{\sim}{\tau}_2) \circ \underset{\sim}{I}(\underset{\sim}{\tau}_3)) \circ \underset{\sim}{I}^{T_c}(\underset{\sim}{\tau}_4)\}^T\right]$$

where T_r, T_c, T stands for rows transposed, columns transposed, complete transposition. This is necessary to have the relations defined on the correct product space. For example, if

$$I \subsetneq U_P \times U_Q$$

then $I^{T_r} \subsetneq U_{\neg P} \times U_Q$.

For the special case of classical two valued logic when $\underset{\sim}{\tau}_1 = \underset{\sim}{\tau}_2 = \underset{\sim}{\tau}_3 = \underset{\sim}{\tau}_4 =$ absolutely true then v(P) = absolutely true so that P is a valid conclusion.

4. (a) $P \supset Q$ is $\underset{\sim}{\tau}$
P is $\underset{\sim}{\tau}_1$

$\therefore v(Q) = \underset{\sim}{\tau}_1 \circ \underset{\sim}{I}(\underset{\sim}{\tau})$

(b) $P \supset Q$ is $\underset{\sim}{\tau}$
Q is $\underset{\sim}{\tau}_1$

$\therefore v(P) = \underset{\sim}{I}(\underset{\sim}{\tau}) \circ \underset{\sim}{\tau}_1$.

These two cases corresponding to Modus Ponens and Modus Tollens were discussed in detail by Baldwin [16] and computational considerations given by Baldwin and Guild [14]. We will illustrate the use of these two forms for modelling in the next section. See Figure 8 for graphical illustration.

For a more detailed treatment of examples, which illustrate transformations to reduce dimensionality problems and comparison of computational details with Wang's algorithm and the resolution principle for binary logic problems, see Baldwin [18]. Example 4 is used to resolve the heap paradox in Baldwin and Guild [20].

The above method has general applicability and in the next section we will show more precisely how it can be used for fuzzy reasoning.

4. Fuzzy Reasoning

We can combine the concepts of T.F.M. and I.T.F.M. dealt with in section 2 with the fuzzy logic of section 3 to give an approach to fuzzy reasoning with propositions containing fuzzy descriptions. We illustrate this with examples below. Such an approach can be used to construct fuzzy algorithms and controllers, to model complex systems and should therefore be a powerful method for building mathematical models of systems which are too complex for precise analysis or which necessarily use fuzzy concepts. Consider that we know the compound proposition

$$[(u \text{ is } \underset{\sim}{p}) \supset (w \text{ is } \underset{\sim}{q})] \text{ is } \underset{\sim}{\tau} \ ; \ \underset{\sim}{p} \subsetneq X \ ; \ \underset{\sim}{q} \subsetneq Y \ ; \ \underset{\sim}{\tau} \subsetneq U$$

We can write this as $P \supset Q$ is $\underset{\sim}{\tau}$.

Suppose further that we are given the data u is $\underset{\sim}{p}'$; $\underset{\sim}{p}' \subsetneq X$

then we can determine v(P) using I.T.F.M. as

$$v(P) = \text{I.T.F.M.}\ (\tilde{p}/\tilde{p}')$$

so that we can write

$$\frac{P \supset Q \text{ is } \tilde{\tau} \qquad P \text{ is I.T.F.M. } (\tilde{p}/\tilde{p}')}{}$$

so that $v(Q) = \text{I.T.F.M. } (\tilde{p}/\tilde{p}') \circ \tilde{I}(\tilde{\tau})$

and thus we can write (w is $\tilde{q}$) is v(Q) and use T.F.M. to obtain w is $\tilde{q}'$ where $\tilde{q}' = \text{T.F.M. } \left(\tilde{q} \mid \text{I.T.F.M. } (\tilde{p}/\tilde{p}') \circ \tilde{I}(\tilde{\tau})\right)$.

We can thus write

$$[(u \text{ is } \tilde{p}) \supset (w \text{ is } \tilde{q})] \text{ is } \tilde{\tau}\ ; \quad \tilde{p}, \tilde{p}' \subseteq X\ ; \quad \tilde{q} \subseteq Y\ ; \quad \tilde{\tau} \subseteq U$$
$$u \text{ is } \tilde{p}'$$

$\therefore$ w is $\tilde{q}'$ where $\tilde{q}' = \text{T.F.M. } \left(\tilde{q} \mid \text{I.T.F.M. } (\tilde{p}/\tilde{p}') \circ \tilde{I}(\tilde{\tau})\right)$; $\tilde{q}' \subseteq Y$.

Thus for brevity we can write this example as

$$\tilde{p} \supset \tilde{q} \text{ is } \tilde{\tau}$$
$$\tilde{p}'$$

$$\tilde{q}' = \text{T.F.M. } [\tilde{q} \mid \text{I.T.F.M. } (\tilde{p}/\tilde{p}') \circ \tilde{I}(\tilde{\tau})].$$

We now give further examples in this abreviated form.

Example 4.2

$$\tilde{p} \supset \tilde{q} \text{ is } \tilde{\tau}\ ; \quad \tilde{p} \subseteq X\ ; \quad \tilde{q}, \tilde{q}' \subseteq Y\ ; \quad \tilde{\tau} \subseteq U$$
$$\tilde{q}'$$

$\therefore$ $\tilde{p}' = \text{T.F.M. } [\tilde{p} \mid \tilde{I}(\tilde{\tau}) \circ \text{I.T.F.M. } (\tilde{q}/\tilde{q}')]$; $\tilde{p}' \subseteq X$

Example 4.3

$$[(\tilde{p} \text{ AND } \tilde{q}) \supset \tilde{r}] \text{ is } \tilde{\tau}\ ; \quad \tilde{p}, \tilde{p}' \subseteq X\ ; \quad \tilde{q} \subseteq Y\ ; \quad \tilde{r}, \tilde{r}' \subseteq Z\ ; \quad \tilde{\tau} \subseteq U$$
$$\tilde{p}', \tilde{r}'$$

$\therefore$ $\tilde{q}' = \text{T.F.M. } [\tilde{q} \mid \text{I.T.F.M. } (\tilde{p}/\tilde{p}') \circ \tilde{C}(\tilde{I}(\tilde{\tau}) \circ \text{I.T.F.M. } (\tilde{r}/\tilde{r}'))]$; $\tilde{q}' \subseteq Y$.

This is an example of a 'mixed input argument' as described by Baldwin [19]. To see this more clearly this example is equivalent to the fuzzy logic problem

(i) (P AND Q) $\supset$ R is $\tilde{\tau}$

(ii) P is $\tilde{\tau}_1$

(iii) R is $\tilde{\tau}_2$

We first use (i) and (iii) to eliminate R using theorem 3.6, i.e. $v(P \text{ AND } Q) = \tilde{I}(\tilde{\tau}) \circ \tilde{\tau}_2$.

We then have the reduced problem

$$P \text{ AND } Q \text{ is } \tilde{I}(\tilde{\tau}) \circ \tilde{\tau}_2$$
$$P \text{ is } \tilde{\tau}_1$$

and again we use theorem 3.6 to eliminate P. Thus

$$v(Q) = \tilde{\tau}_1 \circ \tilde{C}\left(\tilde{I}(\tilde{\tau}) \circ \tilde{\tau}_2\right).$$

Example 4.4

$$[(\tilde{p} \supset \tilde{r}) \text{ OR } (\tilde{q} \supset \tilde{r})] \text{ is } \tilde{\tau}\ ; \quad \tilde{p}, \tilde{p}' \subseteq X\ ; \quad \tilde{q} \subseteq Y\ ; \quad \tilde{r}, \tilde{r}' \subseteq Z\ ; \quad \tilde{\tau} \subseteq U$$
$$\tilde{p}', \tilde{r}'$$

$$v(\tilde{r} \mid \tilde{r}') = \text{I.T.F.M. } (\tilde{r}/\tilde{r}')$$

$$v(\underset{\sim}{p} \supset \underset{\sim}{r}/\underset{\sim}{p}',\underset{\sim}{r}') = \{\text{I.T.F.M.}\ (\underset{\sim}{p}/\underset{\sim}{p}') \circ \underset{\sim}{I}(e_\eta) \circ v(\underset{\sim}{r}/\underset{\sim}{r}')\} \text{ i.e. a function of } \eta$$
$$v(\underset{\sim}{q} \supset \underset{\sim}{r}) = v(\underset{\sim}{p} \supset \underset{\sim}{r}/\underset{\sim}{p}',\underset{\sim}{r}') \circ \underset{\sim}{D}(\underset{\sim}{\tau})$$
$$v(\underset{\sim}{q}) = \underset{\sim}{I}\{v(\underset{\sim}{q} \supset \underset{\sim}{r})\} \circ v(\underset{\sim}{r}/\underset{\sim}{r}')$$
$$\underset{\sim}{q}' = \text{T.F.M.}\ \{\underset{\sim}{q}/v(\underset{\sim}{q})\}$$

Example 4.5

The computational details for this example can be found in [15] and has been computed using FUZLOG, a general purpose approximate reasoning computer program.

(Popular Cricketer ⊃ Good Cricketer) is very very true
(Good Cricketer ≡ (Good Bowler OR Good Batsman) is very very true
(Bradman was a popular cricketer) is extremely true
(Bradman was a good bowler) is false

(Bradman was a good cricketer) is $\underset{\sim}{\tau}_1$
(Bradman was a good batsman) is $\underset{\sim}{\tau}_2$

Figure 9 shows results for $\underset{\sim}{\tau}_1$ and $\underset{\sim}{\tau}_2$ and the particular definition of very very true and extremely true used.

5. Applications

The method of approximate reasoning based upon the theories of fuzzy sets and fuzzy logic will play an important part in the analysis and modelling of complex systems. Simulation methods as described by Forrestor [21] combined with modelling using fuzzy sets and fuzzy logic could provide a powerful tool for evaluating the dynamic consequences of economic, political and social policies. This need for dealing with multi-loop nonlinear feedback social systems is discussed by Gabor [22]. The human mind is capable of specifying the components of complex systems and even the relations between any two of them, bit by bit, but it cannot embrace the whole simultaneously. This motivates not only the need for computers and simulation methods but also a fuzzy logic to simulate human reasoning.

To date the approach described in this paper has been applied to the ergonomic design of a ship control room [23] and is being applied to military strategy design, robotic concept learning, pattern recognition and adaptive forecasting methods. The method gives very good results for Mamdani's fuzzy control problem [24] and work is now beginning on its application to the synthesis of controllers for management and engineering systems. Much research remains to be done in the application of this approach to data processing, concept definition, fuzzy algorithms and programs as discussed by Zadeh in [2]. The fields of artificial intelligence, operational research, control theory and systems theory and the social sciences are rich in problems which have a need for the application of fuzzy reasoning. In decision theory, for example, the approach of utility theory is often difficult to apply. The goals of a decision maker are multi dimensional and conflict is an essential element in any realistic model. This results in a necessary compromise between the various conflicting demands. Fuzzy logic with its (T.F.M.) procedure can be used to model such phenomena as this. Comparison of various decision policies is not as straight forward as when non fuzzy mathematics is used. For a discussion of this and related matters see [27].

6. Conclusion

A fuzzy logic based upon truth value restrictions is proposed and a method of analysis for this logic useful for theorem proving and reasoning with fuzzy statements and loose concepts developed.

Much work needs to be done with this approach to gain experience in its application to large scale reasoning problems. With the limited experience to date the approach appears to be useful for many areas of application including data processing, analysis and control of complex systems, decision theory and simulation methods.

References

[1] Zadeh L.A., (June 1965), "Fuzzy Sets", Information and Control, Vol.8, pp 338-353.

[2] Zadeh L.A., (January 1973), "Outline of a new approach to the analysis of complex systems and decision processes", IEEE trans. on Systems, Man and Cybernetics, SMC-3, pp 28-44.

[3] Zadeh L.A., (1975), "Calculus of fuzzy restrictions" in Zadeh et al, Eds., "Fuzzy Sets and their applications to cognitive and decision processes", New York: Academic Press, pp 1-39.

[4] Zadeh L.A., (1975), "Fuzzy Logic and approximate reasoning", Synthese, Vol.30, pp 407-428.

[5] Russell B., (1923), "Vagueness", Australian Journal of Psychology and Philosophy 1, pp 84-92.

[6] Bellman R.E., (1971), "Mathematics, Systems and Society", F.E.K. Report, Stockholm.

[7] Popper K.R., (1959), "The Logic of Scientific Discovery" Hutchinson of London.

[8] Kaufmann A., (1975), "Introduction to the theory of fuzzy subsets", Vol. 1, New York: Academic Press.

[9] Zadeh L.A., (1968), "Probability Measures of Fuzzy Events", Maths. Anal. and Appl. 23, pp 421-427.

[10] De Luca A., Termini S., (1972), "A definition of a non-probabilistic entropy in the setting of Fuzzy Sets", Information and Control, Vol.20, pp 301-312.

[11] Baldwin J.F. and Pilsworth B., (1978), "Fuzzy truth definition of possibility measure for decision classification", EM/FS/12 University of Bristol Internal Report, Engineering Mathematics Department; presented at the Workshop of Fuzzy Reasoning, Queen Mary College, London, 15th September 1978. To appear in Proceedings thereof in Int.J.Man-Machine Studies (1979).

[12] Bellman R.E. and Zadeh L.A., (1970) "Decision making in a Fuzzy environment" Management Science, Vol.17, No. 4, pp B141-B164.

[13] Gaines B.R., (1976), "Foundations of Fuzzy Reasoning", Int.Journal of Man-Machine Systems, 8, pp 623-668.

[14] Baldwin J.F. and Guild N.C.F., (1978), "Feasible algorithms for approximate reasoning using a fuzzy logic", Internal report University of Bristol Engineering Mathematics Department EM/FS/8. Submitted for publication.

[15] Baldwin J.F. and Guild N.C.F., (1978), "Fuzlog : A computer program for fuzzy reasoning", University of Bristol, Engineering Mathematics Research Report EM/FS/17. Submitted for publication.

[16] Baldwin J.F., (1978), "A new approach to approximate reasoning using a fuzzy logic". To appear in Fuzzy sets and systems.

[17] Baldwin J.F., (1978), "A model of fuzzy reasoning and fuzzy logic", University of Bristol, Engineering Mathematics Department Research Report EM/FS/10. Submitted for publication.

[18] Baldwin J.R., (1978), "Fuzzy logic and fuzzy reasoning". To appear in Proceedings of Workshop of Fuzzy Reasoning held at Queen Mary College (September 1978) in Int.J.Man-Machine Studies (1979).

[19] Baldwin J.F., (1978), "Fuzzy logic and approximate reasoning for mixed input arguments", University of Bristol, Engineering Mathematics Research Report EM/FS/4. Submitted for publication.

[20] Baldwin J.F., and Guild N.F.C., (1978), "The resolution of two paradoxes by approximate reasoning using a fuzzy logic", University of Bristol, Engineering Mathematics Department Research Report EM/FS/11. Submitted for publication.

[21] Forrester Jay W., (1971), "World Dynamics", Wright-Allen Press.

[22] Gabor D., (1971), "The Mature Society", Secker and Warburg, London.

[23] Baldwin J.F. and Guild N.C.F., (1978), "A model for multi-criterial decision-making using fuzzy logic". To appear in Proceedings of Workshop of Fuzzy Reasoning held at Queen Mary College (September 1978) in Int. J.Man-Machine Studies (1979).

[24] Mamdani E.H. and Assilian S., (1975), "An experiment in linguistic synthesis with a fuzzy logic controller", Int. J. for Man-Machine Studies 7, pp1-13.

[25] Baldwin J.F. and Pilsworth B., (1978), "A theory of fuzzy probability", University of Bristol, Engineering Mathematics Department Research Report EM/FS/15.

[26] Zadeh L.A., (1978), "Fuzzy sets as a basis for a theory of possibility", Fuzzy Sets and Systems, Vol. 1, pp 3-28.

[27] Baldwin J.F. and Guild N.C.F., (1978), "Comparison of fuzzy sets on the same decision space". To appear in Int.J. of Fuzzy Sets and Systems (1979).

[28] Baldwin J.F. and Pilsworth B. (1978), "Axiomatic approach to implication for approximate reasoning using a fuzzy logic". To appear in Int.J. Fuzzy Sets and Systems.

[29] Ragade R.K. (1977), "A Mathematical Model of Approximate Communication in Information Systems" in the Proceedings of the Annual North American Meeting of the Society for General Systems Research, Denver Co. pp 334 - 346.

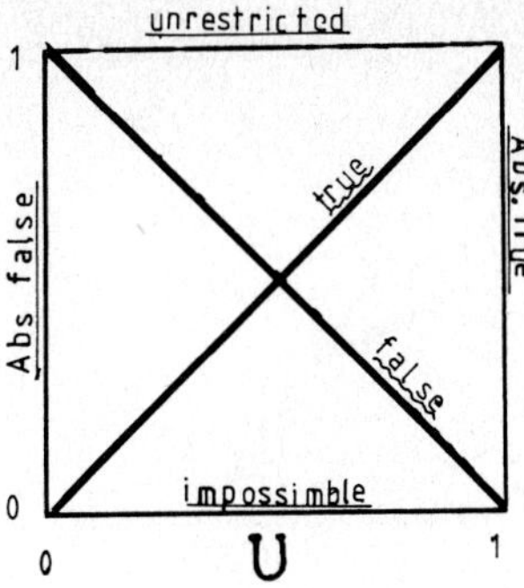

FIGURE 1

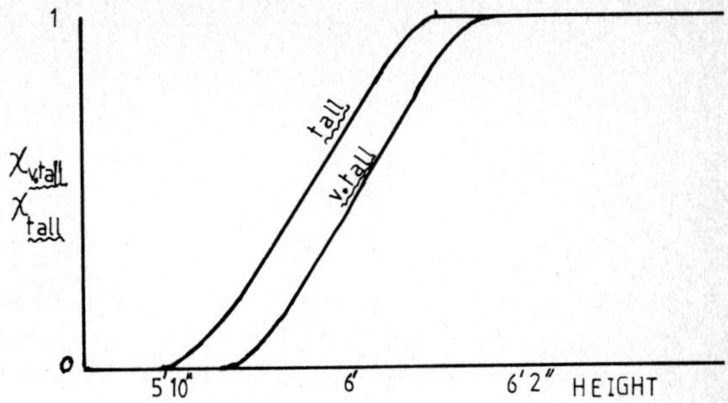

FIGURE 2

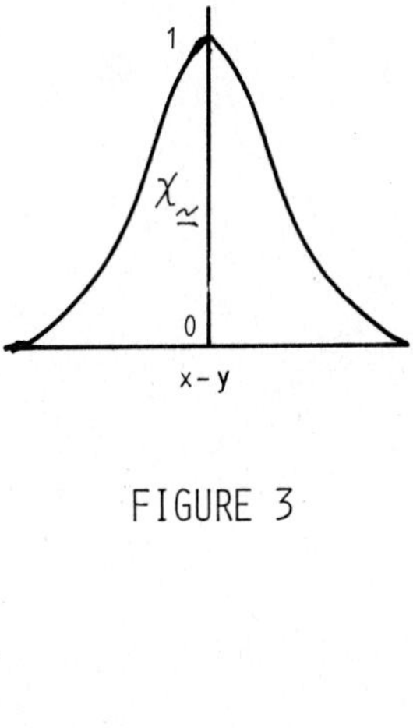

FIGURE 3

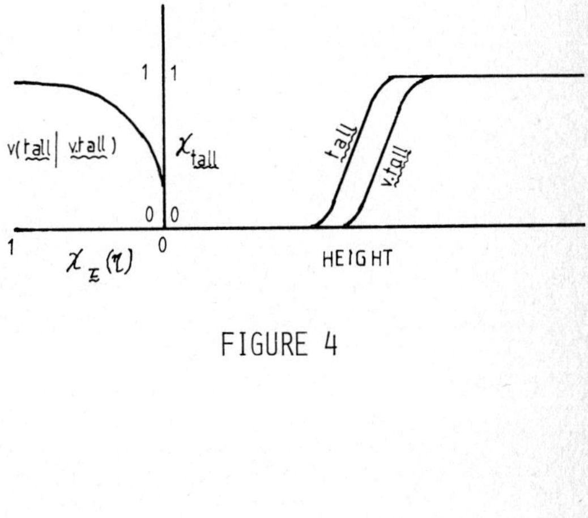

FIGURE 4

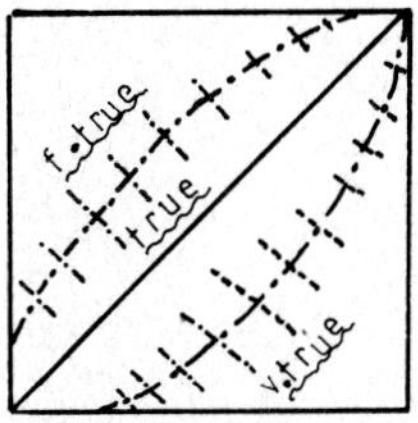

FIGURE 5

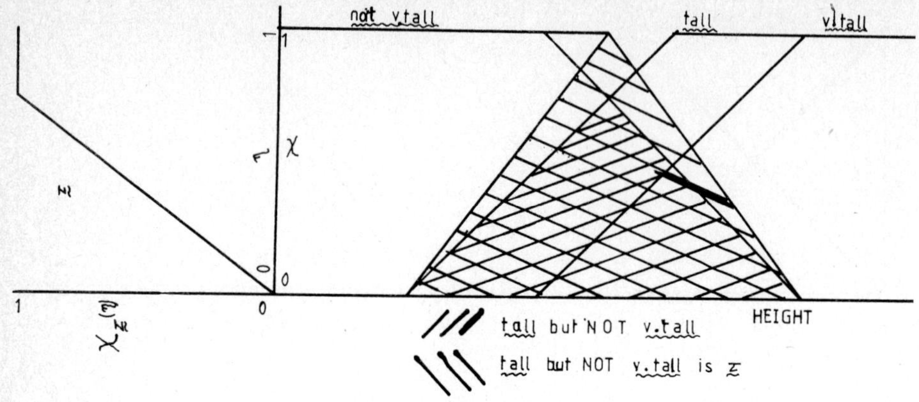

FIGURE 6

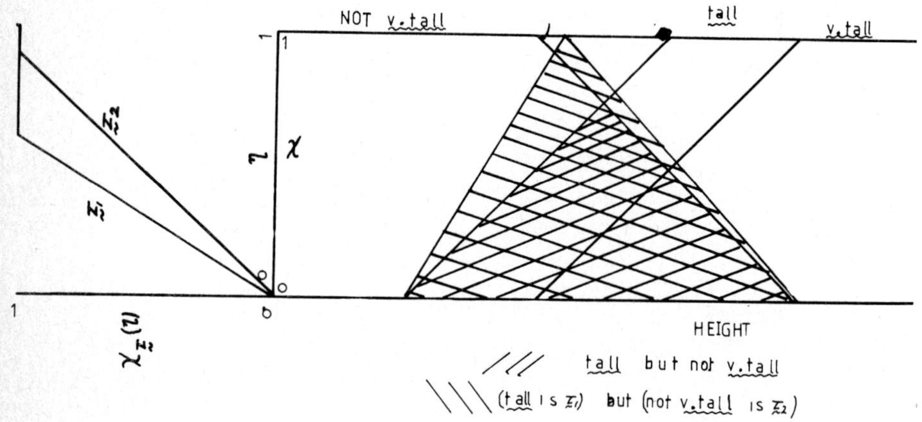

FIGURE 7

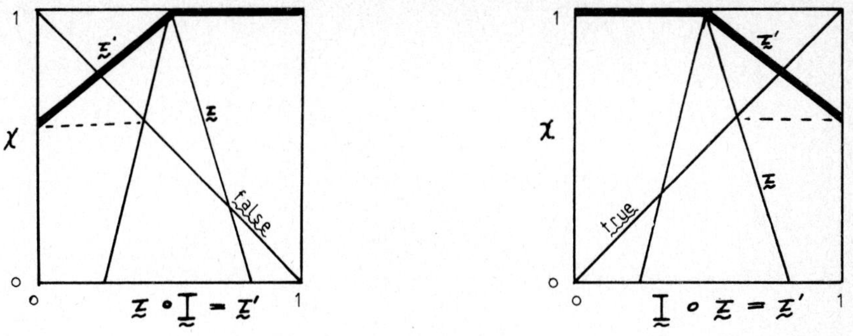

FIGURE 8

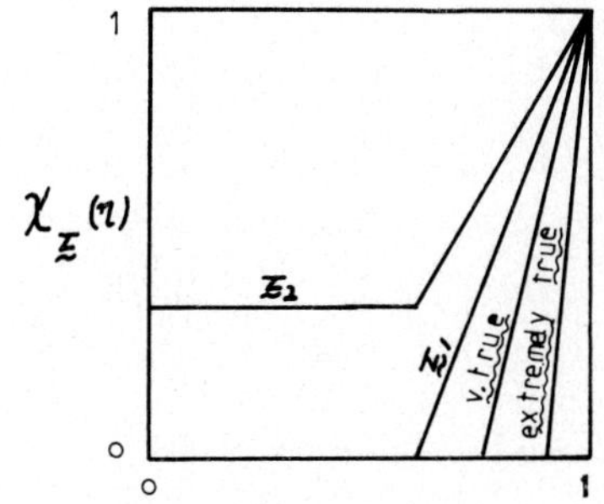

FIGURE 9

ADVANCES IN FUZZY SET THEORY AND APPLICATIONS
M.M. Gupta, R.K. Ragade, R.R. Yager (editors)
© *North-Holland Publishing Company, 1979*

SOME METHODS OF FUZZY REASONING

Masaharu MIZUMOTO*, Satoru FUKAMI**, and Kokichi TANAKA***

* Department of Management Engineering
Osaka Electro-Communication University
Neyagawa, Osaka 572, Japan

** Yokosuka Electrical Communication Laboratory
Nippon Telegraph and Telephone Public Corporation
Yokosuka, Kanagawa 238-03, Japan

*** Department of Information and Computer Sciences
Osaka University
Toyonaka, Osaka 560, Japan

L.A. Zadeh and E.H. Mamdani proposed the methods for fuzzy reasoning in which the antecedent involves a fuzzy conditional proposition "If x is A then y is B" with A and B being fuzzy concepts.
This paper points out that the consequences inferred by their methods do not always fit our intuitions, and suggests some new methods which fit our intuitions under several criteria such as modus ponens and modus tollens.

INTRODUCTION

In much of human reasoning, the form of reasoning is approximate rather than exact as in the statement "A red plum is ripe and this plum is more or less red. Then this plum is more or less ripe." L.A. Zadeh [3] and E.H. Mamdani [1] suggested methods for such a fuzzy reasoning in which antecedent involves a fuzzy conditional proposition "If x is A then y is B", where A and B are fuzzy concepts. In this paper we point out that the consequences inferred by their methods do not always fit our intuitions, and we suggest several improved methods which fit our intuitions under several criteria such as modus ponens and modus tollens. Moreover, the problems of syllogism and contrapositive proposition are discussed using our new methods.

FUZZY CONDITIONAL INFERENCES

We shall consider the following form of inference in which a fuzzy conditional proposition is contained:

$$
\begin{array}{ll}
\text{Ant 1:} & \text{If } x \text{ is } A \text{ then } y \text{ is } B. \\
\text{Ant 2:} & x \text{ is } A'. \\
\hline
\text{Cons:} & y \text{ is } B'.
\end{array}
\qquad (1)
$$

where x and y are the names of objects, and A, A', B and B' are the labels of fuzzy sets in universes of discourse U, U, V and V, respectively.
An example of this form of inference is the following.

$$
\begin{array}{l}
\text{If a tomato is } \underline{red} \text{ then the tomato is } \underline{ripe}. \\
\text{This tomato is } \underline{very\ red}. \\
\hline
\text{This tomato is } \underline{very\ ripe}.
\end{array}
$$

From now on, we call this form of inference as "fuzzy conditional inference."

For this form of inference, Zadeh suggested the following methods [3]. In (1), Ant 1 translates into the following fuzzy relation R_m or R_a, that is,

$$R_m = (A \times B) \cup (\neg A \times V) , \quad (2)$$

$$R_a = (\neg A \times V) \oplus (U \times B) . \quad (3)$$

Mamdani [1] also proposed the following method:

$$R_c = A \times B , \quad (4)$$

where x, U, 7 and ⊕ denote cartesian product, union, complement and bounded-sum*, respectively. Ant 2 in (1) translates into the unary fuzzy relation (that is, fuzzy set):

$$R = A'. \quad (5)$$

Then the consequence B' in Cons of (1) can be obtained by the composition "o" of A' and R_m (or R_a, R_c), that is,

---

* A fuzzy set A in a universe of discourse U is characterized by a membership function μ_A which takes the values in the unit interval [0, 1], i.e.,

$$\mu_A : U \longrightarrow [0, 1] \quad (A.1)$$

and is represented as

$$A = \int_U \mu_A(u)/u . \quad (A.2)$$

In particular, when U is a finite set, we may represent A as

$$A = \mu_A(u_1)/u_1 + \mu_A(u_2)/u_2 + \dots + \mu_A(u_n)/u_n , \quad (A.3)$$

where + is the union of $\mu_A(u_i)/u_i$'s.

Let A and B be fuzzy sets in U, then we have

Union: $$A \cup B = \int \mu_A(u) \vee \mu_B(u) / u , \quad (A.4)$$

Complement: $$\neg A = \int 1 - \mu_A(u) / u , \quad (A.5)$$

Bounded-Sum: $$A \oplus B = \int 1 \wedge (\mu_A(u) + \mu_B(u)) / u , \quad (A.6)$$

where ∨, ∧ and + stand for max, min and arithmetic sum, respectively.

If A and B are fuzzy sets in U and V, respectively, then the cartesian product of A and B is defined by

Cartesian Product: $$A \times B = \int \mu_A(u) \wedge \mu_B(v) / (u, v) . \quad (A.7)$$

A fuzzy relation R from U to V is a fuzzy set in U x V defined by

$$R = \int \mu_R(u, v) / (u, v) . \quad (A.8)$$

Let R and S be fuzzy relations in U x W and W x V, respectively, then the composition of R and S is a fuzzy relation in U x V given by

Composition: $$R \circ S = \int \vee_w \big(\mu_R(u, w) \wedge \mu_S(w, v)\big) / (u, v). \quad (A.9)$$

If R is a unary fuzzy relation (that is, a fuzzy set) over W, then the composition of R and S is defined as

$$R \circ S = \int \vee_w \big(\mu_R(w) \wedge \mu_S(w, v)\big) / v . \quad (A.10)$$

$$B'_m = A' \circ ((A \times B) \cup (\neg A \times V)), \quad (6)$$

$$B'_a = A' \circ ((\neg A \times V) \oplus (U \times B)), \quad (7)$$

$$B'_c = A' \circ (A \times B). \quad (8)$$

On the above form of inferences, according to our intuitions it seems that the relations between A' in Ant 2 and B' in Cons in (1) ought to be satisfied as shown in Table I.

Table I Relations between Ant 2 and Cons under Ant 1

	Ant 2	Cons
Relation I (modus ponens)	x is A	y is B
Relation II-1	x is <u>very</u> A	y is <u>very</u> B
Relation II-2	x is <u>very</u> A	y is B
Relation III	x is <u>more or less</u> A	y is <u>more or less</u> B
Relation IV-1	x is <u>not</u> A	y is <u>unknown</u>
Relation IV-2	x is <u>not</u> A	y is <u>not</u> B
Relation V (modus tollens)	y is <u>not</u> B	x is <u>not</u> A

Relation I corresponds to modus ponens. Relation II-2 is inconsistent with Relation II-1, but in Ant 1 if there is not a strong casual relation between "x is A" and "y is B", the satisfaction of Relation II-2 will be permitted. Relation III is a natural one. Relation IV-1 asserts that when x is not A, any information about y can not be deduced from Ant 1. Thus this relation may be thought to be quite natural, since the fuzzy conditional proposition "If x is A then y is B" does not make any assertion when x is not A. The satisfaction of Relation IV-2 is demanded when the fuzzy proposition "If x is A then y is B" means tacitly the proposition "If x is A then y is B else y is not B." Relation V corresponds to modus tollens in which the form of inference is

Ant 1: If x is A then y is B.
Ant 2: y is not B. (9)

Cons: x is not A.

ZADEH'S AND MAMDANI'S METHODS FOR FUZZY CONDITIONAL INFERENCE

We shall show in this section that Zadeh's methods do not satisfy the relations except Relation IV-1 and that Mamdani's method does not satisfy the relations except Relation I and II-2.

Now, we shall show what will B'_m, B'_a and B'_c be when A' in (6)-(8) is equal to A, very A (= A^2), more or less A (= $A^{0.5}$) or not A (= $\neg A$), where fuzzy sets A in U and B in V are given as in Fig.1 and Fig.2, respectively*.

---

* A^2 and $A^{0.5}$ can be used to approximate the effect of the linguistic mofifiers very and more or less, and are defined as

$$\text{very } A = A^2 = \int \mu_A(u)^2/u \ ; \quad \text{more or less } A = A^{0.5} = \int \sqrt{\mu_A(u)}/u.$$

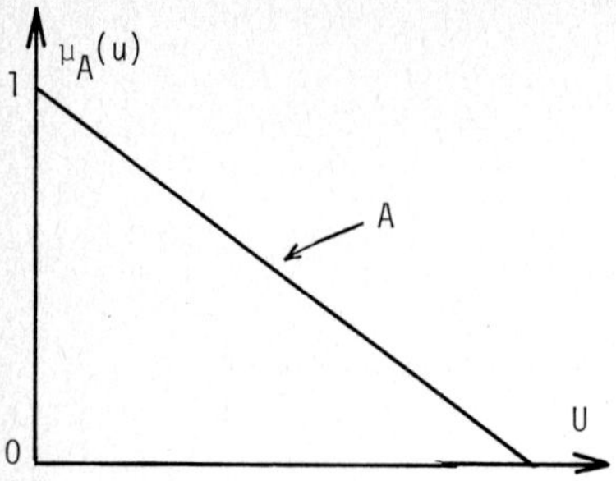

Fig.1 Fuzzy Set A in U.

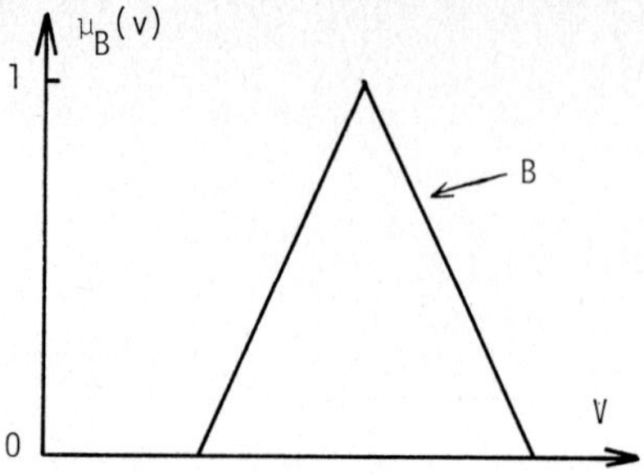

Fig.2 Fuzzy Set B in V.

[I] The Case of Maximin Rule (R_m):

Let A' be A, then B'_m becomes as follows using (6).

$$B'_m = A \circ ((A \times B) \cup (\neg A \times V))$$
$$= \int_U \mu_A(u)/u \circ \int_{U\times V} \big((\mu_A(u)\wedge\mu_B(v)) \vee (1-\mu_A(u))\big)/(u,v)$$
$$= \int_V \bigvee_{u\in U} \big(\mu_A(u) \wedge [(\mu_A(u)\wedge\mu_B(v)) \vee (1-\mu_A(u))]\big) / v . \qquad (10)$$

Now, let

$$S_m(\mu_A(u)) = \mu_A(u) \wedge [(\mu_A(u)\wedge\mu_B(v)) \vee (1-\mu_A(u))] . \qquad (11)$$

The value $S_m(\mu_A(u))$ with a parameter $\mu_B(v)$ is shown in Fig.3. If $\mu_B(v) = 0.3$, then $S_m(\mu_A(u))$ becomes what is shown by the dotted line '----'. In Fig.1 $\mu_A(u)$ takes all the values in the unit interval [0, 1] according to u varying all over U. Thus, the membership function of B'_m of (10) can be obtained as

$$\mu_{B'_m}(v) = \bigvee_{u\in U} S_m(\mu_A(u)) = \begin{cases} \mu_B(v) & \ldots\ldots \ \mu_B(v) \geq 0.5 , \\ 0.5 & \ldots\ldots \ \mu_B(v) \leq 0.5 , \end{cases} \qquad (12)$$

and is shown in Fig.4. From this, $B'_m \neq B$ is obtained and thus it is found that the Relation I in Table I is not satisfied.

Second, suppose A' = <u>very</u> A (= A^2), then

$$B'_m = A^2 \circ ((A \times B) \cup (\neg A \times V))$$
$$= \int_U \mu_A(u)^2/u \circ \int_{U\times V} \big((\mu_A(u)\wedge\mu_B(v)) \vee (1-\mu_A(u))\big)/(u,v)$$
$$= \int_V \bigvee_{u\in U} \big(\mu_A(u)^2 \wedge [(\mu_A(u)\wedge\mu_B(v)) \vee (1-\mu_A(u))]\big) / v . \qquad (13)$$

Now, let

$$S'_m(\mu_A(u)) = \mu_A(u)^2 \wedge [(\mu_A(u)\wedge\mu_B(v)) \vee (1-\mu_A(u))] . \qquad (14)$$

When $\mu_B(v) = 0.3$, the value of $S'_m(\mu_A(u))$ is indicated by the line '— - — -' in Fig.3. Thus, the membership function of B'_m of (13) will be obtained as

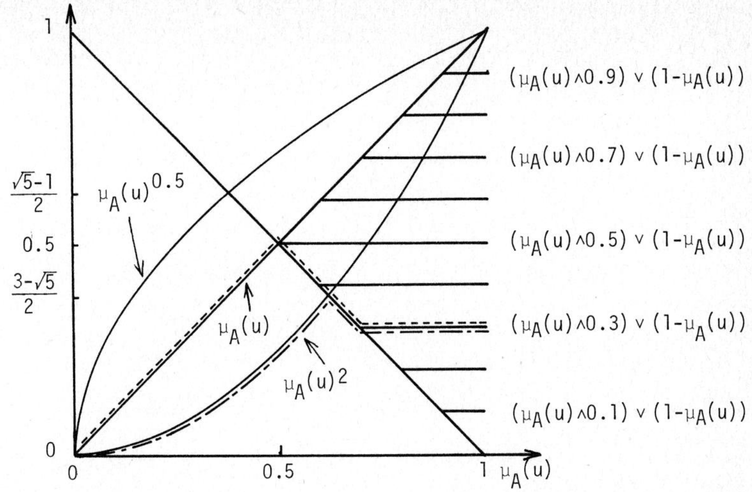

Fig. 3 Diagram of $(\mu_A(u) \wedge \mu_B(v)) \vee (1-\mu_A(u))$, $\mu_A(u)$, $\mu_A(u)^2$ and $\mu_A(u)^{0.5}$.

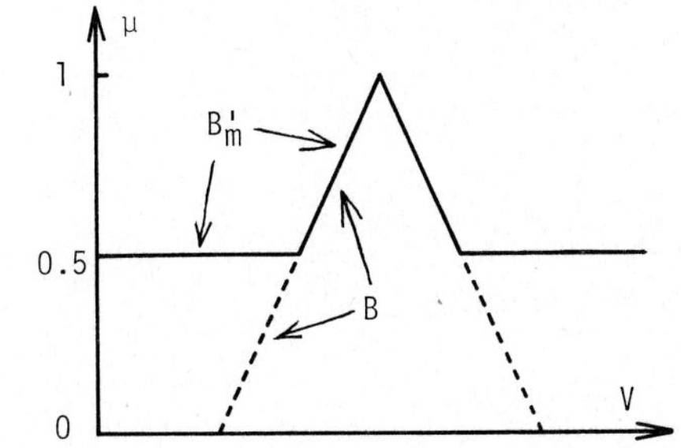

Fig.4 B'_m of (12) when A' = A.

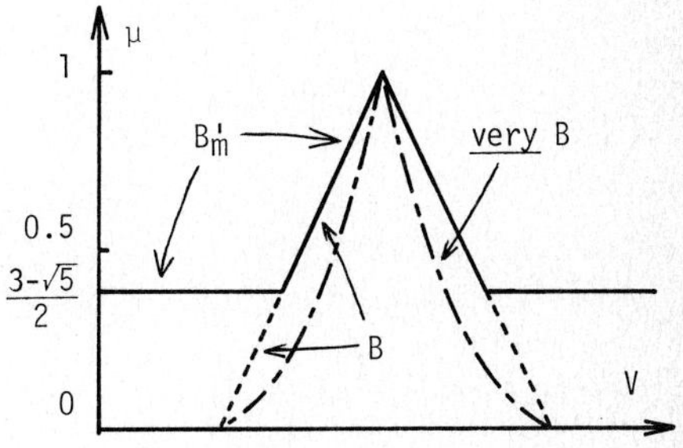

Fig.5 B'_m of (15) when A' = very A.

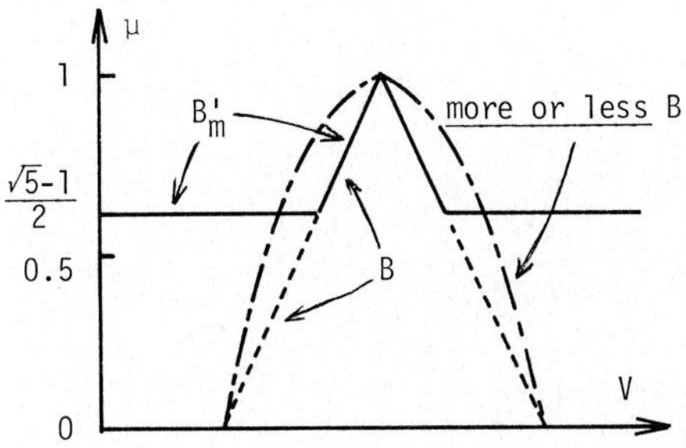

Fig.6 B'_m of (16) when A' = more or less A.

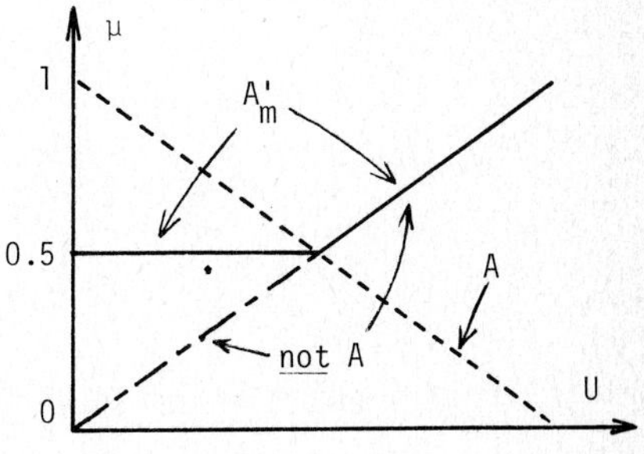

Fig.7 A'_m of (20) when B' = not B

$$\mu_{B'_m}(v) = \bigvee_{u \in U} S'_m(\mu_A(u)) = \begin{cases} \mu_B(v) & \cdots\; \mu_B(v) \geq \dfrac{3-\sqrt{5}}{2}, \\ \dfrac{3-\sqrt{5}}{2} & \cdots\; \mu_B(v) \leq \dfrac{3-\sqrt{5}}{2}. \end{cases} \tag{15}$$

This membership function $\mu_{B'_m}(v)$ is shown in Fig.5, which indicates that

$$B'_m \neq \underline{\text{very}}\ B, \quad B$$

and that both Relation II-1 and Relation II-2 in Table I are not satisfied.

Third, when $A' =$ more or less A ($= A^{0.5}$), the membership function of B'_m is given as

$$\mu_{B'_m}(v) = \begin{cases} \mu_B(v) & \cdots\; \mu_B(v) \geq \dfrac{\sqrt{5}-1}{2}, \\ \dfrac{\sqrt{5}-1}{2} & \cdots\; \mu_B(v) \leq \dfrac{\sqrt{5}-1}{2}, \end{cases} \tag{16}$$

and is shown in Fig.6. From this figure $B'_m \neq$ more or less B is obtained and thus Relation III is not satisfied.

Fourth, we shall show that Relation IV-1 is satisfied when $A' =$ not A. Let $A' =$ not A ($= \neg A$), then we have from (6)

$$\begin{aligned} B'_m &= \neg A \circ ((A \times B) \cup (\neg A \times V)) \\ &= \int_U 1-\mu_A(u)/u \;\circ\; \int_{U\times V} \big((\mu_A(u)\wedge\mu_B(v)) \vee (1-\mu_A(u))\big)/(u,v) \\ &= \int_V \bigvee_{u\in U} \big((1-\mu_A(u)) \wedge [(\mu_A(u)\wedge\mu_B(v)) \vee (1-\mu_A(u))]\big) / v\,. \end{aligned} \tag{17}$$

From Fig.1 there exists $u \in U$ which makes $\mu_A(u) = 0$, so that

$$\begin{aligned} (17) &= \int_V 1 \wedge [(0 \wedge \mu_B(v)) \vee 1] / v = \int_V 1/v = V \\ &= \underline{\text{unknown}}\,. \end{aligned} \tag{18}$$

This shows that Relation IV-1 is satisfied.

Finally, we shall consider Relation V (modus tollens) in Table I. When $B' =$ not B, we can have A'_m as

$$\begin{aligned} A'_m &= ((A \times B) \cup (\neg A \times V)) \circ (\neg B) \\ &= \int_{U\times V} \big((\mu_A(u)\wedge\mu_B(v)) \vee (1-\mu_A(u))\big)/(u,v) \;\circ\; \int_V 1-\mu_B(v)/v \\ &= \int_U \bigvee_{v\in V} \big([(\mu_A(u)\wedge\mu_B(v)) \vee (1-\mu_A(u))] \wedge (1-\mu_B(v))\big) / u\,. \end{aligned} \tag{19}$$

From (19) the membership function of A'_m is obtained by

$$\mu_{A'_m}(u) = \begin{cases} 0.5 & \cdots\; \mu_A(u) \geq 0.5, \\ 1-\mu_A(u) & \cdots\; \mu_A(u) \leq 0.5. \end{cases} \tag{20}$$

Thus it is found that Relation V is not satisfied (see Fig.7).

[II] The Case of Arithmetic Rule (R_a):

In the same way as shown in the maximin rule R_m of [I], we shall indicate that the relations except Relation IV-1 do not hold in the case of arithmetic rule R_a of (3).

Suppose that $A' = A^{\alpha}$ ($\alpha > 0$) as a general case, then the consequence B'_a is obtained from (7) as follows.

$$\begin{aligned} B'_a &= A^{\alpha} \circ ((\neg A \times V) \oplus (U \times B)) \\ &= \int_U \mu_A(u)^{\alpha}/u \ \circ \ \int_{U\times V} \big(1 \wedge (1 - \mu_A(u) + \mu_B(v))\big)/(u,v) \\ &= \int_V \bigvee_{u\in U} \big(\mu_A(u)^{\alpha} \wedge [1 \wedge (1 - \mu_A(u) + \mu_B(v))]\big) \, / \, v \, . \end{aligned} \tag{21}$$

Now, let

$$S_a(\mu_A(u), \alpha) = \mu_A(u)^{\alpha} \wedge [1 \wedge (1 - \mu_A(u) + \mu_B(v))] \, . \tag{22}$$

From Fig.1, $\mu_A(u)$ takes all the values in [0, 1] according to u varying over U. Then from Fig. 8, when $\alpha = 1$,

$$\bigvee_{u\in U} S_a(\mu_A(u), 1) = \frac{1 + \mu_B(v)}{2} \, .$$

Hence we have B'_a of (21) at $\alpha = 1$, that is, $A' = A$, as follows.

$$B'_a = \int_V \frac{1 + \mu_B(v)}{2} / v \, , \tag{23}$$

and the membership function of B'_a is shown in Fig.9. From this figure, $B'_a \neq B$ and thus it is shown that Relation I in Table I is not satisfied.

When $\alpha = 2$, that is, $A' =$ very A ($= A^2$), we have from (22) and Fig.8

$$\bigvee_{u\in U} S_a(\mu_A(u), 2) = \frac{3 + 2\mu_B(v) - \sqrt{5 + 4\mu_B(v)}}{2} \, .$$

Thus, B'_a of (21) at $\alpha = 2$, that is, $A' =$ very A, is obtained as follows.

$$B'_a = \int_V \frac{3 + 2\mu_B(v) - \sqrt{5 + 4\mu_B(v)}}{2} / v \, . \tag{24}$$

Hence the membership functions of B'_a, very B and B are shown in Fig.10 and we have

$B'_a \neq$ very B, B.

This shows that Relation II-1 and Relation II-2 are not satisfied.

When $\alpha = 0.5$, that is, $A' =$ more or less A ($= A^{0.5}$), we can have B'_a as

$$B'_a = \int_V \frac{-1 + \sqrt{5 + 4\mu_B(v)}}{2} / v \, . \tag{25}$$

The membership function of B'_a is shown in Fig.11, which indicates that Relation III is not satisfied, that is, we have $B'_a \neq$ more or less B.

When $A' =$ not A ($= \neg A$), it is shown that Relation IV-1 is satisfied.

$$\begin{aligned} B'_a &= (\neg A) \circ ((\neg A \times V) \oplus (U \times B)) \\ &= \int_V \bigvee_{u\in U} \big((1 - \mu_A(u)) \wedge [1 \wedge (1 - \mu_A(u) + \mu_B(v))]\big) \, / \, v \end{aligned}$$

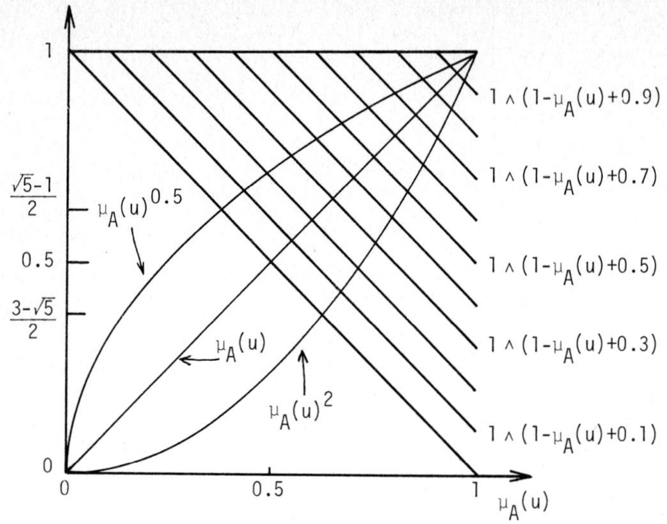

Fig. 8 Diagram of $1 \wedge (1 - \mu_A(u) + \mu_B(v))$, $\mu_A(u)$, $\mu_A(u)^2$ and $\mu_A(u)^{0.5}$.

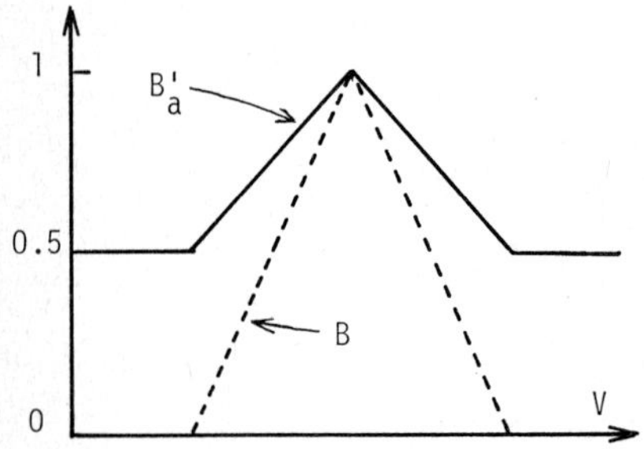

Fig. 9 B'_a of (23) when A' = A.

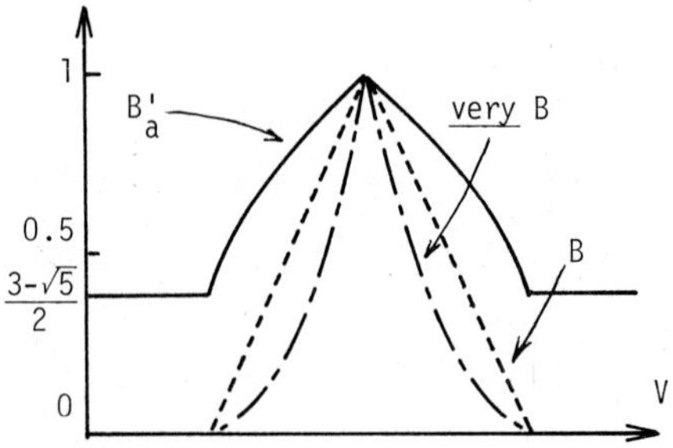

Fig.10 B'_a of (24) when A' = <u>very</u> A.

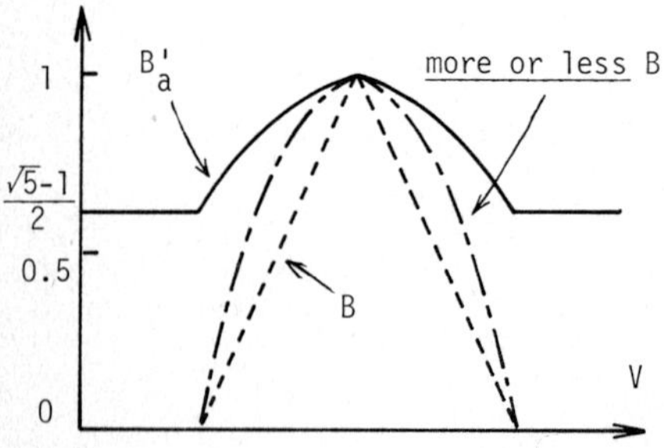

Fig.11 B'_a of (25) when A' = <u>more or less</u> A.

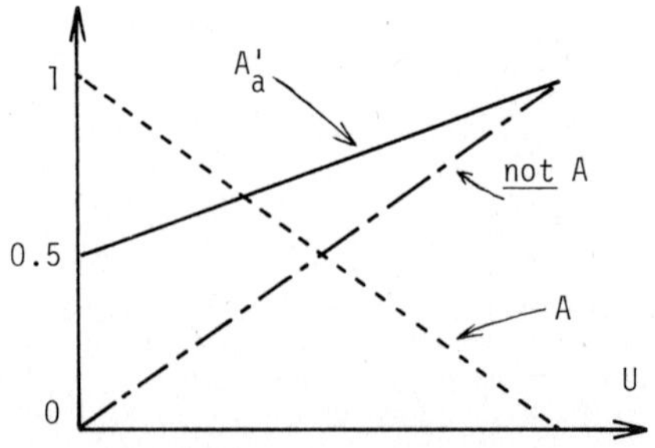

Fig.12 A'_a of (27) when B' = <u>not</u> B.

$$= \int_V 1 \wedge [1 \wedge (1 + \mu_B(v))]/v \quad \ldots \text{ existence of } u \text{ such as } \mu_A(u)=0$$

$$= \int_V 1/v = V$$

$$= \underline{\text{unknown}} \; . \qquad (26)$$

This shows that Relation IV-1 is satisfied. Note that this relation can not be satisfied if $\mu_A(u) > 0$ for all $u \in U$ as in the case of R_s.

As for the modus tollens (Relation V in Table I), let $B' = \underline{\text{not}}\ B\ (= 7B)$, then we can obtain A'_a by the following.

$$A'_a = ((7A \times V) \oplus (U \times B)) \circ (7B)$$

$$= \int_U \bigvee_{v \in V} \left([1 \wedge (1 - \mu_A(u) + \mu_B(v))] \wedge (1 - \mu_B(v))\right) / u$$

$$= \int_U 1 - \frac{\mu_A(u)}{2}/u \; . \qquad (27)$$

Thus, $A'_a \neq \underline{\text{not}}\ A$, which shows that Relation V is not satisfied (see Fig.12).

[III] <u>The Case of Mini Operation Rule (R_c):</u>

We shall indicate that the relations except Relation I and Relation II-2 do not hold in the case of mini operation rule R_c of (4).

Suppose $A' = A^{\alpha}$, then from (8) we obtain that

$$B'_c = A^{\alpha} \circ (A \times B)$$

$$= \int_U \mu_A(u)^{\alpha}/u \;\circ\; \int_{U \times V} (\mu_A(u) \wedge \mu_B(v))/(u, v)$$

$$= \int_V \bigvee_{u \in U} \left(\mu_A(u)^{\alpha} \wedge (\mu_A(u) \wedge \mu_B(v))\right) / v \; . \qquad (28)$$

Since there exists $u \in U$ which makes $\mu_A(u) = 1$ from Fig.1, (28) will be

$$(28) = \int_V 1 \wedge (1 \wedge \mu_B(v))/v = \int_V \mu_B(v)/v$$

$$= B \; . \qquad (29)$$

This shows that Relations I and II-2 are satisfied, but Relations II-1 and III are not satisfied.

Second, let $A' = \underline{\text{not}}\ A\ (= 7A)$, then

$$B'_c = (7A) \circ (A \times B)$$

$$= \int_V \bigvee_{u \in U} \left((1 - \mu_A(u)) \wedge (\mu_A(u) \wedge \mu_B(v))\right) / v \; .$$

Thus, the membership function of B'_c will be given as

$$\mu_{B'_c}(v) = \begin{cases} 0.5 & \ldots\ldots \quad \mu_B(v) \geq 0.5 \; , \\ \mu_B(v) & \ldots\ldots \quad \mu_B(v) \leq 0.5 \; . \end{cases} \qquad (30)$$

This shows that Relations IV-1 and IV-2 are not satisfied.

As to the modus tollens of Relation V, let B' = not B (= 7B), then we have A'_c as

$$A'_c = (A \times B) \circ (7B)$$
$$= \int_U \bigvee_{v \in V} \big((\mu_A(u) \wedge \mu_B(v)) \wedge (1 - \mu_B(v))\big) / u .$$

The membership function of A'_c is given by

$$\mu_{A'_c}(u) = \begin{cases} 0.5 & \ldots\ldots \quad \mu_A(u) \geq 0.5 , \\ \mu_A(u) & \ldots\ldots \quad \mu_A(u) \leq 0.5 . \end{cases} \tag{31}$$

Thus, Relation V is not satisfied under R_c.

It is interesting to note that when A' = unknown (= U), we have

$$\underline{\text{unknown}} \circ R_c = B .$$

This consequence can not be accepted according to our intuitions.

Above discussions show that using the methods (Zadeh's methods R_m, R_a and Mamdani's method R_c), almost all the relations in Table I can not be satisfied and it may be clear that the consequences inferred by these methods do not always fit our intuitions.

NEW METHODS FOR FUZZY CONDITIONAL INFERENCE

In this section we shall introduce several new methods for fuzzy conditional inference which satisfy to almost all the relations in Table I.

Let fuzzy sets A and B in Ant 1 of (1) be as follows.

$$A = \int_U \mu_A(u)/u; \qquad B = \int_V \mu_B(v)/v,$$

where U and V may be discrete.

Suppose that $\mu_A(u)$ and $\mu_B(v)$ satisfy the following conditions†.

$$\text{(i)} \quad \{ \mu_A(u) \mid u \in U \} \supseteqq \{ \mu_B(v) \mid v \in V \}. \tag{32}$$

$$\text{(ii)} \quad \exists\, u \in U \;\; \mu_A(u) = 0; \quad \exists\, u' \in U \;\; \mu_A(u') = 1. \tag{33}$$

$$\text{(iii)} \quad \exists\, v \in V \;\; \mu_B(v) = 0; \quad \exists\, v' \in V \;\; \mu_B(v') = 1. \tag{34}$$

[IV] Method R_s Satisfying Relations I, II-1, III, IV-1 and V:

If Ant 1 of (1) translates into the following fuzzy relation R_s:

$$R_s = A \times V \underset{s}{\longrightarrow} U \times B$$

$$= \int_{U \times V} (\mu_A(u) \underset{s}{\longrightarrow} \mu_B(v))/(u, v) , \tag{35}$$

† It is noted that we have discussed the methods by Zadeh and Mamdani under the same conditions.

where

$$\mu_A(u) \underset{s}{\longrightarrow} \mu_B(v) = \begin{cases} 1 & \ldots\ldots \quad \mu_A(u) \leq \mu_B(v), \\ 0 & \ldots\ldots \quad \mu_A(u) > \mu_B(v), \end{cases} \tag{36}$$

then the consequence B'_s is obtained by

$$B'_s = A' \circ R_s = A' \circ (A \times V \underset{s}{\longrightarrow} U \times B). \tag{37}$$

Note that the definition of (36) is based on the implication in S_{aleph} logic system [2].

Using this method we shall first show that Relations I, II-1, III and IV-1 are satisfied under the assumptions of (32)-(34).

As a general case, suppose $A' = A^{\alpha}$ ($\alpha > 0$), then (37) will be

$$\begin{aligned} B'_s &= A^{\alpha} \circ (A \times V \underset{s}{\longrightarrow} U \times B) \\ &= \int_U \mu_A(u)^{\alpha}/u \quad \circ \int_{U \times V} (\mu_A(u) \underset{s}{\longrightarrow} \mu_B(v))/(u,v) \\ &= \int_V \bigvee_{u \in U} \left(\mu_A(u)^{\alpha} \wedge (\mu_A(u) \underset{s}{\longrightarrow} \mu_B(v))\right) / v . \end{aligned} \tag{38}$$

Here, for each v in V, we can obtain two sets U_1 and U_2 such that

$$U_1 = \{ u \mid \mu_A(u) \leq \mu_B(v)\}, \tag{39}$$

$$U_2 = \{ u \mid \mu_A(u) > \mu_B(v)\}, \tag{40}$$

with $U = U_1 \cup U_2$ and $U_1 \cap U_2 = \Phi$.

Then from (38) we can obtain that

$$\begin{aligned} (38) &= \int_V \Big(\bigvee_{u \in U_1} [\mu_A(u)^{\alpha} \wedge (\mu_A(u) \underset{s}{\longrightarrow} \mu_B(v))] \ \vee \bigvee_{u \in U_2} [\mu_A(u)^{\alpha} \wedge (\mu_A(u) \underset{s}{\longrightarrow} \mu_B(v))]\Big) / v \\ &= \int_V \Big(\bigvee_{u \in U_1} [\mu_A(u)^{\alpha} \wedge 1] \ \vee \bigvee_{u \in U_2} [\mu_A(u)^{\alpha} \wedge 0]\Big) / v \\ &= \int_V \bigvee_{u \in U_1} \mu_A(u)^{\alpha} / v \\ &= \int_V \mu_B(v)^{\alpha}/v \qquad \ldots\ldots \quad \text{from (32) and (39)} \\ &= B^{\alpha} . \end{aligned} \tag{41}$$

This shows that when $\alpha = 1$ ($A' = A$), $\alpha = 2$ (A' = <u>very</u> A) and $\alpha = 0.5$ (A' = <u>more or less</u> A), Relations I, II-1 and III are satisfied, respectively.

Next, let A' = <u>not</u> A ($= \neg A$), then (37) becomes

$$\begin{aligned} B'_s &= (\neg A) \circ (A \times V \underset{s}{\longrightarrow} U \times B) \\ &= \int_V \bigvee_{u \in U} \left((1 - \mu_A(u)) \wedge (\mu_A(u) \underset{s}{\longrightarrow} \mu_B(v))\right) / v. \end{aligned} \tag{42}$$

From the assumption (33) there exists u in U which makes $\mu_A(u) = 0$. Thus

$$\bigvee_{u \in U} \left((1 - \mu_A(u)) \wedge (\mu_A(u) \underset{s}{\longrightarrow} \mu_B(v))\right) = 1.$$

Therefore,

$$(42) = \int_V 1/v = V = \underline{\text{unknown}}.$$

This indicates the satisfaction of Relation IV-1.

Finally, we shall investigate Relation V (modus tollens) under R_s.

Suppose that $\mu_A(u)$ and $\mu_B(v)$ of fuzzy sets A and B satisfy the following condition instead of the condition (32).

$$\text{(i')} \quad \{ \mu_A(u) \mid u \in U \} \subseteqq \{ \mu_B(v) \mid v \in V \} . \tag{43}$$

When B' = not B (= ¬B), A'_s is given as

$$A'_s = (A \times V \underset{s}{\longrightarrow} U \times B) \circ (\neg B)$$

$$= \int_U \bigvee_{v \in V} \left((\mu_A(u) \underset{s}{\longrightarrow} \mu_B(v)) \wedge (1 - \mu_B(v)) \right) / u . \tag{44}$$

Here, for each u ε U, we can have two sets V_1 and V_2 such that

$$V_1 = \{ v \mid \mu_A(u) \le \mu_B(v) \} , \tag{45}$$

$$V_2 = \{ v \mid \mu_A(u) > \mu_B(v) \} . \tag{46}$$

Then (44) will be

$$(44) = \int_U \bigvee_{v \in V_1} (1 - \mu_B(v)) / u$$

$$= \int_U (1 - \mu_A(u))/u \quad \text{...... from (43) and (45)}$$

$$= \neg A . \tag{47}$$

This shows that A'_s = not A (= ¬A) when B' = not B (= ¬B), so that Relation V is satisfied.

Note: In this discussion of Relation V (modus tollens) under R_s, we have used the condition (43) rather than (32). Thus, if we introduce the following condition (i") satisfying both (32) and (43), then R_s can satisfy all the Relations I, II-1, III, IV-1 and V at the same time.

$$\text{(i")} \quad \{ \mu_A(u) \mid u \in U \} = \{ \mu_B(v) \mid v \in V \} . \tag{48}$$

(Example 1) Let †

$$U = V = 0 + 1 + 2 + 3 + 4 + 5 + 6 + 7 + 8 + 9 + 10, \tag{49}$$

$$A = \underline{\text{small}} = 1/0 + 0.8/1 + 0.6/2 + 0.2/3, \tag{50}$$

$$B = \underline{\text{middle}} = 0.2/2 + 0.6/3 + 0.8/4 + 1/5 + 0.8/6 + 0.6/7 + 0.2/8 . \tag{51}$$

Then, using R_s of (35), the fuzzy conditional proposition:

If x is small then y is middle

translates into

$$R_s = \underline{\text{small}} \times V \underset{s}{\longrightarrow} U \times \underline{\text{middle}}$$

---

† It is noted that the fuzzy sets A and B satisfy the conditions (48), (33) and (34).

$$
= \begin{array}{c|ccccccccccc|}
 & 0 & 1 & 2 & 3 & 4 & 5 & 6 & 7 & 8 & 9 & 10 \\
\hline
0 & 0 & 0 & 0 & 0 & 0 & 1 & 0 & 0 & 0 & 0 & 0 \\
1 & 0 & 0 & 0 & 0 & 1 & 1 & 1 & 0 & 0 & 0 & 0 \\
2 & 0 & 0 & 0 & 1 & 1 & 1 & 1 & 1 & 0 & 0 & 0 \\
3 & 0 & 0 & 1 & 1 & 1 & 1 & 1 & 1 & 1 & 0 & 0 \\
4 & 1 & 1 & 1 & 1 & 1 & 1 & 1 & 1 & 1 & 1 & 1 \\
5 & 1 & 1 & 1 & 1 & 1 & 1 & 1 & 1 & 1 & 1 & 1 \\
6 & 1 & 1 & 1 & 1 & 1 & 1 & 1 & 1 & 1 & 1 & 1 \\
7 & 1 & 1 & 1 & 1 & 1 & 1 & 1 & 1 & 1 & 1 & 1 \\
8 & 1 & 1 & 1 & 1 & 1 & 1 & 1 & 1 & 1 & 1 & 1 \\
9 & 1 & 1 & 1 & 1 & 1 & 1 & 1 & 1 & 1 & 1 & 1 \\
10 & 1 & 1 & 1 & 1 & 1 & 1 & 1 & 1 & 1 & 1 & 1
\end{array} .
$$

Thus, we can obtain the consequence B'_S of (37) as follows when A' is small, very small, more or less small or not small, where it is assumed that

very small = small^2 = 1/0 + 0.64/1 + 0.36/2 + 0.04/3,

more or less small = small$^{0.5}$ = 1/0 + 0.89/1 + 0.77/2 + 0.45/3,

not small = $\neg$small = 0.2/1 + 0.4/2 + 0.8/3 + 1/(4 + 5 + ... + 10),

very middle = 0.04/2 + 0.36/3 + 0.64/4 + 1/5 + 0.64/6 + 0.36/7 + 0.04/8,

more or less middle = 0.45/2 + 0.77/3 + 0.89/4 + 1/5 + 0.89/6 + 0.77/7 + 0.45/8,

not middle = 1/0 + 1/1 + 0.8/2 + 0.4/3 + 0.2/4 + 0/5 + 0.2/6 + 0.4/7 + 0.8/8 + 1/9 + 1/10,

unknown = 1/0 + 1/1 + 1/2 + ... + 1/10 (= 0 + 1 + 2 + ... + 10).

The consequence B'_S will be

(i) small o R_S = 0.2/2 + 0.6/3 + 0.8/4 + 1/5 + 0.8/6 + 0.6/7 + 0.2/8
= middle .

(ii) very small o R_S = 0.04/2 + 0.36/3 + 0.64/4 + 1/5 + 0.64/6 + 0.36/7 + 0.04/8
= very middle .

(iii) more or less small o R_S
= 0.45/2 + 0.77/3 + 0.89/4 + 1/5 + 0.89/6 + 0.77/7 + 0.45/8
= more or less middle.

(iv) not small o R_S = 1/0 + 1/1 + 1/2 + ... + 1/10
= unknown.

As for the modus tollens of Relation V, we have

(v) R_S o not middle = 0.2/1 + 0.4/2 + 0.8/3 + 1/(4 + 5 + ... + 10)
= not small.

Stated in English, these inferences may be expressed as follows.

i) If x is <u>small</u> then y is <u>middle</u>.
x is <u>small</u>.

---

y is <u>middle</u>.

ii) If x is <u>small</u> then y is <u>middle</u>.
x is <u>very small</u>.

---

y is <u>very middle</u>.

iii) If x is <u>small</u> then y is <u>middle</u>.
x is <u>more or less small</u>.

---

y is <u>more or less middle</u>.

iv) If x is <u>small</u> then y is <u>middle</u>.
x is <u>not small</u>.

---

y is <u>unknown</u>.

v) If x is <u>small</u> then y is <u>middle</u>.
y is <u>not middle</u>.

---

x is <u>not small</u>.

[V] <u>Method R_g Satisfying Relations I, II-2, III and IV-1</u>:

If Ant 1 of (1) translates into the following fuzzy relation R_g:

$$R_g = A \times V \xrightarrow[g]{} U \times B$$
$$= \int_{U\times V} (\mu_A(u) \xrightarrow[g]{} \mu_B(v))/(u,v), \tag{52}$$

where

$$\mu_A(u) \xrightarrow[g]{} \mu_B(v) = \begin{cases} 1 & \ldots\ldots \mu_A(u) \le \mu_B(v), \\ \mu_B(v) & \ldots\ldots \mu_A(u) > \mu_B(v), \end{cases} \tag{53}$$

then the consequence B'_g is obtained by

$$B'_g = A' \circ R_g = A' \circ (A \times V \xrightarrow[g]{} U \times B). \tag{54}$$

It is noted that the definition (53) is from Gödel's definition of the implication in G_{aleph} logic system [2].

As a general case, let $A' = A^{\alpha}$ ($\alpha > 0$), then (54) becomes

$$R'_g = A^{\alpha} \circ (A \times V \xrightarrow[g]{} U \times B)$$
$$= \int_V \bigvee_{u\in U} \left\{\mu_A(u)^{\alpha} \wedge (\mu_A(u) \xrightarrow[g]{} \mu_B(v))\right\} / v$$
$$= \int_V \left\{ \bigvee_{u\in U_1} (\mu_A(u)^{\alpha} \wedge 1) \vee \bigvee_{u\in U_2} (\mu_A(u)^{\alpha} \wedge \mu_B(v))\right\} / v \quad \ldots \text{ from (39) and (40)}$$

$$= \int_V \left(\mu_B(v)^{\alpha} \vee [(\bigvee_{u\varepsilon U_2} \mu_A(u)^{\alpha}) \wedge \mu_B(v)]\right)/v \quad \text{... from (32) and (39)}$$

$$= \int_V \left(\mu_B(v)^{\alpha} \vee [1 \wedge \mu_B(v)]\right)/v \quad \text{..... from (33) and (40)}$$

$$= \int_V (\mu_B(v)^{\alpha} \vee \mu_B(v))/v$$

$$= \begin{cases} B^{\alpha} & \text{......} \quad \alpha \leq 1 , \\ B & \text{......} \quad \alpha \geq 1 . \end{cases} \tag{55}$$

This shows that when $\alpha = 1$, $\alpha = 2$ and $\alpha = 0.5$, Relations I, II-2 and III are satisfied, respectively.

The satisfaction of Relation IV-1 can be shown in the similar way as in the case of R_S.

Finally we shall consider Relation V (modus tollens) under R_g and show that R_g does not satisfy it.

Let B' = not B (= 7B), then A'_g is obtained as

$$A'_g = (A \times V \xrightarrow[g]{} U \times B) \circ (7B)$$

$$= \int_U \bigvee_{v\varepsilon V} \left((\mu_A(u) \xrightarrow[g]{} \mu_B(v)) \wedge (1 - \mu_B(v))\right) / u$$

$$= \int_U \left(\bigvee_{v\varepsilon V_1} (1 - \mu_B(v)) \vee \bigvee_{v\varepsilon V_2} [\mu_B(v) \wedge (1 - \mu_B(v))]\right) / u \quad \text{... from (45) and (46)}$$

$$= \int_U \left((1 - \mu_A(u)) \vee \bigvee_{v\varepsilon V_2} [\mu_B(v) \wedge (1 - \mu_B(v))]\right)/u \quad \text{... from (43) and (45).}$$

If the membership functions $\mu_A(u)$ and $\mu_B(v)$ are continuous functions as in Fig.1 and 2, then the membership function of A'_g will be given by

$$\mu_{A'_g}(u) = \begin{cases} 0.5 & \text{......} \quad \mu_A(u) \geq 0.5 , \\ 1 - \mu_A(u) & \text{......} \quad \mu_A(u) \leq 0.5 . \end{cases} \tag{56}$$

Anyway, it is found that $A'_g \neq 7A$ under R_g when B' = 7B, so that R_g does not satisfy Relation V.

(Example 2) R_g is obtained from (52) as follows, where small and middle are the same as (50) and (51), respectively.

$$R_g = \underline{\text{small}} \times V \xrightarrow[g]{} U \times \underline{\text{middle}} .$$

Then we have

(i) small o R_g = middle.

(ii) very small o R_g = middle.

(iii) more or less small o R_g = more or less middle.

(iv) not small o R_g = unknown.

(v) R_g o <u>not middle</u> = 0.4/(0 + 1 + 2) + 0.8/3 + 1/(4 + 5 + ... + 10)
$\neq$ <u>not small</u>.

where R_g is given by the matrix form as

R_g =

	0	1	2	3	4	5	6	7	8	9	10
0	0	0	.2	.6	.8	1	.8	.6	.2	0	0
1	0	0	.2	.6	1	1	1	.6	.2	0	0
2	0	0	.2	1	1	1	1	1	.2	0	0
3	0	0	1	1	1	1	1	1	1	0	0
4	1	1	1	1	1	1	1	1	1	1	1
5	1	1	1	1	1	1	1	1	1	1	1
6	1	1	1	1	1	1	1	1	1	1	1
7	1	1	1	1	1	1	1	1	1	1	1
8	1	1	1	1	1	1	1	1	1	1	1
9	1	1	1	1	1	1	1	1	1	1	1
10	1	1	1	1	1	1	1	1	1	1	1

[VI] <u>Method R_{sg} Satisfying Relations I, II-1, III, IV-2 and V:</u>

If Ant 1 of (1) translates into the following fuzzy relation R_{sg}:

$$R_{sg} = (A \times V \xrightarrow[s]{} U \times B) \cap (\neg A \times V \xrightarrow[g]{} U \times \neg B), \tag{57}$$

then we can show that under the conditions of (48), (33) and (34)

$$B'_{sg} = A^{\alpha} \circ ((A \times V \xrightarrow[s]{} U \times B) \cap (\neg A \times V \xrightarrow[g]{} U \times \neg B))$$
$$= \int_V \bigvee_{u \in U} \left(\mu_A(u)^{\alpha} \wedge (\mu_A(u) \xrightarrow[s]{} \mu_B(v)) \wedge ((1 - \mu_A(u)) \xrightarrow[g]{} (1 - \mu_B(v)))\right) / v$$
$$= B^{\alpha}. \tag{58}$$

Similarly we can obtain that

$$(\neg A) \circ R_{sg} = \neg B , \tag{59}$$

$$R_{sg} \circ (\neg B) = \neg A . \tag{60}$$

Therefore, R_{sg} satisfies Relations I, II-1, III, IV-2 and V.

<u>(Example 3)</u> Using the same fuzzy sets <u>small</u> (50) and <u>middle</u> (51), R_{sg} becomes

$$R_{sg} = (\underline{small} \times V \xrightarrow[s]{} U \times \underline{middle}) \cap (\neg\underline{small} \times V \xrightarrow[g]{} U \times \neg\underline{middle}).$$

(i) <u>small</u> o R_{sg} = <u>middle</u>.

(ii) <u>very small</u> o R_{sg} = <u>very middle</u>.

(iii) <u>more or less small</u> o R_{sg} = <u>more or less middle</u>.

(iv) <u>not small</u> o R_{sg} = <u>not middle</u>.

(v) R_{sg} o <u>not middle</u> = <u>not small</u>.

where R_{sg} in the form of matrix is given by

$$R_{sg} = \begin{array}{c} \\ 0 \\ 1 \\ 2 \\ 3 \\ 4 \\ 5 \\ 6 \\ 7 \\ 8 \\ 9 \\ 10 \end{array} \begin{array}{c} \begin{array}{ccccccccccc} 0 & 1 & 2 & 3 & 4 & 5 & 6 & 7 & 8 & 9 & 10 \end{array} \\ \left[\begin{array}{ccccccccccc} 0 & 0 & 0 & 0 & 0 & 1 & 0 & 0 & 0 & 0 & 0 \\ 0 & 0 & 0 & 0 & 1 & 0 & 1 & 0 & 0 & 0 & 0 \\ 0 & 0 & 0 & 1 & .2 & 0 & .2 & 1 & 0 & 0 & 0 \\ 0 & 0 & 1 & .4 & .2 & 0 & .2 & .4 & 1 & 0 & 0 \\ 1 & 1 & .8 & .4 & .2 & 0 & .2 & .4 & .8 & 1 & 1 \\ 1 & 1 & .8 & .4 & .2 & 0 & .2 & .4 & .8 & 1 & 1 \\ 1 & 1 & .8 & .4 & .2 & 0 & .2 & .4 & .8 & 1 & 1 \\ 1 & 1 & .8 & .4 & .2 & 0 & .2 & .4 & .8 & 1 & 1 \\ 1 & 1 & .8 & .4 & .2 & 0 & .2 & .4 & .8 & 1 & 1 \\ 1 & 1 & .8 & .4 & .2 & 0 & .2 & .4 & .8 & 1 & 1 \\ 1 & 1 & .8 & .4 & .2 & 0 & .2 & .4 & .8 & 1 & 1 \end{array}\right] \end{array} .$$

[VII] Method R_{gg} Satisfying Relations I, II-2, III and IV-2:

Ant 1 of (1) which is translated into the following fuzzy relation R_{gg}:

$$R_{gg} = (A \times V \xrightarrow[g]{} U \times B) \cap (\neg A \times V \xrightarrow[g]{} U \times \neg B) \tag{61}$$

satisfies the following.

$$A^{\alpha} \circ R_{gg} = \begin{cases} B^{\alpha} & \ldots\ldots \quad \alpha \leq 1 , \\ B & \ldots\ldots \quad \alpha \geq 1 . \end{cases} \tag{62}$$

$$(\neg A) \circ R_{gg} = \neg B. \tag{63}$$

$$R_{gg} \circ (\neg B) = A'_{gg} , \tag{64}$$

where

$$\mu_{A'_{gg}}(u) = \begin{cases} 0.5 & \ldots \quad \mu_A(u) \geq 0.5 , \\ 1 - \mu_A(u) & \ldots \quad \mu_A(u) \leq 0.5 . \end{cases} \tag{65}$$

with the assumption that μ_A and μ_B are continuous.

Hence, R_{gg} is shown to satisfy Relations I, II-2, III and IV-2.

(Example 4)

$$R_{gg} = (\underline{\text{small}} \times V \xrightarrow[g]{} U \times \underline{\text{middle}}) \cap (\neg\underline{\text{small}} \times V \xrightarrow[g]{} U \times \neg\underline{\text{middle}}).$$

(i) small $\circ\, R_{gg}$ = middle.

(ii) very small $\circ\, R_{gg}$ = middle.

(iii) more or less small $\circ\, R_{gg}$ = more or less middle.

(iv) not small $\circ\, R_{gg}$ = not middle.

(v) $R_{gg} \circ$ not middle $= 0.4/(0 + 1 + 2) + 0.8/3 + 1/(4 + 5 + \ldots + 10)$

$\neq$ not small.

where R_{gg} is given by the following.

$$
R_{gg} = \begin{array}{c} \\ 0 \\ 1 \\ 2 \\ 3 \\ 4 \\ 5 \\ 6 \\ 7 \\ 8 \\ 9 \\ 10 \end{array}
\begin{array}{c} \begin{array}{ccccccccccc} 0 & 1 & 2 & 3 & 4 & 5 & 6 & 7 & 8 & 9 & 10 \end{array} \\
\left(\begin{array}{ccccccccccc}
0 & 0 & .2 & .6 & .8 & 1 & .8 & .6 & .2 & 0 & 0 \\
0 & 0 & .2 & .6 & 1 & 0 & 1 & .6 & .2 & 0 & 0 \\
0 & 0 & .2 & 1 & .2 & 0 & .2 & 1 & .2 & 0 & 0 \\
0 & 0 & 1 & .4 & .2 & 0 & .2 & .4 & 1 & 0 & 0 \\
1 & 1 & .8 & .4 & .2 & 0 & .2 & .4 & .8 & 1 & 1 \\
1 & 1 & .8 & .4 & .2 & 0 & .2 & .4 & .8 & 1 & 1 \\
1 & 1 & .8 & .4 & .2 & 0 & .2 & .4 & .8 & 1 & 1 \\
1 & 1 & .8 & .4 & .2 & 0 & .2 & .4 & .8 & 1 & 1 \\
1 & 1 & .8 & .4 & .2 & 0 & .2 & .4 & .8 & 1 & 1 \\
1 & 1 & .8 & .4 & .2 & 0 & .2 & .4 & .8 & 1 & 1 \\
1 & 1 & .8 & .4 & .2 & 0 & .2 & .4 & .8 & 1 & 1
\end{array}\right) \end{array} .
$$

The satisfaction or failure of each relation in Table I under each method is summarized in Table II.

Table II Satisfaction of Each Relation in Table I

	Ant 2	Cons	R_m	R_a	R_c	R_s	R_g	R_{sg}	R_{gg}
Relation I (modus ponens)	A	B	X	X	O	O	O	O	O
Relation II-1	very A	very B	X	X	X	O	X	O	X
Relation II-2	very A	B	X	X	O	X	O	X	O
Relation III	more or less A	more or less B	X	X	X	O	O	O	O
Relation IV-1	not A	unknown	O	O	X	O	O	X	X
Relation IV-2	not A	not B	X	X	X	X	X	O	O
Relation V (modus tollens)	not B	not A	X	X	X	O	X	O	X

SOME PROPERTIES OF R_s AND R_g

In this section we shall describe some interesting properties of R_s defined in (35) and R_g defined in (52). Note that the fuzzy relations R_m and R_a defined by Zadeh do not have these properties and the fuzzy relation R_c defined by Mamdani has only the Property 2.

[Property 1] Let fuzzy conditional propositions P_1, P_2 and P_3 be given as

$$P_1 = \text{If } x \text{ is A then } y \text{ is B},$$
$$P_2 = \text{If } y \text{ is B then } z \text{ is C},$$
$$P_3 = \text{If } x \text{ is A then } z \text{ is C},$$

where A, B and C are fuzzy sets in U, V, and W, respectively.

Let

$$R_s(A, B) = A \times V \xrightarrow[s]{} U \times B ,$$

$$R_s(B, C) = B \times W \xrightarrow[s]{} V \times C ,$$

$$R_s(A, C) = A \times W \xrightarrow[s]{} U \times C ,$$

be fuzzy relations which are translated, respectively, from P_1, P_2 and P_3 using (35) and let

$$R_g(A, B) = A \times V \xrightarrow[g]{} U \times B ,$$

$$R_g(B, C) = B \times W \xrightarrow[g]{} V \times C ,$$

$$R_g(A, C) = A \times W \xrightarrow[g]{} U \times C ,$$

be fuzzy relations translated from P_1, P_2 and P_3 by the use of (52). Then, under the following conditions, that is,

$$\{ \mu_A(u) \mid u \in U \} \supseteqq \{ \mu_B(v) \mid v \in V \} \supseteqq \{ \mu_C(w) \mid w \in W \}. \quad (66)$$

$$\exists\, u \in U \quad \mu_A(u) = 0; \qquad \exists\, u' \in U \quad \mu_A(u') = 1 . \quad (67)$$

$$\exists\, v \in V \quad \mu_B(v) = 0; \qquad \exists\, v' \in V \quad \mu_B(v') = 1 . \quad (68)$$

$$\exists\, w \in W \quad \mu_C(w) = 0; \qquad \exists\, w' \in W \quad \mu_C(w') = 1 . \quad (69)$$

the following syllogisms hold.

$$R_s(A, C) = R_s(A, B) \circ R_s(B, C) . \quad (70)$$

$$R_g(A, C) = R_g(A, B) \circ R_g(B, C) . \quad (71)$$

(Example 5) Let fuzzy sets A, B and C be given as

$$A = 1/1 + 0.8/2 + 0.6/3 + 0.4/4 + 0.2/5,$$
$$B = 0.2/4 + 0.4/5 + 0.8/6 + 1/7,$$
$$C = 0.4/2 + 0.8/3 + 1/4 + 0.8/5 + 0.2/6,$$

with $U = V = W = 1 + 2 + 3 + 4 + 5 + 6 + 7$. Then $R_g(A, B)$ and $R_g(B, C)$ are given as

$R_g(A, B) = A \times V \xrightarrow[g]{} U \times B$ =

	1	2	3	4	5	6	7
1	0	0	0	.2	.4	.8	1
2	0	0	0	.2	.4	1	1
3	0	0	0	.2	.4	1	1
4	0	0	0	.2	1	1	1
5	0	0	0	1	1	1	1
6	1	1	1	1	1	1	1
7	1	1	1	1	1	1	1

$R_g(B, C) = B \times W \xrightarrow[g]{} V \times C$ =

	1	2	3	4	5	6	7
1	1	1	1	1	1	1	1
2	1	1	1	1	1	1	1
3	1	1	1	1	1	1	1
4	0	1	1	1	1	1	0
5	0	1	1	1	1	.2	0
6	0	.4	1	1	1	.2	0
7	0	.4	.8	1	.8	.2	0

Thus, the composition of $R_g(A, B)$ and $R_g(B, C)$ leads to

$$R_g(A, B) \circ R_g(B, C) = R_g(A, C) ,$$

where $R_g(A, C)$ is given as follows.

$$R_g(A, C) = A \times W \xrightarrow[g]{} U \times C = \begin{array}{c} \\ 1 \\ 2 \\ 3 \\ 4 \\ 5 \\ 6 \\ 7 \end{array} \begin{array}{c} \begin{array}{ccccccc} 1 & 2 & 3 & 4 & 5 & 6 & 7 \end{array} \\ \left(\begin{array}{ccccccc} 0 & .4 & .8 & 1 & .8 & .2 & 0 \\ 0 & .4 & 1 & 1 & 1 & .2 & 0 \\ 0 & .4 & 1 & 1 & 1 & .2 & 0 \\ 0 & 1 & 1 & 1 & 1 & .2 & 0 \\ 0 & 1 & 1 & 1 & 1 & 1 & 0 \\ 1 & 1 & 1 & 1 & 1 & 1 & 1 \\ 1 & 1 & 1 & 1 & 1 & 1 & 1 \end{array} \right) \end{array} .$$

This shows the satisfaction of (71).

[<u>Property 2</u>] For the fuzzy conditional proposition P_1:

$$P_1 = \text{If } x \text{ is } A \text{ then } y \text{ is } B,$$

and its contrapositive proposition P_2:

$$P_2 = \text{If } y \text{ is } \underline{\text{not}}\ B \text{ then } x \text{ is } \underline{\text{not}}\ A,$$

let $R_s(A, B)$ and $R_s(\neg B, \neg A)$ be fuzzy relations which are translated from P_1 and P_2, respectively, using (35). Then the following equality holds.

$$R_s(\neg B, \neg A) = \tilde{R}_s(A, B) , \tag{72}$$

where $\tilde{R}_s(A, B)$ denotes the inverse relation of $R_s(A, B)$.

Note that R_g does not satisfy this Property 2.

CONCLUSION

In this paper we pointed out that the methods by Zadeh and Mamdani for the fuzzy conditional inference do not give the consequences which fit our intuitions, and gave several new methods which fit our intuitions under several criteria such as modus ponens and modus tollens.

The inference form treated here is simple, so the formalization of inference methods for more complicated forms of inference such that the propositions have their truth values or contain the proposition "If ... then ... else ..." will be future subjects.

REFERENCES

[1] Mamdani, E.H. (1977). Application of fuzzy logic to approximate reasoning using linguistic systems. <u>IEEE Trans. on Computer</u>, <u>c-26</u>, 1182-1191.

[2] Rescher, N. (1969). <u>Many Valued Logic</u>. New York: McGraw-Hill.

[3] Zadeh, L.A. (1975). Calculus of fuzzy restrictions, in "<u>Fuzzy Sets and Their Applications to Cognitive and Decision Processes</u>" (eds. L.A. Zadeh, K.S. Fu, K. Tanaka and M. Shimura). New York: Academic Press, pp. 1-39.

ADVANCES IN FUZZY SET THEORY AND APPLICATIONS
M.M. Gupta, R.K. Ragade, R.R. Yager (editors)

AN APPROACH TO FUZZY REASONING METHOD

Yahachiro TSUKAMOTO

Department of Control Engineering, Tokyo Institute of Technology, Tokyo, Japan

1. INTRODUCTION

Recently, L.A.Zadeh has suggested that the essence and power of human reasoning is in its capability to grasp and use inexact concepts directly. With a perspective that the use of fuzzy logic and linguistic approach will provide a new framework for the understanding of such a human reasoning process, he has developed in detail a model for fuzzy reasoning in approximate terms |1,2,3|. One of the most remarkable findings may be the compositional rule of inference formulated by L.A.Zadeh |4|. Moreover the method of solution by E.Sanchez |5| for the inverse problem of fuzzy relation equations is also taken as a rather complicated case of fuzzy reasoning. However, here we do not touch upon these problems, but only propose some new rules of inference corresponding to *modus ponens*, *modus tollens*, etc., in the situation in which fuzzy propositions and fuzzy truth values are used. Concerning the rules of inference using implication, we shall attain to the different kinds of rules from the formulation given by L.A.Zadeh |4|. The present methods will enable us to carry out some interesting fuzzy reasonings together with the technique of truth qualification. The process of fuzzy reasoning is illustrated in detail by considering two examples.

First it deserves to note that the term "fuzzy logic" has been used variously in the literature. According to the nice classification by B.R.Gaines |6|, "fuzzy logic" in this paper is used as a basis for linguistic reasoning with the vague statements using fuzzy set theory for the fuzzification of logical structure. Specifically, we shall consider a fuzzy logic stemming from the fuzzification of Łukasiewicz's infinite-valued logic.

2. PRELIMINARIES

In order to approach to fuzzy reasoning methods without failure, it is useful to consider in advance the general properties which should be satisfied by the rule of inference in a fuzzy version. First we give the definitions concerning special linguistic truth values and the relationships between them.

[Definition 1] Let τ be a linguistic truth value characterized by the membership function as

$$\mu_\tau : V \to [0, 1], \tag{1}$$

where V denotes the range of the ordinary numerical truth value used in infinite-valued logic, that is,

$$V \triangleq [0, 1]. \tag{2}$$

Then, " τ is possibly true" and " τ is possibly false" are defined by, respectively,

$$\mu_\tau(1) = 1 \quad \text{and} \quad \mu_\tau(0) = 1. \tag{3}$$

[Definition 2] Let by $\underline{P}$ and $\underline{Q}$ are denoted two different linguistic truth values of fuzzy propositions, P and Q. Then, "$\underline{P}$ is more true than $\underline{Q}$" or "$\underline{Q}$ is more false than $\underline{P}$" are defined by

$$\underline{P} \sqcup \underline{Q} = \underline{P}, \tag{4}$$

where $\sqcup$ denotes join-composition.

Let us assume that the linguistic truth value of implication, $\underline{P \rightarrow Q}$, is always possibly true. Then, by introspection, we set the following postulations as:

(1) The more true $\underline{P}$ or $\underline{P \rightarrow Q}$ are, the more true is $\underline{Q}$,
(2) If $\underline{P}$ is more false than *unknown*, $\underline{P}$ is *unknown*,
(3) The more false $\underline{Q}$ is, or the more true $\underline{P \rightarrow Q}$ is, the more false is $\underline{P}$,
(4) If $\underline{Q}$ is more true than *unknown*, $\underline{P}$ is *unknown*,

where *unknown* is a linguistic truth value such that the values of its membership function are 1 for all $v \in V$.

In the above, (1) and (2) are concerned with *modus ponens*, and (3) and (4) with *modus tollens* in approximate reasoning. The rules of inference, what-so-ever their forms are, should be as such ones satisfying the above postulations.

In what follows, in addition to the extension principle, truth qualification and the operations of fuzzy sets of type 1 developed by L.A.Zadeh, we shall use the following special linguistic truth values as;

$$\textit{true} = \int_{v \in V} v/v \tag{5}$$

$$\textit{false} = \int_{v \in V} 1-v/v \tag{6}$$

$$\textit{completely true} = \int_{v \in [0,\, 1)} 0/v \; + 1/1 \tag{7}$$

$$\textit{completely false} = 1/0 + \int_{v \in (0,\, 1]} 0/v \tag{8}$$

where $\int$ and + denote union. In the above, *completely true* and *completely false* are corresponding to TRUE and FALSE respectively in binary logic.

3. FUZZIFICATION OF $Ł_{\text{Aleph-1}}$

Let f be a mapping from $X \times Y$ to Z and let A and B be fuzzy subsets of X and Y respectively. Then, by the application of the extension principle to f, the image of AxB under f is a fuzzy subset of Z, and it is described as

$$f(A,B) = \int_Z \mu_{A\times B}(x,y) \,/\, f(x,y) \tag{9}$$

or, equivalently, for each $z \in Z$,

$$\mu_{f(A,B)}(z) = \begin{cases} \sup_{(x,y) \in f^{-1}(z)} (\mu_A(x) \wedge \mu_B(y)); & \exists (x,y) \in f^{-1}(z) \\ 0 & ; \text{if there is no } (x,y) \text{ such that } z=f(x,y), \end{cases} \quad (10)$$

where AxB denotes Cartesian product of A and B, which is a fuzzy subset of X x Y.

In this study, we shall apply the extension principle also to a case in which

$$F: X \times Y \to 2^Z,$$

where 2^Z denotes power set of Z. When this is the case, F^{-1} is defined by

$$F^{-1} : Z \to 2^{X \times Y} \quad (11)$$

$$F^{-1}(z) = \{(x,y) \mid \exists (x,y) \in X \times Y,\ z \in F(x,y)\}. \quad (12)$$

Let us denote an ordinary numerical truth value of a proposition, P, by /P/ or p(small letter). Then, the truth rules in L_1 (abbreviated from $Ł_{Aleph-1}$) are described as follows;

$$/\neg P/ = 1 - /P/ \quad (13)$$

$$/P \text{ AND } Q/ = /P/ \wedge /Q/ \quad (14)$$

$$/P \text{ OR } Q/ = /P/ \vee /Q/ \quad (15)$$

$$/P \to Q/ = (1 - /P/ + /Q/) \wedge 1 \quad (16)$$

where $\neg$, AND, OR and $\to$ denote negation, conjuction, disjunction and implication, respectively, and $a \wedge b$ stands for min{a, b} and $a \vee b$ for max{a, b}. The truth functions corresponding to the above expressions are written as, respectively,

$$f_1(p) = 1 - p \quad (13)'$$

$$f_2(p,q) = p \wedge q \quad (14)'$$

$$f_3(p,q) = p \vee q \quad (15)'$$

$$f_4(p,q) = (1 - p + q) \wedge 1, \quad (16)'$$

where f_1 is a mapping from V to V and f_2 to f_4 are functions from V x V to V.

Let $\underline{P}$ and $\underline{Q}$ be linguistic truth values given as

$$\underline{P} = \int_V \mu_{\underline{P}}(p)/p \quad \text{and} \quad \underline{Q} = \int_V \mu_{\underline{Q}}(q)/q. \quad (17)$$

Then, by the application of the extension principle to $f_i (i=1,..,4)$, we obtain for each $r \in V$

$$\mu_{\underline{\neg P}}(r) = \sup_{p \in f_1^{-1}(r)} (\mu_{\underline{P}}(p)) \quad (18)$$

$$\mu_{\underline{R_i}}(r) = \sup_{(p,q) \in f_i^{-1}(r)} (\mu_{\underline{P} \times \underline{Q}}(p,q)), \quad i=2,3,4 \quad (19)$$

,where $R_2 \triangleq$ P AND Q, $R_3 \triangleq$ P OR Q and $R_4 \triangleq P \to Q$.

The above representations are only analytical ones for logical connectives in a fuzzy logic from the fuzzification of L_1. It has been discussed by L.A.Zadeh |4|, C.V.Negoita & D.A.Ralescu |7|, or M.Sugeno |8|, that a fuzzy subset can be described by means of infinite nonfuzzy interval-valued sets, that is, its α level-sets. Therefore, in order to derive the linguistic truth value of the resulting proposition through an operation for some propositions, we only have to consider how its α level-sets for all $\alpha \in (0, 1]$ are obtained.

Let us first consider the derivation of α level-sets of $\neg\underline{P}$. From Eq. (18), we can readily obtain

$$(\neg\underline{P})_\alpha = \{ r \mid f_1^{-1}(r) \in \underline{P}_\alpha \} \qquad \text{for } \forall\alpha \in (0, 1]. \tag{20}$$

Since f_1 is one-to-one correspondence between V and V, we can directly write the membership function for negation as, for each $r \in V$,

$$\mu_{\neg\underline{P}}(r) = \mu_{\underline{P}}(1-r). \tag{21}$$

Let us now consider the cases for conjunction, disjunction and implication. Let f and R be generic names for respectively f_i and R_i (i = 2, 3, 4) in Eq.(19). If $\mu_{\underline{R}}(r) \geqq \alpha$, then we have

$$\sup_{(p,q) \in f^{-1}(r)} (\mu_{\underline{P}}(p) \wedge \mu_{\underline{Q}}(q)) \geqq \alpha \tag{22}$$

,that is,

$$\exists (p,q) \in f^{-1}(r) \; ; \; ((\mu_{\underline{P}}(p) \geqq \alpha) \text{ and } (\mu_{\underline{Q}}(q) \geqq \alpha)), \tag{23}$$

where

$$f^{-1}(r) = \{ (p,q) \in V \times V \mid \exists (p,q),\ r = f(p,q) \}. \tag{24}$$

Thus, it follows that, for $\forall\alpha \in (0, 1]$,

$$\mu_{\underline{R}}(r) \geqq \alpha \leftrightarrow (\exists p, \exists q \; ; \; p \in \underline{P}_\alpha, q \in \underline{Q}_\alpha,\ r = f(p,q)). \tag{25}$$

Consequently, in order to obtain $\underline{R}_\alpha$ for given $\underline{P}_\alpha$ and $\underline{Q}_\alpha$, we only have to find the interval-valued sets as;

$$\underline{R}_\alpha = \{r \mid \exists p, \exists q \; ; \; p \in \underline{P}_\alpha,\ q \in \underline{Q}_\alpha,\ r = f(p,q)\},\ \alpha \in (0, 1] \tag{26}$$

Now let $\overline{\underline{P}}_\alpha$ and $\overline{\underline{Q}}_\alpha$ be cylindrical extensions defined as

$$\overline{\underline{P}}_\alpha \triangleq \underline{P}_\alpha \times V \quad \text{and} \quad \overline{\underline{Q}}_\alpha \triangleq V \times \underline{Q}_\alpha. \tag{27}$$

Then we can write the above condition, Eq.(26), simply as,

$$\underline{R}_\alpha = \{r \mid f^{-1}(r) \cap (\overline{\underline{P}}_\alpha \cap \overline{\underline{Q}}_\alpha) \neq \phi \}, \tag{28}$$

where ϕ denotes empty set.

It is convenient to use the following denotations for the comparison among two interval-valued sets.

$$1 - [p_1, p_2] \triangleq [1-p_2, 1-p_1]$$

$$[p_1, p_2] \wedge [q_1, q_2] \triangleq [p_1 \wedge q_1, p_2 \wedge q_2]$$

$$[p_1, p_2] \vee [q_1, q_2] \triangleq [p_1 \vee q_1, p_2 \vee q_2]$$

$$[p_1, p_2] \boxplus [q_1, q_2] \triangleq [(p_1 + q_1) \wedge 1, (p_2 + q_2) \wedge 1].$$

Let us assume that $\underline{P}_\alpha$ and $\underline{Q}_\alpha$ are given by subsets of V as

$$\underline{P}_\alpha = [p_1, p_2] \text{ and } \underline{Q}_\alpha = [q_1, q_2] \text{ for } \forall\alpha \in (0, 1]. \quad (29)$$

Then we obtain the following proposition concerned with the computational methods of the logical connectives in a fuzzy logic.

[Proposition 1] Let P and Q be fuzzy propositions whose truthness are described by the linguistic truth values denoted by $\underline{P}$ and $\underline{Q}$, respectively. If α level-sets of $\underline{P}$ and $\underline{Q}$ satisfy the above assumption, Eq.(29), then the logical connectives in fuzzy logic are calculated as follows; for $\forall\alpha \in (0, 1]$,

$$\text{Negation} \quad R1 \triangleq \neg P \quad : \underline{R1}_\alpha = 1 - \underline{P}_\alpha \quad (30)$$

$$\text{Conjunction} \quad R2 \triangleq P \text{ AND } Q : \underline{R2}_\alpha = \underline{P}_\alpha \wedge \underline{Q}_\alpha \quad (31)$$

$$\text{Disjunction} \quad R3 \triangleq P \text{ OR } Q : \underline{R3}_\alpha = \underline{P}_\alpha \vee \underline{Q}_\alpha \quad (32)$$

$$\text{Implication} \quad R4 \triangleq P \to Q \quad : \underline{R4}_\alpha = (\underline{\neg P})_\alpha \boxplus \underline{Q}_\alpha. \quad (33)$$

[Proof] As for Negation, the above result, Eq.(30), can be directly derived from Eq.(21). The inverse images defined by Eq.(24) are respectively expressed as; for each $r \in V$,

$$f_2^{-1}(r) = \{(p,q) \mid (p=r,\ q \geqq r) \text{ or } (p \geqq r,\ q=r)\} \quad (34)$$

$$f_3^{-1}(r) = \{(p,q) \mid (p=r,\ q \leqq r) \text{ or } (p \leqq r,\ q=r)\} \quad (35)$$

$$f_4^{-1}(r) = \begin{cases} \{(p,q) \mid q \geqq p\}, & \text{if } r = 1 \\ \{(p,q) \mid r = 1-p+q\}, & \text{if } r \neq 1. \end{cases} \quad (36)$$

By solving Eq.(28), one may obtain the results, Eqs.(31), (32) and (33). For instance, Eq.(31) is also induced by the graphical illustration as shown in Fig.1 where the bold line denotes the area of $f_2^{-1}(r)$ for a given r, and the rectangular area stands for that of $(\underline{P}_\alpha \cap \underline{Q}_\alpha)$. Q.E.D.

Note that the assumption made in Proposition 1 means that $\underline{P}$ and $\underline{Q}$ are normal and convex fuzzy subsets of V. In effect, if $\underline{P}$ is normal, we have

$$\sup_{p \in V} \mu_{\underline{P}}(p) = 1 \quad (37)$$

,and, consequently, there exists a non-empty subset $\underline{P}_\alpha$ for $\forall\alpha \in (0, 1]$. Further, if $\underline{P}$ is convex, we have

$$\mu_{\underline{P}}(y) \geqq (\mu_{\underline{P}}(x) \wedge \mu_{\underline{P}}(z))$$

$$\text{for } \forall y : y \in [x, z]. \quad (38)$$

The above shows that $\underline{P}_\alpha$ can be described by $[p_1, p_2]$ $(0 \leqq p_1 \leqq p_2 \leqq 1)$ as given in Eq.(29).
This holds for $\underline{Q}$.
If $\underline{P}$ or $\underline{Q}$ are not convex, then we shall have, more generally,

$$\underline{P} = \bigcup_i \bar{p}_i, \quad \underline{Q} = \bigcup_j \bar{q}_j \quad (39)$$

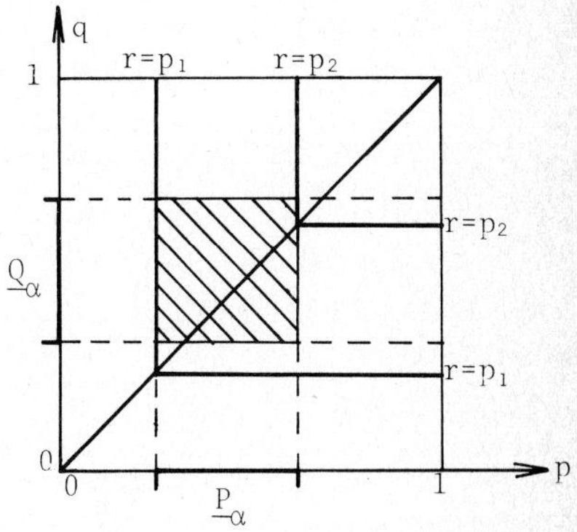

Fig.1 Graphical Illustration for Proof of Eq.(31)

,where

$$\bar{p}_i \triangleq [p_{2i-1}, p_{2i}], \quad p_{2i} < p_{2i+1}, \quad i = 1,\dots,(m-1)$$

$$\bar{q}_j \triangleq [q_{2j-1}, q_{2j}], \quad q_{2j} < q_{2j+1}, \quad j = 1,\dots,(n-1).$$

Then, it is easily verified that Eqs.(30),...,(33) can be rewritten as follows; for $\forall \alpha \in (0, 1]$,

$$\underline{R1}_\alpha = \bigcup_i (1 - \bar{p}_i) \qquad (30)' \qquad \underline{R2}_\alpha = \bigcup_i \bigcup_j (\bar{p}_i \wedge \bar{q}_j) \qquad (31)'$$

$$\underline{R3}_\alpha = \bigcup_i \bigcup_j (\bar{p}_i \vee \bar{q}_j) \qquad (32)' \qquad \underline{R3}_\alpha = \bigcup_i \bigcup_j ((1-\bar{p}_i) \boxplus \bar{q}_j). \qquad (33)'$$

4. RULES OF INFERENCE USING FUZZY ASSERTIONS

The rule of deduction called *modus ponens* in a traditional exact reasoning is formally written as

$$\frac{P \quad P \to Q}{Q} \tag{40}$$

,that is, the deduction of the consequent is valid if the antecedent and implication are valid. Here we consider a case where P or P→Q are given as fuzzy propositions and their truth status are characterized by linguistic truth values. In case of such a fuzzy reasoning, *modus ponens* is in general written as

$$\frac{P_2 \quad P_1 \to Q_1}{Q_2} \tag{41}$$

where $P_1 \triangleq$ "X is A_1", $P_2 \triangleq$ "X is A_2", $Q_1 \triangleq$ "Y is B_1" and $Q_2 \triangleq$ "Y is B_2", where X and Y are the names of objects, and A_1, A_2, B_1 and B_2 are fuzzy subsets of the respective universe of discourse.
Then, the process of the above reasoning is decomposed into the following three stages as:
(1) to find $\underline{P}_1$, given P_1 and P_2,
(2) to determine $\underline{Q}_1$ using $\underline{P}_1$ and $\underline{P_1 \to Q_1}$
(3) to find Q_2 using $\underline{Q}_1$ and Q_1.
In the above, the last and first stages are carried out by the truth qualification and its converse, respectively. Here we only consider the second process, that is, a method to determine the structure of the form as

$$\underline{Q} = \underline{Q}(\underline{P}, \underline{P \to Q}). \tag{42}$$

Let us start from considering the above problem by using numerical truth values. The truth function for implication in L_1 is defined by

$$r = (1 - p + q) \wedge 1.$$

In the above, if r and p are specified in advance, we may consider the following correspondence as;

$$F : V \times V \to 2^V \tag{43}$$

$$F(p,r) = \begin{cases} [p, 1] & \text{if } r = 1 \\ p + r - 1 & \text{if } 0 \le p + r - 1 < 1,\ r \ne 1 \\ \phi & \text{otherwise} \end{cases} \tag{44}$$

Further, the inverse of this correspondence is defined as;

$$F^{-1}(q) \triangleq \{ (p,r) \in V \times V \mid q \in F(p,r)\} \quad \text{for each } q \in V . \tag{44}$$

Now let $\underline{P}$ and $\underline{P \to Q}$ be given by linguistic truth values. By the application of the extension principle to F, we can readily obtain

$$\underline{Q}_\alpha = \{ q \mid F^{-1}(q) \cap (\overline{\underline{P}}_\alpha \cap \overline{\underline{R}}_\alpha) \neq \phi \} \quad \text{for each } \alpha \in (0, 1] \tag{45}$$

where

$$\overline{\underline{P}}_\alpha \triangleq \underline{P}_\alpha \times V \quad \text{and} \quad \overline{\underline{R}}_\alpha \triangleq V \times \underline{R}_\alpha . \tag{46}$$

If $\underline{P}_\alpha$ and $\underline{R}_\alpha$ are given as the following interval-valued sets respectively as; for each $\alpha \in (0, 1]$,

$$\underline{P}_\alpha = [p_1, p_2] \quad \text{and} \quad \underline{R}_\alpha = [r_1, r_2] \tag{47}$$

,then we obtain by Eq.(45), for each $\alpha \in (0, 1]$,

$$\underline{Q}_\alpha = \begin{cases} [p_1+r_1-1 \vee 0, \ 1] & \text{if } r_2=1 \\ [p_1+r_1-1 \vee 0, \ p_2+r_2-1] & \text{if } p_2+r_2 \geqq 1, \ r_2 \neq 1 \\ \phi & \text{if } 0 \leqq p_2+r_2 < 1 . \end{cases} \tag{48}$$

Here we define a dummy proposition denoted by Π.

[Definition 3] Π is a dummy proposition whose linguistic truth value is always *unknown*.

Thus we reach the following proposition concerning the fuzzified case of *modus ponens*.

[Proposition 2] Fuzzy *modus ponens*
Suppose that the truthness of fuzzy propositions P and "P→Q" are given by the linguistic truth values. If $\underline{P}$ is convex and normal and if $\underline{R}$ ($\triangleq \underline{P \to Q}$) is possibly true and convex, then $\underline{Q}$ is deduced as

$$Q = (\neg (R \to (\neg P))) \text{ OR } \Pi, \tag{49}$$

where negation, disjunction and implication are subject to the operations defined in Proposition 1.

[Proof] By assumption, we can write as, for each $\alpha \in (0, 1]$,

$$\underline{P}_\alpha = [p_1, p_2] \quad \text{and} \quad \underline{R}_\alpha = [r, 1] \quad (0 \leqq r \leqq 1)$$

and, by definition, we have

$$\underline{\Pi}_\alpha = [0, 1] \quad \text{for } \forall \alpha \in (0, 1].$$

By setting $r_2=1$ and $r_1=r$ in Eq.(48), we obtain readily

$$\underline{Q}_\alpha = [p_1 + r - 1 \vee 0 , \ 1], \tag{50}$$

which is the same as α level-set of the right side of Eq.(49) for $\forall \alpha \in (0, 1]$.
Q.E.D.

From the above, we can obtain the followings.

[Lemma 1] If $\underline{P}$ is possibly true and convex, so is $\underline{Q}$ deduced by fuzzy *modus ponens*.

[Proof] Since

$$p_1(\alpha)|_{\alpha=1} = 1 \quad \text{and} \quad r(\alpha)|_{\alpha=1} = 1,$$

we have

$$\{1\} \in \underline{Q}_\alpha|_{\alpha=1} ,$$

which shows that $\underline{Q}$ is possibly true. On the other hand, since $p_1(\alpha)$ and $r(\alpha)$ are monotonously non-decreasing with respect to α, so is $q_1(\alpha)$.
Q.E.D.

[Lemma 2] If $\underline{P}$ is possibly false, then $\underline{Q}$ deduced by fuzzy *modus ponens* always becomes *unknown*.

[Proof] By assumption and Eq.(50), we have

$$\underline{Q}_{\alpha}|_{\alpha=1} = [0,1],$$

which leads directly to

$$\underline{Q} = \int_V 1/q \quad (\triangleq \textit{unknown}). \qquad \text{Q.E.D.}$$

It should be noticed that such rule of inference as in Proposition 2 satisfies both the postulations (1) and (2) stated before.

The similar extension as in the above is possible for a deductive rule based on *modus tollens* in a traditional logic. However, for saving space, we only state the result in the fuzzified case.

[Proposition 3] fuzzy *modus tollens*
Suppose that the truthness of Q and "P→Q" are given by the linguistic truth values. If $\underline{Q}$ is convex and normal and if $\underline{R}$ ($\triangleq \underline{P \to Q}$) is possibly true and convex, then $\underline{P}$ is deduced as

$$P = (R \to Q) \text{ AND } \Pi \tag{51}$$

where the operations of implication and conjunction are subject to the definitions in Proposition 1.

[Lemma 3] If $\underline{Q}$ is possibly false and convex, so is $\underline{P}$ deduced by fuzzy *modus tollens*.

[Lemma 4] If $\underline{Q}$ is possibly true, then $\underline{P}$ by fuzzy *modus tollens* becomes *unknown* even if $\underline{R}$ takes on a high truth status.

One may easily find that the above method of reasoning satisfies the postulations (3) and (4) stated before.

Finally let us consider the fuzzified case of inference by the use of equivalence. In L_1, it holds that if

$$r \leqq /P \leftrightarrow Q/ \leqq 1, \tag{52}$$

then

$$(p+r-1) \vee 0 \leqq q \leqq (p-r+1) \wedge 1, \tag{53}$$

where p and q is the numerical truth values of P and Q respectively. So we may consider the following correspondence as;

$$G : V \times V \to 2^V$$

$$G(p,r) = \{ q \in V \mid q \in [p+r-1 \vee 0, \ p-r+1 \wedge 1] \}. \tag{54}$$

[Proposition 4]
Suppose that the truthness of P and P↔Q are given by the linguistic truth values. If $\underline{P}$ is convex and normal and if $\underline{R}$ ($\triangleq \underline{P \leftrightarrow Q}$) is possibly true and convex, then $\underline{Q}$ is deduced as follows;

$$\underline{Q} = \underline{Q1} \textit{ and } \underline{Q2} \tag{55}$$

where

$$\underline{Q1} = \underline{(\neg (R \to (\neg P))) \text{ OR } \Pi}, \quad \underline{Q2} = \underline{(R \to P) \text{ AND } \Pi}$$

and *and* stands for intersection of fuzzy sets of type 1.

The above proposition can be derived from the application of the extension principle to G in Eq.(54). The detail proof is to be referred to Y.Tsukamoto |9|.

In the propositions described above, the normality of the linguistic truth value of the antecedent is not always necessary. When $\underline{P}$ is not normal, we only have to consider α level-sets of the linguistic truth value of the consequent for α such that $\underline{P}_{\alpha} \neq \phi$. Then we can write Proposition 2 to Proposition 4 as follows; for $\forall \alpha$; $\underline{P}_{\alpha} \neq \phi$,

$$\underset{=}{Q} = \underset{=}{Q}(\underset{=}{P}, \underset{=}{R}),\ R \triangleq \text{"}P \to Q\text{"} : \underset{=}{Q}_\alpha = [p_1(\alpha)+r(\alpha)-1 \vee 0,\ 1] \quad (56)$$

$$\underset{=}{Q} = \underset{=}{Q}(\underset{=}{P}, \underset{=}{R}),\ R \triangleq \text{"}Q \to P\text{"} : \underset{=}{Q}_\alpha = [0,\ p_2(\alpha)-r(\alpha)+1 \wedge 1] \quad (57)$$

$$\underset{=}{Q} = \underset{=}{Q}(\underset{=}{P}, \underset{=}{R}),\ R \triangleq \text{"}P \leftrightarrow Q\text{"} : \underset{=}{Q}_\alpha = [p_1(\alpha)+r(\alpha)-1 \vee 0,\ p_2(\alpha)-r(\alpha)+1 \wedge 1] \quad (58)$$

Note that the direction of implication in (57) is opposite to that in Proposition 3.
On the other hand, the assumption that the linguistic truth value of implication itself is possibly true is highly necessary.

5. COMPARISON WITH OTHER LOGICAL SYSTEMS

First it deserves to note that the present reasoning methods can be always reduced to the forms of reasoning by non-fuzzy logic. Let us consider the case of binary logic. The possible truth values in binary logic are TRUE and FALSE which are corresponding to *completely true* and *completely false*, respectively, in linguistic truth values. Let us denote them simply by T and F. If $\underset{=}{R}$ is T, Proposition 2 to Proposition 4 can be rewritten as follows;

$$\text{Prop.2} : Q = P \text{ OR } \Pi \quad (\text{Eq.}(56)) \quad (59)$$

$$\text{Prop.3} : Q = P \text{ AND } \Pi \quad (\text{Eq.}(57)) \quad (60)$$

$$\text{Prop.4} : Q = P \quad (\text{Eq.}(58)). \quad (61)$$

Thus, $\underset{=}{Q}$ deduced by *modus ponens* becomes T or {T,F} according as $\underset{=}{P}$ takes on T or F. Furthermore, $\underset{=}{Q}$ deduced by *modus tollens* becomes {T,F} or F according as $\underset{=}{P}$ takes on T or F. These results are coincident with those from the use of binary logic. This holds for the inference by the use of equivalence. Moreover, it is easily shown by considering a special value of α that the present reasoning method can be reduced to the form of Łukasiewicz infinite-valued logic.

Secondly let us compare the present method with ones from the fuzzification of some other infinite-valued logics. Specifically we discuss the fuzzy version of *modus ponens* and consider the following three different definitions of implication as; for $R \triangleq$ "$P \to Q$",

$$(1)\ \text{Dienes' VSS} : r = (1 - p) \vee q \quad (62)$$

$$(2)\ \text{Goguen } |10| : r = q/p \wedge 1 \quad (q/0 \triangleq 1) \quad (63)$$

$$(3)\ \text{Gödel } |11| : r = \begin{cases} 1 & \text{if } q \geqq p \\ q & \text{if } q < p \end{cases} \quad (64)$$

One may carry out the fuzzification of the implication defined above and can derive the respective rule of inference corresponding to *modus ponens* in the similar way as in the preceding section. Under the assumption that $\underset{=}{R}$ is possibly true and convex, we can reach the following results as; for $\forall\alpha \in (0, 1]$,

$$(1)\ \underset{=}{Q}_\alpha = \begin{cases} [0, 1] & \text{if } p_1+r_1 \leqq 1 \\ [r_1, 1] & \text{otherwise} \end{cases} \quad (65)$$

$$(2)\ \underset{=}{Q}_\alpha = [p_1 \cdot r_1, 1] \quad (66)$$

$$(3)\ \underset{=}{Q}_\alpha = [p_1 \wedge r_1, 1] \quad (67)$$

where · denotes arithmetic multiplication. These are corresponding to Eq.(56). The result, Eq.(65), from the fuzzification of Dienes' Variant Standard Sequence shows that it does not the postulation (1) stated before, say, if $\underset{=}{R}$ is *completely true*, the truthness of the consequent is also *completely true* as long as $\underset{=}{P}$ is more true than *unknown*. So this case must be out of consideration on fuzzy reasoning using fuzzy implication. The other results are more or less satisfactory in view of the postulations (1) and (2).

Among the lower limits in Eqs.(56),(66) and (67), one may find an order as

$$(p_1+r_1-1)\vee 0 \leqq p_1\cdot r_1 \leqq (p_1\wedge r_1), \tag{68}$$

which shows that the reasoning method from the fuzzification of L_1 is the most safely among the above three possible methods. The most safely does not necessarily means the best, but, generally stating, fuzzy reasoning should be moderate in nature. This is an essential reason why we have adopted L_1 as an underlying logic of fuzzy logic.

Finally it may be very interesting to note that the above three forms can be expressed by the following one form as;

$$Q = (P \text{ } \mathit{AND} \text{ } R) \text{ OR } \Pi \tag{69}$$

where *AND* stands for the fuzzified cases of bold conjunction in L_1, interactive conjunction in Goguen and non-interactive conjunction in Gödel respectively. In the above, bold conjuction has been defined by R.Giles |12|.

6. ILLUSTRATIVE EXAMPLES

To make clear the reasoning process proposed here, let us consider two problems, a reasoning process using fuzzy *modus ponens* and the logical paradox called *falakros* (the bald man).

Problem 1. Let P_1 and Q_1 be fuzzy propositions as

$$P_1 \triangleq \text{"}u_1 \text{ is } \mathit{small}\text{"} \quad \text{and} \quad Q_1 \triangleq \text{"}u_2 \text{ is } \mathit{large}\text{"}, \tag{70}$$

where *small* and *large* are fuzzy subsets of U ($\triangleq$[1, 9]) and these meanings are assumed to be given by the membership functions S* and L* respectively as shown in Fig.2. Further suppose that we have the following vague knowledge concerning the relationship between the above two propositions as;

$$R \triangleq \text{"}P_1 \rightarrow Q_1\text{"} \quad \text{with } \underline{R} \text{ as } \mathit{very\ very\ true}. \tag{71}$$

Let us now assume that we have received the information as

$$P_1 \triangleq \text{"}u_1 \text{ is } \mathit{very\ small}\text{"} \tag{72}$$

where *very small* is a fuzzy set characterized by the membership function S as shown in Fig.2. Then, what can we say about u_2 ?

This problem is resoluted as follows. First let L be a fuzzy subset of U, that is, our solution to be found is expressed as "u_2 is L". According to the reasoning processes stated at the beginning of the section 4, the relative linguistic truth value of P_1 to reference proposition P_2 is obtained by the following equation as;

$$\underline{P}_1(S) = \int_V \mu_S(\mu_{S^*}^{-1}(v))/v \tag{73}$$

The above calculation is so called converse of truth qualification (see E.Sanchez |13|). Subsequently, applying Proposition 2 to $\underline{P}_1$ and $\underline{R}$, one may obtain $\underline{Q}_1$ as shown in the left side of Fig.2. Finally, by means of truth qualifiaction, we have

$$L = \int_U \mu_{\underline{Q}_1}(\mu_{L^*}(u))/u \tag{74}$$

Thus, the meaning of the fuzzy set L to be found is given by the dotted line in Fig.2. Now we may say about u_2 as follows;

"u_2 is *large*".

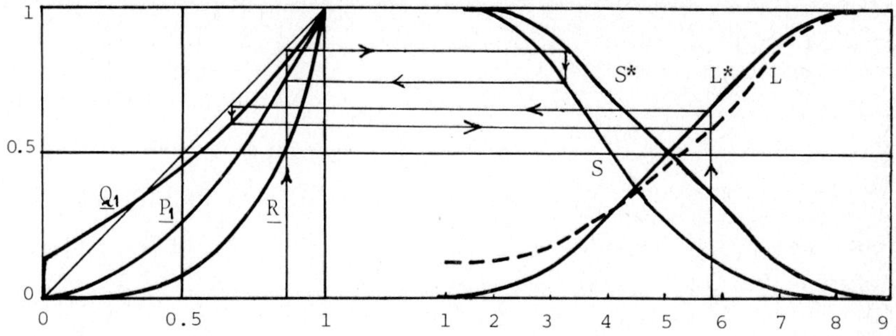

Fig.2 Method of Solution to Problem 1

Such a process of fuzzy reasoning as illustrated in the above may be described by the use of a diagram as shown in Fig.3, where Tq and Tq^{-1} denote truth qualification and its converse respectively, and F.*m.p.* stands for fuzzy *modus ponens*.

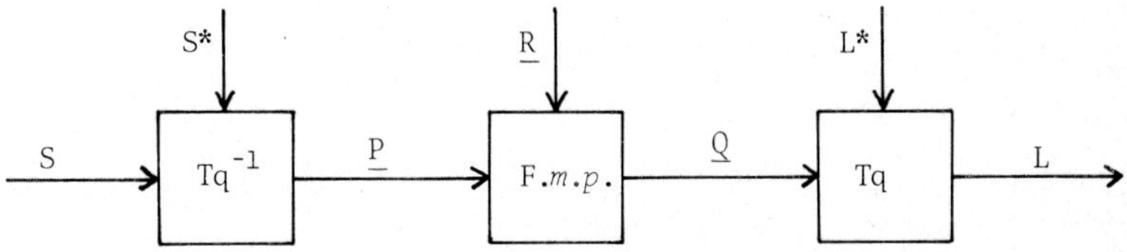

Fig.3 Diagram for a Fuzzy Reasoning Process

If S and (S*,R,L*) in the above example could be taken as an input variable and structural paramerters respectively, then L would be taken as its resulting output variable. So it will be convenient to use the following denotation for representing such a reasoning process as;

$$L = \Psi\ (S;\ S^*,\underline{R},L^*) \tag{75}$$

For instance, if S in the above problem is given as "*rather small*", "*large*" and "*around* 4", then one may obtain the outputs as "*large rather than small*", "*unknown*" and "*not very small*", respectively. The second case in the above represents that we can not say any statement concerning u_2, in other word, "u_2 is all possible".

Let us now discuss the bald man as Problem 2. J.A.Goguen |10| first suggested a measure of validity for deduction which decreases as the number of applications of *modus ponens* increases. Furthermore, B.R. Gaines |14| asserted that in fuzzy reasoning 'we should pay for each application of *modus ponens*, and has shown a resolution of the ancient logical paradox of the bald man by using infinite-valued logic. In order to resolve the paradox in the context of the present reasoning method, we only have to say that "a person with only one more hair than a bald man is still bald" is *almost completely true*. Then, the bald man is resoluted in the following way.

Let P_n be a statement as "a person who has n hairs is bald". Then, the statement under discussion can be described as follows;

$$\forall n, \; R_n \triangleq P_{n-1} \rightarrow P_n \; . \tag{76}$$

This formulation was given by B.R.Gaines |14|. Now we can use the two kinds of knowledges as

$$\underline{P}_0 = \textit{completely true} \tag{77}$$

and

$$\forall n, \; \underline{R}_n = \int_V v^m/v, \qquad m : \text{a large positive number.} \tag{78}$$

Provided that $\underline{P}_{n-1}$ is given, we obtain by the application of fuzzy *modus ponens* to Eq.(76), for $\forall \alpha \in (0, 1]$

$$(\underline{P}_n)_\alpha = [\underline{p}_{n-1}(\alpha) + r(\alpha) - 1 \vee 0, \; 1], \tag{79}$$

where

$$(\underline{R}_n)_\alpha = [r(\alpha), \; 1] \quad \text{and} \quad (\underline{P}_n)_\alpha = [\underline{p}_n(\alpha), \; \overline{p}_n(\alpha)]. \tag{80}$$

By induction on n,

$$(\underline{P}_n)_\alpha = [(1 - n(1 - r(\alpha))) \vee 0, \; 1] \qquad \text{for } \forall \alpha \in (0, 1]. \tag{81}$$

Since, by Eq.(78),

$$r(\alpha) = \sqrt[m]{\alpha} \; , \tag{82}$$

we obtain

$$\underline{p}_n(\alpha) = (1 - n(1 - \sqrt[m]{\alpha})) \vee 0, \quad n = 1, 2, \text{-----} \; . \tag{83}$$

Consequently,

$$\forall v \in V \; ; \; \mu_{\underline{P}_n}(v) \geqq (1 - \frac{1}{n})^m. \tag{84}$$

Thus,

$$\forall v \in V \; ; \; \mu_{\underline{P}_n}(v) \rightarrow 1 \quad (n \rightarrow \infty), \tag{85}$$

that is, $\underline{P}_n$ approaches to *unknown* as n increases.

The above shows that a large number of applications of *modus ponens* will result on *unknown* as to the truthness of the consequent even if the initial premise is *completely true*. This may be a reasonable result as one deduced by using only the knowledge of the truthness of P_0 and R_n.

7. CONCLUDING REMARKS

First of all, it should be noticed that the deductive reasoning in the use of the present methods is carried out by assigning a fuzzy set to a predicate adjective in the fuzzy proposition to be deduced. The general properties of the reasoning by using implication, say, if *modus ponens*, the property that fuzziness increases in the conclusion as the truthness of the premise decreases, will appear in the resulting fuzzy set. This can be found in the reasoning process as shown in the first example in the preceding section, and this point may be essentially important for fuzzy reasoning methods.

In this study we have not stated the applicability of the present method to practical problems, yet Y.Tsukamoto et al. |15| has applied it to the construction of a new algorithm of fuzzy control.

Finally it should be said that we have not necessarily a unified fuzzy logic together with linguistic approach as the foundation of fuzzy reasoning using everyday life languages, and we do not know yet even whether we should have only one unified fuzzy logic or not. Much more attention should be payed to what are meant by the connectives used in natural languages since the validity of the definitions of logical connectives in fuzzy logic are depending heavily on their semantics.

ACKNOWLEDGEMENTS

The author is grateful to Prof.T.Terano and Dr.M.Sugeno for their valuable advices during the process of this study. He is also owed much to the existing studies by Prof.L.A.Zadeh, Prof.E.Sanchez, Prof.B.R.Gaines and many other pioneering workers on general fuzzy problems. He wishes to express his thanks to all of them.

REFERENCES

|1| L.A.Zadeh, K.S.Fu, K.Tanaka and M.Shimura: Fuzzy sets and their application to cognitive and decision processes, New York, Academic Press (1975).

|2| L.A.Zadeh: Fuzzy logic and approximate reasoning, Synthese, 30 (1975) 407-428.

|3| L.A.Zadeh: An fuzzy-algorithmic approach to the definition of complex and imprecise concept, Int. J. Man-Machine Studies, 8 (1976) 249-291.

|4| L.A.Zadeh: The concept of a linguistic variable and its application to approximate reasoning, Information Sciences, 8, 199-249, 8, 301-357, 9, 43-80 (1975).

|5| E.Sanchez: Resolution of composite fuzzy relational equations, Information & Control, 30 (1976) 38-48.

|6| B.R.Gaines: Fuzzy and probability uncertainty logic, Information & Control, 38 (1978) 154-169.

|7| C.V.Negoita and D.A.Ralescu: Applications of fuzzy sets to systems analysis, Birkhauser, Stuttgart (1975).

|8| M.Sugeno: Fuzzy systems with underlying deterministic systems, 4-th Meeting of the Europian Working Group for Fuzzy Sets, Stockholm (1976).

|9| Y.Tsukamoto: Fuzzy logic based on Łukasiewicz logic and its applications to diagnosis and control, Doctor thesis, Tokyo Institute of Technology (1979).

|10| J.A.Goguen: The logic of inexact concepts, Synthese, 19 (1968-1969) 323-373.

|11| N.Rescher: Many-valued logic, New York, McGraw-Hill (1969).

|12| R.Giles: Łukasiewicz logic and fuzzy set theory, Int. J. Man-Machine Studies, 6 (1976) 313-327.

|13| E.Sanchez: On the truth qualification in natural languages, Proc. Int. Conf. on Cybernetics & Society, 2, Tokyo, Japan (1978) 1233-1236.

|14| B.R.Gaines: Foundations of fuzzy reasoning, Int. J. Man-Machine Studies, 6 (1976) 313-327.

|15| Y.Tsukamoto, T.Takagi and M.Sugeno: Fuzzification of $Ł_{\text{Aleph-1}}$ and its application to control, Proc. Int. Conf. on Cybernetics & Society, 2, Tokyo, Japan (1978) 1217-1221.

****

PART TWO

THEORY

ADVANCES IN FUZZY SET THEORY AND APPLICATIONS
M.M. Gupta, R.K. Ragade, R.R. Yager (editors)

SOME PROPERTIES OF FUZZY NUMBERS

Masaharu MIZUMOTO* and Kokichi TANAKA**

* Department of Management Engineering
Osaka Electro-Communication University
Neyagawa, Osaka 572, Japan

** Department of Information and Computer Sciences
Osaka University
Toyonaka, Osaka 560, Japan

A fuzzy number is a fuzzy set in the real line and its operations of +, -, x and ÷ can be defined by using the extension principle. This paper investigates the algebraic properties of fuzzy numbers under the four arithmetic operations of +, -, x and ÷. Furthermore, the ordering of fuzzy numbers is introduced and some properties of fuzzy numbers under join ($\sqcup$) and meet ($\sqcap$) are discussed.

INTRODUCTION

Recently, L.A. Zadeh proposed the interesting concept of the extension principle by which a binary operation defined on a set X may be extended to fuzzy sets in X, and defined the operations for fuzzy sets of type 2 [2, 3] and fuzzy numbers [1, 3].

In this paper we discuss the algebraic properties of fuzzy numbers, which are fuzzy sets in the real line, under the four arithmetic operations, namely, +, -, x and ÷ which are defined by the extension principle [3]. First, as for the convexity of fuzzy numbers, the fuzzy numbers obtained by applying the operations of +, - and x to convex fuzzy numbers are also convex fuzzy numbers, though the convexity can not be preserved in general if ÷ is applied to convex fuzzy numbers. Second, the convex fuzzy numbers do not form such algebraic structures as a ring and a field, since the distributive law is not satisfied and there exist no inverse fuzzy numbers under + and x. Third, the positive convex fuzzy numbers defined over the positive real line, however, satisfy the distributive law and hence form a commutative semiring. And fourth, the ordering of fuzzy numbers is introduced and the properties of fuzzy numbers under the join and the meet combined with the four arithmetic operations are investigated.

FUZZY NUMBERS

We shall briefly review some of the basic definitions relating to fuzzy numbers and their operations of +, -, x and ÷.

<u>Fuzzy Numbers</u>: A <u>fuzzy number</u> A in the real line R is a fuzzy set characterized by a membership function μ_A as

$$\mu_A : R \longrightarrow [0, 1]. \tag{1}$$

A fuzzy number A is expressed as

$$A = \int_{x \in R} \mu_A(x)/x , \tag{2}$$

with the understanding that $\mu_A(x) \in [0,1]$ represents the grade of membership of x in A and $\int$ denotes the union of $\mu_A(x)/x$'s.

(Example 1) A fuzzy number $\underset{\sim}{2}$ which denotes "about 2" will be given as

$$\underset{\sim}{2} = \int_1^2 x - 1/x + \int_2^3 3 - x/x \tag{3}$$

and can be illustrated by the dotted line in Fig. 2, where + stands for the union.

Convex Fuzzy Numbers: A fuzzy number A in R is said to be convex if for any real numbers x, y, z ε R with $x \le y \le z$,

$$\mu_A(y) \ge \mu_A(x) \wedge \mu_A(z) \qquad (4)$$

with $\wedge$ standing for min. A fuzzy number A is called normal if the following holds.

$$\max_x \mu_A(x) = 1. \qquad (5)$$

A fuzzy number which is normal and convex is referred to as a normal convex fuzzy number.

(Example 2) Fuzzy numbers as shown in Fig. 2 are all normal convex fuzzy numbers. Fig.1 gives various kinds of fuzzy numbers, in which it is noted that the fuzzy number A_2 is not convex because the support (defined in (9)) of A_2 is discrete, that is, A_2 does not satisfy (4).

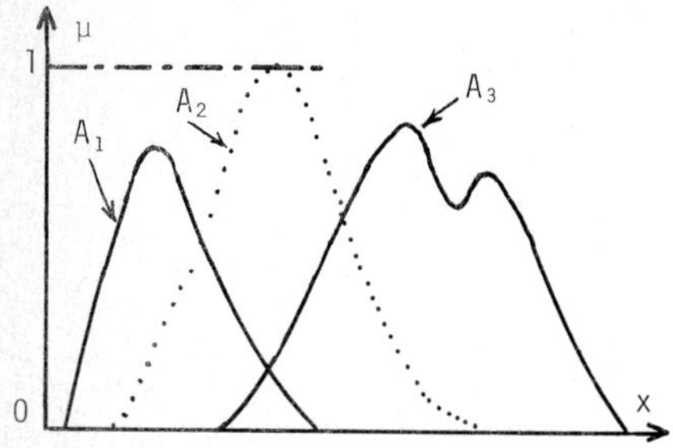

Fig.1 Various Kinds of Fuzzy Numbers (A_1: convex, A_2: normal non-convex, A_3: non-convex).

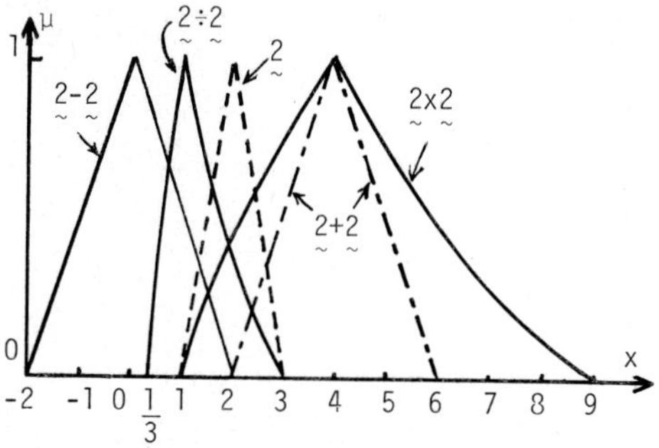

Fig.2 Fuzzy Numbers $\tilde{2}$, $\tilde{2}+\tilde{2}$, $\tilde{2}-\tilde{2}$, $\tilde{2}\times\tilde{2}$, $\tilde{2}\div\tilde{2}$.

Level Sets: An α-level set of a fuzzy number A is a non-fuzzy set denoted by A_α and is defined by

$$A_\alpha = \{ x \mid \mu_A(x) \ge \alpha\}, \quad 0 < \alpha \le 1 . \qquad (6)$$

It is easy to show that

$$\alpha_1 \le \alpha_2 \implies A_{\alpha_1} \supseteq A_{\alpha_2} . \qquad (7)$$

If two fuzzy numbers A and B are equal, that is, $\mu_A(x) = \mu_B(x)$ for all x ε R, then we can obtain $A_\alpha = B_\alpha$ for any α, and vice versa. Let fuzzy number A be convex, A_α is a convex set (or an interval) in R, and vice versa. A fuzzy number A may be decomposed into its level sets through the resolution identity [3].

$$A = \int_0^1 \alpha A_\alpha , \qquad (8)$$

where αA_α is the product of a scalar α with the set A_α and $\int$ is the union of A_α's, with α ranging from 0 to 1.

Support: The support Γ_A of a fuzzy number A is defined, as a special case of level set, by the following.

$$\Gamma_A = \{ x \mid \mu_A(x) > 0\} . \qquad (9)$$

Extension Principle: Let A and B be fuzzy numbers in R and let * be a binary operation defined in R. Then the operation * can be extended to the fuzzy numbers A and B by the defining relation (the extension principle).

$$A * B = \int_{x,y\varepsilon R} (\mu_A(x) \wedge \mu_B(y)) \,/\, (x * y), \qquad (10)$$

where $\wedge$ stands for min.

In (10) let the binary operation * be replaced by the ordinary four arithmetic operations of +, -, x and ÷, then the four arithmetic operations over fuzzy numbers are defined by the following.

Four Arithmetic Operations for Fuzzy Numbers: Let A and B be fuzzy numbers in R, we have from (10)

$$A + B = \int (\mu_A(x) \wedge \mu_B(y)) / (x + y) , \tag{11}$$

$$A - B = \int (\mu_A(x) \wedge \mu_B(y)) / (x - y) , \tag{12}$$

$$A \times B = \int (\mu_A(x) \wedge \mu_B(y)) / (x \times y) , \tag{13}$$

$$A \div B = \int (\mu_A(x) \wedge \mu_B(y)) / (x \div y) . \tag{14}$$

The membership functions of these fuzzy numbers are obtained by

$$\mu_{A+B}(a) = \bigvee_{x+y=a} (\mu_A(x) \wedge \mu_B(y))$$

$$= \bigvee_{x} (\mu_A(x) \wedge \mu_B(a - x)), \tag{15}$$

$$\mu_{A-B}(a) = \bigvee_{x} (\mu_A(x) \wedge \mu_B(x - a)), \tag{16}$$

$$\mu_{AxB}(a) = \bigvee_{x(\neq 0)} (\mu_A(x) \wedge \mu_B(\frac{a}{x})), \tag{17}$$

$$\mu_{A\div B}(a) = \bigvee_{x} (\mu_A(x) \wedge \mu_B(\frac{x}{a})) \tag{18}$$

$$= \bigvee_{y} (\mu_A(ay) \wedge \mu_B(y)) . \tag{19}$$

Although these definitions are useful for any fuzzy numbers, it will be more convenient to convex fuzzy numbers to use the concept of α-level sets of fuzzy numbers.

Let A_α and B_α be α-level sets of convex fuzzy numbers A and B, respectively, then the α-level sets are intervals in R, which are special convex fuzzy numbers whose grades are unity at x belonging to A_α and zero elsewhere. Let the α-level set of, say, the sum A + B of A and B be denoted by $(A + B)_\alpha$, we can obtain

$$(A + B)_\alpha = A_\alpha + B_\alpha . \tag{20}$$

In other words, the α-level set $(A + B)_\alpha$ is the sum of the α-level sets A_α and B_α. Thus, using the resolution identity (8), we can express A + B as

$$A + B = \int_0^1 \alpha(A + B)_\alpha = \int_0^1 \alpha(A_\alpha + B_\alpha) \cdot \tag{21}$$

In a similar fashion, we can obtain A - B, A x B and A ÷ B as follows.

$$A - B = \int_0^1 \alpha(A_\alpha - B_\alpha), \tag{22}$$

$$A \times B = \int_0^1 \alpha(A_\alpha \times B_\alpha), \tag{23}$$

$$A \div B = \int_0^1 \alpha(A_\alpha \div B_\alpha) . \tag{24}$$

(Example 3) For the convex fuzzy number $\underset{\sim}{2}$ given by (3), the fuzzy numbers $\underset{\sim}{2} + \underset{\sim}{2}$, $\underset{\sim}{2} - \underset{\sim}{2}$, $\underset{\sim}{2} \times \underset{\sim}{2}$ and $\underset{\sim}{2} \div \underset{\sim}{2}$ are depicted in Fig.2 and are expressed as

$$\underset{\sim}{2} + \underset{\sim}{2} = \int_{2}^{4} \frac{x}{2} - 1/x \; + \int_{4}^{6} 3 - \frac{x}{2}/x , \tag{25}$$

$$\underset{\sim}{2} - \underset{\sim}{2} = \int_{-2}^{0} \frac{x}{2} + 1/x \; + \int_{0}^{2} 1 - \frac{x}{2}/x , \tag{26}$$

$$\underset{\sim}{2} \times \underset{\sim}{2} = \int_{1}^{4} \sqrt{x} - 1/x \; + \int_{4}^{9} 3 - \sqrt{x}/x , \tag{27}$$

$$\underset{\sim}{2} \div \underset{\sim}{2} = \int_{\frac{1}{3}}^{1} 3 - \frac{4}{x+1}/x \; + \int_{1}^{3} \frac{4}{x+1} - 1/x . \tag{28}$$

ALGEBRAIC PROPERTIES OF FUZZY NUMBERS

This section discusses the algebraic properties of fuzzy numbers under the operations of +, -, x and ÷. We shall begin with the convexity of fuzzy numbers under these operations.

[Theorem 1] If A and B are convex fuzzy numbers in the real line R, then A + B, A - B and A x B are also convex fuzzy numbers.

Proof: In general, let M_1, M_2, N_1 and N_2 be intervals in R and let $M_1 \subseteq M_2$ and $N_1 \subseteq N_2$, then we can obtain that

$$M_1 + N_1 \subseteq M_2 + N_2 \;; \quad M_1 \times N_1 \subseteq M_2 \times N_2$$

and that $M_i + N_i$ and $M_i \times N_i$ $(i = 1, 2)$ also become intervals in R. For each $0 < \alpha \le 1$, the α-level sets A_α and B_α of convex fuzzy numbers A and B are convex sets (or intervals) in R. Thus, for any α_1 and α_2 with $0 < \alpha_1 \le \alpha_2 \le 1$, the relations $A_{\alpha_2} \subseteq A_{\alpha_1}$ and $B_{\alpha_2} \subseteq B_{\alpha_1}$ are derived from (7) and hence $A_{\alpha_2} + B_{\alpha_2} \subseteq A_{\alpha_1} + B_{\alpha_1}$ and $A_{\alpha_2} \times B_{\alpha_2} \subseteq A_{\alpha_1} \times B_{\alpha_1}$ are obtained, which leads to $(A + B)_{\alpha_2} \subseteq (A + B)_{\alpha_1}$ and $(A \times B)_{\alpha_2} \subseteq (A \times B)_{\alpha_1}$. Furthermore, $(A + B)_{\alpha_i}$ and $(A \times B)_{\alpha_i}$ are intervals (or convex sets) for each α_i $(i = 1, 2)$. Thus, fuzzy numbers A + B and A x B are shown to be convex fuzzy numbers.

Next, we shall prove the convexity of A - B. Let - B be defined by 0 - B, then the membership function of - B will be expressed as

$$\mu_{-B}(x) = \mu_B(-x), \quad x \in R \tag{29}$$

and - B can be easily shown to be convex if B is convex. Thus, A - B is proved to be convex since A - B is represented as A + (-B). Q.E.D.

It should be noted that for discrete fuzzy numbers, the convexity of A + B, A - B and A x B does not hold in general even if A and B are in the shape of "convex" like A_2 in Fig.1.

In order to discuss the convexity of fuzzy numbers under ÷, we shall define a special fuzzy number called positive, negative or zero fuzzy number.

Positive Fuzzy Numbers: A fuzzy number A is said to be positive if $0 < a_1 \le a_2$ holds for the support $\Gamma_A = [a_1, a_2]$ of A, that is, Γ_A is in the positive real line. Similarly, A is called negative if $a_1 \le a_2 < 0$ and zero if $a_1 \le 0 \le a_2$.

(Example 4) Fig.2 shows that the fuzzy number $\underset{\sim}{2} - \underset{\sim}{2}$ is a zero fuzzy number and the other fuzzy numbers are all positive.

[Lemma 2] If B is a zero convex fuzzy number, then $\frac{1}{B}$ (= 1 ÷ B) is not a convex fuzzy number.

Proof: The fuzzy number $\frac{1}{B}$ will be defined by the membership function

$$\mu_{\frac{1}{B}}(x) = \mu_B(\tfrac{1}{x}) , \quad x \in R \tag{30}$$

by using (14). Thus, for example, if B is a zero convex fuzzy number depicted in Fig.3 and is expressed by

$$B = \int_{-1}^{1} \frac{x+1}{2}/x + \int_{1}^{2} 2 - x/x ,$$

then the application of (30) to B yields

$$\frac{1}{B} = \int_{-\infty}^{-1} \frac{1}{2}(\frac{1}{x} + 1)/x + \int_{\frac{1}{2}}^{1} 2 - \frac{1}{x}/x + \int_{1}^{\infty} \frac{1}{2}(\frac{1}{x} + 1)/x$$

and thus $\frac{1}{B}$ is not a convex fuzzy number (see Fig.3). Q.E.D.

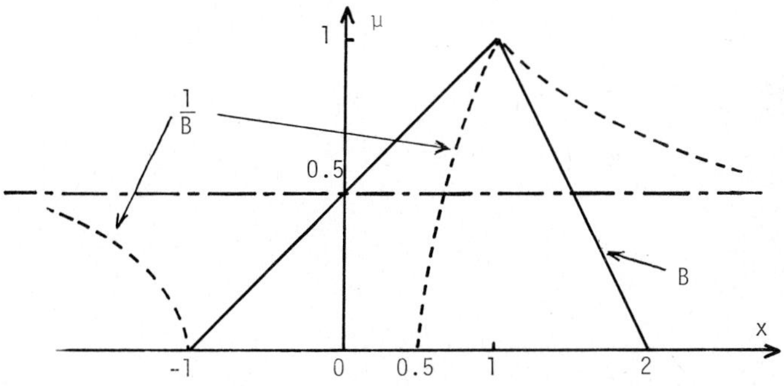

Fig.3 $\frac{1}{B}$ for the Zero Convex Fuzzy Number B.

[Theorem 3] Let A and B be convex fuzzy numbers, then A ÷ B is not, in general, a convex fuzzy number.

In this theorem, however, if B is not a zero fuzzy number but a positive (or negative) fuzzy number, the convexity will be reserved.

[Theorem 4] If A is a convex fuzzy number and B is a positive (or negative) convex fuzzy number, then A ÷ B is a convex fuzzy number.

Proof: It will be sufficient to prove that $\frac{1}{B}$ is convex if B is positive convex, since A ÷ B can be represented as A x $(\frac{1}{B})$. Let x, y, z be real numbers such that $0 < x \le y \le z$, then $0 < \frac{1}{z} \le \frac{1}{y} \le \frac{1}{x}$ holds. Thus, we can have $\mu_B(\frac{1}{y}) \ge \mu_B(\frac{1}{z}) \wedge \mu_B(\frac{1}{x})$ in virtue of the convexity of B. Using (30) we can write $\mu_{\frac{1}{B}}(y) \ge \mu_{\frac{1}{B}}(z) \wedge \mu_{\frac{1}{B}}(x)$, which leads to the convexity of $\frac{1}{B}$.

The normality of fuzzy numbers can be easily shown by the following.

[Theorem 5] If A and B are normal fuzzy numbers, then A + B, A - B, A x B and A ÷ B are also normal.

Note. For two fuzzy numbers A and B, if the one is convex and the other is non-convex, then the execution results of A and B under +, -, x and ÷ may be convex or non-convex. We shall show this by the example.

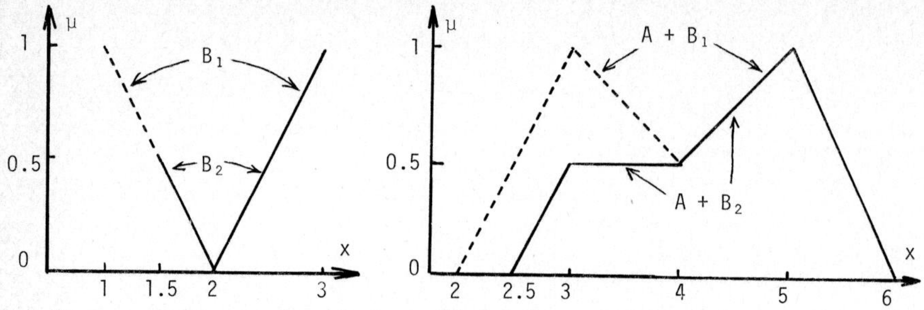

(a) Non-convex fuzzy numbers B_1, B_2. (b) $A+B_1$(non-convex) and $A+B_2$(convex).

Fig.4 Diagram of Example 5.

(Example 5) Let A be the convex fuzzy number $\underset{\sim}{2}$ given by (3) and B_1 be a non-convex fuzzy number such as (see Fig.4)

$$B_1 = \int_1^2 2 - x/x + \int_2^3 x - 2/x \ .$$

Then we have $A + B_1$ as

$$A + B_1 = \int_2^3 x - 2/x + \int_3^4 \frac{5 - x}{2}/x + \int_4^5 \frac{x - 3}{2}/x + \int_5^6 6 - x/x$$

which indicates that $A + B_1$ is non-convex (see Fig.4). On the contrary, let B_2 be also a non-convex fuzzy number such as

$$B_2 = \int_{1.5}^2 2 - x/x + \int_2^3 x - 2/x \ ,$$

then $A + B_2$ is given by

$$A + B_2 = \int_{2.5}^3 x - 2.5/x + \int_3^4 0.5/x + \int_4^5 \frac{x - 3}{2}/x + \int_5^6 6 - x/x \ .$$

This shows that $A + B_2$ is convex.

We shall next investigate the algebraic properties of fuzzy numbers under +, -, x and ÷. As is well-known, the family of real numbers forms a field under the ordinary operations + and x. Convex fuzzy numbers, however, are shown not to have their inverses and not to satisfy the distributive law. So the family of convex fuzzy numbers (needless to say, arbitrary fuzzy numbers) does not form the algebraic structures such as a ring and a field. On the contrary, positive convex fuzzy numbers defined in the positive real line satisfy the distributive law and thus they form a commutative semiring with zero and unity.

[Theorem 6] For any fuzzy numbers A, B and C, we have

$$\left.\begin{aligned}(A + B) + C &= A + (B + C)\\ (A \times B) \times C &= A \times (B \times C)\end{aligned}\right\} \text{(associative laws)} \qquad (31)$$

$$\left.\begin{aligned}A + B &= B + A\\ A \times B &= B \times A\end{aligned}\right\} \text{(commutative laws)} \qquad (32)$$

$$\left.\begin{array}{l} A + 0 = A \\ A \times 1 = A \end{array}\right\} \qquad \text{(identity laws)} \tag{33}$$

where 0 and 1 are zero and unity, respectively, in the ordinary sense.

[Theorem 7] For any fuzzy number A, there exist no inverse fuzzy numbers A' and A" under + and x, respectively, such that

$$A + A' = 0 , \tag{34}$$

$$A \times A'' = 1 . \tag{35}$$

Proof: Assume that A is arbitrary fuzzy number and A' satisfies (34) for A. It follows from (11) that

$$A + A' = \int (\mu_A(x) \wedge \mu_{A'}(y)) \; / \; (x + y) = 1/0, \tag{36}$$

where 1/0 is a fuzzy number which means a zero 0 in the ordinary sense.

[a] The case where A is not normal: It is immediately shown that (36) can not be satisfied for any A'.

[b] The case where A is normal: For x and y (= -x) satisfying x + y = 0, it is necessary to satisfy from (36)

$$\bigvee_x \; [\mu_A(x) \wedge \mu_{A'}(-x)] = 1$$

and hence to satisfy

$$\mu_{A'}(-x_o) = \mu_A(x_o) = 1 \tag{37}$$

for some $x_o \in R$. On the other hand, for x and y with $x + y \neq 0$, we must have by (36)

$$\mu_A(x) \wedge \mu_{A'}(y) = 0 . \tag{38}$$

Since $\mu_{A'}(-x_o) = 1$ holds from (37), $\mu_A(x) = 0$ must hold for all x such that $x + (-x_o) \neq 0$ in view of (38). This is contrary to the assumption that A is arbitrary normal fuzzy number. Thus, it has been proved that there does not exist an inverse fuzzy number for A under +. The same holds for the case of the operation x. Q.E.D.

It is noted that if A is reduced to a real number, -A and $\frac{1}{A}$ are the inverses of A under + and x, respectively.

[Corollary 8] For -A in (29) and $\frac{1}{A}$ in (30) of a fuzzy number A, we have in general

$$A + (-A) \neq 0 , \tag{39}$$

$$A \times (\tfrac{1}{A}) \neq 1 . \tag{40}$$

Proof: This is obvious from $\underset{\sim}{2} - \underset{\sim}{2}$ (= $\underset{\sim}{2} + (-\underset{\sim}{2})$) and $\underset{\sim}{2} \div \underset{\sim}{2}$ (= $\underset{\sim}{2} \times (\frac{1}{\underset{\sim}{2}})$) in Fig.2.

[Theorem 9] When A, B and C are any fuzzy numbers, the following does not hold in general.

$$A \times (B + C) = (A \times B) + (A \times C) \qquad \text{(distributive law)} \tag{41}$$

The same is true for the case where A, B and C are normal convex fuzzy numbers.

Proof: It will be sufficient to show the example of normal convex fuzzy numbers which do not satisfy (41). Now, suppose that A, B and C are normal convex fuzzy numbers such that

$$A = \int_2^3 x - 2/x \; + \; \int_3^4 4 - x/x \; , \tag{42}$$

$$B = \int_{1}^{2} 1/x \ , \tag{43}$$

$$C = \int_{-1}^{1} \frac{1}{2}(x + 1)/x \ . \tag{44}$$

Then

$$A \times (B + C) = \int_{0}^{6} \frac{\sqrt{4 + 2x} - 2}{2}/x \ + \int_{6}^{9} 1/x \ + \int_{9}^{12} 4 - \frac{x}{3}/x \ ,$$

$$(A \times B) + (A \times C) = \int_{-2}^{2\cdot 5} \frac{5 - \sqrt{21 - 2x}}{2} /x + \int_{2\cdot 5}^{6} \frac{\sqrt{4 + 2x} - 2}{2} /x + \int_{6}^{9} 1/x + \int_{9}^{12} 4 - \frac{x}{3}/x \ .$$

Thus the distributive law (41) does not hold for the normal convex fuzzy numbers A, B and C.
Q.E.D.

From Theorems 7 and 9, we can find that (normal) convex fuzzy numbers (needless to say, arbitrary fuzzy numbers) do not satisfy the distributive law and do not have their inverses. Therefore, the family of (normal) convex fuzzy numbers does not form such algebraic structures as a ring and a field.

In the next theorem, however, the distributive law is shown to be satisfied for the positive convex fuzzy numbers.

[Theorem 10] The distributive law of (41) is satisfied for the positive convex fuzzy numbers A, B and C.

Proof: Let α-level sets of positive convex fuzzy numbers A, B and C be $A_\alpha = [a_1, a_2]$, $B_\alpha = [b_1, b_2]$ and $C_\alpha = [c_1, c_2]$, respectively, then each level set is an interval in R and $0 < a_1 \le a_2$, $0 < b_1 \le b_2$ and $0 < c_1 \le c_2$ hold. Thus, for each $0 < \alpha \le 1$,

$$\begin{aligned}[A \times (B + C)]_\alpha &= A_\alpha \times (B_\alpha + C_\alpha) \\ &= [a_1, a_2] \times ([b_1, b_2] + [c_1, c_2]) \\ &= [a_1, a_2] \times [b_1+c_1, b_2+c_2] \\ &= [a_1(b_1+c_1), a_2(b_2+c_2)] \quad \ldots a_i, b_i, c_i > 0 \ .\end{aligned}$$

The right hand member of (41) will be

$$\begin{aligned}[(A \times B) + (A \times C)]_\alpha &= (A_\alpha \times B_\alpha) + (A_\alpha \times C_\alpha) \\ &= ([a_1, a_2] \times [b_1, b_2]) + ([a_1, a_2] + [c_1, c_2]) \\ &= [a_1b_1, a_2b_2] + [a_1c_1, a_2c_2] \quad \ldots a_i, b_i, c_i > 0 \\ &= [a_1b_1+a_1c_1, a_2b_2+a_2c_2] \\ &= [a_1(b_1+c_1), a_2(b_2+c_2)] \\ &= [A \times (B + C)]_\alpha \ .\end{aligned}$$

Thus, using the resolution identity of (8), we can obtain A x (B + C) = (A x B) + (A x C).
Q.E.D.

Note that when α-level set is an empty set Φ, the following holds.

$$A_\alpha + \Phi = \Phi \ ; \qquad A_\alpha \times \Phi = \Phi \ .$$

[Theorem 11] The family of positive convex fuzzy numbers forms a commutative semiring with zero 0 and unity 1† under + and x.

Proof: Positive convex fuzzy numbers are closed under + and x (Theorem 1), and associative (31), commutative (32) and distributive (Theorem 10), and have zero 0 and unity 1 (33) under + and x. Q.E.D.

In Theorem 10, fuzzy numbers A, B and C are assumed to be positive convex, that is, their α-level sets are all positive intervals. It will be, however, found that the following identity (45) can hold even for the case where α-level sets are not positive intervals. Table I summarizes this fact, where the symbols +, 0 and - mean positive, zero and negative intervals, respectively. As an illustration,

$$A_\alpha \times (B_\alpha + C_\alpha) = (A_\alpha \times B_\alpha) + (A_\alpha \times C_\alpha). \tag{45}$$

let $A_\alpha = [-a_1, -a_2]$ be a negative interval and $B_\alpha = [b_1, b_2]$ and $C_\alpha = [c_1, c_2]$ be positive intervals, then we can have

$$\begin{aligned} A_\alpha \times (B_\alpha + C_\alpha) &= [-a_1, -a_2] \times [b_1+c_1, b_2+c_2] \\ &= [-a_1(b_2+c_2), -a_2(b_1+c_1)], \end{aligned}$$

$$\begin{aligned} (A_\alpha \times B_\alpha) + (A_\alpha \times C_\alpha) &= [-a_1b_2, -a_2b_1] + [-a_1c_2, -a_2c_1] \\ &= [-a_1(b_2+c_2), -a_2(b_1+c_1)] = A_\alpha \times (B_\alpha + C_\alpha), \end{aligned}$$

which indicates the satisfaction of (45).

Table I. The Combination of A_α, B_α and C_α Satisfying $A_\alpha \times (B_\alpha + C_\alpha) = (A_\alpha \times B_\alpha) + (A_\alpha \times C_\alpha)$
(+: positive interval, 0: zero interval, -: negative interval)

A_α	+	+	+	0	0	-	-	-
B_α	+	0	-	+	-	+	0	-
C_α	+	0	-	+	-	+	0	-

† The algebraic system R = <R; +, x> with addition + and multiplication x is called a commutative semiring with zero 0 and unity 1 if it satisfies these laws:

(i) Closure property:

$$a, b \in R \Longrightarrow a + b, a \times b \in R .$$

(ii) Associative laws:

$$(a + b) + c = a + (b + c); \quad (a \times b) \times c = a \times (b \times c) .$$

(iii) Commutative laws:

$$a + b = b + a; \quad a \times b = b \times a .$$

(iv) Distributive law:

$$a \times (b + c) = (a \times b) + (a \times c) .$$

(v) Existence of identities: There exist a zero 0 and a unity 1 such that

$$a + 0 = a; \quad a \times 1 = a .$$

If this system R also satisfies the following, R is a field.

(vi) Existence of inverses: There exist a' and a'' for each a such that

$$a + a' = 0; \quad a \times a'' = 1.$$

Therefore, if for each $\alpha \in (0, 1]$, A_α, B_α and C_α of convex fuzzy numbers A, B and C satisfy either of the conditions of Table I, then the distributive law of (41) is shown to be satisfied.

(Example 6) If A, B and C are all negative convex fuzzy numbers, then their α-level sets, which are negative intervals, satisfy the condition in Table I. Thus, the negative convex fuzzy numbers are shown to satisfy the distributive law (41). As another example, let A, B and C be convex fuzzy numbers of (42), (43) and (44), respectively, then $(A_\alpha, B_\alpha, C_\alpha) = (+, +, 0)$ at $\alpha \le 0.5$ does not satisfy the condition of Table I. Thus, these convex fuzzy numbers A, B and C can not satisfy the distributive law (41) as shown in the proof of Theorem 9. However, if changed A with C, then $(A_\alpha, B_\alpha, C_\alpha) = (0, +, +)$ at $\alpha \le 0.5$ and $(A_\alpha, B_\alpha, C_\alpha) = (+, +, +)$ at $\alpha > 0.5$. Hence, this case satisfies the distributive law (41).

From this example it follows that negative convex fuzzy numbers satisfy the distributive law. However, negative convex fuzzy numbers never form a commutative semiring unlike the case of positive convex fuzzy numbers. The reason is that negative convex fuzzy numbers are not closed under x, that is, A x B becomes positive when A and B are negative fuzzy numbers. Table II and III show which intervals $A_\alpha + B_\alpha$ and $A_\alpha \times B_\alpha$ can take.

Table II Intervals of $A_\alpha + B_\alpha$

A_α \ B_α	-	0	+
-	-	-, 0	0, ±
0	-, 0	0	0, +
+	0, ±	0, +	+

Table III Intervals of $A_\alpha \times B_\alpha$

A_α \ B_α	-	0	+
-	+	0	-
0	0	0	0
+	-	0	+

ORDERING OF FUZZY NUMBERS

This section introduces order relations, join and meet for fuzzy numbers, and discusses the algebraic properties of fuzzy numbers under these operations combined with the four arithmetic operations +, -, x and ÷.

The ordering, join and meet of fuzzy numbers can be defined in a similar way as those of fuzzy grades [2, 3] which are fuzzy sets in the unit interval [0, 1].

Join and Meet: Join ($\sqcup$) and meet ($\sqcap$) of fuzzy numbers A and B are defined as follows by using the extension principle (10).

$$A \sqcup B = \int (\mu_A(x) \wedge \mu_B(y)) \,/\, (x \vee y) , \tag{46}$$

$$A \sqcap B = \int (\mu_A(x) \wedge \mu_B(y)) \,/\, (x \wedge y) , \tag{47}$$

where $\vee$ stands for max and $\wedge$ for min.

(Example 7) If $A = \underset{\sim}{2} + \underset{\sim}{2}$ and $B = \underset{\sim}{2} \times \underset{\sim}{2}$ are fuzzy numbers given in (25) and (27), respectively, then

$$A \sqcup B = \int_2^4 \frac{x}{2} - 1/x \;+\; \int_4^9 3 - \sqrt{x}/x , \tag{48}$$

$$A \sqcap B = \int_1^4 \sqrt{x} - 1/x \; + \; \int_4^6 3 - \frac{x}{2}/x \; . \qquad (49)$$

Order Relations: Two kinds of order relations, namely, $\underset{\bullet}{\sqsubseteq}$ and $\overset{\bullet}{\sqsubseteq}$, for fuzzy numbers A and B can be defined by

$$A \underset{\bullet}{\sqsubseteq} B \iff A \sqcap B = A \, , \qquad (50)$$

$$A \overset{\bullet}{\sqsubseteq} B \iff A \sqcup B = B \, . \qquad (51)$$

The following properties of fuzzy numbers under join ($\sqcup$), meet ($\sqcap$) and order relations ($\underset{\bullet}{\sqsubseteq}$, $\overset{\bullet}{\sqsubseteq}$) are obtained (cf. [2]).

[Theorem 12] Arbitrary fuzzy numbers satisfy idempotent laws, commutative laws and associative laws under the operations of join ($\sqcup$) and meet ($\sqcap$). Thus, they constitute a partially ordered set under the order relation ($\overset{\bullet}{\sqsubseteq}$). The same is true of the order relation ($\underset{\bullet}{\sqsubseteq}$). In general, however, we have $\overset{\bullet}{\sqsubseteq} \neq \underset{\bullet}{\sqsubseteq}$.

[Theorem 13] Convex fuzzy numbers also satisfy distributive laws and are closed under $\sqcup$ and $\sqcap$. Thus, they form a commutative semiring under $\sqcup$ and $\sqcap$, but do not form a lattice because they do not satisfy absorption laws under $\sqcup$ and $\sqcap$.

[Theorem 14] Normal convex fuzzy numbers are closed and also satisfy absorption laws under $\sqcup$ and $\sqcap$. Therefore, they form a distributive lattice under $\sqcup$ and $\sqcap$. Consequently, order relation $\underset{\bullet}{\sqsubseteq}$ becomes coincident with $\overset{\bullet}{\sqsubseteq}$ and hence the order relation for normal convex fuzzy numbers can be defined as

$$A \sqsubseteq B \iff A \sqcap B = A \iff A \sqcup B = B \qquad (52)$$

(Example 8) Let $A = \underset{\sim}{2} \div \underset{\sim}{2}$ and $B = \underset{\sim}{2}$ be normal convex fuzzy numbers in (28) and (3), respectively, then $(\underset{\sim}{2} \div \underset{\sim}{2}) \sqcup \underset{\sim}{2} = \underset{\sim}{2}$ and $(\underset{\sim}{2} \div \underset{\sim}{2}) \sqcap \underset{\sim}{2} = (\underset{\sim}{2} \div \underset{\sim}{2})$ are obtained from (50) and (51). Thus, we can have $(\underset{\sim}{2} \div \underset{\sim}{2}) \sqsubseteq \underset{\sim}{2}$ by (52). But, for $A = \underset{\sim}{2} + \underset{\sim}{2}$ and $B = \underset{\sim}{2} \times \underset{\sim}{2}$, it follows from (48) and (49) that $A \sqcap B \neq A$ and $A \sqcup B \neq B$. So $\underset{\sim}{2} + \underset{\sim}{2}$ and $\underset{\sim}{2} \times \underset{\sim}{2}$ are incomparable.

Next we shall obtain the algebraic properties of fuzzy numbers under the operations of $\sqcup$ and $\sqcap$ combined with the operations of +, -, x and ÷.

[Theorem 15] Let A, B and C be convex fuzzy numbers, then the following properties are obtained.

$$A + (B \sqcup C) = (A + B) \sqcup (A + C) \qquad (53)$$

$$A + (B \sqcap C) = (A + B) \sqcap (A + C) \qquad (54)$$

$$A - (B \sqcup C) = (A - B) \sqcap (A - C) \qquad (55)$$

$$A - (B \sqcap C) = (A - B) \sqcup (A - C) \qquad (56)$$

$$(B \sqcup C) - A = (B - A) \sqcup (C - A) \qquad (57)$$

$$(B \sqcap C) - A = (B - A) \sqcap (C - A) \qquad (58)$$

[Theorem 16] If A, B and C are positive convex fuzzy numbers, then we have

$$A \times (B \sqcup C) = (A \times B) \sqcup (A \times C) \qquad (59)$$

$$A \times (B \sqcap C) = (A \times B) \sqcap (A \times C) \qquad (60)$$

$$A \div (B \sqcup C) = (A \div B) \sqcap (A \div C) \qquad (61)$$

$$A \div (B \sqcap C) = (A \div B) \sqcup (A \div C) \qquad (62)$$

$$(B \sqcup C) \div A = (B \div A) \sqcup (C \div A) \qquad (63)$$

$$(B \sqcap C) \div A = (B \div A) \sqcap (C \div A) \qquad (64)$$

It is interesting to note in the above two theorems that if A is convex, then Eqs. (53) - (64) do hold even if B and C are non-convex and that, conversely, if A is non-convex, they do not hold in general even if B and C are convex.

(Example 9) Let A be a positive non-convex fuzzy number and let B and C be positive convex fuzzy numbers such that

$$A = \int_{0}^{0.5} 1 - 2x/x \;+\; \int_{0.5}^{1} 2x - 1/x \,,$$

$$B = \int_{0}^{0.5} 2x/x \;+\; \int_{0.5}^{1} 1/x \,,$$

$$C = \int_{0}^{0.5} 2x/x \;+\; \int_{0.5}^{1} 2(1 - x)/x \,.$$

Then

$$A \times (B \sqcap C) = \int_{0}^{\frac{3}{16}} \frac{3 - \sqrt{1 + 16x}}{2}\Big/x + \int_{\frac{3}{16}}^{0.5} \frac{\sqrt{1 + 16x} - 1}{2}\Big/x + \int_{0.5}^{1} 2(1 - x)/x \,,$$

$$(A \times B) \sqcap (A \times C) = \int_{0}^{u_o} 1 - 2x/x + \int_{u_o}^{0.5} \frac{\sqrt{1 + 16x} - 1}{2}\Big/x + \int_{0.5}^{1} 2(1 - x)/x \,,$$

where $u_o = \dfrac{5 - \sqrt{17}}{4}$. Thus (60) is not satisfied when A is non-convex.

[Theorem 17] Convex fuzzy numbers form a commutative semiring with zero $+\infty$ and unity 0 under $\sqcap$ (as addition) and + (as multiplication). Similarly, they also form a commutative semiring with zero $-\infty$ and unity 0 under $\sqcup$ and +. Positive convex fuzzy numbers form a commutative semiring with zero $+\infty$ and unity 1 under $\sqcap$ and x. The same holds under $\sqcup$ and x, where zero is $-\infty$ and unity is 1.

CONCLUSION

In this paper we have investigated the algebraic properties of fuzzy numbers in the real line under the four arithmetic operations and under join and meet.

The concept of fuzzy numbers will find interesting applications to such fields as decision making and system science, where imprecise data in the real line are usually treated.

REFERENCES

[1] Chang, C.L. (1975). Interpretation and execution of fuzzy programs, in "Fuzzy Sets and Their Applications to Cognitive and Decision Processes" (eds. L.A. Zadeh, K.S. Fu, K. Tanaka & M. Shimura). Academic Press, New York. pp.151-170.

[2] Mizumoto, M. & Tanaka, K. (1976). Some properties of fuzzy sets of type 2. Information and Control, 31, 312-340.

[3] Zadeh, L.A. (1975). The concept of a linguistic variable and its application to approximate reasoning (I), (II), (III). Information Sciences, 8, 199-249; 8, 301-357; 9, 43-80.

ADVANCES IN FUZZY SET THEORY AND APPLICATIONS
M.M. Gupta, R.K. Ragade, R.R. Yager (editors)

FUZZY VARIABLES IN A RANDOM ENVIRONMENT

Steven NAHMIAS

Department of Industrial Engineering, University of Pittsburgh
(Visiting: Departments of Operations Research and Industrial Engineering, Stanford University, 1978-1979 academic year)

This paper treats the problem of modelling an environment in which imprecision arises from both fuzziness and randomness simultaneously. We demonstrate that when one assumes a fuzzy set is defined as its membership function, a serious inconsistency arises when the fuzzy sets become crisp. However by using the pattern space approach (developed earlier) and the concept of a fuzzy variable, the inconsistency is no longer present. Computation of the expectation of the fuzzy random variable is considered for a variety of classes of membership functions as well as for arbitrary membership functions which assume values on a discrete set. A brief discussion of fuzzy interference and its relationship to the sample survey problem is included.

I. Introduction

The purpose of this study is to explore a problem in which imprecision arises from both fuzziness and randomness simultaneously. Our goal will be not only to develop a model which can suitably accommodate both fuzziness and randomness, but to give further evidence that the fuzzy variable definition proposed by Nahmias [2] remains consistent when randomness is introduced while other approaches to defining fuzziness do not.

There are undoubtably many ways that both fuzziness and randomness can occur simultaneously. For example, we can speak of probabilities of fuzzy events. An example of this type would be to determine the probability that the weather will be pleasant tomorrow. Zadeh [6] has considered this particular aspect of the problem and extended the concept of a probability measure to fuzzy events in a very natural fashion. Let $(\Omega, \mathfrak{F}, P)$ be a probability space. Then it is well known that for any event $A \in \mathfrak{F}$, $P(A) = E[I_A(\omega)]$ where $I_A : \Omega \rightarrow R$ is the random variable on Ω given by the indicator function of the event A. Since a fuzzy event is identified by its membership function, μ_A, we may think of μ_A as a random variable on Ω as well when A is a fuzzy event. It follows by analogy that we would define $P(A) = E[\mu_A(\omega)]$.

We will consider an entirely different situation in which both fuzziness and randomness arise simultaneously. Fuzziness presents a model for describing subjective opinion. We may then think of a variety of fuzzy quantities which would represent the opinions of several different individuals, or more generally, a population of individuals. Consider now performing an experiment whose outcome would correspond to soliciting the opinion of one individual chosen at random. Hence we wish to model the situation where an experiment is performed whose outcome is a fuzzy quantity (that is, a fuzzy set).

We now come to a fundamental problem. What, precisely, is a fuzzy set? Zadeh [5] somewhat sidesteps the issue in his original paper by saying that a fuzzy set A is <u>characterized</u> by a membership function $\mu_A(x)$. This is similar to saying that a bicycle is characterized by two wheels, handlebars, a chain, etc. which never tells you what a bicycle is. (A better definition would be: a bicycle is a <u>vehicle</u> which is characterized by ...). Later authors however have been unambiguous in defining a fuzzy set as the membership function (see Goguen [1] and Negoita and Ralescu [3]). In fact, Negoita and Ralescu have developed an interesting mathematical structure of classes of membership functions based on this definition.

Nahmias [2] takes an entirely different approach. By constructing a mathematical structure analogous to the sample space of probability, he introduces the fuzzy variable (i.e., fuzzy set) which turns out to be analogous to the random variable. We will demonstrate that if one assumes a fuzzy set is defined by its membership function, then one obtains a definition for the expected value which is not consistent with that which is known to be true for crisp (i.e., non-fuzzy) sets. This inconsistency is not present, however, when the pattern space model of fuzziness is used.

II. Defining Fuzzy Sets as Membership Functions

In this section we will assume that a fuzzy set is defined as its membership function and develop a model in which an experiment is performed whose outcome is a fuzzy quantity. We will show that the results of this approach do not remain consistent with known results from probability when each of the fuzzy sets reduce to single points. Let $(\Omega, \mathfrak{F}, P)$ be a probability space. We may think of each point $\omega \in \Omega$ corresponding to an outcome of the experiment. Suppose that $\Phi = (\mu_1, \mu_2, \ldots, \mu_n)$ are some finite collection of membership functions each of which is defined on R_1 , the real line. Note that we are specializing to fuzzy sets on the line so that for each $1 \le i \le n$, $\mu_i : R_1 \to [0,1]$.

Let Y be a mapping from $\Omega \times R_1$ onto Φ and further assume that Y is discrete. In other words, for each $\omega \in \Omega$, $Y(\omega, \cdot)$ represents the outcome of the experiment which corresponds to one of the membership functions in Φ .

Since Y assumes only finitely many values, we may partition Ω into n mutually exclusive and exhaustive sets $(A_1,\ldots,A_n)$ such that $Y(\omega,t)$ equals $\mu_i(t)$ for each $\omega \in A_i$, $1 \le i \le n$ and $t \in R_1$. It follows that Y may be given the representation

$$Y(\omega,t) = \sum_{i=1}^{n} \mu_i(t) I_{A_i}(\omega) \quad . \tag{1}$$

Suppose that we define $p_i = P(A_i)$. Then what we are saying is that there are exactly n possible outcomes of the sampling experiment which correspond to the n membership functions $(\mu_1,\ldots,\mu_n)$ each of which occur with respective probabilities $(p_1,\ldots,p_n)$. In order to further clarify the situation, consider the following simple example. Two individuals have differing opinions on some issue and their opinions are represented by the membership functions μ_1 and μ_2. The experiment consists of flipping a coin and soliciting the opinion (i.e., the membership function) of the first individual if the outcome is heads and the second if it is tails. For this example $p_1 = p_2 = 0.5$.

We next consider the expected value associated with this experiment. From (1) it follows that $EY(t)$ will be a membership function and will be given by

$$EY(t) = \sum_{i=1}^{n} \mu_i(t) E[I_{A_i}(\omega)] = \sum_{i=1}^{n} \mu_i(t) p_i \quad . \tag{2}$$

On the surface, this appears to be quite reasonable. Since the individual outcomes of the experiment are membership functions, it is only natural that the expectation should also be a membership function, and that it is computed as the weighted sum of the individual membership functions comprising all of the possible outcomes of the experiment. Suppose however that the opinions of the individuals are not fuzzy but are crisp. In that case there would be real numbers t_i, $1 \le i \le n$, such that

$$\mu_i(t_i) = 1$$

$$\mu_i(t) = 0 \text{ if } t \neq t_i \quad .$$

To see why this is so, consider a specific example. Suppose each individual in a group is asked what salary each thinks his job performance merits. If there is no ambiguity regarding the opinions, then there will be one dollar amount that has grade of membership one in the fuzzy sets corresponding to each of their opinions.

In this case, the mapping Y should reduce to an ordinary random variable which assumes the values $(t_1,\ldots,t_n)$ with respective probabilities $(p_1,\ldots,p_n)$.

The mean of Y should be given by $\Sigma_{i=1}^{n} t_i p_i$, or equivalently, the mean of Y should correspond to the membership function of the non-fuzzy set consisting of the single point $\Sigma_{i=1}^{n} t_i p_i$. However from (2) this is clearly not what results. The mean obtained is the membership function

$$EY(t) = \begin{cases} p_i & \text{at } t = t_i \\ 0 & \text{elsewhere} \end{cases}$$

which is the fuzzy set whose membership function is pictured in Figure 1. Hence we see that defining a fuzzy set as its membership function results in a model which does not remain consistent with known results for crisp sets. The membership function pictured in Figure 1 corresponds to a fuzzy rather than a crisp set even though no fuzziness is present in the experiment. We may then conclude that defining a fuzzy set as its membership function is not correct.

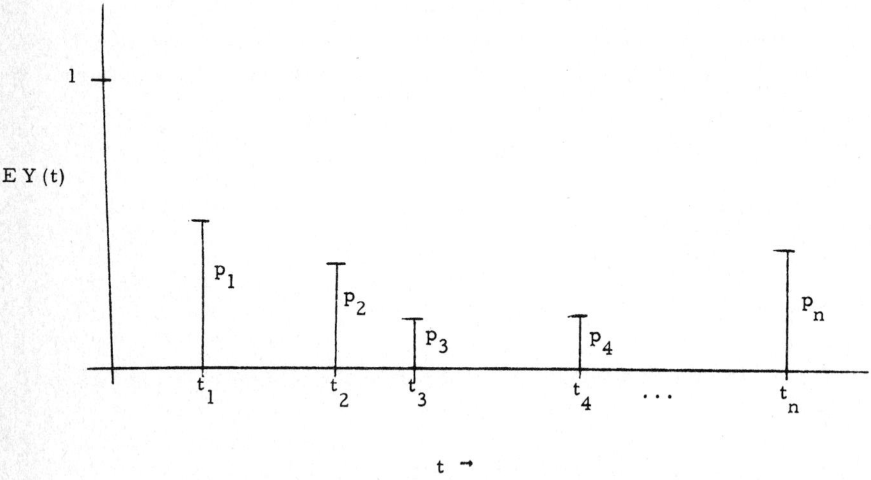

FIGURE 1. The expectation of an experiment in which individual outcomes are non-fuzzy and a fuzzy set is defined as its membership function.

III. The Pattern Space Model

Defining a fuzzy set as its membership function is incorrect for essentially the same reason that it is incorrect to define a random variable as its probability density. When we talk about fuzzy sets on the real line (in this case the so-called reference set corresponds to R_1), the points on the line need to correspond to the range of a function rather than the domain. One finds the same phenomenon occurring in probability theory: points on the line correspond to the range of a random variable rather than the domain. In probability there is a natural domain for a random variable, namely the sample space corresponding to the set of all possible outcomes of an experiment. In the fuzzy model, however the domain of the fuzzy variable does not have such an obvious physical interpretation.

The domain of the fuzzy variable will be labelled the pattern space. The pattern space is an abstract space whose physical significance will depend upon the particular situation one is modelling. Let us consider the case where a computer is attempting to interpret hand-written messages. For this example, the pattern space would correspond to the set of all possible symbols the computer might encounter. The fuzzy variable which maps this space to R_1 essentially imposes a linear ordering on the pattern space.

The pattern space model was originally reported in reference [2]. A pattern space is a triple $(\Gamma, \mathcal{G}, \sigma)$ where $\gamma \in \Gamma$ are arbitrary points, $\mathcal{G}$ is the discrete topology on Γ, and σ (called the scale) is a real valued set function on $\mathcal{G}$ with the properties:

(1) $\sigma(\phi) = 0$, $\sigma(\Gamma) = 1$.

(2) For any arbitrary collection of sets A_α of $\mathcal{G}$ (finite, countable, or uncountable)

$$\sigma(\bigcup_\alpha A_\alpha) = \sup_\alpha \sigma(A_\alpha) \quad .$$

In addition, two sets $A, B \in \mathcal{G}$ are said to be unrelated if $\sigma(A \cap B) = \min[\sigma(A), \sigma(B)]$.

The idea of a fuzzy set on the line is now replaced by a fuzzy variable, X , which is a real valued function defined on Γ . The fuzzy variable plays the same role in our structure as does the random variable in probability. The membership function, $\mu(t)$, is obtained as the extension of the scale to the line via the fuzzy variable. That is,

$$\mu(t) = \sigma\{\gamma : X(\gamma) = t\} \quad .$$

To see that the concepts of the pattern space and scale are not vacuous, let $\mu(t)$ be the membership function for any "normal" fuzzy set on the line: that is, any real valued function whose range is the unit interval and whose supremum is one. Suppose $\Gamma = R_1$ and let $\mathcal{G}$ be the discrete topology on the line. Define for any set of real numbers $A \in \mathcal{G}$:

$$\sigma(A) = \sup_{t \in A} \mu(t) \quad .$$

If A_α represent an arbitrary collection of elements of $\mathcal{G}$, then

$$\sigma(\cup A_\alpha) = \sup_{t \in \cup A_\alpha} \mu(t)$$

$$= \sup_{\alpha} \sup_{t \in A_\alpha} \mu(t)$$

$$= \sup_{\alpha} \sigma(A_\alpha) \quad \text{as required}$$

We may think of σ assigning a weight to each set which corresponds to the largest grade of membership of those points comprising the set. (Hence the appropriateness of the term "scale".)

Suppose further that $\mu(t)$ is continuous and monotonically increasing. This would be the case if, for example, it corresponded to the fuzzy variable "large numbers". Then it is clear that for any sets of real numbers $A, B \in \mathcal{G}$ such that $A \cap B \neq \phi$, A and B will be unrelated. [This will not necessarily be the case, however, if $\mu(t)$ is not monotonic.]

For $\Gamma = R_1$, the fuzzy variable on Γ will correspond simply to the identity mapping. We should point out that not all pattern spaces need to be isomorphic to R_1. When elements of the pattern space are not linearly ordered, the somewhat more general structure developed in [2] is appropriate.

A collection of fuzzy variables $(X_1, \ldots, X_n)$ is said to be unrelated if the sets $\{X_1 = x_1\}, \{X_2 = x_2\}, \ldots, \{X_n = x_n\}$ are unrelated for all real numbers $x_1, x_2, \ldots, x_n$. In what follows we will assume all fuzzy variables are unrelated. The analogy between probability and fuzzy theory is presented in Table 1.

To help clarify how the mechanics of this approach would work, let us again consider the problem of a computer interpreting hand-written script. Let us consider the portion of the pattern space corresponding to those symbols which might be interpreted as the letter "a". Each of these symbols might be labelled as points γ_i in Γ as follows:

γ_1	γ_2	γ_3	γ_4	γ_5	γ_6	γ_7	γ_8

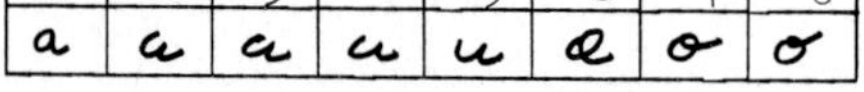

Suppose the fuzzy variable, X, maps each point γ_i to the real number i, that is

$$X(\gamma_i) = i \quad \text{for} \quad 1 \leq i \leq 8 \quad .$$

Table 1

The probability theory/fuzzy theory analogy

Probability theory	Fuzzy theory
Sample space Ω	Pattern space Γ
σ – algebra of events, $\mathfrak{F}$	Discrete topology of subsets, $\mathcal{G}$
Probability P defined on $\mathfrak{F}$ (a special type of measure)	Scale σ defined on $\mathcal{G}$ (a special type of capacity)
Independence of events $P(A \cap B) = P(A) \cdot P(B)$	Unrelatedness of sets $\sigma(A \cap B) = \min [\sigma(A), \sigma(B)]$
Union of countable collection of disjoint sets $P\left(\bigcup_{i=1}^{\infty} A_i\right) = \sum_{i=1}^{\infty} P(A_i)$	Union of arbitrary collection of sets $\sigma\left(\bigcup_{\alpha} A_\alpha\right) = \sup_\alpha \sigma(A_\alpha)$
Random variable: real valued measurable function on Ω	Fuzzy variable: real valued function on Γ
The probability distribution is obtained as the extension of P to the Borel sets on the line. The probability density is the derivative of the distribution function.	The membership function is obtained as the extension of σ to single point sets on the line.

Further suppose the scale, σ, makes the following assignment:

$\sigma(\gamma_1)$	$\sigma(\gamma_2)$	$\sigma(\gamma_3)$	$\sigma(\gamma_4)$	$\sigma(\gamma_5)$	$\sigma(\gamma_6)$	$\sigma(\gamma_7)$	$\sigma(\gamma_8)$
.90	.80	.70	.65	.40	.55	.35	.10

Then the membership function of X, μ_X, is obtained by the rule $\mu_X(i) = \sigma(\gamma_i)$ which specifies the grade of membership associated with the collection of symbols.

That the computer may be reading other symbols in addition to the letter "a" can be modelled by considering a distinct fuzzy variable. Let Y be associated with the letter "u" say. Then Y might be defined as follows:

$$Y(\gamma_1) = 8 \qquad Y(\gamma_5) = 1$$
$$Y(\gamma_2) = 7 \qquad Y(\gamma_6) = 8$$
$$Y(\gamma_3) = 4 \qquad Y(\gamma_7) = 8$$
$$Y(\gamma_4) = 3 \qquad Y(\gamma_8) = 8$$

and σ remains the same.

Note that this assignment gives that the symbol corresponding say, to γ_5, has grade of membership of .40 in the set of objects that are interpreted as the letter "a" , and grade of membership .90 in the set of objects interpreted as the letter "u" . (It should be noted that for this example the fuzzy variables X and Y are related as one might expect.)

We will need some results from [2]. Let X and Y be two unrelated fuzzy variables on $(\Gamma, \mathcal{G}, \sigma)$ with membership functions μ_X and μ_Y. Then

(i) If $Z = X + Y$, then $\mu_Z(z) = \sup_x \min [\mu_X(x), \mu_Y(z-x)]$

(ii) If $g: R_1 \to R_1$ and $Z = g(X)$, then

$$\mu_Z(z) = \sup_{x \in g^{-1}(z)} \mu_X(x) \quad [= \mu_X(g^{-1}(z)) \text{ if } g \text{ is invertible}]$$

Using (i) and (ii) one can demonstrate that various classes of fuzzy variables are closed under addition and scalar multiplication. The following results were proven in [2]:

(iii) Let $(X_1, \ldots, X_n)$ be a collection of n unrelated fuzzy variables whose membership functions are of the form

$$\mu_i(t) = \exp\{-[(t - a_i)/b_i]^2\}$$

where $(a_1, \ldots, a_n)$ are an arbitrary set of scalars and $(b_1, \ldots, b_n)$ are a set of arbitrary non-negative scalars. Define

$$Z = \sum_{i=1}^{n} \alpha_i X_i \quad \text{where } \alpha_i \geq 0 \text{ and } \sum_{i=1}^{n} \alpha_i = 1 .$$

Then

$$\mu_Z(z) = \exp\{-[(z - a)/b]^2\}$$

where

$$a = \sum_{i=1}^{n} \alpha_i a_i \quad \text{and} \quad b = \sum_{i=1}^{n} \alpha_i b_i .$$

(iv) Let $(X_1,\ldots,X_n)$ be a collection of n unrelated fuzzy variables whose membership functions are of the form

$$\mu_i(t) = \begin{cases} (t/\lambda_i r)^r \exp(r - t/\lambda_i) & \text{for } t \geq 0 \\ 0 & \text{for } t < 0 \end{cases}$$

where $(\lambda_1,\ldots,\lambda_n)$ are a set of arbitrary positive scalars, and r is a fixed positive scalar. Then

$$Z = \sum_{i=1}^{n} \alpha_i X_i \quad \text{where } \alpha_i \geq 0 \text{ and } \sum_{i=1}^{n} \alpha_i = 1$$

has a membership function of the same form as each of the X_i with parameter

$$\lambda = \sum_{i=1}^{n} \alpha_i \lambda_i$$

(r is assumed to be fixed).

It is interesting to note that both (iii) and (iv) are analogous to similar results for convolutions of normal and gamma random variables. However, when adding fuzzy variables, the shape of the membership function remains preserved even in cases where a similar result does not hold for random variables. We give the following result without proof. (The proof is similar to those of (iii) and (iv) which can be found in [2].)

(v) Let $(X_1,\ldots,X_n)$ be n unrelated fuzzy variables whose membership functions are given by

$$\mu_i(t) = \begin{cases} 1 & \text{for } a_i \leq t \leq b_i \qquad 1 \leq i \leq n \\ 0 & \text{otherwise} \end{cases}$$

Then if

$$Z = \sum_{i=1}^{n} \alpha_i X_i \quad \text{where } \alpha_i \geq 0 \text{ and } \sum_{i=1}^{n} \alpha_i = 1 \quad ,$$

$$\mu_Z(z) = \begin{cases} 1 & \text{if } z = \sum_{i=1}^{n} x_i \text{ for some } \alpha_i a_i \leq x_i \leq \alpha_i b_i \\ 0 & \text{otherwise} \end{cases}$$

Notice that in (v), the fuzzy variables are crisp in that they correspond unambiguously to closed intervals on the line. Their sum, Z , is simply the pointwise algebraic sum of these intervals (rescaled by the constants α_i) as one would expect. It is interesting to note that (v) does not have an analogy in probability. The convolution of uniform densities is not uniform. Undoubtedly, there are numerous other classes of fuzzy variables that are also closed under addition.

In many respects the notion of a fuzzy variable has been hinted at in previous or concurrent work. Result (i), for example, is the same as the extension principle discussed by Zadeh [7] while (ii) is the same as the definition of the transformation of a fuzzy set given by Zadeh [5]. In addition, the concept of a fuzzy number discussed by Mizumoto and Tanaka [1a] is a special case of a fuzzy variable. The important point is that (i) and (ii) are derived as a consequence of the pattern space model rather than simply stated as definitions. A rigorous and well structured framework allows us to derive unambiguously a calculus for performing algebraic transformations on fuzzy quantities.

IV. Fuzzy Variables in a Random Environment

We will now show how the fuzzy variable approach can be used to model an environment in which both fuzziness and randomness are present and further demonstrate that the inconsistency which arises by assuming a fuzzy set is defined as its membership function will no longer occur.

Let $(\Gamma, \mathcal{G}, \sigma)$ be a pattern space and suppose that $(X_1, \ldots, X_n)$ are n unrelated fuzzy variables defined on Γ . Further, suppose that $(\Omega, \mathfrak{F}, P)$ is a probability space. A discrete fuzzy random variable Y is a mapping on $\Omega \times \Gamma$ and has the representation

$$Y(\omega, \gamma) = \sum_{i=1}^{n} X_i(\gamma) I_{A_i}(\omega) \tag{3}$$

where $(A_1, \ldots, A_n)$ form a partition of Ω . If $p_i = P(A_i)$ then we may think of Y as taking on the value $X_i(\gamma)$ with probability p_i . In this sense Y describes an experiment whose outcome is one of the fuzzy variables $(X_1, \ldots, X_n)$. It follows from (3) that EY will be a fuzzy variable and will have the representation

$$EY(\gamma) = \sum_{i=1}^{n} X_i(\gamma) E[I_{A_i}(\omega)] = \sum_{i=1}^{n} X_i(\gamma) \cdot p_i \tag{4}$$

The membership function of the fuzzy variable EY , which may be denoted μ_{EY} , can be obtained by applying the rules (i) and (ii) for addition and scalar multiplication of fuzzy variables.

Let us consider the particular case where each of the X_i are non-fuzzy and reduce to single points on the line. In this case $X_i(\gamma) = t_i$ for all $\gamma \in \Gamma$ and $1 \leq i \leq n$ where $(t_1,\ldots,t_n)$ are some set of real numbers. It now follows from (4) that

$$EY(\gamma) = \sum_{i=1}^{n} t_i p_i \quad \text{for all } \gamma \in \Gamma \tag{5}$$

which is precisely the expectation of a random variable whose n possible outcomes are $(t_1,\ldots,t_n)$ with probabilities $(p_1,\ldots,p_n)$. We can also derive (5) by following the rules (i) and (ii). The membership function of X_i would be

$$\mu_{X_i}(x) = \begin{cases} 1 & \text{if } x = t_i \\ 0 & \text{otherwise,} \end{cases}$$

so that by applying (ii) the membership function of $p_i X_i$ is

$$\mu_{p_i X_i}(x) = \begin{cases} 1 & \text{if } x = p_i t_1 \\ 0 & \text{otherwise.} \end{cases}$$

Now by applying (i), one obtains

$$\mu_{p_1 X_1 + p_2 X_2}(x) = \begin{cases} 1 & \text{if } x = p_1 t_1 + p_2 t_2 \\ 0 & \text{otherwise,} \end{cases}$$

so that continued application of (i) gives an alternative representation of (5):

$$\mu_{EY}(x) = \begin{cases} 1 & \text{if } x = \sum_{i=1}^{n} p_i t_i \\ 0 & \text{otherwise.} \end{cases}$$

Hence we see that by assuming that the outcomes of the experiment are fuzzy variables rather than membership functions, we obtain a definition of the expectation which remains consistent when the fuzzy variables become non-fuzzy and Y reduces to an ordinary random variable.

The membership function of Y (not of EY) will be a random element whose outcomes correspond to the membership functions of the fuzzy variables $(X_1,\ldots,X_n)$. For fixed $\omega \in \Omega$, $Y(\omega,\cdot)$ is a fuzzy variable on $(\Gamma,\mathcal{G},\sigma)$ whose membership function is given by

$$\mu_{Y(\omega,\cdot)}(x) = \sigma\{\gamma : Y(\omega,\gamma) = t\} \quad .$$

For all $\omega \in A_i$, $Y(\omega,\cdot) = X_i(\cdot)$ which gives

$$\mu_{Y(\omega,\cdot)}(t) = \sigma\{\gamma : X_i(\gamma) = t\} = \mu_{X_i}(t)$$

for $\omega \in A_i$. It follows that

$$\mu_Y(t) = \sum_{i=1}^{n} \mu_{X_i}(t) I_{A_i}(\omega) \quad .$$

That is, the membership function of Y will correspond to the membership function of X_i with probability p_i .

In general, we would be interested in computing $\mu_{EY}(t)$. When the membership functions of $(X_1,\ldots,X_n)$ are of general forms, computing μ_{EY} can be cumbersome. However, by utilizing the results (iii), (iv), (v) above, this computation may be done directly if one is willing to assume the form of the membership functions of $(X_1,\ldots,X_n)$ is one of those given above. We have the following results: (the proofs are straightforward)

<u>Theorem</u>. Let Y be a discrete fuzzy random variable that assumes values $(X_1,\ldots,X_n)$ with probabilities $(p_1,\ldots,p_n)$. Then

(a)
$$\mu_{EY}(t) = \exp\{-[(t-a)/b]^2\}$$

where

$$a = \sum_{i=1}^{n} p_i a_i \quad \text{and} \quad b = \sum_{i=1}^{n} p_i b_i$$

if $(X_1,\ldots,X_n)$ are as given in (iii) above.

(b)
$$\mu_{EY}(t) = \begin{cases} (t/\lambda r)^r \exp(r - t/\lambda) & \text{for } t \geq 0 \\ 0 & \text{for } t < 0 \end{cases}$$

where

$$\lambda = \sum_{i=1}^{n} p_i \lambda_i$$

if $(X_1,\ldots,X_n)$ are as given in (iv) above.

(c)
$$\mu_{EY}(t) = \begin{cases} 1 & \text{if } t = \sum_{i=1}^{n} x_i \text{ for some } a_i p_i \leq x_i \leq b_i p_i \\ 0 & \text{otherwise} \end{cases}$$

if $(X_1,\ldots,X_n)$ are given as in (v) above.

V. Computation of EY for Empirical Membership Functions

In trying to use concepts of fuzziness in actual problems, a procedure must be employed for inferring or estimating the membership functions corresponding to individual opinion. (Saaty [4] has developed an approach to this problem based on analysis of a pairwise comparison matrix.) We will call the result of this estimation an empirical membership function. It may not necessarily be true that these empirical membership functions can be closely approximated by one of the continuous forms described above.

Suppose that X_i is a fuzzy variable with membership function $\mu_{X_i}(t)$ which is estimated at some discrete set of points $(t_1, t_2, \ldots, t_k)$. With no loss in generality, assume that each t_i corresponds to a non-negative integer. (More generally, we could assume that each t_i was a member of some discrete set which would be isomorphic to the positive integers.) We will adopt the notational convention that $[u]$ is the closest integer to u, rounding up at .5.

Now, suppose that Y is a fuzzy random variable whose outcomes are $(X_1, \ldots, X_n)$ with respective probabilities $(p_1, \ldots, p_n)$ and assume that the membership functions $\mu_{X_i}(t)$ are specified for a set of integer points of the argument t. The computation of the membership function of the fuzzy variable EY would have to be carried out by the following procedure:

(1) Compute the membership function of $Z_1 = p_1X_1 + p_2X_2$ by the rule

$$\mu_{Z_1}(t) = \max_{u=0,1,2,\ldots} \min \left\{ \mu_{X_1}([u/p_1]) \, , \, \mu_{X_2}([(t-u)/p_2]) \right\} .$$

(2) Compute the membership function of $Z_2 = Z_1 + p_3X_3$ by the rule

$$\mu_{Z_2}(t) = \max_{u=0,1,2,\ldots} \min \left\{ \mu_{Z_1}(u) \, , \, \mu_{X_3}([(t-u)/p_3]) \right\} .$$

(3) Continue with this procedure until the membership function of $Z_n = EY$ is obtained.

Example. In order to illustrate how this procedure would be used, consider the following simple example. Alumni of colleges A and B are asked their opinion regarding the number of points each thinks college A's football team will score at the upcoming game. We suppose that there is some ambiguity regarding their opinions so that it is necessary to assume the opinion of each is a fuzzy variable rather than a real number. The following empirical membership functions are estimated:

$$\mu_{X_1}(t) = \begin{cases} .4 & \text{for } t = 12, 13, 14 \\ .8 & \text{for } t = 15, 16, 17, 18 \\ .2 & \text{for } t = 19, 20 \\ 0 & \text{otherwise,} \end{cases}$$

$$\mu_{X_2}(t) = \begin{cases} .6 & \text{for } t = 0, 1, 2, 3, 4 \\ .3 & \text{for } t = 5, 6, 7, 8 \\ 0 & \text{otherwise,} \end{cases}$$

Now suppose a fair coin is flipped and the opinion of the alumnus from college $A(X_1)$ is picked if heads are obtained and the opinion of the alumnus from college $B(X_2)$ is picked if tails are obtained. The outcome of this experiment may be represented by the fuzzy random variable Y whose expectation is $EY = .5X_1 + .5X_2$. Following the rules outlined above we obtain the membership function of EY as

$$\mu_{E(Y)}(t) = \begin{cases} .4 & \text{for } t = 6, 7 \\ .6 & \text{for } t = 8, 9, 10, 11 \\ .3 & \text{for } t = 12 \\ .2 & \text{for } t = 13, 14 \\ 0 & \text{otherwise.} \end{cases}$$

VI. Fuzzy Inference and the Sample Survey Problem

The sample survey problem may be stated as follows: It is desired to estimate the characteristics of a population of individuals whose opinions exhibit a pattern of statistical variation as one samples different individuals. If we are willing to assume the outcome of the sampling experiment corresponds to a real number, then the problem of inferring the characteristics of the population reduces the classical problem of statistical inference.

In order to account for the ambiguity of individual opinion, a more precise model would treat the realization of a particular sample point as a fuzzy variable rather than simply as a real number. Prior to sampling, each observation would be considered to be a fuzzy random variable, Y . Hence the random sample would correspond to a sequence of fuzzy random variables $(Y_1, Y_2, \ldots, Y_n)$. A typical inference problem would be to estimate the membership function of the true mean EY . In this case, a natural choice might be the membership function of $\overline{Y} = (1/n) \Sigma_{i=1}^{n} Y_i$.

In order to make the problem tractable, one could assume that each membership function in the population was of the form $\mu(t;\theta)$ where θ is a real variable. The statistical variation in the sample could then be characterized by a probability distribution on θ which itself may have one or more unknown parameters. The estimation of these parameters (and hence the estimation of θ as well) could then be accomplished by classical means.

To be specific, suppose that the membership function of each individual in the population is given by

$$\mu(t;\theta) = e^{-\left[\frac{(t-\theta)}{b}\right]^2}$$

where b is a known constant. Further, suppose that the probability distribution on θ, say $f(\theta)$, is normal with unknown mean μ and unknown standard deviation σ. That is

$$f(\theta) = \frac{1}{\sigma\sqrt{2\pi}} e^{-\frac{1}{2}\left(\frac{\theta-\mu}{\sigma}\right)^2} \qquad -\infty < \theta < +\infty$$

In this case, the outcome of the sampling experiment would correspond to a sequence of n observations on the random variable θ, say $(\theta_1, \theta_2, \ldots, \theta_n)$. Estimators for μ and σ may now be obtained by classical means which could then be used to estimate functionals of the fuzzy random variable Y. For example, a reasonable estimator for the membership function of EY would be $\mu(t;\bar{\theta})$ where $\bar{\theta} = (1/n)\,\Sigma_{i=1}^{n}\,\theta_i$.

This example was intended for illustrative purposes only. The general problem of fuzzy inference is extremely complex. Estimators would be functions of $(Y_1, \ldots, Y_n)$ which implies that they also would be fuzzy random variables. We have not developed a theoretical basis for dealing with such a problem. However, the sample survey problem is a fundamental one which might ultimately be addressed by techniques such as those alluded to here.

VII. Concluding Remarks

Our goal in this discussion has been to tentatively explore an approach for dealing with the simultaneous phenomenon of fuzziness and randomness. One of the important features of our results is they provide further evidence that the pattern space model of fuzziness appears to be better than other approaches which have been suggested.

Although the concept of fuzziness is intuitively appealing, it remains to be seen whether or not this concept will ever be used in solving actual problems.

The notion that the strengths of individual opinion can be equated with degrees of membership is reminiscent of the notion of the existence of cardinal utilities. The existence of cardinal utilities has never been established.

There are also other problems. Perhaps the most salient is that there appears to be little physical verification for many of the results or definitions which have been expounded in the hundreds of papers published in this area. Ultimately, the test of the notion of fuzziness will be whether or not it can predict measurable phenomena.

Another problem is the development of a consistent and coherent theory. One of the unanswered questions in the area of fuzzy modelling is the physical basis of the minimum operator in the definition of unrelatedness (or non-interactiveness according to Zadeh [7]). In fact, most of the definitions one finds in the literature are the result of taking analogous results from probability and substituting the supremum operator for the integral operator and minimum operator for the product. Little physical evidence seems to exist to justify these analogies.

Our comments are meant to be more cautionary rather than negative. Much work needs to be done before fuzzy theory can be compared to probability theory either in its level of rigor or its potential usefulness. In time, many of the problems alluded to above may be resolved, and accurate quantitative modelling of subjective phenomenon will be the result.

References

[1] Goguen, J. A., "L-Fuzzy Sets," Journal of Mathematical Analysis and Applications, 18, 145-174 (1967).

[1a] Mizumoto, M. and Tanaka, K., "Algebraic Properties of Fuzzy Numbers," Faculty of Engineering Science, Osaka University, Osaka, Japan (1978).

[2] Nahmias, S., "Fuzzy Variables," Journal of Fuzzy Sets and Systems, 1, 97-110 (1978).

[3] Negoita, C. V. and Ralescu, D. A., Applications of Fuzzy Sets to Systems Analysis, John A. Wiley, New York (1975) (translated from the Rumanian).

[4] Saaty, T. L., "Exploring the Interface Between Hierarchies, Multiple Objectives and Fuzzy Sets," Journal of Fuzzy Sets and Systems, 1, (1977).

[5] Zadeh, L. A., "Fuzzy Sets," Information and Control, 8, 338-353 (1965).

[6] Zadeh, L. A., "Probability Measures of Fuzzy Events," Journal of Mathematical Analysis and Applications, 10, 421-427 (1968).

[7] Zadeh, L. A., "The Concept of a Linguistic Variable and Its Application to Approximate Reasoning," Information Science, 8, 199-251 (1975).

ADVANCES IN FUZZY SET THEORY AND APPLICATIONS
M.M. Gupta, R.K. Ragade, R.R. Yager (editors)

ON FUZZY STATISTICS

Abraham KANDEL

Associate Professor and Director of Computer Science
Department of Mathematics
Florida State University
Tallahassee, Florida 32306 U.S.A.

Preface

Every change is confronted with resistance. The acceptance of a new idea involves a spiritual and mental effort, and its transformation to real-life applications requires some revolution in the standard way of thinking.

This work is dedicated to Professor Lotfi A. Zadeh, the man behind the power of fuzziness, whose original ideas have made this work possible.

Abstract

Averages, we all must recoginize, are not natural realities. Indeed, they are not reality at all, but artificial constructs of the human mind which have been created, paradoxically, to help us understand parts of reality. Here an attempt is made to extend the axiomatic basis of probability theory, utilizing the concepts of multiple valued events and fuzzy expected values, in order to derive a more natural way of expressing the central tendency of a data-set. Through this model we can use the notion of subjectiveness to achieve qualitative and quantitative fuzzy methods which have possibility of applications in a variety of fields in science and engineering.

1. Motivation

Ordinarily, imprecision and indeterminacy are considered to be statistical random characteristics, and are taken into account by the methods of probability theory. In real situations, a frequent source of imprecision is not only the presence of random variables, but the impossibility, in principle, of operating with exact data as a result of the complexity of the system, imprecision of the constraints and objectives. At the same time, classes of objects appear in the problems that do not have clear boundaries; the imprecision of such classes is expressed in the possibility that an element does not only belong or not belong to a certain class but intermediate grades of membership are also possible.

Intuitively, a similarity is felt between the concepts of fuzziness and probability. The problems in which they are used are similar or coincide. There are problems in which indeterminacy is encountered, due to random factors, inexact knowledge, or the theoretical impossibility or lack of necessity to obtain exact solutions. The similarity is also underscored by the fact that the intervals of variation of the membership grade of fuzzy sets $\chi \varepsilon [0,1]$ coincide. However, between the concepts of fuzziness and probability there are also essential differences.

Probability is an <u>objective</u> characteristic; the conclusions of probability theory can, in general, be <u>tested by experience</u>.

The membership grade is <u>subjective</u>, although it is natural to assign a lower membership grade to an event that, considered from the aspect of probability, would have a lower probability of occurence. The fact that the assignment of a membership function of a fuzzy set is "nonstatistical" does not mean that we cannot use probability distribution functions in assigning membership functions. As a matter of fact, a careful examination of the variables of fuzzy sets reveals that they may be classified into two types: statistical and nonstatistical. The variable "magnitude of x" is an example of the former type. However, if one considers the "class of tall men", the "height of a man" can be considered to be a statistical variable. In this case, for instance, if a man is under 5 feet tall we would not call him a tall man by "everybody's" standard, and we should assign to him a low grade of membership in the class of tall men. Similarly, if a man is over 7 feet, he certainly deserves a high grade of membership in the class of tall men.

The operations applied in the theory of fuzzy sets [8]-[11], (finding maxima and minima) have a special form and differ from operations applied in probability.

It is only possible to judge the effectiveness of the methods of fuzzy set theory when a sufficient number of functioning devices shall have been constructed on the basis of this theory.

It is well known that probability theory can be constructed as an axiomatic theory, based on three axioms, coinciding with corresponding axioms of measure theory. Here an attempt is made to generalize to fuzzy sets certain positions of the axiomatic probability theory, utilizing the concept of fuzzy event, in a different way from Zadeh's extension [9].

The motivation for the development of fuzzy statistics is its philosophical and conceptual relation to subjective probability. In the subjective view, probability represents the degree of belief that a given person has in a given event on the basis of given evidence. This view (called also personalistic or judgmental probability) can be best described by an early article by James Bernoulli [1] who defines probability as degree of confidence in a proposition of whose truth we cannot be certain. His "degree of confidence" is identified with the probability of an event and depends on the knowledge that the individual has at his disposal. Thus, it varies from individual to individual and can be best described as the art of guessing (Ars conjectandi).

The difficulty in applying subjective probability stems from the vagueness associated with judgements made via subjective analysis. The postulates of subjective probability cannot be applied, as is widely recognized, to many interesting and useful theories of modern science which are inexact. Subjective probability can be regarded as a personal way of treating objective views in that they are concerned with individual judgment. These judgment, however, are not additive since human behavior is often contradictory to the assumption of subjective probability, that an individual is using additive measures in his criteria for evaluation [6]-[7].

We shall remove the restrictive device known as additivity and formulate the basic structure of a fuzzy statistic to be applied as our analytic model of investigation of imprecise data.

2. Compatibility and Fuzzy Expectation

In the classical approach a probability system is a triple (Ω, S, P), where Ω is an arbitrary set (the sample space which includes all possible outcomes), S is a set of events, and P is a real valued function defined for each $A \in S$ such that:

1) $0 \leq P(A) \leq 1 \quad \forall A \in S$

2) $P(\Omega) = 1$

3) If $A_1, A_2, \ldots$ is any sequence of pairwise disjoint sets in S, then

$$P(\bigcup_n A_n) = \sum_n P(A_n).$$

A function P satisfying the above three rules is called a probability measure [2] and the elements of S are referred to as events. Property (3) is known as countable additivity. First, an event can be inexact a fuzzy event in the sense that it belongs to a fuzzy set. Secondly, even if A is a non-fuzzy event (well-defined), P(A) may be ill-defined (e.g., vague evaluations of exact events, "soft" predictions).

At this point we would like to discuss the exhaustive power of the probability system. Clearly, for finite spaces, P(A) of any event A is the sum of the probabilities of all sample points in it. We have no argument about the probability measure itself or about the calculus by which mathematical probabilities may be manipulated. However, our alternative measure and calculus will eventually eliminate the exhaustivity involved in the classical structure, which restricts every sample point into a well defined set.

Interestingly enough, we are not the only ones who do not agree with classical probability. There have been, and continue to be, divergent opinions among theorists about the nature of the relations between the mathematics of probability and the events to which it is applied. There are, in fact, so many divergent views, special cases, and subtle philosophical questions that a brief summary must neccessarily ignore most of the fine points.

The frequency view most often used is based on an experiment in which trials are made and a record is kept of proportion of trials whose outcome is the desired one to the total number of trials. The probability of the event is taken to be the limiting value of the proportion as the number of trials increases without bound. If z is one of a set of exclusive and exhaustive outcomes of trials and n_z is the number of times it has occurred in n trials, the probability of z is

$$p(z) \triangleq \lim_{n\to\infty} \frac{n_z}{n}$$

if the limit exists. If it does not, p(z) is undefined. It is never possible to assign an exact probability on this basis unless the set of possible trials is exhausted, n being finite, since we can take only a finite amount of data. But we can hypothesize until further evidence suggests otherwise. The principal difficulty with the frequency view of probability is that it makes no sense for situations that happen only once, since it defines probability in terms of collections or sequences of events. From this viewpoint, the probability of a unique event, such as heads on a particular toss of a coin, must either be undefined or else be zero or one, depending on which way the coin falls. Similarly a pure frequency view must reject a priori probabilities which are not based on some number of trials but on collateral evidence, such as the number of sides of a die.

Still another view has been advanced that probability is a measure of the degree to which one statement is confirmed by other statements. This is called logical probability and is defined even if the statements themselves are false. It refers to relations among the statements rather than to relations among events. The logical view is plausible when situations can be reduced to a set of cases to which the evidence points equally.

The emerging consensus among decision theorists is a view of probability that frankly admits a subjective component. It takes into account that there is an element of human judgment even in the seemingly most objective procedures for determining quantitative probabilities, and it does not require that there be only one correct value unless the evidence logically entails it. The essence of this subjective or personal view is that probability is intimately related to human decision making, reflecting a person's degree of belief that the event in question will actually occur. Degree of belief, in this context, is interpreted as the extent to which the belief would contribute to a disposition to act rather than as an intensity of feeling.

A subjective element can be seen in the frequency view in the necessity to hypothesize a limiting relative frequency and to reasess the probability if the evidence indicates it. This amounts to a personal decision. When should we reassess the probability? It is assumed that when evidence is available it will be used to update probabilities, and, in essence, quantify the subjective evaluation involved. In order to do so we use the term "grade of membership" as the implication of the quantity which depends on a parameter involving human subjective evaluation.

It is interesting to note that Sugeno and Terano [6]-[7], have used fuzzy integrals over the range [0,1] to express their subjective evaluation of fuzzy objects. They use this concept in order to achieve identification of human characteristics as well as to apply it to macroscopic optimization using conditional models and to the evaluation of fuzzy objects.

Definition 2.1: Let B be a Borel field (σ-algebra) of subsets of a sample space Ω. A set function $\mu(\cdot)$ defined on B is called a fuzzy measure if it has the following properties:

1) $\mu(\Phi) = 0$ (Φ is the empty set)

2) $\mu(\Omega) = 1$

3) If $\alpha, \beta \in B$ with $\alpha \subset \beta$, then $\mu(\alpha) \leq \mu(\beta)$

4) If $\{\alpha_j | 1 \leq j < \infty\}$ is a monotone sequence,

   then $\lim_{j\to\infty}[\mu(\alpha_j)] = \mu[\lim_{j\to\infty}(\alpha_j)]$.

Clearly Φ, $\Omega \in B$; also if $\alpha_j \in B$ and $\{\alpha_j | 1 \leq j < \infty\}$ is a monotonic sequence then $\lim_{j\to\infty}(\alpha_j) \in B$. In the above definition (1) and (2) mean that the fuzzy measure is bounded and non-negative, (3) means monotonicity (in a similar way to countable additive measures used in probability), and (4) means continuity. It should be noted that if Ω is a finite set, then the continuity requirement can be deleted.

Definition 2.2: (Ω,B,μ) is a fuzzy measure space; its analog in probability is (Ω,S,P). $\mu(\cdot)$ is the fuzzy measure of (Ω,B).

One particular example of a fuzzy measure is P, the probability measure.

Let χ be the membership function of fuzzy set α. Namely, $\chi(x)$ is the grade of membership of x in α. Let $\chi : \Omega \to [0,1]$ and $\xi_T = \{x | \chi(x) \geq T\}$. The function χ is called a B-measurable function if $\xi_T \in B$ $\forall T \in [0,1]$.

Note: Using the Borel field as the domain of the fuzzy measures enables us to extend the notion of measurability of functions from the theory of Lebesgue measures.

Definition 2.3: Let χ be a B-measurable function. The fuzzy expected value (FEV) of χ over a fuzzy set α with respect to the measure $\mu(\cdot)$ is defined as

$$\sup_{T\varepsilon[0,1]} \{\min[T, \mu(\xi_T)]\}$$

where $\xi_T \subseteq \alpha$

Note: The B-measurable function χ is called the compatibility function. In the case of linguistic variables [10] the numerical variables whose values constitute what may be called the base variable, describe a fuzzy restriction on the meaning of the linguistic variable. This fuzzy restriction (which is clearly subjective) on the values of the base variable is characterized by the compatibility function χ which associates with each value of the base variable a number in the interval [0,1], representing its compatibility with the fuzzy restriction.

If the composition established in Definition 2.3 under the operation "Max-min" seems like a pessimistic evaluation of the FEV, does a similar composition using the operation "min-Max" act as an optimistic model?

The answer to the above question lies in the following theorem.

Theorem 2.1: Let $\chi:\Omega\to[0,1]$ and

$$\xi_T = \{x|\chi(x) \geq T\} \varepsilon B.$$

Then

$$\sup_{T \varepsilon [0,1]} \{\min[T, \mu(\xi_T)]\} = \inf_{T \varepsilon [0,1]} \{\mathrm{Max}[T, \mu(\xi_T)]\},$$

where μ is a fuzzy measure.

Proof: Let $\sup_{T \varepsilon [0,1]} \{\min[T, \mu(\xi_T)]\} = H.$

Then

$$H = \mathrm{Max}\, \{\sup_{T \varepsilon [0,H]} \{\min[T, \mu(\xi_T)]\}, \sup_{T \varepsilon [H,1]} \{\min[T, \mu(\xi_T)]\}\}.$$

Since $\mu(\xi_T)$ increases as T decreases we have

$$\sup_{T \varepsilon [0,H]} \{\min[T, \mu(\xi_T))]\} \leq H$$

and $$\sup_{T \varepsilon [H,1]} \{\min[T, \mu(\xi_T))]\} \leq \sup_{T \varepsilon [0,H]} \{\min[T, \mu(\xi_T)]\}.$$

Thus $$\sup_{T \varepsilon [0,H]} \min [T, \mu(\xi_T)]\} = H$$

and $$\mu(\xi_H) \geq H, \mu(\xi_H+) = \mu\{x|\chi(x) > H\} \leq H.$$

Assume

$$\operatorname*{Inf}_{T \in [0,1]} \{\mathrm{Max}[T, \mu(\xi_T)]\} = R$$

such that $R \neq H$.

Then $$R = \min \{\operatorname*{Inf}_{T \in [0,R]} \{\mathrm{Max}[T, \mu(\xi_T)]\}, \operatorname*{Inf}_{T \in [R,1]} \{\mathrm{Max}[T,\mu(\xi_T)]\}\} ,$$

since $$\operatorname*{Inf}_{T \in [R,1]} \{\mathrm{Max}[T, \mu(\xi_T)]\} \geq R ;$$

clearly $$\operatorname*{Inf}_{T \in [0,R]} \{\mathrm{Max}[T, \mu(\xi_T)]\} \geq \operatorname*{Inf}_{T \in [R,1]} \{\mathrm{Max}[T, \mu(\xi_T)]\} ,$$

and thus we define $$\operatorname*{Inf}_{T \in [R,1]} \{\mathrm{Max}[T, \mu(\xi_T)]\} = R.$$

Since the turning point between $\mu(\xi_T)$ and T is at H where

$$\mu(\xi_H) \geq H \quad \text{and} \quad \mu(\xi_H+) \leq H$$

it is quite obvious that R and H must be identical.

Namely

$$\operatorname*{Sup}_{T \in [0,1]} \{\min[T, \mu(\xi_T)]\} = \operatorname*{Inf}_{T \in [0,1]} \{\mathrm{Max}[T, \mu(\xi_T)]\} = H.$$

Q.E.D.

Theorem 2.1 can also be derived via a set of interesting lemmas. We feel that the following is not just a repetition of a proof but that the lemmas give more insight to the properties of the FEV.

Lemma 2.1: For every $T_1, T_2 \in [0,1]$

$$\min[T_1,\mu(\xi_{T_1})] \leq \mathrm{Max}[T_2,\mu(\xi_{T_2})].$$

Proof: Case 1: $T_1 \leq T_2$. In this case

$$\min[T_1,\mu(\xi_{T_1})] \leq T_1 \leq T_2 \leq \mathrm{Max}[T_2,\mu(\xi_{T_2})].$$

Case 2: $T_2 < T_1$. In this case $\mu(\xi_{T_1}) \leq \mu(\xi_{T_2})$.

Hence

$$\min[T_1,\mu(\xi_{T_1})] \leq \mu(\xi_{T_1}) \leq \mu(\xi_{T_2}) \leq \mathrm{Max}[T_2,\mu(\xi_{T_2})]$$

from which the lemma follows. Q.E.D.

Lemma 2.2: $$\sup_{T \in [0,1]} \{\min T,\mu(\xi_T)]\} \leq$$

$$\leq \inf_{T \in [0,1]} \{\max T,\mu(\xi_T)]\}.$$

Proof: By Lemma 2.1, every member of $\{\min[T,\mu(\xi_T)], T \in [0,1]\}$ is less than every member of $\{\max[T,\mu(\xi_T)], T \in [0,1]\}$. The lemma follows.

Lemma 2.3: There is no real number s such that

$$\sup_{T \in [0,1]} \{\min[T,\mu(\xi_T)]\} < s < \inf_{T \in [0,1]} \{\max[T,\mu(\xi_T)]\}.$$

Proof: Suppose such an s exists. We distinguish two cases.

Case 1: $\mu(\xi_s) \leq s$

Case 2: $\mu(\xi_s) > s$

Each case leads to a contradiction.

Case 1:

$$s < \inf_{T \in [0,1]} \{\max[T,\mu(\xi_T)]\} \leq \max[s,\mu(\xi_s)] = s.$$

Case 2:

$$s = \min[s,\mu(\xi_s)] \leq \sup_{T \in [0,1]} \{\min[T,\mu(\xi_T)]\} < s.$$

Since both cases lead to a contradiction, the lemma follows. Q.E.D.

Lemma 2.4: It is not true that

$$\sup_{T \in [0,1]} \{\min[T,\mu(\xi_T)]\} < \inf_{T \in [0,1]} \{\max[T,\mu(\xi_T)]\}.$$

Proof: If this inequality held, there would be a real number between the left and the right hand members, contradicting Lemma 2.3. Q.E.D.

Hence, Theorem 2.1 follows from Lemmas 2.2 and 2.4.

Since the FEV is defined with no use of addition; subtraction, multiplication, or division, it is not to be expected that it will satisfy identities involving such operations nor operators derived from them such as derivatives. Thus the FEV is not additive or linear. But it does satisfy powerful identities involving arbitrary monotone functions.

Clearly Definition 2.3 can be extended to any real interval $[a,b]$ by extending T and μ to this interval, under the same transformation that χ undergoes. Hence, we can define the FEV of an extended grade of membership function $\eta \varepsilon [a,b]$, as

$$\mathrm{FEV}(\eta) \triangleq \sup_{T^* \varepsilon [0,1]} \{\min[T^*, \mu^*(\xi_{T^*})]\}$$

where $\xi_{T^*} = \{x | \eta(x) \geq T^*\}$ and $\mu^* \varepsilon [a,b]$.

In particular, if C is a constant and we want to evaluate $\mathrm{FEV}(\chi + C)$ from some compatibility function χ, then the constant C acts as a factor to scale the compatibility curve and thus to scale the FEV and to evaluate $C + \mathrm{FEV}(\chi)$. Thus the value of $\mathrm{FEV}(\chi + C)$ will be in the interval $[C, C + 1]$ if $\chi \varepsilon [0,1]$. Similarly, if C is a positive constant, the $\mathrm{FEV}(C \cdot \chi)$ is basically a normalization of the compatibility curve χ by a factor C which acts as a normalization factor on the entire $\mathrm{FEV}(\chi)$. Thus we have the following results.

Proposition 2.1: Let C be a constant and $\chi \to [0,1]$. Then $\mathrm{FEV}(\chi + C) = \mathrm{FEV}(\chi) + C$

Proposition 2.2: Let $C \geq 0$ be a constant and $\chi \to [0,1]$. Then $\mathrm{FEV}(C \cdot \chi) = C \cdot \mathrm{FEV}(\chi)$.

These last results provide some insight into the nature of the FEV as a tool for partitioning of a population into "above FEV" and "below FEV" which is independent of the compatibility curves, up to a monotone function. Thus the FEV is more of a typical value of the population.

In the following section we shall illustrate and obtain the relations between the FEV and the standard probabilistic expectation on one hand, and between the FEV and measures of central tendency.

3. FEV and Measures of Central Tendency

An average is a value which is typical of a set of data. Since such typical values tend to lie centrally within a set of data arranged according to magnitude, averages are also known as measures of central tendency. Several types of averages can be defined, the most common being the arithmetic mean, the median, the mode, the geometric mean and the harmonic mean.

Before discussing the relations of the FEV to the mean and to the median we shall illustrate the use of the concept via the following example.

Example 3.1: Using the base variable AGE, we have x people of age α and y people of age β where α and β correspond to the compatible values α_T and β_T, respectively. Assume

$$0 \leq \alpha_T \leq \beta_T \leq 1$$

and $\mu(\xi_T) = \mu\{w | \chi(w) \geq T\}$. In this example $\mu(\cdot)$ acts as a standard probabilistic measure.

Clearly

$$\beta_T < T \leq 1 \rightarrow \mu(\xi_T) = 0$$

$$\alpha_T < T \leq \beta_T \rightarrow \mu(\xi_T) = \frac{y}{x+y}$$

$$0 \leq T \leq \alpha_T \rightarrow \mu(\xi_T) = 1$$

Thus

$$\text{FEV} = \begin{cases} \alpha_T & \text{if} \quad \alpha_T \geq \frac{y}{x+y} \\ \beta_T & \text{if} \quad \beta_T \leq \frac{y}{x+y} \\ \frac{y}{x+y} & \text{if} \quad \alpha_T \leq \frac{y}{x+y} \leq \beta_T \end{cases}$$

In what follows we shall construct a combinatorial scheme to generalize the above example.

Assume a finite set of data points where there are $n + 1$ distinct levels of compatibility such that

$$0 \leq \alpha_1 \leq \alpha_2 \leq \ldots \leq \alpha_{n+1} \leq 1 ,$$

which imply n distinct levels of fuzzy measure $\mu(\xi_T)$, excluding 0 and 1. Clearly the two sets representing

$$\{\alpha_i\}_{i=1}^{n+1} \quad \text{and} \quad \{\mu_j(\xi_T)\}_{j=1}^{n}$$

are in increasing order and we trivially exclude any permutation in each of the sets. Thus the set representing the union of these two sets has $(2n + 1)$ elements; in order to find all possible arrangements of these elements we can view the problem of finding the number of arrangements as follows:

Find γ , the number of n-arrangements of $(2n + 1)$ objects, of which exactly $(n + 1)$ are alike of one kind and exactly n are alike of the second kind.

We present now our main claim.

<u>Theorem 3.1</u>: The median (middle value) of the set of $2n + 1$ numbers, representing $\{\alpha_i\}_{i=1}^{n+1}$ and $\{\mu_j(\xi_T)\}_{j=1}^{n}$ as described above, where $0 \leq \alpha_1 \leq \alpha_2 \leq \ldots \alpha_{n+1} \leq 1$ for a finite n , arranged in order of magnitude (i.e., in an array left to right), is the FEV.

<u>Proof</u>: In order to find the FEV we have to compare α_i's , $i > 1$, with the proper $\mu_i(\xi_T)$, take the minimum of these two and then the Maximum over all minimum values. Clearly $\min(\alpha_1,1) = \alpha_1$. It should be noted that α_1 cannot appear to the right of the middle of the array since it has to be followed by at least all the

members of the sequence $\{\alpha_k\}_{k=2}^{n+1}$. It is important to note that the values of the two sequences are independent but not their place in the array. This is due to the fact that if $\alpha_j \geq \alpha_i$ then $\mu_j(\xi_T) \leq \mu_i(\xi_T)$. Because of that and because the array is arranged in an increasing order, the first part of the array, including n numbers (n + 1 left to right and excluding α_1) is compared with the second part of the array, including the last n numbers (right to left). Since we take the minimum of every comparison, the results must lie in the first n + 1 numbers of the array. Out of these n + 1 numbers we are interested in the maximum, which is obviously the number in the (n + 1)<u>th</u> place since the array is increasing in order. This number is the <u>median</u> of the array. Q.E.D.

<u>Example 3.2</u>: Let A be our tested population such that

x people are of age α (compatible value α_T) ,

y people are of age β (compatible value β_T) ,

and z people are of age γ (compatible value γ_T)

where $x + y + z = A$, $1 \geq \gamma_T \geq \beta_T \geq \alpha_T \geq 0$, and $\mu(\xi_T) = \mu\{w|\chi(w) \geq T\} = \frac{|\xi_T|}{A}$

Since n = 2 there are 10 different arrangements as given below; the FEV's can easily be verified graphically and thus will be omitted here.

1) $\alpha_T, \beta_T, \gamma_T, \frac{z}{A}, \frac{z+y}{A}$; FEV = γ_T.

2) $\frac{z}{A}, \frac{z+y}{A}, \alpha_T, \beta_T, \gamma_T$; FEV = α_T.

3) $\alpha_T, \frac{z}{A}, \beta_T, \gamma_T, \frac{z+y}{A}$;

4) $\alpha_T, \frac{z}{A}, \beta_T, \frac{z+y}{A}, \gamma_T$;

5) $\frac{z}{A}, \alpha_T, \beta_T, \frac{z+y}{A}, \gamma_T$;

6) $\frac{z}{A}, \alpha_T, \beta_T, \gamma_T, \frac{z+y}{A}$;

(3)–(6): FEV = β_T.

7) $\alpha_T, \beta_T, \frac{z}{A}, \frac{z+y}{A}, \gamma_T$;

8) $\alpha_T, \beta_T, \frac{z}{A}, \gamma_T, \frac{z+y}{A}$;

(7)–(8): FEV = $\frac{z}{A}$

9) $\alpha_T, \frac{z}{A}, \frac{z+y}{A}, \beta_T, \gamma_T$;

10) $\frac{z}{A}, \alpha_T, \frac{z+y}{A}, \beta_T, \gamma_T$;

(9)–(10): FEV = $\frac{z+y}{A}$.

<u>Example 3.3</u>: Using the base variable "Hourly Wages" let us assume a given population and given a subjective compatibility curve such that

1 person is making \$3.00 $\rightarrow \alpha_1 = 0.40$

3 persons are making \$4.00 $\rightarrow \alpha_2 = 0.50$

4 persons are making \$4.20 → $\alpha_3 = 0.55$

2 persons are making \$4.50 → $\alpha_4 = 0.60$

2 persons are making \$10.00 → $\alpha_5 = 1.00$

Arranging the sequences $\{\alpha_i\}_{i=1}^{5}$ and $\{\mu_j(\xi_T)\}_{j=1}^{4}$ where $\mu(\xi_T) = \mu\{w|\chi(w) \geq T\} = \frac{|\xi_T|}{12}$ in an array we get the values

$$\left\{\frac{1}{6}, \frac{1}{3}, \frac{2}{5}, \frac{1}{2}, \frac{11}{20}, \frac{3}{5}, \frac{2}{3}, \frac{11}{12}, 1\right\}.$$

Both the FEV and the median are $\frac{11}{20}$. The probabilistic expected value (mean) is 0.61 (using the same compatibility curve on [0,1] interval).

In this example the FEV gives a better indication of the average hourly wage than the mean, by not being affected by the extreme value of \$10.00.

It should be noted that a different population distribution might give a FEV which is different from the median.

Example 3.4: The hourly wages of five people are \$2.20, \$2.50, \$2.70, \$3.50, and \$10.00; using the same compatibility curve of Example 3.3, these wages give the following sequence of α_i's, $1 \leq i \leq 5$:

$$0.25,\ 0.3,\ 0.35,\ 0.45,\ 1.$$

Namely, the median is 0.35 (\$2.70), the FEV is 0.4 (\$3.00), and the mean is 0.47 (\$3.80).

Since the median of a data set is quantile that, like the expected value, acts as a "center" for a given distribution, we claim that the first $\mu_j(\xi_T)$ in the array (from left to right), which is greater or equal to 1/2 indicates the respective α_j as the median of the data set. If no such $\mu_j(\xi_T)$ exists then α_1 is the median. We shall denote the above specified $\mu_j(\xi_T)$ and α_j as μ_{median} and α_{median}, respectively. Clearly, if $\mu_{median} \geq$ FEV then $\alpha_{median} \leq$ FEV and if $\mu_{median} <$ FEV then $\alpha_{median} >$ FEV; since $\mu_{median} \geq \frac{1}{2}$, the mean as obtained is closer to the FEV than to the median. Namely

$$|\text{ Mean-FEV }| \leq |\text{ Mean-Median }|,$$

or using the empirical result for unimodal frequency curves which are moderately skewed (asymmetrical), we have

$$|\text{ Mean-FEV }| \leq \frac{1}{3}|\text{ Mean-Mode }|.$$

Similar results can be derived for grouped data. It should be pointed out that different compatibility curves with the same fuzzy measure and the same frequency distribution might yield a different FEV. This not so when standard median is used, since the FEV is also a function of the subjective evaluation.

We will now formulate an easier way to obtain the value of the FEV.

Definition 3.1: A value $T \in [0,1]$ is a subtypical value of a function ξ if $T \leq \mu(\xi_T)$, supertypical if $T \geq \mu(\xi_T)$, and typical if $T = \mu(\xi_T)$. Note that a typical value is both subtypical and supertypical.

Theorem 3.2: If a number T_1 is a subtypical value of two functions, ξ and η , and if for every $T \geq T_1$ $\xi_T = \eta_T$, then $FEV(\xi) = FEV(\eta)$.

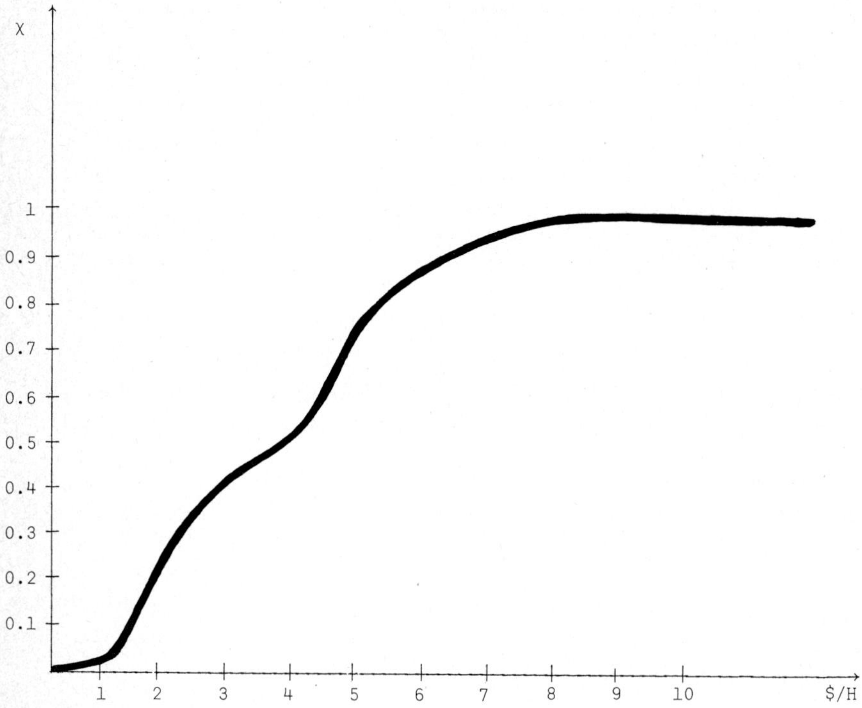

FIGURE 3.1 Compatibility curve for hourly wages

Proof: For $T < T_1$ we have

$$T < T_1 \le \mu(\xi_{T_1}) \le \mu(\xi_T)$$

and

$$T < T_1 \le \mu(\eta_{T_1}) \le \mu(\eta_T).$$

Thus

$$\min[T,\mu(\xi_T)] = T$$

and

$$\min[T,\mu(\eta_T)] = T$$

so that

$$\min[T,\mu(\xi_T)] = \min[T,\mu(\eta_T)].$$

For $T \ge T_1$ it is given that $\xi_T = \eta_T$ and therefore

$$\mu(\xi_T) = \mu(\eta_T)$$

and

$$\min[T,\mu(\xi_T)] = \min[T,\mu(\eta_T)].$$

Hence

$$\underset{T\varepsilon[0,1]}{\mathrm{Sup}}\{\min[T,\mu(\xi_T)]\} = \underset{T\varepsilon[0,1]}{\mathrm{Sup}}\{\min[T,\mu(\eta_T)]\}$$

and the result follows. Q.E.D.

Theorem 3.2 implies that values of a function that are below a subtypical value may be ignored in calculating the FEV. Similarly, we show in Theorem 3.3, which is the dual of Theorem 3.2, that values of a function that are above a supertypical value may be ignored in calculating the FEV.

Theorem 3.3: If a number T_2 is a supertypical value of two functions, ξ and η, and if for every $T \le T_2$ $\xi_T = \eta_T$, then $FEV(\eta) = FEV(\xi)$.

Proof: For $T > T_2$ we have

$$T > T_2 \geq \mu(\xi_{T_2}) \geq \mu(\xi_T)$$

and

$$T > T_2 \geq \mu(\eta_{T_2}) \geq \mu(\eta_T).$$

Thus

$$\text{Max}[T,\mu(\xi_T)] = \text{Max}[T,\mu(\eta_T)].$$

For $T \leq T_2$ it is given that $\xi_T = \eta_T$ and therefore

$$\mu(\xi_T) = \mu(\eta_T)$$

and

$$\text{Max}[T,\mu(\xi_T)] = \text{Max}[T,\mu(\eta_T)].$$

Hence

$$\underset{T\varepsilon[0,1]}{\text{Inf}}\{\text{Max}[T,\mu(\xi_T)]\} = \underset{T\varepsilon[0.1]}{\text{Inf}}\{\text{Max}[T,\mu(\eta_T)]\}$$

and by use of Theorem 2.1 the result follows. Q.E.D.

Now let the domain of a function χ be the union of a finite number of subspaces $K = \{s_1,s_2,\ldots,s_n\}$ such that χ: $K \to [0.1]$. If we consider a fuzzy measure space $(K, 2^K, \mu)$ we can write the FEV of χ as

$$\text{FEV}(\chi) = \underset{K'\,\varepsilon\,2^K}{\text{Max}}\{\text{min}[\underset{s\,\varepsilon\,K'}{\text{min}}\,\chi(s),\ \mu(K')]\}.$$

We assume that $\chi(s_i) \leq \chi(s_{i+1})$ for $1 \leq i \leq n-1$. If this is not the case then rearrangement of $\chi(s_i)$ is necessary. Then the following holds.

<u>Theorem 3.4</u>: The FEV of χ in $(K, 2^k, \mu)$ can be written as

$$\underset{i}{\text{Max}}\{\text{min}[\chi(s_i),\mu(K_i)]\}$$

where $K_i = \{s_i,s_{i+1},\ldots,s_n\}$, $1 \leq i \leq n$.

Proof: Let $\chi(s_i) = \min_{s\epsilon K'} \chi(s)$. Then

$$\operatorname*{Max}_{K'\epsilon 2^K}\{\min[\min_{s\epsilon K'} \chi(s),\mu(K')]\} \le \operatorname*{Max}_i\{\min[\chi(s_i),\mu(K_i)]\},$$

and since $\{K_i | 1 \le i \le n < 2^K\}$, the reverse inequality holds too. The equality follows. Q.E.D.

Even though the power set 2^K has 2^n members, Theorem 3.4 calls for a monotone sequence of subsets of K such that $K_1 > K_2 > \ldots > K_n$ so that we can find the FEV by calculating $\min[\chi(s_i),\mu(K_i)]$ at n points at most.

Theorem 3.5: The FEV of χ in $(K, 2^K, \mu)$ can be written as

$$\min[\chi(s_j),\mu(K_j)] \quad \text{iff} \quad \chi(s_{j-1}) \le \mu(K_j) \le \chi(s_j)$$

or

$$\mu(K_j) > \chi(s_j) \ge \mu(K_{j+1}).$$

Proof: a) necessity: Clearly since $\mu(K_i)$ is a monotonic decreasing function and $\chi(s_i)$ monotonic increasing, the function

$$\min[\chi(s_i), \mu(K_i)]$$

has a single peak for some i. Thus it is necessary that

$$\min[\chi(s_{j-1}),\mu(K_{j-1})] \le \min[\chi(s_j),\mu(K_j)] \ge \min[\chi(s_{j+1}),\mu(K_{j+1})].$$

b) sufficiency: Assume $\chi(s_j) \ge \mu(K_j)$. Then we have

$$\chi(s_{j-1}) \le \mu(K_i) \le \chi(s_j)$$

from the above inequality.

If we assume

$$\chi(s_j) < \mu(K_j)$$

then

$$\mu(K_j) > \chi(s_j) > \mu(K_{j+1}).$$

Hence the proof is complete. Q.E.D.

Thus there is no need to evaluate $\min[\chi(s_j),\mu(K_j)]$ for all i but only at point j fulfilling the requirements of Theorem 3.5. Moreover, since χ is a known function there is a need to evaluate $\mu(K_i)$ for three different points only.

Clearly this technique, as far as complexity of computations is concerned, has advantages over classical evaluations of probabilistic expected values.

Another relation between the FEV, when $\mu(\cdot)$ is taken to be additive (probability measure), and the probabilistic expected value can be established via the following theorem.

Theorem 3.6: Let (Ω, B, p) be a probability space and let $\chi:\Omega\to[0.1]$ be a B-measurable function. Then

$$|\Delta| = |\{\int_\Omega \chi(x)dp - \operatorname*{Sup}_{T\varepsilon[0,1]} \{\min[T,p(\xi_T)\}| \le 1/4$$

where $\xi_T = \{x|\chi(x) \ge T\}$

Proof: Let $\hat{\chi}(x) \triangleq \operatorname{Sup}[\chi(x)]$ and $\underset{\wedge}{\chi}(x) \triangleq \inf[\chi(x)]$. Clearly $\Omega-\xi_T = \{x|\chi(x) < T\}$ and therefore

$$\int_\Omega \chi(x)dp = \int_{\Omega-\xi_T} \chi(x)dp + \int_{\xi_T} \chi(x)dp$$

$$\le \int_{\Omega-\xi_T} T\,dp + \int_{\xi_T} \chi(x)dp$$

$$\le T.p(\Omega-\xi_T) + \hat{\chi}(x).p(\xi_T)$$

$$= T.[1-p(\xi_T)] + \hat{\chi}(x)\ p(\xi_T)$$

$$= T + p(\xi_T)\cdot[\hat{\chi}(x) - T].$$

Similarly,

$$\int_\Omega \chi(x)dp \ge \int_{\xi_T} T\,dp + \int_{\Omega-\xi_T} \chi(x)dp$$

$$\ge T.p(\xi_T) + \underset{\wedge}{\chi}(x).p(\Omega-\xi_T)$$

$$= T.p(\xi_T) + \underset{\wedge}{\chi}(x).[1-p(\xi_T)]$$

$$= \underset{\wedge}{\chi}(x) + p(\xi_T)\cdot[T - \underset{\wedge}{\chi}(x)].$$

Define the quatity H by

$$H \triangleq \operatorname*{Sup}_{T\varepsilon[0,1]} \{\min[T,p(\xi_T)]\}$$

Then

$$H = \operatorname{Max}\{ \sup_{T\varepsilon[0,H]} \{\min[T,p(\xi_T)]\}, \sup_{T\varepsilon[H,1]} \{\min[T,p(\xi_T)]\}\};$$

clearly

$$\sup_{T\varepsilon[0,H]} \{\min[T,p(\xi_T)]\} \le H$$

since $p(\xi_T)$ increases as T decreases, and therefore it is trivial that

$$\sup_{T\varepsilon[H,1]} \{\min[T,p(\xi_T)]\} \le \sup_{T\varepsilon[0,H]} \{\min[T,p(\xi_T)]\}$$

and thus

$$\sup_{T\varepsilon[0,H]} \{\min[T,p(\xi_T)]\} = H ,$$

and

$$p(\xi_H) \ge H.$$

Similarly

$$p(\xi_H+) \le H \quad \text{since}$$

$$\xi_H+ = \{x|\chi(x) > H\} \quad \text{and}$$

$$\sup_{T\varepsilon[H,1]} \{\min[T,p(\xi_T)]\} \le \sup_{T\varepsilon[0,H]} \{\min[T,p(\xi_T)]\}.$$

Hence,

$$\lim_{T\to H^+} \left| \int_\Omega \chi(x)dp - \sup_T\{\min[T,p(\xi_T)]\} \right| \le$$

$$\le H^+ + p(\xi_H+)\cdot[\hat{\chi}(x) - H^+] - H \le$$

$$\le p(\xi_H+)\cdot[\hat{\chi}(x) - H^+]$$

and

$$\lim_{T\to H^-} \left| \int_\Omega \chi(x)dp - \sup_T\{\min[T,p(\xi_T)]\} \right| \ge$$

$$\ge \underset{\wedge}{\chi}(x) + p(\xi_H-)\cdot[H^- - \underset{\wedge}{\chi}(x)] - H \ge$$

$$\ge [1 - p(\xi_H-)]\cdot[\underset{\wedge}{\chi}(x) - H^-].$$

Hence,

$$[1-p(\xi_H-)]\cdot[\underset{\wedge}{\chi}(x) - H^-] \le \Delta \le p(\xi_H+)\cdot[\hat{\chi}(x) - H^+].$$

Clearly

$$\hat{\chi}(x) = 1 \quad \text{and} \quad \underset{\wedge}{\chi}(x) = 0.$$

Thus

$$H^-\cdot[p(\xi_H-) - 1] \le \Delta \le (1-H^+)\cdot p(\xi_H+)$$

and hence in the limit

$$|\Delta| \le H(1-H).$$

$$\frac{d[H(1 - H)]}{dH} = 1-2H = 0 \quad \text{implies} \quad H = 1/2 ,$$

and $|\Delta| \le 1/4$. Q.E.D.

This means that by a subjective evaluation of grades of membership via the FEV, using a probabilistic measure, we obtain an average value different by no more than a 1/4 from the classical probabilistic expected value.

4. Conclusions and Remarks

The problem of selecting a proper measure is related to selecting a criterion for optimality condition. Clearly a function of the form $\mu(\cdot)|\mu(A) = \sup_{x\varepsilon A}[\chi(x)]$ satisfies the conditions of a fuzzy measure.

To summarize this point, we are confronted with an identification problem; namely, we have to find an optimal set function $\mu(\cdot)$, among the set of all set functions satisfying the fuzzy measures $\hat{\mu}(\cdot)$ such that a certain optimality criterion is satisfied, e.g., find $\mu(\cdot)$ such that $\varepsilon_{op} = \min_{\mu\varepsilon\hat{\mu}(\cdot)} \varepsilon$. It should be pointed out, however, that even in the cases where the fuzzy measure used was a probability measure, the concept of FEV had advantages over the classical methods of "average evaluation".

As a last remark, in this paper on fuzzy statistics, it is interesting to note that regarding the logic of uncertainty, a formal propositional calculus has been developed by Gaines [3], in which statements consisting of truth-values in the continuous interval [0,1] have been assigned to propositions in a lattice. These truth-values form a continuous, order-preserving valuation of the lattice giving a Basic Probability Logic (BPL) which satisfies the usual definitions of a probability logic or probability over a language except for the Law of the Excluded Middle (LEM).

A common semantics for the BPL and all its derivatives can be given in terms of the binary responses of a population. Hence fuzzy logic and probability logic have common formal and semantic foundations. Defining the population further brings out the distinctions normally made between "degree of membership" and "probability", not now as absolute divisions but rather, for example, as the distinction between "physical-frequentist" and "subjective" interpretations of probability. For example the "population" may be one of physical events, or of people giving opinions, or of decision-making elements (neurons).

To quote Gaines [3]:

> ...it is suggested that regardless of its correctness in particular circumstances the advantages of fuzzy logic may be seen in its strong truth-functionality (TF). It is unique amongst the various logics of uncertainty in not requiring memory of the structures of two propositions (in terms of the lattice generators) when computing connectives involving them. This is not only of practical importance but may also explain the role of fuzzy logic in modelling human reasoning where short-term memory capabilities are notoriously weak.

In conclusion, Gaines [3] shows that a non-functional BPL provides a formal foundation for a general logic of uncertainty encompassing both fuzzy probability logics. LEM is consistent with weaker forms of TF leading to partially fuzzy logics with Max-min connectives between primitive propositions, logics of statistical independence, mutual exclusion, etc. Gaines' model also clarifies the distinction between fuzzy "degree of membership" and conventional "probability", showing it to be one of detailed semantic interpretation rather than one of logic or basic semantics.

Thus, the wider framework for logics of uncertainty described by Gaines [3] establishes a close link between fuzzy logic and probability theory, to the mutual advantage of both fields. It also makes clear the unique computational advantage of fuzzy logic derived from its strong truth-functionality, which is so much emphasized.

5. References

[1] Bernoulli, J. (1713), Ars Conjectandi, Basel.

[2] Breiman, L. (1968), Probability, Addison Wesley, Reading Maps.

[3] Gaines, B.R. (1975), "General fuzzy logics", EES-MMS-FUZZ-75, University of Essex, U.K.

[4] Kandel, A. (1978), "Fuzzy statistics and forecast evaluation", IEEE Trans. on Systems, Man and Cybernetics, SMC-8, No. 5, pp. 396-401.

[5] Kandel, A. and W.J. Byatt (1978), "Fuzzy sets, fuzzy algebra, and fuzzy statistics", Proceedings of the IEEE, 66, No. 12, pp. 1619-1639.

[6] Sugeno, M., and T. Terano (1974), "An approach to the identification of human characteristics by applying fuzzy integrals", Proc. 3rd IFAC Symp. on Identification and System Parameter Estimation, Hague.

[7] Sugeno, M., and T. Terano (1975), "Analytical representation of fuzzy systems", in Special Interest Discussion on Fuzzy Automata and Decision Processes, 6th IFAC World Congress, Boston, Mass.

[8] Zadeh, L.A. (1965), "Fuzzy Sets", Information and Control, 8, pp. 338-353.

[9] Zadeh, L.A. (1968), "Probability Measures of fuzzy events", J. Math. Anal. & App., 10, pp. 421-427.

[10] Zadeh, L.A. (1976), "A fuzzy-algorithmic approach to the definition of complex or imprecise concepts", in Bossel et.al., Systems Theory in the Social Sciences, Birkhauser Verlag, Basel, pp. 202-282.

[11] Zadeh, L.A. (1978), "Fuzzy sets as a basis for a theory of possibility", Fuzzy Sets and Systems, 1, 1, pp. 3-28.

ADVANCES IN FUZZY SET THEORY AND APPLICATIONS
M.M. Gupta, R.K. Ragade, R.R. Yager (editors)

EXTENDED FUZZY EXPRESSION OF PROBABILISTIC SETS

Kaoru HIROTA

Department of Information Science, Sagami Institute of Technology,
Tsujido-Nishikaigan, 1-1-25 Fujisawa-city, Kanagawa 251, Japan

The concept of probabilistic sets is proposed first in terms of a probabilistic expression. Another mutually equivalent expression, called an extended fuzzy expression of probabilistic sets, is also introduced. One interpretation is given theoretically on the relationship between fuzziness and probability.

1. INTRODUCTION

A concept of probabilistic sets has been proposed by using both probability and fuzzy theory (Hirota 1975). A probabilistic set A on a total space X is defined by a defining function $\mu_A(x,\omega)$,

$$\begin{aligned} \mu_A:\ & X\times\Omega \longrightarrow [0,1], \\ & \ \ \omega \qquad\qquad\ \ \omega \\ & (x,\omega) \longmapsto \mu_A(x,\omega) \end{aligned} \tag{1}$$

where $\mu_A(x,\cdot)$ is a measurable function on a parameter space (Ω,B,P). A pair $(\mu_A(x,\omega),P(\cdot))$ is called a probabilistic expression of the probabilistic set A. It is the main aim of this work to deal with the relationship between probability theory and fuzzy theory. Moment analysis shows that a countable family of functions $\{m_A{}^n(x)\}_{n=1}^{\infty}$ provides the same information as $(\mu_A(x,\omega),P(\cdot))$ under the following conditions;

1) $n \geqq m \ \rightarrow\ 1 \geqq m_A{}^m(x) \geqq m_A{}^n(x) \geqq 0,$ (2)

2) $\phi_A(x,t) \underset{\Delta}{=} \sum_{n=0}^{\infty} (i^n/n!)\cdot m_A{}^n(x)\cdot t^n \quad (m_A{}^o(x) \equiv 1),$ (3)

   is a positive definite function of $t\varepsilon R$ for each $x\varepsilon X$.

3) $\psi_A(x,t) \underset{\Delta}{=} \phi_A(x,-it),$ (4)

   is a monotonically non-decreasing function of t for each x.

A family $\{m_A{}^n(x)\}_{n=1}^{\infty}$ is called an extended fuzzy expression of the probabilistic set A; where $m_A{}^n(x)$ is called a n-th monitor of A, especially $m_A{}^1(x)$ is called a membership function of A and $v_A(x) \underset{\Delta}{=} m_A{}^2(x)-(m_A{}^1(x))^2$ $(\geqq 0)$ is called a vagueness function of A. It can be shown theoretically that both the membership function and the vagueness function express almost all informations of the probabilistic set A. It will also be clarified that the information given by probability is same as the information given by the membership function, the vagueness function and so on.

2. PROBABILISTIC EXPRESSION OF PROBABILISTIC SETS

2-1. DEFINITION OF PROBABILISTIC SETS

In decision making theory and in pattern recognition, there are a few real problems remaining unsolved, i.e. ambiguity of objects and subjectivity of observers. Lately these problems have become of general interests, and they have been studied by many researchers among whom the fuzzy concept by L.A.Zadeh (Zadeh 1965) is especially excellent. However, there are few carefully thought-out investigations by paying attention to the inherent and special characteristics of decision making and pattern recognition. Realizing this point early, we commenced a study of this subject and recently the investigation has been summarized as a concept of probabilistic sets (Hirota 1975).

A set of objects that we want to discuss will be called a *total space* and be denoted by

$$X = \{x\}. \tag{5}$$

A *fuzzy set* A on X, as is well known, is defined by a *membership function* m_A,

$$m_A: X \longrightarrow [0,1]. \tag{6}$$

This [0,1]-quantization is apparently a good idea, but the situation in general is so complicated that all the problems of ambiguity and subjectivity are not completely solved by this [0,1]-idea alone. The evaluation given by observers might not be determined uniquely in [0,1]-interval. Hence we shall introduce a probability space (Ω, B, P), called a *parameter space*, whose element represents a standard of judgement of observers. It is assumed that if a standard $\omega(\varepsilon\Omega)$ is fixed, then the degree of ambiguity of considered objects (i.e. elements of X) can be definitely determined. A set of all degrees of ambiguity, denoted by (Ω_c, B_c), will be called a *characteristic space*. We usually adopt ([0,1],Borel sets) as the characteristic space. A (B,B_c)-measurable function $\mu(\omega)$ will be called a *characteristic variable*, and a set of characteristic variables will be denoted by M,

$$M \underset{\Delta}{=} \{ \mu \mid \mu : \Omega \longrightarrow \Omega_c \quad (B,B_c)\text{-mble.} \}. \tag{7}$$

We can consider a quotient space $M/\equiv$ by the following equivalence relation $\equiv$,

$$\mu_1 \equiv \mu_2 \underset{def}{\leftrightarrow} \mu_1(\omega) = \mu_2(\omega) \quad \text{for P a.e. } \omega\varepsilon\Omega. \tag{8}$$

A *probabilistic set* A on X is defined by a $M/\equiv$-valued mapping χ_A on X,

$$\chi_A: X \longrightarrow M/\equiv. \tag{9}$$

A family of all probabilistic sets on X is denoted by $\mathcal{P}(X)$.

In order to understand a physical meaning, it will be easy to consider the following mapping μ_A instead of χ_A,

$$\begin{array}{ccc} \mu_A: X\times\Omega & \longrightarrow & [0,1]. \\ \omega & & \omega \\ (x,\omega) & \longmapsto & \mu_A(x,\omega) \end{array} \tag{10}$$

Both χ_A and μ_A can be considered the same, and each of them is called a *defining function* of the probabilistic set A. However, in the definition of probabilistic sets in terms of μ_A's, we must consider the equivalence relation $\equiv$; i.e. μ_A and $\mu_{A'}$ must be regarded as the same one if there exists $E(\varepsilon B)$ for each x such that

$$P(E) = 1, \tag{11}$$

$$\mu_A(x,\omega) = \mu_{A'}(x,\omega) \quad \text{for all } \omega\varepsilon E. \tag{12}$$

Here, we shall also use the following brief notation,

$$\mu_A(x,\omega) = \mu_{A'}(x,\omega) \text{ for all } x\varepsilon X \text{ and P a.e. } \omega\varepsilon\Omega. \tag{13}$$

We shall mainly use $\mu_A(x,\omega)$ as the definition of the probabilistic set A in the following.

It is clear from the abovestated definition that the probabilistic set A is expressed completely by both the defining function $\mu_A(x,\omega)$ and the probability measure $P(\omega)$ (exactly speaking, by both $\mu_A(x,\omega)$ and (Ω,B,P)). A pair $(\mu_A(x,\omega), P(\omega))$ will be called a *probabilistic expression* of the probabilistic set A.

2-2. FUNDAMENTAL OPERATIONS OF PROBABILISTIC SETS

We shall introduce several operations in $\mathcal{P}(X)$. For arbitrary two probabilistic sets A and B, whose defining functions are $\mu_A(x,\omega)$ and $\mu_B(x,\omega)$ respectively, A is said to be included in B $(A \subset B)$ if for each $x(\varepsilon X)$ there exists $E(\varepsilon B)$ which satisfies,

$$P(E) = 1, \tag{14}$$

$$\mu_A(x,\omega) \leqq \mu_B(x,\omega) \quad \text{for all } \omega\varepsilon E. \tag{15}$$

In this situation we shall use a brief notation as follows:

$$\mu_A(x,\omega) \leqq \mu_B(x,\omega) \quad \text{for all } x\varepsilon X \text{ and P a.e.}\omega\varepsilon\Omega. \tag{16}$$

Remark; If both $A \subset B$ and $B \subset A$ are satisfied, A and B are regarded as the same. (Consider the equivalence relation described in a previous section.)

It will be easily confirmed that the inclusion-relation $\subset$ provides a partial order in $P(X)$. Hence, $(P(X),\subset)$ constitutes a *poset* (*p*artially *o*rdered *set*). Moreover it can be verified from a lattice-theoretical viewpoint that the family of probabilistic sets $(P(X),\subset)$ constitutes a *complete pseudo-Boolean algebra* (Hirota 1977).

Remark; A pseudo-Boolean algebra is a relative-complemented lattice with a minimum element and is a subclass of distributive lattices. In this case the minimum element is a *null set* ϕ defined by

$$\mu_\phi(x,\omega) = 0 \text{ for all } x\varepsilon X \text{ and P a.e. } \omega\varepsilon\Omega. \tag{17}$$

For example, a "union" of $\{A_\gamma\}_{\gamma\varepsilon\Gamma}$($\Gamma$:possibly infinite) is given by the following procedure. (Here, the union is defined, of course, as a supremum with respect to the partial order $\subset$.): Let $\mu_{A_\gamma}(x,\omega)$ be a defining function of A_γ for all $\gamma\varepsilon\Gamma$. Put

$$a(x) \underset{\Delta}{=} \sup\{\int_\Omega \max\{\mu_{A_{\gamma i}}(x,\omega) \mid 1 \leqq i \leqq n\}\cdot dP(\omega) \mid n\varepsilon N(\text{natural numbers}), \gamma_i\varepsilon\Gamma\}, \tag{18}$$

(of course $0 \leqq a(x) \leqq 1$).

Choose a countably infinite subsequence $\{\max\{\mu_{A_{\gamma i}}(x,\omega) \mid 1 \leqq i \leqq n_j\} \mid n_j\varepsilon N, \gamma_i\varepsilon\Gamma\}_{j=1}^{\infty}$ such that

$$\lim_{j\to\infty} \int_\Omega \max\{\mu_{A_{\gamma i}}(x,\omega) \mid 1 \leqq i \leqq n_j\} dP(\omega) = a(x). \tag{19}$$

(Note that this procedure is possible for each fixed $x\varepsilon X$.) Then the union $\bigcup A_\gamma$ is given by

$$\mu_{\cup A_\gamma}(x,\omega) \underset{\Delta}{=} \sup\{\max\{\mu_{A_{\gamma i}}(x,\omega) \mid 1 \leqq i \leqq n_j\} \mid 1 \leqq j < \infty\}. \tag{20}$$

An "intersection" of $\{A_\gamma\}_{\gamma\varepsilon\Gamma}$, which is denoted by $\bigcap A_\gamma$, is a dual concept of the union $\bigcup A_\gamma$. (Exchange the simbols "max" and "sup" appeared in (8)(9)(10) for "min" and "inf" respectively.) Some other useful operations in $P(X)$ can also be defined. However all of them are omitted here, except a λ-*sum* of A and B ($0 \leqq \lambda \leqq 1$) defined by

$$\mu_{A \underset{\lambda}{+} B}(x,\omega) \underset{\Delta}{=} \lambda\cdot\mu_A(x,\omega) + (1-\lambda)\cdot\mu_B(x,\omega). \tag{21}$$

Although the abovestated procedure of union is rather complicated, it can be simplified in the case that the index set Γ is at most countably infinite. For instance, the union of A and B may be defined by

$$\mu_{A\cup B}(x,\omega) = \max\{\mu_A(x,\omega), \mu_B(x,\omega)\}, \tag{22}$$

for each $x\varepsilon X$ and each $\omega\varepsilon\Omega$, and the union of $\{A_n\}_{n=1}^{\infty}$ may be defined by

$$\mu_{\cup A_n}(x,\omega) = \sup\{\mu_{A_n}(x,\omega) \mid 1 \leqq n < \infty\}, \tag{23}$$

for each $x\varepsilon X$ and each $\omega\varepsilon\Omega$. The complexity in a general case arises from the fact that M (cf.(7)) is not closed by more than countably infinite operations (Hirota 1977).

3. EXTENDED FUZZY EXPRESSION OF PROBABILISTIC SETS

3-1. SUMMARY OF RESULTS

It was mentioned in a previous section that the information about a probabilistic set A was expressed completely by a pair $(\mu_A(x,\omega), P(\omega))$ which was called a probabilistic expression of the probabilistic set A. The main purpose of the present section is to introduce another mutually equivalent expression of probabilistic sets, called an extended fuzzy expression.

The extended fuzzy expression of A is given by a countably infinite set of functions (called "monitors") $\{m_A{}^n(x)\}_{n=1}^{\infty}$. The moment analysis shows that the main information is concentrated on lower monitors such as $m_A{}^1(x)$ (called a "membership function" of A) and $v_A(x) \underset{\Delta}{=} m_A{}^2(x) - (m_A{}^1(x))^2$ (called a "vagueness function" of A).

It is sufficient practically to consider both the membership function and the vagueness function. From the possibility of two mutually equivalent expressions of probabilistic sets, we can draw an interesting conclusion to the "fuzzy vs probability" controversy: The classical fuzzy concept, i.e. the notion of membership functions alone introduced by L.A.Zadeh (1965), is not sufficient. However, if other concepts such as vagueness functions are taken into consideration in addition to the notion of membership functions, then the modified fuzzy concepts, called the extended fuzzy expression, provides the same information as the notion of probability. Hence, the equality between the notion of "(modified) fuzzy" and that of "probability" is confirmed theoretically. Moreover, by using membership functions, vagueness functions and higher monitors successively, we can expect to obtain useful results which are different from the results given by probabilistic approaches.

We shall also mention about the extended fuzzy expression of plural probabilistic sets and the fundamental operations of probabilistic sets in terms of extended fuzzy expression.

3-2. EXTENDED FUZZY EXPRESSION OF A SINGLE PROBABILISTIC SET

Let $(\mu_A(x,\omega),P(\omega))$ be a *probabilistic expression* of a probabilistic set A on X. Another mutually equivalent expression, called an *extended fuzzy expression*, will be given in the following.

Consider the following induced measure $\Phi_A(x,\cdot)$ on (Ω_c,B_c) (=([0,1],Borel Sets)) for each fixed $x(\varepsilon X)$,

$$\Phi_A(x,E) \underset{\Delta}{=} P(\{\omega \mid \mu_A(x,\omega)\varepsilon E\}) \qquad \text{for all } E\varepsilon B_c. \tag{24}$$

If we do not consider the structure of Ω (since the important point is not each parameter $\omega\varepsilon\Omega$, but the evaluation value $\mu_A(x,\omega)$ in Ω_c), then we may consider that $\Phi_A(x,\cdot)$ has the same information as $(\mu_A(x,\omega),P(\omega))$. We can consider the following transformation from $\Phi_A(x,\cdot)$ to $\phi_A(x,t)$ (where $t\varepsilon R$(real numbers)) for each fixed $x\varepsilon X$,

$$\phi_A(x,t) \underset{\Delta}{=} \int_0^1 \exp(it\alpha)\cdot d\Phi_A(x,\alpha). \tag{25}$$

It is clear from the Levy-Haviland inversion formula (cf. Haviland 1935) that the transformation (25) has an inverse. Hence it can be concluded that $\phi_A(x,t)$ has the same information as $\Phi_A(x,\cdot)$. Here $\phi_A(x,t)$ satisfies the following proposition. (Proofs of propositions in the following are given in the appendix.)

[Prop.1]

$$\left.\frac{\partial^n \phi_A(x,t)}{\partial t^n}\right|_{t=0} = i^n\cdot E[\mu_A(x,\cdot)^n], \tag{26}$$

$$\text{where } E[\mu_A(x,\cdot)^n] \underset{\Delta}{=} \int_\Omega \mu_A(x,\omega)^n\cdot dP(\omega) \tag{27}$$

$$= \int_0^1 \alpha^n\cdot d\Phi_A(x,\alpha). \tag{28}$$

$$\phi_A(x,t) = \sum_{n=o}^{\infty} \frac{i^n}{n!} E[\mu_A(x,\cdot)^n]\cdot t^n. \tag{29}$$

Since each n-th moment $E[\mu_A(x,\cdot)^n]$ is a function of $x\varepsilon X$, we shall use the following notations,

$$m_A{}^n(x) \underset{\Delta}{=} E[\mu_A(x,\cdot)^n], \tag{30}$$

$$\phi_A(x,t) = \sum_{n=o}^{\infty} \frac{i^n}{n!} m_A{}^n(x)\cdot t^n. \tag{31}$$

Then we have the following proposition.

[Prop. 2]

1) $n \geqq m \quad \rightarrow \quad 1 = m_A{}^o(x) \geqq m_A{}^m(x) \geqq m_A{}^n(x) \geqq 0.$ (32)

2) For each fixed $x\varepsilon X$, $\phi_A(x,t)$ (31) is a positive definite function of $t\varepsilon R$ in Bochner's sense; i.e. for each $n\varepsilon N$, $t_1,t_2,\ldots,t_n \varepsilon R$, $z_1,z_2,\ldots,z_n \varepsilon C$(complex numbers), we have,

$$\sum_{i,j=1}^{n} z_i \cdot \phi_A(x,t_i-t_j)\cdot \overline{z_j} = (\ldots z_i \ldots)\begin{pmatrix} & \vdots & \\ \ldots & \phi_A(x,t_i-t_j) & \ldots \\ & \vdots & \end{pmatrix}\overline{\begin{pmatrix} \vdots \\ z_j \\ \vdots \end{pmatrix}} \geqq 0. \quad (33)$$

3) If we consider the following function $\psi_A(x,t)$,

$$\psi_A(x,t) \underset{\Delta}{=} \phi_A(x,-it), \quad (34)$$

then $\psi_A(x,t)$ is a monotone non-decreasing function of t for each fixed $x \varepsilon X$. ($\psi_A(x,t)$ is called a *moment generating function*.)

[Prop.3]

Let $\{m_A{}^n(x)\}_{n=1}^{\infty}$ be a countable set of functions which satisfy three conditions 1)2)3) mentioned in Prop.2. Then we can constitute uniquely a probability measure $\Phi_A(x,\cdot)$ on (Ω_c, B_c) (=([0,1],Borel sets)) for each $x\varepsilon X$.

The noteworthy point from above propositions is summarized as the following theorem 1.

[Theorem 1]

The probabilistic expression $(\mu_A(x,\omega),P(\omega))$ of the probabilistic set A has another equivalent expression $\{m_A{}^n(x)\}_{n=o}^{\infty}$ where

1) $n \geqq m \rightarrow 1 = m_A{}^o(x) \geqq m_A{}^m(x) \geqq m_A{}^n(x) \geqq 0$, (35)

2) For each fixed $x\varepsilon X$,

$$\phi_A(x,t) \underset{\Delta}{=} \sum_{n=o}^{\infty} \frac{i^n}{n!} m_A{}^n(x)\cdot t^n, \quad (36)$$

is a positive definite function of $t\varepsilon R$.

3) For each fixed $x\varepsilon X$,

$$\psi_A(x,t) \underset{\Delta}{=} \phi_A(x,-it) = \sum_{n=o}^{\infty} \frac{1}{n!} m_A{}^n(x)\cdot t^n, \quad (37)$$

is a monotone non-decreasing function of t.

[Def. 1]

A set of countably infinite functions $\{m_A{}^n(x)\}_{n=o}^{\infty}$ with three conditions 1)2)3) in theorem 1 is called an *extended fuzzy expression* of the probabilistic set A. The function $m_A{}^n(x)$ is called a n-th *monitor* of A; especially $m_A{}^1(x)$ is called a *membership function* of A, and a function $v_A(x) \underset{\Delta}{=} m_A{}^2(x) - (m_A{}^1(x))^2$ is called a *vagueness function* of A.

Remark; It must be noted that main information of the extended fuzzy expression is concentrated on lower monitors such as the membership function and the vagueness function since we have (35) and

$$\lim_{n\to\infty} \int_{\Omega}(\mu_A(x,\omega)-E[\mu_A(x,\cdot)])^n \cdot dP(\omega) = \lim_{n\to\infty} M_0^n[\mu_A(x,\cdot)] = 0 \quad \text{for all } x\varepsilon X, \quad (38)$$

(cf. Hirota 1977).

Many discussions have been done on fuzzy concepts since L.A.Zadeh presented his paper [6]. Some anti-fuzzy scientists have been persisting that the fuzziness is not a new concept and that so-called fuzzy concepts can be derived by probability and two-valued logic. Whereas other fuzzy scientists have been insisting that the fuzziness is one thing and the (probabilistic) randomness is another. From the facts described above, we may conclude that,

1) There are two possible approaches to the problem of ambiguity and subjectivity, i.e. a probabilistic expression and an extended fuzzy expression.
2) The two approaches are mutually equivalent from a theoretical viewpoint. The most important point, however, is that there may exist several differences between the two approaches in applications. In probabilistic expression, the basic idea is to give a probability distribution $\Phi_A(x,\cdot)$, whereas, in extended fuzzy expression, it is to provide a membership function, a vagueness function and so on. Hence it may be possible to obtain different kinds of useful results accord-

ing to each approach.

3) It was shown that although the notion of membership function (by L.A.Zadeh 1965) provided an important information, it was not sufficient. We must consider a vagueness function and other higher monitors in order to get a good approximation. However, it was also shown that the important information was concentrated on lower monitors such as a membership function and a vagueness function.

3-3. EXTENDED FUZZY EXPRESSION OF PLURAL PROBABILISTIC SETS

At first we shall deal with a case of two probabilistic sets. Let A and B be two probabilistic sets on X whose defining functions are $\mu_A(x,\omega)$ and $\mu_B(x,\omega)$ respectively. A triplet $(\mu_A(x,\omega),\mu_B(x,\omega),P(\omega))$ is called a *probabilistic expression* of A and B. We shall derive another expression, called an *extended fuzzy expression* of A and B, in the following.

We can introduce a probability measure $\Phi_{A,B}(x,\cdot)$ on $(\Omega_c\times\Omega_c,B_c\times B_c)(=([0,1]^2$, Borel sets)) by

$$\Phi_{A,B}(x,E\times E')= P(\{\omega|\mu_A(x,\omega)\varepsilon E\}\bigcap\{\omega|\mu_B(x,\omega)\varepsilon E'\}) \quad \text{for all } E,E'\ \varepsilon\ B_C. \tag{39}$$

This probability measure $\Phi_{A,B}(x,\cdot)$ provides the same information as the probabilistic expression. (The reason has already mentioned in the case of a single probabilistic set.) It is easy to see that moment relations are preserved, i.e.

$$\int_0^1\int_0^1\alpha^n\cdot\beta^m\cdot d\Phi_{A,B}(x,(\alpha,\beta))= \int_\Omega\mu_A(x,\omega)^n\cdot\mu_B(x,\omega)^m\cdot dP(\omega)$$
$$\underset{\Delta}{=} E[\mu_A(x,\cdot)^n\cdot\mu_B(x,\cdot)^m]. \tag{40}$$

We can consider a transformation from $\Phi_{A,B}(x,\cdot)$ to $\phi_{A,B}(x,(s,t))$ by

$$\phi_{A,B}(x,(s,t))\underset{\Delta}{=} \int_0^1\int_0^1\exp(i(s\alpha+t\beta))\cdot d\Phi_{A,B}(x,(\alpha,\beta)). \tag{41}$$

[Prop.4]

The transformation (41) has an inverse, i.e.

$$\int_0^1\int_0^1\chi(\alpha;a_1,a_2)\cdot\chi(\beta;b_1,b_2)\cdot d\Phi_{A,B}(x,(\alpha,\beta))$$
$$= \lim_{T\to\infty} \left(\frac{1}{2\pi}\right)^2\int_{-T}^{T}\int_{-T}^{T} \frac{e^{-ia_2s}-e^{-ia_1s}}{-is}\cdot\frac{e^{-b_2t}-e^{-ib_1t}}{-it}\,\phi_{A,B}(x,(s,t))\cdot dsdt, \tag{42}$$

where $0\leqq a_1<a_2\leqq 1$, $0\leqq b_1<b_2\leqq 1$,
$$\chi(\alpha;a_1,a_2)=\begin{cases} 1 & a_1<\alpha<a_2 \\ 1/2 & \alpha=a_1 \text{ or } a_2 \\ 0 & \text{otherwise.}\end{cases} \tag{43}$$

We shall give several comments. Let $F_{A,B}(x,(\alpha,\beta))$, defined by

$$F_{A,B}(x,(\alpha,\beta))\underset{\Delta}{=} \Phi_{A,B}(x,(-\infty,\alpha]\times(-\infty,\beta]), \tag{44}$$

be a *cumulative distribution function*. Then we have

1) $F_{A,B}(x,(\alpha,\beta)) \nearrow$ with respect to α and β, (45)

2) $\lim_{\alpha,\beta\to-\infty} F_{A,B}(x,(\alpha,\beta))= 0$, $\lim_{\alpha,\beta\to+\infty} F_{A,B}(x,(\alpha,\beta))= 1$, (46)

3) $\lim_{h,k\to+0} F_{A,B}(X,(\alpha+h,\beta+k))= F_{A,B}(x,(\alpha,\beta))$ (right continuous). (47)

Conversely, it is well-known that a probability measure $\Phi_{A,B}(x,\cdot)$ can be uniquily constructed by $F_{A,B}(x,(\alpha,\beta))$ which satisfies (45)(46)(47). Therefore there exists an one to one correspondence between $\Phi_{A,B}(x,\cdot)$ and $F_{A,B}(x,(\alpha,\beta))$. If $\Phi_{A,B}(x,\cdot)$ is absolutely continuous and can be expressed by

$$\Phi_{A,B}(x,E)= \iint_E f_{A,B}(x,(\alpha,\beta))d\alpha d\beta \quad \text{for all } E\varepsilon B_c\times B_c, \tag{48}$$

where $f_{A,B}(x,(\alpha,\beta))$ is a non-negative, Baire function of α and β with an integral value 1, then $f_{A,B}(x,(\alpha,\beta))$ is said to be a *probability density function* (p.d.f.) of $\Phi_{A,B}(x,\cdot)$. If there exists the p.d.f. $f_{A,B}(x,(\alpha,\beta))$ of $\Phi_{A,B}(x,\cdot)$, then we have

$$F_{A,B}(x,(\alpha,\beta))= \int_{-\infty}^{\alpha}\int_{-\infty}^{\beta}f_{A,B}(x,(\alpha,\beta))\cdot d\alpha d\beta. \tag{49}$$

[Prop. 5]

If $\phi_{A,B}(x,(s,t))$ belongs to $L^1(R^2)$ as a function of (s,t), then $\Phi_{A,B}(x,\cdot)$ has a continuous p.d.f. $f_{A,B}(x,(\alpha,\beta))$ and the following relation holds,

$$f_{A,B}(x,(\alpha,\beta))=\left(\frac{1}{2\pi}\right)^2\int_{-\infty}^{\infty}\int_{-\infty}^{\infty}e^{-i(\alpha s+\beta t)}\cdot\phi_{A,B}(x,(s,t))\cdot dsdt. \quad (50)$$

Let us return to the main problem. We have the following propositions about the function $\phi_{A,B}(x,(s,t))$(41).

[Prop. 6]

There exists a partial derivative of $\phi_{A,B}(x,(s,t))$ of arbitrary order with respect to s and t, and we have

$$\left.\frac{\partial^{n+m}\phi_{A,B}(X,(s,t))}{\partial s^n\ \partial t^m}\right|_{s=t=0}=i^{n+m}\cdot E[\mu_A(x,\cdot)^n\cdot\mu_B(x,\cdot)^m]. \quad (51)$$

$$\phi_{A,B}(x,(s,t))=\sum_{n=0}^{\infty}\frac{1}{n!}\sum_{r=0}^{n}i^n\binom{n}{r}\cdot E[\mu_A(x,\cdot)^r\cdot\mu_B(x,\cdot)^{n-r}]\cdot s^r t^{n-r}. \quad (52)$$

[Prop. 7]

We shall use the following notations in the same manner as the case of a single probabilistic set,

$$m_{A,B}^{n,m}(x)\underset{\Delta}{=}E[\mu_A(x,\cdot)^n\cdot\mu_B(x,\cdot)^m], \quad (53)$$

$$\phi_{A,B}(x,(s,t))\underset{\Delta}{=}\sum_{n=0}^{\infty}\frac{1}{n!}\sum_{r=0}^{n}i^n\binom{n}{r}\cdot m_{A,B}^{r,n-r}(x)\cdot s^r\cdot t^{n-r}. \quad (54)$$

Then we have

1) $1=m_{A,B}^{0,0}(x)\geqq m_{A,B}^{n,m}(x)\geqq m_{A,B}^{n',m'}(x)\geqq 0\quad (n'\geqq n,m'\geqq m)$, (55)

2) $\psi_{A,B}(x,(s,t))\underset{\Delta}{=}\phi_{A,B}(x,(-is,-it))$, (56)

is a monotone non-decreasing function of (s,t) for each $x(\varepsilon X)$.

3) For each fixed $x\varepsilon X$, $\phi_{A,B}(x,(s,t))$ is a positive definite function of (s,t); i.e. for each $n\varepsilon N$, $(s_1,t_1),\ldots,(s_n,t_n)\varepsilon R$, $z_1,\ldots,z_n\varepsilon C$, we have

$$\sum_{i,j=1}^{n}z_i\cdot\phi_{A,B}(x,(s_i-s_j,t_i-t_j))\cdot\overline{z_j}$$

$$=(\ldots,z_i,\ldots)\begin{pmatrix} & \vdots & \\ \ldots, & \phi_{A,B}(x,(s_i-s_j,t_i-t_j)), & \ldots \\ & \vdots & \end{pmatrix}\overline{\begin{pmatrix}\vdots\\ z_j\\ \vdots\end{pmatrix}}\geqq 0. \quad (57)$$

[Prop. 8]

Let $\{m_{A,B}^{n,m}(x)\}_{n,m=o}^{\infty}$ be a countably infinite function-matrix with three properties 1)2)3) described in Prop.7. Then we can constitute uniquely a probability measure $\Phi_{A,B}(x,\cdot)$ on $(\Omega_c\times\Omega_c,B_c\times B_c)(=([0,1]^2,\text{Borel sets}))$.

With these propositions, the following theorem 2 is demonstrated.

[Theorem 2]

The probabilistic expression $(\mu_A(x,\omega),\mu_B(x,\omega),P(\omega))$ of two probabilistic sets A and B has another equivalent expression $\{m_{A,B}^{n,m}(x)\}_{n,m=o}^{\infty}$ where

1) $1=m_{A,B}^{o,o}(x)\geqq m_{A,B}^{n,m}(x)\geqq m_{A,B}^{n',m'}(x)\geqq 0\quad (n'\geqq n,m'\geqq m)$, (58)

2) For each fixed $x\varepsilon X$,

$$\phi_{A,B}(x,(s,t))\underset{\Delta}{=}\sum_{n=o}^{\infty}\frac{1}{n!}\sum_{r=0}^{n}i^n\binom{n}{r}\cdot m_{A,B}^{r,n-r}(x)\cdot s^r\cdot t^{n-r}, \quad (59)$$

is a positive definite function of (s,t).

3) For each fixed $x\varepsilon X$,

$$\psi_{A,B}(x,(s,t))\underset{\Delta}{=}\phi_{A,B}(x,(-is,-it)), \quad (60)$$

is a monotone non-decreasing function of (s,t).

[Def. 2]

A countably infinite function-matrix $\{m_{A,B}^{n,m}(x)\}_{n,m=o}^{\infty}$ with three conditions 1)2)3) in theorem 2 is called an *extended fuzzy expression* of the probabilistic sets A and B. The function $m^{n,m}_{A,B}(x)$ is called a *(n,m)-th monitor* of A and B; especially, $m_A(x) \underset{\Delta}{=} m^{1,0}_{A,B}(x)$ is called a *membership function* of A, $m_B(x) \underset{\Delta}{=} m^{0,1}_{A,B}(x)$ is called a *membership function* of B, $v_A(x) \underset{\Delta}{=} m^{2,0}_{A,B}(x) - (m^{1,0}_{A,B}(x))^2$ is called a *vagueness function* of A, $v_B(x) \underset{\Delta}{=} m^{0,2}_{A,B}(x) - (m^{0,1}_{A,B}(x))^2$ is called a *vagueness function* of B, $v_{A,B}(x) \underset{\Delta}{=} m^{1,1}_{A,B}(x) - m^{1,0}_{A,B}(x) \cdot m^{0,1}_{A,B}(x)$ is called a *co-vagueness function* of A and B.

In the extended fuzzy expression of A and B, the important information is concentrated on lower monitors. In most applications, it will be sufficient practically to give the 1-st and the 2-nd monitors, i.e.

$$\begin{pmatrix} 1 & m_A(x) & v_A(x) \\ m_B(x) & v_{A,B}(x) & \times \\ v_B(x) & \times & \times \end{pmatrix}. \tag{61}$$

We have been discussing about the extended fuzzy expression of two probabilistic sets. It will be easily verified to expand this notion into the case of more than two (in general n) probabilistic sets. Moreover, it is possible to develop this extended fuzzy expression to the case of infinite (possibly non-countably infinite) probabilistic sets: Let $\{A_\gamma\}_{\gamma\varepsilon\Gamma}$ be a family of probabilistic sets (Γ: possibly infinite), whose *probabilistic expression* is given by $(\{\mu_{A_\gamma}(x,\omega)\}_{\gamma\varepsilon\Gamma}, P(\omega))$. Then its *extended fuzzy expression* is given as follows. Let $I=\{\gamma_1,\gamma_2,\ldots,\gamma_m\}$ be an arbitrary, finite subset of Γ. An extended fuzzy expression of $\{A_\gamma\}_{\gamma\varepsilon I}$ will be given by a set of monitors $\{m^{n_1,n_2,\ldots,n_m}_{A_{\gamma 1},A_{\gamma 2},\ldots,A_{\gamma m}}(x)\}^{\infty}_{n_1,n_2,\ldots,n_m=o}$. Consider a class of such sets of monitors for all finite subsets I's of Γ. Then the class is called an *extended fuzzy expression* of $\{A_\gamma\}_{\gamma\varepsilon\Gamma}$.

3-4. OPERATIONS OF PROBABILISTIC SETS IN TERMS OF EXTENDED FUZZY EXPRESSION

We defined several operations of probabilistic sets such as union (20), intersection, and λ-sum (21). However, all of these operations were defined in terms of defining functions. It is also possible to define these operations in terms of monitors.

Let $\{m^{n,m}_{A,B}(x)\}^{\infty}_{n,m=o}$ be a set of monitors of two probabilistic sets A and B. Then for each $x\varepsilon X$, we can constitute uniquely a probability measure $\Phi_{A,B}(x,\cdot)$ on [0,1] by Prop.8. By using this probability measure, we can define monitors of various binary operations of A and B as follows:

1) the union of A and B ($A \cup B$)

For each fixed $x\varepsilon X$, we can constitute a probability measure $\Phi_{A\cup B}(x,\cdot)$ on ([0,1], Borel sets) by

$$\Phi_{A\cup B}(x,(a,b]) \underset{\Delta}{=} \Phi_{A,B}(x,E), \tag{62}$$

$$\text{where } E = \{(\alpha,\beta) \mid a < \max(\alpha,\beta) \leqq b, \quad \alpha,\beta\varepsilon[0,1]\}. \tag{63}$$

Then the n-th monitor of $A \cup B$ is given by

$$m^n_{A\cup B}(x) = \int_0^1 \alpha^n \cdot d\Phi_{A\cup B}(x,\alpha). \tag{64}$$

2) the intersection of A and B ($A \cap B$)

$$\Phi_{A\cap B}(x,(a,b]) \underset{\Delta}{=} \Phi_{A,B}(x,E), \tag{65}$$

$$\text{where } E = \{(\alpha,\beta) \mid a < \min(\alpha,\beta) \leqq b, \quad \alpha,\beta\varepsilon[0,1]\}. \tag{66}$$

$$m^n_{A\cap B}(x) = \int_0^1 \alpha^n \cdot d\Phi_{A\cap B}(x,\alpha). \tag{67}$$

3) the λ-sum of A and B ($A \underset{\lambda}{+} B$)

$$\Phi_{A\underset{\lambda}{+}B}(x,(a,b]) = \Phi_{A,B}(x,E), \tag{68}$$

$$\text{where } E = \{(\alpha,\beta) \mid a < \lambda\alpha+(1-\lambda)\beta \leqq b, \quad \alpha,\beta\varepsilon[0,1]\}. \tag{69}$$

$$m^n_{A\underset{\lambda}{+}B}(x) = \int_0^1 \alpha^n \cdot d\Phi_{A\underset{\lambda}{+}B}(x,\alpha). \tag{70}$$

4) the algebraic sum of A and B ($A \oplus B$) (cf. Hirota 1977)

$$\Phi_{A\oplus B}(x,(a,b]) \overline{\overline{\Delta}}\ \Phi_{A,B}(x,E), \quad (71)$$

$$\text{where } E=\{(\alpha,\beta)\mid a< \alpha+\beta-\alpha\beta \leqq b, \quad \alpha,\beta\varepsilon[0,1]\}. \quad (72)$$

$$m^n_{A\oplus B}(x)= \int_0^1 \alpha^n \cdot d\Phi_{A\oplus B}(x,\alpha). \quad (73)$$

5) the algebraic product of A and B (A·B)

$$\Phi_{A\cdot B}(x,(a,b]) \overline{\overline{\Delta}}\ \Phi_{A,B}(x,E), \quad (74)$$

$$\text{where } E=\{(\alpha,\beta)\mid a< \alpha\beta \leqq b, \quad \alpha,\beta\varepsilon[0,1]\}. \quad (75)$$

$$m^n_{A\cdot B}(x)= \int_0^1 \alpha^n \cdot d\Phi_{A\cdot B}(x,\alpha). \quad (76)$$

Other operations can be defined in almost the same manner.

[Example]

Let the total space X be a set of real numbers, A be a probabilistic set "numbers nearly equal to 1" on X and B be a probabilistic set "numbers nearly equal to -1" on X. Here the probabilistic expression of A and B are given, for example, by

$$\omega=(\xi,\eta)\varepsilon\ \Omega= [0,1]^2, \quad (77)$$

$$p(\xi,\eta)= 1 \quad (\text{uniform distribution on } [0,1]^2), \quad (78)$$

$$\mu_A(x,(\xi,\eta))= \{\min(1,(x-1)^2)\cdot\xi+ \max(1,2-(x-1)^2)/2\} /\{1+(x-1)^2\}, \quad (79)$$

$$\mu_B(x,(\xi,\eta))= \{\min(1,(x+1)^2)\cdot\eta+ \max(1,2-(x+1)^2)/2\} /\{1+(x+1)^2\}, \quad (80)$$

(cf. Fig (a)). In the extended fuzzy expression of A and B, membership functions are given by (cf. Fig (b)),

$$m_A(x)= 1/\{1+(x-1)^2\}, \quad (81)$$

$$m_B(x)= 1/\{1+(x+1)^2\}, \quad (82)$$

vagueness functions are given by,

$$v_A(x)= [\min\{m_A(x),(1-m_A(x))\}]^2/12, \quad (83)$$

$$v_B(x)= [\min\{m_B(x),(1-m_B(x))\}]^2/12, \quad (84)$$

$$v_{AB}(x)= 0, \quad (85)$$

and so on. The membership function of $A\cup B$ (i.e. "numbers nearly equal to 1 or -1") and the vagueness function of $A\cup B$ are given as in Fig (c) and (d) respectively; where ① is a result by using only membership functions (81) and (82), ② is a result by using both membership functions (81)(82) and vagueness functions (83)(84)(85), and ③ is a result by using all monitors (or equivalently by using the probabilistic expression (77)(78)(79)(80)). To summarize our interpretation of the results, we can explain that

1) The membership function $m_{A\cup B}(x)$ in Fig (c) provides the first approximation of probabilistic set $A\cup B$. The result ① (which is the same result as a classical fuzzy operation by Zadeh) has a continuous but non-smooth (i.e. non-differentiable) point (see x=0). It will be natural to expect a smooth curve like ③ as the first approximation of "numbers nearly equal to 1 or -1". The result ② provides a good approximation of ③.
2) From a viewpoint of the vagueness function $v_{A\cup B}(x)$ in Fig (d), the result ① of classical fuzzy operation has no "vagueness" ($v_{A\cup B}(x)= 0$) but it seems to be unnatural. Since the vagueness function provides the second information and since it indicates a disordered degree of judgements, we would like to expect a curve like ③. And ② also gives a fairly good approximation of ③.
3) Although information given by the membership function alone (see ① in Fig (c)(d)) is not sufficient, almost all insufficient information can be expressed by the vagueness function (compare ② with ③).

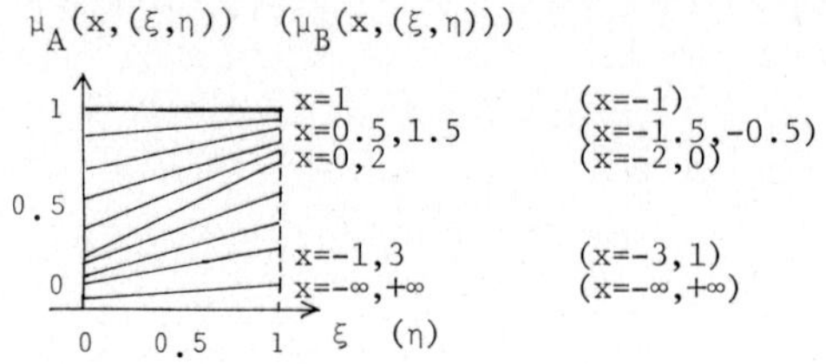

(a) defining functions of A and B. (Probabilistic Expression)

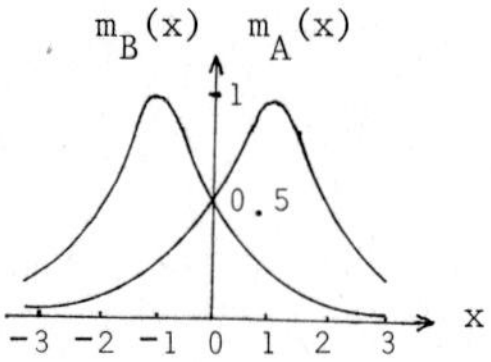

(b) membership functions of A and B. (Extended Fuzzy Expression)

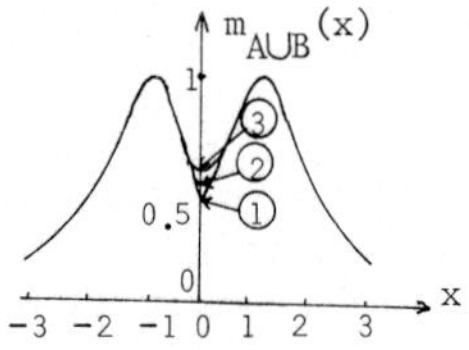

(c) the membership function of $A \cup B$.

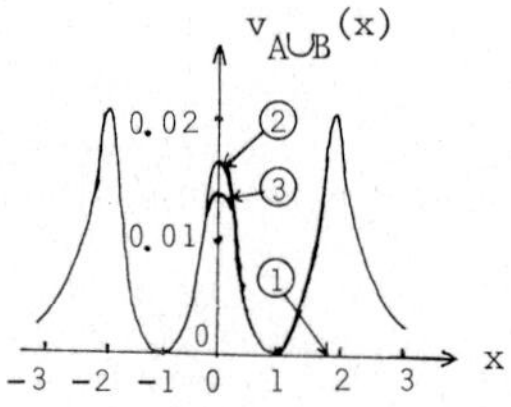

(d) the vagueness function of $A \cup B$.

Fig.1 Probabilistic sets; A="numbers nearly equal to 1", B="numbers nearly equal to -1", $A \cup B$="numbers nearly equal to 1 or -1".

4. CONCLUSION

The concept of probabilistic sets has been introduced as an analytical method for the problem of ambiguity and subjectivity in pattern recognition and decision making theory. The summary of the idea is as follows: All the things we can interfere are expressed by concepts of probability and the cases we can not intervene are unified by using fuzzy concepts. In this work, however, only mathematical aspects of probabilistic sets were discussed.

We proposed two mutually equivalent expressions of probabilistic sets, i.e. the probabilistic expression and the extended fuzzy expression. Equality between a concept of probability and that of fuzzy theory is also assured theoretically. But we also indicated that the study of fuzziness enables us to obtain useful results which are different from the results given by probabilistic approaches. In fact, several applications have been studied by us in the field of pattern recognition. We will be glad if our idea is any help to the people concerned. (A part of this work was supported by scientific research fund from the Yamada Foundation in Japan. The title is "Probabilistic Set Theory and its Application to Pattern Recognition".)

REFERENCES:

[1] Halmos,P.R.,Measure Theory, Van Nostrand Reinhold Comp.,p.110, New York (1950)
[2] Haviland, On the inversion formula for Fourier Stieltjes transforms in more than one dimension, Amer. J. Math. 57, (1935)
[3] Hirota,K., Kakuritsu-Shugoron to sono Oyo-rei (Probabilistic sets and its applications),The Behaviormetric Society of Japan 3-rd conference, pp.24-27,(in Japanese),(1975)
[4] ditto, Concepts of Probabilistic Sets, Proc. of IEEE, 77CH 1269-0SC, pp.1361-1366 (1977)
[5] Yoshida,K., Functional Analysis (3-rd Ed.), p.346, Springer-Verlag, (1971)
[6] Zadeh,L.A., Fuzzy Sets, Inf. and Control 8, pp.338-353, (1965)
[7] Hirota,K. & Iijima,T., A decision making model - A new approach based on the concepts of probabilistic sets -, Proc. of IEEE, 78 CH 1306-0SMC, pp.1348-1353, (1978)
[8] ditto, The bounded variation quantity (B.V.Q.) and its application to feature extractions, Proc. of 4-th Int. Joint Conf. on Pattern Recognition, pp.456-461, (1978)

APPENDIX - Proofs of Propositions -

[Prop.1],[Prop.2],[Prop.3]

Proofs of these propositions are omitted, since they are included in the following proofs.

[Prop.4]

It will be sufficient to show a one dimensional case, i.e.

$$\int_0^1 \chi(\alpha;a_1,a_2)\cdot d\Phi(\alpha)= \lim_{T\to\infty} \frac{1}{2\pi} \int_{-T}^{T} \frac{\exp(-ia_2s)-\exp(-ia_1s)}{-is}\cdot\phi(s)\cdot ds. \qquad \text{(A-1)}$$

(Here we use simplified notations.) By applying the Fubini's theorem, we have

$$\begin{aligned} F(T) &\underset{\Delta}{=} \int_{-T}^{T} \frac{e^{-ia_2s}-e^{-ia_1s}}{-is}\,\phi(s)ds= \int_{-T}^{T}\left(\int_{a_1}^{a_2} e^{-ixs}dx\right)\left(\int_0^1 e^{i\alpha s}d\Phi(\alpha)\right)ds \\ &= \int_0^1 d\Phi(\alpha)\int_{a_1}^{a_2}dx \int_{-T}^{T} e^{is(\alpha-x)}ds= 2\int_0^1 d\Phi(\alpha)\int_{a_1}^{a_2}\frac{\sin T(x-\alpha)}{x-\alpha}\,dx \\ &= 2\int_0^1 d\Phi(\alpha)\int_{T(a_1-\alpha)}^{T(a_2-\alpha)} \frac{\sin u}{u}\,du= 2\int_0^1\{G(T(a_2-\alpha))-G(T(a_1-\alpha))\}d\Phi(\alpha), \end{aligned} \qquad \text{(A-2)}$$

where $G(x)= \int_0^x \frac{\sin u}{u}\,du.$ (A-3)

Since $G(x)$ is continuous and since $\lim_{x\to\infty} G(x)= \pi/2$, $\lim_{x\to-\infty} G(x)= -\pi/2$, $G(x)$ is bounded and the right hand side of (A-2) is also bounded. Hence we can apply the Lebesgue's dominated convergence theorem (Halmos 1950), and we have

$$\lim_{T\to\infty} F(T)= 2\int_0^1 \lim_{T\to\infty}\{G(T(a_2-\alpha))-G(T(a_1-\alpha))\}d\Phi(\alpha)= 2\pi\int_0^1\chi(\alpha;a_1,a_2)d\Phi(\alpha). \qquad \text{(A-4)}$$

[Prop. 5]

The proof is omitted, since this result is well-known in probability theory.

[Prop. 6]

$$\frac{\partial \phi_{A,B}(x,(s,t))}{\partial s} = \lim_{h\to 0} \frac{\phi_{A,B}(x,(s+h,t)) - \phi_{A,B}(x,(s,t))}{h}$$

$$= \lim_{h\to 0} \iint \frac{\exp(i((s+h)\alpha+t\beta))-\exp(i(s\alpha+t\beta))}{h} d\Phi_{A,B}(x,(\alpha,\beta))$$

$$= \lim_{h\to 0} \iint \frac{e^{ih\alpha}-1}{h} e^{i(s\alpha+t\beta)} \cdot d\Phi_{A,B}(x,(\alpha,\beta))$$

$$= \iint \lim_{h\to 0} \frac{e^{ih\alpha}-1}{h} e^{i(s\alpha+t\beta)} \cdot d\Phi_{A,B}(x,(\alpha,\beta)) \quad \text{(cf. Lebesgue's dominated convergence theorem)}$$

$$= i \iint \alpha e^{i(s\alpha+t\beta)} \cdot d\Phi_{A,B}(x,(\alpha,\beta)). \tag{A-5}$$

$$\lim_{s,t\to 0} \frac{\partial \phi_{A,B}(x,(s,t))}{\partial s} = i \iint \lim_{s,t\to 0} \alpha\cdot e^{i(s\alpha+t\beta)} d\Phi_{A,B}(x,(\alpha,\beta))$$

$$= i \iint \alpha d\Phi_{A,B}(x,(\alpha,\beta)) = i\cdot E[\mu_A(x,\cdot)]. \tag{A-6}$$

We can obtain (51) almost in the same manner. By applying the Taylor's theorem of a function of two variables, we have

$$\phi_{A,B}(x,(s,t)) = \sum_{n=o}^{N} \frac{1}{n!} \sum_{r=o}^{n} i^n \binom{n}{r} E[\mu_A(x,\cdot)^r \mu_B(x,\cdot)^{n-r}]\cdot s^r t^{n-r} + R^N_{A,B}(x,(s,t)), \tag{A-7}$$

where the remainder $R^N_{A,B}(x,(s,t))$ is given by

$$R^N_{A,B}(x,(s,t)) = \frac{1}{(N+1)!} \sum_{r=o}^{N+1} \binom{N+1}{r} \frac{\partial^{N+1}\phi_{A,B}(x,(s,t))}{\partial s^r \partial t^{N+1-r}}\Bigg|_{s=\theta s,\, t=\theta t} \cdot s^r t^{N+1-r} \quad (0< \theta< 1)$$

$$= \frac{1}{(N+1)!} \sum_{r=o}^{N+1} \binom{N+1}{r} i^{N+1} \iint \alpha^r \beta^{N+1-r} \cdot e^{i\theta(s\alpha+t\beta)} \cdot d\Phi_{A,B}(x,(\alpha,\beta))\cdot s^r t^{N+1-r}. \tag{A-8}$$

$$|R^N_{A,B}(x,(s,t))| \leqq \frac{1}{(N+1)!} \sum_{r=o}^{N+1} \binom{N+1}{r} \iint |\alpha|^r |\beta|^{N+1-r} \cdot d\Phi_{A,B}(x,(\alpha,\beta)) |s|^r |t|^{N+1-r}$$

$$= \frac{1}{(N+1)!} \sum_{r=o}^{N+1} \binom{N+1}{r} E[\mu_A(x,\cdot)^r \mu_B(x,\cdot)^{N+1-r}] |s|^r |t|^{N+1-r}$$

$$\leqq \frac{1}{(N+1)!} \sum_{r=o}^{N+1} \binom{N+1}{r} |s|^r |t|^{N+1-r}$$

$$\leqq \frac{(2C)^{N+1}}{(N+1)!} \quad \text{for all } (s,t)\varepsilon \{(s,t)|\ s^2+t^2 \leqq C^2\}$$

$$\longrightarrow 0 \quad \text{uniformly as } N\to +\infty. \tag{A-9}$$

Therefore, we obtain (52).

[Prop. 7]

1) & 2)

It is clear from the following equations,

$$m^{n,m}_{A,B}(x) = \int_\Omega \mu_A(x,\omega)^n \mu_B(x,\omega)^m dP(\omega), \tag{A-10}$$

$$\psi_{A,B}(x,(s,t)) = \int_0^1\int_0^1 \exp(s\alpha+t\beta)\ d\Phi_{A,B}(x,(\alpha,\beta)). \tag{A-11}$$

3) Since

$$\phi_{A,B}(x,(s,t)) = \int_0^1\int_0^1 \exp(i(s\alpha+t\beta)) d\Phi_{A,B}(x,(\alpha,\beta)), \tag{A-12}$$

we have

$$\sum_{i,j} z_i\cdot\phi_{A,B}(x,(s_i-s_j,t_i-t_j))\cdot\overline{z_j}$$

$$= \iint \sum_{i,j} z_i\cdot\exp(i((s_i-s_j)\alpha+(t_i-t_j)\beta)\cdot\overline{z_j}\ d\Phi_{A,B}(x,(\alpha,\beta))$$

$$= \iint \sum_{i,j} z_i \cdot \exp(i(s_i\alpha+t_i\beta)) \cdot \exp(-i(s_j\alpha+t_j\beta)) \cdot \overline{z_j} \cdot d\Phi_{A,B}(x,(\alpha,\beta))$$

$$= \iint \Big|\sum_i z_i \cdot \exp(i(s_i\alpha+t_i\beta))\Big|^2 d\Phi_{A,B}(x,(\alpha,\beta)) \geqq 0. \quad \text{(A-13)}$$

[Prop.8]

1) Since $\phi_{A,B}(x,(s,t))$ defined by (54) is continuous at $(s,t)=(0,0)$ for each x and is a positive definite function, we can define a measure $\Phi_{A,B}(x,\cdot)$ on R^2 (cf. the Bochner's theorem (e.g. Yoshida 1971)) and we have

$$\phi_{A,B}(x,(s,t)) = \int_{-\infty}^{\infty}\int_{-\infty}^{\infty} e^{i(s\alpha+t\beta)} d\Phi_{A,B}(x,(\alpha,\beta)). \quad \text{(A-14)}$$

Moreover, the measure $\Phi_{A,B}(x,\cdot)$ becomes a probability measure on R^2, since we have

$$\phi_{A,B}(x,(0,0)) = m^{o,o}_{A,B}(x) = 1 = \int_{-\infty}^{\infty}\int_{-\infty}^{\infty} d\Phi_{A,B}(x,(\alpha,\beta)). \quad \text{(A-15)}$$

2) We shall prove the uniqueness of $\Phi_{A,B}(x,\cdot)$. It is sufficient to show $\Phi_{A,B}(x,\cdot) = \Phi'_{A,B}(x,\cdot)$ from the assumption of

$$\phi_{A,B}(x,(s,t)) = \iint \exp(i(s\alpha+t\beta)) d\Phi_{A,B}(x,(\alpha,\beta))$$

$$= \iint \exp(i(s\alpha+t\beta)) d\Phi'_{A,B}(x,(\alpha,\beta)). \quad \text{(A-16)}$$

Consider the following probability measure $\Psi(\cdot)$ on R^2,

$$\Psi(\cdot) \underset{\Delta}{=} \frac{1}{2}\{\Phi_{A,B}(x,\cdot)+\Phi'_{A,B}(x,\cdot)\}, \quad \text{(A-17)}$$

and the following hyperplanes on R^2 (i.e. lines)

$$H_1(a) = \{(a,\beta) \mid \beta\varepsilon(-\infty,\infty)\}, \quad \text{(A-18)}$$

$$H_2(a) = \{(\alpha,a) \mid \alpha\varepsilon(-\infty,\infty)\}. \quad \text{(A-19)}$$

Since a set of a's such that

$$\Psi(H_i(a)) > 0 \quad i=1 \text{ or } 2, \quad \text{(A-20)}$$

is at most countable, we shall write it by

$$D \underset{\Delta}{=} \{a_{i,n} \mid i=1,2,\ n=1,2,\dots\}. \quad \text{(A-21)}$$

Then we have

$$\Phi_{A,B}(x,H_i(a)) = \Phi'_{A,B}(x,H_i(a)) = 0 \quad i=1,2,\ a\varepsilon D^c. \quad \text{(A-22)}$$

If we choose an interval I,

$$I = \{(\alpha,\beta) \mid a_1 < \alpha \leqq a_2,\ b_1 < \beta \leqq b_2\}, \quad \text{(A-23)}$$

such that $a_1,a_2,b_1,b_2 \varepsilon D^c$, then we have, by (A-16) and the Levy-Haviland inversion formula (Haviland 1935),

$$\Phi_{A,B}(x,I) = \Phi_{A,B}(x,I^a) = \Phi_{A,B}(x,I^i) = \Phi'_{A,B}(x,I^a) = \Phi'_{A,B}(x,I^i) = \Phi'_{A,B}(x,I). \quad \text{(A-24)}$$

Let an interval I_0 be

$$I_0 \underset{\Delta}{=} (-\infty,\alpha_0]\times(-\infty,\beta_0], \quad \text{(A-25)}$$

where (α_0,β_0) (εR^2) is arbitrarily fixed. Then we have

$$\Phi_{A,B}(x,I_0) = \Phi'_{A,B}(x,I_0). \quad \text{(A-26)}$$

(Since D^c is dense, we can select a countable sequence of intervals $\{I_n\}_{n=1}^{\infty}$ such that

$$I_n = (\alpha_n^1,\alpha_n^2]\times(\beta_n^1,\beta_n^2] \quad \text{where } \alpha_n^1,\alpha_n^2,\beta_n^1,\beta_n^2 \varepsilon D^c, \quad \text{(A-27)}$$

$$I_n \nearrow I_0. \quad \text{(A-28)}$$

Here,

$$\Phi_{A,B}(\hat{x},I_n) = \Phi'_{A,B}(x,I_n) \quad n=1,2,\dots \quad , \quad \text{(A-29)}$$

is valid by (A-24). Considering the monotone-property of probability measure, we have (A-26).)

Therefore, we have $\Phi_{A,B}(x,\cdot) = \Phi'_{A,B}(x,\cdot)$. (Note the abovementioned remark that there exists an one to one correspondence between the cumulative distribution function and the probability measure.)

3) Finally, we shall prove that $\Phi_{A,B}(x,\cdot)$ is a probability measure on $[0,1]^2$, i.e.

$$\Phi_{A,B}(x,[0,1]\times[0,1])= 1. \tag{A-30}$$

Let a be an arbitrary, positive number which is greater than 1, and put

$$U_a \underset{\Delta}{=} (-\infty,-a)\times(-\infty,\infty)\bigcup(a,\infty)\times(-\infty,\infty). \tag{A-31}$$

Since the following equations hold,

$$m_{A,B}^{2n,o}(x)= \int_{-\infty}^{\infty}\int_{-\infty}^{\infty}\alpha^{2n}d\Phi_{A,B}(x,(\alpha,\beta))\geqq \iint_{U_a}\alpha^{2n}d\Phi_{A,B}(x,(\alpha,\beta))$$
$$\geqq a^{2n}\Phi_{A,B}(x,U_a), \tag{A-32}$$

$$1\geqq m_{A,B}^{2n,o}(x)\searrow\geqq 0, \tag{A-33}$$

we have

$$\Phi_{A,B}(x,U_a)= 0 \quad \text{for all } a> 1. \tag{A-34}$$

Considering the fact that $\Phi_{A,B}(x,\cdot)$ is monotone, we have

$$\Phi_{A,B}(x,U_1)= 0. \tag{A-35}$$

We can also verify that

$$\Phi_{A,B}(x,V_1)= 0, \tag{A-36}$$

$$\text{where } V_1= (-\infty,\infty)\times(-\infty,-1)\bigcup(-\infty,\infty)\times(1,\infty), \tag{A-37}$$

in almost the same manner as (A-35). Hence, from (A-35) and (A-36), we have

$$\Phi_{A,B}(x,[-1,1]\times[-1,1])= 1, \tag{A-38}$$

i.e. $\Phi_{A,B}(x,\cdot)$ becomes a probability measure on $[-1,1]^2$.

In the next place, put

$$U_a= \{(\alpha,\beta)\mid \alpha< a,\ \beta\varepsilon(-\infty,\infty)\} \quad \text{where } a\,\varepsilon\,[-1,0). \tag{A-39}$$

If we substitute t=0 in (56) and (A-14), then we have

$$\psi_{A,B}(x,(s,0))= \phi_{A,B}(x,(-is,0))=\int_{-1}^{1}\int_{-1}^{1}\exp(s\alpha)\cdot d\Phi_{A,B}(x,(\alpha,\beta))$$
$$\geqq \int_{-1}^{a}\int_{-1}^{1}\exp(s\alpha)\cdot d\Phi_{A,B}(x,(\alpha,\beta))$$
$$\geqq \exp(s\alpha)\cdot\Phi_{A,B}(x,U_a) \quad \text{for } s< 0. \tag{A-40}$$

If $\Phi_{A,B}(x,U_a)> 0$, then the right hand side of (A-40) tends to $+\infty$ as s tends to $-\infty$. This contradicts the monotone-assumption of $\psi_{A,B}(x,(s,t))$. Hence we have

$$\Phi_{A,B}(x,U_a)= 0 \quad \text{for all } a\varepsilon[-1,0). \tag{A-41}$$

Since the measure $\Phi_{A,B}(x,\cdot)$ is monotone, we have

$$\Phi_{A,B}(x,U_o)= 0. \tag{A-42}$$

We can also conclude that

$$\Phi_{A,B}(x,V_o)= 0, \tag{A-43}$$

$$\text{where } V_o= \{ (\alpha,\beta) \mid \alpha\varepsilon(-\infty,\infty),\ \beta< 0\}. \tag{A-44}$$

Considering (A-42) and (A-43), we can obtain the desired result,

$$\Phi_{A,B}(x,[0,1]\times[0,1])= 1, \tag{A-45}$$

i.e. $\Phi_{A,B}(x,\cdot)$ is a probability measure on $[0,1]^2$.

ADVANCES IN FUZZY SET THEORY AND APPLICATIONS
M.M. Gupta, R.K. Ragade, R.R. Yager (editors)

POSSIBILISTICALLY DEPENDENT VARIABLES AND A GENERAL THEORY OF FUZZY SETS [1)]

Ellen Hisdal

Institute of Informatics, University of Oslo, Box 1080, Blindern, Oslo 3,Norway.

The theory of possibilities has opened up a new era in the theory of fuzzy sets. For multidimensional universes there exist both joint and conditional fuzzy subsets of the mathematical attribute universe. All these subsets must be specified for a complete description of a verbal proposition. The definition of noninteraction is restricted to noninteractive relations only. It is shown that noninteraction does not necessarily imply independence of the possibilistic variables. Furthermore it is suggested that the AND connective should consistently be represented by the operation of intersection which is valid for the fuzzy set system within which one operates. Previously this would always have resulted in a noninteractive relation for the conjunctive proposition containing the AND. This is no longer the case when one makes use of conditional possibilities. It is suggested that there exists a general theory of fuzzy sets of which the theories of 1) nonfuzzy sets, 2) probability and 3) max-min fuzzy sets are special cases. Each special case or fuzzy sets system is characterized by its allowed range for the values of the grades of membership and by the definition of the operations of union and intersection.

1. INTRODUCTION

The theory of sets was introduced by George Boole in the middle of the 19-th century in his book "An Investigation of the Laws of Thought on which are Founded the Mathematical Theories of Logic and Probabilities"[B1]. Every set (usually called "class" by Boole) refers to a "universe of discourse which is in the strictest sense the ultimate subject of the discourse"[B1p.42]. The office of a descriptive term of a set is, according to Boole, to raise in the mind the conception of all beings or objects which exist within the supposed universe of discourse and to which the particular description is applicable. E.g. "Sometimes in discoursing of men we imply (without expressing the limitation) that it is of men only under certain circumstances and conditions that we speak, as of men in the vigour of life". A term like 'tall men' will then refer to the set or class of tall men contained in the universe of discourse of 'men in the vigour of life'.

The two fundamental operations in Boole's theory are the operation of union and intersection between sets, for which he uses the + and × signs respectively[B1p.33]. From these fundamental concepts of sets and the operations between them Boole builds up his beautiful theory of mathematical logic.

In 1965 Boole's theory received an enormous extension through the advent of Zadeh's theory of fuzzy sets[Z1],[Z2],[Z3], similar to the extension of Newtonian mechanics introduced by the theory of relativity. However in the case of mechanics, Newton's early theory is sufficient to explain the mechanical phenomena of everyday life. The opposite is true in the case of the logical theories. It is the new and extended theory of fuzzy sets which allows us to express the processes of everyday logic and approximate reasoning in mathematical terms. This sometimes gives rise to opposition to the theory of fuzzy sets from people who think that the approximate reasoning used by humans in everyday life (e.g. for purposes of decision making) is inferior to the exact reasoning of Boolean logic, and that therefore the mathematics associated with such approximate reasoning must be simpler than the mathematics of exact logic. The misunderstanding is propably due to the fact that we are not conscious of the intermediate steps in the information processing in our brains, and it is therefore very easy to underestimate the great complexity of this process. We do know however that humans are still much superior to computers in making decisions based on inexact information. The processing of such information is more difficult than that of exact one, and it is therefore not surprising that it needs a more elaborate tool.

The fuzzy period itself can again be divided into two. The extensions of Boole's theory introduced in the first period from 1965-1977 may be compared to the extensions introduced into Newtonian mechanics by the theory of special relativity. While Boole's sets are sharply bounded regions within the universe of discourse, fuzzy sets are unsharp regions within this universe. Every element in the universe of discourse has a grade of membership in the range from 0 to 1 in a specified fuzzy set. Thus the universe of discourse plays a more prominent role in the new theory. E.g. to perform the operations of union and intersection between fuzzy sets we must usually take into account all elements of the universe of discourse. The union (and similarly the intersection) of two fuzzy sets is a new fuzzy set such that a given element of the universe of discourse has a grade of membership which is a function of the grades of membership of the corresponding elements in the component fuzzy sets. The definition for the union which was introduced in the first period is the supremum (denoted by max or $\vee$) over the grades of membership of the component sets. The definition for the intersection is the infimum (denoted by min or $\wedge$). These definitions should now possibly be considered as default definitions only.

The first period is also characterized by the great emphasis on the role of linguistic variables. Although Boole already emphasized the close connection between logic and language[B1pp.27-30], this connection has become much more evident through the approximate reasoning of fuzzy set logic[Z4],[Z5].

The second fuzzy period started with Zadeh's paper "Fuzzy Sets as a Basis for a Theory of Possibility"[Z6]. Here the concept of possibility is defined as practically synonymous with grade of membership. E.g. if the grade of membership of 60 years in the fuzzy set OLD is 0.5, then the possibility that Age(John) = 60 years, given that 'John is old', is 0.5. Zadeh also discusses the conceptual difference between possibilities and probabilities.

For one-dimensional universes of discourse the concept of possibility does not introduce anything new into the mathematical structure of fuzzy set theory. For multidimensional universes it turned out that the new concept was very fruitful for clarification and refinement of the theory of the first period. Furthermore the concept was an aid in seeing the theory of fuzzy sets in a broader perspective. Thus we can speak about a general theory of fuzzy sets of which 1) Boole's theory of nonfuzzy sets, 2) the theory of probability and 3) the theory of the special or max-min fuzzy sets of the first fuzzy period are special cases.

In a one-dimensional universe a fuzzy set is represented by a unary fuzzy relation whose entries are the grades of membership or possibilities. E.g. the fuzzy set OLD may have the subjective possibility distribution $\Pi(u)$, partly specified in Table 1.1, for the variable U = Age,

Table 1.1
Unary fuzzy relation or possibility distribution representing OLD.

u in years	...45	55	65	.....	95
$\Pi(u)$	... 0	0.1	0.6	.....	1.0

For multidimensional universes the theory of fuzzy sets operates with multidimensional fuzzy relations which specify the grade of membership or possibility of each point in the universe. An example of a fuzzy relation R which describes the fuzzy set HUMAN in the attribute universe UxV = WeightxHeight is given in Table 1.2

Table 1.2

Binary fuzzy relation R or possibility distribution representing HUMAN.

u in kg \ v in cm	155	165	...	225
⋮				
55	1	1		0.1
65	0.7	1		0.3
⋮				
95	0.1	0.2		0.9

Such relations have been used to infer a fuzzy value of e.g. v (or possibility distribution $\Pi(v)$) from a fuzzy value of u by taking the composition of $\Pi(u)$ with R [Z2],

$$\Pi(v) = \Pi(u) \circ R .$$

The composition is the matrix "product" in which addition is replaced by the max operation and multiplication by the min operation.

Here however the second fuzzy set period has brought with it important innovations. For a 2-dimensional universe UxV we must specify three fuzzy subsets of this universe in order to have a complete description of a verbal proposition like "John is human" or "John is fat". These three necessary fuzzy subsets are the joint possibility distribution $\Pi(u,v)$ and the two conditional distributions $\Pi(v|u)$ and $\Pi(u|v)$. It turns out however that there are many cases in which the numerical values of the possibilities of the three distributions are equal. (One important such case are nonfuzzy relations.) In these cases the description of the first period, which used only one relation, is sufficient.

The next section discusses these questions as well as the difference between noninteraction and independence of possibilistic variables. It is a slightly modified version of paper[H1]. In section 3 we suggest that there exists a more general theory of fuzzy sets which allows other operations than max and min for union and intersection respectively. The theory of probability can be formulated as a special case of such a general fuzzy set system. In this interpretation probabilities are grades of membership in fuzzy sets. The resulting mathematical formulas are the same as in the traditional theory of probability. However the fuzzy set formulation, in which probabilities are an integral part of the definition of a set, gives a more direct connection with everyday language and logic. It may therefore be intuitively more appealing to many people. The AND connective is an illustration of the connection with ordinary language. It is discussed in section 4.

The special max-min fuzzy sets are discussed in section 5. Both their definition and some of their resulting properties lie, loosely speaking, somewhere in between Boolean sets and probabilistic sets. In reality they represent however a completely new system whose mathematical properties are especially well adapted to represent the linguistic concept of possibility.

We remark that the question of the existence of general fuzzy set systems with other operations than max and min for union and intersection respectively has received quite a bit of attention in the literature. Good lists of references can be found in Gaines'[G1] and Yager's[Y1] papers. Yager shows that there exists a infinite family of operations with the desirable properties of union and intersection. However neither Yager's paper nor the other references deal with the subject of conditional and joint possibilities. So we do not yet know whether the suggested operations can be adapted to a theory which allows for possibilistic dependence in the case of multidimensional universes.

2. CONDITIONAL POSSIBILITIES INDEPENDENCE AND NONINTERACTION

2.1 Introduction. In his paper on "Fuzzy sets as a basis for a theory of possibility" Zadeh[Z6] proposes a tentative expression for a conditional possibility distribution function. He remarks that in some applications it may be appropriate to modify this expression. Recently Nguyen[N1] has derived a formula for the conditional possibilities from the joint (and the marginal) possibilities. Here we proceed in the reverse direction and compute the joint possibilities from the conditional and the marginal ones.

Zadeh tentatively equates the conditional possibilities to the joint ones. Our results show that if the joint possibilities are given, then the conditional possibilities may always be set equal to the joint ones. However, usually there will exist also other solutions for the conditional possibilities. It depends on the particular problem at hand which solution we decide to use. The interrelations between the different possibility distribution functions are derived in Section 2.2. In Section 2.3 we show that fuzzy relations representing conditional possibility distributions should be used in connection with Zadeh's compositional rule of inference. Zadeh's examples of fuzzy relations used for purposes of inference are shown to be either conditional possibility distributions, or joint distributions of a special structure which is such that the solution for the conditional distribution is unique and equal to the joint one. In Section 2.4 we show that we must distinguish between noninteraction and independence of fuzzy or possibilistic variables, while the corresponding relations in the theory of probability are equivalent. The noninteractive AND relation, used for purposes of inference, is discussed. In Section 2.5 we give examples of different possibility distributions in connection with a character recognition problem.

A few remarks on terminology and notation are in order. We use the following three expressions to denote the same concept:(1) a fuzzy subset of a universe of discourse U;(2) a fuzzy value of a variable X which takes on values in U;(3) a possibility distribution $\Pi_X(U)$.

Let $U=\{u_i\}$ be a universe of discourse, u denoting its generic element. X is a variable which takes on values in U. To simplify the notation we will, for the most part, denote the possibility distribution $\Pi_X(u)$ for X by $\Pi(u)$. Similarly we will largely omit the X subscripts in the joint and the conditional distributions.

When we have two universes of discourse U_1,U_2, then the joint possibility distribution $\Pi(u_1,u_2)$ and the two conditional distributions $\Pi(u_2|u_1)$, $\Pi(u_1|u_2)$ are all binary fuzzy relations in the universe $U=U_1\times U_2$.

2.2 Conditional, marginal and joint possibilities. Let $X=(X_1,X_2)$ be a binary fuzzy variable taking on values in $U=U_1\times U_2$, and let $\Pi_{X_2|X_1}(u_2|u_1)=\Pi(u_2|u_1)$ be the conditional possibility distribution of X_2, given that X_1 is assigned a specific value from the universe U_1. Let $\Pi_{X_1}(u_1)=\Pi(u_1)$ be the marginal possibility for X_1. u_1,u_2 are the generic elements of U_1,U_2 respectively.

We define the conditional and joint possibilities in the following way.
Conditional possibility:

$$\Pi(X_2=u_2|X_1=u_1)=\text{possibility that } X_2 \text{ is } u_2 \text{ given that } X_1 \text{ has the value } u_1. \tag{2.1a}$$

Joint possibility:

$$\Pi(X_1=u_1,X_2=u_2)=\text{possibility that } (X_1=u_1 \text{ AND that } (X_2=u_2 \text{ given that } X_1 \text{ has the value } u_1)). \tag{2.1b}$$

With Zadeh's default definition for AND[Z7p.34] we obtain the following equation for the joint possibilities,

$$\Pi(u_1,u_2)=\Pi(u_1)\wedge\Pi(u_2|u_1). \tag{2.2}$$

In words: The joint possibility for the event u_1,u_2 equals the minimum of the prior (i.e.the marginal) possibility for u_1, and the conditional possibility for u_2, given u_1. Similarly

$$\Pi(u_1,u_2)=\Pi(u_2)\wedge\Pi(u_1|u_2). \tag{2.3}$$

The derivation of Eq.(2.2) from Eq(2.1b) is discussed on more detail in Sec. 4.

The marginal possibility distributions are given, according to Zadeh[Z6 sec.3.1]by the following equations which are analogous to those in the theory of probability:

$$\Pi(u_1)=\bigvee_{u_2}\Pi(u_1,u_2), \tag{2.4}$$

$$\Pi(u_2)=\bigvee_{u_1}\Pi(u_1,u_2), \tag{2.5}$$

where v denotes max.

From Eqs.(2.2) and (2.3) it follows that

$$\Pi(u_2|u_1)=\begin{cases}\Pi(u_1,u_2) & \text{for } \Pi(u_1)>\Pi(u_1,u_2),\\ [\Pi(u_1,u_2),1] & \text{for } \Pi(u_1)=\Pi(u_1,u_2),\end{cases} \tag{2.6}$$

$$\Pi(u_1|u_2)=\begin{cases}\Pi(u_1,u_2) & \text{for } \Pi(u_2)>\Pi(u_1,u_2),\\ [\Pi(u_1,u_2),1] & \text{for } \Pi(u_2)=\Pi(u_1,u_2).\end{cases} \tag{2.7}$$

The lower line on the right hand sides of these equations denotes that the conditional possibility can have any value in the interval $[\Pi(u_1,u_2),1]$.

Eq.(2.6) shows that the conditional distribution $\Pi(u_2|u_1)$ is determined uniquely from the joint and the marginal ones only in the subdomain for which $\Pi(u_1,u_2) < \Pi(u_1)$. In this subdomain the conditional distribution equals the joint one. For some distributions (see rule 4 below and example,Table 2.1) this subdomain is equal to the whole $U=U_1\times U_2$ domain of the possibility distributions. In this case Zadeh's tentative expression for the conditional possibilities will hold everywhere. In many other cases it will hold for a large part of the U domain.

Although Eqs.(2.6),(2.7) have some similarities with Nguyen's result, the lower part on the right hand side of our equation leaves more latitude for the values of the conditional possibilities. Furthermore Nguyen has different subdomains for which the upper and lower results are valid.

Substituting from Eq.(2.2) into Eq.(2.4) we obtain

$$\Pi(u_1)=\bigvee_{u_2}(\Pi(u_1)\wedge\Pi(u_2|u_1)). \tag{2.8}$$

From (2.8) it follows that

$$\bigvee_{u_2}\Pi(u_2|u_1)\geq\Pi(u_1)=\bigvee_{u_2}\Pi(u_1,u_2). \tag{2.9}$$

Similarly

$$\bigvee_{u_1}\Pi(u_1|u_2)\geq\Pi(u_2)=\bigvee_{u_2}\Pi(u_1,u_2). \tag{2.10}$$

In words: A given row in the $\Pi(u_2|u_1)$ table must contain at least one entry which is bigger than or equal to $\Pi(u_1)$. A given column in the $\Pi(u_1|u_2)$ table must contain at least one entry which is bigger than or equal to $\Pi(u_2)$. (We assume that rectangular tables for both the joint and the two conditional possibilities are set up in such a way that u_1 varies vertically and u_2 horizontally. See tables of Sec. 2.5.)

In Section 2.3 we show that relations representing conditional possibility distributions should be used for purposes of inference. In practice it is however often most natural to set up the joint distributions from a few prior premises. In other cases it may be expedient to set up some entries of the joint distribution and some entries of the conditional distribution from the prior premises, and to derive the other entries. It therefore becomes an important practical question how to derive the different distributions from one another. In the following we summarize some rules for such derivations. Examples are given in Section 2.5.

Rule 1. The conditional and the marginal distribution cannot be chosen independently of each other but must satisfy the constraints of Eqs.(2.9),(2.10).
Rule 2. When a conditional and the corresponding marginal distribution are given, then the joint distribution is uniquely determined by Eq.(2.2) or (2.3).
Rule 3. When the joint distribution (and therefore also the marginal ones) is given, then a possible solution for the conditional distributions is that they are equal to the joint distribution. In general this solution is not unique. For entries in a given row of the joint distribution which are equal to the supremum over that row, the corresponding entries in $\Pi(u_2|u_1)$ may have any value in the closed interval $[\Pi(u_1,u_2),1]$. A similar statement holds for the columns of the joint distribution and $\Pi(u_1|u_2)$. The remaining entries must equal those of the joint distribution. Rule 3 follows from Eqs.(2.6),(2.7).
Rule 4. If the joint distribution contains a 1 entry in a given row, then the entries in that row of the $\Pi(u_1|u_2)$ table are unique and equal to the row of the joint distribution. A similar result holds for columns and the $\Pi(u_1|u_2)$ table. The rule follows from rule 3.
Rule 5. If a given $\Pi(u_1,u_2)=1$, then $\Pi(u_1)=\Pi(u_2)=\Pi(u_2|u_1)=\Pi(u_1|u_2)=1$. The rule follows from Eqs.(2.4)-(2.7).
Rule 6. If $\Pi(u_1,u_2)=0$ and $\Pi(u_1)\neq 0$, then $\Pi(u_2|u_1)=0$, and similarly when we exchange the indices 1,2. The rule follows from Eqs.(2.6),(2.7).
Rule 7. The following relations hold always: $\Pi(u_1,u_2)\leq\Pi(u_1)$;$\Pi(u_1,u_2)\leq\Pi(u_2)$; and therefore $\Pi(u_1,u_2)\leq\Pi(u_1)\wedge\Pi(u_2)$. The first two relations follow from Eqs.(2.2),(2.3).
Rule 8. If for a given pair of u_1,u_2 values $\Pi(u_2|u_1)\leq\Pi(u_1)$ and $\Pi(u_1|u_2)\leq\Pi(u_2)$, then $\Pi(u_2|u_1)=\Pi(u_1|u_2)=\Pi(u_1,u_2)$. The rule follows from Eqs.(2.2),(2.3).
Rule 9. When X_2 is possibilistically independent of X_1 (see Eq.(2.18)) and X_1 is possibilistically independent of X_2 then the largest $\Pi(u_1)$ equals the largest $\Pi(u_2)$. The rule follows from the constraints (2.9),(2.10).

An important special case of rule 4 occurs for nonfuzzy relations whose entries are either 0 or 1. Furthermore one usually assumes for such relations that all the marginal possibilities are equal to 1. (A marginal possibility of 0 implies that the particular element does not belong to the universe of discourse.) It then follows from rule 4 that all the conditional possibilities are equal to the corresponding joint ones. This explains why a single relation is a sufficient specification of a property in the multidimensional, nonfuzzy case.

2.3 The compositional rule of inference. In the theory of fuzzy sets, binary relations play an important role for purposes of inference and approximate reasoning. Suppose that we are given a fuzzy relation R from U_1 to U_2, and a specific (fuzzy or non-fuzzy) value of u_1 (i.e. an arbitrary possibility distribution $\Pi(u_1)$). According to Zadeh's compositional rule of inference [Z7pp.20,21],[Z6.Eq.(3.43)], the relation allows us to infer the following value of u_2, induced by u_1,

$$\Pi(u_2)=\Pi(u_1)\circ R. \qquad (2.11)$$

where $\circ$ denotes the composition. The right hand side of Eq.(2.11) is thus the max-min matrix product of the row vector $\Pi(u_1)$ and the matrix representing R.

Before Zadeh's recent paper on possibility distributions [Z6], our attention had not been called to the fact that there exist joint and conditional possibility distributions. In Section 2.2 we have shown that when the joint possibility distribution is given, then there exists a solution for the conditional possibilities such that they are equal to the joint ones, but this solution is not always unique. Indeed in Section 2.5 we show some examples where it is reasonable to choose conditional distributions which are (for some entries of the tables) different from the joint distribution.

We must therefore now ask ourselves whether we should use the joint possibility distribution or one of the conditional distributions for R when we wish to infer a (fuzzy) value uf u_2 from a (fuzzy) value of u_1 according to Eq.(2.11). This question is easily answered by assuming that u_1 has a non fuzzy value, i.e. we assume that the ith element of the row vector $\Pi(u_1)$ equals 1, all the other elements being 0. In this case Eq.(2.11) yields for $\Pi(u_2)$ the ith row of R. Now the ith row of $\Pi(u_2|u_1)$ is the possibility distribution of u_2, assuming that u_1 has the non-fuzzy value $u_1=u_{1i}$. This row therefore equals the fuzzy value (i.e. the possibility distribution) of u_2 which we can infer from the given value of u_1. It is thus $\Pi(u_2|u_1)$, the conditional possibility distribution of u_2 given u_1, that we must use for R in Eq.(2.11) when we wish to infer a value of u_2 from a value of u_1. The suggestion that conditional fuzzy subsets should be used for inference was already put forth by Kaufmann[K1] in the first period. However Kaufmann does not use the concept of joint and marginal fuzzy sets and of their connection with conditional ones as given by Eqs.(2.2),(2.3).

The examples of fuzzy relations used by Zadeh to illustrate the compositional rule of inference are of different types.

The first type of relations is automatically a conditional possibility distribution because it represents a conditional statement of the "IF A THEN B ELSE C" type[Z7 p.12]. Similarly when Zadeh[Z8],[Z4] sets up relational tableaus for purposes of pattern recognition, he uses a table which lists the truth value that an object belong to a certain class, when the measurements of the object have given values. Again we have a conditional possibility distribution here.

Another type of relation which is often used by Zadeh represents approximately "Equal" (see e.g. Zadeh[Z7p.20] and Table 2.1 of the present paper). This relation has only 1 entries along the main diagonal. If we assume that Zadeh's table represents the joint distribution, then it follows from rule 4 of Section 2.2 that the solution for the conditional distribution is unique and equal to the joint distribution. Therefore Zadeh's use of this relation for purposes of inference is again correct.

Finally Zadeh often uses the non-interactive AND relation for purposes of inference (e.g.Zadeh[Z7 pp.30,34]):

$$R = \Pi(u_1) \wedge \Pi(u_2) \tag{2.12}$$

At first sight this may seem like a paradoxical situation for two reasons. In the first place Eq.(2.12) is symmetrical in u_1 and u_2. It is therefore more reasonable to assume that it represents a joint distribution and not a conditional one.However, we have seen that there always exists a solution for the conditional possibility distribution such that it is equal to the joint one. So at least in some cases it should be correct to use Eq.(2.12) also for the relation representing the conditional possibilities. But here another apparent paradox comes in. Assuming that Eq. (2.12) represents the joint distribution, then it certainly has a striking similarity to the analogous equation in the theory of probability, $P(u_1,u_2)=P(u_1)\cdot P(u_2)$, which is valid if and only if u_1 and u_2 are independent. But if u_1 and u_2 are independent, it is not possible to draw inferences about u_2 from a given value of u_1. How can Zadeh then use a relation of type (2.12) for purposes of inference ? Section 2.4 is devoted to the resolution of this apparent paradox.

<u>2.4. Independence and non-interaction</u>. The law of compound probabilities [F1,

chapter V, Eq.(1.5)],

$$P(u_1,u_2) = P(u_1) \cdot P(u_2|u_1) \tag{2.13}$$

is the probabilistic equivalent of the linguistic statement, Eq.(2.1b), and of our possibilistic equation (2.2).

In the theory of probability the random variable X_2 is defined to be statistically independent of X_1 when for all values of u_2,

$$P(u_2|u_1) = P(u_2) = \text{const.independent of } u_1. \tag{2.14}$$

It then follows from Eq.(2.13) that

$$P(u_1,u_2) = P(u_1) \cdot P(u_2) \quad \text{for all } u_1,u_2 \tag{2.15}$$

when X_1 and X_2 are independent.

Since we can exchange the indices 1,2 in Eq.(2.13), we have in general

$$P(u_1,u_2) = P(u_2) \cdot P(u_1|u_2) \text{ for alle } u_1,u_2, \tag{2.16}$$

and it follows from (2.15) that

$$P(u_1|u_2) = P(u_1) \quad \text{for all } u_1,u_2. \tag{2.17}$$

Eq.(2.17) tells us that X_1 is independent of X_2 when X_2 is independent of X_1. Thus Eqs.(2.14),(2.15) and (2.17) are equivalent statements in the theory of probability. We shall now show that the analogous equations in the theory of possibility are not necessarily equivalent. This is due to the special properties of the min operator which replaces the multiplication operator of the theory of probability.

In analogy with Eq.(2.14) we define the fuzzy variable X_2 to be possibilistically independent of X_1 when for all values of u_2

$$\Pi(u_2|u_1) = \Pi(u_2) = \text{const.independent of } u_1. \tag{2.18}$$

Substituting this value of $\Pi(u_2|u_1)$ into Eq. (2.2), it follows that for X_2 independent of X_1

$$\Pi(u_1,u_2) = \Pi(u_1) \wedge \Pi(u_2) \quad \text{for all } u_1,u_2. \tag{2.19}$$

When Eq.(2.19) holds, then the variables X_1,X_2 are defined to be non-interactive according to Zadeh[Z6p.15]. We can therefore conclude that when the fuzzy variable X_2 is possibilistically independent of X_1, then the joint distribution for X_1,X_2 is non-interactive. However, the reverse statement does not neseccarily hold.

Assume that the joint distribution for X_1,X_2 is non-interactive, and therefore that Eq.(2.19) holds. But according to Eq. (2.2), the joint possibilities are always equal to

$$\Pi(u_1,u_2) = \Pi(u_1) \wedge \Pi(u_2|u_1). \tag{2.20}$$

From Eqs.(2.19) and (2.20) it follows that when the joint distribution is non-interactive, and $\Pi(u_1)$ and $\Pi(u_2)$ are given, then

$$\Pi(u_2|u_1) = \begin{cases} \Pi(u_2) & \text{for } \Pi(u_2) < \Pi(u_1). \\ [\Pi(u_1),1] & \text{for } \Pi(u_2) \geqq \Pi(u_1). \end{cases} \tag{2.21}$$

Again the lower line on the right hand side denotes that $\Pi(u_2|u_1)$ may assume any value in the interval $[\Pi(u_1),1]$. The value $\Pi(u_2)$ is included in this interval.

Comparing Eqs.(2.18) and (2.21) we notice that when the joint distribution is non-interactive (as defined by Eq.(2.19))then there exists a solution for the conditional possibilities which is such that X_1 and X_2 are possibilistically independent. In general this solution will not be unique. We can therefore have non-interaction between two fuzzy variables which are possibilistically dependent.

We now return to the non-interactive AND relation. We saw in Section 2.3 that when a relation is used for purposes of inference, the relation must be a conditional possibility distribution. For a non-interactive AND used for purposes or inferring a fuzzy value of u_2 from a fuzzy value of u_1 we must therefore have

$$\Pi(u_2|u_1)=\Pi(u_1) \wedge \Pi(u_2) \text{ for all } u_1,u_2. \tag{2.22}$$

Notice that the left hand side of this equation is now a conditional and not the joint distribution.

Substituting this value of the conditional possibility into Eq.(2.20) we obtain

$$\Pi(u_1,u_2)=\Pi(u_1) \wedge (\Pi(u_1) \wedge \Pi(u_2))=\Pi(u_1) \wedge \Pi(u_2). \tag{2.23}$$

We conclude that when a non-interactive conditional AND relation of the type of (2.22) holds, then the joint distribution is non-interactive according to the definition of Eq.(2.19). But we saw that this does not necessarily mean that X_2 is independent of X_1. Indeed it will be only in very special cases that Eq.(2.22)for the conditional possibilities will result in possibilistic independence as defined by (2.18). We can therefore conclude that a non-interactive AND of the type of Eq.(2.22) used by Zadeh for purposes of inference[Z7 p,20] is correct.

When a non-interactive AND relation can be used to infer a fuzzy value of u_2 from a fuzzy value of u_1, then Eq.(2.22) must hold and therefore also Eq.(2.23). From this it does not necessarily follow that we must use the conditional distribution $\Pi(u_1) \wedge \Pi(u_2)$ when we wish to infer a value of u_1 from a value of u_2. In general we have that

$$\Pi(u_1,u_2)=\Pi(u_1|u_2) \wedge \Pi(u_2). \tag{2.24}$$

In our case $\Pi(u_1,u_2)=\Pi(u_2|u_1)$ according to Eqs.(2.22),(2.23) and it follows from Eq.(2.24) that

$$\Pi(u_2|u_1)=\Pi(u_1|u_2) \wedge \Pi(u_2). \tag{2.25}$$

Solving Eq.(2.25) for $\Pi(u_1|u_2)$ and using Eq.(2.22) we obtain

$$\Pi(u_1|u_2)=\begin{cases}\Pi(u_2|u_1) & \text{for } \Pi(u_1)<\Pi(u_2),\\ [\Pi(u_2|u_1),1] & \text{for } \Pi(u_1)\geq\Pi(u_2).\end{cases} \tag{2.26}$$

In the following we summarize the results of this section.

Non-interaction is a much less stringent condition than possibilistic independence. When X_2 is possibilistically independent of X_1, then the joint distribution is non-interactive. It does not necessarily follow that X_1 is independent of X_2. When the joint distribution is non-interactive, the variables may be possibilistically dependent.

We must distinguish between the case of a non-interactive joint distribution, Eq. (2.19), and the case of a non-interactive conditional distribution, Eq.(2.22). When one of the conditional distributions is non-interactive, then the joint distribution is non-interactive. It does not necessarily follow that the other conditional distribution is non-interactive.

When a conditional distribution is non-interactive, the variables will usually be possibilistically dependent.

2.5 Examples. To illustrate the different possibility distributions we use an example from the automatic recognition of handwritten letters. In the structural recognition scheme of Hisdal et al.[H2] letters like capital U,V,J are divided into a left and a right stroke at their lowest point. The description of these letters is: 2 "essentially vertical" strokes connected at the bottom. In addition to this structural description, we have information about: u_1=height of left stroke and u_2=height of right stroke. We have thus a universe of discourse $U=U_1 \times U_2$ for the two height attributes of the letters belonging to the structural family "Two strokes connected at bottom". For purposes of illustration we let each universe consist of three possible height values $U_1 = U_2 = \{1,2,3\}$ (in units of 3 mm). Each of the three possibility distributions $\Pi(u_1,u_2)$,$\Pi(u_2|u_1)$,$\Pi(u_1|u_2)$ represents a fuzzy subset of the universe $U = U_1 \times U_2$.

Table 2.1 for the joint possibility distribution of the letter class "U or V" was set up directly from a few prior premises. E.g. that letters with equal height of the two strokes have a possibility 1 to belong to class "U or V", while a difference of 2 units between the heights was assigned a possibility 0 to belong to this class. The same table can also be used for a fuzzy description of the figure "Square", with u_1 = height, u_2 = width, or to describe the fuzzy relation "Equal" between the universes U_1,U_2(see Zadeh[Z9p.20]). Since all the entries on the main diagonal are equal to 1, it follows from rule 4 of Section 2.2 that the derived tables for the conditional possibilities are identical with the joint possibility table in this case.

Table 2.1

Possibility distributions for "Equal" or for letter class "U or V" or for the figure "Square". This table was set up for the joint possibilities $\Pi(u_1,u_2)$. With the aid of rule 4 of Section 2.2 it follows that the same table describes the conditional possibilities $\Pi(u_2|u_1)$ and $\Pi(u_1|u_2)$.u_1=height of left stroke; u_2=height of right stroke

	u_2 =	1	2	3	$\Pi(u_1)$
u_1 = 1		1	0.5	0	1
2		0.5	1	0.5	1
3		0	0.5	1	1
$\Pi(u_2)$		1	1	1	

Tables2.2 describes the letter class "Capital U or V", written by a given person with a large handwriting. They can also describe a "Rather large square". The entries marked by a star were assumed a priori, the other entries being derived from them. Because we now assume that we are presented only with letters written by the specific person with the large handwriting, the diagonal elements of the joint possibility table were assigned the shown uneven distribution. For the conditional possibilities all the diagonal elements were assigned the value 1, because we assume that if the left stroke is short, the person will also write a short right stroke as long as she wishes to write a U or V. Furthermore we made the prior assumption that the two conditional tables are transposes of each other so that we have a symmetry with respect to the two strokes.

The value 0.1 for $\Pi(u_1{=}1)$ and $\Pi(u_2{=}1)$ was derived from Eqs.(2.2),(2.3) using the u_1=1, u_2=1 entries of Tables 2.2a and 2.2b. $\Pi(u_1{=}2)$ and $\Pi(u_2{=}2)$ were derived from the prior assumption that the diagonal element in the given row (column) of $\Pi(u_1,u_2)$ is the largest one. The off-diagonal 0.1 entries in Table 2.2a are now derived from Tables 2.2b,c and Eqs.(2.2),(2.3). The 0.1 entries in Tables 2.2b,2.2c are derived from Eqs.(2.6),(2.7). We see that also in this example the derived entry for the conditional possibility has a unique value.

The entries for which the joint possibilities differ from the conditional ones are the diagonal elements and the (1,2) and (2,1) entries for Tables 2.2b, 2.2c respectively. The reason that the conditional possibilities can be larger than the joint ones is that once we assume that an event with a very low possibility has occurred, e.g. u_1=1, then the writer, who is aware of the fact that the reader has to distinguish between U's and J's, may see to it that the right stroke will also be short as long as she wishes to write a U.

Tables 2.2

Possibility distributions for letter class "Capital U or V" with u_1 =height of left stroke; u_2=height of right stroke. The distributions can also describe a "Rather large square" with u_1=height, u_2=width. The entries without a star are computed from the entries marked by a star.

Table 2.2a $\Pi(u_1,u_2)$.

	u_2 = 1	2	3	$\Pi(u_1)$
u_1 = 1	0.1*	0.1	0	0.1
2	0.1	0.9*	0.5	0.9
3	0	0.5	1*	1
$\Pi(u_2)$	0.1	0.9	1	

Table 2.2b $\Pi(u_2|u_1)$.

	u_2 = 1	2	3	$\Pi(u_1)$
u_1 = 1	1*	0.5*	0*	0.1
2	0.1	1*	0.5*	0.9
3	0*	0.5*	1*	1
$\Pi(u_2)$	0.1	0.9	1	

Table 2.2c $\Pi(u_1|u_2)$.

	u_2 = 1	2	3	$\Pi(u_1)$
u_1 = 1	1*	0.1	0*	0.1
2	0.5	1*	0.5*	0.9
3	0*	0.5*	1*	1
$\Pi(u_2)$	0.1	0.9	1	

Table 2.3 shows possibility distributions for letter class "Capital J" written by the same person. In this case we set up assumed values for the joint possibilities. The conditional possibility tables were derived from the rules of Section 2.2 and from Eqs.(2.6),(2.7). In each of the two conditional tables there are now two non-unique entries. E.g. for u_1=2, u_2=3 we have that $\Pi(u_2|u_1)$ may have any value in the interval [0.4,1]. We must now make an additional prior assumption in order to fix a unique value for this entry. Since Table 2.3 refers to the case where we are presented only with letters J, we make the prior assumption that given that u_1=2, the best choice that is left for the writer is that u_2 assume its maximum value of 3. We therefore assign the possibility 1 to this and the other three interval-valued entries in Tables 2.3b, 2.3c.

This example also explains the result of Eq.(2.6) which says that only those conditional possibilities $\Pi(u_2|u_1)$ can be different from the joint ones which correspond to the supremum of the joint possibilities over a given row; while the other conditional possibilities in that row must necessarily be equal to the joint ones. What is the reason for the discrimination against the other conditional possibilities ?

It seems that the concept of possibilities makes it possible to work with processes that are not completely random but have a certain degree of directed control. If such control is present at all, then the occurrence of a combination of u_1, u_2 values with a small marginal possibility for u_1 will have the effect that the writer will try to compensate for this faulty feature. If she does so at all, she will choose the best possible value for u_2, namely the one with the largest joint possibility. It is therefore only for the biggest entries in the row of the joint possibilities, that the conditional possibilities may exceed them.

Tables 2.3

Possibilty distributions for letter class "Capital J" with u_1=height of left stroke, u_2=height of right stroke. The tables can also describe a "Somewhat thin rectangle in horizontal position" with u_1=height, u_2=width. The entries without a star are computed from the entries marked by a star.

Table 2.3a $\Pi(u_1,u_2)$.

	u_2 =	1	2	3	$\Pi(u_1)$
u_1 = 1		0.1*	0.7*	1*	1
2		0*	0.1*	0.4*	0.4
3		0*	0*	0.1*	0.1
$\Pi(u_2)$		0.1	0.7	1	

Table 2.3b $\Pi(u_2 \mid u_1)$.

	u_2 =	1	2	3	$\Pi(u_1)$
u_1 = 1		0.1	0.7	1	1
2		0	0.1	[~~0.4~~,1]	0.4
3		0	0	[~~0.4~~,1]	0.1
$\Pi(u_2)$		0.1	0.7	1	

Table 2.3c $\Pi(u_1 \mid u_2)$.

	u_2 =	1	2	3	$\Pi(u_1)$
u_1 = 1		[~~0.4~~,1]	[~~0.7~~,1]	1	1
2		0	0.1	0.4	0.4
3		0	0	0.1	0.1
$\Pi(u_2)$		0.1	0.7	1	

Suppose now that we are asked to classify letters written by the person to whom Tables 2.2, 2.3 apply. For simplicity we assume that the person writes only the capital letters U,V,J and we must decide whether a given letter belongs to class "Capital U or V" or to class "Capital J".

We can now proceed by two slightly different methods, both of which use a universe of dicourse $U \times V = U_1 \times U_2 \times V$. U_1 represents the height of the left stroke, U_2 the height of the right stroke and V gives us information about the membership grade of the object in a letter class.

In the first method, which is the one suggested by Zadeh[Z8,Z4] we set up one 3-dimensional relation for each class. Let Y denote the variable which takes on values in V. Y can take on linguistic truth values which are answers to the question: Given the (possibly fuzzy) value of $X = X_1, X_2$, how true is it that the object belongs to the class represented by the relation ? With Zadeh's method we have one relation for class (U or V) and one relation for class J. The relations represent $\Pi_{Y|X}(v|u)$, the conditional possibility for Y given X. The value of X is obtained from measurements performed on the particular object. The object is classified as belonging to the class of the relation which results in the highest truth value.

Tables 2.4

Class conditional possibilities and classification table. u_1 = height of left stroke, u_2 = height of right stroke.

Table 2.4a. $\Pi(u|v)$ with $u=u_1,u_2$ representing the features or measurements of the object, and v the class to which it belongs. v_1=class "U or V", v_2=class "J". The entries of the two columns are the same as those of Tables 2.2a, 2.3a respectively. $\Pi(v_1),\Pi(v_2)$ are assumed to be 1. It then follows that $\Pi(u,v)$ is represented by the same table, and $\Pi(u)$ can be found from Eq.(2.4).

	v = U or V	J	$\Pi(u)$
$u=u_1,u_2$ = 1,1	0.1	0.1	0.1
2,1	0.1	0	0.1
3,1	0	0	0
1,2	0.1	0.7	0.7
2,2	0.9	0.1	0.9
3,2	0.5	0	0.5
1,3	0	1	1
2,3	0.5	0.4	0.5
3,3	1	0.1	1
$\Pi(v)$	1	1	

Table 2.4 b. $\Pi(v|u)$, the a posteriori possibilities for the class v when the measurements u of the object are known. The table is derived from Table 2.4a using Eq.(2.6). The object is classified as belonging to the column with the highest entry (marked by +).

	v = U or V	J	$\Pi(u)$
$u=u_1,u_2$ = 1,1	[0.1,1]	[0.1,1]	0.1
2,1	[0.1,1]+	0	0.1
3,1	[0.1]	[0.1]	0
1,2	0.1	[0.7,1]+	0.7
2,2	[0.9,1]+	0.1	0.9
3,2	[0.5,1]+	0	0.5
1,3	0	1+	1
2,3	[0.5,1]+	0.4	0.5
3,3	1+	0.1	1
$\Pi(v)$	1	1	

Table 2.4. A slight variation of Zadeh's method for pattern classification is to let the elements of the universe V be the possible classes. In our case, V will have two elements: v_1=class(U or V), v_2=class J. The entries in the relational tableau are now the grades of membership of the object in the class represented by the given column. Expressed in another way, the entries are the conditional possibilities $\Pi(v|u)$ for class v when the two features of the object, $u=u_1,u_2$, are

given. (See Table 2.4b.) The possibilities $\Pi(v|u)$ may also be called aposteriori possibilities for class v when the features or measurements of the object are given; as contrasted with $\Pi(v)$, the apriori possibilities of class v before we are acquainted with the measurements of the object.

The object is classified according to the column (class) with the highest conditional possibility (see + signs in Table 2.4b). When the measurements of the object do not coincide exactly with any of the u values of the table, then we assign a fuzzy value to $X=X_1,X_2$ and "interpolate" (see[Z8]) by taking the composition of the measured fuzzy X with the relation of Table 2.4b. This way we obtain the aposteriori possibilities $\Pi(v_1|u),\Pi(v_2|u)$ respectively.

Table 2.4a lists $\Pi(u|v)=\Pi(u_1,u_2|v)$, the possibilities for a measurement $u=u_1,u_2$ when the class v is given. Since Table 2.2a represents the joint possibilities for u_1,u_2 under the assumption that the object belongs to letter class "Capital U or V", the entries in the "U or V" column of Table 2.4a are identical with the entries of Table 2.2a. Similarly the entries of Table 2.3a for the letter class "Capital J" are repeated in the J column of Table 2.4a.

We now make the assumption that both letter classes have a priori or marginal possibilities $\Pi(v_1)=\Pi(v_2)=1$. From Eq.(2.3) (with u_1 in that equation being replaced by $u=u_1,u_2$ and u_2 by v) we find that Table 2.4a also represents the joint possibilities $\Pi(u,v)=\Pi(u_1,u_2,v)$.

For purposes of classification we are, however, interested in $\Pi(v|u)$, the aposteriori possibilities for the class v when the features or measurements $u=u_1,u_2$ are given. Using Eq.(2.6) and the joint possibilities of Table 2.4a, we obtain Table 2.4b for the a posteriori possibilities of class v. Again an entry like [0.1,1] indicates that $\Pi(v|u)$ may have any value in this interval unless we fix the value with the aid of an additional prior assumption. The biggest entry in each row is marked by a + to indicate the decision made by the classification scheme for an object with measurements $u=u_1,u_2$. The decision for $u_1,u_2=2,3$ is somewhat doubtful and may indicate that the 0.4 entry in Table 2.3a is too small. When the two entries in a row are equal, then the object cannot be identified.

We remark that Tables 2.2-2.4 were set up solely for the purpose of illustrating relations for possibilistically dependent variables. In a practical case one might find it worthwhile to reduce the two measurements u_1,u_2, to a single feature,e.g. the ratio u_2/u_1 of the lengths of the two strokes.

2.6. Cloncluding remarks. We have pointed out some similarities and some differences between the mathematical representations of probabilities and possibilities. The conceptual differences have been pointed out by Zadeh[Z6].

In practice there are many instances of statistical systems, e.g. in pattern recognition, where the probabilities are known only so inaccurately that the performance of the system is seriously impaired (see e.g.Duda and Hart[D1]).

In the case of a fuzzy or possibilistic system on the other hand, rough numerical values may often be sufficient. In many cases it is easy to assign reasonable numerical values to the possibilities (e.g. the value 1), without having to perform a large statistical experiment. Because of the min-max operations, the system may be less sensitive to variations in these values than the corresponding statistical system.

Thus inference with the aid of fuzzy set theory, or equivalently with the aid of a possibilistic description, lies somewhere midways between the uncertainty inherent in probabilistic reasoning and the rigid TRUE-FALSE reasoning of Boolean algebra.

It would be of great value if experimental workers would try all these three descriptions on different practical cases so that we can learn which of them is most suitable for the analysis of a given type of practical problem.

3. A GENERAL THEORY OF FUZZY SETS

At this point we cannot help but note a similarity in form between certain formulas of the theory of possibility and analogous formulas in the theory of probability. Thus Zadeh's equations (2.4), (2.5) for the marginal possibilities, (2.11) for the compositional rule of inference, and our Eqs.(2.2),(2.3) for the joint possibilities, have their equivalents in the theory of probability when we replace the max operation by + and the min operation by ·. That there exist a certain analogy between possibilities and probabilities was already pointed out by Zadeh in his original paper on possibilities[Z6]. On the other hand we also know that the theory of fuzzy sets and relations is a generalization of Boole's theory of nonfuzzy sets.

These considerations suggest that there exists a more general theory of fuzzy sets of which 1) the theory of nonfuzzy sets, 2) the theory of probability and 3) Zadeh's special theory of the max-min fuzzy sets are special cases.

The generalizations which such a theory must introduce into the classical theories of nonfuzzy sets and of probability are of two types. The first generalization concerns the grades of membership of the elements of a set. Instead of allowing only the values 0 or 1 as in Boole's theory, Zadeh's theory allows values for the grades of membership in the whole real interval [0,1]. The second generalization concerns the allowed operations between the elements of different sets. These may be either + and · or max and min. Yager's paper[Y1] shows that for one-dimensional universes there exist other operations which lead to a consistent theory. It is quite possible that his results can be generalized to the case of possibilistically dependent variables.

Table 3.1. The three special cases of generalized fuzzy set systems. The definitions of the three systems are summarized above the double line. Some consequences of the definitions are given below the double line.

	(1) nonfuzzy Boolean sets	(2) probabilistic sets	(3) Zadeh's special max-min sets
range for values of grades of membership	0 or 1	[0,1]	[0,1]
special constraints		$\sum_u \Pi(u) = 1$	
union (op.on gr. of m.s.)	max	+	max
intersection(op.on gr.of m.s.)	min	·	min
conditional possibilities equal to joint possibilities?	yes	no	often
what possibilities can be used for inference?	conditional or joint	conditional	conditional (often joint)
particularization of joint possibilities allowed?	yes	no	often

The definitions of the three types of fuzzy set systems are summarized in Table 3.1 above the double line. Theories 1) and 3) use the operation of max between corresponding elements of a set for union and the operation of min for intersection;

while theory 2) uses + and · respectively. Theories 2) and 3) operate with the extended range [0,1] for the values of the grades of membership. In this interpretation probabilities are grades of membership (or possibilities) in a fuzzy set system which uses + and · for union and intersection respectively, and which has the additional constraint that the grades of membership of a given fuzzy set must add up to 1. This serious constraint exists neither in Zadeh's max-min sets nor in Boole's nonfuzzy sets. For probabilities it is quite essential both because of the interpretation of probabilities as limiting cases of frequencies, and in order to keep the operation of union from yielding grades of membership which exceed 1.

We will from now on use the symbols ⓤ and ⓘ for the albegraic operations between grades of membership corresponding to union and intersection respectively. ⓤ and ⓘ are the operations of addition and multiplication respectively in the theory of probability. In theories 1) and 3) they are the operations of max and min.

4. THE AND CONNECTIVE

Let P_A, P_B be two propositions, and let A,B be the sets induced by these propositions. In Boolean logic the set induced by the proposition P_A AND P_B is the intersection of (the cylindrical extensions of) A and B. This definition of the conjunctive composition has been retained by Zadeh for fuzzy propositions and sets as a default definition (using the definition of intersection which is valid for fuzzy sets). When nothing else is specified, then the joint possibilities induced by "P_A AND P_B" are

$$\Pi_{A,B}(u,v) = \Pi_A(u) \wedge \Pi_B(v) . \tag{4.1}$$

This AND is called noninteractive by Zadeh because the induced set is a noninteractive relation. But he leaves open the possibility of other types of AND within his max-min system (i.e.the system where the max operation is used for union and the min operation for intersection). These other AND's are called "noninteractive" by Zadeh[Z7p.34],[Z10p.32],[Z3p.36]. In ref.[Z2] he mentions the possibility of a noninteractive AND which uses the multiplication operation instead of min in Eq. (4.1).

With the developments of the second fuzzy set period we have however the means of retaining the definition of AND from the Boolean period and still obtaining joint fuzzy sets whose possibilities differ from the right hand side of Eq.(4.1).

In the first place, as explained in the previous section, we can have generalized fuzzy set systems (e.g. a probabilistic system) which use operations other than max and min for union and intersection respectively. We therefore define the elements of a generalized noninteractive relation by the equation

$$\Pi_{A,B}(u,v) = \Pi_A(u) \; ⓘ \; \Pi_B(v). \tag{4.2}$$

An example of a proposition which induces the possibility distribution (4.2) is "John is light and tall", where "light" is represented by the fuzzy set A and "tall" by B. Such a proposition implies that we assume that weight and height are independent attributes either because this is so in reality or because we have no information about the dependence of height upon weight.

When such information is available, then the true joint distribution is obtained in all cases from the proposition:

> The possibility that (u,v) has the value (u_i,v_j) is equal to the possibility that ($u=u_i$ AND that ($v=v_j$ given that $u=u_i$)). (4.3)

We have here a conjunctive composition connecting the fuzzy set A_u (with possibilities $\Pi(u)$) and the fuzzy set $B_{v|u}$ (with possibilities $\Pi(v|u)$). By using for the

AND the standard intersection operation which is valid for the particular fuzzy set system within which we operate, we opbtain

$$\Pi(u,v) = \Pi(u) \ \textcircled{i} \ \Pi(v|u) \ . \qquad (4.4)$$

This equation is generally valid within any fuzzy set system. It will always give the true joint distribution. We have thus justified equation (2.2). It is the form which (4.4) assumes in a max-min system. (See also appendix.)

An example of a linguistic proposition which induces a possibility distribution of the form of (4.4) is "John is light and (tall considering his weight)".

A possible objection to our suggestion that probabilities are a special case of possibilities might be the following. A fundamental law in Boolean logic is that the intersection of a set with itself must yield the same set. This law has been retained by Bellman and Giertz as a requirement for all types of fuzzy set systems [B2]. However in the above form the law is not satisfied by probabilistic fuzzy sets which use multiplication for $\textcircled{i}$.

The solution to this paradox is to use a careful formulation of the logical statement containing the AND connective. A conjunctive proposition always connects two propositions concerning the value of two (possibly fuzzy) variables respectively. In the present case these two variables happen to be the same. But this is no excuse for not using the general formulation of Eqs.(4.3),(4.4). This formulation should always be used in the case of possibilistic dependence.

The question "What is the possibility that u is u_i and that u is u_i" does not make much sense logically. But the question "What is the possiblity that u is u_i AND that (u is u_i given that u is u_i)" is logically satisfactory although redundant. Its mathematical representation is given by the generally valid Eq.(4.4) with v=u.

Now $\Pi(u=u_i|u=u_i) = 1$, and the complete matrix of $\Pi(v|u)$ is the unity matrix in the present case of complete identity and dependence between u and v. Eq.(4.4) becomes therefore

$$\Pi(u=u_i,u=u_i) = \Pi(u=u_i). \qquad (4.5)$$

The right hand side of this equation represents the original fuzzy set as required.

In the following we summarize our suggestion concerning the AND connective. The conjunctive composition containing the AND should always be represented by the intersection between (the cylindrical extensions of) the fuzzy sets on each side of the AND. The joint possibilities induced by the conjunctive proposition may then have values which differ from $\Pi(u) \wedge \Pi(v)$ for two reasons. Reason 1 is that we may operate within a fuzzy set system in which the operation for intersection is different from the min operation. Reason 2 is that the possibility distribution induced by one of the component propositions is a conditional and not a marginal one.

Because AND is now always represented by the same operation within a given system, it might be well to discard the term "noninteractive AND" altogether, and to talk only about "noninteractive relations" whose elements are defined by (4.2).

We should like to point out that Zadeh's definition of noninteractive and interactive AND's, corresponding to different operations within the max-min system, was indispensable in the first period in which the difference between (and even the existence of) marginal, conditional and joint possibilities was not clear. Similarly Bellman and Giertz's requirement that the intersection of a set with itself must yield the same set was a natural one in the first period in which the concept of possibilistic variables and dependence did not exist. This requirments has been dropped by Yager[Y1]with the result that he has found a countably infinite family of $\textcircled{u}$ and $\textcircled{i}$ operations.

Although we have suggested here that different operations of $\textcircled{i}$ should not be used within the same fuzzy set system, there does exist a context in which different systems may be "mixed", namely the concept of the possibility of a possibility [H3],[H4]. Here the first possibility may be a probability, and we have then the concept of the probability of a fuzzy set or event which was defined by Zadeh already in the first period[Z11].

5. THE MAX-MIN FUZZY SETS

Boole's book [B1] is divided into two parts. The first part consists of the theory of sets and logic, the second part treats the theory of probability. In section 3 we showed that each of these parts fits into its particular niche of a generalized theory of fuzzy sets.

Table 3.1 shows that the max-min fuzzy sets lie somewhere in between the two classical systems. They have taken one of their properties, the allowed range for the grades of membership, from the theory of probability. While the operations of union and intersection are those of the theory of nonfuzzy sets. As a result there are several features of the max-min fuzzy sets which lie in between the corresponding features in the two classical theories. Some of these features are listed under the double line. Zadeh's intriguing concept of particularization is especially well suited to demonstrate the two limiting aspects of max-min fuzzy sets [H5]. But just as a quantum mechanical system is neither a pure particle nor a pure wave, so the max-min fuzzy sets represent a completely new system.

APPENDIX. TRANSLATION OF A LINGUISTIC DESCRIPTION INTO MATHEMATICAL LANGUAGE.

In the following we present a more exact formulation of how a description in ordinary language translates into a possibility distribution. Such translation is discussed by Zadeh,[Z12p.412]. Here we extend the discussion to the case of possibilistically dependent variables. As an example we use the universe U with elements u_i for the attribute Weight, and the universe V with elements v_j for Height.

We shall say that the description "light" in ordinary language induces a fuzzy subset of U with grades of membership $\mu_{light}(u)$, see first row of Table A1. A similar statement holds for "tall" in the second row. The notation for conditional fuzzy subsets is shown in the next two rows, and for joint subsets in the last row.

From Table A1 we can infer Table A2 which lists the translation of the linguistic descriptions into possibility distributions.

Table A1.

Label or description in ordinary language	Notation for grades of membership of induced fuzzy subset	Universe
light	$\mu_{light}(u)$	U
tall	$\mu_{tall}(v)$	V
light considering its height	$\mu_{light\|Height}(u\|v)$	U×V
tall considering its weight	$\mu_{tall\|Weight}(v\|u)$	U×V
light and tall	$\mu_{light,tall}(u,v)$	U×V

Table A2.

Label or description in ordinary language	Translation into mathematical language in terms of possibility distributions
light	$\Pi(u=u_i \mid \text{light}) = \mu_{light}(u_i)$
tall	$\Pi(v=v_j \mid \text{tall}) = \mu_{tall}(v_j)$
light considering its height	$\Pi[(u=u_i \mid v=v_j) \mid (\text{light} \mid \text{Height})] = \mu_{light \mid Height}(u_i \mid v_j)$
tall considering its weight	$\Pi[(v=v_j \mid u=u_i) \mid (\text{tall} \mid \text{Weight})] = \mu_{tall \mid Weight}(u_i \mid v_j)$
light and tall	$\Pi[(u=u_i, v=v_j) \mid (\text{light AND tall})] =$
(with implied dependence between Weight and Height)	$\Pi(u_i \mid \text{light})$ ⓘ $\Pi[(v_j \mid u_i) \mid (\text{tall} \mid \text{Weight})] =$ $\Pi(v_j \mid \text{tall})$ ⓘ $\Pi[(u_i \mid v_j) \mid (\text{light} \mid \text{Height})]$

We remark that our translation of the conjunctive composition (containing AND) is quite parallel to Zadeh's translation in the noninteractive case [Z12p.425]. However we replace one of the marginal distributions by a conditional one in the case of possibilistic dependence. ⓘ denotes the operation for intersection.

References

B1. G.Boole, An Investigation of the Laws of Thought on which are Founded the Mathematical Theories of Logic and Probabilities (Dover Publications,Inc.New York). (Original Edition Macmillan 1854.)

B2. R.Bellman and M.Giertz, On the Analytic Formalism of the Theory of Fuzzy Sets, Information Sciences 5 (1973) 149-156.

D1. R.O. Duda and P.E. Hart, Pattern Classification and Scene Analysis (Wiley, New York, 1973) 44, 68, 86, 92-95.

F1. W.Feller, An Introduction to Probability Theory and Its Applications, Vol.1 (Wiley, New York, 1968) 3rd.ed.

G1. B.R.Gaines, Foundations of Fuzzy Reasoning, Int. J.Man-Machine Studies, 8, 623-668, 1976.

H1. E.Hisdal, Conditional Possibilities, Independence and Noninteraction, Fuzzy Sets and Systems I (1978) 283-297.

H2 E.Hisdal, N.Christophersen, T.Ericson, A.Løkketangen and S.Vestøl, Structural recognition of handwriting, Proc. 3rd Int.Joint Conf. on Pattern Recognition, IEEE catalog number 76CH1140-3C (November 1976).

H3. E.Hisdal, Conditional and Joint Possibilities of Higher Order, ISBN 82-90230-37-0 Res.Rep. No 42, Institute of Informatics, University of Oslo, Boks 1080, Blindern, Oslo 3, Norway.

H4. E.Hisdal, Concrete and Mathematical Sets - Measures of Second Order Possibilities, ISBN 82-90239-38-9 Res.Rep.No 43, Institute of Informatics, University of Oslo, Boks 1080, Blindern, Oslo 3, Norway.

H5. E.Hisdal, Particularization - The Theory of Fuzzy Sets versus Classical Theories ISBN 82-90230-39-7 No 44, Institute of Informatics, University of Oslo, Boks 1080, Blindern, Oslo 3, Norway.

K1. A.Kaufmann, Introduction to the Theory of Fuzzy Subsets I (Academic Press,New York, 1975).

N1. N.T.Nguyen, On Conditional Possibility Distributions, Fuzzy Sets and Systems I (1978) 299-309.

Y1. R.R.Yager, On a General Class of Fuzzy Connectives, Tech.Report.No RRY 78-18, Iona College, New Rochelle, N.Y. 10801.

Z1. L.A.Zadeh, Fuzzy Sets, Inform. Control 8 (1965) 338-353.

Z2. L.A.Zadeh, Outline of a New Approach to the Analysis of Complex Systems and Decision Processes, IEEE Transactions on Systems, Man and Cybernetics SMC-3 (1973) 28-44.
Z3. L.A.Zadeh, Theory of Fuzzy Sets, Memorandum M77-1, Electronics Research Laboratory, University of California, Berkeley (1977).
Z4. L.A.Zadeh, A Fuzzy-Algorithmic Approach to the Definition of Complex or Imprecise Concepts, Int. J. Man-Machine Studies 8 (1976) 249-291.
Z5. L.A.Zadeh, Theory of Approximate Reasoning (AR), Memorandum M77/58, Electronics Research Laboratory, University of California, Berkeley (1977).
Z6. L.A.Zadeh, Fuzzy Sets as a Basis for a Theory of Possibility, Fuzzy Sets and Systems I (1978) 3-28.
Z7. L.A.Zadeh, "Calculus of Fuzzy Restrictions", in *Fuzzy Sets and Their Applications to Cognitive and Decision Processes,* L.A.Zadeh, K.S.Fu, K.Tanaka, M.Shimura (eds.), Academic Press, New York, 1975.
Z8. L.A.Zadeh, Fuzzy Sets and their Application to Pattern Classification and Scene Analysis, Memorandum M 607, Electronics Research Laboratory, University of California, Berkeley (1976).
Z9. L.A.Zadeh, Fuzzy Logic and Approximate Reasoning, Memorandum No. ERL-M479, Electronics Research Laboratory, Univ. of California, Berkeley, CA (12 Nov. 1974).
Z10.R.E.Bellman and L.A.Zadeh, Local and Fuzzy Logics, Memorandom M584, Electronics Research Laboratory, University of California, Berkeley (1976).
Z11.L.A.Zadeh, Probability Measures of Fuzzy Events, J.Math.An.Appl.23 (1968) 421-427.
Z12.L.A.Zadeh, PRUF - A Meaning Representation Language for Natural Languages, Int. J.Man-Machine Studies 10 (1978) 395-460.

[1] This paper is dedicated to Professor Lotfi A. Zadeh.

ADVANCES IN FUZZY SET THEORY AND APPLICATIONS
M.M. Gupta, R.K. Ragade, R.R. Yager (editors)

TOWARD A CALCULUS OF THE MATHEMATICAL NOTION OF POSSIBILITY

Hung T. NGUYEN

Department of Mathematics and Statistics
University of Massachusetts
Amherst, MA 01003, U.S.A.

Abstract This paper is concerned with the study of the notion of Possibility measures recently introduced by L.A. Zadeh for the analysis of situations in which the uncertainty is not statistical in nature. The intuitive notion of Possibility is presented as a set-function which is a special case of a Precapacity of Choquet. This set-function is extended to linguistic variables using general formulation of Optimization under elastic constraints. The notion of non-interaction of fuzzy variables is discussed in the symmetric and non-symmetric cases. In the symmetric case, we justify this notion by using product space representation and the concept of non-compensation. We also point out relations between Possibility measures and fuzzy measures, belief functions and probability measures. We proceed to give a definition of conditional Possibility distributions which is consistent with the non-interaction in the symmetric case. We study various properties of conditional distributions, especially relations between marginal, joint and conditional Possibility distributions. This study is not dealt under very general conditions, but only in a simple and readily understandable form.

Keywords possibility measures, non-interaction, fuzzy variables, conditional possibility distributions.

I. INTRODUCTION

This paper presents a study of the mathematical notion of possibility introduced by L.A. Zadeh [1]. This notion of uncertainty is useful for the analysis of various problems in which the uncertainty is not statistical in nature, especially in humanistic systems.

Starting from the intuitive meaning of the notion of possibility, we consider possibility measures as set-functions satisfying an appropriate axiom. These measures are extended to fuzzy sets to cover the case of linguistic variables, and this will be done by using the formulation of optimization under elastic constraints. Some fundamental concepts related to possibility measures, such as possibility distributions, non-interaction of fuzzy variables, joint and marginal possibility distributions are discussed in detail. The problem of conditioning in possibility theory is investigated and various properties of conditional possibility distributions are derived.

II. POSSIBILITY MEASURES

If U is a set, then $P(U)$ [resp. $\underset{\sim}{P}(U)$] denotes the collection of all subsets (resp. fuzzy subsets) of U. The symbols $\wedge$ and $\vee$ stand for infimum and supremum respectively. We start with the concept of possibility introduced in [1].

Definition 1. A possibility measure on U is a mapping from $\mathcal{P}(U)$ to the unit interval and satisfying the following axiom:

$$\hat{\Pi}\left(\bigcup_I A_i\right) = \bigvee_I \hat{\Pi}(A_i), \quad \text{for any index set I.} \qquad (1)$$

Remarks:

(i) $\hat{\Pi}(U) \leq 1$

(ii) Denote by $\emptyset$ the empty set, $\hat{\Pi}(\emptyset) = 0$ by convention.

(iii) As a consequence of (1), $A \subset B \Rightarrow \hat{\Pi}(A) \leq \hat{\Pi}(B)$.

(iv) $\hat{\Pi}$ is not additive, but has the property similar to that of the Hausdorff dimension:

$$\hat{\Pi}(A \cup B) = \mathrm{Max}\{\hat{\Pi}(A), \hat{\Pi}(B)\}$$

for all sets A and B, disjoint or not. More precisely $\hat{\Pi}$ is a strong precapacity [2] and formally $-\mathrm{Log}\hat{\Pi}(\cdot)$ is a generalized Information measure in the sense of Kampé de Fériet [3].

(v) For a study of the relationship between possibility measures, probability measures, fuzzy measures [4] and belief functions [5], see [6]. In particular (U finite)

a. By (ii), (iii) and (iv) a possibility measure $\hat{\Pi}$ is also a fuzzy measure if $\hat{\Pi}(U) = 1$.

b. The Dirac measure is the only possibility measure which is also a belief function, and a probability measure.

(vi) Since $\hat{\Pi}$ is defined on $\mathcal{P}(U)$, there is no need to talk about the notion of measurability in possibility theory.

Definition 2. A possibility distribution Π on U is a mapping from U to [0,1].

Remarks:

(i) In applications, a possibility distribution is some specific fuzzy subset of U.

(ii) A possibility distribution Π is a "density" of a possibility measure $\hat{\Pi}$, in the sense that:

$$\hat{\Pi}(A) = \bigvee_A \Pi(u) \qquad (2)$$

and conversely, the possibility distribution Π associated with the possibility measure $\hat{\Pi}$ is given by

$$\Pi(u) = \hat{\Pi}(\{u\}).$$

Note that $\Pi(\cdot) \not\equiv 0$. In practical situations, it is more convenient to start with a possibility distribution.

Definition 3. A fuzzy variable X on U is a variable taking values in U following a possibility distribution Π_X on U.

Remarks:

(i) Since $\Pi_X \not\equiv 0$, it is possible to consider $\Pi_X(u)$ as the possibility that X takes the value u, in symbol, Poss(X=u), for each $u \in U$.

(ii) The quantity $\hat{\Pi}_X(A) = \sup_A \Pi_X(u)$, $A \subset U$, is interpreted as Poss $(X \in A)$.

With this interpretation, the expression (2) is clear as far as the intuitive meaning of the notion of possibility is concerned.

Now assume that X is a linguistic variable [7], i.e. a variable taking values in $\mathcal{P}(U)$, it is necessary to extend the notion of possibility measure to $\mathcal{P}(U)$.

Proposition 1. Let X be a fuzzy variable on U with possibility distribution Π_X, and $\hat{\Pi}_X$ the associated possibility measure. Then:

$$A \in \mathcal{P}(U),\ \hat{\Pi}_X(A) = \text{Poss}(X \text{ is } A) = \bigvee_U [\Pi_X(u) \wedge \mu_A(u)] \tag{3}$$

where μ_A is the membership function of A.

Proof: The extension of $\hat{\Pi}_X$ to $\mathcal{P}(U)$ can be expressed formally as

$$\hat{\Pi}_X(A) = \sup_A \Pi_X(u). \tag{4}$$

The meaning of (4), when A is a fuzzy subset, has been defined and investigated in [8] as the problem of optimization of bounded real-valued functions under elastic constraints. With the notation in [8], let f_A be the restriction of Π_X to A, associated with the maximizing set $M(\Pi_X)$, i.e.

$$f_A(u) = \Pi_X(u) \wedge \Phi_{(\Pi_X,A)}(u)$$

where $$\Phi_{(\Pi_X,A)}(u) = \alpha(\Pi_X)\mu_A(u) + \beta(\Pi_X)$$

with $$\alpha(\Pi_X) = \{\bigvee_U \Pi_X(u)\} \vee 0 - \{\bigwedge_U \Pi_X(u)\} \wedge 0$$

and $$\beta(\Pi_X) = \{\bigvee_U \Pi_X(u)\} \wedge 0 + \{\bigwedge_U \Pi_X(u)\} \wedge 0.$$

Thus $$\hat{\Pi}_X(A) = \bigvee_{S_A} f_A(u)$$

where $$S_A = \{u \in U: \mu_A(u) \neq 0\}$$

But $$\alpha(\Pi_X) = \bigvee_U \Pi_X(u) = 1 \qquad (\Pi_X \text{ normal})$$

and $$\beta(\Pi_X) = 0$$

therefore $$\hat{\Pi}_X(A) = \bigvee_{S_A} [\Pi_X(u) \wedge \mu_A(u)] = \bigvee_U [\Pi_X(u) \wedge \mu_A(u)].$$

Remark: When the uncertainty in some situations is due to both randomness and fuzziness, the concept of possibility distributions can be used as subjective measures. As an illustration, let X be a random variable defined on a probability space (Ω,A,P) and taking values in a finite set $\{a_1,\ldots,a_n\}$. Suppose that the probabilities $P(X = A)$, $i=1,\ldots,n$ are not known exactly. With partial information, we may consider the fuzzy variable $Y_k = P(X = a_i)$ on the interval $[0,1]$ with some possibility distribution Π_{Y_i}, $i=1,\ldots,n$. In this case, X is regarded as having a linguistic probability distribution [7],[9], with linguistic expectation given by

$$EX = \sum_{i=1}^{n} a_i \Pi_{Y_i} . \tag{5}$$

EX is a fuzzy number with membership function μ_{EX} given by (using the extension principle [9]):

$$\mu_{EX}(z) = \begin{cases} \bigvee\limits_{(x_1,\ldots,x_n) \in [0,1]^n} [\bigwedge\limits_{i=1}^{n} \Pi_{Y_i}(x_i)] \\ x_1+\ldots+x_n = 1 \\ \sum\limits_{i=1}^{n} a_i x_i = z \end{cases}$$

The linguistic expectation (5) can be used in the following situation: Let Y be a linguistic random variable, defined on (Ω,A,P) with value in $\{\mu_1,\ldots,\mu_n\}$ where μ_i, $i=1,\ldots,n$, are fuzzy numbers. Set $p_i = P(Y = \mu_i)$, then

$EY = \sum_{i=1}^{n} p_i \mu_i$ is a fuzzy number characterized by

$$\mu_{EY}(z) = \begin{cases} \bigvee\limits_{(u_1,\ldots,u_n)} [\bigwedge\limits_{i=1}^{n} \mu_i(u_i)] \\ \sum\limits_{i=1}^{n} p_i u_i = z \end{cases}$$

In particular, if μ_i are real numbers, say

$$\mu_i(t) = \begin{cases} 1 & \text{if } t = t_i \\ 0 & \text{if } t \neq t_i \end{cases} \quad \text{for all } i$$

then (6) reduces to:

$$\mu_{EY}(z) = \begin{cases} 1 & \text{if } \sum\limits_{i=1}^{n} p_i t_i = z \\ 0 & \text{otherwise} \end{cases}$$

i.e. (6) is a generalization of the ordinary expectation.

III. NON-INTERACTION

We consider first the notion of non-interaction of two fuzzy variables X and Y on U and V respectively. This notion is symmetric in the sense that if X is non-interactive with respect to Y, then Y is also non-interactive with respect to X. Therefore we can say simply that X and Y are non-interactive. After, we shall consider a weaker, non-symmetric notion called weak non-interaction.

Intuitively, two fuzzy variables X and Y, on U and V respectively, are non-interactive if knowing that X = u with $\Pi_X(u)$, we still have total freedom to assign any value v to Y with $\Pi_Y(v)$ and vice-versa. Therefore:

$$\left.\begin{aligned} &\mathrm{Poss}[(X,Y) \in \{u\} \times V] = \mathrm{Poss}\ (X = u) \\ &\mathrm{Poss}[(X,Y) \in U \times \{v\}] = \mathrm{Poss}\ (Y = v) \end{aligned}\right\} \tag{7}$$

These relations imply that Π_X and Π_Y are marginal possibility distributions, i.e.

$$\left.\begin{aligned} \Pi_X(u) &= \sup_{v\in V} \mathrm{Poss}[X=u,Y=v] = \sup_{v\in V} \Pi_{(X,Y)}(u,v) \\ \Pi_X(v) &= \sup_{u\in U} \mathrm{Poss}[X=u,Y=v] = \sup_{u\in U} \Pi_{(X,Y)}(u,v) \end{aligned}\right\} \tag{8}$$

This follows simply from the property of the possibility measure $\hat{\Pi}_{(X,Y)}$ in the product space U × V.

<u>Remark</u>: We will discuss in detail the notion of marginal possibility distributions in section IV.

In [10], the notion of non-interaction is defined as follows:

<u>Definition 4</u>. Two fuzzy variables X and Y are said to be non-interactive with respect to Π_X, Π_Y and $\Pi_{(X,Y)}$, if:

$$\Pi_{(X,Y)}(u,v) = \Pi_X(u) \wedge \Pi_Y(v), \quad \forall(u,v) \in U \times V. \tag{9}$$

The rational for identifying non-interaction with set-intersection is provided by the following lemma [10].

<u>Lemma 1</u>. If f: [0,1] × [0,1] → [0,1] such that:

a) f is continuous in both arguments

b) f is monotone non-decreasing in both arguments

c) $f(x,0) = f(0,y), \quad \forall x,y \in [0,1]$

d) $f(x,x) = x, \quad \forall x \in [0,1]$

e) [property of non-compensation]: For all $x \in [0,1]$, there does not exist $\alpha,\beta \in [0,1]$ such that

$$\alpha > x,\ \beta < x\ (\text{or } \alpha < x,\ \beta > x) \quad \text{and} \quad f(\alpha,\beta) = f(x,x).$$

Then $f(x,y) = x \wedge y, \quad \forall x,y \in [0,1]$.

The proof of this lemma can be found in [10]. We shall proceed, following the work of J. Kampé de Fériet [11], to examine all conditions of lemma 1, starting from relations (7).

For simplicity, we will assume that Π_X, Π_Y are surjective and normal (i.e. $\operatorname{Sup}_U \Pi_X(u) = \operatorname{Sup}_V \Pi_Y(v) = 1$).

Remark: It is clear that if Π_X and Π_Y are normal, then:

(i) $\Pi_X(u) = \operatorname{Sup}_{v \in V}[\Pi_X(u) \wedge \Pi_Y(v)]$

(ii) $\Pi_{(X,Y)}$ is normal, if in addition, Π_X and Π_Y are marginals.

It is also intuitive to express the non-interaction notion in terms of possibility measures. The relations (7) become consequences of the following conditions.

For any $A \subset U$ and $B \subset V$:

$$\begin{aligned} \operatorname{Poss}[(X,Y) \in A \times V] &= \operatorname{Poss}(X \in A) \\ \operatorname{Poss}[(X,Y) \in U \times B] &= \operatorname{Poss}(Y \in B) \end{aligned} \tag{10}$$

From now on we assume that:

$$\hat{\Pi}_{(X,Y)}(A \times B) = f[\hat{\Pi}_X(A), \hat{\Pi}_Y(B)] \tag{11}$$

which implies that $\Pi_{(X,Y)}(u,v) = f[\Pi_X(u), \Pi_Y(v)]$.

Proposition 2. We have

$$f(x,0) = f(0,y) = 0, \quad \forall x,y \in [0,1].$$

Proof: From (8), $\Pi_Y(v) = \operatorname{Sup}_U \Pi_{(X,Y)}(u,v), \quad \forall v \in V$. If $\Pi_Y(v_\circ) = 0$ and $\Pi_X(u_\circ) = x$, then:

$$f(x,0) = \Pi_{(X,Y)}[\Pi_X(u_\circ), \Pi_Y(v\ )] \le \Pi_Y(v_\circ) = 0.$$

Proposition 3. $f(\cdot,\cdot)$ is monotone non-decreasing in both arguments.

Proof: Assume that $\Pi_X(\hat{u}) \le \Pi_X(u^1)$. Let $\hat{x} = \Pi_X(\hat{u})$ and $x^1 = \Pi_X(u^1)$, and

$A = \Pi_X^{-1}([0,\hat{x}]) \subset B = \Pi_X^{-1}([0,x^1])$. Since $A \times \{v\} \subset B \times \{v\} \Longrightarrow$

$\hat{\Pi}_{(X,Y)}(A \times \{v\}) \le \hat{\Pi}_{(X,Y)}(B \times \{v\})$ but by (11) $\hat{\Pi}_{(X,Y)}(A \times \{v\}) = f[\hat{\Pi}_X(A), \Pi_Y(v)] =$

$f(\hat{x},y) \le f[\hat{\Pi}_X(B), \Pi_Y(v)] = f(x^1,y)$.

Proposition 4. $f(\cdot,\cdot)$ is continuous in both arguments.

Proof: The continuity of $f(x,y)$ is an immediate consequence of the properties of possibility measures $\hat{\Pi}_X$, $\hat{\Pi}_Y$ and $\hat{\Pi}_{(X,Y)}$.

Proposition 5. Assume that

(i) For each $x \in [0,1]$, $f(x,\cdot)$ takes all values in $[0,x]$

(ii) $f[f(x,y),z] = f[x,f(y,z)]$

(iii) f satisfies the non-compensation property. Then $\forall x \in [0,1]$, $f(x,x) = x$.

Proof: By the marginal distribution property, we have $f(x,y) \leq x \wedge y$, $\forall x,y \in [0,1]$. Therefore $f(x,x) \leq x$. If $f(x,x) < x$, then take $\alpha = f(x,x)$, so $\alpha < x$. By (i), there exists $\beta \in [0,1]$ such that $f(x,\beta) = x$. By (ii), $f[f(x,x);\beta] = f(\alpha,\beta) = f[x,f(x,\beta)] = f(x,x)$. But $f(x,\beta) = x > f(x,x) \Rightarrow \beta > x$ since f is increasing, therefore (iii) is violated.

Remarks:

(a) Since $f(x,y) \leq x \Rightarrow$ the range of $f(x,\cdot) \subset [0,x]$.

(b) The associativity (i) follows from $u \times v \times w = (u \times v) \times w = u \times (v \times w)$.

(c) If $f(x,x) = x$, $\forall x \in [0,1]$, then the non-compensation property is satisfied.

IV. CONDITIONING

IV.-1 Marginal distributions. Let (X,Y) be a fuzzy variable on $U \times V$ with joint possibility distribution $\Pi_{(X,Y)}$.

Definition 4 [7]. The marginal possibility distributions Π_X and Π_Y are defined by:

$$\begin{cases} \Pi_X(u) = \bigvee_V \Pi_{(X,Y)}(u,v) \\ \\ \Pi_Y(v) = \bigvee_U \Pi_{(X,Y)}(u,v) \end{cases} \tag{12}$$

Remark: If Π_X and Π_Y are normal, then:

$$\Pi_X(u) = \bigvee_V [\Pi_X(u) \wedge \Pi_Y(v)]$$

therefore, if X and Y are non-interactive, the relations (12) are consistent.

Proposition 6. If Π_X and Π_Y are marginal possibility distributions of $\Pi_{(X,Y)}$, then:

$$\begin{cases} \Pi_X(u) = \bigvee_{v \in V} \Pi_{X/Y}(u/v) \\ \\ \Pi_Y(v) = \bigvee_{u \in U} \Pi_{Y/X}(v/u) \end{cases} \tag{13}$$

where $\Pi_{X/Y}$ and $\Pi_{Y/X}$ are conditional possibility distributions [12].

Proof: Recall that [12] (see also section IV-2 below):

$$\Pi_{X/Y}(u/v) = \begin{cases} \Pi_{(X,Y)}(u,v) & \text{if } \Pi_X(u) \le \Pi_Y(v) \\ \Pi_{(X,Y)}(u,v)\dfrac{\Pi_X(u)}{\Pi_Y(v)} & \text{if } \Pi_X(u) > \Pi_Y(v) \end{cases} \tag{14}$$

From this explicit definition:

$$\Pi_{(X,Y)}(u,v) \le \Pi_{X/Y}(u/v) \le \Pi_X(u)$$

therefore: $$\bigvee_{v\in V} \Pi_{(X,Y)}(u,v) \le \bigvee_{v\in V} \Pi_{X/Y}(u/v) \le \Pi_X(u).$$

But $$\Pi_X(u) = \bigvee_{v\in V} \Pi_{(X,Y)}(u,v) \Rightarrow \Pi_X(u) = \bigvee_{v\in V} \Pi_{X/Y}(u/v).$$

Proposition 7. $\Pi_X(u) = \Pi_{X/Y}(u/\hat{v})$ where $\hat{v}$ is solution of $\Pi_{Y/X}(v/u) = 1$.

Proof: From (14) we have the following relation:

$$\Pi_{X/Y}(u/v)\Pi_Y(v) = \Pi_{Y/X}(v/u)\Pi_X(u) \tag{15}$$

Now let $\hat{v}$ such that $\Pi_{Y/X}(\hat{v}/u) = 1$. From (13): $\Pi_Y(\hat{v}) = \bigvee_{u^1\in U} \Pi_{Y/X}(\hat{v}/u^1) \Rightarrow$

$\Rightarrow \Pi_Y(\hat{v}) = 1$. Thus, for $v = \hat{v}$, the relation (15) becomes:

$$\Pi_{X/Y}(u/\hat{v})\Pi_Y(\hat{v}) = \Pi_{Y/X}(\hat{v}/u)\Pi_X(u) \Rightarrow \Pi_{X/Y}(u/\hat{v}) = \Pi_X(u).$$

IV.-2 Conditional possibility distributions. Let X and Y be two random variables with joint density $f(x,y)$ and marginal densities $f_X(x), f_Y(y)$.

Denote by $f_{X/Y}$ the conditional density of X given Y. Then:

(i) $$f_{X/Y}(x/y)f_Y(y) = f_{Y/X}(y/x)f_X(x) \tag{16}$$

(ii) $$f_{X/Y}(x/y) = \begin{cases} \psi(x,y) & \text{if } f_X(x) \le f_Y(y) \\ \psi(x,y)\dfrac{f_X(x)}{f_Y(y)} & \text{if } f_X(x) > f_Y(y) \end{cases} \tag{17}$$

where $\psi(x,y) = f_{X/Y}(x/y) \wedge f_{Y/X}(y/x)$. Note that, in general, $\psi(x,y) \ne f(x,y)$.

Now let X and Y be two fuzzy variables on U and V respectively. To define the conditional possibility distribution of X given Y, in symbol $\Pi_{X/Y}(u/v)$, we start with the following assumptions:

a) Π_X and Π_Y are marginal distributions of $\Pi_{(X,Y)}$.

b) Π_X and Π_Y are normal.

Definition 6 [12].

$$\Pi_{X/Y}(u/v) = \Pi_{(X,Y)}(u,v)\left[1 \vee \frac{\Pi_X(u)}{\Pi_Y(v)}\right], \quad \Pi_Y(v) \neq 0 \tag{18}$$

Remarks:

(i) This is a condensed form of (14).

(ii) a) and b) $\Rightarrow \Pi_{(X,Y)}$ is normal.

(iii) $0 \leq \Pi_{X/Y}(u/v) \leq 1$ by a).

(iv) b) $\Rightarrow \Pi_X(u) = \operatorname{Sup}_{v \in V}[\Pi_X(u) \wedge \Pi_Y(v)]$, $\forall u \in U$.

(v) In [12], the relation (18) was derived from a functional equation compatible with the non-interaction, i.e.

If $$\Pi_{(X,Y)}(u,v) = \Pi_X(u) \wedge \Pi_Y(v)$$

then $$\begin{cases} \Pi_{X/Y}(u/v) = \Pi_X(u) \\ \Pi_{Y/X}(v/u) = \Pi_Y(v). \end{cases}$$

(vi) It is proved in [12] that:

$$\Pi_X(u) = \bigvee_{v \in V} [\Pi_{X/Y}(u/v) \wedge \Pi_Y(v)].$$

(vii) It is clear that:

$$\Pi_{X/Y}(u/v)\Pi_Y(v) = \Pi_{Y/X}(v/u)\Pi_X(u).$$

We get the same relation as (16) since $\Pi_{X/Y}$ has the same form as $f_{X/Y}$ (see relation (17)).

Recently, Zadeh has suggested that conditional possibility distributions have to be defined such that:

$$\Pi_{X/Y}(u/v) \wedge \Pi_{Y/X}(v/u) = \Pi_{(X,Y)}(u,v). \tag{19}$$

It is easy to see that (19) is satisfied for conditional possibility distributions defined by (18).

Proposition 8. Given $\Pi_{X/Y}$ and $\Pi_{Y/X}$, define:

$$\begin{cases} \Pi_X(u) = \bigvee_{v\in V} \Pi_{X/Y}(u/v) \\ \Pi_Y(v) = \bigvee_{u\in U} \Pi_{Y/X}(v/u) \end{cases}$$

Then
$$\Pi_{X/Y}(u/v)\Pi_Y(v) = \Pi_{Y/X}(v/u)\Pi_X(v) \tag{20}$$

is a necessary and sufficient condition for:

$$\Pi_{X/Y}(u/v) = \Pi_{(X,Y)}(u,v)\left[1 \vee \frac{\Pi_X(u)}{\Pi_Y(v)}\right] \tag{21}$$

where $\Pi_{(X,Y)}(u,v) = \Pi_{X/Y}(u/v) \wedge \Pi_{Y/X}(v/u)$.

Proof:

a) Sufficiency Assume (20)

If $\Pi_X(u) \le \Pi_Y(v) \Rightarrow \Pi_{X/Y}(u/v) \le \Pi_{Y/X}(v/u)$ by (20),

therefore: $\Pi_{(X,Y)}(u,v) = \Pi_{X/Y}(u/v)$.

If $\Pi_X(u) > \Pi_Y(v) \Rightarrow \Pi_{X/Y}(u/v) > \Pi_{Y/X}(v/u)$ by (20),

therefore: $\Pi_{(X,Y)}(u,v) = \Pi_{Y/X}(v/u)$.

But from (20),
$$\Pi_{X/Y}(u/v) = \Pi_{Y/X}(v/u)\frac{\Pi_X(u)}{\Pi_Y(v)}$$
$$= \Pi_{(X,Y)}(u,v)\frac{\Pi_X(u)}{\Pi_Y(v)}.$$

b) Necessity Assume (21)

If
$$\Pi_X(u) \le \Pi_Y(v) \Rightarrow \begin{cases} \Pi_{X/Y}(u/v) = \Pi_{(X,Y)}(u,v) \\ \Pi_{Y/X}(v/u) = \Pi_{(X,Y)}(u,v)\dfrac{\Pi_Y(v)}{\Pi_X(u)} \end{cases} \Rightarrow (20).$$

If $\Pi_X(u) > \Pi_Y(v)$, by symmetry, we also get (20).

V. WEAK NON-INTERACTION

As far as we are concerned with the notion of non-interaction defined in Section III, a consistent definition of conditional possibility distribution is the one proposed in Section IV. Note that, in this approach, we have defined conditional distributions from joint distributions.

Now suppose that, given two fuzzy variables X and Y, it is easier to assign subjectively $\Pi_{X/Y}$ and $\Pi_{Y/X}$ simply as two possibility distributions. Based on the investigation in Section IV above, there are good grounds to believe that

$$\Pi_{(X,Y)}(u,v) = \Pi_{X/Y}(u/v) \wedge \Pi_{Y/X}(v/u) \tag{22}$$

which can be taken as a rule to infer joint distribution from conditional ones.

We say that X is weakly non-interactive with respect to Y if

$$\Pi_{X/Y}(u/v) = \Pi_X(u)$$

where
$$\Pi_X(u) = \bigvee_V \Pi_{X/Y}(u/v).$$

This notion of weak non-interaction is obviously non-symmetric. We say that X and Y are weakly non-interactive if X is weakly non-interactive with respect to Y and vice versa.

We always have:

$$\Pi_{(X,Y)}(u,v) \leq \Pi_X(u) \wedge \Pi_{Y/X}(v/u) \leq \Pi_X(u) \wedge \Pi_Y(v).$$

If X and Y are weakly non-interactive, then:

$$\Pi_{(X,Y)}(u,v) = \Pi_X(u) \wedge \Pi_Y(v).$$

If X is weakly non-interactive with respect to Y, then:

$$\Pi_{(X,Y)}(u,v) = \Pi_X(u) \wedge \Pi_{Y/X}(v/u) \tag{23}$$

(but the converse is not necessarily true). It might happen that X and Y are not weakly non-interactive, but:

$$\Pi_{(X,Y)}(u,v) = \Pi_X(u) \wedge \Pi_{Y/X}(v/u) = \Pi_Y(v) \wedge \Pi_{X/Y}(u/v).$$

Note that the infimum operation creates the non-symmetry. It is interesting to investigate this weak non-interaction and develop an appropriate formulation for conditional possibility distributions.

VI. CONCLUDING REMARKS

This paper touches upon only some fundamental concepts in the theory possibility. Since the introduction of fuzzy sets, applications and calculus using fuzzy concepts have been widely developed. The calculus of possibility seems important in human decision-making problems. This direction has to be explored in order to provide rigorous basis for inference from humanistic systems. For example, for the problem of possibility qualification in natural languages discussed in [1], it is interesting to extend not only the domain of a possibility measure but also its range. More specifically, a possibility measure $\tilde{\Pi}$ on U can be regarded as a mapping from $\underset{\sim}{P}(U)$ to $\underset{\sim}{P}[0,1]$. To meet the needs in general situations, when we face randomness and fuzziness, it is necessary to consider fuzzy variables having linguistic probability or possibility distributions.

REFERENCES

[1] L. A. Zadeh, Fuzzy sets as a basis for a theory of possibility, Fuzzy Sets and Systems, Vol. 1, No. 1 (1978), 3-28.

[2] H. T. Nguyen, Sur les mesures d'information de type Inf. Lecture Notes in Math., No. 398 (1974), 62-75, Springer-Verlag.

[3] J. Kampé de Fériet, Mesure de l'Information fournie par un évènement, Coll. Inter. C.N.R.S. 186, Paris (1970), 191-221.

[4] M. Sugeno, Theory of fuzzy integrals and its applications, Thesis, Tokyo Inst. of Tech., Tokyo (1974).

[5] G. Shafer, A mathematical theory of evidence, Princeton University Press (1976).

[6] G. Banon, Distinction entre plusieurs sous-ensembles de mesures floues, Note interne LAAS No. 78I11 (1978) Toulouse.

[7] L. A. Zadeh, The concept of a linguistic variable and its application to approximate reasoning, Inf. Sci., 8 (1975), 119-249, 301-357, 9 (1975), 43-80.

[8] H. T. Nguyen, Some mathematical tools for linguistic probabilities, Proceedings IEEE on Decision and Control (1977), 1345-1350.

[9] H. T. Nguyen, On fuzziness and linguistic probabilities, J. Math Anal. and Appl. Vol. 61, No. 3 (1977), 658-671.

[10] R. E. Bellman, L. A. Zadeh, Local and fuzzy logics, Modern uses of multiple-valued logics, D. Epstein, ed. Dordrecht, D. Reidel (1977).

[11] J. Kampé de Fériet, Remarks on the definition of a probability on a product-space with given marginal probabilities, J. Math and Phy. Sci., 6 (1972), 129-132.

[12] H. T. Nguyen, On conditional possibility distributions, Fuzzy sets and Systems, 1 (1978), 299-309.

ADVANCES IN FUZZY SET THEORY AND APPLICATIONS
M.M. Gupta, R.K. Ragade, R.R. Yager (editors)

ON THE THEORY OF FUZZY SWITCHING MECHANISMS (FSM's)*

Abraham KANDEL

Associate Professor and Director of Computer Science
Department of Mathematics
Florida State University
Tallahassee, Florida 32306 U.S.A.

Abstract

The theory of fuzzy switching mechanisms (FSM's) described in this paper is related both to the theory of fuzzy sets and to the treatment of switching circuits in the binary world. In this paper we are concerned with the study of such imprecise functions, their properties, and possible applications. The enumeration of the number of distinct fuzzy switching functions will ba addressed as well as minimization and simplification procedures.

1. Introduction

Let $X = \{x\}$ denote a space of objects. Then a fuzzy set (Zadeh [1]) A in X is a set of ordered pairs $A = \{(x, \chi_A(x))\}$, $x \in X$ where $\chi_A(x)$ is termed the grade membership of x in A. We shall assume for simplicity that $\chi_A(x)$ is a number in the interval [0,1], with grades 1 and 0 representing respectively, full membership and nonmembership in a fuzzy set. In the sequel, the term "fuzzy variable" will replace the term "membership grade" of a variable in a set.

Definition 1: A fuzzy algebra is the system $\mathcal{Z} = \langle Z,+,*,-\rangle$ where Z has at least two distinct fuzzy variables, and $\forall x,y,z, \in Z$, system $\mathcal{Z}$ satisfies the following set of axioms:

(1) Idempotency: $x + x = x$ $\qquad x*x = x$

(2) Commutativity: $x + y = y + x$ $\qquad x*y = y*x$

(3) Associativity: $(x + y) + z = x + (y + z)$ $\qquad (x*y)*z = x*(y*z)$

(4) Absorption: $x + (x*y) = x$ $\qquad x*(x + y) = x$

(5) Distributivity: $x + (y*z) = (x + y)*(x + z)$ $\qquad x*(y + z) = x*y + x*z$

(6) Complement: If $x \in Z$ then there is a unique complement $\bar{x}$ of x such that $\bar{x} \in Z$ and $\bar{\bar{x}} = x$.

(7) Identities: $(\exists! e_+)(\forall x)$ such that $x + e_+ = e_+ + x = x$
$(\exists! e^*)(\forall x)$ such that $x*e_* = e_**x = x$

(8) De-Morgan Laws: $\overline{x + y} = \bar{x}*\bar{y}$ $\qquad \overline{x*y} = \bar{x} + \bar{y}$

The system is a distributive lattice with existence of unique identities under + and *. It is noted that a Boolean algebra is a complemented distributive lattice with existence of unique identities under + and *. However, for every element x in Boolean algebra there exists a unique complement, $\bar{x}$, such

*This paper is dedicated to Professor Raymond T. Yeh

that $x\bar{x} = 0$ and $x + \bar{x} = 1$, which is not so in fuzzy algebra. Hence, every Boolean algebra is a fuzzy algebra, but not vice versa.

In this case we will use a particular fuzzy algebra defined by the system

$$\mathcal{Z} = \langle [0,1],+,*,- \rangle,$$

where +, *, and - are interpreted as Max, min, and complement ($\bar{x} = 1 - x$, $\forall x, x \in [0,1]$), respectively. The unique identities e_+ and e_* are 0 and 1, respectively.

Conventionally, we shall drop the * symbol, i.e., $x*y$ will be written as xy.

We can now define fuzzy forms, generated by $x_1,\ldots,x_n$, recursively as follows:

a) The numbers 0 and 1 are fuzzy forms.

b) A fuzzy variable x_i is a fuzzy form.

c) If A is a fuzzy form, then $\bar{A}$ is a fuzzy form.

d) If A and B are fuzzy forms, then A + B and AB are fuzzy forms.

e) The only fuzzy forms are those given by rules (a)-(d).

As in two valued logic, we shall merge the concepts of fuzzy switching functions and fuzzy forms, to an extent by representing the mapping by fuzzy forms.

The grade membership $\chi(S)$ of a form S is uniquely determined through the following rules:

1) $\chi(S) = 0$ if $S = 0$;

2) $\chi(S) = 1$ if $S = 1$;

3) $\chi(S) = \chi(x_i)$ if $S = x_i$;

4) $\chi(S) = 1 - \chi(A)$ if $S = \bar{A}$

5) $\chi(S) = \min[\chi(A),\chi(B)]$ if $S = AB$; and

6) $\chi(S) = \text{Max}[\chi(A),\chi(B)]$ if $S = A + B$.

In order to determine consistency of forms in fuzzy algebra we shall expand the forms into disjunctive and conjunctive forms, using the following definitions:

A literal is a variable x_i , or $\bar{x}_i$, the complement of x_i.

A clause is a disjunction of one or more literals.

A phrase is a conjunction of one or more literals.

A form S is said to be in disjunctive normal form if $S = P_1 + P_2 + \cdots + P_m$, $m \geq 1$ and every P_i , $1 \leq i \leq m$, is a phrase.

A form S is said to be in conjunctive normal form if $S = C_1C_2,\ldots,C_k$, $k \geq 1$ and every C_j, $1 \leq j \leq k$, is a clause

It can be easily seen that forms in fuzzy logic can be expressed in disjunctive and conjunctive normal forms, in a similar way to two-valued logic.

2. Fuzzy Functions and FSM's

A lattice L is defined to be a partially ordered set with the property that every two elements of the poset have a greatest lower bound (glb) and a least upper bound (lub). (The glb and the lub of any two elements may be found by taking the intersection and the union, respectively, of the two elements.) The poset is usually represented diagramatically for simple posets by placing the elements of the set at the vertices of a graph; lines between the vertices connect two elements to their glb and the lub.

An atom of a lattice is any element of that lattice which "covers" the minimal element(s) of the lattice -- i.e., there is no element between the minimal element and any atom. More precisely, let a be an element of L. Then a is an atom iff $a \neq 0$ and for all elements x of L , either $ax = 0$ or $ax = a$. In other words, the minimum of an atom and any other element of the lattice must be either that atom or else zero.

The lattice representing fuzzy switching functions over n variables may be constructed in a manner similar to that used in constructing a Boolean lattice, although the process is much more tedious because the number of vertices increase at a much greater rate.

Proposition 1: The unique atom of the lattice representing all fuzzy functions of n variables is the minterm $x_1\bar{x}_1x_2\bar{x}_2\ldots x_n\bar{x}_n$.

Proof: We will first show that the minterm $x_1\bar{x}_1x_2\bar{x}_2\ldots x_n\bar{x}_n$ is indeed an atom. By the definition of the atom, for all other elements q of the fuzzy lattice, $qx_1\bar{x}_1\ldots x_n\bar{x}_n$ must be either 0 or the atom itself. Obviously if $q = 0$, then the product will be 0 ; if q is non-zero, then the product must be the term $x_1\bar{x}_1\ldots x_n\bar{x}_n$ again by the rules of minimum and maximum (alternatively, $x_1\bar{x}_1\ldots x_n\bar{x}_n$ must contain all the factors of q ; therefore, their product will be the atom itself).

To show the uniqueness of this atom, consider the following consequences of the basic definition of an atom of a lattice: if a and b are atoms, then if $ab \neq 0$, $a = b$. We have shown that there exists at least one atom; let it be a. The above proof demonstrates that for $b \neq 0$, $ab = a$. Since $a \neq 0$, $ab \neq 0$, so $a = b$. Therefore, any other atom possible turns out to be the same atom.

From the uniqueness of this atom we can conclude that the diagram of the lattice for fuzzy function of n variables has the above fundamental phrase directly above (distance 1 from) the "bottom" of the graph (zero).

The size of the lattice - the number of vertices which it has - is a direct measurement of the number of fuzzy functions of n variables that exist. Before being able to specify the lattice for a generalized case (instead of the specific cases given above), it will be necessary to determine exactly how many unique fuzzy switching functions (FSF's) there are.

3. Fuzzy Primes Implicants and Fuzzy Forms

In a disjunctive normal form, each phrase corresponds to a logic gate and each literal to an input line. The ratio between the cost of a logic gate and the cost of an input line will depend on the type of gates used in the realization. However, practically, the cost of an additional input line on an already existing gate, will be several times less than the cost of an additional logic gate. On this basis, the elimination of gates will be the primary objective of the minimization process, leading to the following definition of a minimal expression.

Definition 2: A disjunctive normal form is regarded as a minimal complexity form if there exists

(1) no other equivalent form involving fewer number phrases, and

(2) no other equivalent form involving the same number of phrases but a smaller total number of literals.

Definition 3: A phrase f_j subsumes another phrase f_ℓ iff f_j contains all the literals of f_ℓ, and thus $\chi(f_j) \le \chi(f_\ell)$. A fuzzy phrase f_k is said to be a fuzzy implicant of F iff $\chi(f_k) \le \chi(F)$ under all possible assignments. A fuzzy implicant f_j is said to be a fuzzy prime implicant of F. (i.e., $\chi(f_j) \le \chi(f_k) \le \chi(F) \leftrightarrow k = j$). It has been shown [Kandel, [3]] that the minimal complexity form must consist of a sum of phrases representing fuzzy prime implications. In order to find the complete set of F.P.I.'s the fuzzy consensus has been defined and utilized as follows:

Theorem 1: let P be a phrase of fuzzy literals from the set $\{x_i\}_{i=1}^n$. A disjunction D of any variable x_k and its complement $\bar{x}_k$, $1 \le k \le n$, can be appended to P without affecting the general value of the phrase iff there exists a variable x_i and its complement $\bar{x}_i$ in P, for some i, $1 \le i \le n$.

Proof: a) Assume x_i and $\bar{x}_i$ in P. ($P = \alpha x_i \beta \bar{x}_i \gamma$, where α, β and γ are conjunctions of literals from the set $\{x_i\}_{i=1}^n$). Obviously, $x_i \bar{x}_i \le 0.5$ and thus, $P \le 0.5$. However, $D \ge 0.5$ and therefore the "if" part is proved.

b) Assume that $x_i \bar{x}_i$ for some i, $1 \le i \le n$ is not in P. Then assign grades of membership larger than 0.5 to all literals of P. Clearly, $P \ge 0.5$. Q.E.D.

Similarly, we can prove the dual theorem.

Theorem 2: Let C be a clause of fuzzy literals from the set $\{x_i\}_{i=1}^n$. A conjunction Q of any variable x_k and its complement $\bar{x}_k$, $1 \le k \le n$, can be appended to C without affecting the general value of the clause iff there exists a variable x_i and its complement $\bar{x}_i$ in C, for some i, $1 \le i \le n$.

In general if F is a conjunction of formulas it can take a value ≤ 0.5 if certain conditions are satisfied. Clearly, if F is of the form $F = [\sum_j x_j \bar{x}_j \gamma_j]\beta$, $1 \le j \le n$ when β and $\{\gamma_j\}_{j=1}^n$ are formulas in $\{x_i\}_{i=1}^n$, then $F \le 0.5$. This is a trivial case where one formula in the conjunction is ≤ 0.5 and thus F is ≤ 0.5. A more general case can be proven as follows.

Theorem 3: Let the set $\{F_i\}_{i=1}^{\omega}$ be a set of fuzzy formulas over $x_1, x_2, \ldots, x_n$, and let F be a conjunction of formulas from this set. A disjunction F_d, of any formula F_k and its complement $\bar{F}_k$, can be appended to or deleted from the conjunction representing F, without affecting the value of F, if there exist functions F_s and F_t in the conjunction representing F such that $\bar{F}_s$ is subsumed by F_t.

Proof: Let F_s and F_t be in the conjunction representing F such that $\bar{F}_s$ is subsumed by F_t. Then $F_s F_t \le 0.5$ since

$$\overline{F_sF_t} = \bar{F}_s + \bar{F}_t = \bar{F}_s + F_t + \bar{F}_t \geq 0.5 \qquad \text{Q.E.D.}$$

Theorem 4: Let the set $\{F_j\}_{j=1}^{\nu}$ be a set of fuzzy formulas over $x_1, x_2, \ldots, x_n$, and let F be a disjunction of formulas from this set.

A conjunction F_c of any formula F_r and its complement $\bar{F}_r$, can be appended to or deleted from the disjunction representing F, without affecting the value of F, if there exist functions F_i and F_m in the disjunction representing F such that F_i is subsumed by $\bar{F}_m$.

Proof: $F_i + F_m \geq 0.5$ since $F_i + F_m = F_i + \bar{F}_m + F_m$. Q.E.D.

It should be noted that these formulas can be generated by combining several subformulas under the rules of fuzzy algebra. Thus the sets $\{F_j\}_{j=1}^{\omega,\nu}$ are compositions or decompositions of a variety of fuzzy forms.

It is obvious that in order to achieve the minimized form of a fuzzy switching function no modifications in the original definition of fuzzy consensus [6] are needed. However, in order to obtain the sum of all fuzzy prime implicants of a fuzzy switching function and not only the essential FPI's the following definition of fuzzy consensus should be used.

Definition 4: Let R and Q be two phrases over the set of fuzzy variables $x_1, x_2, \ldots x_n$. The fuzzy consensus of R and Q, written $R\psi Q$, is defined to be the set of phrases $\{R_iQ_i\}$, where $R = x_iR_i$ and $Q = \bar{x}_iQ_i$ (or $R = \bar{x}_iR_i$ and $Q = x_iQ_i$) and $x_i \in \{x_1, x_2, \ldots, x_n\}$, if the phrase R_iQ_i includes the conjunction $x_j\bar{x}_j$ for at least one j, $j \in \{1,2,\ldots,n\}$. If the phrase R_iQ_i does not include $x_j\bar{x}_j$ for any j, $j \in \{1,2,\ldots,n\}$, then

$$R\psi Q = \{R_iQ_ix_j\bar{x}_j \mid j = 1,2,\ldots,n\},\ x_i \in \{x_1, x_2, \ldots, x_n\}.$$

if none of the above occurs then we say that

$$R\psi Q = 0$$

The phrases added whenever $x_j\bar{x}_j \not\subset R_iQ_i$ are not needed fuzzy prime implicants. This can be seen from the following proposition.

Proposition 2: Let $R = x_iR_i$ and $Q = \bar{x}_iQ_i$ (or $R = \bar{x}_iR_i$ and $Q = x_iQ_i$) and R_iQ_i does not include $x_j\bar{x}_j$ for any j, $j \in \{1,2,\ldots,n\}$. Then

$$R + Q + (R\psi Q) = R + Q$$

Proof:
$$\begin{aligned} R+Q+(R\psi Q) &= x_iR_i + \bar{x}_iQ_i + \{R_iQ_ix_j\bar{x}_j \mid j = 1,2,\ldots,n\} = \\ &= x_iR_i + \bar{x}_iQ_i + \{R_iQ_ix_j\bar{x}_j(x_i + \bar{x}_i) \mid j = 1,2,\ldots,n\} = \\ &= x_iR_i + \bar{x}_iQ_i + \{x_iR_iQ_ix_j\bar{x}_j + \bar{x}_iQ_iR_ix_j\bar{x}_j \mid j = 1,2,\ldots,n\} = \end{aligned}$$

$= x_i R_i + \bar{x}_i Q_i = R + Q.$ Q.E.D.

We shall define two kinds of fuzzy phrases. The first kind, to which we shall refer as type-1 phrase, are phrases which contain a conjunction of the form $x_j \bar{x}_j$ for at least one j, $j \in \{1,2,\ldots,n\}$. Otherwise we refer to the phrase as a type-2 phrase. Clearly a type-1 phrase cannot be subsumed by type-2 phrases. However, they can subsume some of them. For the case where members of the set $\{R_i Q_i\}$ do not include a conjunction of the form $x_j \bar{x}_j$ for at least one j, $j \in \{1,2,\ldots,n\}$, two situations must be checked:

a) R and Q are both type-2 phrases. Since $\{R_i Q_i x_j \bar{x}_j \mid j = 1,2,\ldots,n\}$ is a set of type-1 phrases covered by $R + Q$, this set is not needed.

b) R is a type-1 phrase and Q is a type-2 phrase. In order for members of $\{R_i Q_i\}$ not to include a conjunction $x_j \bar{x}_j$ for any j, $j \in \{1,2,\ldots,n\}$, R must be of the form $\alpha x_i \bar{x}_i \beta$ and Q must be of the form $\gamma x_i \delta$ (or $\gamma' x_i \delta'$) where the phrase $\alpha x_i \beta\gamma\delta$ (or $\alpha \bar{x}_i \beta\gamma'\delta'$) is a type-2 phrase. Thus

$$R_i = \alpha x_i \beta \quad (\text{or} \quad \alpha \bar{x}_i \beta)$$

$$Q_i = \gamma\delta \quad (\text{or} \quad \gamma'\delta')$$

and obviously the set $\{R_i Q_i x_j \bar{x}_j \mid j = 1,2,\ldots,n\}$ is covered by Q and is not needed.

<u>Example 1</u>: Let $f(x_1,x_2,x_3) = x_1 x_2 + \bar{x}_2 x_3$.
If the variables are Boolean then the Boolean consensus will add the phrase $x_1 x_3$. However in the fuzzy case

$$x_j \bar{x}_j \not\leq x_1 x_3, \quad j \in \{1,2,3\}$$

and thus

$$x_1 x_2 \,\psi\, \bar{x}_2 x_3 = x_1 x_3 (x_1 \bar{x}_1 + x_2 \bar{x}_2 + x_3 \bar{x}_3) = x_1 \bar{x}_1 x_3 + x_1 x_2 \bar{x}_2 x_3 + x_1 x_3 \bar{x}_3.$$

The phrase $x_1 x_2 \bar{x}_2 x_3$ is not a fuzzy prime implicant since it subsumes both $x_1 x_2$ and $\bar{x}_2 x_3$. The phrases $x_1 \bar{x}_1 x_3$ and $x_1 x_3 \bar{x}_3$ are, however, fuzzy prime implicants, even though not essentials.

<u>Example 2</u>: $f(x_1,x_2) = x_1 x_2 + x_1 \bar{x}_2$
The terms $x_1 x_2$ and $x_1 \bar{x}_2$ are obviously fuzzy prime implicants of $f(x_1,x_2)$. The consensus as originally defined [3] will produce $x_1 x_2 \,\psi\, x_1 x_2 = 0$. However, the term $x_1 \bar{x}_1$ can be added to $f(x_1,x_2)$ without changing the function and it subsumes no other fuzzy implicant of $f(x_1,x_2)$. Hence

$$f(x_1,x_2) = x_1 x_2 + x_1 \bar{x}_2 + x_1 \bar{x}_1$$

where this is the fuzzy prime implicant representation of the function. Clearly,

$$x_1\bar{x}_1 = x_1\bar{x}_1x_2 + x_1\bar{x}_1\bar{x}_2$$

and thus

$$f(x_1,x_2) = x_1x_2 + x_1\bar{x}_2 + x_1\bar{x}_1x_2 + x_1\bar{x}_1\bar{x}_2$$

and $f_{min}(x_1,x_2) = x_1x_2 + x_1\bar{x}_2$. Namely $x_1\bar{x}_1$ is not an <u>essential</u> fuzzy prime implicant.

<u>Proposition 3</u>: Let $f(x_1,x_2,\ldots,x_n) = \prod_k(x_k + \bar{x}_k + \sigma_k)$ where σ_k is an arbitrary function in $x_1,x_2,\ldots,x_n$ and $k \in \{1,2,\ldots,n\}$. Then $f(x_1,x_2,\ldots,x_n) =$ $= \prod_k(x_k + \bar{x}_k + \sigma_k) + \sum_j x_j\bar{x}_j\gamma_j$ where γ_j is also some arbitrary function in $x_1,x_2,\ldots,x_n$ and $j \in \{1,2,\ldots,n\}$.

<u>Proof</u>: $\prod_k(x_k + \bar{x}_k + \sigma_k) = \min_k[\text{Max}(x_k,\bar{x}_k,\sigma_k)] \geq 0.5.$

$\sum_j x_j\bar{x}_j\gamma_j = \text{Max}_j[\min(x_j,\bar{x}_j,\sigma_j)] \leq 0.5.$ Q.E.D.

It is interesting to note that the modified definition of fuzzy consensus might add fuzzy prime implicants which include variables not even presented in the original fuzzy switching function. For example, let

$$f(x_1,x_2) = x_1 + \bar{x}_1 .$$

Then

$$f(x_1,x_2) = x_1 + \bar{x}_1 + x_2\bar{x}_2$$

where all three phrases are fuzzy prime implicants. However,

$$x_2\bar{x}_2 = x_2\bar{x}_2(x_1 + \bar{x}_1) = x_2\bar{x}_2x_1 + x_2\bar{x}_2\bar{x}_1$$

and these two phrases are included in x_1 and $\bar{x}_1$, respectively. Hence

$$f_{min}(x_1,x_2) = x_1 + \bar{x}_1$$

and $x_2\bar{x}_2$ is not an essential fuzzy prime implicant of the function.

4. Minimization of FSF's

Based on Definitions 2 and 4 the following theorems can be used to obtain a minimized form of a FSF.

Theorem 5: Let R, Q, and W each represent a phrase. If $W \in R \psi Q$, then $R + Q \supseteq W$.

Proof: Since $W \in R \psi Q$ by definition W includes the conjunction $x_j \bar{x}_j$ for at least one j, $j \in \{1,2,\ldots,n\}$ or otherwise $W = 0$. If $W = 0$, clearly $R + Q \supseteq W$. If not so, four cases must be checked.

Case 1: $R \supseteq W$ and $Q \supseteq W$

Case 2: $R \supseteq W$ and $Q \not\supseteq W$

Case 3: $R \not\supseteq W$ and $Q \supseteq W$

Case 4: $R \not\supseteq W$ and $Q \not\supseteq W$

In the first three cases it is clear that $R + Q \supseteq W$. We now shall prove the fourth case, namely, $W \in R \psi Q$, $R \not\supseteq W$, and $Q \not\supseteq W$.

Let $R|W$ be the product of those literals of R that are not present in W and $Q|W$ be the product of those literals of Q which are not present in W. For example, if $R = \bar{x}_1$, $Q = x_1 x_2 \bar{x}_2$, and $W = x_2 \bar{x}_2$, then $R|W = \bar{x}_1$ and $Q|W = x_1$. Then, since $W \in R \psi Q$, it must be true that $R|W = (\overline{Q|W})$, and since $R|W$ and $Q|W$ are both phrases, it follows that $R|W$ must be a single literal, say x_i, and $Q|W$ must be $\bar{x}_i$.

Since $W \in R \psi Q$, by Theorem 3

$$W = W(x_i + \bar{x}_i) = Wx_i + W\bar{x}_i .$$

However,

$$R + Q \supseteq Wx_i$$

and

$$R + Q \supseteq W\bar{x}_i$$

and thus, $R + Q \supseteq W$ since neither R nor Q can contain any literals other than x_i that are not present in W. Q.E.D.

The set of axioms of fuzzy algebra and Theorems 1 - 4 form the basis for the method of fuzzy iterated consensus and minimization of fuzzy functions. It will be shown in the following that the successive addition of fuzzy consensus phrases to a sum of products expression and the removal of phrases that are included in other phrases $(x + xy = x)$ will result in an expression that represents the function as the sum of all its F.P.I.'s.

Theorem 6: A sum-of-products expression $F = P_1 + P_2 + \ldots + P_r$ for the function $F(x_1,x_2,\ldots,x_n)$ is the sum of F.P.I.'s of $F(x_1,x_2,\ldots,x_n)$ if and only if:

1) No phrase subsumes any other phrase, $P_j \not\subseteq P_i$ for any i and j, $i \neq j$, i, $j \in \{1,2,\ldots,r\}$.

2) The fuzzy consensus of any two phrases, $P_i \psi P_j$, either does not exist ($P_i \psi P_j = 0$) or every phrase that belongs to the set describing $P_i \psi P_j$ subsumes some other phrase from the set $\{P_k\}_{k=1}^{r}$.

Proof: It is necessary to show that the fact that $F = P_1 + P_2 + \cdots + P_r$ is not the sum of all FPI's of $f(x_1,x_2,\ldots,x_n)$ implies that either $P_j \subseteq P_i$ for some i and j or that some phrase from the fuzzy consensus $P_i \psi P_j$ exists which is not included in some other phrase from the set $\{P_k\}_{k=1}^{r}$.

It will now be assumed that $F = P_1 + P_2 + \cdots + P_r$ is a sum of products expression but not the sum of all FPI's of $f(x_1,x_2,\ldots,x_n)$. It will then be shown that there must be some P_i and P_j such that $P_j \subseteq P_i$ of that same phrase from the set describing $P_i \psi P_j$ exists for some i and j and there is no P_k that includes this phrase.

There are two possible reasons for F not being the sum of all FPI's. Either some of the P_j are not FPI's or some FPI's of $f(x_1,x_2,\ldots,x_n)$ are missing from F.

If P_j is not an FPI, then there must be an FPI T of $f(x_1,x_2,\ldots,x_n)$ that includes P_j, $P_j \subseteq T$.

If T occurs as one of the phrases, say $T = P_i$, then it follows that $P_j \subseteq P_i$. If T does not occur as one of the P_i, then this is the situation where at least one of the FPI's is missing from F.

It will thus be assumed next that there is some FPI T of $f(x_1,x_2,\ldots,x_n)$ that is missing from F. Since T is an FPI and is not identical with any of the P_i it follows that $T \nsubseteq P_i$, $\forall i$, $i = 1,2,\ldots,r$.

Since T must be covered by the representation of the function, either $T \subseteq P_i$ for some i, or T is covered by a set of phrases, generated with respect to Theorem 3. However, $F \nsubseteq P_i$, $\forall i$, and therefore T must include at least one conjunction of the form $x_j\bar{x}_j$ for some j, $j \in \{1,2,\ldots,n\}$, in order for T to be covered in the representation of the function and to be an FPI of $f(x_1,x_2,\ldots,x_n)$.

It may be possible to add some literals to T, forming a phrase T' which still has the property that $T' \nsubseteq P_i$, $\forall i$. In general there may be several phrases satisfying the requirements placed on T'. In this case T' is defined as one of those phrases which satisfies the requirements given and contains as many literals as any other phrase satisfying these requirements. In case it is not possible to join an additional variable to T without having the resulting phrase included in one of the P_i, T' is defined to be T itself.

In summary T' satisfies the following conditions.

Condition 1: T' contains no variables other than $x_1,x_2,\ldots,x_n$.

Condition 2: $T' \subseteq T$.

Condition 3: $T' \nsubseteq P_i$, for any i, $i = 1,2,\ldots,r$.

Condition 4: No phrase exists having more literals than T' and satisfying Conditions 1 and 2.

It follows from the definition of T' that one of the variables will be missing from T'. If T' had all the set of variables appearing in it, Condition 3 cannot be satisfied in the definition of T'.

Next, the phrases formed by joining the missing varibale, say x_k, and its complement $\bar{x}_k$, to T', $T'x_k$, and $T'\bar{x}_k$ will be considered.

Unless $T'x_k$ and $T'\bar{x}_k$ fail to satisfy Condition 3, it is clear that T' could not have satisfied Condition 4 as originally assumed. Thus $T'x_k$ and $T'\bar{x}_k$ must violate Condition 3 and there must be some P_1 and P_s such that $T'x_k \subseteq P_1$ and $T'\bar{x}_k \subseteq P_s$. From the fact that $T'x_k \subseteq P_1$ and $T' \nsubseteq P_1$, it follows that the literal x_k must appear in P_1 so that it must be possible to express P_1 as $P_1 = x_k Q_1$. Similarly, it must be possible to express P_s as $P_s = \bar{x}_k Q_s$. It follows from the fact that $T'x_k \subseteq x_k Q_1 = P_1$, that $T' \subseteq Q_i$ since neither Q_1 nor T' contains x_k, and similarly that $T' \subseteq Q_s$ so that $T' \subseteq Q_1 Q_s$ and thus the phrase $Q_1 Q_s$ is nonzero. Clearly, $P_1 \psi P_s$ is not empty since at least $Q_1 Q_s$ belongs to the set describing $P_1 \psi P_s$. This is due to the fact that $Q_1 Q_s$ includes $x_j \bar{x}_j$ for some j since $T' \subseteq Q_1 Q_s$.

All that remains to be shown is that there exist no P_i such that $Q_1 Q_s \subseteq P_i$. This is not possible, for if $Q_1 Q_s \subseteq P_i$, then because $T' \subseteq Q_1 Q_s$, it follows that $T' \subseteq P_i$, which contradicts Condition 3. This argument can be repeated for every phrase in the set describing $P_1 \psi P_s$. In order to complete the proof of the theorem, it is necessary to show that if $F = P_1 + P_2 + \cdots + P_r$ is the sum of all the FPI's of $f(x_1,x_2,\ldots,x_n)$, then: 1) no phrase includes any other phrase, $P_j \nsubseteq P_i$ for any i and j, $i \neq j$, i, $j \in \{1,2,\ldots,r\}$; and 2) the fuzzy consensus of any two phrases $P_i \psi P_j$ either does not exist $(P_i \psi P_j = 0)$ or every phrase that belongs to the set describing $P_i \psi P_j$ is included in some other phrase from $\{P_k\}_{k=1}^{r}$.

Each of the P_i is an FPI by definition and it follows from the definition of an FPI that it is not included in any other FPI of the same function. If $P_i \psi P_j$ exists, it is a set of phrases which is included in $f(x_1,x_2,\ldots,x_n)$. By the definition of an FPI, any phrase which is included in $f(x_1,x_2,\ldots,x_n)$ must be included in some FPI of $f(x_1,x_2,\ldots,x_n)$, and thus every phrase that belongs to the set describing $P_i \psi P_j$ must be included in some other phrase from $\{P_k\}_{k=1}^{r}$. Q.E.D.

Evidently the theorem

$$xy + \bar{x}z + yz = xy + \bar{x}z$$

which is the basic Boolean consensus, does not hold in fuzzy logic. (Just take, for example, $\mu(x) = 0.7$, $\mu(y) = 0.8$, and $\mu(z) = 0.9$.) However, the following theorem is true in fuzzy logic.

Based on Theorem 6 an algorithm that converts the function from its disjunctive normal form to a sum of FPI's has been devised.

Algorithm 1:

Input: The set of phrases representing $F(x_1,x_2,\ldots,x_n)$.

Output: Set of FPI's of $F(x_1,x_2,\ldots,x_n)$.

Step 1: Compare each phrase with every other phrase in the expression, and remove any phrase which subsumes any other phrase.

Step 2: Add the fuzzy consensus phrases which do not subsume some other phrases.

The process is iteratively repeated and it terminates when all possible fuzzy consensus operations have been performed. The remaining phrases include all the essential FPI's of $F(x_1,x_2,\ldots,x_n)$.

The minimization algorithm which is based on the above theorem may be broken down into three steps, namely:

1. Compute the fuzzy prime implicants of the function.
2. Remove the essential fuzzy prime implicants.
3. Find a minimal complexity cover of the remainder.

In this paper we propose a novel way to implement such a minimization algorithm, where each fuzzy literal is represented by a double-bit notation or a decimal equivalent

$$x_i \rightarrow 10 \quad (2)$$

$$\bar{x}_i \rightarrow 01 \quad (1)$$

$$x_i\bar{x}_i \rightarrow 11 \quad (3)$$

and 00 (0) if x_i does not appear in the phrase.

Given a fuzzy function F, expressed as a disjunction of phrases, none of which subsume any other phrase of F; table K is then constructed with one row for each phrase of F, such that row a_i contains the variables of phrase i of F. Also, table K is to be constructed with one column for each variable of F.

Example 3: Given a fuzzy function over three variables,

$$F(x_1,x_2,x_3) = x_1x_2\bar{x}_2x_3 + \bar{x}_1x_2\bar{x}_2\bar{x}_3 + x_1\bar{x}_1x_2x_3 + \bar{x}_1\bar{x}_2x_3$$

The binary representation will be:

$$F(x_1,x_2,x_3) = 101110 + 011101 + 111010 + 010110$$

or

$$F(x_1,x_2,x_3) = 2,3,2 + 1,3,1 + 3,2,2 + 1,1,2$$

Construction of Table K yields

	x_1	x_2	x_3
$x_1x_2\bar{x}_2x_3$ →	2	3	2
$\bar{x}_1x_2\bar{x}_2\bar{x}_3$ →	1	3	1
$x_1\bar{x}_1x_2x_3$ →	3	2	2
$\bar{x}_1\bar{x}_2x_3$ →	1	1	2

The following is a minimization algorithm which is simple to apply.

Algorithm 2: Given a fuzzy switching F function represented as a disjunction of phrases.

Step 1: Remove every phrase in F which subsumes any other phrase in F.

Step 2: Construct table K with the remaining phrases of F.
(The following is performed for every row in Table K.)

Step 3: Find 1 or 2 in the row.

Step 4: Find 2 or 1 , respectively in a column.

Step 5: Find at least one 3 in the two rows where the pair were found. If one cannot be found then try to find another pair of 1 and 2 in a column in the two rows.

Step 6: Set up a new table, K_1, where the following results will be placed in the next free row.

(i) Place a 0 in the same column where the 1, 2 couple found in Step 4, is placed.

(ii) A 3 should be placed in the column detected from Step 5.

(iii) In any other column place a 3 if one member of the pair is 3. If one member of the pair is 0 place the other value. When the column consists of 1 and 2 place a 3 in the new row.

Step 7: Remove every phrase of F which subsumes the new phrase. If there is none to remove then the new phrase shall be removed.

Step 8: Check for every phrase in Table K_1 to verify that the two original rows, which created the new phrase, have not been removed and, then remove the new phrase from K_1.

Now the phrases remaining in Tables K and K_1 are a minimal fuzzy function.

Example 4:

$$F(x_1,x_2) = x_1 + \bar{x}_1x_2\bar{x}_2$$

$$F(x_1,x_2) \rightarrow 1000 + 0111$$

$$F(x_1,x_2) \rightarrow 2,0 + 1,3$$

$$K = \begin{array}{|c|c|l} \hline 2 & 0 & \surd \\ \hline 1 & 3 & x \\ \hline \end{array} \qquad \left\{ \begin{array}{l} \text{this row subsumes} \\ \text{the row in } K_1 \end{array} \right\}$$

$$K_1 = \begin{array}{|c|c|l} \hline 0 & 3 & \surd \\ \hline \end{array}$$

$$F(x_1,x_2) \rightarrow 2,0 + 0,3$$

$$F(x_1,x_2) \rightarrow 1000 + 0011$$

$$F(x_1,x_2) = x_1 + x_2\bar{x}_2$$

Example 5:

$$F(x_1,x_2,x_3,x_4) = x_1x_2x_3\bar{x}_3x_4 + \bar{x}_1x_2x_3\bar{x}_3 + \bar{x}_2x_3\bar{x}_3 + x_3\bar{x}_3\bar{x}_4$$

$$F(x_1,x_2,x_3,x_4) \rightarrow 10101110 + 01101100 + 00011100 + 00001101$$

$$F(x_1,x_2,x_3,x_4) \rightarrow 2,2,3,2 + 1,2,3,0 + 0,1,3,0 + 0,0,3,1$$

K =

x_1	x_2	x_3	x_4	
2	2	3	2	x
1	2	3	0	x
0	1	3	0	x
0	0	3	1	x

K_1 =

x_1	x_2	x_3	x_4	
2	2	3	0	√ x
2	0	3	2	x
0	2	3	2	x
1	0	3	0	√ x
0	2	3	0	√ x
0	0	3	0	√

$$F(x_1,x_2,x_3,x_4) \rightarrow 0,0,3,0$$

$$F(x_1,x_2,x_3,x_4) \rightarrow 00001100$$

$$F(x_1,x_2,x_3,x_4) = x_3\bar{x}_3$$

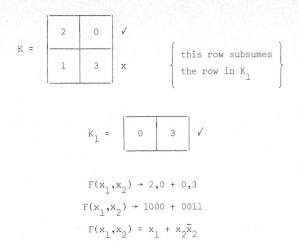

It should be noted that the advantages of the algorithm lie basically in the theoretical importance of "fuzzy consensus" and its practical use for simplification of fuzzy espressions as well as in the simplicity by which the algorithm is executed.

5. On the Enumeration of FSF's

In this section we shall discuss the problem of the enumeration of distinct fuzzy switching functions.

From combinatorial arguments it has been shown [2] that there are 2^{4^n} possible fuzzy functions of n variables. However, most of these are simply non-minimized forms of a relatively few number of unique fuzzy functions. The problem can be appreciated best by considering the number of functions of two variables that exist: for Boolean functions there are 2^{2^2}, or 16 ; for fuzzy functions though, there are 2^{4^2}, or 65,536. It is obviously impossible to enumerate them by hand, which in the case of Boolean functions is quite reasonable for small n.

Kameda and Sadeh [7] have shown that there are $\binom{n}{k}$ ways of selecting k out of n variables and each variable may be either complemented or left uncomplemented. Thus the number of type-2 phrases containing k literals is $\binom{n}{k}2^k$. It follows that the total number of type-2 phrases is

$$\sum_{k=1}^{n} \binom{n}{k}2^k = (1+2)^n - 1 = 3^n - 1.$$

As for the fundamental phrases of type-1, each such phrase contains, for each k, either x_k only, $\bar{x}_k$ only, or $x_k\bar{x}_k$. Thus there are 3^n such phrases. Since 2^n of them are simple, altogether there are $2 \cdot 3^n - 2^n - 1$ fundamental phrases. Clearly, any function can be defined by disjunction of a nonempty subset of the set of all fundamental phrases. Therefore there are at most $2^{2 \cdot 3^n - 2^n - 1} - 1$ distinct fuzzy switching functions.

Following Kameda and Sadeh [7] we call a set of phrases "independent" iff none of the phrases in the set implies any other in the same set.

We use this definition in order to generate an independence relation R^* between the fundamental phrases. This relation can be graphically represented as an upper diagonal binary matrix showing the independence relation of the $2 \cdot 3^n - 2^n - 1$ elements.

Thus the number of distinct fuzzy switching functions will be given by M_n where

$$M_n = 2 \cdot 3^n - 2^n - 1 + \sum_{i=1}^{MAX} t\ell_i$$

and $t\ell_i$ is transitivity level i on the matrix, and MAX will denote the maximum level of transitivity for this particular n, given by the maximum cover fraction discussed by Schwede [8] and Schwede and Kandel [9].

Converting M_n to an explicit function of n will ever, it should be noted that for $n = 2$, we get directly that $M_2 = 82$ where the Kameda and Sadeh [7] technique yields an upper bound of 8192 and a lower bound of 49.

6. Applications

A 1976 bibliography compiled by Kandel and Davis [10] lists 571 references on fuzzy sets and their applications; more recently, Gaines and Kohout [11] list an extended bibliography of fuzzy systems and closely related subjects and a 1979 updated version appears in this volume. Many papers in these bibliographies are devoted to the subject of fuzzy switching logic (SWLOG in [11]) and a recent monograph is dealing mainly with this topic (Kandel and Lee [12]). The enormous growth in research dealing with fuzziness since L. A. Zadeh's original paper in 1965 (Zadeh [1]), is easily explained: models based on conventional set theory are not always suitable whenever imprecision and inexactness arise. It is clear that there is much more work in this area than any of us, involved in the field of fuzzy set theory and its applications, have realized, and that the subfield of fuzzy switching functions and their applicability is both important and promising as far as handling scientific models using approximate information.

Many important notions, fuzzy in nature may be assigned a precise meaning using the concept of the membership function of a fuzzy set and thereby making them amenable to mathematical analysis as well as to engineering applications. One such example is the notion of a fuzzy switching function introduced to analyze hazard detection in binary systems (Kandel [12]). The main objective of the present proposal is to provide for the first time a formulation of fuzzy switching theory and its applications, general enough to encompass all pieces of existing formulations and yet suitable enough for dealing with problems of practical interest.

At this time we have a clear definition of fuzzy switching functions (Kandel [13]), basic notions of their simplification, which are quite different from classical minimizations of Boolean functions (Kandel [3]-[6]), some primitive knowledge of their disjunctive decomposition (Kandel [14]), and some basic assertions regarding fuzzy chains (Kandel and Yelowitz [15]), fuzzy maps (Schwede and Kandel [9]), and fuzzy matrices (Kandel [16]).

We know very little on the applicability to problems of practical interest. The only evidence to that is a preliminary study relating fuzzy switching functions to the problem of hazard detection in binary systems (Kandel [2]) and the many features fuzzy functions and their properties have in common with the design of branching questionnaires and the problems related to the efficient design of branching questionnaires as illustrated by Zadeh in [17].

We feel that in many scientific models the amount of information is determined by the amount of the uncertainty - or, more exactly, it is determined by the amount by which the uncertainty has been reduced; namely, we can measure information as the decrease of uncertainty. The concept of information itself has been implicit in many areas of science, both as a substantive concept important in its own right and as consonant concept which is ancillary to the entire structure of science. Thus, models using approximate information, especially in pattern recognition and classification, feature selection, branching questionnaires, optimal encoding and optimal design of decision tables, and in fault/transient analysis of systems have many practical implications. These models can be successfully used utilizing the concept of fuzzy switching functions and their properties.

7. References

[1] Zadeh, L.A.: Fuzzy sets, Information and Control. 8, pp. 338-353, 1965.

[2] Kandel, A.: Applications of fuzzy logic to the detection of static hazards in combinational switching systems, Int. J. of Computer and Information Sciences. pp. 129-139, 1974.

[3] Kandel, A.: On the minimization of fuzzy functions, IEEE Trans. on Computers. C-22, pp. 828-832, 1973.

[4] Kandel, A.: On the minimization of imcompletely specified fuzzy functions, Information and Control. pp. 141-153, 1974.

[5] Kandel, A.: Inexact switching logic, IEEE Trans. on Systems, Man, and Cybernetics. SMC-6, pp. 215-219, 1976.

[6] Kandel, A.: A note on the simplification of fuzzy switching functions, Information Sciences. 13, pp. 91-94, 1977.

[7] Kameda, T. and Sadeh, E.: Bounds on the number of fuzzy functions, Information and Control. 35, pp. 139-145, 1977.

[8] Schwede, G.W.: N-variable fuzzy maps with application to disjunctive decomposition of fuzzy functions, Proc. 6th Int. Symp. on Multiple-Valued Logic. pp. 203-218, Logan, Utah, 1976.

[9] Schwede, G.W., and Kandel, A.: Fuzzy maps, IEEE Trans. on Systems, Man, and Cybernetics, SMC-7, pp. 619-674, 1977.

[10] Kandel, A. and Davis, H.A.: The first fuzzy decade, New Mexico Institute of Mining and Technology, Socorro, New Mexico, CSR 140, 1976.

[11] Gaines, B.R. and Kohout, L.J.: "The fuzzy decade: a bibliography of fuzzy systems and closely related topics", in Fuzzy Automata and Decision Processes, M.M. Gupta, G.M. Saridis, B.R. Gaines, Editors. North Holland, N.Y., 1977.

[12] Kandel, A. and S.C. Lee: Fuzzy Switching and Automata: Theory and Applications. New York; Crane, Russak, and Co., 1979.

[13] Kandel, A.: On the properties of fuzzy switching functions, J. of Cybernetics. 4, pp. 119-126, 1974. Also presented as U.S.-Japan Seminar on Fuzzy Set Theory, Berkeley, CA, July, 1974.

[14] Kandel, A.: On the decomposition of fuzzy functions, IEEE Trans. on Computers. C-25, pp. 1124-1130, 1976.

[15] Kandel, A. and Yelowitz, L.: Fuzzy chains, IEEE Trans. on Systems, Man, and Cybernetics. SMC-4, pp. 472-475, 1974.

[16] Kandel, A.: Properties of fuzzy matrices and their applications to hierarchical sturctures, Conf. Record of the 9th Asilomar Conf. on Circuits, Systems, and Computers. pp. 531-538, Pacific Grove, CA, 1975.

[17] Zadeh, L.A.: A fuzzy-algorithmic approach to the definition of complex or imprecise concepts, Int. J. Man-Maching Studies. 8, pp. 249-291, 1976.

ADVANCES IN FUZZY SET THEORY AND APPLICATIONS
M.M. Gupta, R.K. Ragade, R.R. Yager (editors)

FUZZY SETS AND LANGUAGES

Wolfgang WECHLER

Sektion Mathematik
Technische Universität Dresden
DDR-8027 Dresden, German Democratic Republic

Abstract. The primary goal of the present paper is the development of a fuzzy set theory based on an appropriate many-valued logic. Thus, we get the advantage that fuzzy sets can be treated in a way syntactically analogouos to usual sets. It turns out that this approach is suitable and very convenient for investigating generalized languages.

1. Introduction

The rapid development of so-called "soft" sciences, whose key notions cannot precisely be determined as in physics for instance, requires a suitable and convenient mathematical representation for inexact concepts. It seems to be a contradiction in itself unless the modifier "inexact" is explained in this context.

In physics any observable is uniquely defined by a related measurement. By a thorough analysis of these measurement processes, the range of validity of the definitions can be found out, within which they are meaningful. Outside, there is not any possibility of defining such observables because either the considered objects are disturbed in such a way during measurement that they lose their identity or systems are so complex that they have an enormous number of observables whose greater part is more or less irrelevant for investigation. In this case an exact description by a large amount of data in a quantitative manner would not be feasible and even not meaningful. Therefore, an optimum has to be found, namely a description which is as exact (in a quantitative manner) as practicable, on the one hand, and as convenient and suitable as possible, on the other.

Let us illustrate the situation in pattern recognition. The first step in solving a problem is the selection of significant characteristics of a considered object. Only, by such reduction, the problem becomes conceivable in general. After that the object in question will be compared with standard ones in order to decide the appropriate class of representatives. But this comparison often possesses an inherent uncertainty related more to ambiguity and indeterminancy than to random variations. Thus the calculus of probability cannot be applied in such cases. Hence it is necessary to build up a mathematical formalism which enables us to describe fuzziness generally.

The theory of fuzzy sets proposed by L.A. Zadeh [11] is a first attempt at constructing a suitable mathematical formalism applicable in areas just mentioned. In analogy to ordinary set theory a syntactical representation of fuzzy sets by formulas of an appropriate logic is necessary. Evidently, such a logic must be many-valued. Its basic features have been derived by J.A. Goguen [3]. There are close connections to the well-known many-valued logics introduced by E.L. Post and J. Łukasiewicz as well as to intuitionistic logic (cf. [4,5]).

Section 2 deals with fuzzy sets. An appropriate many-valued logic called fuzzy logic is developed in Section 2.1. This development is only done to such an extent that we are able to generalize the notion of a model of a (first order) theory. Considering fuzzy models of a theory for the relation of membership we are in a position to introduce fuzzy sets in Section 2.2.

In Section 3 we apply the formalism of fuzzy sets to generalize basic notions of the theory of languages and grammars. Using syntactical description of fuzzy sets we get a strong analogy between ordinary and generalized concepts which shows the advantage of the developed formalism.

In order to see how fuzzy languages are related to usual languages, so-called R-regular languages are investigated in Section 4. These languages are obtained from fuzzy languages by means of cut-points. The following problems arise: How to characterize a family of languages to be a family of R-regular languages and under which condition a family of R-regular languages equals the family of regular languages.

2. Fuzzy Sets

Many of well-known applications of fuzzy sets in various fields like pattern recognition, artificial intelligence, automata and language theory etc. have shown fuzzy concepts and methods to be useful. Hence, it is desirable to have a rigorous treatment of a calculus of fuzzy sets. Its goal should be a self-contained universe governed by the fundamental membership relation ε. Such an approach has to be a "set theory" in a certain sense, that means the universe in question must be determined by a fuzzy interpretation of an appropriate ε-theory. The necessary framework has to be a generalized logic.

2.1. Fuzzy Logic

At first we recall some basic concepts in order to be able to formulate as general an algebraic structure on the set of truth values as possible. A semiring R is a set with two binary operations: addition + and multiplication $\cdot$ such that (R,+) is a commutative monoid with 0 as unit element. With respect to the multiplication $(R,\cdot)$ is a monoid with 1 as unit and 0 as zero element. Both operations are connected by the usual distributive laws. R is said to be a partially ordered semiring (hereafter abbreviated po-semiring) if R is a partially ordered set with 0 as bottom element such that both operations are monotone.

Examples. 1) The simplest po-semiring is the so-called Boolean semiring $\mathbb{B} = \{0,1\}$ with $0 < 1$ and $1 + 1 = 1$. Note that the remaining rules of addition and multiplication are forced by the axioms.
2) The semiring $\mathbb{N}$ of all natural numbers including 0 with respect to usual ordering and operations.
3) Any distributive lattice with 0 and 1 can be considered as a po-semiring.

Definition 1. An algebraic structure $(T,\cap,+,\cdot,\rightarrow)$ with four binary operations called intersection, sum, product, and residuation is said to be a truth value algebra, or shortly TV-algebra, if the following axioms are satisfied:

(1) $(T,\cap)$ is a $\cap$-semilattice.

(2) $(T,+,\cdot)$ is a po-semiring with respect to the induced partial order of (1).

(3) $a \to b = 1$ if and only if $a \leq b$.

(4) $a \leq a'$ implies $a' \to b \leq a \to b$.

(5) $b \leq b'$ implies $a \to b \leq a \to b'$.

Examples. 1) Every Boolean lattice $(L,\cap,\cup)$ can be considered as a TV-algebra $(L,\cap,+,\cdot,\to)$ by putting $a + b = a \cup b$, $a \cdot b = a \cap b$ and $a \to b = \bar{a} \cup b$, where $\bar{a}$ is the complement of a.
2) The structure $\mathrm{Ł} = ([0,1],\cap,+,\cdot,\to)$ with $a \cap b = \min\{a,b\}$, $a + b = \max\{a,b\}$, $a \cdot b = \max\{a + b - 1, 0\}$ and $a \to b = \min\{1, 1 - a + b\}$ is a TV-algebra. Ł is the algebraic structure of truth values for the many-valued Łukasiewicz logic.
3) With each po-semiring $(R,+,\cdot)$, a TV-algebra $(R,\cap,+,\cdot,\to)$ may be associated by setting

$$a \cap b = \begin{cases} \inf\{a,b\} & \text{if } \inf\{a,b\} \text{ exists} \\ 0 & \text{otherwise} \end{cases}$$

and

$$a \to b = \begin{cases} 1 & \text{if } a \leq b \\ 0 & \text{otherwise.} \end{cases}$$

Hence a TV-algebra $(R,\cap,+,\cdot,\to)$ is called R-algebra whenever the possible values of $\to$ are only 0 and 1.

Now let $(T,\cap,+,\cdot,\to)$ be a fixed TV-algebra. We are going to introduce the fuzzy propositional calculus on a set V of propositional variables. Its so-called F-propositions are constructed from symbols of the following kinds:

propositional variables v of V,

logical constants $\underline{a}$ for each truth value a of T,

logical connectives $\wedge, \nabla, \Delta, \Rightarrow$,

parantheses.

Definition 2. The set of all F-propositions is inductively defined by the following rules:

(1) Any proposition variable v of V is an F-proposition.

(2) Any logical constant $\underline{a}$ $(a \in T)$ is an F-proposition.

(3) If α and β are F-propositions, so are $(\alpha) \wedge (\beta)$, $(\alpha) \nabla (\beta)$, $(\alpha) \Delta (\beta)$, and $(\alpha) \Rightarrow (\beta)$.

There are no F-propositions other than those constructed from the atomic ones described in (1) and (2), or obtained by means of successive applications of the rule (3).

The use of parantheses ensures a unique representation of compound F-propositions. In order to avoid an excessive number of parantheses we adopt the usual convention for omitting them.

The following derived connectives are useful in the sequel:

$(-\alpha) := (\alpha \Rightarrow \underline{0})$

$(\alpha \Leftrightarrow \beta) := ((\alpha \Rightarrow \beta) \wedge (\beta \Rightarrow \alpha))$.

They are called negation and biimplication, resp.

F-propositions are interpreted as follows. Let f be a mapping from V into T. Then f can uniquely be extended to a mapping "val_f" from the set of all F-propositions into T by setting

(1) $val_f(v) = f(v)$ for all $v \in V$

(2) $val_f(\underline{a}) = a$ for all $a \in T$

(3) $val_f(\alpha \wedge \beta) = val_f(\alpha) \cap val_f(\beta)$

$val_f(\alpha \nabla \beta) = val_f(\alpha) + val_f(\beta)$

$val_f(\alpha \Delta \beta) = val_f(\alpha) \cdot val_f(\beta)$

$val_f(\alpha \Rightarrow \beta) = val_f(\alpha) \rightarrow val_f(\beta)$

for all F-propositions α and β.

Thus semantic interpretation of the fuzzy propositional calculus in a TV-algebra $(T, \cap, +, \cdot, \rightarrow)$ is essentially determined by the following correspondence

logical connectice		operation in T	
$\wedge$	conjunction	$\cap$	intersection
∇	disjunction	+	addition
Δ	context	$\cdot$	multiplication
=>	implication	$\rightarrow$	residuation

For a given f the truth value $val_f(\alpha)$ of an F-proposition α can be computed in a finite number of steps in analogy to the usual truth table method.

Next, the <u>predicate calculus</u> (of first order) shall be generalized to such an extent that the definition of fuzzy models of first order theories becomes possible. Let P be a set of predicate symbols divided into disjoint subsets P_n, $n \geq 1$, where P_n contains the predicate symbols of arity n. Atomic formulas $\pi(x_1,\ldots,x_n)$ are obtained by filling the empty places of a predicate symbol π of P_n with individual variables $x_1,\ldots,x_n$ of V. The set of all atomic formulas is denoted by P(V). We are now in a position to define the set of all F-formulas using the following kinds of symbols

individual symbols v of V,

logical constants $\underline{a}$ ($a \in T$),

predicate symbols π of P,

logical connectives $\wedge, \triangledown, \Delta, \Rightarrow$,

quantifiers $\bigwedge$, $\exists$, $\forall$,

parantheses.

<u>Definition 3</u>. The set of all <u>F-formulas</u> is inductively defined by the following rules:

(1) Any individual variable v of V is an F-formula.

(2) Any logical constant $\underline{a}$ ($a \in T$) is an F-formula.

(3) Any atomic formula $\pi(x_1,\ldots,x_n)$ of P(V) is an F-formula.

(4) If α and β are F-formulas, so are $(\alpha) \wedge (\beta)$, $(\alpha) \triangledown (\beta)$, $(\alpha)\ \Delta\ (\beta)$, and $(\alpha) \Rightarrow (\beta)$.

(5) If α is an F-formula and x is an individual variable of V, then $\bigwedge x(\alpha)$, $\exists x(\alpha)$, and $\forall x(\alpha)$ are F-formulas.

There are no F-formulas other than the elementary ones described in (1), (2), and (3), or obtained by means of successive applications of the rule (4) and (5).

The usual convention for omitting parantheses is adopted. We say that an individual variable in an F-formula α that is not written just after a quantifier is bound in α if it occurs in α within the scope of a quantifier and is identical with the variable just after the quantifier. An individual variable which occurs in an F-formula α and which does not occur just after a quantifier is called free if it is not bound. The set of free variables of α is denoted by var(α).

In order to interpret F-formulas we have to think of the elements of V as names of objects. Therefore, let M be an arbitrary set (of objects) and let f be a mapping from V into M. In this context the predicate symbols have to be considered as T-valued relations among those objects, whereby an n-ary T-valued relation on M is a mapping from M^n into T. We call an algebraic structure

$$\underline{M} = (M, \{\pi_M \mid \pi \in P\})$$

with $\pi_M: M^n \longrightarrow T$ if $\pi \in P_n$, $n \geqq 1$, a <u>fuzzy model</u> (of the theory with an empty set of axioms).

<u>Definition 4.</u> A <u>fuzzy interpretation</u> with respect to a given mapping f from V into M and $\underline{M} = (M, \{\pi_M \mid \pi \in P\})$ is a mapping "$val_{f,M}$" from the set of all F-formulas into T defined by

(1) $val_{f,M}(v) = f(v)$ for $v \in V$

(2) $val_{f,M}(\underline{a}) = a$ for $a \in T$

(3) $val_{f,M}(\pi(x_1,\ldots,x_n)) = \pi_M(f(x_1),\ldots,f(x_n))$
for $\pi \in P_n$, $x_1,\ldots,x_n \in V$, $n \geqq 1$

(4) $val_{f,M}(\alpha \wedge \beta) = val_{f,M}(\alpha) \cap val_{f,M}(\beta)$
$val_{f,M}(\alpha \triangledown \beta) = val_{f,M}(\alpha) + val_{f,M}(\beta)$
$val_{f,M}(\alpha \vartriangle \beta) = val_{f,M}(\alpha) \cdot val_{f,M}(\beta)$
$val_{f,M}(\alpha \Rrightarrow \beta) = val_{f,M}(\alpha) \rightarrow val_{f,M}(\beta)$
for all F-formulas α and β

(5) Let α be an F-formula with a quantification depth less than or equal to n. Assume that $val_{f,M}$ is already a fuzzy interpretation for all F-formulas with a quantification depth less than or equal to n, then put

$$val_{f,M}(\bigwedge x(\alpha)) = \begin{cases} \bigcap_{m \in M} \lambda m \; val_{f[x/m],M}(\alpha) & \text{if it exists} \\ 0 & \text{otherwise} \end{cases}$$

$$val_{f,M}(\exists x(\alpha)) = \begin{cases} \sum_{m \in M} \lambda m \; val_{f[x/m],M}(\alpha) & \text{if it exists} \\ 0 & \text{otherwise} \end{cases}$$

$$\mathrm{val}_{f,M}(\forall x(\alpha)) = \begin{cases} \prod_{m \in M} \lambda m\, \mathrm{val}_{f[x/m],M}(\alpha) & \text{if it exists} \\ 0 & \text{otherwise.} \end{cases}$$

Thereby, we use the notation of the λ-calculus, that means $\lambda m\ \mathrm{val}_{f[x/m],M}(\alpha)$ is considered as a function of m provided f, x, M and α are fixed. As usually, f[x/m] denotes the function which maps x onto m and which is identical with f for all x' of V − {x}. Now, the presented correspondence between logical connectives and operations may be extended as follows

quantifier	infinite operation
$\bigwedge$ universal quantifier	$\bigcap$ intersection
$\exists$ existential quantifier	$\sum$ sum
$\forall$ context quantifier	$\prod$ product

For a given fuzzy model $\underline{M}$, a fuzzy interpretation is uniquely determined by the mapping f from V into M. Thus, we can define

$$|{=}{=}_{\underline{M}}\ \alpha = \bigcap\{\mathrm{val}_{f,M}(\alpha) \mid f \text{ arbitrary mapping from V into M}\}.$$

An F-formula α is called <u>valid</u> in $\underline{M}$ if $|{=}{=}_{\underline{M}}\ \alpha = 1$ holds. Let α be an F-formula with $\mathrm{var}(\alpha) = \{x_1,\dots,x_n\}$ and let f be a mapping from V into M satisfying $f(x_i) = m_i$ for $i = 1,\dots,n$. Then there exists a mapping from M^n into T assigning $\|\alpha(m_1,\dots,m_n)\| := \mathrm{val}_{f,M}(\alpha)$ to each n-tuple $(m_1,\dots,m_n)$ of M^n.

It can easily be seen that

$$|{=}{=}_{\underline{M}}\ \alpha = \bigcap_{m_1,\dots,m_n \in M} \|\alpha(m_1,\dots,m_n)\|$$

holds for any F-formula α with $\mathrm{var}(\alpha) = \{x_1,\dots,x_n\}$.

A (first order) <u>theory</u> $\mathcal{T}$ is given by a set P of predicate symbols and a chosen subset of F-formulas Ax whose elements are called mathematical axioms. We say that $\underline{M} = (M,\{\pi_M \mid \pi \in P\})$ is a fuzzy model of $\mathcal{T}$ if there exists a fuzzy interpretation such that each α of Ax is valid in $\underline{M}$.

2.2. Fuzzy Models of the ε-Theory

The aim of this section is to define fuzzy models of the ε-theory whose predicate symbols are ε (membership) and ϱ (equality). As usual, instead of $\varepsilon(x,y)$ and $\varrho(x,y)$ we write $x \in y$ and $x = y$, resp. The F-formulas $-(x \in y)$ and $-(x = y)$ will be abbreviated as $x \notin y$ and $x \neq y$, resp.

Definition 5. A theory $\mathcal{T}$ with the predicate symbols ε and ϱ is called ε-theory if Ax consists of the following mathematical axioms

(α_1) $\bigwedge x(x = x)$

(α_2) $\bigwedge x \bigwedge y(x = y \Rightarrow y = x)$

(α_3) $\bigwedge x \bigwedge y \bigwedge z(x = y \Rightarrow (y = z \Rightarrow x = z))$

(α_4) $\bigwedge x \bigwedge y \bigwedge z(x = y \Rightarrow (x \in y \Rightarrow y \in z))$

(α_5) $\bigwedge x \bigwedge y \bigwedge z(x = y \Leftrightarrow (z \in x \Leftrightarrow z \in y))$

(α_6) $\bigwedge x(x \notin x)$.

Consequently, a fuzzy model of the ε-theory is an algebraic structure $\underline{M} = (M,\{\varepsilon_M,\varrho_M\})$ equipped with two mappings ε_M and ϱ_M from $M \times M$ into a TV-algebra such that $\alpha_1,\dots,\alpha_6$ are valid. Throughout this section the underlying TV-algebra is assumed to be an R-algebra $(R,\cap,+,\cdot,\rightarrow)$. The related fuzzy models are called R-fuzzy models.

At first, Zadeh's approach will be reformulated in our framework. R^X denotes the set of all mappings from a given set X into R. An R-fuzzy model of the ε-theory can be constructed as follows:

$$\underline{M}_1 = (X \cup R^X,\{\varepsilon_{M_1},\varrho_{M_1}\}),$$

where

$$\varepsilon_{M_1}(x,y) = \begin{cases} y(x) & \text{if } y \text{ belongs to } R^X \text{ and } x \text{ belongs to } X \\ 0 & \text{otherwise} \end{cases}$$

and

$$\varrho_{M_1}(x,y) = \begin{cases} 1 & \text{if } x \text{ equals } y \\ 0 & \text{otherwise.} \end{cases}$$

Obviously, the elements of $\underline{M}_1$ that belongs to R^X are the fuzzy sets in the sense of L.A. Zadeh [11].

But this model is too restrictive to be used for generalizing set-theoretic operations in general. Therefore, our next model is based on a set U with the following properties:

(1) If X is a set of U, then each x of X is in U, the power set of X is in U, and $\bigcup X$ is in U.

(2) For elements x and y in U the ordered pair (x,y), the set $\{x,y\}$, and the cartesian product $x \times y$ are in U

(3) If f is a mapping from X onto Y, where X is in U and $Y \subset U$, then Y is in U.

(4) $\mathbb{N}$ and R are in U.

These closure properties for U ensure that any of the standard operation of set theory applied to elements of U will always produce elements of U (cf. [1]).

Now hold U fixed. We put

$$F = \{A \mid A \text{ is a mapping from a set } X \text{ of } U \text{ into } R\}.$$

Because of the closure properties F is a subset of U.

<u>Theorem 1</u>. $\underline{M} = (M,\{\varepsilon_M,\varrho_M\})$ is an R-fuzzy model of the ε-theory if ε_M and ϱ_M are defined by

$$\varepsilon_M(x,y) = \begin{cases} 1 & \text{if } x \text{ is in } y \\ y(x) & \text{if } y \text{ is in } U \text{ and } x \text{ belongs to } \mathrm{dom}(y) \\ 0 & \text{otherwise} \end{cases}$$

and

$$\varrho_M(x,y) = \begin{cases} 1 & \text{if } x \text{ equals } y \\ 0 & \text{otherwise.} \end{cases}$$

The proof is straightforward and can be found in [10].

Instead of $\varepsilon_M(x,y)$ and $\varrho_M(x,y)$ we will write $x \mathrel{\dot\in} y$ and $x \mathrel{\dot=} y$, resp. Occasionally, the point may be omitted if no confusion can arise. The elements of $\underline{M}$ belonging to F are called <u>R-fuzzy sets</u>. An R-fuzzy set A is said to be crisp if either $x \mathrel{\dot\in} A = 0$ or $x \mathrel{\dot\in} A = 1$ for all x of U.

In order to have a syntactical description of R-fuzzy sets we need the following <u>comprehension principle</u> [10]: Let α be an F-formula with one free variable. Assume that $\|\alpha(x)\|$ is defined in $\underline{M}$ for all x of U. If an R-fuzzy set X of $\underline{M}$ exists such that $\|\alpha(x)\| \neq 0$ if and only if x belongs to X, then there is an R-fuzzy set A in $\underline{M}$ satisfying $x \mathrel{\dot\in} A = \|\alpha(x)\|$ for all x of U. In that case we will write

$$A = \{x \mid \alpha(x)\} \quad \text{or} \quad A = \{x \in X \mid \alpha(x)\}.$$

Now we are in a position to introduce set-theoretic operations in a syntactical way. Let A and B be R-fuzzy sets. For instance, the sum A + B and the intersection A ∩ B are defined by setting

$$A + B = \{x \mid x \in A \vee x \in B\} \quad \text{and} \quad A \cap B = \{x \mid x \in A \wedge x \in B\}.$$

The inclusion ⊂ is defined as usual

$$A \subset B = \| \exists x(x \in A \Rightarrow x \in B)\|.$$

Consequently, the power set of an R-fuzzy set could be introduced. But for the application in Section 3, a modified power set is needed:

$$S_R A = \{A' \mid x \in A' \neq 0 \Rightarrow x \in A \neq 0\}.$$

We say $S_R A$ is the set of all R-fuzzy sets with support A, whereby the support $|A|$ of an arbitrary R-fuzzy set A is defined by $|A| = \{x \mid x \in A \neq 0\}$. Hence, $S_R A = \{A' \mid |A'| \subset |A|\}$. Especially, $S_R^f A$ denotes the set of all R-fuzzy sets A' with a finite support which is included in the support of A.

The cartesian product A × B is syntactically defined by

$$A \times B = \{x \mid \exists a \exists b(x = (a,b) \wedge (a \in A) \wedge (b \in B))\}.$$

By definition, we get $(A \times B)(a,b) = A(a) \cap B(b)$ for all pairs (a,b) of $|A| \times |B|$.

Next, the notion of a (binary) relation can be generalized. We call ϱ an R-fuzzy relation between A and B if the support of ϱ is included in $|A| \times |B|$. The set of all R-fuzzy relations between A and B is denoted by $Rel_R(A,B)$. Accordingly, $Rel_R^f(A,B)$ denotes the set of all R-fuzzy relations between A and B that have a finite support. In order to define the product $\varrho \cdot \sigma$ of R-fuzzy relations ϱ of $Rel_R(A,B)$, σ of $Rel_R(B,C)$ in analogy to the usual product we have to restrict the R-fuzzy set B. If B has a finite support, then put

$$\varrho \cdot \sigma = \{x \mid \exists a \exists b \exists c(x = (a,c) \Delta ((a,b) \in \varrho) \Delta ((b,c) \in \sigma))\}.$$

By definition, $\varrho \cdot \sigma$ is a mapping from $|A| \times |C|$ into R given by

$$(\varrho \cdot \sigma)(a,b) = \sum_{b \in |B|} \varrho(a,b) \cdot \sigma(b,c)$$

for all pairs (a,c) of $|A| \times |C|$. Since $|B|$ is assumed to be finite the summation is well-defined.

3. Context-free and Regular R-fuzzy Languages

The aim of this section is to show the advantage of our formalism in dealing with generalized languages. Let X be a finite set. X^* denotes the free monoid generated by X with the empty word e as unit. An R-fuzzy language over X is an R-fuzzy set with support X^*.

Since from the syntactical point of view the developed theory of R-fuzzy sets is very similar to set theory the investigation of R-fuzzy languages reduces to that of usual languages. In other words the basic concepts can be generalized in a straightforward way and, consequently, the known methods are applicable in solving related problems. R-fuzzy languages may be considered as formal power series But using the calculus of formal power series (cf. [6]) we are force to think in a way different from the set-theoretic approach. This, however, is not necessary for our formalism. Thus, it is more suitable and convenient.

In the sequel we focus our attention to the fixed point characterization of context-free and regular R-fuzzy languages. A more detaile study of other problems (algebraic characterization and acceptors) is done in [10].

At first, basic structures on the set $S_R X^*$ of all R-fuzzy languages over X have to be introduced. Let L and L' be R-fuzzy languages over X. The product $L \cdot L'$ is defined by

$$L \cdot L' = \{w \in X^* \mid \exists u \; \exists v (w = uv \wedge u \in L \wedge v \in L')\}.$$

By the definition of a fuzzy interpretation we get

$$\|\alpha(w)\| = \sum_{u,v \in X^*} \|w = uv\| \cdot \|u \in L\| \cdot \|v \in L'\| = \sum_{w = uv} L(u) \cdot L'(v).$$

Since each word w over X possesses only a finite number of factorizations w = uv, the summation is well-defined. Hence

$$(L \cdot L')(w) = \sum_{w = uv} L(u) \cdot L'(v) \quad \text{for all} \quad w \in X^*.$$

Evidently, $S_R X^*$ forms a semiring with respect to addition and multiplication.

To deal with infinite sums, which do not always exist, a metric on $S_R X^*$ is introduced:

$$d(L,L') = \begin{cases} 0 & \text{if } L = L' \\ 2^{-\min\{|w| \mid L(w) \neq L'(w)\}} & \text{if } L \neq L' \end{cases}, \quad L,L' \in S_R X^*.$$

In [10] it is shown that $S_R X^*$ is a complete metric space. Assume that L is an R-fuzzy language with $e \notin L$. Then

$$L^* = \lim_{n\to\infty} \sum_{i=0}^{n} L^i \quad \text{with} \quad L^0 = \{e\},\ L^{i+1} = L^i \cdot L \text{ for } i \geq 1$$

exists. L* is called the iteration of L.

A context-free <u>R-fuzzy grammar</u> G is a triple G = (X,N,π) consisting of two disjoint finite sets X and N (of terminals and non-terminals, resp.) and an R-fuzzy relation $\pi \in \mathrm{Rel}^f_R(N,V^*)$, where $V = X \cup N$. In particular, G is called regular if $\pi \in \mathrm{Rel}^f_R(N,X^*\cdot(N \cup \{e\}))$. As usual, we extend π to an R-fuzzy relation ==> of $\mathrm{Rel}_R(V^*,V^*)$ by

$$v_1 \Longrightarrow v_2 = \begin{cases} \pi(\alpha,v) & \text{if there are words } u,u',v \text{ of } V^* \text{ and } \alpha \text{ of } N \text{ such that } v_1 = u\alpha u' \text{ and } v_2 = uvu' \\ 0 & \text{otherwise.} \end{cases}$$

The reflexive and transitive closure ==>* is defined by

$$\Longrightarrow^* = \lim_{n\to\infty} \sum_{i=0}^{n} \Longrightarrow^i .$$

But ==>* does not always exist. In order to ensure the existence, a general assumption has to be stated: π can be decomposed into $\pi = \pi_0 + \pi_1$ so that $\pi_0 \cdot \pi_0 = \emptyset$ and $\pi_1 \cdot (N \cup \{e\}) = \emptyset$. Now put

$$L(G,\alpha) = \{w \in X^* \mid \alpha \Longrightarrow^* w\}.$$

L(G,α) is called the R-fuzzy language generated by G with respect to α.

<u>Definition 6</u>. An <u>R-fuzzy language</u> L over X is said to be <u>context-free</u> if there is a context-free R-fuzzy grammar G = (X,N,π) such that L = L(G,α) for some α of N. In particular, L is called <u>regular</u> if G is regular.

Let G = (X,N,π) be a context-free grammar with $N = \{\alpha_1,\dots,\alpha_n\}$. With G, we associate an operator Φ_G on $(S_R X^*)^n$ defined by

$$\Phi_G(\underline{Z})_i = \{w \in X^* \mid \exists v(\pi(\alpha_i,v) \wedge w \in v[\underline{\alpha}/\underline{Z}])\}$$

for $\underline{Z} = (Z_1,\dots,Z_n)$ of $(S_R X^*)^n$. The substitution $v[\underline{\alpha}/\underline{Z}]$ of $\underline{\alpha} = (\alpha_1,\dots,\alpha_n)$ by $\underline{Z}$ in v of V* is defined by induction over the length of v:

(1) $x[\underline{\alpha}/\underline{Z}] = \{x\}$ and $\alpha_i[\underline{\alpha}/\underline{Z}] = Z_i$

(2) $(vx)[\underline{\alpha}/\underline{Z}] = v[\underline{\alpha}/\underline{Z}]\cdot\{x\}$ and $(v\alpha_i)[\underline{\alpha}/\underline{Z}] = v[\underline{\alpha}/\underline{Z}]\cdot Z_i$.

In [10] it has been proved that Φ_G is contractive for any context-free resp. regular grammar G (fulfilling the general assumption). Thus, by Banach's fixed point principle Φ_G has a unique fixed point which is equal to $(L(G,\alpha))_{\alpha\in N}$. Therefore, we get

Theorem 2. An R-fuzzy language is context-free resp. regular if and only if it is a component of the fixed point of the mapping associated to a context-free resp. regular grammar.

4. R-regular Languages

In this section we show how R-fuzzy languages may be compared. For that reason cut-point languages are introduced. The underlying R-algebra is regarded as a po-semiring throughout this section. Thus, an R-fuzzy language L over X is a mapping from X* into R. Let c be an arbitrary element of R. We call $L_c = \{w \in X^* \mid L(w) \geqq c\}$ a cut-point language of L. Now, R-fuzzy languages can be compared for different po-semirings R by means of the associated cut-point languages.

Definition 7. A subset $A \subset X^*$ is called an R-context-free resp. R-regular language if there is a context-free resp. regular R-fuzzy language L such that $A = L_c$ for some c of R.

In the sequel we restrict ourselves on R-regular languages. The first problem arises: Which languages are R-regular?

Theorem 3 (cf. [9]). Any language is R-regular.

The proof of this theorem is based on the fact that any language A over X can be accepted by the following (possible infinite) deterministic automaton $\mathcal{A}_d = (Z, z_o, Z_f, d)$ over X, where $Z = \{w^{-1}A \mid w \in X^*$ and $w^{-1}A \neq \emptyset\}$, $z_o = A$, $Z_f = \{w^{-1}A \mid w \in A\}$, and $d(x, w^{-1}A) = (wx)^{-1}A$ for $x \in X$ and $w \in X^*$ (cf. [2]). Thereby, $w^{-1}A = \{v \in X^* \mid wv \in A\}$ is the quotion of A with respect to $w \in X^*$. Now, we "simulate" $\mathcal{A}_d$ by a finite R-fuzzy automaton (cf. [2,8]) $\mathcal{A} = (S, S_o, S_f, m)$ with two internal states, i.e. $S = \{s_o, s_f\}$. S_o and S_f are R-fuzzy sets with support S defined by $S_o(s) = 1$ if and only if $s = s_o$ and $S_o(s) = 0$ if and only if $s = s_f$, and $S_f(s) = r$ if and only if $s = s_o$ and $S_f(s) = 1$ if and only if $s = s_f$, where $r = 0$ if $e \notin A$ and $r = e$ if $e \in A$. The transition function m is a mapping from X into all 2×2 matrices with coefficients of R defined by

$$m(x) = \begin{vmatrix} 0 & x^{-1}A \\ 0 & x \end{vmatrix} \qquad \text{for each } x \in X.$$

It remains to define the appropriate po-semiring R by

$$R = \{0,1,\infty\} \cup \{w^{-1}A \mid w \in X^*\} \cup X^*$$

whose operations and partial order are given as follows

(1) 0 is zero and 1 is unit element.

(2) $r + r' = \infty$ for $r, r' \in R - \{0\}$.

(3) $r \cdot r' = \begin{cases} rr' & \text{if } r, r' \in X^* \\ (wr')^{-1}A & \text{if } r = w^{-1}A \text{ for some } w \text{ of } X^* \text{ and } r' \in X^* \\ \infty & \text{otherwise.} \end{cases}$

(4) 0 is bottom and ∞ is top element.

(5) $v \leqq (wv)^{-1}A$ if and only if $w \in A$ for all $v, w \in X^*$.

An easy calculation shows

$$m(e) = \begin{vmatrix} 1 & 0 \\ 0 & 1 \end{vmatrix} \quad \text{and} \quad m(w) = \begin{vmatrix} 0 & w^{-1}A \\ 0 & w \end{vmatrix} \qquad \text{for } w \in X^* - \{e\}.$$

Consequently, $\mathfrak{A}$ accepts the following regular R-fuzzy language L with $L(e) = e$ if $e \in A$ resp. $L(e) = 0$ if $e \notin A$ and $L(w) = w^{-1}A$ for each non-empty word w of X^*. Since $A = L_e$, A is an R-regular language which proves the theorem.

In addition, it can be shown that any family of languages is inculded in a family of R-regular languages for a certain po-semiring R. This leads to the next problem: Which families of languages are families of R-regular languages?

Theorem 4 [9,10]. A family if languages is a family of R-regular languages, where R is a po-semiring with a top element, if and only if it is closed under inverse gsm mapping, intersection and union with regular languages.

Obviously, the family of all regular languages is a family of R-regular languages if we put for R the Boolean semiring $\mathbb{B}$. But also the family of all stochastic languages is a family of R-regular languages if R is choosen to be the po-semiring of all positive real numbers (cf. [7]).

References

[1] Cohen, P.M. (1965): Universal Algabra.
Harper and Row, Publ., New York, Evaston and London

[2] Eilenberg, S. (1974): Automata, Languages, and Machines Vol.A
Academic Press
New York and London

[3] Goguen, J.A. (1968/69): The logic of inexact concepts.
Synthese 19, 325-373

[4] Rasiowa, H. (1974): An Algebraic Approach to Non-Classical Logics.
North-Holland Publ. Comp.
Amsterdam and London

[5] Rescher, N. (1969): Many-Valued Logic.
Mc Graw-Hill
New York

[6] Salomaa, A. and Soittola, M. (1978): Automata-Theoretic Aspects of Formal Power Series.
Springer-Verlag
New York, Heidelberg, Berlin

[7] Turakainen, P. (1969): Generalized automata and stochastic languages.
Proc. Amer. Math. Soc. 21, 303-309

[8] Wechler, W. and Dimitrov, V. (1974): R-fuzzy automata.
In "Information Processing '74", North-Holland Publ. Comp.
Amsterdam

[9] Wechler, W. (1977): Families of R-fuzzy languages.
Lecture Notes in Computer Science Vol. 56, 117-186

[10] Wechler, W. (1978): The Concept of Fuzziness in Automata and Language Theory.
Akademie-Verlag
Berlin

[11] Zadeh, L.A. (1965): Fuzzy sets.
Information and Control 8, 338-353

ADVANCES IN FUZZY SET THEORY AND APPLICATIONS
M.M. Gupta, R.K. Ragade, R.R. Yager (editors)

DECISION-MAKING UNDER FUZZINESS(*)

Didier DUBOIS
Laboratoire I.M.A.G.
B.P. 53
38041 GRENONLE - Cédex
FRANCE

Henri PRADE
Langages et Systèmes informatiques
Université Paul Sabatier
TOULOUSE
FRANCE

ABSTRACT :

In this paper decision-making is viewed as choosing an action in a given situation, so as to maximise a utility function. Several cases are solved depending on where fuzziness lies. Lastly, a problem is solved when the actions have unequal feasibility. Most of the proposed methods are computationally tractable, owing to results in fuzzy algebra (i.e. elementary operations extended to fuzzy sets of the real line) which are first reviewed.

1. INTRODUCTION :

Uncertainty is a very familiar phenomenon in decision-making analysis. Until recently, Decision Theory put forward only probabilistic models to deal with the lack of knowledge about utility values, subjective preferences and diagnosis of situations [12].

It was thus implicity assumed that uncertainty was always a consequence of randomness.

However, the evaluation of alternatives is often also pervaded with fuzziness, especially when subjective factors are involved. Fuzziness, in the sense of Zadeh [15], means a lack of precise boundaries for some considered subsets of a given universe. For example, we deal with fuzziness when a linguistic value A (e.g. tall, i.e. A is a fuzzy set of the real line $\mathbb{R}$) is assigned to a given variable V (e.g. size), which is then called a linguistic variable [19]. This fuzzy set A assigned to the linguistic variable V, may also be viewed sometimes as a fuzzy restriction on the possible non fuzzy values that the real variable v (on which the linguistic one V is based) may take : V = A is understood as $\pi(v = a) = \mu_A(a)$, i.e. the possibility for v = a is equal to the grade of membership of a in A. A is then equivalent to the possibility distribution π [20]. A proposition such as "Jack is tall" involves a fuzzy set "tall" over a universe of heights. "Tall" is a linguistic value assigned to the linguistic variable V = height of Jack. The proposition "Jack is tall" translates into a possibility distribution $\pi = \mu_{tall}$ on Jack's height. Here "tall" acts as a fuzzy restriction on the possible numerical values of this height. Note that precise grades of membership are usually out of reach. Anyway, the knowledge of the rough shape of μ_A is usually sufficient to actually apply fuzzy sets theory (see [6] IV chap.1).

---

* This work was carried out while the authors were visiting scholars, respectively at :
- the School of Electrical Engineering. Purdue University, 47907 IN.
- the Stanford Art. Int. Lab. ; Stanford University, 94305 CA.

They were supported by scholarships of the "Institut de Recherches en Informatique et Automatique" Rocquencourt. 78150 LE CHESNAY - France.

Bellman an Zadeh [2] were the first to model fuzziness in the framework of multicriteria decision-making. They pointed out the fact that in a fuzzy environment, the difference between criteria and constraints vanishes. Another kind of decision-making problem can be stated as : find the best strategy (or action) in a given situation, so as to maximize a given utility function.

This paper proposes solutions to this problem in different cases, according to where fuzziness lies. Jain [7], [8], [9] has been the first to investigate the cases when the situation, the utility function or both are fuzzy. His computation methods were discussed in [3] . We give here a new method, already sketched in [6], which seems more satisfactory for the intuition. Moreover we consider also the cases when the situation is only known through classical or even linguistic probabilities, or when there is a feasibility constraint on the actions, expressed by a possibility distribution. Some results in fuzzy algebra, which are needed for practical computation, are first recalled. These results detailed in [4], [5], [6] include the works by Nahmias[1], Mizumoto Tanaka[2] and provide simple methods for effective computation in all cases.

2. BACKGROUND : some results in fuzzy algebra :

2.1. Definition

A fuzzy number $\tilde{m}$ is a fuzzy set of the real line $\mathbb{R}$ which is :

- continuous, i.e. its membership function $\mu_{\tilde{m}}$ is continuous (except perhaps for a finite number of points),
- normalized, i.e. $\exists x \in \mathbb{R}$, $\mu_{\tilde{m}}(x) = 1$
- convex i.e. $\forall x, z \in \mathbb{R}$, $\forall y \in [x,z]$, $\mu_{\tilde{m}}(y) \geq \min(\mu_{\tilde{m}}(x), \mu_{\tilde{m}}(z))$

  NB . If $\exists$ a,b, $a \neq b$ such that $\forall x \in [a,b]$, $\mu_{\tilde{m}}(x) = 1$,

  we shall speak of fuzzy interval rather than of fuzzy number.

A fuzzy number may be interpreted as a fuzzy value or as a possibility distribution on the non fuzzy value of a variable.

2.2. L - R representation of a fuzzy number :

Let L and R be two continuous bell-shaped functions from to [0,1] such that :

- L (0) = R (0) = 1
- L (-x) = L (x) ; R (-x) = R (x)
- L and R are non-increasing on [0, +∞]
- L (+∞) = R (+∞) = 0

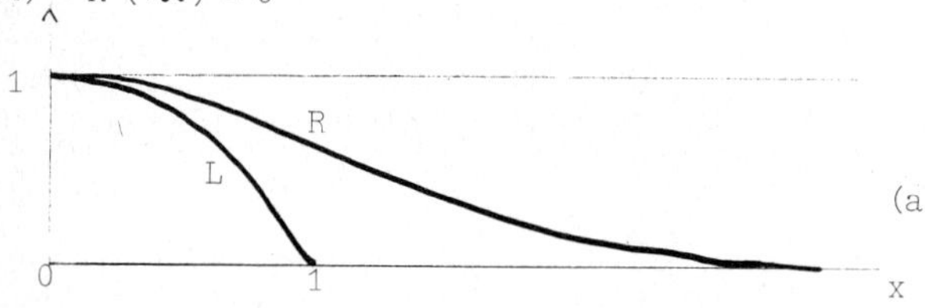

(a) reference functions

---

1) Fuzzy sets and Systems Vol. 1, n° 2, 1978, pp. 97-111

2) Systems Computers Control. Vol. 7, n° 5, 1976, pp. 73-81

Any membership function $\mu_{\tilde{m}}$ of a fuzzy number can be represented by :

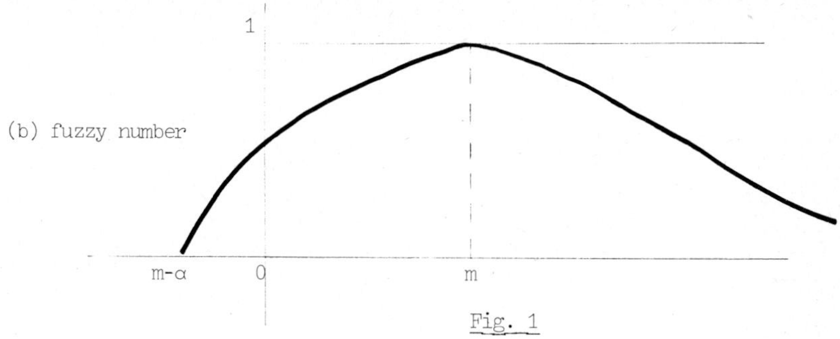

Fig. 1

$$\mu_{\tilde{m}}(x) = \begin{cases} L(\frac{m-x}{\alpha}) & \text{for } x \leq m \\ R(\frac{x-m}{\beta}) & \text{for } x \geq m \end{cases}$$

where $\mu_{\tilde{m}}(m) = 1$ and $(\alpha, \beta) \in \mathbb{R}_+^2$ Symbollically, it will be written :

$\tilde{m} = (m, \alpha, \beta)_{LR}$.

m is the mean value of $\tilde{m}$, i.e. $\mu_{\tilde{m}}(m) = 1$ ($\tilde{m}$ is "approximately m"); α and β are the left and right spreads of $\tilde{m}$;
L and R, the left and the right reference functions.
The attractivences of such a representation will appear more clearly when performing addition on fuzzy numbers.

2.3. Extension principle

For the sake of simplicity this principle is now stated for real functions of two real variable. Let f be such a function and let A and B be two fuzzy sets of R. The extension principle (Zadeh [19]) allows us to define the image of A and B through F, say f (A, B) whose membership function is :

$$\mu_{f(A,B)}(z) = \sup_{x,y \in f^{-1}(z)} \min(\mu_A(x), \mu_B(y)) \qquad (1)$$

where $f^{-1}(z) = \{(x,y) \in \mathbb{R}^2 | f(x,y) = z\}$.

This principle can be interpreted as follows : the possibility for the pair (A,B) to be represented by (x,y) is $\mu_{AxB}(x,y) = \min(\mu_A(x),\mu_B(y))$. AxB is the cartesian product of A and B.
The possibility for f(A,B) to be represented by z is the greatest possibility value for a pair (x,y) in the converse image of z, $f^{-1}(z)$, to be in AxB. Note that whenever $f^{-1}(z) = \emptyset$, $\mu_{f(A,B)}(z) = 0$.

It is easy to show that (see Dubois Prade [5], [6])

$$f(A \cup B, C) = f(A,C) \cup f(B,C) \quad \forall A,B,C \in \tilde{\mathcal{P}}(\mathbb{R}) \qquad (2)$$

$$hgt(f(A,B)) = \min(hgt(A), hgt(B)) \quad \text{(flattening effect)} \qquad (3)$$

$\tilde{\mathcal{P}}(\mathbb{R})$ denotes the set of fuzzy sets of $\mathbb{R}$

Jain [7] assumes finite support fuzzy sets and uses a different extension principle :

He replaces sup by a probabilistic sum $\tilde{+}$ ($u \tilde{+} v = u + v - uv$), so that the membership value $\mu_{f(A_1 A_2)}(z)$ depends upon the cardinality of the set $\{(x,y) \in S(A_1) \times S(A_2) \mid z = f(x,y)\}$. Particularly, it can be shown (cf [3]) that when this set is not finite $f(A_1, A_2)$ is a classical set. In this paper we consider a "possibilistic" extension principle, and the supports of the fuzzy sets are allowed to be infinite sets.

Remark :

The support $S(A)$ of a fuzzy set A is $S(A) = \{x \in X \mid \mu_A(x) > 0\}$.

2.4. Addition of fuzzy numbers

Let $\tilde{m}$ and $\tilde{n}$ be two fuzzy numbers. Using Zadeh's extension principle [7], the membership function of the extended sum $\tilde{m} \oplus \tilde{n}$ is :

$$\mu_{\tilde{m} \oplus \tilde{n}}(t) = \sup_{t=x+y} \min(\mu_{\tilde{m}}(x), \mu_{\tilde{n}}(y)) \qquad (4)$$

let u and v be two real variables which have respectively $\mu_{\tilde{m}}$ and $\mu_{\tilde{n}}$ as possibility distributions. The possibility of having $u = x$ and $v = y$ is equal to $\min(\mu_{\tilde{m}}(x), \mu_{\tilde{n}}(y))$. (1) means that the possibility that $u+v = t$ is equal to the upper bound of the possibilities that $u = x$ and $v = y$, when $x+y = t$.

(4) can be written :

$$\mu_{\tilde{m}\oplus\tilde{n}}(t) = \sup_{x} \min(\mu_{\tilde{m}}(x), \mu_{\tilde{n}}(t-x)) \qquad (5)$$

suppose $\tilde{m} = (m, \alpha, \beta)_{LR}$ and $\tilde{n} = (n, \gamma, \delta)_{LR}$

$\mu_{\tilde{n}}(t-x)$ is the membership function of a number of type RL which is $(t-n, \delta, \gamma)_{RL}$. Because of the convexity and continuity of $\tilde{m}$ and $\tilde{n}$, the upper bound in (2) is obtained for :

$$\mu_{\tilde{m}}(x) = \mu_{\tilde{n}}(t-x) \qquad \text{(see figure 2)}$$

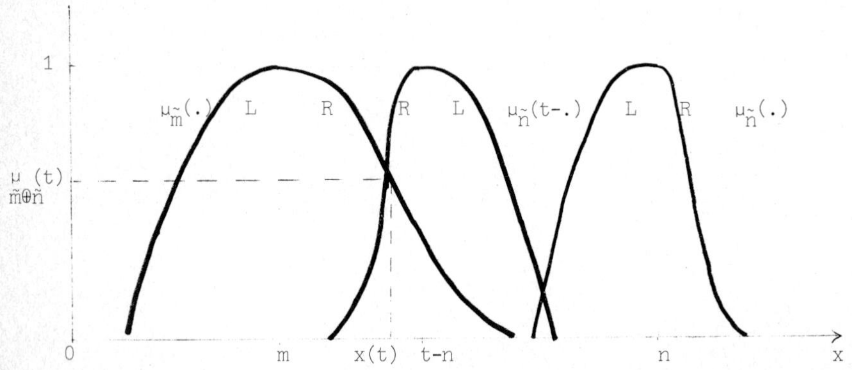

Figure 2

i.e. if $t-x \geq n$ and $x \geq m$ (which implies $t \geq m+n$) :

$$R\left(\frac{x(t)-m}{\beta}\right) = R\left(\frac{t-x(t)-n}{\delta}\right)$$

or
$$\frac{x(t)-m}{\beta} = \frac{t-x(t)-n}{\delta}$$

or
$$x(t) = \frac{(t-n)\beta + m\delta}{\beta+\delta}$$

and
$$\mu_{\tilde{m}\oplus\tilde{n}}(t) = R\left(\frac{\frac{\beta(t-n)+m\delta}{\beta+\delta} - m}{\beta}\right) = R\left(\frac{t-(m+n)}{\beta+\delta}\right)$$

it can similarly be shown that $\mu_{m\oplus n}(t) = L\left(\frac{m+n-t}{\alpha+\gamma}\right)$ when $t \leq m+n$

hence we have the formula :

$$(m, \alpha, \beta)_{LR} \oplus (n, \gamma, \delta)_{LR} = (m+n, \alpha+\gamma, \beta+\delta)_{LR}$$

N.B. Nahmias (op.cit.) obtained this result for triangular membership functions only. Actually, it is much more general.

Even more generally $(m, \alpha, \beta)_{LR} \oplus (n, \gamma, \delta)_{L'R'} = (m+n, 1,1)_{L'',R''}$ with $L'' = (\alpha L^{-1} + \gamma L^{-1})^{-1}$, $R'' = (\beta L^{-1} + \delta L^{-1})^{-1}$. (See [6]).

2.5. Multiplication of a fuzzy number by a scalar :

Let λ be a positive real number and $\tilde{m}$ be a fuzzy number. the membership function of $\lambda.\tilde{m}$ is, using the extension principle [17] :

$$\mu_{\lambda.\tilde{m}}(t) = \sup_{t=\lambda x} \mu_{\tilde{m}}(x) = \mu_{\tilde{m}}\left(\frac{t}{\lambda}\right)$$

For $\lambda = -1$ we get the opposite of $\tilde{m}$, namely $-\tilde{m}$

Using the L R representation of $\tilde{m}$:

$$\mu_{\lambda.\tilde{m}}(t) = \begin{cases} L\left(\frac{\lambda m - t}{\lambda\alpha}\right) & \text{for } t \leq \lambda m \\ R\left(\frac{t-\lambda m}{\lambda\beta}\right) & \text{for } t \geq \lambda m \end{cases} \qquad (\lambda > 0)$$

i.e.

if $\lambda > 0$ $\quad \lambda\tilde{m} = \lambda(m, \alpha, \beta)_{LR} = (\lambda m, \lambda\alpha, \lambda\beta)_{LR}$

if $\lambda < 0$ $\quad \lambda\tilde{m} = \lambda(m, \alpha, \beta)_{LR} = (\lambda m, -\lambda\beta, -\lambda\alpha)_{RL}$

2.6. Extended max and min for fuzzy numbers :

When comparing fuzzy numbers $\tilde{m}$ and $\tilde{n}$; two different questions may arise :

Q1 : is there a least upper bound (resp. a greatest lower bound) for the pair $(\tilde{m}, \tilde{n})$ in the sense of the canonical order of $\mathbb{R}$? More precisely, and in other words, what is the possibility distribution of the real variable max (u,v) (resp. min(u,v)) when the possibility distributions of u and v are respectively given by $\mu_{\tilde{m}}$ and $\mu_{\tilde{n}}$?

Q2 : what is the possibility that the value x of the real variable u (whose possibility distribution is $\mu_{\tilde{m}}$) is greater than or equal to the value y of the real variable v (whose possibility distribution is $\mu_{\tilde{n}}$) ? Or, roughly speaking, what is the truth value of the assertion "$\tilde{m}$ is greater than $\tilde{n}$" ?

The answer to the first question is a fuzzy set of the real line $\mathbb{R}$ (or if we prefer a possibility distribution) ; the answer to the second one is just a real number of the interval [0,1]. We deal with the former question in this section and with the latter in the next one.

Using Zadeh's extension principle, the membership function of the extended max of $\tilde{m}$ and $\tilde{n}$, namely $\tilde{\max}$ $(\tilde{m}, \tilde{n})$ is :

$$\mu_{\tilde{\max}(\tilde{m},\tilde{n})}(t) = \sup_{t=\max(x,y)} \min(\mu_{\tilde{m}}(x), \mu_{\tilde{n}}(y)) \qquad (7)$$

Thus, $\tilde{\max}$ is commutative and associative.

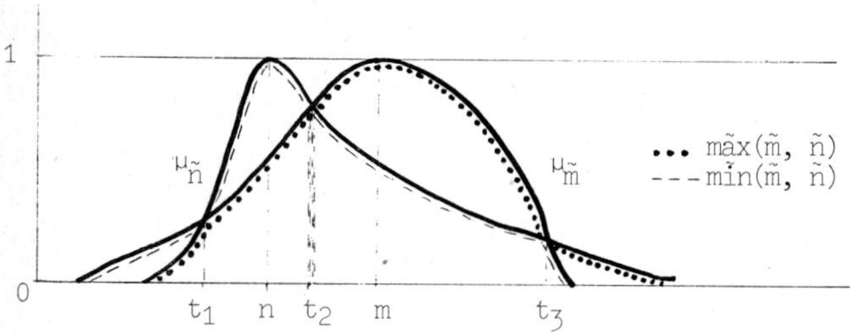

Figure 3

Let us prove, for $\tilde{m}$ and $\tilde{n}$ pictured on figure 3, that $\tilde{\max}(\tilde{m},\tilde{n})$ is the fuzzy number whose membership function appears in dotted line :

. for $t \le t_1$, the least upper bound in (7) is reached for y=t and $x \le t$.
Hence $\mu_{\tilde{\max}(\tilde{m},\tilde{n})}(t) = \mu_{\tilde{n}}(t)$ if $t \le t_1$

. for $t_1 \le t \le t_2$, the least upper bound in (3) is reached for x=t and $t_1 \le y \le t_2$.

Hence $\mu_{\max(\tilde{m},\tilde{n})}(t) = \mu_{\tilde{m}}(t)$ if $t_1 \le t \le t_2$

. for $t_2 \le t \le t_3$, $\mu_{\tilde{m}}(t) \ge \mu_{\tilde{n}}(t)$ and $\exists(x,y)$ such that $\mu_{\tilde{n}}(y) = 1$ and $t = \max(x,y)$. Here x=t, y=n.

Hence $\mu_{\tilde{\max}(\tilde{m},\tilde{n})}(t) = \mu_{\tilde{m}}(t)$ if $t_2 \le t \le t_3$

. for $t \ge t_3$, $\mu_{\tilde{m}}(t) \le \mu_{\tilde{n}}(t)$ and the least upper bound in (7) is reached for y=t and $x \le t_3$.

Hence $\mu_{\tilde{\max}(\tilde{m},\tilde{n})}(t) = \mu_{\tilde{n}}(t)$ if $t \ge t_3$

Q.E.D.

So, an important conclusion is that $\tilde{\max}(\tilde{m},\tilde{n})$ is a fuzzy number different from $\tilde{m}$ or $\tilde{n}$. Also worth noticing is that the calculation of $\tilde{\max}(\tilde{m},\tilde{n})$ is rather easy, at least when $\tilde{m}$ and $\tilde{n}$ are fuzzy numbers (i.e. convex and normalized) . $\tilde{\min}$ is defined and calculated similarly to $\tilde{\max}$ [4] [5] [6]. (See figure 3).
$\tilde{\min}$ and $\tilde{\max}$ are idempotent, associative, commutative, mutually distributive ; they

satisfy the following equality :

$$\tilde{\max}(M,N) \oplus \tilde{\min}(M,N) = M \oplus N, \forall M, N \text{ fuzzy numbers} \quad (8)$$

These properties and others are studied in Mizumoto, Tanaka [10] and Dubois Prade [4], [5], [6].

Proofs pertaining to $\tilde{m} \oplus \tilde{n}$, and $\tilde{\max}(\tilde{m},\tilde{n})$ were given here in order to have a paper as self-contained as possible and also because the proof used here for $\tilde{m} \oplus \tilde{n}$ may be somewhat generalized for the problem of the "interactive sum" which we will have to deal with in part 5.3. However, a different and very general approach which encompasses most of the usual n-ary non interactive operations (for instance addition, subtraction, multiplication, division, exponentiation, max, min, power elevation, ... and many others) has been first proposed in [4] refined and extended to any kind of piecewise continuous fuzzy sets of the real line in [5] and [6].

In [4], [5], [6], exact or approximate formulas, which use the L R representation and are thus easy to perform , are provided for all the usual operations on fuzzy numbers. A general algorithm, which performs operations on possibly non convex or / and non normalized fuzzy sets of $\mathbb{R}$ is also presented in [5], [6].

The basic idea of this algorithm is a discretization of the valuation set [0,1] instead of a discretization of the supports of the fuzzy numbers which some authors use (for instance Jain [8]).

2.7. Is $\tilde{m}$ greater than $\tilde{n}$?

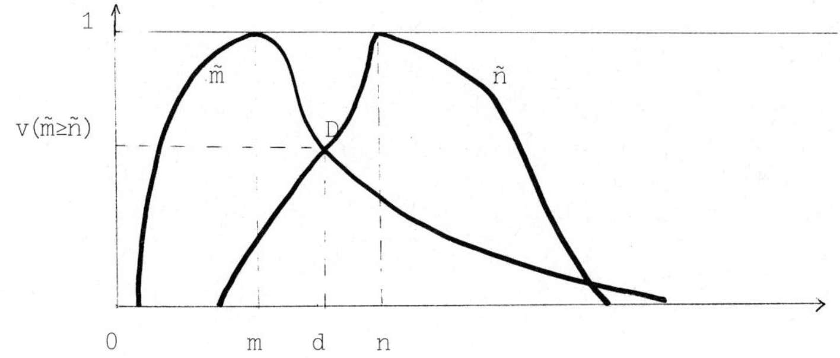

Figure 4

The possibility that $\tilde{m}$ is greater than (or equal to) $\tilde{n}$, symbolically denoted $v(\tilde{m} \geq \tilde{n})$(truth value of the statement "$\tilde{m}$ is greater than or equal to $\tilde{n}$"), is

$$v(\tilde{m} \geq \tilde{n}) = \sup_{x \geq y} \min (\mu_{\tilde{m}}(x), \mu_{\tilde{n}}(y)) \quad (9)$$

i.e. it is the least upper bound of the possibilities that u = x, v = y and x ≥ y, where $\mu_{\tilde{m}}$ and $\mu_{\tilde{n}}$ are respectively the possibility distributions of the real variables u and v.

In the case of fig. 4

$v(\tilde{n} \geq \tilde{m}) = 1$ because $\mu_{\tilde{m}}(m) = 1$, $\mu_{\tilde{n}}(n) = 1$ and $n > m$

and it is easy to check that :

$v(\tilde{m} \geq \tilde{n})$ = height of $\tilde{m} \cap \tilde{n}$ = ordinate of D. (fig. 4). (10)

(which is general for fuzzy numbers as soon as the mean value n of $\tilde{n}$ is greater than or equal to m, the mean value of $\tilde{m}$).

When $\tilde{m} = (m, \alpha, \beta)_{RL}$ and $\tilde{n} = (n, \gamma, \delta)_{LR}$ the ordinate of D is given by the equation :

$$L\left(\frac{n-d}{\gamma}\right) = L\left(\frac{m-d}{\beta}\right) = \mu_{\tilde{m}}(d)$$

i.e. $v(\tilde{m} \geq \tilde{n}) = L\left(\frac{n-m}{\beta+\gamma}\right)$ if $n \geq m$ (11)

The answer a to the question "can $\tilde{m}$ be greater than $\tilde{n}$?" is fuzzy and can be represented as a fuzzy set of the universe {yes, no} :

$$a = v(\tilde{m} \geq \tilde{n}) / \text{yes} + v(\tilde{n} \geq \tilde{m}) / \text{no}$$

Note that there may be a complete ambiguity of the answer, when,

$$v(\tilde{m} \geq \tilde{n}) = v(\tilde{n} \geq \tilde{m}) = 1$$

N.B.1. When $\tilde{m}$ and $\tilde{n}$ are no more convex or normalized, the use of formula (9) can be made easier by noticing that the only values of x an y to be considered are those which correspond to the maxima of $\mu_{\tilde{m}}$ and $\mu_{\tilde{n}}$.

Let $m_1 \ldots m_k$ (resp : $n_1 \ldots n_l$) be the local maxima of the continuous fuzzy set of $\mathbb{R}$, $\tilde{m}$ (resp : $\tilde{n}$). The following formula holds :

$$v(\tilde{m} \geq \tilde{n}) = \max(\text{hgt}(\tilde{m} \cap \tilde{n}), \max_{\substack{i=1,k \\ j=1,l \\ m_i \geq n_j}} \min(\mu_{\tilde{m}}(m_i), \mu_{\tilde{n}}(n_j))) \quad (12)$$

(see the Annex.1 for the proof) "hgt" means "height of a fuzzy set".

N.B.2 : the possibility for $\tilde{m}$ to be greater than k fuzzy sets $\tilde{m}_i$ of $\mathbb{R}$ can be defined by :

$$v(\tilde{m} \geq \tilde{m}_1, \tilde{m}_2 \ldots, \tilde{m}_k) = v((\tilde{m} \geq \tilde{m}_1) \text{ and } (\tilde{m} \geq \tilde{m}_2) \text{ and } \ldots \text{ and } (\tilde{m} \geq \tilde{m}_k))$$

$$= \min_{i=1,k} v(\tilde{m} \geq \tilde{m}_i)$$

Baas and Kwakernaak [1] have proposed an extension of (9)

$$v'(\tilde{m} \geq \tilde{m}_1, \tilde{m}_2 \ldots, \tilde{m}_k) = \sup_{\substack{t \geq t_1 \\ \vdots \\ t \geq t_k}} \min(\mu_{\tilde{m}}(t), \mu_{\tilde{m}_1}(t_1), \ldots \mu_{\tilde{m}_k}(t_k)) \quad (13)$$

$v' = v$ when the fuzzy sets involved are fuzzy numbers.

N.B.3 : $\tilde{m}$ could be considered as greater than $\tilde{n}$ as soon as $\widetilde{\max}(\tilde{m}, \tilde{n}) = \tilde{m}$ (or equivalently $\widetilde{\min}(\tilde{m}, \tilde{n}) = \tilde{n}$ owing to (8). However, $\tilde{m}$ and $\tilde{n}$ can be very close to each other with $\widetilde{\max}(\tilde{m}, \tilde{n}) = \tilde{m}$ being true. Thus $v(\tilde{m} \geq \tilde{n})$ and $v(\tilde{n} \geq \tilde{m})$ are a better index to assess the possibility of proximity for the highest-membership parts of $\tilde{m}$ and $\tilde{n}$, $\widetilde{\max}$ is interesting whenever $\widetilde{\max}(\tilde{m},\tilde{n}) \neq \tilde{m}$ or $\tilde{n}$ since it means that, taking into account lower membership grades will not yield the same conclusions as considering only the highest ones (see fig.3)

2.8. Interactivity

In the definition of the extension principle (formula 1) the domain over which the membership function of $A_1 \times A_2$ is to be maximized is $f^{-1}(z) = \{(x,y) \mid f(x,y)=z\}$ This no longer true when there is some dependency between the variables x and y. This dependency can be represented by a fuzzy relation R on X x X. The extension principle can there be changed into :

$$(14) \quad \mu_{f(A,B)}(z) = \sup_{(x,y)\in f^{-1}(z)} \min(\mu_A(x), \mu_B(y), \mu_R(x,y))$$

Particularly when R is a non fuzzy domain, the function $\mu_{A\times B}$ is to be maximized over $f^{-1}(z) \cap R$. The variables x and y are then said to be interactive (Zadeh [19]) so is the operation f.

3 - FUZZY SITUATION

3.1 Statement of the problem :

Let $X = (x_1, x_2, \ldots, x_M)$ be a set of possible states of the considered system or universe. X describes the various possible situations, assumed uncontrollable which may be encountered. A given number N of actions or alternatives $a_1, a_2, \ldots, a_N$ exist, which can be performed.

The utility of performing a_j when the state of the universe is x_i is u_{ij}. u_{ij} is assumed to be a real number. When the universe is in the state x_{i_o}, the best action is a_{j_o} such that $u_{i_o j_o} = \max_{j=1,N} u_{i_o j}$

N.B. : There is no objection to assume X as a continuum. In the paper, X is a discrete finite set only for the sake of clarity.

3.2 Fuzzy state :

Now the state of the universe is a fuzzy set F of X, whose membership function is $\mu_F : X \to [0,1]$. This situation can happen if the state is a linguistic variable on X, or when the universe is too complex, so that its state can only be roughly perceived and approximately measured.

The extension principle allows us to induce for each action a_j a fuzzy utility U_j which reflects the lack of well-defined knowledge of the state :

$$U_j = \sum_{i=1}^{M} \mu_F(x_i) / u_{ij}$$

This notation must be understood as : U_j is a fuzzy set of the real line whose membership function is μ_{U_j} such that $\mu_{U_j}(u_{ij}) = \mu_F(x_i)$.

The possibility distribution on the actual value of the state induces a possibility distribution on the value of the utility of performing a given action a_j.

Now, $\tilde{\max}$ can be used to calculate the fuzzy value U of the "greatest utility" in the situation described by F : $U = \tilde{\max}_{j=1,N} U_j$

First note that U_j induced by F, is not necessarily convex ; however U_j is normalized as soon as F is. Second, if X is discrete (and has not a lot of elements), the support (included in $\mathbb{R}$) of U_j will be discrete and the result of the operation $\tilde{\max}$ will be calculated using (7), directly. But if X has too many elements, it may be interesting to "approximate" each U_j by a continuous fuzzy set before performing $\tilde{\max}$ although the result obtained by rediscretizing U may be slightly different from the result which would be given by a direct use of forrula (7) (see [6]). Moreover if X is a continuum equipped with a metric, it will be assumed that the mapping $x \to U_j(x)$ (utility of performing a_j when the state is x) is continuous, then if μ_F is continuous, $\mu_{U_j(F)}$ will be too ; it is defined by $\mu_{U_j(F)}(u) = \sup_{x \in U_j^{-1}(u)} \mu_F(x)$

As pointed out in 2.4, U is generally different from each U_j and thus does not correspond to a particular well-defined alternative.
In fact, strictly speaking, U corresponds to a fuzzy alternative $\tilde{a}$ (a fuzzy set of $\{a_1, a_2, \ldots a_N\}$) - viewing U as a fuzzy set of the U_j's (parts of which U is made of) ; the degree to which a_j belongs to $\tilde{a}$ is equal to the greatest ordinate of the parts of U_j actually present in U.

We may prefer to have a scalar measure of the dominance $d(a_j)$ of an action a_j over the other ones. Using formula (12),

$$d(a_j) = \min_{k=1,N} v(U_j \geq U_k) \quad \text{because} \quad v(U_j \geq U_j) = 1$$

we may as well use formula (13).

3.3 Examples

Example 1 : $X = \{x_1, x_2, x_3\}$ $A = \{a_1, a_2, a_3\}$

The utilities are given by the matrix :

$$U = [U_{ij}] = \begin{bmatrix} 7 & 4 & 2 \\ 3 & 6 & 4 \\ 3 & 4 & 8 \end{bmatrix}$$

The fuzzy state in $F = 0.8/x_1 + 1/x_2 + 0.5/x_3$

Hence : $U_1 = 0.8/7 + 1/3$ ×

$U_2 = 0.8/4 + 1/6$ □

$U_3 = 0.8/2 + 1/4 + 0.5/8.$ •

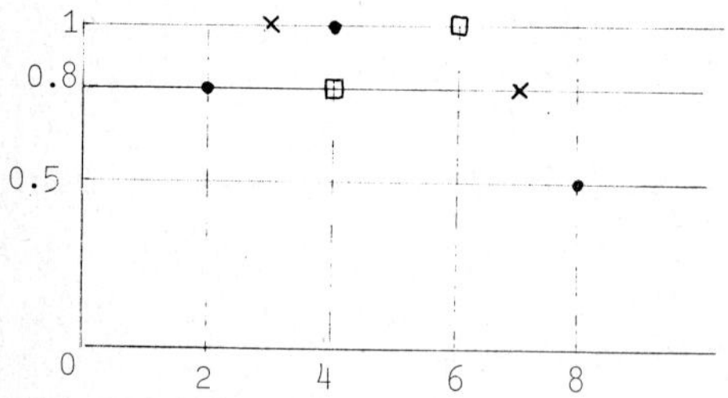

Figure 5

From the definition :

$$v(U_1 \geq U_2) = \max_{u_1 \geq u_2} \min(\mu_{U_1}(u_1), \mu_{U_2}(u_2))$$

$$= \max(\min(\mu_{U_1}(7), \mu_{U_2}(4)), \min(\mu_{U_1}(7), \mu_{U_2}(6))$$

$$= \max(\min(0.8, 0.8), \min(0.8, 1)) = 0.8$$

By applying (12) and the definition of "v", it is easy to check that :

$$d(a_1) = v(U_1 \geq U_2, U_3) = \min(0.8, 0.8) = 0.8$$
$$d(a_2) = v(U_2 \geq U_1, U_3) = \min(1,1) = 1$$
$$d(a_3) = v(U_3 \geq U_1, U_2) = \min(1, 0.8) = 0.8$$

Thus, a_2 can be considered as the best action. Moreover $v(U_1 \geq U_3) = 0.8$ $v(U_3 \geq U_1) = 1$: a_3 is the second best action. However, the dominance of a_2 over $\{a_1, a_3\}$ and that of a_3 over a_1 is not very strong.

Lastly, $\tilde{\max}(U_1, U_2, U_3) = 0.8/4 + 1/6 + 0.8/7 + 0.5/8 \notin \{ U_1, U_2, U_3 \}$

For instance, the membership value of 4 in $\tilde{\max}(U_1, U_2, U_3)$ is

$$\mu(4) = \max_{4=\max(u_1, u_2, u_3)} \min(\mu_{U_1}(u_1), \mu_{U_2}(u_2), \mu_{U_3}(u_3))$$
$$= \min(\mu_{U_1}(3), \mu_{U_2}(4), \mu_{U_3}(4)) = 0.8$$

It corresponds to a fuzzy best action $\tilde{a} = 0.8/a_1 + 1/a_2 + 0.5/a_3$

Both rankings are different because the index "v" favours high membership grades and the $\tilde{\max}$ takes into account large utility values. Generally a ranking using $\tilde{\max}$ may be riskier than using the index "v".

N.B. : a ranking based on $\tilde{\min}$ would involve small utility values for discarding bad actions.

Example 2 :

$X = [0,10]$ $A = (a_1, a_2, a_3)$

and $u_1(x) = \max(0,8-x)$; $u_2(x) = x \leq 5$; $u_2(x) = 10-x$; $x \geq 5$.

$u_3(x) = \frac{2x}{3}$

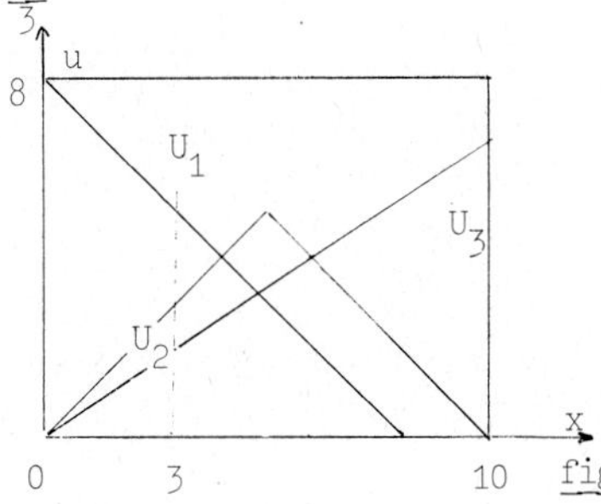

fig. 6

F = "approximately 3" = $(3,3,3)_{LL}$
with $L(x) = \max(1 - |x|, 0)$

To calculate the fuzzy utilities, we apply the extension principle ; for instance

$$\mu_{U_1}(u) = \sup_{u=u_1(x)} \mu_F(x) = \mu_F(8-u) \quad \text{for } u \in [0,8]$$

since $\mu_F(x) = L\left(\frac{3-x}{3}\right)$ we get $\mu_{U_1}(u) = L\left(\frac{u-5}{3}\right)$.

Similarly $\mu_{U_2}(u) = \max(L(\frac{u-3}{3}), L(\frac{7-u}{3})) = L\left(\frac{u-3}{3}\right)$ for $u \in [0,5]$

$= 0$ for $u \geq 5$.

$$\mu_{U_3}(u) = L\left(\frac{u-2}{2}\right)$$

That is to say $U_1 = (5, 3, 3)_{LL}$ $U_2 = \tilde{\min}(5, (3,3,3)_{LL})$ $U_3 = (2,2,2)_{LL}$

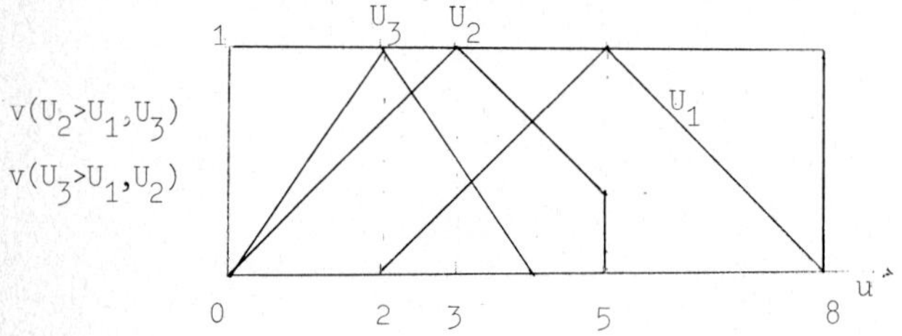

Fig. 7

Here $\tilde{\max}(U_1, U_2, U_3) = U_1$; a_1 is obviously the best alternative.

We have also : $v(U_1 \geq U_2, U_3) = 1 > v(U_2 \geq U_1, U_3) > v(U_3 \geq U_1, U_2)$

A is ranked as :1) a_1 2) a_2 3) a_3

Here, in spite of the fuzziness of the state, the ranking of alternative is not blurred, and is the same as in the non fuzzy case (x=3).

3.4 Comments :

It is obvious that if the evaluation of the alternatives is too blurred with fuzziness, a rational choice is not possible. More information is then needed to get more precise evaluations. But, even with fuzzy evaluations, there are situations where a rational decision is still possible : it occurs whenever U is equal or approximately equal to one of the U_i's (and different from the others) the corresponding $d(a_i)$ is equal to 1 (if the U_j's are normalized) and the other $d(a_j)$ are very different from 1. (Some threshold may be used).

Perhaps, the most appealing feature of the fuzzy approach here is to point out the limit of the arbitrariness in the decision process.

4 - FUZZY UTILITY :

4.1 Non fuzzy state : if the utilities are only vaguely known, and their values linguistically assigned, the u_{ij}'s are fuzzy numbers, say $\tilde{u}_{ij}$. Fuzzy utilities can be also considered when there is a tolerance interval on the utility values, without precise boundaries.

Choosing the best alternative, in the state x_{i_o}, is the problem of ranking the fuzzy numbers $\tilde{u}_{i_o j}$: this can be done as discussed in 3.2, with the results of 2.5 and 2.6.

4.2 Fuzzy state F :

This is the combined problem. Now the extension principle allows us to assign the membership value $\mu_F(x_i)$ to each fuzzy utility value $\tilde{u}_{ij}$. This induces a twice fuzzy utility $\tilde{U}_j$ for action a_j, defined as

$$\tilde{U}_j = \sum_{i=1}^{M} \mu_F(x_i) / \tilde{u}_{ij} \qquad j = 1,N$$

$\tilde{U}_j$ is a fuzzy set of fuzzy sets of the real line, i.e. a level 2 fuzzy set [15].

A way of dealing with level 2 fuzzy sets is to reduce them to ordinary fuzzy sets through the process of "s-fuzzification" (Zadeh [16]).
First define a fuzzification kernel as a mapping K from $\mathbb{R}$ to the set of fuzzy sets of $\mathbb{R}$, say $\tilde{\mathcal{P}}(\mathbb{R})$.

$$r \in \mathbb{R} \mapsto K(r) \in \tilde{\mathcal{P}}(\mathbb{R})$$

The fuzziness of any fuzzy set A of $\mathbb{R}$ can be increased by blurring the elements of its support, using a fuzzification kernel K, so that each r becomes a fuzzy set of numbers clustered around r. A is changed into the fuzzy set

$$F(A,K) = \bigcup_r \mu_A(r).K(r)$$

the membership function is $\mu_{F(A,K)}(x) = \sup_r \mu_A(r).\mu_{K(r)}(x)$

this transformation is called s-fuzzification of A (s=support).
Now considering the $\tilde{u}_{ij}$'s as deriving from a fuzzification kernel i.e. the utility values are the result of blurring non fuzzy utility values u_{ij}, we may state :

$$U_j = \bigcup_{i=1,M} \mu_F(x_i)\tilde{u}_{ij}, \text{ i.e. } U_j \text{ is the result of s-fuzzifying the fuzzy state F.}$$

Hence we obtain $\mu_{U_j}(t) = \max_{i=1,M} (\mu_F(x_i) \cdot \mu_{\tilde{u}_{ij}}(t)), \forall t \in \mathbb{R}$ (14)

When the $\tilde{u}_{ij}$'s are not fuzzy, this formula gives back the result of 3.

Once the $\tilde{U}_j$'s have been built, their comparison can be performed as in 3.2.

N.B. The choice of the product as a combination operator in formula (14) is not the only possible one. For instance, "min" could be used. The rationale underlying the product is that the membership values $\mu_F(x_i)$ and $\mu_{\tilde{U}_{ij}}(t)$ cannot be compared. Interpreting the height of a fuzzy set as its suitability to restrict the values of a variable, $\mu_F(x_i)$ acts as a modifier of the suitability of $\tilde{u}_{ij}$, which then depends on the possibility for x_i to be the actual state of nature.

Example :

On example 1 in paragraph 3.2 we blur the utility matrix into the fuzzy matrix :

$$\begin{bmatrix} (7,4) & (4,2) & (2,1) \\ (3,2) & (6,3) & (4,2) \\ (3,1) & (4,2) & (8,4) \end{bmatrix}$$

where each utility value is a symmetrical fuzzy number $(u_{ij}, \alpha, \alpha)_{LL} = (u_{ij}, \alpha)$ with $L(x) = 1 \dot{-} |x|$. The fuzzy state is still $F = 0.8/x_1 + 1/x_2 + 0.5/x_3$

Applying (14) we get

$$\mu_{U_1}(u) = \max(0.8\ L\ (\tfrac{7-u}{4}),\ L\ (\tfrac{3-u}{2}),\ 0.5\ L\ (3-u))$$
$$= \max(0.8\ L\ (\tfrac{7-u}{4}),\ L\ (\tfrac{3-u}{2}))$$

$$\mu_{U_2}(u) = \max\ [0.8.L(\tfrac{4-u}{2}),\ L(\tfrac{6-u}{3})]$$

$$\mu_{U_3}(u) = \max\ [0.8.L(2-u),\ L(\tfrac{4-u}{2}),\ 0.5\ L(\tfrac{8-u}{4})]$$

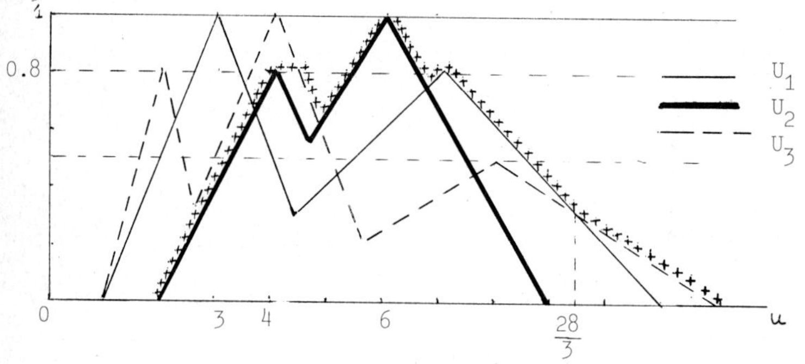

Figure 7

Let us calculate $v(U_i \geq U_j)$ $i,j = 1,3$, using the result in appendix 1. (see the table)

$v(U_i \geq U_j)$	1	2	3
1	1	0.8	0.8
2	1	1	1
3	1	0.8	1

Obviously, the fuzziness of the utilities does not affect the analysis of the decision. This is due to the fact that the index "v" considers the highest membership values, mainly.

To build $\tilde{\max}\ (U_1, U_2, U_3)$, we use the following procedure : each multimodal U_1 is decomposed into a union of convex (here, triangular) fuzzy sets i.e.

$U_1 = V_{21} \cup V_{11}$, $U_2 = V_{12} \cup V_{22}$, $U_3 = V_{13} \cup V_{23} \cup V_{33}$. Then it is easy to calculate $\tilde{\max}\ (V_{i1}, V_{j2}, V_{k3})$ for $(i,j,k) \in \{1,2\} \times \{1,2\} \times \{1,2,3\}$, applying the flattening effect (formula (3)). $\tilde{\max}$ (U_1, U_2, U_3)is the union of the resulting fuzzy sets, due to (2). It is pictured in +++ line on the above figure 7. The fuzzy action is now $a = 1/a_2 + 0.8/a_1 + 0.8/a_3$. There is a slight change with the non fuzzy utility case, due to the shape of U_3 in the upper neighborhood of 4.

However, we could interpret the shape of $\tilde{\max}(U_1, U_2, U_3)$ as follows :

- choosing a_3 as the best action is to hope that is utility will be greater than $\frac{28}{3}$ (see fig. 7), which is not very possible (membership values are below $\frac{1}{3}$).
- choosing a_2 as the best action is the safest decision but the most possible utility is only 6.
- choosing a_1 as the best action is quite risky since it may have a utility value which is only close to 3. However, there is a non negligible possibility (0.8) for a_1 to be better than a_2, by reaching 7 as a utility value.

A similar analysis could also be carried out on figure 5.
Hence, the blurring of the U_j's does not affect too much the decision analysis. However, would the U_j's be fuzzier, involving constant membership intervals, the conclusions could be greatly altered.

5 - PROBABILISTIC SITUATION

5.1. The non fuzzy case :

Now assume the state of the universe is only known in probability, i.e. there is a probability distribution p over x, such that

$$\sum_{i=1}^{M} p(x_i) = 1$$

The decision problem when the utility function is not fuzzy is solved by assigning to each action a_j the non fuzzy pay-off U_j,

$$U_j = \sum_{i=1}^{M} p(x_i)\, u_{ij}$$

the best action is a_{j_o} such that : $U_{j_o} = \max_{j=1,N} U_j$

5.2. The fuzzy case : fuzzy utilites :

If the $\tilde{u}_{ij}$'s are fuzzy, the same formula holds provided that we use the extended addition between real fuzzy numbers : $U_j = p(x_1)\, \tilde{u}_{1j} \oplus p(x_2).\tilde{u}_{2j} \oplus \ldots \oplus p(x_M).\tilde{u}_{Mj}$

Using formulas of sections 2.3 and 2.4 for LR represented fuzzy numbers the computation of the U_j is very easy. Since we are mainly interested in a qualitative modelling of the fuzzy numbers, it is not an important limitation to use the same references L and R for all the fuzzy numbers. But, if the fuzzy numbers are not of the same LR type, an easy-to-implement algorithm can be found in [5] or [6]. The ranking of the U_j's can be done as discussed in 3.2, 3.3, 3.4.

Example :

$p(x_1) = 0.2 \qquad p(x_2) = 0.5 \qquad p(x_3) = 0.3$

The fuzzy utility matrix is the one of section 4.2. Denoting (u_{ij},α) the fuzzy number $(u_{ij}, \alpha, \alpha)_{LL}$ and applying the addition and scalar product formulae (2.3) we get

$$U_1 = (7 \times 0.2 + 3 \times 0.5 + 3 \times 0.3,\ 4 \times 0.2 + 2 \times 0.5 + 1 \times 0.3) = (3.8,\ 2.1)$$

Similarly $U_2 = (5\ ,\ 2.5)$ $U_3 = (4.8,\ 2.4)$

It is easy to see that $\tilde{\max}(U_1, U_2, U_3) = U_2$. But although a_2 is still the best action a_3 is not so bad a choice since

$$V(U_3 > U_2) = L\left(\frac{5-4.8}{2.4+2.5}\right)$$

by (11). With $L = \max(0,1-|.|)$, $v(U_3 > U_2) = \frac{47}{49}$, close to 1.

5.3. Fuzzy probabilities. Non fuzzy utilities :

5.3.1. Statement of the problem

We assume now the probability distribution over the state space x is assigned linguistically, i.e. x_i can be thought, for instance, "very probable", where "very probable" is a fuzzy set of [0,1] . Let $\tilde{p}(x_i)$ be the linguistic probability assigned to x_i. The utility of an action a_j is not given by :

$$U_j = \tilde{p}(x_1).u_{1j} \oplus \tilde{p}(x_2).u_{2j} \oplus \ldots \oplus \tilde{p}(x_M).u_{Mj}$$

(where the u_{ij}'s are assumed non-fuzzy), because the variables p_i which are fuzzily restricted by the $\tilde{p}(x_i)$ are not independent :

they are related to each other through the orthogonality condition $\sum_{i=1}^{M} p_i = 1$;

in other words they are interactive (see 2.8). More specifically the membership function of U_j is defined by :

$$\mu_{U_j}(t) = \sup_{\substack{t=\sum_{i=1}^{M} p_i u_{ij} \\ \sum_{i=1}^{M} p_i = 1}} \ \min_{i=1,M} \ \mu_{\tilde{p}(x_i)}(p_i) \qquad (15)$$

The constraint $\sum_{i=1}^{M} p_i = 1$ makes the calculation of μ_{U_j} more tricky. However, when M = 2 the result is easily obtained since the two equations

$$\left\{\begin{array}{l} t = p_1.u_{1j} + p_2\, u_{2j} \\ 1 = p_1 + p_2 \end{array}\right\}$$ uniquely define p_1 and p_2 when $u_{1j} \neq u_{2j}$;

Hence, $\mu_{U_j}(t) = \min\left(\mu_{\tilde{p}(x_1)}\left(\frac{t - u_{2j}}{u_{1j}-u_{2j}}\right),\ \mu_{\tilde{p}(x_2)}\left(\frac{u_{1j} - t}{u_{1j}-u_{2j}}\right)\right)$

When $u_{1j} = u_{2j}$, we get :

$$\mu_{U_j}(t) = \mu_{\tilde{p}(x_1) \oplus \tilde{p}(x_2)}(1) \qquad \text{for } t = u_{1j}$$

$$\mu_{U_j}(t) = 0 \qquad \text{otherwise}$$

These formulae yield the classical result if the $\tilde{p}(x_i)$'s are just real numbers. However, note that U_j is a normalized fuzzy set as soon as the mean values of $\tilde{p}(x_1)$ and $\tilde{p}(x_2)$, say $\bar{p}_1$ and $\bar{p}_2$, are such that $\bar{p}_1 + \bar{p}_2 = 1$.
More specifically we have

$$\mu_{U_j}(\bar{p}_1 u_{1j} + \bar{p}_2 u_{2j}) = 1$$

More generally, if a M-tuple $(\tilde{p}(x_1), \ldots, \tilde{p}(x_M))$ of (linguistic) fuzzy probabilities satisfies $\sum_{i=1}^{M} \bar{p}_i = 1$, where $\bar{p}_i$ is such that $\mu_{\tilde{p}(x_i)}(\bar{p}_i) = 1$, then the fuzzy utility U_j given by formula (15)is normalized and we have :

$$\mu_{U_j}\left(\sum_{i=1}^{M} u_{ij}\, \bar{p}_i\right) = 1$$

5.3.2. The three-state problem

We now consider the case $M = 3$; for convenience the following notation is adopted $a = u_{1j}$, $b = u_{2j}$, $c = u_{3j}$, together with the assumption $a < b < c$. We must calculate the interactive sum :

$$\mu_{U_j}(t) = \sup_{p_1,p_2,p_3} \min(\mu_{\tilde{p}(x_1)}(p_1), \mu_{\tilde{p}(x_2)}(p_2), \mu_{\tilde{p}(x_3)}(p_3))$$

under the constraints $t = ap_1 + bp_2 + cp_3$

$$1 = p_1 + p_2 + p_3$$

It is equivalent to :

(16) $$\mu_{U_j}(t) = \sup_{p_1} \min\left(\mu_{\tilde{p}(x_1)}(p_1) , \mu_{\tilde{p}(x_2)}\left[\frac{t-c+(c-a)p_1}{b-c}\right] , \mu_{\tilde{p}(x_3)}\left[\frac{t-b+(b-a)p_1}{c-b}\right]\right)$$

In order to calculate U_j, assume now $\tilde{p}(x_i)$ is the LL-fuzzy number $(\bar{p}_i, \hat{p}_i, \check{p}_i)_{LL}$ for all $i = 1,3$. Note that the membership functions of $\tilde{p}(x_2)$ and $\tilde{p}(x_3)$ have been transformed by an affinity followed by a translation, and thus remain LL-fuzzy numbers. Now denoting $\mu_{\tilde{p}}$, $\mu_{\tilde{q}(t)}$ and $\mu_{\tilde{r}(t)}$ the membership functions :

$$\mu_{\tilde{p}(x_1)}, \mu_{\tilde{p}(x_2)}\left[\frac{t-c+(c-a).}{b-c}\right] , \mu_{\tilde{p}(x_3)}\left[\frac{t-b+(b-a).}{c-b}\right]$$

we get : $$\tilde{q}(t) = \left(\frac{(b-c)\bar{p}_2 - (t-c)}{c-a} , \left(\frac{c-b}{c-a}\right) \check{p}_2, \left(\frac{c-b}{c-a}\right) \hat{p}_2\right)$$

$$\tilde{r}(t) = \left(\frac{(c-b)\bar{p}_3 - (t-b)}{b-a}, \left(\frac{c-b}{b-a}\right) \hat{p}_3, \left(\frac{c-b}{b-a}\right) \check{p}_3 \right)$$

Obviously, for $t = \bar{t} = \bar{p}_1 a + \bar{p}_2 b + \bar{p}_3 c$, $\tilde{p}$, $\tilde{q}(t)$ and $\tilde{r}(t)$ have the same mean value $\bar{p}_1$. The fuzzy utility U_j is such that :

$$\mu_{U_j}(t) = \sup_{p_1} \min(\mu_{\tilde{p}}(p_1), \mu_{\tilde{q}(t)}(p_1), \mu_{\tilde{r}(t)}(p_1))$$

It can be checked (see Annex 2) that for any fuzzy numbers $\tilde{p}$, $\tilde{q}(t)$, $\tilde{r}(t)$,

$$\mu_{U_j}(t) = \min(\mathrm{hgt}(\tilde{p} \cap \tilde{q}(t)), \mathrm{hgt}(\tilde{q}(t) \cap \tilde{r}(t)), \mathrm{hgt}(\tilde{r}(t) \cap \tilde{p})) \quad (17)$$

This result is intuitive by looking at fig. 8.

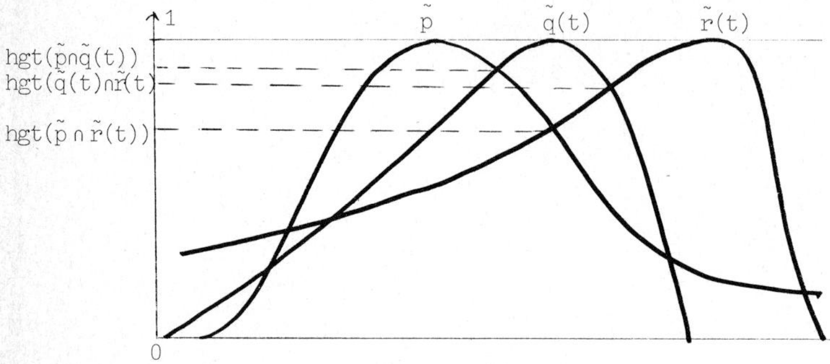

Figure 8

The height of the intersection of two LL fuzzy numbers is very easy to calculate since from 2.7 :

$$\mathrm{hgt}((m, \alpha, \beta)_{LL} \cap (n, \gamma, \delta)_{LL}) = L\left(\frac{n-m}{\beta+\gamma}\right) \quad \text{for } n \geq m.$$

For instance, when $\frac{(b-c)\bar{p}_2 - (t-c)}{c-a} \geq \bar{p}_1$, i.e. $t \leq \bar{t}$, it comes :

$$\mathrm{hgt}(\tilde{p} \cap \tilde{q}(t)) = L\left(\frac{\frac{(b-c)\bar{p}_2 - (t-c)}{(c-a)} - \bar{p}_1}{\check{p}_1 + \left(\frac{c-b}{c-a}\right) \check{p}_2} \right) = L\left(\frac{\bar{t} - t}{(c-a)\check{p}_1 + (c-b)\check{p}_2}\right)$$

and when $t \geq \bar{t}$ $\mathrm{hgt}(\tilde{p} \cap \tilde{q}(t)) = L\left(\frac{t - \bar{t}}{(c-a)\hat{p}_1 + (c-b)\hat{p}_2}\right)$

It is the membership function of a LL fuzzy number, say Npq :

$$\mathrm{Npq} = (\bar{t}, (c-a)\check{p}_1 + (c-b)\check{p}_2, (c-a)\hat{p}_1 + (c-b)\hat{p}_2)_{LL}$$

Similarly, for $\mathrm{hgt}(\tilde{q}(t) \cap \tilde{r}(t))$ and $\mathrm{hgt}(\tilde{r}(t) \cap \tilde{p})$ respectively, we get :

$$\mathrm{Nqr} = (\bar{t}, (b-a)\check{p}_1 + (c-b)\hat{p}_3, (b-a)\hat{p}_1 + (c-b)\check{p}_3)_{LL}$$

$$\mathrm{Nrp} = (\bar{t}, (b-a)\hat{p}_2 + (c-a)\hat{p}_3, (b-a)\check{p}_2 + (c-a)\check{p}_3)_{LL}$$

So that $\mu_{U_j}(t)$, due to the common mean value $\bar{t}$ of the three terms in (17) is also a LL-fuzzy-number membership function : $U_j = (\bar{t}, \bar{\alpha}, \bar{\beta})_{LL}$ with :

$$(18) \quad \bar{\alpha} = \min((c-a)\check{p}_1 + (c-b)\check{p}_2, (b-a)\check{p}_1 + (c-b)\hat{p}_3, (b-a)\hat{p}_2 + (c-a)\hat{p}_3)$$

$\bar{\beta}$ is the same as $\bar{\alpha}$, exchanging "∧" and "∨".

Note that the interactive sum just calculated has generally the same mean value as the non interactive sum $a.\tilde{p}(x_1) \oplus b.\tilde{p}(x_2) \oplus c.\tilde{p}(x_3)$.

N.B. When two of the coefficients a, b, c are equal, say $b = c > a$, formula ( ) becomes :

$$\mu_{U_j}(t) = \sup_{p_2, p_3} \min(\mu_{\tilde{p}(x_1)}(\frac{b-t}{b-a}), \mu_{\tilde{p}(x_2)}(p_2), \mu_{\tilde{p}(x_3)}(p_3))$$

subject to $p_2 + p_3 = \frac{t-a}{b-a} \equiv z$ for short.

i.e. $$\mu_{U_j}(t) = \min(\mu_{\tilde{p}(x_1)}(\frac{b-t}{b-a}), \sup_{p_2+p_3=z} \min(\mu_{\tilde{p}(x_2)}(p_2), \mu_{\tilde{p}(x_3)}(p_3)))$$

$$= \min(\mu_{\tilde{p}(x_1)}(\frac{b-t}{b-a}), \mu_{\tilde{p}(x_2)\oplus\tilde{p}(x_3)}(\frac{t-a}{b-a}))$$

U_j is still a LL-fuzzy number $(\bar{t}, \bar{\alpha}, \bar{\beta})_{LL}$ with :

$$\bar{t} = a\bar{p}_1 + b(1-\bar{p}_1)$$

$$\bar{\alpha} = (b-a) . \min(\check{p}_1, \hat{p}_2+\hat{p}_3)$$

$$\bar{\beta} = (b-a) . \min(\hat{p}_1, \check{p}_2+\check{p}_3)$$

When $a = b = c$, U_j is found similarly to the case $M = 2$, $u_{1j} = u_{2j}$.

5.3.3. Example

Let us calculate U_3, from example 1 of 3.2, when $u_{13} = 2$, $u_{23} = 4$, $u_{33}=8$ and the $\tilde{p}(x_i)$'s are symmetrical LL fuzzy numbers (0.2, 0.1), (0.5, 0.2), (0.3, 0.1) with the convention $(\bar{p}_1, \check{p}_1, \check{p}_1)_{LL} = (\bar{p}_1, \check{p}_1)$

We get $\tilde{p} = (0.2, 0.1)$; $\tilde{q}(t) = (\frac{b-t}{6}, \frac{0.4}{3})$; $\tilde{r}(t) = (\frac{5.2-t}{2}, 0.2)$

Then $hgt(\tilde{p} \cap \tilde{q}(t)) = L(\frac{t-4.8}{1.4})$

$hgt(\tilde{r}(t) \cap \tilde{q}(t)) = L(t-4.8)$

$hgt(\tilde{p} \cap \tilde{r}(t)) = L(\frac{t-4.8}{0.6})$

All three intersection values are LL-fuzzy number membership functions with the same mean value $\bar{t} = 4.8$. U_3 is the fuzzy number having the least spread, that is :

$$U_3 = (4.8, 0.6)$$

Note that the result of the non-interactive sum is $(4.8, 1.8) \neq U_3$.

NB.1. From the above analysis it seems that a similar approach could be carried out for M greater than 3. More specifically formulae (17) and (18) could be extended.

NB.2. When both utilities and probabilities are fuzzy numbers U_j is defined by :

$$\mu_{U_j}(t) = \sup_{(p_1 \ldots p_M ; u_{1j} \ldots u_{Mj})} \min_{i=1,M} (\mu_{\tilde{u}_{ij}}(u_{ij}), \mu_{\tilde{p}(x_i)}(p_i))$$

subject to : $t = p_1 u_{1j} + \ldots + p_M u_{Mj}$

$1 = p_1 + \ldots p_M$

which seems much more difficult to calculate except by a direct application of the formula. This problem has been already considered by Watson, Weiss and Donnell [13] in a somewhat restrictive context (N = M = 2).

6 - ACTIONS WITH UNEQUAL FEASIBILITY :

We assume now that there is a constraint on the realizability of the actions. This means that one may have a very good utility but it is difficult to perform the corresponding action for some reason. For instance, if the action consists in buying new clothes, the most expensive ones will be together the strongest or the most beautiful, but also the most difficult to buy.

This constraint is assumed fuzzy, i.e. each action is assigned a degree of possibility π_j, so that the set of possible actions is a fuzzy set G of $A = \{a_1 \ldots a_N\}$. When neither the state nor the utility is fuzzy, each action a_j is valued by $U_j = u_{i_0 j}$ where it is supposed that x_{i_0} is the state of the universe.

Denoting f the function $A \to R : a_j \to u_{i_0 j} = U_j$

the best action in the one which maximizes f over the fuzzy domain G.

The problem of maximizing a function over a fuzzy domain has already been solved. Several methods are available.

A first method is to consider the maximization of f as a fuzzy objective ; the corresponding fuzzy set is the maximizing set of f (Zadeh [18]), say M such that (for f positive): $\mu_M(a) = \dfrac{f(a)}{\sup f} \quad \forall\, a \in A.$

the best action a^* is then defined, as done in Bellman and Zadeh [2], by maximizing the membership function of the decision set $D = M \cap G$:

$$\mu_D(a^*) = \sup_{a_j \in A} \min(\mu_M(a_j), \pi_j)$$

An algorithm for determining a^* when A is a continuum and a metric space can be found in H. Tanaka et al. [13].

Another approach stems from the idea that the maximum of a function over a fuzzy domain is fuzzy, and can be defined through the extension principle (see 1.2).

Denoting $C_\alpha = \{a_j \in A \mid \pi_j \geq \alpha\}$, we extend the function

$$: \{C_\alpha \mid \alpha \in]0,1]\} \to \mathbb{R}^+$$

$$C_\alpha \mapsto \sup_{a \in C_\alpha} f(a)$$

π is viewed as the fuzzy set of α cuts $\sum_{\alpha \in]0,1]} \alpha / C_\alpha$

and the fuzzy maximal value of f on the fuzzy domain defined by π is nothing but $\varphi(\pi)$ such that :

$$\mu_{\varphi(\pi)}(r) = \sup_{r = \sup_{a \in C_\alpha} f(a)} \alpha$$

Denoting $A(\alpha) = \{a_j \in A \mid f(a_j) = \sup_{a \in C_\alpha} f(a)\}$, the support of the fuzzy set Ω of optimal actions is $S(\Omega) = \bigcup_{\alpha \in]0,1]} A(\alpha)$ and $\forall a \in S(\Omega)$, $\mu_\Omega(a) = \sup_{a \in A(\alpha)} \alpha$

This approach, which includes the first one ($a^* \in S(\Omega)$ generally) was proposed by Orlowski [12]. More details can be found in Dubois-Prade [6].

CONCLUDING REMARKS :

Decision-making when fuzziness blurs the set of alternatives seems to be no more a fiction. Several simple methods have been described for various situations ; most of them do not require too much computation, and are intuitively appealing.

Further work can be developped. Firstly, the situations of 6 can be extended to fuzzy states, fuzzy utilities, which require a method to maximize a fuzzy function over a fuzzy domain. This task can be actually carried out. Secondly, more complex models can be investigated : the actions can be evaluted from several points of view and a multicriteria analysis can be considered.

A general survey of existing decision-making methods with fuzzy sets encompassing other approaches (and related problems) than the one presented here, can be found in [6].

REFERENCES

(1) - BAAS S.M., KWAKERNAAK H., "Rating and ranking of Multiple-Aspect Alternatives Using Fuzzy sets" Automatica 13 (1977) 47-58.

(2) - BELLMAN R.E., ZADEH L.A., "Decision-making in a fuzzy environment", Mgmt. SCI., 17 n°4 (Déc. 1970) B-141, B-164.

(3) - DUBOIS D., PRADE H., "Comment on 'Tolerance Analysis Using Fuzzy Sets' and 'A Procedure for Multiple Aspect Decision-making'". Int.J. of Systems-Science. 9 n° 3 (1978), 357-360.

(4) - DUBOIS D., PRADE H., " Operations on Fuzzy Numbers" Int. J. of Systems-Science, 9 n°6 (1978), 613-626.

(5) - DUBOIS D., PRADE H., "Fuzzy real Algebra : Some Results 'Purdue University Memorandum TR-EE78 - 13, Part. A (Feb. 1978). To appear in Int. J. for Fuzzy Sets and Systems.

(6) - DUBOIS D., PRADE H., "Fuzzy Sets and Systems" Academic Press, Inc to appear 1979.

(7) - JAIN R., "Decision-making in the presence of fuzzy variables", IEEE Trans. S.M.C. 6, n° 10, (1976), 698-703.

(8) - JAIN R., "A procedure for multiple aspect decision-making" Int. J. Systems Science 8, n°1, (1977), 1-7.

(9) - JAIN R., "Decision-making in the presence of fuzziness and uncertainty", In Proc. IEEE. Conf. on Decision and Control, New Orleans, (1978), 1318-1323.

(10)- MIZUMOTO M., TANAKA K., "Some properties of fuzzy sets of type 2", Inf. and Cont. 31, (1976), 312-340.

(11) - ORLOWSKY S.A., "On programming with Fuzzy Constraints Sets" Kybernetes 6, (1977), 197-201.

(12) - RAIFFA & KEENEY, "Decision-Making under multiple objectives" (Wiley, 1976, New York).

(13) - TANAKA H. OKUDA T., ASAI K., "On Fuzzy Mathematical programming", J. of Cybernetics 3, n° 4, (1973), 37-46.

(14) - WATSON S.R., WEISS J.J., DONNELL M., "Fuzzy decision analysis", IEEE Trans. on S.M.C. 9, n°1, (1979), 1-9.

(15) - ZADEH L.A., "Fuzzy Sets", Inf. & Cont. 8, (1965), 338-353.

(16) - ZADEH L.A., " Quantitative Fuzzy Semantics", Inf. Sci. 3, (1971), 159-176.

(17) - ZADEH L.A., "A Fuzzy set Theoretic Interpretation of Linguistic Hedges", J. of Cybernetics Vol. 2 n° 3, (1972), 4-34.

(18) - ZADEH L.A., "On fuzzy algorithms", Memo UCB-ERL, M-325, Berkeley, (1972).

(19) - ZADEH L.A., "The Concept of a Linguistic Variable and its Application to Approximate Reasoning", Part 1 : Inf. Sci. Vol. 8, 199-249 -
Part 2 : Inf. Sci. Vol 8, 301-357 -
Part 3 : Inf. Sci. Vol 9, (1975), 43-80.

(20) - ZADEH L.A., "Fuzzy sets as a basis for a theory of possibility", Int. J. for "Fuzzy sets ans Systems", 1 n°1, (1978), 3-28.

ANNEX 1

Proposition

Let $\tilde{m}$ be a continuous fuzzy set of $\mathbb{R}$ with k local maxima $m_1, m_2, \ldots, m_k$ and $\tilde{n}$ another continuous fuzzy set of $\mathbb{R}$ with l local maxima $n_1, n_2, \ldots, n_l$.
Then :

$$v(\tilde{m} \geq \tilde{n}) = \max(\mathrm{hgt}(\tilde{m} \cap \tilde{n}), \max_{\substack{i=1,k \\ j=1,l \\ m_i \geq n_j}} \min(\mu_{\tilde{m}}(m_i), \mu_{\tilde{n}}(n_j))$$

Proof

$$v(\tilde{m} \geq \tilde{n}) = \sup_{x,y:x \geq y} \min(\mu_{\tilde{m}}(x), \mu_{\tilde{n}}(y))$$

* $k=l=1$: the proposition obviously holds

* $v(\tilde{m} \geq \tilde{n}) \geq \mathrm{hgt}(\tilde{m} \cap \tilde{n})$ obviously holds

* Let $I_1 \ldots I_k$ (resp. : $I_1 \ldots I_l$) be disjoint intervals such that $\bigcup_{i=1,k} I_i = \mathbb{R}$ (resp : $\bigcup_{i=1,l} I_j = \mathbb{R}$) and m_i(resp:n_j) is a global maximum over I_i (resp:I_j) (i.e. for instance $\tilde{m}$ is convex over I_i(resp:I_j); we have :

$$v(\tilde{m} \geq \tilde{n}) = \max_{\substack{i=1,k\\ j=1,l}} \sup_{\substack{(x,y)\in I_i x I_j\\ x \geq y}} \min(\mu_{\tilde{m}}(x), \mu_{\tilde{n}}(y))$$

for a given pair (I_i, I_j) we compare two convex fuzzy sets of $\mathbb{R}$, $\tilde{m}|_{I_i}$ and $\tilde{n}|_{I_j}$.

So, $v(\tilde{m}|_{I_i} \geq \tilde{n}|_{I_j}) = hgt(\tilde{m}|_{I_i} \cap \tilde{n}|_{I_j})$ if $m_i \leq n_j$

$= \min(\mu_{\tilde{m}}(m_i), \mu_{\tilde{n}}(n_j))$ if $m_i \geq n_j$

Hence $v(\tilde{m} \geq \tilde{n}) = \max [\max_{i,j:m_i \geq n_j} hgt(\tilde{m}|_{I_i} \cap \tilde{n}|_{I_j}), \max_{i,j:m_i \leq n_j} \min(\mu_{\tilde{m}}(m_i), \mu_{\tilde{n}}(n_j))]$

But since $\max_{\substack{i,j\\ m_i \leq n_j}} hgt(\tilde{m}|_{I_i} \cap \tilde{n}|_{I_j}) \leq hgt(\tilde{m} \cap \tilde{n})$, the proposition holds

Q.E.D.

ANNEX 2

Proposition

Let $\{A_i \mid i=1,3\}$ be three fuzzy numbers such that for all i

μ_{A_i} is strictly in creasing on $(-\infty, a_i)$ and strictly decreasing on

$[a_i, +\infty)$, with $\mu_{A_i}(a_i) = 1$.

Then (19) $hgt(A_1 \cap A_2 \cap A_3) = \min(hgt(A_1 \cap A_2), hgt(A_1 \cap A_3), hgt(A_2 \cap A_3))$

Proof

* $hgt(A_1 \cap A_2 \cap A_3) \leq hgt(A_i \cap A_j)$, $\forall$ ij (20)

  Hence $hgt(A_1 \cap A_2 \cap A_3) \leq \min(hgt(A_1 \cap A_2), hgt(A_1 \cap A_3), hgt(A_2 \cap A_3)$

* Assume now that $hgt(A_1 \cap A_2)$ is the least term of the three, and that $a_1 < a_2$.

  These assumptions do not restrict the generality of the proof.

  As a consequence, and calling x_{12} the point such that $\mu_{A_1 \cap A_2}(x_{12}) = hgt(A_1 \cap A_2)$

  We can state $a_1 < x_{12} < a_2$.

Now we prove that $\mu_{A_3}(x_{12}) \geq hgt(A_1 \cap A_2)$.

for suppose $\mu_{A_3}(x_{12}) < hgt(A_1 \cap A_2)$. Assume $x_{12} \leq a_3$

Hence μ_{A_3} is increasing on $[x_{12}, a_3]$ and $a_1 < a_3$

Hence μ_{A_1} is decreasing on $[x_{12}, a_3]$

Since $\mu_{A_1}(x_{12}) > \mu_{A_3}(x_{12})$ we infer that the maximum of $\mu_{A_1 \cap A_3}$ is reached for $x_{13} > x_{12}$, that is $\mu_{A_1}(x_{13}) < \mu_{A_1}(x_{12})$

Hence $hgt(A_1 \cap A_3) < hgt(A_1 \cap A_2)$ which contradicts the fact that $hgt(A_1 \cap A_2)$ is the least term on the left side of (19).

Assuming $x_{12} > a_3$ would lead to a similar result.

* Now, since $\mu_{A_3}(x_{12}) \geq hgt(A_1 \cap A_2)$ we get

$$hgt(A_1 \cap A_2 \cap A_3) \leq \min(\mu_{A_1}(x_{12}), \mu_{A_2}(x_{12}), \mu_{A_3}(x_{12}))$$

by applying (20). But from the very definition of $hgt(A_1 \cap A_2 \cap A_3)$ we know that $hgt(A_1 \cap A_2 \cap A_3) \geq \min(\mu_{A_1}(x), \mu_{A_2}(x), \mu_{A_3}(x)), \forall x \in \mathbb{R}$

Hence (19) holds.

QED

ADVANCES IN FUZZY SET THEORY AND APPLICATIONS
M.M. Gupta, R.K. Ragade, R.R. Yager (editors)

Fuzzy Information and Decision in Statistical Model

H. Tanaka*, T. Okuda** and K. Asai*

* Department of Industrial Engineering, University of Osaka Prefecture, Osaka 591, Japan.
** Department of Industrial Management, Osaka Institute of Technology, Osaka 535, Japan.

This paper deals with some aspects on decision-making in a vague environment where both, fuzziness and randomness are to be treated as fuzzy events. Fuzziness can be included in a formal model by defining a fuzzy set that expresses the subjective judgement about an imprecise meaning of an object of the decision problem under consideration. Our main goal is to generalize the Bayes-Approach to the fuzzy case where the state space, the action space and the information space are a set of fuzzy events.

1. Introduction

In usual decision problems, it is customary to treat vagueness by the use of probability theory, but it has become obvious that there are two different types of vagueness;randomness and fuzziness. Fuzziness means the imprecision of meaning of an object, and randomness is the uncertainty of occurrence of an object. For example, the statement "the demand will increase about 10% with a probability of 0.9" is vague because of randomness which is expressed by "a probability of 0.9" and also because of fuzziness which is expressed by "about 10%". It is well-known from the fuzzy set theory that the meaning of the words "about 10%" can be precisely defined by a fuzzy set, which depends on subjective judgement. From the above point of view, many management problems, where human aspects have to be included, would contain both randomness and fuzziness.

The concept of the probability of fuzzy events might be very useful for treating a decision problem which contains such a fuzzy statement as that mentioned above. Then, let us briefly introduce the definition of the probability of fuzzy events due to Zadeh[1]. Let S be a set of non-fuzzy events $\{s_1, \cdots, s_n\}$ with probabilities $P(s_k)$, respectively. A fuzzy event $\underset{\sim}{F}$ is a fuzzy set whose membership function is denoted by $\chi_F(\cdot)$. The probability of the fuzzy event $\underset{\sim}{F}$ is defined as follows :

$$P(\underset{\sim}{F}) = \sum_{k=1}^{n} \chi_F(s_k) P(s_k).$$

With this definition, the fuzzy statement "the demand will increase about 10% with a probability of 0.9" can be expressed by a mathematical model

$$\sum_{k=1}^{n} \chi_F(s_k) P(s_k) = 0.9$$

where F denotes the meaning of "about 10%". If the fuzzy statement is viewed as a kind of information, it means that we have had prior fuzzy information.

With the above view, a statistical model of decision-making in a fuzzy environment have been defined as a specific formulation of fuzzy decision problems[2]~[5]. First, the Fuzzified Bayes Formula is derived to generalize the Bayes-Approach to the fuzzy case where the information space is a set of fuzzy events. Next, a mechanism of obtaining a fuzzy information which is called a fuzzy observation system is discussed.

Finally, a fuzzy decision problem is analyzed to decide an optimal fuzzy action and it is shown that the value of fuzzy information is positive in a decision problem. A few numerical examples are given to illustrate our approach.

In this approach, the emphases are followings :
(i) Even a fuzzy information can be treated from the same aspect as in the statistical decision analysis.
(ii) A fuzzy information source is rather as valuable as an exact information source is. Further, a fuzzy information is easier obtainable than an exact information.
(iii) Our intuitive concept of decision might be well reflected in our model of fuzzy decision problems.

2. Fuzzy Information in a Statistical Decision Problem

First we will describe an usual statistical decision problem. A statistical decision ploblem is generally represented in the form of a quadruple $<S, D, P(s_k), U(d_j, s_k)>$, where $S=\{s_1, \cdots, s_n\}$ is a set of states of nature, $D=\{d_1, \cdots, d_r\}$ is a set of actions, $P(s_k)$ is a prior probability of s_k and $U(\cdot, \cdot)$ is a utility function on $S\times D$. In this problem, we have the indefiniteness concerning which states of nature will occur. This uncertainty is indicated by prior probability $P(s_k)(k=1, \cdots, n)$ which is also called a prior information. $U(d_j, s_k)$ is an utility of selecting d_j when a state s_k occurs. If the state which will occur were known with certainty beforehand, say s_i, it is reasonable to select the action which yields the greatest utility among the column s_i. When we know only probabilities $P(s_k)$ of s_k, the optimal action that we should select is defined as follows.

DEFINITION 1. An expected utility of action d_j is

$$U(d_i) = \sum_k U(d_i, s_k)P(s_k). \tag{1}$$

An optimal decision can be defined as an action d^0 which maximizes $U(d_i)$, i.e.

$$U(d^0) \equiv \max_i U(d_i). \tag{2}$$

Now, let us assume that there is an information source $X=\{x_1, \cdots, x_m\}$ from which an information x_j can be obtained with the conditional probability $f(x_j|s_k)$ if s_k is tuue state. Here conditional probabilities $f(x_j|s_k)(j=1,\cdots,m, k=1,\cdots,n)$ are given. This structure of getting an information x_j is called a probabilistic information system denoted by X. A given information can be used to choose one action. In other words, it is to associate a given information x_j with such action that maximizes the expected utility with posterior probability $P(s_k|x_j)$.

DEFINITION 2. An expected utility of action d_j after having an information x_j is

$$U(d_i) = \sum_k U(d_i, s_k)P(s_k|x_j). \tag{3}$$

An optimal decision $d^0_{x_j}$ is defined as

$$U(d^0_{x_j}|x_j) \equiv \max_i U(d_i|x_j). \tag{4}$$

As well-known, the posterior probability $P(s_k|x_j)$ is calculated by the bayes formula

$$P(s_k|x_j) = \frac{f(x_j|s_k)P(s_k)}{f(x_j)} \tag{5}$$

where $f(x_j) = \sum_i f(x_j|s_i)P(s_i)$.

The posterior probability $P(s_k|x_j)$ might play an important role for finding out an optimal decision given an information. This case is called posterior analysis.

Now, we consider the case that is called preposterior analysis. This means that we estimate an information system before we obtain some information x_j from the information system.

DEFINITION 3. The expected utility for having the probabilistic information system X is

$$U(d_X^0) = \sum_j U(d_{x_j}^0 \mid x_j) f(x_j). \tag{6}$$

DEFINITION 4. The worth of probabilistic information system X is

$$V(X) = U(d_X^0) - U(d^0). \tag{7}$$

Since $V(X)$ is non-negative, the increase of utility due to the information system is called as the worth of the information system. Furthermore, we introduce following definitions.

DEFINITION 5. The entropy of S given an information x_j as a criterion of uncertainty is defined by

$$H(S \mid x_j) = -\sum_k P(s_k \mid x_j) \log P(s_k \mid x_j). \tag{8}$$

By averaging (8) over the set X, we have

$$H(S \mid X) = \sum_j H(S \mid x_j) f(x_j) \tag{9}$$

which is called the entropy of S after having the probabilistic information system X. The quantity of information system X is defined by

$$I(X) = H(S) - H(S \mid X) \tag{10}$$

where $H(S)$ corresponds to the case without the information x_j in (8).

Now using there definitions, we consider fuzzy informations.

DEFINITION 6. A fuzzy information $\underset{\sim}{M}$ is a fuzzy set on X, which is given by the membership function $\chi_{\underset{\sim}{M}} : X \to [0, 1]$. A set of fuzzy informations $\mathscr{M} = \{\underset{\sim}{M}_1, \cdots, \underset{\sim}{M}_q\}$ is called a fuzzy information system.

An example of fuzzy information is the expression "Being in the vicinity of x_j" in Fig. 1. Many of informations which we can get are a kind of fuzzy information. Hence,

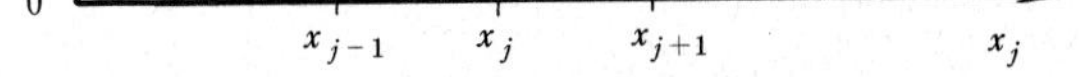

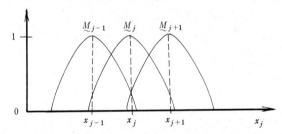

Fig. 1. Fuzzy information $M_j \equiv$ "Being in the vicinity of x_j".

it will be needed to deal with fuzzy informations. First, the bayes formula is modified to obtain the posterior probability $P(s_k \mid \underset{\sim}{M})$ given a fuzzy information $\underset{\sim}{M}$. We will call it a fuzzified bayes formula. The fuzzy information system considered here is illustrated in Fig. 2 where the fuzzy observation system has a set of fuzzy informations $\mathscr{M} = \{\underset{\sim}{M}_1, \cdots, \underset{\sim}{M}_q\}$ like Fig. 1 as outputs. We receive a fuzzy information $\underset{\sim}{M}_m$ out of $\mathscr{M}$ instead of an exact information x_m. The mechanism of obtaining a fuzzy information $\underset{\sim}{M}_m$ in Fig. 2 is discussed later.

The concept of the probability of a fuzzy event might be very useful for constructing a fuzzified bayes formula. Then, let us briefly introduce the definition of probability of a fuzzy event. Let S be a set of non-fuzzy events $\{s_1, \cdots, s_n\}$ with probabilities $P(s_k)$. A fuzzy event $\underset{\sim}{F}$ is a fuzzy set whose membership function is $\chi_F(\cdot)$.

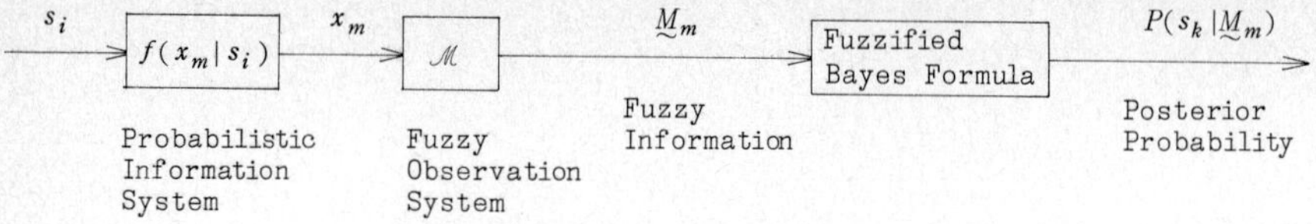

Fig. 2. Fuzzy information system.

DEFINITION 7. The probability of the fuzzy event $\underset{\sim}{F}$ is defined as

$$P(\underset{\sim}{F}) = \sum_k \chi_F(s_k)P(s_k). \tag{11}$$

Analogously to the well-known bayes formula with the definition of the probability of fuzzy event, we can define the posterior probability of s_k given a fuzzy information $\underset{\sim}{M}$ as follows :

$$P(s_k|\underset{\sim}{M}) = \frac{\sum_j f(x_j|s_k)\chi_M(x_j)P(s_k)}{P(\underset{\sim}{M})} \tag{12}$$

where $P(\underset{\sim}{M}) = \sum_j \chi_M(x_j)f(x_j)$.

Equivalently, we have

$$P(s_k|\underset{\sim}{M}) = \sum_j^m P(s_k|x_j)\frac{\chi_M(x_j)f(x_j)}{P(\underset{\sim}{M})}. \tag{13}$$

We can see from eqn.(13) that $P(s_k|\underset{\sim}{M})$ is a linear combination of all $P(s_k|x_j)$ in the weighting parameter

$$\chi(x_j|\underset{\sim}{M}) = \chi_M(x_j)f(x_j)/P(\underset{\sim}{M}). \tag{14}$$

This parameter can be regarded as the posterior probability of x_j given fuzzy information $\underset{\sim}{M}$ from similarity to the bayes formura.

PROPOSITION 1. (i) $\sum_k P(s_k|\underset{\sim}{M}) = 1$.

(ii) If the information $\underset{\sim}{M}$ is quite fuzzy in the sense that $\chi_M(x_j)$ is constant for all j, we have

$$P(s_k|\underset{\sim}{M}) = P(s_k).$$

This fact means that this information $\underset{\sim}{M}$ gives us nothing.

(iii) If the information $\underset{\sim}{M}$ is non-fuzzy, that is $\chi_M(x_j)=1$ and $\chi_M(x_j)=0$ for all $i \neq j$, this formulation turns out to be the usual bayes formula.

When we have two fuzzy informations $\underset{\sim}{M}_1$ and $\underset{\sim}{M}_2$, the posterior probability can be written in the form of

$$P(s_k|\underset{\sim}{M}_1, \underset{\sim}{M}_2) = \frac{\sum_{j=1}^m \sum_{j'=1}^m \chi_{M_1}(x_j)\chi_{M_2}(x'_j)f(x_j|s_k)f(x'_j|s_k)P(s_k)}{P(\underset{\sim}{M}_1)P(\underset{\sim}{M}_2)} \tag{15}$$

where

$$P(\underset{\sim}{M}_i) = \sum_{j=1}^m \chi_{M_i}(x_j)f(x_j) \quad \text{for } i=1,2 \tag{16}$$

and it is assumed throughout this paper that each occurrence of x_j is independent, i.e.

$$f(x_j, x'_j|s_k) = f(x_j|s_k)\cdot f(x'_j|s_k). \tag{17}$$

Similarly, we can write $P(s_k | \underset{\sim}{M_1} : \underset{\sim}{M_2})$ with the prior probability $P(s_k | M_1)$ as follows:

$$P(s_k | \underset{\sim}{M_1} : \underset{\sim}{M_2}) = \frac{\sum_{j=1}^{m} f(x_j | s_k) \chi_{M_2}(x_j) P(s_k | \underset{\sim}{M_1})}{P(\underset{\sim}{M_2})} \tag{18}$$

but it is clear that

$$P(s_k | \underset{\sim}{M_1}, \underset{\sim}{M_2}) = P(s_k | \underset{\sim}{M_1} : \underset{\sim}{M_2}). \tag{19}$$

From the above, it is seen that a fuzzy information consisting of two fuzzy informations $\underset{\sim}{M_1}$ and $\underset{\sim}{M_2}$ on $X \times X$ is defined as

$$\underset{\sim}{M_1} \cdot \underset{\sim}{M_2} \Longleftrightarrow \chi_{M_1}(\cdot) \cdot \chi_{M_2}(\cdot). \tag{20}$$

In the context of fuzzy language, the following is defined:

$$\overbrace{\text{very} \cdots \text{very}}^{n \text{ times}} \underset{\sim}{M} = \underset{\sim}{M}^n \Longleftrightarrow \chi_M^n(\cdot). \tag{21}$$

It should be noted that the posterior probability $P(s_k | \underset{\sim}{M}^2)$ is generally not equal to $P(s_k | \underset{\sim}{M}, \underset{\sim}{M})$, because the system of fuzzified bayes formula is connected with the probabilistic information system like Fig.2. On the other hand, we will discuss the case that the probabilistic information system is perfect, that is, the matrix $[f(x_j | s_k)]$ is identity matrix I with $S = X$. It follows from eqn.(12) that

$$P(s_k | \overbrace{\underset{\sim}{M}, \cdots, \underset{\sim}{M}}^{n \text{ times}}) = \frac{\chi_M^n(s_k) P(s_k)}{\sum_{k=1}^{n} \chi_M^n(s_k) P(s_k)}. \tag{22}$$

Hence we have

$$P(s_k | \overbrace{\underset{\sim}{M}, \cdots, \underset{\sim}{M}}^{n \text{ times}}) = P(s_k | \underset{\sim}{M}^n). \tag{23}$$

In this case, it can be said that receiving a fuzzy information n times is just equivalent to getting a fuzzy information $\overbrace{\text{very} \cdots \text{very}}^{n \text{ times}} \underset{\sim}{M}$. If the fuzzy information $\underset{\sim}{M}$ is defined as $\chi_M(x_j) = 1$ for only one j and $\chi_M(x_l) \neq 1$ for all $l \neq j$ like Fig.1, it follows from eqn.(22) that

$$\lim_{n \to \infty} P(s_k | \overbrace{M, \cdots, M}^{n \text{ times}}) = \begin{cases} 1 \text{ for } k = j \\ 0, \text{ otherwise}. \end{cases} \tag{24}$$

DEFINITION 8. The conditional entropy of S given a fuzzy information $\underset{\sim}{M}$ can be defined by

$$H(S | \underset{\sim}{M}) = \sum_{k=1}^{n} P(s_k | \underset{\sim}{M}) \log P(s_k | \underset{\sim}{M}). \tag{25}$$

In addition to the condition that the probabilistic information system is perfect, let us here assume that we have no prior information with regard to states of nature, that is, prior probabilities are uniformly distributed. Then, we have the following proposition.

PROPOSITION 2. $H(S | \underset{\sim}{M}) - H(S | \underset{\sim}{M}^2) \geq 0.$ (26)

This proposition can easily be proved as follows:

$$H(\widetilde{S} | \underset{\sim}{M}) - H(\widetilde{S} | \underset{\sim}{M}^2) \geq -\sum_{k=1}^{n} \frac{\chi_M(s_k)}{C_1} \log \frac{\chi_M(s_k)}{C_1}$$

$$+\sum_{k=1}^{n} \frac{\chi_M^2(s_k)}{C_2} \log \frac{\chi_M^2(s_k)}{C_1}$$

$$=\sum_{k=1}^{n} \frac{\chi_M(s_k)}{C_1 C_2}(-C_2+\chi_M(s_k)C_1) \log \frac{\chi_M(s_k)C_1}{C_2} \geq 0,$$

where

$$C_1=\sum_{k=1}^{n} \chi_M(s_k) \text{ and } C_2=\sum_{k=1}^{n} \chi_M^2(s_k).$$

This means that the fuzzy information "very $\underset{\sim}{M}$" is more useful than the fuzzy information $\underset{\sim}{M}$ in a sense of the entropy. This situation is illustrated in Fig.3.

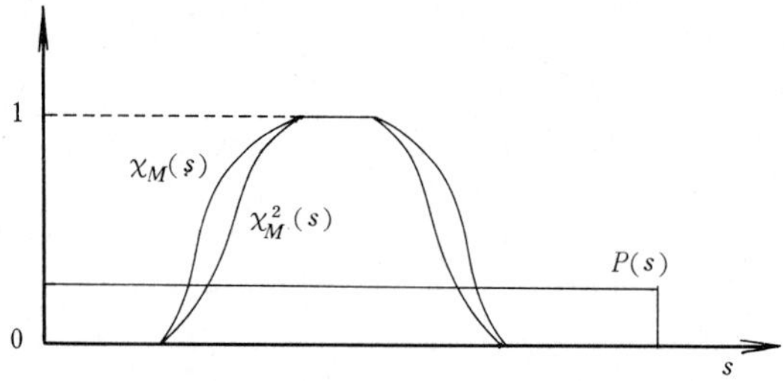

Fig.3. Illustration of $H(S\,|\,\underset{\sim}{M}) \geq H(S\,|\,\underset{\sim}{M}^2)$.

EXAMPLE 1.

The set D of action is assumed to be identical with S and the utility function is $U(d_i, s_k)=1$ for $i=k$ and $U(d_i, s_k)=0$ for $i \neq k$. Let us also assume that the information system is perfect, that is the matrix $[f(x_j\,|\,s_k)]$ is the identity matrix I with $S=X$. This example is a decision problem for guessing the actual state of nature given a fuzzy information $\underset{\sim}{M}$. Here a die game is assumed and the given fuzzy information $\underset{\sim}{M}$, which is expressed by a fuzzy statement "the spot on the die is middle" is shown as a membership function in Table 1. The probability of occurence of s_i will be 1/6 for all i. Then we have

(i) $P(s_1\,|\,\underset{\sim}{M}) = P(s_6\,|\,\underset{\sim}{M}) = 0$, $P(s_2\,|\,\underset{\sim}{M}) = P(s_5\,|\,\underset{\sim}{M}) = \frac{1}{6}$,

$P(s_3\,|\,\underset{\sim}{M}) = P(s_4\,|\,\underset{\sim}{M}) = \frac{1}{3}$.

(ii) The expected value without informations $U(d^0) = \frac{1}{6}$.

(iii) The expected value after having the fuzzy information $\underset{\sim}{M}$

$U(d_M^0) = \max_k U(d_i, s_k)\, P(s_k\,|\,\underset{\sim}{M}) = \frac{1}{3}$.

Table 1. Membership functions of $\underset{\sim}{M}$ and $\underset{\sim}{M}^2$.

S	1	2	3	4	5	6
$\underset{\sim}{M}$	0	0.5	1	1	0.5	0
$\underset{\sim}{M}^2$	0	0.25	1	1	0.25	0

As regarding to the fuzzy information "very M"$=\underset{\sim}{M}^2$ whose membership function is $\chi_{M^2}(\cdot)=\chi_M^2(\cdot)$, we have

(i) $P(s_1\,|\,\underset{\sim}{M}^2) = P(s_6\,|\,\underset{\sim}{M}^2) = 0$, $P(s_2\,|\,\underset{\sim}{M}^2) = P(s_5\,|\,\underset{\sim}{M}^2) = \frac{1}{10}$,

$P(s_3\,|\,\underset{\sim}{M}^2) = P(s_4\,|\,\underset{\sim}{M}^2) = \frac{2}{5}$.

(ii) The expected value after having the fuzzy information $\underset{\sim}{M}^2$

$$U(d^0_{\underset{\sim}{M}^2}) = \max_i \sum_k U(d_i, s_k) P(s_k | \underset{\sim}{M}^2) = \frac{2}{5}.$$

Comparing the fuzzy information $\underset{\sim}{M}$ with $\underset{\sim}{M}^2$, we have

$$U(d^0_{\underset{\sim}{M}}) \leq U(d^0_{\underset{\sim}{M}^2}), \quad H(S | \underset{\sim}{M}) \geq H(S | \underset{\sim}{M}^2).$$

Then $\underset{\sim}{M}^2$ is more informative than M.

Until now, we have not discussed the fuzzy observation system. In other words, the mechanism of obtaining fuzzy information $\underset{\sim}{M}_m$ coming from a true state s_k was not discussed. As mentioned before, it may be natural that the probability of occurrence of the fuzzy information $\underset{\sim}{M}_m$ is defined by

$$P(\underset{\sim}{M}_m) = \sum_j \chi_{M_m}(x_j) f(x_j).$$

But we have to define the relation between the input x_j and the output M_m in Fig.2. Let us, therefore, regard the relation as follows.

If x_j occurs as the output of a probabilistic information system, the fuzzy observation system lets us know a fuzzy information $\underset{\sim}{M}_m$ with the probability of occurrence of $\underset{\sim}{M}_m$ which is equal to $\chi_{M_m}(x_j)$. Hence, in what follows, it is assumed that $\mathscr{M}$ is orthogonal, that is $\sum_m \chi_{M_m}(x_j) = 1$, for all j. Following the definition of the probability of fuzzy events, we have

$$f(\underset{\sim}{M}_m | s_k) = \sum_j \chi_{M_m}(x_j) f(x_j | s_k). \tag{27}$$

Here, let us regard $\{\chi_{M_1}(x_j), \cdots, \chi_{M_q}(x_j)\}$ as kinds of probabilities with which we get fuzzy information. For example, in Fig.4, if x_1 and x_3 occur, our assumption means that we can get the fuzzy information $\underset{\sim}{M}_1$ and $\underset{\sim}{M}_2$ with probability one. If x_2 occurs,

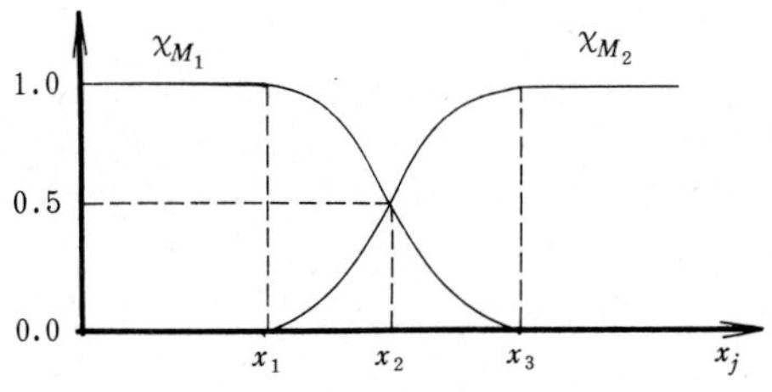

Fig.4. Explanation of membership function $\chi_{M_i}(x_j)$.

fuzzy information which we can get from the fuzzy information system is $\underset{\sim}{M}_1$ or $\underset{\sim}{M}_2$ with the probability of 0.5. Hence, we can interpret $[\chi_{M_m}(x_j)]$ as a fuzzified observation matrix. Fig.5 explains the mechanism for obtaining fuzzy information $\underset{\sim}{M}_m$ coming from a true state s_k.

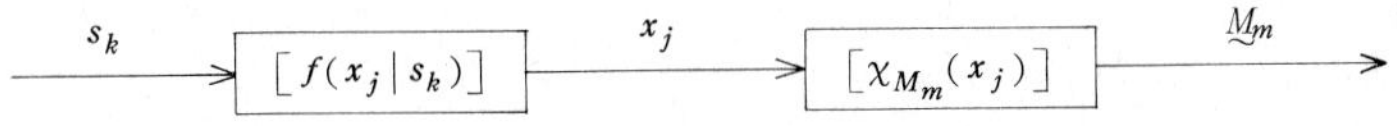

Fig.5. Mechanism for obtaining fuzzy information $\underset{\sim}{M}_m$.

Using this interpretation of the fuzzy observation, we can make the following proposition.

PROPOSITION 3.

If $P_r[\underset{\sim}{M}m\,;\, f(\underset{\sim}{M}m\,|\,s_h) = f(\underset{\sim}{M}m\,|\,s_g)\,|\,s_k] \neq 1$, for $h \neq g$, $\forall m$, $\forall h$, $\forall g$ and $\forall k$ (28)

holds, we get for $\forall \mathcal{E} > 0$ when s_k is a true state,

$$\lim_{m\to\infty} P_r[P(s_k\,|\,\underset{\sim}{M}_1, \cdots, \underset{\sim}{M}m) \geq 1 - \mathcal{E}\,|\,s_k] = 1$$

and (29)

$$\lim_{m\to\infty} P_r[P(s_h\,|\,\underset{\sim}{M}_1, \cdots, \underset{\sim}{M}m) \leq \mathcal{E}\,|\,s_k] = 1, \text{ for } h \neq k.$$

The proof is analogous to the proof in the non-fuzzy case by using the fuzzy information system $[f(x_j\,|\,s_k)]\cdot[\chi_{M_m}(x_j)]$ instead of the usual probabilistic information system $[f(x_j\,|\,s_k)]$. This proposision implies that if we get an infinite sequence of fuzzy informations, we can find out which state of nature will be true. Although informations are fuzzy, those are available for improving decision-making.

DEFINITION 9. The worth of fuzzy information system $\mathcal{M}$ is

$$V(\mathcal{M}) = U(d^0_{\mathcal{M}}) - U(d^0) \tag{30}$$

where

$$U(d^0_{\mathcal{M}}) = \sum_m U(d^0_{\underset{\sim}{M}m}\,|\,\underset{\sim}{M}m)\,P(\underset{\sim}{M}m) \tag{31}$$

and

$$U(d^0_{\underset{\sim}{M}m}\,|\,\underset{\sim}{M}m) = \max_i \sum_k U(d_i, s_k)\,P(s_k\,|\,\underset{\sim}{M}m). \tag{32}$$

EXAMPLE 2.

We consider a numerical example of a decision problem which has

$$S = \{s_1, s_2\},\ D = \{d_1, d_2\},\ X = \{x_1, \cdots, x_8\} \text{ and } \mathcal{M} = \{\underset{\sim}{M}_1, \underset{\sim}{M}_2, \underset{\sim}{M}_3\}.$$

The utility function is given in Table 2 and the conditional probability $f(x_j\,|\,s_i)$ and the membership function $\chi_{M_i}(\cdot)$ are shown in Table 3. It is assumed that we have no prior information, i.e. $P(s_1) = P(s_2) = 0.5$. Then we have in the cases:

(i) No information system : $U(d^0) = 1$

(ii) Exact information system X: $U(d^0_X) = 1.875$

$$V(X) = U(d^0_X) - U(d^0) = 0.875$$

(iii) Fuzzy information system $\mathcal{M}$: $U(d^0_{\mathcal{M}}) = 1.808$

$$V(\mathcal{M}) = U(d^0_{\mathcal{M}}) - U(d^0) = 0.808$$

Table 2. Utility $U(d_i, s_k)$

	s_1	s_2
d_1	4	−2
d_2	−1	2

Table 3. Membership function $\chi_{M_i}(\cdot)$ and $[f(x_j\,|\,s_k)]$

χ_{M_1}	1	1	0.5	0	0	0	0	0
χ_{M_2}	0	0	0.5	1	1	0.5	0	0
χ_{M_3}	0	0	0	0	0	0.5	1	1
X	x_1	x_2	x_3	x_4	x_5	x_6	x_7	x_8
s_1	0	0.05	0.1	0.1	0.2	0.4	0.1	0.05
s_2	0.05	0.1	0.4	0.2	0.1	0.1	0.05	0

It is seen from this example that the value of fuzzy information system is not much different from the value of the exact information system, although $V(X)$ is larger than $V(\mathscr{M})$. Therefore it might be understood that having the exact information system at a great cost is not so valuable, as compared with having the fuzzy information system. In other words, it seems that in many decision problems a fuzzy information system is rather as valuable as the exact information system is.

3. Fuzzy Decision Problem

The fuzzy decision problem is to decide a basic policy in a fuzzy environment. In general a basic policy might be regarded as a fuzzy action. The structure of such a fuzzy decision problem can be illustrated in Fig.6. In usual decision problem there are many elements—a lot of states, feasible actions and available informations. Hence, the utilities for all states and all actions can hardly be analyzed because of insufficient data, its high cost and its limited time. On the other hand, a decision maker in top management is generally not concerned with the detail of each element in a decision problem. He wants to decide roughly what actions should be considered as a direction of policy. Therefore, we try to aggregate states of nature $\{s_i\}$ and actions $\{d_j\}$ by means of what we intend to concern. Hence the state space and action space are divided into a number of parts by means of statements of interest like $\{\underset{\sim}{F}_1, \underset{\sim}{F}_2, \underset{\sim}{F}_3\}$ and $\{\underset{\sim}{A}_1, \underset{\sim}{A}_2, \underset{\sim}{A}_3\}$ (see Fig.6). As regards the information, there are many cases where the obtained information is fuzzy.

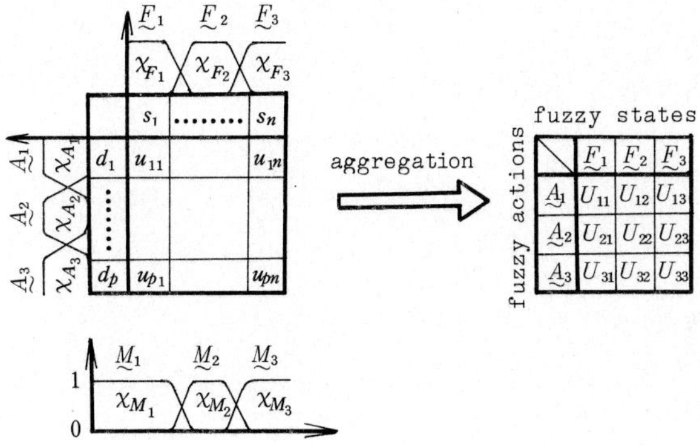

Fig.6. Structure of fuzzy decision problem.

DEFINITION 10. A fuzzy decision problem is represented in the form of a quadruple $\langle \mathscr{F}, \mathscr{A}, P, U \rangle$, where $\mathscr{F}=\{\underset{\sim}{F}_1, \cdots, \underset{\sim}{F}_r\}$ satisfying the orthogonal condition is a set of fuzzy states which are fuzzy sets on S, $\mathscr{A}=\{\underset{\sim}{A}_1, \cdots, \underset{\sim}{A}_l\}$ is a set of fuzzy actions which are fuzzy sets on D, $U(\cdot,\cdot)$ is a utility function on $\mathscr{A}\times\mathscr{F}$ and $P(\cdot)$ is a prior probability on S.

In order to deal with fuzzy information system $\mathscr{M}$ and probabilistic information system X in a fuzzy decision problem, the bayes formula concerned with fuzzy states has to be induced. From the definition of probability of fuzzy events and usual bayes formula, the following fuzzified bayes formula can be defined.

(i) The posterior probability of fuzzy state $\underset{\sim}{F}_k$ given an exact information x_j is

$$P(\underset{\sim}{F}_k \mid x_j) = \frac{\sum_i \chi_{F_k}(s_i) f(x_j \mid s_i) P(s_i)}{f(x_j)} \tag{33}$$

where $f(x_j) = \sum_i f(x_j | s_i) P(s_i)$.

(ii) The posterior probability of fuzzy state $\underset{\sim}{F}_k$ given a fuzzy information $\underset{\sim}{M}_m$ is

$$P(\underset{\sim}{F}_k | \underset{\sim}{M}_m) = \frac{\sum_i \sum_j \chi_{F_k}(s_i) \chi_{M_m}(x_j) f(x_j | s_i) P(s_i)}{\sum_j \chi_{M_m}(x_j) f(x_j)} . \tag{34}$$

Now, using the above definition, the expected utility of a fuzzy action can be defined as follows.

DEFINITION 11. The expected utility of fuzzy action $U(\underset{\sim}{A}_i)$ is

$$U(\underset{\sim}{A}_i) = \sum_j U(\underset{\sim}{A}_i, \underset{\sim}{F}_j) P(\underset{\sim}{F}_j), \tag{35}$$

and the optimal fuzzy action $\underset{\sim}{A}^0$ is defined by

$$U(\underset{\sim}{A}^0) \equiv \max_i U(\underset{\sim}{A}_i). \tag{36}$$

Thus, the fuzzy decision problem on $\mathscr{A} \times \mathscr{F}$ can easily be analyzed because of the smaller number of elements $|\mathscr{A}| \times |\mathscr{F}|$ than $|D| \times |S|$, where $|\ |$ denotes the number of elements in the set. Here a utility function $U(\underset{\sim}{A}_i, \underset{\sim}{F}_j)$ is subjectively or roughly defined on $\mathscr{A} \times \mathscr{F}$. Hence, without analyzing exact utilities we can decide an optimal decision $\underset{\sim}{A}^0$ from our formulation as just a direction of policy. The direction with the rough restriction given from the higher level may be shown by $\underset{\sim}{A}^0$. Then, a decision problem at the lower level can be reformulated on a limited subset of actions $A^* = \{d_j | \chi_{A^0}(d_j) \geqq \alpha^*\}$, where α^* denotes a satisfactory threshold.

From the similarity to Definition 11, the expect utility of fuzzy action $\underset{\sim}{A}_i$ given an exact information x_j is

$$U(\underset{\sim}{A}_i | x_j) = \sum_k U(\underset{\sim}{A}_i, \underset{\sim}{F}_k) P(\underset{\sim}{F}_k | x_j). \tag{37}$$

The optimal fuzzy action $\underset{\sim}{A}^0_{x_j}$ is defined by

$$U(A^0_{x_j} | x_j) \equiv \max_i U(\underset{\sim}{A}_i | x_j). \tag{38}$$

The total expected utility can be represented as

$$U(\underset{\sim}{A}^0_X) = \sum_j U(\underset{\sim}{A}^0_{x_j} | x_j) f(x_j) \tag{39}$$

under the condition that we will be able to have the probabilistic information system X. Thus, the worth of probabilistic information system X can be defined as follows:

$$V(X) = U(\underset{\sim}{A}^0_X) - U(\underset{\sim}{A}^0). \tag{40}$$

Note that this value is calculated until the information will be obtained.

Next, let us consider the case that the probabilistic information system is perfect. This information, as mentioned in the preceding section, is one which lets us exactly know a true state s_i and this information system is denoted by X_∞. If a true state $s_k \in S$ is known from the information system X_∞, the expected utility of fuzzy action $\underset{\sim}{A}_i$ given the true state s_k becomes

$$U(\underset{\sim}{A}_i | s_k) = \sum_j U(\underset{\sim}{A}_i, \underset{\sim}{F}_j) \chi_{F_j}(s_k) \tag{41}$$

where it should be noted that $P(\underset{\sim}{F}_j | s_k) = \chi_{F_j}(s_k)$. Therefore, the total expected utility for having the information system X_∞ is

$$U(\underset{\sim}{A}^0_{X_\infty}) = \sum_k U(\underset{\sim}{A}^0_{s_k} | s_k) P(s_k) \tag{42}$$

where $\underset{\sim}{A}^0_{s_k}$ is defined by

$$U(\underset{\sim}{A}^0_{s_k} | s_k) \equiv \max_i U(\underset{\sim}{A}_i | s_k). \tag{43}$$

Thus, the worth of probabilistic perfect information system X_∞ can be defined as follows :

$$V(X_\infty) = U(\underset{\sim}{A}^0_{X_\infty}) - U(\underset{\sim}{A}^0). \tag{44}$$

Now, assume that a fuzzy information $\underset{\sim}{M}_m$ is obtained from a fuzzy information system $\mathscr{M} = \{\underset{\sim}{M}_1, \cdots, \underset{\sim}{M}_q\}$.

DEFINITION 12. The exected utility of fuzzy action $\underset{\sim}{A}_i$ given a fuzzy information $\underset{\sim}{M}_m$ is

$$U(\underset{\sim}{A}_i \mid \underset{\sim}{M}_m) = \sum_j U(\underset{\sim}{A}_i, \underset{\sim}{F}_j) P(\underset{\sim}{F}_j \mid \underset{\sim}{M}_m) \tag{45}$$

and the optimal fuzzy action $\underset{\sim}{A}^0_{M_m}$ is defined by

$$U(\underset{\sim}{A}^0_{M_m} \mid \underset{\sim}{M}_m) \equiv \max_i U(\underset{\sim}{A}_i \mid \underset{\sim}{M}_m). \tag{46}$$

From above Definition 12 , the worth of fuzzy information system $\mathscr{M}$ can be defined as follows :

$$V(\mathscr{M}) = U(\underset{\sim}{A}^0_{\mathscr{M}}) - U(\underset{\sim}{A}^0) \tag{47}$$

where

$$U(\underset{\sim}{A}^0_{\mathscr{M}}) = \sum_m U(\underset{\sim}{A}^0_{M_m} \mid \underset{\sim}{M}_m) P(\underset{\sim}{M}_m). \tag{48}$$

Finally, let us define a fuzzy perfect information. This information is one which let us know what fuzzy state occurs with probability one, and this information system is denoted by $\mathscr{M}_\infty$. If $P(\underset{\sim}{F}_j) = 1$ is known from $\mathscr{M}_\infty$, the expected utility of $\underset{\sim}{A}_i$ becomes $U(\underset{\sim}{A}_i \mid \underset{\sim}{F}_j) = U(\underset{\sim}{A}_i, \underset{\sim}{F}_j)$. The optimal fuzzy action $\underset{\sim}{A}^0_{F_j}$ can be defined by $U(\underset{\sim}{A}^0_{F_j} \mid \underset{\sim}{F}_j) \equiv \max_i U(\underset{\sim}{A}_i \mid \underset{\sim}{F}_j)$. Thus, the total expected utility for having the fuzzy perfect information system $\mathscr{M}_\infty$ is

$$U(\underset{\sim}{A}^0_{\mathscr{M}_\infty}) = \sum_j U(\underset{\sim}{A}^0_{F_j} \mid \underset{\sim}{F}_j) P(\underset{\sim}{F}_j) \tag{49}$$

and the worth of the information system $\mathscr{M}_\infty$ can be defined as follows :

$$V(\mathscr{M}_\infty) = U(\underset{\sim}{A}^0_{\mathscr{M}_\infty}) - U(\underset{\sim}{A}^0). \tag{50}$$

THEOREM 1. The following relation holds:

$$V(\mathscr{M}_\infty) \geqq V(X_\infty) \geqq V(X) \geqq V(\mathscr{M}) \geqq 0.$$

Proof. First, let us prove $V(\mathscr{M}_\infty) \geqq V(X_\infty)$:

$$\begin{aligned} U(\underset{\sim}{A}^0_{X_\infty}) &= \sum_k U(\underset{\sim}{A}^0_{s_k} \mid s_k) P(s_k) \\ &\leqq \sum_k \{\sum_j \max_i U(\underset{\sim}{A}_i, \underset{\sim}{F}_j) \chi_{F_j}(s_k)\} P(s_k) \\ &= \sum_j U(\underset{\sim}{A}^0_{F_j} \mid \underset{\sim}{F}_j) P(F_j) = U(\underset{\sim}{A}^0_{\mathscr{M}_\infty}). \end{aligned}$$

The relation $V(X_\infty) \geqq V(X)$ can be proved in the same way as the usual statistical decision theory. Next, let us prove $V(X) \geqq V(\mathscr{M})$:

$$\begin{aligned} U(\underset{\sim}{A}^0_{\mathscr{M}}) &= \sum_m \left[\max_i \sum_j U(\underset{\sim}{A}_i, \underset{\sim}{F}_j) \{\sum_k P(\underset{\sim}{F}_j \mid x_k) \chi_{M_m}(x_k) f(x_k)\} \right] \\ &\leqq \sum_m \left[\sum_k \{\max_i \sum_j U(\underset{\sim}{A}_i, \underset{\sim}{F}_j) P(\underset{\sim}{F}_j \mid x_k)\} \chi_{M_m}(x_k) f(x_k) \right] \\ &= \sum_k U(\underset{\sim}{A}^0_{x_k} \mid x_k) f(x_k) = U(\underset{\sim}{A}^0_X). \end{aligned}$$

Finally, let us prove $V(\mathscr{M}) \geqq 0$:

$$U(\underset{\sim}{A}^0_{\mathcal{M}}) = \sum_k U(\underset{\sim}{A}^0_{M_k} | M_k) P(M_k)$$

$$\geq \max_i \sum_j U(\underset{\sim}{A}_i, \underset{\sim}{F}_j) \{ \sum_k P(\underset{\sim}{F}_j, \underset{\sim}{M}_k) \}$$

$$= U(\underset{\sim}{A}^0).$$

In the theorem 1, note that $V(\mathcal{M}) = 0$ occurs in the case that χ_{M_m} is constant or the random variables x and s are mutually independent.

It might be said that the relation $V(X) \geqq V(\mathcal{M})$ is caused by the fact that the information system $\mathcal{M}$ has fuzziness in addition to randomness in the information system X. On the other hand, the relation $V(\mathcal{M}_\infty) \geqq V(X_\infty)$ is caused by the fact that the uncertainty shown by the probability $\{P(F_i)\}$ still remains even when the true state s_i is known.

EXAMPLE 3.

To explain the fuzzy decision problem constructed above, first of all, we will give some comments on the project selection problem at the higher level. Since funds may be obtained from within or outside the firm, the available funds must be very fuzzy at this stage. Also, all projects may hardly be analyzed because of insufficient data, its high cost and its limited time. Therefore, as the decision of the higher level, it is necessary to determine how large scale project should be approximately considered. As regarding the information, the higher level is not interested in the detail about the national economy, but rather in approximate states of business activity which are implied by the data.

Now, let us describe our problem in the above situation. Assume that a firm has been successful enough to be in a position to expand the operations. Three courses of action are considered:

(1) choose the small scale project ($\underset{\sim}{A}_1$),
(2) choose the middle scale project ($\underset{\sim}{A}_2$),
(3) choose the large scale project ($\underset{\sim}{A}_3$).

The meanings of "small", "middle" and "large" are given by fuzzy sets on the action space. The utility from each of these investments depends on the national economy— economic growth, inflation and stagnation. For the sake of simplification, let us take the state of natures $S = \{s_1, \cdots, s_{26}\}$ as the rate of economic growth. The utility for each of the alternatives has been developed for three interested levels of business activity:

(1) low($\underset{\sim}{F}_1$), (2) medium ($\underset{\sim}{F}_2$), (3) high($\underset{\sim}{F}_3$),

whose meanings are defined by fuzzy sets on S. Thus, three fuzzy states of nature are considered. It is also assumed that the utility $U(\underset{\sim}{A}_i, \underset{\sim}{F}_j)$ for each $\underset{\sim}{A}_i$ and $\underset{\sim}{F}_j$ can be approximately estimated as listed in Table 4. The membership functions of $\underset{\sim}{F}_1$, $\underset{\sim}{F}_2$ and $\underset{\sim}{F}_3$, and the prior probabilities $P(s_i)$ are shown in Table 5 respectively.

Let $X = \{x_1, \cdots, x_{26}\}$ be the set of rates of increase of the national gross investments in next term. The conditional probabilities $f(x_j | s_i)$ are shown in Table 6. This information is probabilistic. By contrast, we consid r a fuzzy information system $\mathcal{M} = \{\underset{\sim}{M}_1, \underset{\sim}{M}_2, \underset{\sim}{M}_3\}$, where

$\underset{\sim}{M}_1$: the rate of increase of the national gross investments is approximately more than 10%,

$\underset{\sim}{M}_2$: the rate is approximately equal to 10%,

$\underset{\sim}{M}_3$: the rate is approximately less than 10%.

Table 4. Utility function $U(\underset{\sim}{A}_i, \underset{\sim}{F}_j)$.

fuzzy state / fuzzy action	$\underset{\sim}{F}_1$	$\underset{\sim}{F}_2$	$\underset{\sim}{F}_3$
$\underset{\sim}{A}_1$	90	110	120
$\underset{\sim}{A}_2$	0	150	200
$\underset{\sim}{A}_3$	−80	100	300

Table 5. Membership functions of $\underset{\sim}{F}_1$, $\underset{\sim}{F}_2$ and $\underset{\sim}{F}_3$, and prior probability P.

s_i	$s_1 \sim s_3$	$s_4 \sim s_6$	s_7	s_8	s_9	s_{10}	s_{11}	$s_{12} \sim s_{14}$	s_{15}	s_{16}	s_{17}	s_{18}	s_{19}	s_{20}	s_{21}	s_{22}	$s_{23} \sim s_{26}$
$\chi_{F_1}(s_i)$	1.00	1.00	0.90	0.80	0.50	0.20	0.00	⋯	⋯	⋯	⋯	⋯	⋯	⋯	⋯	⋯	0.00
$\chi_{F_2}(s_i)$	0.00	0.00	0.10	0.20	0.50	0.80	1.00	1.00	1.00	0.90	0.50	0.20	0.10	0.00	⋯	⋯	0.00
$\chi_{F_3}(s_i)$	0.00	⋯	⋯	⋯	⋯	⋯	⋯	⋯	0.00	0.10	0.50	0.80	0.90	1.00	⋯	⋯	1.00
$P(s_i)$	0.03	0.04	⋯	⋯	⋯	⋯	0.04	0.05	0.04	⋯	⋯	⋯	⋯	⋯	⋯	0.04	0.03

Table 6. Conditional probability $f(x_j \mid s_i)$.

s_i \ x_j	x_1	x_2	x_3	x_4	x_5	⋯	x_{22}	x_{23}	x_{24}	x_{25}	x_{26}
s_1	0.70	0.10	0.10	0.05	0.05						
s_2^2	0.10	0.70	0.10	0.05	0.05						
s_3	0.05	0.10	0.70	0.10	0.05				0		
⋮											
s_{24}			0				0.05	0.10	0.70	0.10	0.05
s_{25}							0.05	0.05	0.10	0.70	0.10
s_{26}							0.05	0.05	0.10	0.10	0.70

Table 7. Membership functions of $\underset{\sim}{M}_1$, $\underset{\sim}{M}_2$ and $\underset{\sim}{M}_3$.

x_j	$x_1 \sim x_6$	x_7	x_8	x_9	x_{10}	x_{11}	$x_{12} \sim x_{14}$	x_{15}	x_{16}	x_{17}	x_{18}	x_{19}	x_{20}	x_{21}	$x_{22} \sim x_{26}$
$\chi_{M_1}(x_j)$	1.0	0.9	0.8	0.7	0.5	0.2	0.0	⋯	⋯	⋯	⋯	⋯	⋯	⋯	0.0
$\chi_{M_2}(x_j)$	0.0	0.1	0.2	0.3	0.5	0.8	1.0	0.9	0.8	0.6	0.4	0.3	0.2	0.1	0.0
$\chi_{M_3}(x_j)$	0.0	⋯	⋯	⋯	⋯	⋯	0.0	0.1	0.2	0.4	0.6	0.7	0.8	0.9	1.0

The membership functions of $\underset{\sim}{M}_1, \underset{\sim}{M}_2$ and $\underset{\sim}{M}_3$, which satisfy the orthogoral condition, are given in Table 7.

Now, let us show the solution of this problem. Given a probabilistic information x_j and a probabilistic perfect information s_i, the results of optimal action and its utility are listed in Table 8. The similar results, given a fuzzy information $\underset{\sim}{M}_j$ and a fuzzy perfect information $\underset{\sim}{F}_i$, are shown in Table 9. By taking an average of utility values in Table 8 and 9 respectively, the exected utility and the worth of information system are obtained as shown in Table 10.

From Table 10, the relation $V(\mathcal{M}_\infty) \geq V(X_\infty) \geq V(X) \geq V(\mathcal{M}) \geq 0$ holds in regard to the worth of information. Comparing $V(X_\infty)$ with $V(X)$ numerically, little difference can be found. Thus, it might be understood that obtaining the probabilistic perfect information at a great cost is not so valuable, as compared with obtaining the probabilistic information。 There is also little diference between the value $V(X)=55.3$ and the value $V(\mathcal{M})=50.0$. This suggests that the fuzzy information is sufficiently valuable compard with the probabilistic information in this fuzzy decision problem. On the other hand, the fuzzy perfect information is more informative than the fuzzy information. It is caused by the fact that our interest is not in S, but in $\mathcal{F}$ which is a set of our concerned events on S. These results will be intuitionally agreed on by all.

Table 8. Utility and optimal decision given a probabilistic information

(a) The case of probabilistic information x_j

x_j	$\underset{\sim}{A}_{x_j}$	$U(\underset{\sim}{A}_{x_j} \mid x_j)$
$x_1 \sim x_5$	$\underset{\sim}{A}_1$	90.0
x_6	$\underset{\sim}{A}_1$	90.4
x_7	$\underset{\sim}{A}_1$	92.3
x_8	$\underset{\sim}{A}_1$	94.8
x_9	$\underset{\sim}{A}_1$	100.1
x_{10}	$\underset{\sim}{A}_2$	116.0
x_{11}	$\underset{\sim}{A}_2$	143.6
x_{12}	$\underset{\sim}{A}_2$	148.8
x_{13}	$\underset{\sim}{A}_2$	150.0
x_{14}	$\underset{\sim}{A}_2$	150.2
x_{15}	$\underset{\sim}{A}_2$	151.7
x_{16}	$\underset{\sim}{A}_2$	157.9
x_{17}	$\underset{\sim}{A}_3$	197.0
x_{18}	$\underset{\sim}{A}_3$	251.0
x_{19}	$\underset{\sim}{A}_3$	277.0
x_{20}	$\underset{\sim}{A}_3$	296.0
x_{21}	$\underset{\sim}{A}_3$	299.0
$x_{22} \sim x_{26}$	$\underset{\sim}{A}_3$	300.0

(b) The case of probabilistic perfect information s_i

s_i	$\underset{\sim}{A}_{s_i}$	$U(\underset{\sim}{A}_{s_i} \mid s_i)$
$s_1 \sim s_6$	$\underset{\sim}{A}_1$	90.0
s_7	$\underset{\sim}{A}_1$	92.0
s_8	$\underset{\sim}{A}_1$	94.0
s_9	$\underset{\sim}{A}_1$	100.0
s_{10}	$\underset{\sim}{A}_2$	120.0
$s_{11} \sim s_{15}$	$\underset{\sim}{A}_2$	150.0
s_{16}	$\underset{\sim}{A}_2$	155.0
s_{17}	$\underset{\sim}{A}_3$	200.0
s_{18}	$\underset{\sim}{A}_3$	260.0
s_{19}	$\underset{\sim}{A}_3$	280.0
$s_{20} \sim s_{26}$	$\underset{\sim}{A}_3$	300.0

Table 9. Utility and optimal decision given a fuzzy information.

(a) The case of fuzzy information $\underset{\sim}{M}_j$

$\underset{\sim}{M}_j$	$\underset{\sim}{A}_{M_j}$	$U(\underset{\sim}{A}_{M_j} \mid \underset{\sim}{M}_j)$
$\underset{\sim}{M}_1$	$\underset{\sim}{A}_1$	93.0
$\underset{\sim}{M}_2$	$\underset{\sim}{A}_2$	148.1
$\underset{\sim}{M}_3$	$\underset{\sim}{A}_3$	281.2

(b) The case of fuzzy perfect information $\underset{\sim}{F}_i$.

$\underset{\sim}{F}_i$	$\underset{\sim}{A}_{F_i}$	$U(\underset{\sim}{A}_{F_i} \mid \underset{\sim}{F}_i)$
$\underset{\sim}{F}_1$	$\underset{\sim}{A}_1$	90.0
$\underset{\sim}{F}_2$	$\underset{\sim}{A}_2$	150.0
$\underset{\sim}{F}_3$	$\underset{\sim}{A}_3$	300.0

Table 10. Expected utility and worth of information.

Information	Expected utility	Worth of information
no information	120.7	—
probabilistic information	176.0	55.3
probabilistic perfect information	177.4	56.7
Fuzzy information	170.7	50.0
Fuzzy perfect information	181.4	60.7

4. Quantity of Fuzzy Information

In this section, the quantity of fuzzy information is defined by using the definition of entropy of fuzzy event $\underset{\sim}{A}$ as follows:

$$H(\underset{\sim}{A}) = -P(\underset{\sim}{A}) \log P(\underset{\sim}{A}) - P(\overline{\underset{\sim}{A}}) \log P(\overline{\underset{\sim}{A}}) \tag{51}$$

where $\overline{\underset{\sim}{A}}$ is defined by $\chi_{\overline{A}} = 1 - \chi_A$. The entropy functions of fuzzy event $\underset{\sim}{A}$ on S given a probabilistic information $x_j \in X$ and a fuzzy information $\underset{\sim}{M}_k \in \mathscr{M}$ are respectively

$$H(\underset{\sim}{A} \mid x_j) = -P(\underset{\sim}{A} \mid x_j) \log P(\underset{\sim}{A} \mid x_j) - P(\overline{\underset{\sim}{A}} \mid x_j) \log P(\overline{\underset{\sim}{A}} \mid x_j) \tag{52}$$

and

$$H(\underset{\sim}{A} \mid \underset{\sim}{M}_k) = -P(\underset{\sim}{A} \mid \underset{\sim}{M}_k) \log P(\underset{\sim}{A} \mid \underset{\sim}{M}_k) - P(\overline{\underset{\sim}{A}} \mid \underset{\sim}{M}_k) \log P(\overline{\underset{\sim}{A}} \mid \underset{\sim}{M}_k). \tag{53}$$

DEFINITION 13. Two kind of the conditional entropy of fuzzy events $\underset{\sim}{A}$ on S given a probabilistic information system X and a fuzzy information system $\mathscr{M}$ are respectively defined by

$$H(\underset{\sim}{A} \mid X) = \sum_j H(\underset{\sim}{A} \mid x_j) f(x_j) \tag{54}$$

and

$$H(\underset{\sim}{A} \mid \mathscr{M}) = \sum_k H(\underset{\sim}{A} \mid \underset{\sim}{M}_k) P(\underset{\sim}{M}_k). \tag{55}$$

Now, let $\mathscr{L}(x_i, y_j)$ be a joint probability of $x_i \in X$ and $y_j \in Y$ given two probabilistic information systems X and Y.

PROPOSITION 4: The following relations hold:

(i) $H(\underset{\sim}{A}) \geqq H(\underset{\sim}{A} \mid X)$,

(ii) $H(\underset{\sim}{A} \mid X, Y) \leqq H(\underset{\sim}{A} \mid X) + H(\underset{\sim}{A} \mid Y)$.

Proof. (i) Noting the relation $P(\underset{\sim}{A}) = \sum_j P(\underset{\sim}{A}, x_j)$, we have

$$-P(\underset{\sim}{A})\log P(\underset{\sim}{A})+\sum_j P(\underset{\sim}{A}|x_j)f(x_j)\log P(\underset{\sim}{A}|x_j)$$
$$=-\sum_j P(\underset{\sim}{A},x_j)\log\frac{P(\underset{\sim}{A})}{P(\underset{\sim}{A}|x_j)}.$$

Then, it follows from $\log x \leqq x-1\ (x>0)$ that

$$H(\underset{\sim}{A})-H(\underset{\sim}{A}|X)\geqq-\sum_j P(\underset{\sim}{A},x_j)\left\{\frac{P(\underset{\sim}{A})}{P(\underset{\sim}{A}|x_j)}-1\right\}$$
$$-\sum_j P(\bar{\underset{\sim}{A}},x_j)\left\{\frac{P(\bar{\underset{\sim}{A}})}{P(\bar{\underset{\sim}{A}}|x_j)}-1\right\}=0.$$

(ii) First, from the equalities $\sum_j P(\underset{\sim}{A}|x_i,y_j)f(y_j|x_i)=P(\underset{\sim}{A}|x_i)$ and $\sum_i P(\underset{\sim}{A}|x_i,y_j)\cdot f(x_i,y_j)=P(\underset{\sim}{A}|y_j)$ we have

$$H(\underset{\sim}{A}|X)=-\sum_i\sum_j P(\underset{\sim}{A}|x_i,y_j)\mathscr{L}(x_i,y_j)\log P(\underset{\sim}{A}|x_i)$$
$$-\sum_i\sum_j P(\bar{\underset{\sim}{A}}|x_i,y_j)\mathscr{L}(x_i,y_j)\log P(\bar{\underset{\sim}{A}}|x_i),$$
$$H(\underset{\sim}{A}|Y)=-\sum_i\sum_j P(\underset{\sim}{A}|x_i,y_j)\mathscr{L}(x_i,y_j)\log P(\underset{\sim}{A}|y_j)$$
$$-\sum_i\sum_j P(\bar{\underset{\sim}{A}}|x_i,y_j)\mathscr{L}(x_i,y_j)\log P(\bar{\underset{\sim}{A}}|y_j).$$

Next, using the inqualities $P(\underset{\sim}{A}|x_i)P(\underset{\sim}{A}|y_j)+P(\bar{\underset{\sim}{A}}|x_i)P(\bar{\underset{\sim}{A}}|y_j)\leqq 1$ it follows that

$$-P(\underset{\sim}{A}|x_i,y_j)\log P(\underset{\sim}{A}|x_i,y_j)-P(\bar{\underset{\sim}{A}}|x_i,y_j)\log P(\bar{\underset{\sim}{A}}|x_i,y_j)$$

$$\leqq -P(\underset{\sim}{A}|x_i,y_j)\log P(\underset{\sim}{A}|y_j)-P(\bar{\underset{\sim}{A}}|x_i,y_j)\log P(\bar{\underset{\sim}{A}}|x_i)P(\bar{\underset{\sim}{A}}|y_j).$$

Then, the following relation holds:

$$H(\underset{\sim}{A}|X,Y)=-\sum_i\sum_j\mathscr{L}(x_i,y_j)\Big[P(\underset{\sim}{A}|x_i,y_j)\log P(\underset{\sim}{A}|x_i,y_j)$$
$$+P(\bar{\underset{\sim}{A}}|x_i,y_j)\log P(\bar{\underset{\sim}{A}}|x_i,y_j)\Big]$$
$$\leqq-\sum_i\sum_j\mathscr{L}(x_i,y_j)\Big[P(\underset{\sim}{A}|x_i,y_j)\log P(\underset{\sim}{A}|x_i)P(\underset{\sim}{A}|y_j)$$
$$+P(\bar{\underset{\sim}{A}}|x_i,y_j)\log P(\bar{\underset{\sim}{A}}|x_i)P(\bar{\underset{\sim}{A}}|y_j)\Big]$$
$$=H(\underset{\sim}{A}|X)+H(\underset{\sim}{A}|Y).$$

For fuzzy events $\underset{\sim}{A}$ on S, $\underset{\sim}{B}$ on X and $\underset{\sim}{C}$ on Y, the following proposition holds.

PROPOSITION 5. The following relations hold for two fuzzy information systems $\mathscr{B}=\{\underset{\sim}{B},\bar{\underset{\sim}{B}}\}$ and $\mathscr{C}=\{\underset{\sim}{C},\bar{\underset{\sim}{C}}\}$:

(i) $H(\underset{\sim}{A})\geqq H(\underset{\sim}{A}|\mathscr{B})$,

(ii) $H(\underset{\sim}{A},\underset{\sim}{B})\leqq H(\underset{\sim}{A}|\mathscr{B})+H(\underset{\sim}{B})$,

(iii) $H(\underset{\sim}{A},\underset{\sim}{B})\leqq H(\underset{\sim}{A})+H(\underset{\sim}{B})$,

(iv) $H(\underset{\sim}{A}|\mathscr{B},\mathscr{C})\leqq H(\underset{\sim}{A}|\mathscr{B})+H(\underset{\sim}{A}|\mathscr{C})$.

Proof. (i) From $P(\underset{\sim}{A}, \underset{\sim}{B}) = \sum_i \sum_j \chi_A(s_i)\chi_B(x_j) f(x_j | s_i) P(s_i)$, the equality $P(\underset{\sim}{A}) = P(\underset{\sim}{A}, \underset{\sim}{B}) + P(\underset{\sim}{A}, \bar{\underset{\sim}{B}})$ holds. Noting the relation $P(\underset{\sim}{A}, \underset{\sim}{B}) = P(\underset{\sim}{A}|\underset{\sim}{B}) P(\underset{\sim}{B})$, we have

$$\begin{aligned}
H(\underset{\sim}{A}) - H(\underset{\sim}{A}|\mathcal{B}) &= - P(\underset{\sim}{A}) \log P(\underset{\sim}{A}) - P(\bar{\underset{\sim}{A}}) \log P(\bar{\underset{\sim}{A}}) \\
&\quad + P(\underset{\sim}{A}|\underset{\sim}{B}) P(\underset{\sim}{B}) \log P(\underset{\sim}{A}|\underset{\sim}{B}) + P(\bar{\underset{\sim}{A}}|\underset{\sim}{B}) P(\underset{\sim}{B}) \log P(\bar{\underset{\sim}{A}}|\underset{\sim}{B}) \\
&\quad + P(\underset{\sim}{A}|\underset{\sim}{B}) P(\bar{\underset{\sim}{B}}) \log P(\underset{\sim}{A}|\bar{\underset{\sim}{B}}) + P(\bar{\underset{\sim}{A}}|\bar{\underset{\sim}{B}}) P(\bar{\underset{\sim}{B}}) \log P(\bar{\underset{\sim}{A}}|\bar{\underset{\sim}{B}}) \\
&= - P(\underset{\sim}{A}, B) \log \frac{P(\underset{\sim}{A})}{P(\underset{\sim}{A}|\underset{\sim}{B})} - P(\underset{\sim}{A}, \bar{\underset{\sim}{B}}) \log \frac{P(\underset{\sim}{A})}{P(\underset{\sim}{A}|\bar{\underset{\sim}{B}})} \\
&\quad - P(\bar{\underset{\sim}{A}}, \underset{\sim}{B}) \log \frac{P(\bar{\underset{\sim}{A}})}{P(\bar{\underset{\sim}{A}}|\underset{\sim}{B})} - P(\bar{\underset{\sim}{A}}, \bar{\underset{\sim}{B}}) \log \frac{P(\bar{\underset{\sim}{A}})}{P(\bar{\underset{\sim}{A}}|\bar{\underset{\sim}{B}})} \\
&\geq - P(\underset{\sim}{A}, \underset{\sim}{B}) \left\{ \frac{P(\underset{\sim}{A})}{P(\underset{\sim}{A}|\underset{\sim}{B})} - 1 \right\} - P(\underset{\sim}{A}, \bar{\underset{\sim}{B}}) \left\{ \frac{P(\underset{\sim}{A})}{P(\underset{\sim}{A}|\bar{\underset{\sim}{B}})} - 1 \right\} \\
&\quad - P(\bar{\underset{\sim}{A}}, \underset{\sim}{B}) \left\{ \frac{P(\bar{\underset{\sim}{A}})}{P(\bar{\underset{\sim}{A}}|\underset{\sim}{B})} - 1 \right\} - P(\bar{\underset{\sim}{A}}, \bar{\underset{\sim}{B}}) \left\{ \frac{P(\bar{\underset{\sim}{A}})}{P(\bar{\underset{\sim}{A}}|\bar{\underset{\sim}{B}})} - 1 \right\} \\
&= 0 .
\end{aligned}$$

(ii) This assertion can be proved from the definitions of $H(\underset{\sim}{B})$ and $H(\underset{\sim}{A}|\underset{\sim}{B})$, noting the equality

$$\begin{aligned}
H(\underset{\sim}{A}, \underset{\sim}{B}) &= -P(\underset{\sim}{A}, \underset{\sim}{B}) \log P(\underset{\sim}{A}, \underset{\sim}{B}) - P(\bar{\underset{\sim}{A}}, \underset{\sim}{B}) \log P(\bar{\underset{\sim}{A}}, \underset{\sim}{B}) \\
&\quad - P(\underset{\sim}{A}, \bar{\underset{\sim}{B}}) \log P(\underset{\sim}{A}, \bar{\underset{\sim}{B}}) - P(\bar{\underset{\sim}{A}}, \bar{\underset{\sim}{B}}) \log P(\bar{\underset{\sim}{A}}, \bar{\underset{\sim}{B}}).
\end{aligned}$$

(iii) This assertion follows from (i) and (ii).

(iv) In view of the proof of (ii) in Proposition 4, it is clear that this assertion holds.

On the basis of the definition of entropy of fuzzy events, we have the following definitions.

DEFINITION 14. The entropy of fuzzy state space $\mathcal{F}$ is

$$H(\mathcal{F}) = - \sum_i \left[P(\underset{\sim}{F}_i) \log P(\underset{\sim}{F}_i) + P(\bar{\underset{\sim}{F}}_i) \log P(\bar{\underset{\sim}{F}}_i) \right].$$

DEFINITION 15. The quantity of probabilistic information system X and the quantity of fuzzy information system $\mathcal{M}$ are respectively

$$I(X) = H(\mathcal{F}) - H(\mathcal{F}|X)$$

and

$$I(\mathcal{M}) = H(\mathcal{F}) - H(\mathcal{F}|\mathcal{M}).$$

The following proposition holds for two probabilistic information systems X and Y, and two fuzzy information systems $\mathcal{M}_X$ on X and $\mathcal{M}_Y$ on Y.

PROPOSITION 6.

(i) $\hat{H}(\mathcal{F}|X, Y) \leq H(\mathcal{F}|X) + H(\mathcal{F}|Y)$,

(ii) $I(X) \geq 0$ $(I(Y) \geq 0)$,

(iii) $H(\mathcal{F}|\mathcal{M}_X, \mathcal{M}_Y) \leq H(\mathcal{F}|\mathcal{M}_X) + H(\mathcal{F}|\mathcal{M}_Y)$,

(iv) $I(\mathcal{M}_i) \geq 0$, $i = X, Y$.

This proposition can be immediately proved from Propositions 4 and 5.

REFERENCES :

[1] L.A. Zadeh, Probability Measures of Fuzzy Events, Journal of Mathematical Analysis and Applications, 23 (1968), 421~427.
[2] T. Okuda, H. Tanaka and K. Asai, A Formulation of Fuzzy Decision Problems with Fuzzy Information using Probability Measures of Fuzzy Events, Information and Conrol, 38(1978), 135~147.
[3] H. Tanaka, T. Okuda and K. Asai, A Formulation of Fuzzy Decision Problems and its Application to an Investment Problem, Kybernetes, 5(1976), 25~30.
[4] H. Tanaka, T. Okuda and K. Asai, On Decision-making in Fuzzy Eveironment(Fuzzy Information and Decision-making), International Journal of Production Research, 15(1977), 623~635.
[5] K. Asai, H. Tanaka and T. Okuda, On Discrimination of Fuzzy States in Probability Space, Kybernetes, 6(1977), 185~192.

ADVANCES IN FUZZY SET THEORY AND APPLICATIONS
M.M. Gupta, R.K. Ragade, R.R. Yager (editors)

ENTROPY AND ENERGY MEASURES OF A FUZZY SET

Aldo DE LUCA and Settimo TERMINI

Laboratorio di Cibernetica del C.N.R., Arco Felice (Napoli), Italy.

A general setting for the study of the entropy and energy measures of a fuzzy set is provided. In the previous framework, after a discussion of the measures already introduced by the authors and the new proposals by Trillas and Riera and Knopfmacher, some new results are proven. After an interpretative analysis of these measures, which takes into account also a comparison with the classical entropy of Statistical Thermodynamics and Information Theory, a final section is devoted to the development of an unforseeability *measure in taking decisions in the setting of fuzzy sets theory. It turns out that the previous measure is strongly related both to the entropy measures and to fuzzy sets themselves allowing an interesting interpretation of membership value in frequentistic terms (of decisions).*

1. INTRODUCTION

One of the first motivations, and one of the main aims of all the research in fuzzy sets theory, is to furnish mathematical models which are able to describe systems or classes of systems which escape traditional analysis. This can be pursued by the two different (but noncontradictory) approaches: using well known pieces of mathematics by interpreting them in a different way and trying to build new *calculi* more apt than the existing ones for the modelling of the reality in its complexity (or better, in practice, by making abstractions which are less strong than the usual ones based ultimately, from a conceptual point of view even if not from a formal one, on the idea that, in principle, everything of which it is worth speaking, in the sciences at least, can be affirmed or denied with certainty).

The concept that a *crisp* model, even if never realized in practice, is a good platonic idea to which to refer has shown to be fruitful in many fields in these last centuries, first of all in physics. However, it seems that it could be a simplification, at least from an epistemological point of view, to consider *fuzzy sets*, that is the association of properties with a certain universe of objects without imposing the condition of the existence of sharp boundaries, as the primary and not yet analyzed information that we have on the world (i.e. on the system we are studying). An analysis of this information can be made at two different levels, the first one regards the relationships that have to exist among these data taken as the unitary bases of our discourse, the *atoms* of (our representation of) the world; and that is what has been done by the people who have tried to develop a *logic* of these levels considered at the same time both real and basic; here we limit ourselves to remembering the papers by Joe Goguen [16] and Robin Giles [14] together with the qualitative but deep analysis of Max Black[3-4] without any pretension of completeness. Another approach is to study the interrelation of these *elementary* bulks of information at another level by considering their mutual relationships not among the single bulks but their *average* effect. The situation is then very similar to the statistical thermodynamics in which one does not consider the single particles, or the relations among them but their average effects thus obtaining other macroscopic quantities such as *pressure, temperature* or *volume* and their mutual relationships. There certainly exists a tie between these two levels and we think that a study of it follows the line of the von Neumann idea of a *logic of information* very similar from a formal point of view to classical thermodynamics. However, we shall not pursue this aspect further here but shall start directly from some of these global measures that have been introduced.

An important, subtle and very delicate problem of the theory is the relationship between *fuzziness* and *probability* and other notions of uncertainty. This problem even if is outside the main topics of the paper can be indirectly clarified also by means of the present discussion and for this reason we shall briefly make some comments here (see also the discussion in [7]). The comparison fuzziness-probability can be made at various different levels, for instance, conceptual, mathematical or axiomatic. In making it however one has to be very careful: in fact, the general methodological discussions and debates on *probability*, made during the last centuries, took into account almost every possible sophisticated variant of the notion of *uncertainty* so that one can consider that a conceptual comparison (at least partial) of fuzziness and probability began long before the birth of fuzzy sets and so one has to trace back in the midst of these debates the ancestors of the present discussions on fuzziness and probability.

H. Nurmi in an interesting paper [20] makes the penetrating linguistic observation that in many European languages the classical term denoting probability had to be literally translated as "verisimilitude" and then since, technically, the term probability is mainly used either in a (tendentially) *frequentistic sense* or (subjectivistically) as *degree of belief*, one had to retract from using these terms (in the previous languages) to denote probability in order to avoid ambiguities. He adds, finally, that since *"the concept of fuzziness comes very close to the concept of verisimilitude,...,for reasons of conceptual ortodoxy one could suggest that the terms used in Finnish, German and Swedish to denote probability- "todennakoinen", "wahrscheinlich", "sannolik" respectively - should rather be used to denote fuzziness"*. Nurmi adds that *"thereby some conceptual confusion might be avoided"* but rightly concludes that *"this suggestion is, of course, unrealistic in view of the strenght of our linguistic habits"*. We would add to the previous remarks that, apart from an historical interest in the study of the interrelation between natural languages and scientific languages on one side and their interaction with the global evolution of society on the other, the previous observations by Nurmi form a strong empirical support for the thesis that scientific concepts and theories are born and grow from informal *explicanda* of everyday language and concepts by making precise and exact definitions and uses *but almost always by strongly limiting the original informal notions involved*. An exception to this appears to be the formalization of "computable" which seems to include every intuitive and qualitative facet of the informal notion but this remains the only example of this type which we know. For every notion that makes a jump from an informal qualitative stage to a quantitative, and possibly formalized, level, one has to trace back the major part of its conceptual ancestors and aspects in order to fully appreciate its significance and the most fruitful potential directions of its formal developments.

From this point of view we would add to the previous remarks of Nurmi the fact that fuzziness is strongly related also to the notion of *vagueness* and under certain aspects, especially in the version of Goguen [16], it represents its first formal *explicatum*. An analysis of the relationship probability-fuzziness aspiring to a certain completeness has then to take into account also the relationship vagueness-probability which has not been, as far as we know, exhaustively explored. A second serious handicap for a brief and simple comparison between fuzziness and probability is the fact that the theory of probability is very well developed from a formal point of view so that many notions have a standard and mathematically rich version (i.e. with a characterization of deep theorems) accepted and shared by the scientific community while the theory of fuzzy sets is in a far less developed stage from the point of view of a theory defined by a corpus of meaningful deep theorems which are considered as an unrenouncible part of the theory itself and so, indirectly, implicitly defining or characterizing it.

A consequence of unbalanced situations of this type is that it is almost always possible to force a given developed theory in order to take into account some new facts provided by a new theory in a still embrionic stage of development so that it really seems that an older and more developed theory can seem superior in comparison with a less developed one. We shall not dwell more on this point which belongs to the theory of scientific progress but only observe that this is the main reason why in such comparative studies one is forced to take into account both the mathematical aspects of the theories and their interpretative conceptual side by passing continually from one level to the other. If this passage is made in a deep way the comparison becomes fruitful and the mathematical richness of one of the two theories becomes useful for the comparison since it shows either that the new theory really can be absorbed, and how, (in this case the differences between the two remain only different interpretations of the same formal structure) or that the mathematical form of the first one shows its incompatibility with the conditions imposed by the interpretative requirements of the second one. For a deep and complete analysis of the comparison probability-fuzziness we refer the reader to the papers by Nurmi [20-21] and we shall conclude the present introductory section by noting that some clarification of this comparison can be obtained by a development of the measures of fuzziness for the following two main reasons; the first is that they provide the tools for the comparison between the two notions at the level of their *quantitative measures*, the second is that, since the measures of fuzziness are related to decision problems, their development can help in the discrimination between concepts of probabilistic or fuzzy nature or, still better, in the discrimination of probabilistic or fuzzy aspects in the same concept.

2. NOTATIONS AND MATHEMATICAL PRELIMINARIES.

Let X be a set and L a poset (i.e. a partially ordered set). An *L-fuzzy set* [15] is any map $f: X \to L$. We denote by $\mathcal{L}(X,L)$ the class of all fuzzy sets defined on X. $\mathcal{L}(X,L)$ can be partially ordered by defining:

$$f \le g \Leftrightarrow \forall x \in X\ [\, f(x) \le g(x)\,] \tag{2.1}$$

If L is a lattice one can induce a lattice structure on $\mathcal{L}(X,L)$ by introducing for all $f, g \in \mathcal{L}(X,L)$ the operations $f \vee g$ and $f \wedge g$ defined point-by-point as:

$$(f \vee g)(x) = \sup\{f(x), g(x)\}, \quad (f \wedge g)(x) = \inf\{f(x), g(x)\}.$$

If we take for L the interval $I = [0,1]$ of the real line an I-fuzzy set will be called simply *fuzzy set* [26], or *generalized characteristic function*, and $\mathcal{L}(X,I)$ will be denoted by $\mathcal{L}(X)$. In this case for all $x \in X$ $(f \vee g)(x) = \max\{f(x), g(x)\}$ and $(f \wedge g)(x) = \min\{f(x), g(x)\}$. Further for any $f \in \mathcal{L}(X)$ one can introduce the so-called *complement* f' defined for all $x \in X$ as: $f'(x) = 1 - f(x)$, having for all $f, g \in \mathcal{L}(X)$:

$$(f')' = f \qquad \textit{Involution law}$$

$$(f \vee g)' = f' \wedge g', \quad (f \wedge g)' = f' \vee g' \qquad \textit{De-Morgan laws}$$

We recall that $\mathcal{L}(X)$ is a *complete distributive but noncomplemented lattice* with respect to the operations $(\vee)$ and $(\wedge)$ [26,9]. These operations and the complement (') extend in the case of fuzzy sets the *union, intersection* and *complement* of classical sets expressed in terms of their characteristic functions.

We observe that other operations can be introduced in $\mathcal{L}(X)$ to extend the classical union and intersection, as for instance $f \oplus g$ and fg defined for all $x \in X$ as

$$(f \oplus g)(x) = f(x) + g(x) - f(x)g(x), \quad (fg)(x) = f(x)g(x).$$

However since these operations are not idempotent $\mathcal{L}(X)$ is not a lattice relative to them. Further also the distributive property is not verified.

$\mathcal{L}(X)$ can be also partially ordered by the relation $\le'$ defined as [24]:

$$f \le' g \Leftrightarrow \forall x \in X [\, f(x) \le g(x) \text{ if } g(x) \le 1/2\, ;\, f(x) \ge g(x) \text{ if } g(x) \ge 1/2\,]. \tag{2.2}$$

If $f \le' g$ and $f \ne g$ then f is said *sharper* than g [8].

In the following for all $a \in [0,1]$ we denote by a the fuzzy set which is constantly equal to a for all $x \in X$. One has that 1 and 0 are the maximum and minimum elements of $\mathcal{L}(X)$ ordered by $\le$. If $\mathcal{L}(X)$ is ordered by $\le'$ then $1/2$ is the maximum element and 0 and 1 minimal elements.

Some noteworthy properties of the orderings $\le$ and $\le'$ of $\mathcal{L}(X)$ which can be easily proven by using definitions (2.1) and (2.2) and the involution and De-Morgan laws, are listed below. For all $f, g, h, k \in \mathcal{L}(X)$ one has:

$$
\begin{array}{ll}
1.\ f \le g \Rightarrow f' \le g'; & 2.\ f \le' g \Rightarrow f' \le' g'; \\
3.\ f \le' g \Rightarrow f \wedge f' \le g \wedge g' \ ,\ f \vee f' \ge g \vee g'; & \\
4.\ f \le g \Rightarrow f \wedge 1/2 \le' g \wedge 1/2 \ ; & \\
5.\ f \le' g\ ,\ h \le' k \Rightarrow f \wedge h \le' g \wedge k \ ; & (2.3) \\
6.\ f \le' g \Rightarrow f \wedge f' \le' g \wedge g'\ ,\ f \vee f' \le' g \vee g' \ ; & \\
7.\ f \le' g \Rightarrow f \le' f \wedge g \le' g \ ,\ f \le' f \vee g \le' g \ . &
\end{array}
$$

3. ENTROPY AND ENERGY MEASURES OF A FUZZY SET.

As we said in the introduction a possible description of fuzzy sets can be done by means of functionals of thermodynamic kind, defined in $\mathcal{L}(X)$ and taking values in the set R of nonnegative real numbers, which give global information about the fuzzy sets. In this section we shall introduce two classes of these functionals which play a substantial different role and that we shall call *energy* and *entropy* respectively. We define these quantities in an axiomatic way and postpone interpretative problems to the next section.

Let us premise the following general definitions. Let $< P, \le, \le', i, o, c >$ be a sextuple where P is a set partially ordered by $\le$ and $\le'$, i and o are the maximum and minimum of P relative to $\le$ and both are minimal elements with respect to $\le'$; c is the maximum element of P relative to $\le'$. Let us denote by M the set of minimal elements of P with respect to $\le'$.

Definition 3.1 - An E-function is any map $e: P \to R$ such that:

i. e is isotone with the order $\le$, i.e. for all $p,q \in P$, $p \le q \Rightarrow e(p) \le e(q)$.
ii. $e(p) = 0$ if and only if $p = o$.

Definition 3.2 -An H-function is any map $h: P \to R$ such that:

i. h is isotone with the order $\le'$, i.e. for all $p,q \in P$, $p \le' q \Rightarrow h(p) \le h(q)$.
ii. $h(p) = 0$ if and only if $p \in M$.

The previous definitions imply that E-functions (*resp.* H-functions) reach their maximum value at $p=i$ (*resp.* $p = c$). We simply denote by E and H respectively the classes of all E-functions and H-functions defined in P. These classes satisfy the following general properties.

1. E and H are disjoint classes.
2. *Additivity*. Let be $e_1, e_2 \in E$, $h_1, h_2 \in H$, $0 < a_1, a_2 \in R$ and $e_1 + e_2$, $h_1 + h_2$ the functions defined for all $p \in P$ as:

   $(e_1 + e_2)(p) = e_1(p) + e_2(p)$, $(h_1 + h_2)(p) = h_1(p) + h_2(p)$.

   One has that $e_1 + e_2 \in E$ and $h_1 + h_2 \in H$.
3. For any $0 < a \in R$, $e \in E$ and $h \in H$ let us define the functions ae, ah as:

   $(ae)(p) = ae(p)$, $(ah)(p) = ah(p)$, for all $p \in P$. One has that $ae \in E$, $ah \in H$.
4. *Convexity property*. As consequence of properties 1. and 2. it follows that for all $e_1, e_2 \in E$, $h_1, h_2 \in H$ and $0 < a_1, a_2 \in R$, $a_1 e_1 + a_2 e_2 \in E$ and $a_1 h_1 + a_2 h_2 \in H$.
5. *Multiplicativity*. Let us introduce in E and H the products $e_1 e_2$, $h_1 h_2$ defined for all $e_1, e_2 \in E$, $h_1, h_2 \in H$ as

   $(e_1 e_2)(p) = e_1(p) e_2(p)$, $(h_1 h_2)(p) = h_1(p) h_2(p)$, for all $p \in P$.

   One has that $e_1 e_2 \in E$ and $h_1 h_2 \in H$.
6. Let u be any nondecreasing map of R to R such that $u(x) = 0$ iff $x = 0$. For all $h \in H$ and $e \in E$ the functions $u \circ h$ and $u \circ e$ defined as:

   $(u \circ h)(p) = u(h(p))$, $(u \circ e)(p) = u(e(p))$, for all $p \in P$

   belong to H and E respectively.

Let us now take as P the set $\mathcal{L}(X)$ of all the fuzzy sets defined in X ordered by the relations $\leq$ and $\leq'$ defined by Eq. (2.1) and Eq. (2.2) respectively. In this case $o = 0$, $i = 1$ and M coincides with the subclass of $\mathcal{L}(X)$ formed by all the Boolean characteristic functions. We consider now the classes of E-functions and H-functions in $\mathcal{L}(X)$ that we still denote by E and H respectively. Since in $\mathcal{L}(X)$ is defined the operation of complementation (') we have as a consequence of $(2.3)_2$ that it holds the property:

7. for any $h \in H$ the function h' defined for all $f \in \mathcal{L}(X)$ as $h'(f) = h(f')$ is still an H-function.

Definition 3.3 - An H-function h is called *symmetric* if $h=h'$.

Definition 3.4 - A function $v : \mathcal{L}(X) \to R$ is a *valuation* on the lattice $\mathcal{L}(X)$ if for all $f,g \in \mathcal{L}(X)$

$$v(f) + v(g) = v(f \vee g) + v(f \wedge g) . \qquad (3.1)$$

8. If $h \in H$ is a valuation then it is symmetric if and only if for all $f \in \mathcal{L}(X)$ $h(f) = h(f \vee f') = h(f \wedge f')$.

Proposition 3.1 -For any $e \in E$ the map $h : \mathcal{L}(X) \to R$ defined as $h(f) = e(f \wedge f')$, for all $f \in \mathcal{L}(X)$, is a symmetric H-function. Vice versa if $h \in H$ the map $e: \mathcal{L}(X) \to R$ defined as $e(f) = h(f \wedge 1/2)$ for all $f \in \mathcal{L}(X)$ is an E-function.

Proof - Let $e \in E$ and $h(f) = e(f \wedge f')$ for all $f \in \mathcal{L}(X)$. From the definition of E-function $e(f \wedge f')=0$ if and only if $f \wedge f' = 0$. But $f \wedge f' = 0$ if and only if f is a classical characteristic function. If $f \leq' g$ then from $(2.3)_3$ it follows that $f \wedge f' \leq g \wedge g'$ so that $h(f) = e(f \wedge f') \leq e(g \wedge g') = h(g)$. Moreover from the definition $h(f) = h(f')$ for all $f \in \mathcal{L}(X)$. Hence $h = h'$.
Let now be $h \in H$ and $e(f) = h(f \wedge 1/2)$ for all $f \in \mathcal{L}(X)$. $h(f \wedge 1/2) = 0$ if and only if $f \wedge 1/2$ is a classical characteristic function. However this can occur if and only if $f = 0$. If $f \leq g$ then from $(2.3)_4$ one has $f \wedge 1/2 \leq' g \wedge 1/2$ so that $e(f) = h(f \wedge 1/2) \leq h(g \wedge 1/2) = e(g)$ □

Let us denote by H' the subclass of H formed by all functions $h \in H$ such that $h(f \wedge f') = h(f)$ for all $f \in \mathcal{L}(X)$. A function of this kind is, for instance, any $h \in H$ which is symmetric and satisfies the valuation property (3.1). It holds the following proposition the proof of which we omit for brevity,

Proposition 3.2 - The maps $u : E \to H$, $v : H \to E$ defined as:

$$(u(e))(f) = e(f \wedge f'), \quad (v(h))(f) = h(f \wedge 1/2)$$

for all $e \in E$, $h \in H$ and $f \in \mathcal{L}(X)$, are such that $uvu = u$ and $vuv = v$. Moreover the restriction v' of v to H' is an injection □

Proposition 3.3 -For any $e \in E$, $h \in H$, $a \in (0,1]$, $b \in (0,1/2]$ the functions e_1, e_2 defined as $e_1(f) = e(af)$, $e_2(f) = h(bf)$, for all $f \in \mathcal{L}(X)$, belong to E .

Proof - If $f \leq g$ then $af \leq ag$, $bf \leq bg \leq 1/2$. This implies $e(af) \leq e(ag)$ and since $bf \leq bg$ one has that $h(bf) \leq h(bg)$. Moreover $e(af) = 0$ if and only if $f = 0$ and $h(bf) = 0$ if and only if bf is a classical characteristic function. Since $bf \leq 1/2$ this implies $f = 0$ □

Let us now give a formal definition of the concepts of *energy* and *entropy* of a fuzzy set. We suppose first that $\text{Card}(X) < +\infty$. The case of an infinite support will be considered later. *By energy and entropy we mean an E-function and an H-function defined on $\mathcal{L}(X)$ respectively.* This definition has been intentionally chosen very weak and general in order to capture in the formal notions only the minimal essential features that intuitively any measure of *amount of membership* and of *amount of fuzziness* must possess. We observe, however, that from one hand more general definitions are possible (for instance by considering in $\mathcal{L}(X)$ order relations different from the natural ones $\leq$ and $\leq'$) and from the other hand it is natural to consider further requirements on the E-functions and H-functions in order to restrict the classes of measures. In the following we shall introduce some of these requirements allowing to characterize classes of entropies and energies. A discussion on the meaning of these requirements will be done in the next section. Let us list some of these general requirements:

A. An energy reaches its maximum *only if* $f = 1$. An entropy reaches its maximum *only if* $f = 1/2$. This property of an entropy measure has been required in the original definition of the entropy of a fuzzy set [8].

B. An energy and an entropy are strictly isotonic maps, i.e.

$$f < g \Rightarrow e(f) < e(g) \quad , \quad f <' g \Rightarrow h(f) < h(g) .$$

Such an assumption obviously implies A.

C. The amount of membership and of fuzziness, is the sum of individual contributions evaluated by means of a function depending only on the single point and on the value of the degree of membership. We shall denote by E_s and H_s these two subclasses of E and H respectively. If $e \in E_s$, $h \in H_s$, $f \in \mathcal{L}(X)^s$ one has then:

$$e(f) = \sum_{x \in X} \Phi(x, f(x)) \qquad h(f) = \sum_{x \in X} \Psi(x, f(x)) \qquad (3.2)$$

where $\Phi, \Psi : X \times I \to R$ are two maps which must satisfy, as a consequence of the definitions of E-functions and H-functions, the properties:
For each $x \in X$,

$$\Phi(x,y) = 0 \text{ iff } y = 0 \ ; \ \Phi(x,y) \text{ is nondecreasing in } [0,1]$$
$$\Psi(x,y) = 0 \text{ iff } y \in \{0,1\} \ ; \Psi(x,y) \text{ is nondecreasing in } [0,1/2] \text{ and nonincreasing in } [1/2,1] . \qquad (3.3)$$

Vice versa one can easily verify that if Φ, Ψ are two maps from $X \times I$ to R satisfying Eq. (3.3) then e and h defined by (3.2) are an energy and an entropy measures respectively.
A remarkable property of the energies and entropies of the classes E_s and H_s are obtained supposing that the functions Φ and Ψ can be factorized as:

$$\Phi(x,y) = w(x)\, \phi(y) \ , \ \Psi(x,y) = w(x)\, \psi(y) \qquad x \in X , y \in I$$

where ϕ and ψ are maps from I to R and w is a positive weight function from X to R. In this case one has:

$$e(f) = \sum_{x \in X} w(x)\, \phi(f(x)) , \qquad h(f) = \sum_{x \in X} w(x)\, \psi(f(x)) \qquad (3.4)$$

where ϕ and ψ have to satisfy the properties:

$$\phi(x) = 0 \text{ iff } x = 0 , \ \phi \text{ nondecreasing in } [0,1] \qquad (3.5)$$

$$\psi(x) = 0 \text{ iff } x \in \{0,1\} \ ; \ \psi \text{ nondecreasing in } [0,1/2] \text{ and nonincreasing in } [1/2,1] . \qquad (3.6)$$

Vice versa for any given positive weight function w if ϕ and ψ are two maps from I to R satisfying the previous requirements then the functions e and h defined by Eq. (3.4) are an energy and an entropy measure respectively. Functions ψ satisfying Eq. (3.6) have been considered by Trillas and Riera [24] and called *entropy N-functions* (*norm functions*). In the following we shall call *energy N-function* a function ϕ verifying Eq. (3.5). The entropies of the previous class have been called *sum-product* entropies by Trillas and Riera [24]. We shall denote by E_{sp} and H_{sp} these classes of energies and entropies.
It is easily seen that an entropy $h \in H_{sp}$ is symmetric if and only if $\psi(x) = \psi(1-x)$ for all $x \in [0,1]$.

D. Since energies and entropies are maps from I^n to R, where $n = \text{Card}(X)$, a natural requirement is that these maps are *continuous* functions with respect to one of the standard *Euclidean metrics*. Let us consider, for instance, as metrics:

$$\rho_1(f,g) = [\sum_{x \in X} [f(x) - g(x)]^2]^{1/2} , \ \rho_2(f,g) = \bigvee_{x \in X} |f(x) - g(x)| .$$

It is easy to verify that an energy $e \in E_{sp}$ (*resp.* entropy $h \in H_{sp}$) *is a continuous function of* $f \in \mathcal{L}(X)$ *with respect to the metrics* ρ_1 *or* ρ_2 *if and only if the corresponding N-function* ϕ *(resp.* ψ *) is continuous in the interval* $[0,1]$.

We shall give now some examples of energies of this subclass E_{sp}. A first example is the so-called *power* of a fuzzy set defined for all $f \in \mathcal{L}(X)$ as:

$$P(f) = \sum_{x \in X} f(x) \tag{3.7}$$

This quantity has been called power since it generalizes in the case of fuzzy sets the concept of *cardinality* of a finite set. Moreover the relation (valuation property)

$$P(f \vee g) = P(f) + P(g) - P(f \wedge g),$$

extends the classical relation of the cardinality of the union of two ordinary finite sets. Other measures of the class are

$$\sum_{x \in X} f^2(x) \quad \text{and} \quad \sum_{x \in X} p(x) f(x),$$

where $p: X \to [0,1]$ is a probability distribution over X. The first quantity when f is a probability distribution over X reduces to the *informational energy* of Onicescu [22] . The second quantity represents the *average membership* of an element to the fuzzy set described by f .

A noteworthy subclass of the continuous entropies of the class H_{sp} is that obtained by imposing that *the norm-functions are symmetric and concave in the interval* $[0,1]$. Let us give some examples of entropies of this class. First of all we recall the *logarithmic entropy* d defined as:

$$d(f) = \sum_{x \in X} S(f(x)) \tag{3.8}$$

for all $f \in \mathcal{L}(X)$, where S is the Shannon function defined for all $x \in [0,1]$ as $S(x) = x \ln 1/x + (1-x) \ln 1/(1-x)$ and assuming that $0 \ln 0 = 0$. Other examples are

$$\sigma(f) = \sum_{x \in X} f(x)(1-f(x)) \tag{3.9}$$

and

$$u(f) = \sum_{x \in X} \min\{f(x), 1-f(x)\} = P(f \wedge f') \tag{3.10}$$

We note that formally $\sigma(f)$ equals the sum of the variances of the random variables $\xi(x)$ which take values 1 and 0 with probabilities $f(x)$ and $1-f(x)$, respectively.

It is easy to verify that if $\mu: I \to R$ is a *continuous strictly concave function* in $(0,1)$ such that $\lim_{x \to 0} \mu(x) = \lim_{x \to 1} \mu(x) = 0$, then the function T defined, for all $x \in [0,1]$ as $T(x) = \mu(x) + \mu(1-x)$ is a *symmetric and continuous N-function satisfying requirements A and B*. Entropies of this kind have been introduced in [5] . We observe that in the case of logarithmic entropy $T = S$; in the case of σ , $T(x) = x(1-x)$, $x \in [0,1]$. On the contrary u does not belong to this class even though its N-function is symmetric, continuous and satisfies the requirements A and B.

Let us now examine some remarkable properties of the entropies of the class H_{sp} whose norm-function ψ is a continuous , symmetric and concave function in the interval $[0,1]$. By making use of Jensen's inequality of concave functions one has:

$$\sum_{x \in X} w(x)\psi(f(x)) \leq \left[\sum_{x \in X} w(x)\right] \psi\left(\sum_{x \in X} w(x) f(x) \Big/ \sum_{x \in X} w(x)\right) \tag{3.11}$$

these relations show that if one keeps the energy $\sum w(x)f(x)$ equal to a constant e the corresponding entropy $\sum w(x)\psi(f(x))$ reaches its maximum value on the fuzzy set taking the constant value $e/\sum w(x)$. In particular when w is a constant the maximum value of the entropy is reached when the *energy (or the power) is equidistributed over* X . We observe that if requirement B is verified then the maximum of the entropy is reached only in this case.

A second property is obtained by observing that still from Jensen's inequality one derives $\psi(x/2) \geq x\,\psi(1/2)$, for all $x \in [0,1]$ and then

$$\psi(x) \geq 2x\,\psi(1/2),\ x \in [0,1/2] \quad \text{and} \quad \psi(1-x) \geq 2(1-x)\,\psi(1/2),\ x \in [1/2,1]$$

Since ψ is symmetric the above relations imply

$$\psi(x) \geq 2 \min\{x, 1-x\}\ \psi(1/2), \quad x \in [0,1]$$

or

$$\psi(x)/\psi(1/2) \geq \min\{x, 1-x\} / (1/2). \qquad (3.12)$$

Let us now consider for any entropy measure h of this class the functional h_o obtained from h by normalizing in any point x the value of the elementary entropy $w(x)\,\psi(f(x))$ to its maximum value $w(x)\,\psi(1/2)$. One has that h_o is still an entropy of the class having for all $f \in \mathcal{L}(X)$

$$h_o(f) = \sum_{x \in X} \psi(f(x))/\psi(1/2) = \sum_{x \in X} \psi_o(f(x))$$

where ψ_o is the N-function defined for all $x \in [0,1]$ as $\psi_o(x) = \psi(x)/\psi(1/2)$. For any $f \in \mathcal{L}(X)$, $0 \leq h_o(f) \leq \text{Card}(X)$ and moreover from (3.12) and (3.10) we have the result that the entropies h_o have a *minimal element* given by u_o where for all $f \in \mathcal{L}(X)$, $u_o(f) = 2 \sum_{x \in X} \min\{f(x), 1-f(x)\} = 2\,P(f \wedge f')$.

Furthermore if for any entropy h we denote by $\bar{h}$ the *relative entropy* h/h_{max} one has that $\bar{h}$ is still an entropy of the class, the value of which is given for all $f \in \mathcal{L}(X)$ by $\sum_{x \in X} q(x)\,\psi_o(f(x))$ with $q(x) = w(x) / \sum_{y \in X} w(y)$. Thus it follows from (3.12) that all relative entropies $\bar{h}$ of the class having a same weight-distribution q, possess a minimal element given, for any $f \in \mathcal{L}(X)$, by $2 \sum_{x \in X} q(x) \min\{f(x), 1-f(x)\}$.

E. Some interesting classes of entropy measures have been recently introduced by Trillas and Riera [24]. For any given weight-function w, N-function ψ satisfying Eq. (3.6) and $f \in \mathcal{L}(X)$, the entropy depends on the values $\psi(f(x))$ in one of the following way:

$$\sum_{x \in X} \min\{w(x), \psi(f(x))\}, \quad \operatorname{Max}_{x \in X}\{w(x)\,\psi(f(x))\}, \quad \operatorname{Max}_{x \in X} \min\{w(x), \psi(f(x))\}.$$

One can easily prove that these functionals are entropies. In a similar way if ϕ is a N-function satisfying Eq. (3.5) one can prove that $\sum_{x \in X} \min\{w(x), \phi(f(x))\}$, $\operatorname{Max}_{x \in X}\{w(x)\,\phi(f(x))\}$ and $\operatorname{Max}_{x \in X} \min\{w(x), \phi(f(x))\}$ are energies. We shall call these energies and entropies *sum-min*, *max-product* and *max-min* respectively. The following general proposition [24] allows to characterize a general class of energies and entropies which includes the previous ones and also the sum-product energies and entropies.

Proposition 3.4 - Let $\otimes$ and θ be two binary operations in R such that θ is commutative and associative. Let us further suppose that for all $x, y, z \in R$

$$x \otimes y = 0 \text{ iff } x = 0 \text{ or } y = 0, \quad x\,\theta\,y = 0 \text{ iff } x = 0 \text{ and } y = 0$$

$$x \leq y \Rightarrow z \otimes x \leq z \otimes y \quad \text{and} \quad z\,\theta\,x \leq z\,\theta\,y.$$

If ϕ and ψ are an energy and an entropy N-function respectively, then the functionals defined for all $f \in \mathcal{L}(X)$ as $\theta_{x \in X}[w(x) \otimes \phi(f(x))]$, $\theta_{x \in X}[w(x) \otimes \psi(f(x))]$ are an energy and an entropy respectively.

Some further general properties of θ-entropies can be found in [24]. We only remark here that in the case of sum-min , max-product and max-min entropies and energies the requirements A and B are not in general satisfied even though one makes the hypothesis that ϕ is strictly increasing in $[0,1]$ and ψ strictly increasing in $[0,1/2]$ and strictly decreasing in $[1/2,1]$.

We shall consider now the case $\text{Card}(X) = +\infty$. Measures of energy and entropy in $\mathcal{L}(X)$ can be introduced in two different ways. One could think at first glance of defining as energy and entropy still an E-function and an H-function in $\mathcal{L}(X)$ respectively. However as we shall see it may occur that one must exclude as measures functionals which are natural extensions to the infinite case of the corresponding finite functionals. For this reason in these cases it is perhaps more meaningful to make some natural changes in the definitions of E-functions and H-functions.

Let us first analyze the case when X is an *infinite denumerable set* and consider the sum-product functionals e and h of the kind:

$$e(f) = \sum_{x \in X} w(x)\, \phi(f(x)) \,, \quad h(f) = \sum_{x \in X} w(x)\, \psi(f(x)) \,, f \in \mathcal{L}(X) \tag{3.13}$$

where ϕ and ψ are an energy and an entropy N-functions respectively. If one makes the hypothesis $\sum w(x) < +\infty$ then e and h are *convergent series* for all $f \in \mathcal{L}(X)$ and satisfy the requirements of an E-function and an H-function respectively. This is also the case of functionals as:

$$\sup_{x \in X} \min \{w(x), \phi(f(x))\} \,, \quad \sup_{x \in X} \min \{w(x), \psi(f(x))\} \,.$$

However in the case $\sum w(x) = +\infty$,as for instance when $w = \text{const.}$, sum-product functionals (3.13) are not always convergent. In other words they are only partially defined in $\mathcal{L}(X)$. This difficulty can be formally avoided by assuming as energies and entropies E-functions and H-functions which are maps of $\mathcal{L}(X)$ to $R + \{+\infty\}$ instead of $\mathcal{L}(X)$ to R.

Let us remark that if ψ is a symmetric and continuous concave function in $[0,1]$ the convergence of $h(f) = \sum \psi(f(x))$ implies that $P(f \wedge f') < +\infty$. A necessary condition of convergence for entropies of this class which is also sufficient in the case of logarithmic entropy, can be found in [11] .

Let us consider now the more general case when X is an infinite nondenumerable set. Following Knopfmacher [17] we suppose that a *totally-finite positive measure* w is defined in a σ-algebra $\mathcal{A}$ of subsets of X . Let us denote by $\mathcal{M}(X)$ the subclass of $\mathcal{L}(X)$ formed by *all* fuzzy sets in X which are *measurable functions* . A natural extension of functionals e and h defined by (3.4) and (3.13) is given by:

$$e(f) = \int \phi(f(x))\, dw(x) \quad , \qquad h(f) = \int \psi(f(x))\, dw(x) \tag{3.14}$$

where $f \in \mathcal{M}(X)$ and ϕ , ψ are an energy and an entropy N-functions respectively. Let us observe that:

1. e and h are defined only on measurable fuzzy sets.
2. For all $f,g \in \mathcal{M}(X)$, $f \leq g \Rightarrow e(f) \leq e(g)$ and $f \leq' g \Rightarrow h(f) \leq h(g)$.
3. $e(f) = 0$ if and only if f is equivalent to *0* ,i.e. f vanishes almost everywhere.
4. $h(f) = 0$ if and only if f is equivalent to a classical characteristic function.

Moreover if one supposes that ϕ is strictly increasing in the interval $[0,1]$ and ψ strictly increasing in $[0,1/2]$ and strictly decreasing in $[1/2,1]$ then $e(f)$ reaches its maximum value only if f is equivalent to *1* and $h(f)$ reaches its maximum only if f is equivalent to *1/2*.

Thus if one assumes that functionals e and h as defined by Eq. (3.14) give energy and entropy measures then one has to modify the definition of E-function and H-function accorging to the above points 1,2,3 and 4.

Other extensions to the infinite case of entropies measures have been done by Trillas and Batle [25] for max-min entropies by using the Sugeno fuzzy integral [23].

In conclusion we recall that a concept of entropy can be introduced also for L-fuzzy sets [10]. The entropy in this case is a *vector* instead of a scalar quantity. Similarly also a vector energy can be defined. We shall not however consider here this subject in detail.

4. INTERPRETATION OF THE ENTROPY AND ENERGY OF A FUZZY SET.

The main aim of this section is to examine some interpretative problems related to the measures introduced in the previous section. Some of the things that we shall discuss here have been partly developed in some of our previous papers; we shall repeat here all that is needed for a clear and self-contained discussion.

Let us begin with the concept of entropy of a fuzzy set. We recall that the classical concept of *entropy* as well as the concept of *information* appeared for the first time in Physics and it was Boltzmann who observed that, in the setting of statistical thermodynamics, the entropy is proportional to the logarithm of the number of microscopical states which are compatible with the given macroscopical state. The entropy then measures the quantity of information that is missing while knowing only the macroscopic description and not the detailed microscopic one.

The second very important field in which the notion of entropy, assumed as a measure of missing information, played a very fundamental role is Shannon's information theory. In both fields, namely statistical thermodynamics and information theory the entropy is strictly related to the theory of probability which has to be interpreted, in order to be useful to model the given situations in a strictly frequentistic sense. We have then on one side a set of alternatives each of which is a very definite *crisp* one and on the other an incomplete description which does not allow one to pick up with certainty just one single alternative.

More precisely if X denotes a finite scheme of events $E_1, E_2, \ldots, E_n$ which may occur with probabilities $p_1, p_2, \ldots, p_n$ ($\sum_{i=1}^{n} p_i = 1$) the entropy H(X) of X is given by

$$H(X) = - \sum_{i=1}^{n} p_i \ln p_i \quad .$$

This quantity measures the *(average) amount of information that has to be supplied in order to have no uncertainty in the prevision of the event which will occur.*

It is well known that many objections have been raised against the identification of *information* in this frequentistic sense with the intuitive notion and some different proposals have been made by many people in order to give different formal explicata of the intuitive notion of information. Here we limit ourselves to remembering the founders of two proposals, Bar-Hillel and Carnap [2] for semantic information and Kolmogorov [19] and Chaitin [6] for structural information related to the computational complexity of a string of symbols.

What we wish to do here instead is not to depart strongly from the usual formal approach of statistical thermodynamics and information theory as is done in these last approaches but, without changing the mathematical approach from the beginning, modify the point of view in the description. If we start from a given level of description of the reality, which is not a detailed, crisp one (which in our case is obtained by means of the fuzzy sets) it is possible to obtain some "information" by means of some global measures or a suitable combination of them ? In other words, if a certain level of description of the reality is not a crisp one, is the only way for obtaining a deeper comprehension of it the one which retraces the origin of the vagueness and imprecision in order to cancel it, at least epistemologically or, on the contrary, is it perfectly legitimate to accept vagueness, in certain cases, as strictly inherent and thus ineliminable if one wants to describe the reality (the systems) without too strong simplifications ? If one answers affermatively to this last question, as we do, it becomes natural to think of passing to an upper level in which some global quantitative measures allow a study of the presence of this intrinsic vagueness in our description. In this sense then our measures of fuzziness can be interpreted as measuring the degree of intrinsic vagueness that belongs to our description; they have nothing to do with the missing information for choosing among some (crisp) alternatives but they say how much the fuzzy sets depart from a classical characteristic function (and from the "crisp" predicates that correspond to an *idealization* of the vague predicate which is formally represented by a certain fuzzy set).

These measures of fuzziness can be interpreted, however, as measures of information in a different sense, namely the following. It often happens that we have to simplify for some purposes the complexity of things as they appear; in other words we are forced to take decisions putting into some categories things that naturally do not perfectly fit them; in making these transformations we change fuzzy sets into other fuzzy (or, eventually, crisp) sets. The previous modification of the fuzziness which has to be measured is a quantity of information which we lost in the transformation. The entropy of a fuzzy set then measures *the total amount of information missing in a fuzzy pattern and which is that that one needs in order to have no uncertainty in the classification of the objects of a given universe*. In the next section we shall analyze in more detail this informational interpretation of the entropy of a fuzzy set with respect to decision processes.

In the problem of taking decisions (which usually implies a drastic reduction of all the nuances that some objects or predicates can possess in order to put them in a certain preassigned number of classes with certain features) or anyway in many real cases we are confronted with situations in which the vagueness modelled by the theory of fuzzy sets is mixed with uncertainty of a real statistical nature. Since the theory of probability gives us a developed calculus of the uncertainty of a statistical type it is important to have some quantity which gives at the same time a measure of the uncertainty due both to the probability and to the fuzziness and which reduces to the probabilistic entropy (*resp.* fuzzy entropy) when the fuzziness (*resp.* uncertainty of probabilistic type) is not present.

The first author to introduce a "probability measure of a fuzzy event" was L.A. Zadeh [27] ; however while his proposal of an entropy measure captures very well the idea of a measure of probability that is smoothed by the fact that we have the supplementary information that the event we are considering is a fuzzy one it has the formal disadvantage that it does not reduce to the classical (non fuzzy) expression when the event in non fuzzy.

In [8] we introduced in a heuristic way a probabilistic weighting $\sum_{x \in X} p(x) S(f(x))$ of the logarithmic entropy $d(f)$. Here we want to propose a more detailed discussion on the previous problem and on the interpretation of the quantity $\Sigma\ p(x)\ S(f(x))$ which is still an entropy measure of f . The main idea underlying any measure of fuzziness is that there is a certain symmetry between belonging , and not belonging, i.e. between the extreme values 0 and 1 that a membership function f can take. This fact is well illustrated by the second requirement of our axiom scheme that implies that the maximum of the entropy is reached when $f = 1/2$ and is the main reason for considering $d(f) = H(f) + H(f')$ (where $H(f) = \Sigma\ f(x) \ln 1/f(x)$) and not $H(f)$ as a measure of fuzziness. That is, in measuring the fuzziness we want to start from the given alternatives which are admitted by the given fuzzy set of which we want to evaluate the fuzziness and among which we could choose a decision which will transform the fuzzy set into a crisp set. (A stronger symmetry is found when, as in the case of d, the entropy is required to be symmetric) . If we have a probabilistic scheme superimposed on the intrinsic fuzziness of the description we have to take into account the effect of the probability not only on f which is the formal explicatum of a certain vague predicate *P* but also on f' which stands for the vague predicate *not P* . So, referring to the commonest entropy measure that has been considered, namely the logarithmic entropy, one has that a measure of both probabilistic and fuzzy aspects of f given a probability distribution $p : X \to (0,1]$ ($\sum_{x \in X} p(x) = 1$), is

$$H_{tot} = -\sum_{x \in X} [\, f(x)\, p(x) \ln (f(x)\, p(x)) + f'(x) p(x) \ln (f'(x)\, p(x)) \,] =$$

$$= -\sum_{x \in X} p(x) \ln p(x) + \sum_{x \in X} p(x)\, S(f(x)) = H(X) + \sum_{x \in X} p(x)\, S(f(x)) .$$

This formula has the interesting feature that when f is a classical characteristic function, i.e. when the fuzziness disappears, it reduces to the classical entropy of a probabilistic scheme. Let us note, however, that any measure of the kind

$$H(X) + \sum_{x \in X} p(x)\ \psi\ (f(x))\ ,\ \text{where}\ \sum_{x \in X} p(x)\ \psi\ (f(x))\ \text{is a sum-product}$$

entropy of f with a weight function $w = p$, satisfies the same property.

In an analogous way if one wants to consider the total uncertainty of fuzzy origin related to the contemporary consideration of two fuzzy sets f and g one has first to list the various alternatives arising from considering both f and g at the same time. If f is the explicatum for the vague predicate *P* and g for the vague predicate *Q* one has then four alternatives $P \wedge Q$, $P \wedge \text{not } Q$, $\text{not } P \wedge Q$, $\text{not } P \wedge \text{not } Q$.

Let us assume a measurement of the previous properties by means of the fuzzy sets fg, fg', $f'g$ and $(fg)'$ respectively. These measures satisfy the condition

$$(fg)(x) + (fg')(x) + (f'g)(x) + (fg)'(x) = 1$$

for all $x \in X$ *(orthogonality condition of properties)*. Thus according to the theory developed in [5] for measures of uncertainty in the case of more (orthogonal) properties, *the total amount of uncertainty* is given (by using the logarithmic measure) by:

$$H(fg) + H(fg') + H(f'g) + H((fg)') = d(f) + d(g)\ .$$

Thus, in this case, *the total amount of uncertainty equals the sum of the entropies of f and g*.

For what concerns the concept of energy of a fuzzy set we recall that some functionals satisfying the axioms of an energy have been introduced in [8] and [5] . Intuitively the energy is a *measure of the amount of membership* to a fuzzy set. This measure can concern either the total amount of membership as in the case of the power of a fuzzy set or the average membership as in the case of the measure $\Sigma\ p(x)\ f(x)$, where p is a probability distribution over X . This last quantity in the case of a continuous support and for a measurable fuzzy set f , reduces to

$$\int f(x)\ dw(x)\ ,$$

(where w is a probability measure) that coincides with the so-called *probability of the fuzzy event* described by f (see, Zadeh [27]). The energy is then both formally and conceptually a quantity independent of the entropy even though, as shown by proposition 3.1 , one can construct an entropy measure starting from an energy, and vice versa.

The importance of the use of both these quantities in decision theory has been stressed in [5] . Let us give here the following simple example. Suppose one performs independent statistical decisions on a fuzzy set f assuming that at any point x the probability p(x) of obtaining the answer 1 is given by f(x) (this will be the case if one refers, for instance, to the class of all coherent decisions of Sec. 5). One can introduce at any point $x \in X$ a random variable f^* assuming the values 1 and 0 with probabilities p(x) and 1- p(x) . The quantity $\Sigma\ f^*(x)$ is then a random variable whose average value is P(f) and whose variance is equal to $\sum_{x \in X} f(x)(1 - f(x)) = \sigma(f)$. Thus in this case it is possible to give a statistical interpretation to the power P(f) and to the entropy σ (f) in terms of the two basic statistical quantities : *average value and variance*.

Further a correspondence between the energy concept of a fuzzy set and the thermodynamic one is given by the property (3.11) of sum-product entropies which are concave functions. In this case, as we have seen, the maximum of the entropy over all fuzzy sets having a constant energy is reached when this is equidistributed. Some further analogies can be found in [5] . We stress here that the possibility of the introduction of energy and entropy measures is a definite advantage of the theory of fuzzy sets for the modelling of incomplete (or inexact) information. We have to take seriously the initial intuition of a certain similarity with thermodynamics and consider it as a kind of paradigm for the study of all those systems whose available knowledge is not the best possible (from a classical point of view).

The real success of thermodynamics lies in the fact that it is based on two *independent* macroscopic quantities which allow the construction of a rich calculus in which we have some other (dependent) quantities which assume particular significance according to the context (the problem that is studied, the boundary conditions, the transformations carried in the system and so on). The main strong limitation of the Shannon Theory of Information is that it has not been possible to introduce some other (independent) functionals besides the entropy. Some efforts have been made in this direction; but as far as we know these have not been developed so as to produce a proper extension of the theory. Here we recall the proposal of Onicescu [22] of considering besides the entropy $H(X)$ the quantity

$$\sum_{x \varepsilon X} p^2(x) \quad ,$$

called *informational energy*. Dumitrescu [13] has proposed considering a counterpart of it in the setting of fuzzy set theory by interpreting it as a measure of certainty. We agree with the final observation of Onicescu [22] that *"L'énergie informationelle peut servir, aussi bien que l'entropie, comme fondement d'une Théorie de l'Information"*, however we think that the ultimate usefulness of measures of information, as well as of fuzziness is either in the fact that they allow the proof of deep non trivial theorems, such as Shannon's (which are the real justification of the importance of the entropy concept in the theory of information) or that they allow one (at least) to outline a mathematical structure with more than one independent quantity by means of which to construct a theory at a global level (of a macroscopical description).

A final discussion that we want to make concerns the requirements that have to be satisfied by the entropy and energy measures and the corresponding subclasses of measures introduced in the previous section. First of all we should say that *the properties which are required to a measure of fuzziness, differently from the probabilistic case, do not uniquely determine the entropy*. The particular class of the functional h and even the particular form of it are strictly related to the context and use of the measure itself. We note that, also in the probabilistic setting, measures of uncertainty different from the one of Shannon have been successfully used. The same can be said for the energy measures. Moreover one can consider problems for which more measures of energy have to be taken into account: for instance if one wants to maximize the entropy keeping constant the power $\Sigma\ f(x)$ and an energy like $\Sigma\ w(x)\ f(x)$ (cf. [5]).

We now briefly discuss some of the previous mathematical assumptions defining the requirements themselves. As regard to the general axioms defining entropy and energy measures, as we said in the previous section, in our original introduction of measures of fuzziness [8] the condition that the entropy reaches its maximum value only if $f = 1/2$ has been imposed. This condition can be further reinforced by requiring the strict isotony of the entropy with respect to the order $\leq'$. However in both cases some interesting subclasses of entropies should be excluded as, for instance, the max-min, sum-min, max-product entropies of Trillas and Riera which, in general, do not verify the previous conditions. This is one of the main reasons for considering the more weak and general formulation of the axioms.

Another general requirement on an entropy measure h is the *symmetry*, i.e. $h = h'$. We have already previously discussed the basic symmetry existing between "belonging" and "not belonging" in the measures of fuzziness. The stronger condition of symmetry has to be required any time one wants that any fuzzy set f is completely indistinguishable from the complement f' by making a measure of entropy. This does not imply, of course, that they cannot be distinguished by a measure of energy. Let us note however, that one can immagine cases in which a vague property and its negation are not in the same setting so that it could be useful to have a different evaluation for the values of the entropies $h(f)$ and $h(f')$.

For what concerns the subclass of sum-product measures we observe that the implicit assumption is the *additivity* of the measures, i.e. the measure of energy or of entropy is the sum of *elementary contributions* due to the single points of the universe X. The additivity is a strong requirement that models the real situations only in some cases , namely when a decision performed on the elements of X does not influence the next ones *(independence)* . This can be, often, a useful approximation of some more complex situations.

The *continuity* of a measure is the mathematical traduction of the requirement that slight variations in the values of the fuzzy set produce slight variations in the entropy and energy measures. This is a quite natural assumption for sum-product entropies and energies; it would appear strange to impose this kind of condition for other classes of measures as, for instance, the max-min entropies. In some case one could be interested in requiring that the entropy measure produces a sort of automatic classification so that it is more useful to have discontinuous functionals which in critical places react to a small variation of the values of the fuzzy set with a jump in the value that they assume.

5. INFORMATIONAL INTERPRETATION OF THE ENTROPY OF A FUZZY SET WITH RESPECT TO DECISION PROCESSES.

In this last section we shall outline some interpretative and formal problems raised by an analysis of *decision processes* in the setting of fuzzy sets theory. We shall find some interesting connection among some *unforseeability* measures of the decisions and the entropy measures discussed in the previous sections. It turns out that the measures considered here include as a particular case a measure of uncertainty in decision-making introduced by Backer [1] and that some properties of this last measure can be extended to the general case. The first part of this section is an extension of section 3 of our paper [12] of which we continue to share the general conceptual setting that will be briefly outlined here.

A main general distinction that can be made in studying the decision processes concerns the way in which one looks at the *prototypes* according to which the decisions are taken and a certain classification is obtained . According to the first choice the *prototypes* , or *paradigms* do exist independently of the process we are considering and also of the decisions that we shall take and the final classification obtained. This kind of decision was called in [12] a process with *fixed paradigms* . According to these rules if certain objects do not share totally the meaningful ones of the abstract prototypes they are forced in one of them. At the end of the decision process one has then transformed any fuzzy set into a classical characteristic function modifying then the membership values and also, in general, the power or the energy of the fuzzy set. On the contrary we have that the paradigms remain untouched by the various decisions and as a consequence of this each decision is independent of the other ones since it refers each time only to the abstract paradigm.

One could however think of different decision techniques in which the various decisions are correlated and the classification obtained correspondently is strictly related to the universe of objects at disposition , i.e. the fuzzy sets we have to classify. One could take into account these qualitative requirements by considering the paradigms of the classification not as platonically existing and then given *a priori* and unchangeable but somethings that will change during the same process of decision and which will be exactly defined only after the process has been ultimated.

An informational interpretation of the entropy of a fuzzy set in the setting of this second type of decision process appears complex even if we think that this framework is more akin to the philosophy underlying fuzzy sets theory.

The informational interpretation of the entropy of a fuzzy set given in the preceding section fits well the first kind of decision process; the entropy in fact measures the departures of a fuzzy set from a classical characteristic function and so it evaluates the total amount of information missing in a fuzzy pattern (with respect to an ideal crisp one) and so also the one needed in order to classify without uncertainty the objects of the considered universe.

Formally, a decision D over $\mathcal{L}(X)$ can be defined as any map $D: X \times I \rightarrow \{0,1\}$ where the values 0 and 1 of the range of D refer to one of the two properties which has to be attributed to the elements of X. Furthermore it is natural to add the condition that $D(x,0) = 0$ and that $D(x,1) = 1$ for all $x \varepsilon X$.

We shall show now that it is possible to give to the entropy of a fuzzy set an interpretation *similar* to that of probabilistic information by making suitable hypotheses on the decisions. We shall precisely refer to the subclass of the *coherent* decisions which are defined as:

Definition 5.1 - A decision D is coherent if $D(x,a) = 1 \Rightarrow D(x,b) = 1$ for all $b > a$ and $x \varepsilon X$.

Proposition 5.1 - A decision D is coherent if and only if

$$D(x,a) = \theta\,[\,a - t(x)\,] \quad , x \varepsilon X \ , a \varepsilon I \tag{5.1}$$

where θ is the step function $\theta(x) = 1$ if $x \geq 0$, $\theta(x) = 0$ if $x < 0$, and t is a threshold map $t : X \to (0,1]$.

Proof - It is clear that a decision D defined by Eq. (5.1) is coherent. On the contrary if $D(x,a)$ is a coherent decision let us set $t(x) = \min\{a \mid D(x,a) = 1\}$ for all $x \varepsilon X$. Since $D(x,1) = 1$ and $D(x,0) = 0$ for all $x \varepsilon X$ one has that $0 < t(x) \leq 1$ and $D(x,a) = \theta\,[\,a - t(x)\,]$ □

Let us explicitly note that a threshold map is a fuzzy set defined in X and that there is a one-to-one correspondence between coherent decisions and threshold maps.

Let h be a sum-product entropy

$$h(f) = \sum_{x \varepsilon X} w(x)\ \psi(f(x)) \ , \quad f \varepsilon \mathcal{L}(X),$$

where ψ is a continuous and symmetric concave N-function satisfying Eq. (3.6). Since ψ is symmetric we can write:

$$h(f) = 1/2 \sum_{x \varepsilon X} w(x)[\psi(f(x)) + \psi(1 - f(x))] \ .$$

Moreover ψ can be always written as $\psi(x) = x\,L(1/x)$ for all $x \varepsilon (0,1]$ where L is a nondecreasing function in the interval $[1,+\infty)$ such that $L(x) = x\ \psi(1/x)$ for $x \varepsilon [1,+\infty)$, $L(1) = \lim_{x \to 0} x\,L(1/x) = 0$. As ψ is symmetric it follows that if $x \leq 1/2$, $L(1/(1-x)) \leq L(1/x)$ and if $x \geq 1/2$, $L(1/(1-x)) \geq L(1/x)$. Further if $x \neq 0$ and $L(1/x) = 0$ then $x = 1$.

The entropy $h(f)$ can be written as:

$$h(f) = 1/2 \sum_{x \varepsilon X} w(x)\ [\,f(x)\,L(1/f(x)) + (1 - f(x))\,L(1/(1-f(x)))\,] \tag{5.2}$$

For any coherent decision D whose threshold map is t let us introduce the functional $I_t : \mathcal{L}(X) \to R$ defined for all $f \varepsilon \mathcal{L}(X)$ as:

$$I_t(f) = 1/2 \sum_{x \varepsilon X} w(x)\{\theta(f(x) - t(x))\,L(1/f(x)) + [\,1 - \theta(f(x) - t(x))]\,L(1/(1-f(x)))\} \tag{5.3}$$

where, conventionally $0\,L(1/0) = 0$.

Let us now list some properties of the functional I_t.

1. $I_t(f) = 0$ if and only if f is a classical characteristic function.

Proof - It is clear from the definition that if f is a classical characteristic function then $I_t(f) = 0$. On the contrary if $I_t(f) = 0$ then for any $x \varepsilon X$ either $f(x) \geq t(x)$ and then $L(1/f(x)) = 0$ or $f(x) < t(x)$ and then $L(1/(1-f(x))) = 0$. In the first case $f(x)$ has to be equal to 1 and in the second $f(x) = 0$ □

2. For all t, $I_t(1/2) = h(1/2)$.

Proof - From the definition (5.3), $I_t(1/2) = 1/2 \sum w(x) L(2) = h(1/2)$ □

For any fuzzy set f and threshold map t let us denote by $q(f,t;x)$ the quantity:

$$q(f,t;x) = \theta[f(x) - t(x)] L(1/f(x)) + [1-\theta(f(x)-t(x))] L(1/(1-f(x))) \tag{5.4}$$

for all $x \in X$.

3. If t and t' are two threshold-maps such that $t' \leq' t$ then $I_{t'}(f) \geq I_t(f)$ for all $f \in \mathcal{L}(X)$.

Proof - If $t' \leq' t$ then for all $x \in X$ either $t'(x) \leq t(x) \leq 1/2$ or $t'(x) \geq t(x) \geq 1/2$. Let x be an element of X and suppose that $t'(x) \leq t(x) \leq 1/2$. If $f(x) < t'(x) \leq t(x) \leq 1/2$ then $q(f,t';x) = q(f,t;x) = L(1/(1-f(x)))$. Similarly if $t'(x) \leq t(x) \leq 1/2 \leq f(x)$ then $q(f,t;x) = q(f,t';x) = L(1/f(x))$. Let us now suppose $1/2 \geq f(x) \geq t'(x)$. This implies $q(f,t';x) = L(1/f(x))$. Now if $f(x) \geq t(x)$ then $q(f,t';x) = q(f,t;x)$. If, on the contrary, $f(x) < t(x)$ then $q(f,t;x) = L(1/(1-f(x)))$. Since $f(x) \leq 1/2$ it follows $L(1/(1-f(x))) \geq L(1/f(x))$ and $q(f,t';x) \geq q(f,t;x)$. Thus in all cases $q(f,t';x) \geq q(f,t;x)$ that implies

$$I_{t'}(f) = 1/2 \sum_{x \in X} w(x)\, q(f,t';x) \geq 1/2 \sum_{x \in X} w(x)\, q(f,t;x) = I_t(f).$$

The case $t'(x) \geq t(x) \geq 1/2$ can be dealt symmetrically □

Since any threshold-map t is such that $t \leq' 1/2$ one derives the following corollary

4. For any threshold-map t one has $I_t(f) \geq I_{1/2}(f)$, for all $f \in \mathcal{L}(X)$.

5. For all $f \in \mathcal{L}(X)$ the average value $\langle I_t(f) \rangle$ of $I_t(f)$ over all coherent decisions equals the entropy $h(f)$, i.e. $\langle I_t(f) \rangle = h(f)$.

Proof - From Eq.s (5.2) and (5.3) it suffices to prove that for any $x \in X$ the average value of $\theta(f(x) - t(x))$ when $t(x)$ varies in the interval $(0,1]$ is just equal to $f(x)$. In fact

$$\langle \theta(f(x) - t(x)) \rangle = \int_0^1 \theta(f(x) - t)\, dt = \int_0^{f(x)} dt = f(x) \qquad \square$$

Let us now refer to the entropy measure σ defined by Eq. (3.9) as

$$\sigma(f) = \sum f(x)(1-f(x)) \quad \text{for all } f \in \mathcal{L}(X).$$

In this case $L(1/x) = 1-x$ and $I_t(f) = \sum \{ \theta(f(x) - t(x))(1-f(x)) +$

$[1 - \theta(f(x) - t(x))] f(x) \} = \sum | f(x) - \theta(f(x) - t(x)) |$.

The quantity $I_t(f) = \sum | f(x) - \theta(f(x) - t(x)) |$, in the case of a constant threshold map t, has been introduced by Backer [1] as a measure of the uncertainty in taking decisions corresponding to the threshold t on the fuzzy set f.

We want, in conclusion, make some comments on the interpretation of the functional $I_t(f)$ defined by Eq. (5.3) and on the results which have been obtained. If we restrict ourselves to considering coherent decisions only, the quantity $I_t(f)$ can be seen as a measure of the *unforseeability* of the decision D corresponding to the threshold-map t in the sense that $I_t(f)$ assumes high values on the decisions for which $f(x) \backsim 1$ and $t(x) > f(x)$ or $f(x) \backsim 0$ and $t(x) \leq f(x)$. On the contrary $I_t(f)$ is minimal when t is constantly equal to 1/2. We have moreover seen that the membership value $f(x)$ can be interpreted at any point $x \in X$ as the *frequency* by which a coherent decision will give the answer 1. The interpretation of the entropy measure of a fuzzy set as a measure of (missing) information in order to take a decision is then quite similar to the one of the probabilistic entropy in the case of a probabilistic scheme. Let us stress that nevertheless the strong analogy a main conceptual difference subsists. In the case of the frequentistic conception of probability the occurrence of an event is something independent of the context, while in the previous decision scheme this notion has been substituted with something strictly related to the decision criteria, i.e. to the context.

REFERENCES

1 .E. Backer, A non statistical type of uncertainty in fuzzy events, in Topics in Information Theory (Second Colloq. Keszthely, Hungary, 1975). Colloquia Mathematica Societatis Janos Bolyai vol.16 pp.53-73, North-Holland, Amsterdam (1977).

2 .Y. Bar-Hillel and R. Carnap, An outline of a theory of semantic information, in "Language and Information", Chapter 15 ,pp.221-274, Addison Wesley (1964)

3 .M.Black, Vagueness, Phil.of Science, 4 ,427-455 (1937)

4 . M.Black, Reasoning with loose concepts, Dialogue,2 ,325-337 (1963)

5 .R.Capocelli and A. de Luca, Fuzzy sets and decision theory, Information and Control, 23 , 446-473 (1973)

6 .G.J. Chaitin, On the length of programs for computing finite binary sequences, A C M J. 13,547-569 (1966)

7 .A.de Luca and S.Termini, Algorithmic aspects in complex systems analysis, Scientia, 106 , 659-671 (1971)

8 .A.de Luca and S.Termini, A definition of a nonprobabilistic entropy in the setting of fuzzy sets theory, Information and Control ,20 , 301-312 (1972)

9 .A.de Luca and S.Termini, Algebraic properties of fuzzy sets, J.of Math.Anal.Appl. 40 , 373-386 (1972)

10 .A.de Luca and S.Termini, Entropy of L-fuzzy sets. Information and Control, 24 ,55-73 (1974)

11 .A.de Luca and S.Termini, On the convergence of entropy measures of a fuzzy set, Kybernetes, 6 , 219-227 (1977)

12 .A.de Luca and S.Termini, Measures of ambiguity in the analysis of complex systems ,Proc.6-th Symposium of MFCS, Lectures Notes in Computer Science vol.53, pp.382-389, Springer Verlag (1977)

13 .D. Dumitrescu, A definition of an informational energy in fuzzy sets theory, Studia Univ. Babes-Bolyai ,Mathematica ,2 ,57-59 (1977)

14 .R.Giles, Lukasiewicz logic and fuzzy sets theory, International J.Man-Machine Studies, 8 ,313-327 (1976)

15 .J.A.Goguen, L-fuzzy sets, J.of Math.Anal.Appl., 18 , 145-174 (1967)

16 .J. A.Goguen, The logic of inexact concepts, Synthese, 19 ,325-373 (1969)

17 .J.Knopfmacher, On measures of fuzziness, J.Math.Anal.Appl.,49, 529-534 (1975)

18 .M.Kokawa, K.Nakamura and M.Oda, Experimental approach to fuzzy simulation of memorizing, forgetting and inference process, in "Fuzzy sets and their applications to cognitive and decision processes" (Zadeh, L., Fu, K., Tanaka, K .and Shimura, M. eds.) pp.409-428. Academic Press, Inc. (1975)

19 .A.N.Kolmogorov, Three approaches to the quantitative definition of information, Problemy Peredachi Informatsii, 1 ,3-11 (1965) (russian) translated in International Journal of Computer Mathematics , 2 , 157-168 (1968)

20 .H.Nurmi, Probability and fuzziness, Sixth Research Conference on Subjective Probability, Utility and Decision Making, Warszawa (1977)

21 .H.Nurmi, Modelling impreciseness in human systems, IV European Meeting on Cybernetics and Systems Research, Linz (1978)

22 .O.Onicescu, Energie informationelle, C.R. Acad.Sc.Paris,263,841-842 (1966)

23 .M Sugeno,Theory of fuzzy integrals and its applications,Ph.D.Thesis,Tokyo Institute of Technology (1974)

24 .E.Trillas and T.Riera,Entropies in finite fuzzy sets,Information Sciences, 15 , 159-168 (1978)

25 .E.Trillas and N.Batle,Sobre la integral de Sugeno,Preprint Departamento de Matematicas,Esquela Tecnica Superior de Arquitectura,Universidad Politecnica de Barcellona (1977)

26 .L.A. Zadeh,Fuzzy sets,Information and Control, 8 ,338-353 (1965)

27 .L.A. Zadeh,Probability measures of fuzzy events, J.Math.Anal.Appl.,23 , 421-427 (1968)

ADVANCES IN FUZZY SET THEORY AND APPLICATIONS
M.M. Gupta, R.K. Ragade, R.R. Yager (editors)

SOLUTION CONCEPTS FOR n-PERSONS FUZZY GAMES

Dan BUTNARIU

Str. Tepeş-Vodă nr.2, bloc V1, scara A, etaj 4, apart. 1

Iaşi-6600,Romania

The problem of finding the best decisional alternative in a given political, economical,... conjuncture can be approached in many ways. One of them consists in modeling the real conjuncture as a game with rules which permit to the "players" to exchange fuzzy informations between them. What means, from the mathematical point of view, the best decisional alternative in such a game ? Now, we are going to discuss several answers for that question. Precisely, we are going to define several concepts of "solution" for n-persond fuzzy games and to prove their consistency.

§0, INTRODUCTION

The present paper is a mathematical approach to the problem of the rational behaviour in n-persons fuzzy games. Since the paper [5] consists in a heuristical discussion on "fuzzy games" and a comparative analysis between fuzzy and non-fuzzy games is there explained, the present paper is dedicated to show different manners to answer the following question:"What is the best possible behaviour of a rational player which is a participant in a n-persons fuzzy game ?".

To answer this question, we remind that a n-persons fuzzy game is defined in [5] to be a set of rules which determine the possible exchange of informations between the participants in the game. So, the enounced question can be formulated in terms of regular exchanges of informations: "What is the best regular exchange of informations in a given n-persons fuzzy game ?". It is clear that the answer to this fundamental question can be given in many ways because it is dependent upon the nature of the rules and the nature of the information which can be exchanged in a dpecific game, but it is also dependent upon the individual and subjective thought of any rational person participant in the game concerning these rules and informations.

However, a clssification of these ways can be made. It is based on the essential distinction which exists between the non-cooperative and the cooperative games, where -- similar as in the classical Theory of Games [16] -- characteristic for non-cooperative fuzzy games is the regular impossibility of transmitting the decisional rights from the individual players to representative decidents so as the (fuzzy) coalitions [1], [5]. If this regular impossiblity does not exists in a fuzzy game, then the game is called cooperative Further we shall explain separately the solution concepts for non-cooperative and cooperative fuzzy games.

In the explanation which we are going to make, we shall use the following notations: U will denote an universe (see [5]) and L will denote the real interval [0,1]; $\underline{N}$ and $\underline{R}$ will denote the sets of natural and real numbers respectively; A, B, A_i,... will denote fuzzy sets into U; if A is a fuzzy set (into U), then A(x) will denote the membership degree of $x \in U$ to A; all the fuzzy sets considered in this paper will be fuzzy sets into U; if A_i, $i \in J$ will be a family of fuzzy sets, then $\bigcup_{i \in J} A_i$ and $\bigcap_{i \in J} A_i$ will denote the union and the intersection of the given family respectively; the inclusion between fuzzy sets will be denoted by $\subset$ and L(A) will be the class of all fuzzy sets included in the fuzzy set A.

Now we are able to remind from [5, §1] several definitions which will be used later. Let A, B, C be fuzzy sets. The product of A with B is the fuzzy set defined by:

$$\begin{aligned}(A \times B)(x) &= A(y).B(z) \quad \text{if } x=(y,z) \in U,\\ &= 0 \quad \text{elsewhere.}\end{aligned}$$

A fuzzy relation between A and B is a fuzzy set from $L(A \times B)$. Let R be a fuzzy relation between A and B and S a fuzzy relation between B and C. Then SoR denote the fuzzy relation between A and C defined by:

$$\begin{aligned}(SoR)(z) &= \sup\{R(x,y).S(y,t);\ y \in U\} \quad \text{if } z=(x,t) \in U\\ &= 0 \quad \text{elsewhere.}\end{aligned}$$

If $X \in L(A)$, then the immage of X by R is the fuzzy set R[X] whose membership function is defined by:

$$R[X](y) = \sup\{X(x).R(x,Y);\ x \in U\} .$$

If $Y \in L(B)$, then the counter-immage of Y by R is the fuzzy set given by:

$$R^{-1}[Y](x)=\sup\{R(x,y).Y(y);\ y\in U\}.$$

There will not be notational differences between the subsets of U and their characteristic functions seen as fuzzy sets into U. So, $\emptyset$ will denote the empty subset of U and the identical 0 fuzzy set simoultaneous.[2]

PART ONE
NON-COOPERATIVE n-PERSONS FUZZY GAMES

§1. The mathematical description

The concept of non-cooperative n-persons fuzzy game with n=2 have been discussed first in [5, §2.5]. The conclusions of this discussion can be clearly extended for n-persons fuzzy games with $n\in \underline{N}$, $n\geq 3$. So, the following mathematical description is meaningful:

DEFINITION 1.1: The concept of non-cooperative n-persons fuzzy game can be formaly described as a set of data $G=(\sum_1, \sum_2,\dots,\sum_n; Y_1,\dots,Y_n; E_1,\dots,E_n)$, where $N= 1,2,\dots,n$ is the set of the players of G and for any k in N the following requirements are fulfilled:

(a) $\sum_k=\{\sigma_1^{(k)},\dots,\sigma_{n(k)}^{(k)}\}$ (with $n(k)\in \underline{N}$) is the set of the pure strategies of the player k in G.

(b) Y_k is an unambigous subset of $\underline{R}^{n(k)}$ whose elements are called strategic compositions of k in G; if $w^k=(w_1^k,\dots,w_{n(k)}^k)\in Y_k$, then w_i^k is called investment (or regular investment) of k for his pure strategy $\sigma_i^{(k)}$; a n-vector $w=(w^1,\dots,w^n)\in Z=\prod_{i\in N} Y_i$ is called strategical choice in G.

(c) $E_k\in L(Z)$ and for any $w=(w^1,\dots,w^n)\in Z$, $E_k(w)$ is the possibility degree of the strategical choice w from the point of view of k.

(d) $Z_k=\prod_{i\in N-\{k\}} Y_i$, $W_k=L(Z_k)\times Y_k$ and the pairs $s_k=(A_k,w^k)$ are called (regular) strategic conceptions of the player k in G. The next axiom holds:

<u>Axiom k</u>: If $A_k\in L(Z_k)$ and $A_k\neq\emptyset$, then $E_k[A_k]\neq\emptyset$, i. e. there exists $s_k=(A_k,w^k)$ in W_k so that $E_k[A_k](w^k)\neq 0$.[3]

The intuitive sense of the notions introduced by the Definition 1.1 have been discussed in [5]. The axiom k states that for any unvoid set of informations concerning the future behaviour of his partners in G, the player k can choose/can elaborate a regular strategic conception in the game to be used as a reply for the (incomplete known) future behaviour of the other players in the game.

DEFINITION 1.2: Let G be a n-persons non-cooperative fuzzy game as described in the previous definition. A <u>play</u> in G is a n-vector $s=(s_1,\dots,s_n)\in W=\prod_{k\in N} W_k$, i. e. a set of individual strategic conceptions which can be put into practice simoultaneously.

The problem of the best exchange of informations in G is equivalent with that of finding "the best play in G" because a play $s=(s_1,\dots$ $\dots, s_n)$ in G is, in fact, a mathematical description for an exchange of informations in G. In truth, the play s consisting in the strategic conceptions $s_i=(A_i,w^i)$, $(i\in N)$ shows (by the fuzzy sets A_i) what informations receive the player i concerning the future behaviour of the other players and (by the w^i's) it shows a possible set of individual decisions which can be determined by this informational conjuncture.[4]

Now we must say what a "best play in G" is from the mathematical point of view. To this end we introduce:

DEFINITION 1.3: Let $s_k=(A_k,w^k)$ and $s'=(A'_k,w'^k)$ be two strategic conceptions of k in G. We say that s_k is better than s'_k and we denote $s_k \succ s'_k$ iff $E_k[A_k](w^k) \geq E_k[A_k](w'^k)$.

In fact, a strategic conception s_k is better than the strategic conception s'_k iff the first is considered by k to be more possible than the other by vertue of the rules of the game.

DEFINITION 1.4: Let $s=(s_1,\dots,s_n)$ and $s'=(s'_1,\dots,s'_n)$ be two plays in G, where $s_i=(A_i,w^i)$ and $s'_i=(A'_i,w'^i)$ for any $i\in N$. We say that the play s is <u>preferable</u> to the play s' iff $s_k \succ s'_k$ for any $k\in N$.

What does this mean ? To answer, let us consider that the players of G have received some information one about the others. Based on this information, each player $k\in N$ can estimate the fuzzy set A_k *in* $L(Z_k)$ and can choose $w^k\in Y_k$ for defining his strategic conception $s_k=(A_k,w^k)$. The strategic conceptions s_k, $k\in N$, rise a play in G. Each component s_k of this play is more or less regular (i. e. more or less possible by vertue of the rules of the game); then it is natural to consider that rational players prefer the more regular (or the more possible or the more statutory ...) strategic conceptions and, consequently, the more regular plays.

Now let us consider that the fuzzy sets A^*_k, $k\in N$, have been defined by the players based on the informations exchanged between them. In this case, each player k N must decide which is the best strategic conception to be put into the practice in such an informational conjuncture. Precisely, they must decide which is the best strategic composition $w^k = w^k_*$ that can be choosen according to the received information. It is natural to consider that rational players will choose w^k_* ($k \in N$) so that $s^*_k = (A^*_k, w^k_*)$ ($k \in N$) form the most preferable play in G, play which can be obtained using the informations given by the A^*_k's. So, we can give the next:

DEFINITION 1.5: A <u>possible solution</u> of G is a play $s^* = (s^*_1, \ldots, s^*_n)$ with $s^*_k = (A^*_k, w^k_*)$, $k \in N$, so that for any play $s = (s_1, \ldots, s_n)$ with $s_k = (A^*_k, w^k)$, $k \in N$, the play s can not be preferable to s, i. e.

$$(1.1) \qquad E_k[A^*_k](w^k_*) \geqslant E_k[A^*_k](w^k), \qquad \forall\, w^k \in Y_k, \quad \forall\, k \in N.$$

Using the previous considerations, it seems to be natural to say that a possible solution of G is one of the best plays in G (from the viewpoint of the regularity, legality,...). Hence the Definition 1.5 shows what a best play in G means.

The rules of a large class of games seem to permit the players to collaborate one with the others without the formation of the coalitions (i. e. without the cooperation) being implicitely possible.[6] The collaboration between players must not be thought as a cooperation but as a special manner of exchanging information which have a maximal plausibility (i. e. information so that there is not doubt about it). It is intuitively clear that this manner of exchanging information does not imply the fact that the decisionso of the players which collaborate must be made by a representative decident so as their coalition. Mathematically, the collaboration of the players in a n-persons fuzzy game G consists in playing only plays of the form $s^* = (s^*_1, \ldots, s^*_n)$, where $s^*_k = (A^*_k, w^k_*)$, $k \in N$, with

$$(1.2) \qquad A^*_k(w^1, \ldots, w^{k-1}, w^{k+1}, \ldots, w^n) = 1 \quad \text{if } w^i = w^i_* \text{ for any } i \in N - \{k\},$$
$$= 0 \quad \text{elsewhere.}$$

DEFINITION 1.6: An <u>equilibrium point</u> of G is a possible solution $s^* = (s^*_1, \ldots, s^*_n)$, where $s^*_k = (A^*_k, w^k_*)$, $k \in N$, is verifying (1.2).

We want know how must a n-persons fuzzy game be in order to have a possible solution and, eventually, an equilibrium point. To this aim we dedicate the next section.

§2. Existence criteria for possible solutions of a n-persons fuzzy game

In this section we shall present two theorems which concern with the existence of the possible solutions for n-persons fuzzy games. These theorems reduce the problem of finding possible solutions for n-persons fuzzy games to fixed point problems for fuzzy relations.

DEFINITION 2.1: Let A be a fuzzy set and let R be in $L(A\times A)$. A <u>fixed point</u> for R is an element w^* U such that $R(w^*,w^*)\geqslant R(w^*,w)$ for any w in U.

In the next we consider G be a n-persons fuzzy game given as in the Definition 1.1. We remind that W is denoting the set of the plays in G.

DEFINITION 2.2: The <u>fuzzy individual preference</u> of the player k in G is the fuzzy relation $E_k^* \in L(W\times W)$ which is defined by:

$$(2.1)\qquad E_k^*(s,\bar{s})=E_k[A_k](\bar{w}^k)\vee E_k[A_k](w^k).\prod_{i\in N} E_i[\bar{A}_i](\bar{w}^i),$$

for any two plays $s=(s_1,\ldots,s_n)$, $\bar{s}=(\bar{s}_1,\ldots,\bar{s}_n)$ with $s_j=(A_j,w^j)$, $\bar{s}_j=(\bar{A}_j,\bar{w}^j)$, $j\in N$.[7]

The fuzzy relation E_k^* describes the measure of the preference of the player k for the play s comparative with $\bar{s}$.

DEFINITION 2.3: We call <u>fuzzy preference</u> in G the fuzzy relation $R\in L(W\times W)$ defined by:

$$(2.2)\qquad R(s,\bar{s})=\prod_{i\in N} E^*(s,\bar{s}),\quad \forall s,\ \bar{s}\in W.$$

$R(s,\bar{s})$ is the degree of the preferability of the play s comparative with $\bar{s}$ in G. Now we can give:

THEOREM 2.4: Let $s^*=(s_1^*,\ldots,s_n^*)$ be a play in G with $s_i^*=(A_i^*,w_*^i)$, i N. If $A_i^*\neq\emptyset$ for any i N, then the following assertions are equivalent:

(I) s^* is a possible solution for G;

(II) s^* is a fixed point for the fuzzy relation R, where R is the fuzzy preference in G.

<u>Proof: (I) implies (II).</u> Let i be in N. Then

$$(2.3)\qquad E_i^*(s^*,s^*)=E_i[A_i^*](w_*^i)\vee E_i^2[A_i^*](w_*^i).\prod_{j\in N-\{i\}} E_j[A_j^*](w_*^j)=E_i[A_i^*](w_*^i)$$

and for any $s=(s_1,\ldots,s_n)\in W$ with $s_j=(A_j,w)$, $j\in N$, we have:

$$(2.4)\qquad E_i^*(s^*,s)=E_i[A_i^*](w^i)\vee E_i[A_i^*](w_*^i).\prod_{j\in N} E_j[A_j](w^j).$$

Since s^* is a possible solution for G, the formula (1.1) holds. Hence we deduce:

$$(2.5)\qquad E_j[A_j^*](w_*^j)\geqslant E_j[A_j^*](w_*^j).\prod_{i\in N} E_i[A_i](w^i)$$

and, by (2.3), results:

$$(2.6)\qquad E_i^*(s^*,\bar{s})=E_i[A_i^*](w_*^i)\geqslant E_i[A_i^*](w_*^i).\prod_{i\in N} E_i[A_i](w^i).$$

Hence, from (2.6) and (2.5) and (1.1) results $E_i^*(s^*,s^*) \geqslant E_i^*(s^*,s)$ for any s W and $i \in N$, i. e.

(2.8) $$R(s^*,s^*)=\prod_{i\in N} E_i^*(s^*,s^*) \geqslant \prod_{i\in N} E_i^*(s^*,s)=R(s^*,s), \quad \forall s\in W.$$

Then s^* is a fixed point for R, i. e. the assertion (II) holds.

<u>(II) implies (I)</u>. Under the hypothesis that (II) holds, the formula (2.3) and the formula (2.4) are true. We must prove that:

(2.8) $$E_i[A_i^*]w_*^i \geqslant E_i[A_i^*](w^i) , \quad \forall w^i \in Y_i, \quad \forall i\in N.$$

To this end we define for any $i\in N$ the play $s^{(i)}=(s_1^{(i)},\ldots,s_n^{(i)})$, where $s_j^{(i)}=(A_j^{(i)},w_{(i)}^j)$ with:

(2.9) $$A_j^{(i)}= A_j \quad \text{if } j\neq i,$$
$$= A_j^* \quad \text{if } j=i;$$

and

(2.10) $$w_{(i)}^j= w_*^j \quad \text{if } j\neq i,$$
$$= w^j \quad \text{if } j=i.$$

Then it is not dificult to see that:

(2.11) $$E_i^*(s^*,s^{(i)})=E_i[A_i^*](w^i)\vee E_i[A_i^*](w^i).E_i[A_i^*](w_*^i).\prod_{j\neq i} E_j[A_j](w^j)$$
$$=E_i[A_i^*](w^i),$$

for any play $s=(s_1,\ldots,s_n)$ in G with $s_j=(A_j,w^j)$, $j\in N$. For any $k\neq i$ we have:

(2.12) $$E_k^*(s^*,s^{(i)})=E_k[A_k^*](w_*^k)\vee E_k[A_k^*](w_*^k).\prod_{j\in N} E_j[A_j^{(i)}](w_{(i)}^j)=E_k[A_k^*](w_*^k).$$

According to (2.11) and (2.12) we obtain:

(2.13) $$R(s^*,s^{(i)})=E_i[A_i^*](w^i).\prod_{k\neq i} E_i[A_k^*](w_*^k)\leqslant R(s^*,s^*)=\prod_{k\in N} E_k\ A_k^*((w_*^k).$$

Hence, we have:

(2.14) $$E_i[A_i^*](w^i).\prod_{k\neq i} E_k[A_k^*](w_*^k) \leqslant \prod_{j\in N} E_j\ A_j^*\ (w_*^j),$$

for any w^i in Y_i and i in N. Then one and only one of the following two situations can holds:

<u>Case 1:</u> For any k in N, $E_k[A_k^*](w_*^k)\neq 0$. Then (2.14) can be simplified by $\prod_{k\neq i} E_k\ A_k^*((w_*^k)$ for any i in N. So we obtain (I) in this special case.

<u>Case 2:</u> There is a k in N so that $E_k[A_k^*](w_*^k)=0$. This implies a contradiction. In truth, by (2.3), we have $R(s^*,s^*)=0$. Hence $R(s^*,s)=0$ for any s in W. Then we have:

(2.15) $$\prod_{j\in N} E_j^*(s^*,s)=0, \quad \forall s\in W.$$

Since by (2.1) we deduce that $E_k^*(s^*,s)=E_k[A_k^*](w^k)$, for any s in W, then (2.15) implies

(2.15') $$E_k[A_k^*](w^k).\prod_{h\neq k} E_h^*(s^*,s)=0, \quad \forall s\in W,$$

where w^k is the component of s_k in s. Now, let $\bar{s}=(\bar{s}_1,\ldots,\bar{s}_n)$ be the play in G, with $\bar{s}=(A_i^*,\bar{w}^i)$ for each $i\neq k$, where $\bar{w}^i$ $(i\neq k)$ have the property: $E_i\ A_i^*\ (\bar{w}^i)\neq 0$. When i=k, $\bar{s}_k=(A_k^*,w^k)$ with w^k from s. Such a play exists according to the axioms i used for $i\neq k$. It is easy to observe that:

(2.16) $$E_h^*(s^*,\bar{s})\geqslant E_h[A_h^*](\bar{w}^h)>0, \quad \forall h\in N-\{k\}.$$

Then the formula (2.15') can be simplified by $\prod_{h\neq k} E_h^*(s^*,\bar{s})$ when $s=\bar{s}$. Hence we have:

(2.17) $$E_k[A_k^*](w^k)=0, \quad \forall w^k\in Y_k.$$

This contradicts the axiom k. So, the case 2 can not hold and the theorem is proved.

The previous theorem is a characterisation of the possible solutions in the game G. Clearly, this characterisation can be used for equilibrium points too, but in this last case we can give another theorem which seem to be more interesting for the practitioners because it leads to numerical methods in finding solutions. To enounce this theorem we denote W^* the set of the plays in G of the form $s^*=(s_1^*,\ldots,s_n^*)$ with $s_k^*=(A_k^*,w_*^k)$, $k\in N$, where A_k^* fulfils (1.2) for each k in N. Such a play is called with perfect information. A play with complete information in G is completely defined by its n-vector $w^*=(w_1^*,\ldots,w_n^*)$ from Z.

DEFINITION 2.5: We call restricted preference in G the fuzzy relation $R^*\in L(Z\times Z)$ whose membership function is $R^*(w,w')=R(s,s')$, where s and s' are the plays with perfect information determined by the vectors w and w' respectively.

Now we can enounce the following result:

THEOREM 2.6: Let $w^*=(w_*^1,\ldots,w_*^n)$ be in Z and let s^* be the play with perfect information determined by w^*. The next two assertions are equivalent:

(III) s^* is an equilibrium point of G.

(IV) w^* is a fixed point for the restricted fuzzy preference R^* of the game G.

The proof of this theorem is similar with that of the Theorem 2.4. We omit it here.

The Theorem 2.6 reduces the problem of finding equilibrium points in a n-persons fuzzy game G to that of finding fixed points for the

Restricted fuzzy preference R^*. Existence criteria for fixed points of a fuzzy relation such as R^* have been explained in [9], [6]. The author have constructed an algorithm to approximate fixed points for fuzzy relations which satisfy a set of restrictive conditions.[8]

§3. Using non-cooperative fuzzy games in modelling real phenomena

The aim of this section is to suggest to the reader a possible way for using non-cooperative n-persons fuzzy games in solving practical problems. Precisely, we shall show how a political phenomenon can be modelled using the the terms of the previous sections.

Let us consider two "players" named 1 an 2 respectively which negociate the simoultaneous reduction of their military investments in a conflictual state existing between them. The negociations of the two players consist in a political dialogue with precisely defined topics which concerns the values and the directions of the military investments for each of them. So, let us suppose that $\sigma_1^{(k)},\ldots,\sigma_{n(k)}^{k}$ are the possible directions of the military investments of the player k.[9] Analysing the resources of the military power of k and the availabilities of this player, an unambigous set Y_k contained in $\underline{R}^{n(k)}$ can be known, whose elements $w^k=(w_1^k,\ldots,w_{n(k)}^k)$ are vectors of values (in a specified currency) which can be simoultaneous investments of k in the directions $\sigma_i^{(k)}$, $i=1,\ldots,n(k)$, respectively. It is clear that a negative investment w_i^k means a decrease of the possibilities of k in the direction named $\sigma_i^{(k)}$.

Accepting that the aim of the political negociations between 1 and 2 is to reduce the intensity of the conflictual state, the players must exchange information concerning their future investments; since each of the two players is intereste in reducing his own economical effort in military problems without a reduction of the efficiency of his own military power relative to that of the other player, then each of them tries to obtain from the other acceptable guaranties so that such and such investments in such and such directions will be or will be not realized. According to the plausibility of these guaranties, the players will decide the values of their investments.[10] The exchanges of guaranties concerning the values and the directions of the investments of the players are governed by objective laws (the rules of the game). It must be clear that the "guaranties" exchanged between the players together with the laws which govern them, define a two-persons fuzzy game with the pure strategies $\sigma_i^{(k)}$

and the regular strategic compositions in Y_k. Since a set of guaranties of the player k, together their degree of plausibility from the point of view of the player p ($\neq$k), define clearly a fuzzy set A_p element in $L(Y_k)$, it seem to be natural to consider that an exchange of informations between the two players is satisfactory represented by a pair (s_1, s_2) where $s_k=(A_p, w^k)$ consists in the fuzzy set of the guaranties and in a regular decision of k based upon these data. Ofcourse, the same set of guaranties of a player k can be viwed in many ways as a fuzzy set $A_p \in L(Y_k)$, this because the degree of plausibility of k's guaranties that he will choose such and such value for an investment in such and such direction is subjectively determined being dependent upon factors such as the psychosis of the fear of agression, the tone of the political declarations... .

The objective factors which define the conflictula state between 1 and 2 permit to the players to know their individual possibilities in each conjuncture of the game. Precisely, they are determining, for any strategic composition w^p from Y_p, a fuzzy set $E_{k,w^p}(w^k)$ contained in $L(Y_k)$, so that $E_{k,w^p}(w^k)$ is the degree of the membership of the strategic composition $w = w^k$ to the set of the replies to the choice w^p of the player p. This means that for any $w^p \in Y_p$ not all the strategic compositions in Y_k are equally feasible, i. e. a w^k in Y_k is less or more regular decision of k as reply to a choosen strategic composition w^p of p according to the rules of the game. So, if p is a player and w^p is in Y_p and $w^p=(w_1^p,\dots,w_{n(p)}^p)$ with an investment w_i^p which shows an excesive increasing of the stocks of atomic bombs of p, then the strategic composition w^k with an investment w_j^k determining a reduction of the military atomic power of k is less possible (regular) than another strategic composition w^k with a great $w_j'^k$, this because each of the players wants to ensure his own security and the rules of the game show that an increased investment w_j^p is a threatening for p.

Now, if we consider the set of data $G=(\sum_1, \sum_2, Y_1, y_2, E_1, E_2)$ where $E_k(w^p, w^k)=E_{k,w^p}(w^k)$, we obtaina fuzzy game in the sense of the Definition 1.1. Searching to obtain guaranties as sure as possible concerning an acceptable future behaviour of his partner, a player of this game will make concessions and will offer guaranties that he will not put into practice menacing investments. Hence it is natural to say that the players will prefer to realize plays which are possible solutions and, eventually, equilibrium points in G. In such a game, an equilibrium point means a reciprocally supervised increasing of the military power. Since the equilibrium point of a fuzzy game (if it exists) is not generally unique, the players can use different criteria in choosing one of them. So, a Pareto-minimal equilibrium point seems to realize the ideal of the security with a minimal economical effort.

PART TWO

COOPERATIVE n-PERSONS FUZZY GAMES

§4. Fuzzy coalitions, payoff functions and core for a cooperative n-persons fuzzy game

The concept of "cooperative game" is intrinsically characterized by rules which describe the ways of exchanging informations between the coalitions. From the heuristical point of view, the concept of fuzzy coalition has been discussed in [5, §2.5]. Mathematicaly, a fuzzy coalition in a n-persons game with the players contained in the set $N=\{1,2,\dots,n\}$ is a fuzzy set $S \quad L(N)$. For any $i \in N$, $S(i)$ is the membership degree of i to S. We consider that any fuzzy coalition S in a n-persons fuzzy game, has the behaviour determined by its power in the game. Having in mind the economical, political, military,.... "games", where the power of a coalition is expressed by the value (in a specified currency) which the coalition can invest to realize its statutary objectives, we introduce the next:

DEFINITION 4.1: A cooperative n-persons fuzzy game is a function v from L(N) to $\underline{R}$ with $v(\emptyset)=0$. For any $S \in L(N)$, $v(S)$ is the power of the coalition S in the game v or the value of S in v. We shall denote G(N) the set of n-persons fuzzy games with the set of players N.

For a coalition $S \in L(N)$, the number $|S|=\sum_{i \in N} S(i)$ will be called cardinal number of S. We denote P(N) the set of unambigous coalitions S from L(N), i. e. the set of the coalitions $S \in L(N)$ with $S(i) \in \{0,1\}$ for any $i \in N$. A cooperative non-fuzzy n-persons game is a function from P(N) to $\underline{R}$ with $v(\emptyset)=0$. We shall denote $G_0(N)$ the set of these games. Let us consider that any fuzzy coalition of $v \in G(N)$ obtains the power v(S) in the game by individual contributions of the players in the game and these contributions must be proportional with the grade of membership of each individual participant in the given game and dependent upon the stipulations of the statute of S (see [5]). Hence we can introduce the next:

DEFINITION 4.2: A payoff function for the game $v \in G(N)$ is a function x from L(N) to $\underline{R}^n$ which satisfies the following conditions:

(a) If $S \in L(N)$, then $\sum_{i \in N} x_i(S).S(i) \geq v(S)$.

(b) If $S \in L(N)$ and $S(i)=0$, then $x_i(S)=0$.

We shall denote F(v) the set of the payoff functions of $v \in G(N)$.

For any $v \in G(N)$, we have $F(v) \neq \emptyset$. In truth, the next function x from L(N) to $\underline{R}^n$ is contained in F(v):

$$(4.1) \qquad x_i(S)=\frac{1}{|S|}.\max(v(S),0) \quad \text{if } S \in L(N),\ S(i)\neq 0,$$
$$=0 \quad \text{if } S \in L(N),\ S(i)=0.$$

If $x \in F(v)$ we think $x_i(S).S(i)$ as a possible contribution of the player i for its participation in the coalition S. The heuristical meaning of the axioms (a)-(b) has been discussed in [7]. The value $\underline{x}(S)=\sum_{i \in N} x_i(S).S(i)$ is the total contribution for S by the payoff function x. It is natural to consider that the players would prefer the payoff functions which imply little contributions for their participation in the coalitions of the game (without diminishing the benefits which they will receive by this participation).

DEFINITION 4.3: Let us consider $v \in G(N)$ and $x, y \in F(v)$. We say that x is <u>preferable</u> to y <u>relative to</u> $S \in L(N)$ iff $\underline{x}(S) < \underline{y}(S)$. If there exists S in L(N) so that x is preferable to y relative to S, then we say that x is <u>preferable</u> to y and we denote $x \succ y$. The set

$$(4.2) \qquad D(v)=\{x \in F(v);\ \ \forall y \in F(v),\ y \not\succ x\},$$

is named <u>core</u> of v.

In an usual way we introduce the next notion:

DEFINITION 4.4: Let v be in G(N) and V be a subset of F(v). We shall say that V is a <u>stable set</u> in v iff the following axioms hold:

(a) If $x, y \in V$, then $x \not\succ y$.

(b) If $x \in F(v)$ and $x \notin V$, then there exists $y \in V$ so that $y \succ x$.

Now we are going to prove the consistency of D(v). To this end the next proposition will be used:

PROPOSITION 4.5: Let x be in F(v). Then x D(v) iff there exists an A in L(N) so that $\underline{x}(A) > v(A)$.

<u>Proof:</u> The direct implication is clear. Conversely, if $\underline{x}(A) > v(A)$, then the function y from L(N) to $\underline{R}^n$ defined by the formula:

$$y_i(S)=x_i(S) \quad \text{if } S\neq A,$$
$$=x_i(A)-(\underline{x}(A)-v(A)).|A|^{-1}.$$

has the property $y \succ x$ because $\underline{y}(A) < \underline{x}(A)$ and the proposition holds.

THEOREM 4.6: Let v be in G(N) and

(4.2') $$W(v)=\{x\in F(v);\ \underline{x}(S)=v(S),\quad \forall S\in L(N)\}.$$

Then we have D(v)=W(v) and D(v)≠∅.

<u>Proof:</u> It is sufficient to show that D(v)=W(v) and W(v)≠∅. First we observe that W(v)⊂ D(v) because the Proposition 4.5 holds. Let us suppose that there exists $x^{(o)}$ in D(v)-W(v). We construct the series of functions $x^{(r)}$ from L(N) to $\underline{R}^n$ for r=1,2,...,n, where

(4.3) $$x_i^{(r)}(A)=x_i^{(r-1)}(A)+(v(A)-\underline{x}^{(r-1)}(A)).(A(i))^{-1} \quad \text{if } i=\min\{j;\ A(j)\neq 0\},$$
$$=x_i^{(o)}(A) \quad \text{elsewhere.}$$

We observe that $x=x^{(n)}$ is an element of W(v) and there is an A in L(N) so that $\underline{x}(A)=v(A)<\underline{x}^{(o)}(A)$ because $x^{(o)}$ is not in W(v). Hence x is preferable to $x^{(o)}$, i. e. is not in D(v) and this is a contradiction. Then $x^{(o)}$ in D(v)-W(v) does not exists and D(v)=W(v). Since for any $x^{(o)}$ in F(v) the recursive construction from (4.3) leads to an $x=x^{(n)}\in W(v)$, it results that W(v)≠∅ because F(v)≠∅. The theorem is completely proved.

Now we are going to explain a theorem which shows the difference existing between the classical concept of core of a non-fuzzy game and our concept of core of a fuzzy game (see also [3]).

THEOREM 4.7: D(v) is the unique unvoid stable set for v∈G(N).

<u>Proof:</u> Let V be a stable set for v∈G(N) and let us suppose that V≠∅. It is clear that V contains D(v) as a subset. Let us consider x to be in V. If we suppose that x is not in D(v), then there is an y in F(v) so that y is preferable to x. By the procedure (4.3) with $x^{(o)}=y$ we obtain $z=x^{(n)}\in W(v)=D(v)$. It is easy to see that there exists a fuzzy coalition A L(N) with $\underline{x}(A)>\underline{y}(A)$, i. e. we have $\underline{x}(A)\geqslant\underline{y}(A)>\underline{z}(A)$. Hence, z is preferable to x and z is in D(v). Consequently, we have that x and z are in V and $z\succ x$. This contradicts the fact that V is a stable set for v. The theorem is proved.

The theory devloped in this section is a reply to the classical theory concerning "stable sets" for games [16] . However, it is not simply an extension for this classical approach because our concept of stable set is a two-dimensional one (see [11]) and it is applied to fuzzy games. Also, the previous concept of core is most powerful than the classical "core" even if it is restricted to a game in $G_o(N)$.

§5. Shapley value of a n-persons cooperative fuzzy game

In this section we shall show that, for a large class of cooperative n-persons fuzzy games, there exist payoff functions which describe rational modalities of contributing to the coalitions when they are formed according to the principles 1. and 2. enounced by L. S. Shapley in [17, §6]. Since the intuitive sense of such a payoff function has been extensively discussed in [7, §4], further on we shall explain the proof of the theorem of existence and unicity for this function attached to a given game. To this end we shall use the next classic result:

SHAPLEY'S LEMMA [17, pp. 31o]: If $v \in G_o(N)$, then v can be written in the form:

$$v=\sum_{S \in P(N)} c_S(v).w_S \ , \tag{5.1}$$

where, for any S in P(N), the coeficient $c_S(v)$ is defined by:

$$c_S(v)=\sum_{T \in P(N),\ T \subset S} (-1)^{|S|-|T|}.v(T) \tag{5.2}$$

and w_S is the game from $G_o(N)$ defined by:

$$\begin{aligned} W_S(T)&=1 \quad \text{if } T \in P(N) \text{ and } S \subset T, \\ &=0 \quad \text{if } T \in P(N) \text{ and } S \not\subset T. \end{aligned} \tag{5.3}$$

Also we shall use the following terms and notations:

DEFINITION 5.1: Let S be a fuzzy coalition. We call t-section of S the set:

$$\begin{aligned} S_t(i)&=1 \quad \text{if } i \in N \text{ and } S(i)=t, \\ &=0 \quad \text{if } i \in N \text{ and } S(i) \neq t \ . \end{aligned} \tag{5.4}$$

We say that the game $v \in G(N)$ is a game with proportional values if for any $S \in L(N)$ we have:

$$v(S)=\sum_{t \in L} v(S_t).t \ .^{11} \tag{5.5}$$

The set of the game with proportional values will be denoted $G_p(N)$. The condition (5.5) means that the contribution of a player to the power of a fuzzy coalition in v is proportional with the participation degree of this player to the given coalition or with the degree of accepting the statute of the coalition by the player. Concerning the games with proportinal values, we can give the following result:

LEMMA 5.2: If $v \in G_p(N)$, then $v=\sum_{T \in P(N)} c_S(v).\underline{w}_T$, where, for any S in P(N), the coeficients $c_S(v)$ are defined in (5.2) and the game $\underline{w}_T$ is defined by:

$$w_T(S)=\sum_{t \in L} w_T(S_t).t. \tag{5.6}$$

Proof: Using Shaypley's Lemma for the sets S_t we obtain:

$$v(S)=\sum_{t\in L} v(S_t).t = \sum_{t\in L} t.\sum_{T\in P(N)} c_T(v).w_T(S_t) =$$

$$\sum_{T\in P(N)} c_T(v).\sum_{t\in L} t.w_T(S_t)=\sum_{T\in P(N)} c_T(v).\underline{w}_T(S).$$

The lemma is proved.

Now, we define over L(N) the following relation:

(*) $S \vdash T$ iff $S_t \subset T_t$ for any $t\in L$, $t\neq 0$.

Using this relation we can give:

DEFINITION 5.3: Let v be in G(N) and S, W be two fuzzy coalitions such that $S\vdash W$. We say that S is a v-carrier for W iff for any V in L(N) we have:

(5.7) $V\vdash W$ implies that $v(V_t \; S_t)=v(V_t)$ for any $t\in L$, $t\neq 0$.

If S is a fuzzy coalition, then we call t-level of S the fuzzy set S_t^+ defined by:

(5.8) $S_t^+(i)=1$ if $i\in N$ and $S(i)\geqslant t$,
$=0$ elsewhere.

Using these notations, it is not difficult to prove:

LEMMA 5.4: For any V and W in L(N) the next assertions are true:

(a) $(W\cup V)_t=(W_t\cap V_t^+)\cup(W_t^+\cap V_t^+)$ for any $t\neq 0$ in L.

(b) $(W\cap V)_1=W_1\cap V_1$.

(c) $W\vdash V$ implies that $(W\cap V)_t=W_t$ for any $t\neq 0$ in L.

(d) $W\in P(N)$ and $V\vdash W$ implies that $V\in P(N)$.

(e) $V\in P(N)$ and $W\in L(N)$ implies that $(W\cup V)_t=W_t\cup V_t$, for any $t\neq 0$.

(f) $S\in P(N)$ ans $S\neq\emptyset$ and $T\in L(N)$ implies that T is a $\underline{w}_S$-carrier in N iff $S\vdash T$ and $T\in P(N)$.

(g) If W is a v-carrier in V, then W_t is a v-carrier in V_t for any $t\neq 0$ in L.

Let S be in L(N) and let π be a permutation of N. We denote πS the fuzzy set $(\pi S)(i)=S(\pi^{-1}(i))$ if $i\in N$ and null elsewhere. If $v\in G(N)$, the π-game v is defined by: $(\pi v)(S)=v(\pi^{-1}(S))$, for any $S\in L(N)$. It is clear that, if $v\in G_p(N)$ or if v $G_o(N)$, then πv is in $G_p(N)$ or in $G_o(N)$ respectively. Hence, we introduce:

DEFINITION 5.5: Let G be a subset of G(N). A Shapley function over G is a function f from G to $(\underline{R}^n)^{L(N)}$ which verifies the axioms:

(S1) If $v\in G$, $W\in L(N)$ and S is a v-carrier in W, then:

(5.9) $\sum_{i\in N} f_i(v)(W).S(i)=v(S)$ and ($W(i)=0$ implies that $f_i(v)(W)=0$).

(S2) If $v\in G$, $S\in L(N)$ and π is a permutation of N, then we have:

(5.10) $f_{\pi(i)}(\pi v)(\pi S)=f_i(v)(S)$.

(S3) If a and b are real numbers and u, $v \in G$, $a.u+b.v \in G$, then

(5.11) $$f(a.u+b.v)=a.f(u)+b.f(v).$$

To prove the existence and the uniquess of the Shapley function over $G=G_p(N)$ we need the next result:

THEOREM 5.6: There exists an unique function f' from $G_o(N)$ to $(\underline{R}^n)^{P(N)}$ so that the axioms (s1)-(S3) with f' instead of f and with P(N) instead of L(N) hold. This function f' can be written in the next form:

(5.12) $$f_i'(v)(W)=\sum_{T \in P_i(W)} \frac{(t-1)!.(w-t)!}{w!}.(v(T)-v(T-\{i\})) \text{ if } i \in W,$$
$$= 0 \qquad \text{else.}$$

where $t=|T|$, $w=|W|$, $P_i(W)=\{T \in P(N);\ T \subset W,\ i \in T\}$. If $S \in P(N)$, then

(5.12') $$f_i'(w_S)(W)=\frac{1}{|S|} \quad \text{if } i \in S \subset W,$$
$$=0 \qquad \text{else.}$$

We omit the laborious proof of this theorem (see [8]). Based on it we can obtain two lemmas which are important for our construction.

LEMMA 5.7: If $G^*(N)=\{\underline{w}_S;\ S \in P(N)\}$, then the function f from $G^*(N)$ to $(R^n)^{L(N)}$ defined by:

(5.13) $$f_i(\underline{w}_S)(W)=f_i'(w_S)(W_t) \text{ if } i \in S \subset W_t \text{ and } t \neq 0,$$
$$= 0 \quad \text{else.}$$

verifies the axioms (S1) and (S2). over $G=G^*(N)$.

Proof: The axiom (S2) is clearly verified. To show that (S1) is also verified we observe that, for any $\underline{w}_S$-carrier T in W L(N), we have:

$$\sum_{i \in N} f_i(\underline{w}_S)(W).T(i)=\sum_{i \in N} f_i(\underline{w}_S)(W).\sum_{t \in L} t.T_t(i)=\sum_{t \in L} t.\sum_{i \in N} f_i'(w_S)(W).T_t$$
$$=\sum_{t \in L} t.w_S(T_t)=\underline{w}_S(T).$$

The Lemma 5.4 is used here. The Lemma 5.7 is proved.

LEMMA 5.8: If f^* is a function from $G^*(N)$ to $(\underline{R}^n)^{L(N)}$ which verifies (S1) and (S2) for u and v in $G^*(N)$, then f^* is exactly the function f given in (5.13).

Proof: Let us consider W in L(N), S in P(N) and i in N. The following cases are possible:

(I). There exists t in L, $t \neq 0$, so that S W_t. Then it is clear that S is a $\underline{w}_S$-carrier in W_t and we must prove that $f_j^*(\underline{w}_S)(W)=f_j'(w_S)(W_t)$ if $j \in S$ and null else. Since S is a $\underline{w}_S$-carrier in W_t, we deduce that $\sum_{i\in N} f_i^*(\underline{w}_S)(W).S(i)=\sum_{i\in N} f_i^*(\underline{w}_S)(W_t).S(i)=\underline{w}_S(S)=1$. Let j be an element of S. Using (S2) we deduce that, for any permutation π of N defined by $\pi(k)=k$ if $k\neq j$, i and $\pi(i)=j$ (where $i\in S$), we have:

$$f_i^*(\underline{w}_S)(W)=f_j^*(\pi\underline{w}_S)(\pi W)=f_j^*(\underline{w}_S)(W) \text{ and } \sum_{i\in N} f_i^*(\underline{w}_S)(W).S(i)=1.$$

Hence $|S|.f_i(\underline{w}_S)(W)=1$, i.e. $f_{?}(\underline{w}_S)(W)=|S|^{-1}=f_i^!(w_S)(W_t)$, where the last equality is a consequence of the Theorem 5.6. Now, if j is not an element of S, we observe that S and $T=S\cup\{j\}$ are w_S-carriers in W_t. Hence we deduce that:

$$\sum_{i\in N} f_i^*(\underline{w}_S)(W).S(i)=\underline{w}_S(W)=t=\underline{w}_S(T)=\sum_{i\in N} f_i^*(\underline{w}_S)(W).T(i),$$

i. e. $f_j(\underline{w}_S)(W)=0=f_i^!(w_S)(W_t)$ because $S=T-\{j\}$ with S and T unambiguous sets.

(II). For any $t\neq 0$ in L, we have S $\;$ W_t. In this case, if $V\vdash W$, then V is a $\underline{w}_S$-carrier in W. Let j be in N. If $W(j)=0$, then $f_j(\underline{w}_S)(W)=0$ by the axiom (S1). If $W(j)=a\neq 0$, then $f_j(\underline{w}_S)(W)=0$ because the axiom (S1) can be applied for the next v-carrier U in W:

$$U(k)=a \text{ if } k=j, \quad =0 \text{ if } k\neq j. \qquad k\in N.$$

Hence the lemma is completely proved.

Now we are going to prove the main result of this section. It give an analitical description of the Shapley function over $G_p(N)$.

THEOREM 5.9: Let f be a function from $G_p(N)$ to the set $(\underline{R}^n)^{L(N)}$. defined by:

$$(5.14)\qquad f_i(v)(T)=\sum_{S\in P(N)} c_S(v).f_i(\underline{w}_S)(T)\ ,$$

where the values $f_i(\underline{w}_S)(T)$ are given in (5.13). Then we have:

$$(5.15)\qquad f_i(v)(W)=\begin{cases}\displaystyle\sum_{T\in P_i(W_r)} \frac{(t-1)!.(w(r)-t)!}{w(r)!}.(v(T)-v(T-\{i\})) & \text{if } W(i)=r\neq 0\\ =0 & \text{else,}\end{cases}$$

where $t=|T|$ and $w(r)=|W_r'|$.

In truth, if $W(i)=0$, then it is clear that $f_i(v)(W)=0$. If $r=W(i)\neq 0$, then

$$f_i(v)(W)=\sum_{S\in P_i(W_r)} c_S(v).f_i(\underline{w}_S)(W_r)=\sum_{\{S;\ i\in S\vdash W_r\}} c_S(v).f_i^!(w_S)(W_r)$$

and according to (5.13) the theorem is proved.

Now we are going to prove the consistency of the concept of Shapley function over $G_p(N)$.

THEOREM 5.10: The function f defined in (5.14) is the unique Shapley function over $G_p(N)$.

Proof: If we prove that f is a Shapley function over $G_p(N)$, then results the oneness because the Lemma 5.8 and the Lemma 5.2. hold. Hence, it remains to prove that f verifies the axioms (S1)-(S3). Since $c_S(v)$ is linear dependent upon v, then the axiom (S3) is clearly verified by f over $G_p(N)$. Also, it is easy to see that (S2) holds. To prove that (S1) is accomplished by f, let us denote T a v-carrier of W. Then, for any $t \neq 0$ in L, T_t is a v-carrier in W_t. Hence, $\sum_{i \in N} f_i(v)(W_t).T_t(i)=v(T)$, for any $t \neq 0$ in L. So, acording to (5.14), we deduce:

$$\sum_{i \in N} f_i(v)(W).T(i)=\sum_{t \in L} t.\sum_{i \in N} f_i(v)(W_t).T_t(i)=\sum_{t \in L} t.v(T_t)=v(T)$$

and the theorem is proved.

Now we can give:

DEFINITION 5.11: If $v \in G_p(N)$, then the function f(v) from L(N) to $\underline{R}^n$ is named Shapley value of v.

THEOREM 5.12: If v is in $G_p(N)$, then the Shapley value f(v) is a payoff function of v. Precisely, the Shapley value is an element of D(v).

Proof: The axiom (b) from the Definition 4.2 is clearly accomplished. The axiom (a) from the same definition is accomplished by f(v) because each W L(N) is a v-carrier in itself. For the same reason, $f(v) \in W(v)=D(v)$ and the theorem is completely proved.

§6. Other concepts of solution for cooperative n-persons fuzzy games

In the literature there are several concepts of solution for n-persons fuzzy games which must be mentioned. First we mention the solution conceptsdefined by J. P. Aubin [1] because they seem to be similar to ours. However, in spite of their names, the "core" and the "value" defined by J. P. Aubin in [1] are essentially different from our "core" and "Shapley value". The essential difference consists in the fact that J. P. Aubin's concepts are unidimensional concepts of solution, but our homonymous concepts are two-dimensional solutions. The sense of the terms "unidimensional" and "two-dimensional" solution for a game can be understood by reading the papers [11], [12]. This two-dimensionality is an essential property for the whole explanation concerning cooperative fuzzy games in the present paper. We think that the Two-dimensional Theory of Games, whose fundations are laied by I. Drăgan in [11] and [12], can be a good basis for a Theory of Fuzzy Games and that the solutions defined above represents some few possibilities from a grate set of alternatives.

The space reserved for the present paper is large but not sufficient to explain another concept of solution for fuzzy games in the cooperative case, namely the concept of "bargaining set". This is an extension of the "generalized bargaining set" described in [10].

Interesting mathematical models, which can be used in a game theoretical approach for economical problems, are presented in [13] [15], [18].

Hoping that our paper have suggested a way to approach the Fuzzy Games Theory, we shall be glad to receive the reader's critical remarks.

ACKNOWLEDGEMENTS

The author is grateful to Professor I. Drăgan for the suggestions and encouragements to approach the topic of this paper. Also the author is grateful to his family, especially to his parents, for the inestimable help in writting the present work.

FOOTNOTES

[1]The last dependence is usual named "preference".

[2]The properties of the immages and of the composition for fuzzy relations have been explained in [18], [14], [15], [5].

[3]If E_k is seen as a fuzzy relation between Z_k and Y_k defined by $E_k(z_k, w^k) = E_k(w^1, \ldots, w^n)$ for any z_k with the components w^j, $j \neq k$, then the notation $E_k[A_k]$ is meaningful representing the immage of A_k by E_k.

[4]According to [5], a strategic conception $s_k = (A_k, w^k)$ of the player k in G, describes a possible manner to behave for the player k in G. For a set of strategic compositions $\{w^i;\ i \in N-\{k\}\}$, $A_k(w^1, \ldots, w^{k-1}, w^{k+1}, \ldots, w^n)$ is the degree of plausability (from k's point of view) of the following proposition which concerns the future behaviour of the players $i \neq k$:

" $\{w^i;\ i \in N-\{k\}\}$ will be the strategic compositions which will be choosen by k's partners in the next play in G".

So, the A_k's in a play of G defines an informational conjuncture in the usual intuitive sense.

[5]A possible solution means a regular, a legal,... one by vertue of the rules of the game in a given informational conjuncture.

[6]Here, a clear distinction between the notions of "collaboration" and "cooperation" must be made.

[7]In the formula (2.1) and in the other analogous formula " $\vee$ " means "max", i. e. $a \vee b = \max(a,b)$ if $a,b \in L$.

[8]This will be the topic of the author's communication for the IFAC Conference (september, 1979).

[9]Such a "dirrection" can be identified as a modality of increasing the military possibilities of the player k such as " producing news types of atomic bombs" or "moving the soldiers in a menacing position for the other player"...

[10]This means that a rational player will not invest his resources in increasing the deffensive potential of his army in such and such dirrection when he is sure that the security in that direction is not menaced.

[11]The last sum is finite because there exists at most n sets S_t which are not empty.

REFERENCES

[1] Aubin, J.P., Coeur et valeur des jeux flous à paiements lateraux, Compter Rendues de l'Acad. Sci. Paris, 279, (1974) 891-894.

[2] Aubin, J.P., Coeur et valeur des jeux flous sans paiements lateraux, idem, 963-966.

[3] Aumann, R.J. and Dreze, J.M, Solutions of cooperative games with coalition structures, Core Discussion Paper No. 7334, Center for Operation Research and Econometrics, Univ. Catholique de Louvaine, (September, 1973).

[4] Aumann, R.J. and Maschler, M., The bargaining set for cooperative games, in: Dresher, M., Shapley, L. S. and Tucker, A.K. (eds) Advances in Games Theory,(Princeton University Press, 1964).

[5] Butnariu, D., Fuzzy games. A description of the concept, Fuzzy Sets and Systems, 1 (1978), 181-192.

[6] Butnariu, D., A fixed point theorem and its application to fuzzy games, to appear in: Revue Roumaine de Math. Pures et Appliquées,

[7] Butnariu, D., Stability and Shapley value for n-persons fuzzy games, to appear in Fuzzy Sets and Systems.

[8] Butnariu, D., The generalized Shapley value for a n-persons game, to appear in Ann. St. Univ. "Al. I. Cuza"-Iaşi.

[9] Butnariu, D., An existence theorem for possible solutions of a two-persons fuzzy game, to appear in Bull. Math. de la Soc. Sci. Math. de la R. S. de Roumanie,

[10] Butnariu, D., Coalitional stability for n-persons games, Communication for the International Symp. on Application of Mathematics in System Theory, Braşov, (December, 1978).

[11] Drăgan, I., A network-flow model for solving cooperative games, Symposia Mathematica, vol. XIX (1976), 215-227.

[12] Drăgan, I., Two-dimensional concepts for solutions of cooperative n-persons games, Seminair Arghiriade, Univ. of Timişoara, Romania, 1976.

[13] Féron, R., Economie d'exchange aleatoire flou, Comptes Rendues de l'Acad. Sci. Paris, 282 (1976), 1379-1382.

[14] Goguen, J.,L-Fuzzy sets, J. Math. Anal. Appl., 18 (1967), 145-174.

[15] Negoiţă, C.V. and Ralescu, D., Application of fuzzy sets to systems analysis,(Birkhauser Verlag, Basel, 1975).

[16] Neumann, J.(von) and Morgenstern, O., Theory of Games and Economic Behaviour,(Princeton University Press, 1970).

[17] Shapley, L.S., A value for n-persons games, in: Tucker, W. and Kuhn, H.W., (eds.), Contributions to the Theory of Games, vol. 2 (Princeton University Press, 1953).

[18] Zimmermann, H.J., Description and optimisation in fuzzy systems, Int.J. General Systems, 2 (1976), 201-215.

[19] Zadeh, L.A., Fuzzy sets, Information and Control, 8 (1965), 338-352.

—///—

ADVANCES IN FUZZY SET THEORY AND APPLICATIONS
M.M. Gupta, R.K. Ragade, R.R. Yager (editors)

FUZZY CONCEPTS: THEIR STRUCTURE AND PROBLEMS OF MEASUREMENT

Maria NOWAKOWSKA

Institute of Philosophy and Sociology
Polish Academy of Sciences
Warszawa, Poland

INTRODUCTION AND SUMMARY

This paper deals with the problem of measurement of fuzzy concepts, that is, of an empirical access to the membership functions of fuzzy sets representing concepts. Two classes of concepts will be considered. Firstly, the analysis will concern the concepts which possess a certain internal structure (as is the case for most concepts in the social sciences, especially those on a higher level of generality). A formalism will be introduced to describe the structure of such concepts, and it will be shown how the theory of psychological tests provides a methodology of construction of measurement tools.

Secondly, the analysis will concern fuzzy concepts connected with classification schemes, especially such schemes for which it is meaningless to speak of the notion of the true category of an object and error of classification. The notions introduced will then be applied to the case of linguistic measurement.

FUZZY SETS

Let X be a fixed set. A fuzzy subset A of X is, by definition, a function $f_A: X \longrightarrow [0,1]$ (see [12]). The function f_A is called a membership function; the relation $f_A(x) = 1$ signifies full membership of x in A, i.e., $x \in A$, $f_A(x) = 0$ signifies no membership, $x \notin A$, while intermediate values $f_A(x)$ represent partial membership.

The basic definitions of the theory of fuzzy sets are: two fuzzy sets A and B are equal, $A = B$, iff their membership functions coincide at every point, i.e., $f_A(x) = f_B(x)$ for all x. Given fuzzy sets A,B, the complement $\bar{A}$ of A, intersection $A \cap B$ and union $A \cup B$, are fuzzy sets with membership functions, respectively, $f_{\bar{A}}(x) = 1 - f_A(x)$, $f_{A \cap B}(x) = \min (f_A(x),f_B(x))$, and $f_{A \cup B}(x) = \max (f_A(x),f_B(x))$. These operations extend, of course, to more than two arguments.

The theory of fuzzy sets was introduced in order to create a formal and precise way of speaking about inexact, vague or fuzzy concepts. The standard examples given in nearly every paper dealing with the theory of fuzzy sets ([12-16], [1],[3]) are: the class of very tall men, of numbers approximately equal 5, of publications containing essential contribution to a domain, etc. (stress indicates the source of fuzziness).

The concept of fuzziness should be distinguished from that of stochasticity (see [1]), but we shall not dwell here on this point.

In the first sections we shall apply the theory of fuzzy sets to concepts in the social sciences. Those concepts, with a sufficiently high generality (also called constructs) possess a certain internal structure. Attempts in this paper will be directed at exploring the relations between the structure of the concept and its description in form of a membership function over a suitably defined set of objects.

We shall be considering an arbitrary class of concepts all pertaining to the same set X of objects. For most psychological concepts, the natural domain will be the set X of persons. For the time being, we shall abstract from the internal structure of the concept, and identify it with a fuzzy set in X.

By definition, two concepts A and B coincide, if their membership functions f_A and f_B are identical. As long as the determination of membership function is left open, the decision whether two concepts coincide or not is purely subjective. The theory of fuzzy sets is irrelevant when one makes this decision. Moreover, even if one could determine the membership functions adequately and without error, it is doubtful that such a stringent definition of identity is really needed. Two concepts may well be identical if their membership functions are sufficiently close one to another. Since the term sufficiently close is fuzzy, it appears that one should "fuzzify" the basic concept of the theory, namely that of identity.

To illustrate the type of application, imagine that two scientists define a concept, bearing the same name or label; they each do it in a slightly different way. Assume that both definitions fit intuition equally well, both are equally useful as tools of developing the theory, etc. The question is: Is the phrasing of the definitions sufficient to declare the two concepts different or not? The theory of fuzzy sets, requiring the identity of membership functions, apart from the difficulty in assessing the values of these functions, appears too restrictive in such cases. This is the motivation for the notions introduced below.

FUZZY IDENTITY

Let $\mathfrak{F}$ be a class of fuzzy subsets of X, some of which represent concepts bearing specific names. For $A,B \in \mathfrak{F}$, let

$$d_1(A,B) = \sup_x |f_A(x) - f_B(x)| .$$

Then d_1 defines a metric in the space $\mathfrak{F}$. Obviously, $A = B$ iff $d_1(A,B) = 0$, the equality between A and B being the identity of fuzzy sets as defined by Zadeh. The function

$$h_=(A,B) = 1 - d_1(A,B)$$

assumes values between 0 and 1, and can, therefore, be treated as a membership function of a fuzzy set in $\mathfrak{F}\times\mathfrak{F}$, i.e., of a fuzzy binary relation in $\mathfrak{F}$. From the construction it is obvious that the relation in question is that of identity of fuzzy sets in X. Alternatively, $h_=$ may also be regarded as a measure of similarity of fuzzy concepts, and the subsequent considerations may be easily rephrased in this terminology.

In the usual way, to every fuzzy concept there corresponds an indexed family of nonfuzzy concepts. In this case, we may define

$$A =_\varepsilon B \quad \text{iff} \quad h_=(A,B) \geqslant \varepsilon \ ;$$

the nonfuzzy relation $=_\varepsilon$ will be referred to as ε-identity. For the extreme case $\varepsilon = 1$, one gets the usual identity.

Another way to define a fuzzy identity is to let

$$d_2(A,B) = \int (f_A(x) - f_B(x))^2 p(x)dx,$$

where p is some probability distribution on X. Then d_2 is a metric which depends on the choice of p. Two fuzzy sets are equivalent if their membership functions differ at most on a set $E \subset X$ with $p(E) = 0$.

The fuzzy identity based on metric d_2 may be defined as above, i.e., by putting $h'_=(A,B) = 1 - d_2(A,B)$, and the nonfuzzy ε-identity is then introduced in the standard way.

It may be argued that, at least in some cases, d_2 yields an intuitively more acceptable concept of fuzzy identity than d_1. Indeed, consider a concept such as <u>tall</u>, as specified by two persons, 1 and 2. Suppose that person 1 assigns to it a membership function f_{tall_1} which increases from 0 at $x < 160$ cm to 1 at values $x > 195$ cm. Imagine now that person 2 has the membership function for the concept <u>tall</u> identical to the first, except that he claims that $f_{tall_2}(x)$ assumes value 1 also at negative x. In other words, both persons agree as to who is tall and who is not tall, except that they differ extremally in their opinions for persons of negative height. Since there are no such persons, one could argue that the concept <u>tall</u> is the same for both persons, despite the fact that $d_1(\underline{tall}_1, \underline{tall}_2) = 1$. Hence, technically speaking, these concepts are extremally different. On the other hand, if $p(x)$ is any probability distribution obtainable from some sampling scheme of persons, then $d_2(\underline{tall}_1, \underline{tall}_2) = 0$.

The metric d_2, hence the fuzzy identity $h'_=$, has another advantage. As will be seen from subsequent considerations, it occasionally may lend itself to estimation procedures.

The main trouble with the concept of ε-identity, whether based on d_1 or d_2, is that it is not transitive. This is a phenomenon largely analogous to that encountered in psychophysics for the concept of indistinguishability. In a chain of pairwise indistinguishable objects, e.g. hues, the extreme members need not be indistinguishable.

In the case under consideration, we have only the transitivity as specified by the following theorem, in which $=_\varepsilon$ may mean ε-identity based on either d_1 or d_2.

THEOREM. <u>If</u> $A =_a B$ <u>and</u> $B =_b C$, <u>then</u> $A =_{a+b-1} C$.

Indeed, let f_A, f_B and f_C be the membership functions of A, B and C. If $A =_a B$, then we have $d_i(A,B) \leq 1 - a$, and similarly, $d_i(B,C) \leq 1 - b$. The assertion now follows from the triangle inequality for metric d_i $(i = 1,2)$.

We have, therefore, the phenomenon of dissipation of identity, or decrease of similarity, as one moves along a chain of incompletely identical objects. If $A =_{0.95} B$ and $B =_{0.90} C$, then $A =_{0.85} C$.

Next, since the metrics d_1 and d_2 are invariant under addition of constants and simultaneous change of signs, we have the obvious

THEOREM. <u>For any</u> $A,B \in \mathfrak{F}$, <u>if</u> $A =_a B$, <u>then</u> $\bar{A} =_a \bar{B}$.

Next, the question to which extent is the identity preserved under fuzzy set-theoretical operations of intersections and unions, is answered by the following theorem.

THEOREM. <u>Let</u> $=_\varepsilon$ <u>denote the</u> ε<u>-identity based on metric</u> d_1. <u>If</u> $A =_a C$ <u>and</u> $B =_b D$, <u>then</u> $A \cap B =_{\min(a,b)} C \cap D$ <u>and</u> $A \cup B =_{\min(a,b)} C \cup D$.

In many instances, one deals with a set of more than two fuzzy concepts, which are to a large extent identical. Because of intransitivity of the introduced relations $=_a$, for the ε-identity of these concepts, it is not sufficient that they can be combined into a chain of pairwise ε-identical concepts. One needs a stronger notion.

DEFINITION. Let $\mathfrak{F}' \subset \mathfrak{F}$ be a class of fuzzy concepts. We say that $\mathfrak{F}'$ forms an ε-<u>identical bundle</u>, if $A =_\varepsilon B$ for all $A,B \in \mathfrak{F}'$. Here again, the ε-identity may be based on either the metric d_1 or d_2.

In practical situations, ε-identical bundles, or simply ε-bundles, appear if a concept bearing the same name is explicated in a number of similar, but not identical ways by different researches. To be able to say that these researchers speak of the same concept, these concepts ought to form an ε-bundle for some ε sufficiently close to 1.

The notion of ε-bundle allows us to simplify checking that two concepts are sufficiently distinct to be accepted as different.

THEOREM. <u>Let</u> $\mathfrak{F}', \mathfrak{F}'' \subset \mathfrak{F}$ <u>be, respectively, an</u> ε_1<u>-bundle and an</u> ε_2<u>-bundle in</u> $\mathfrak{F}$. <u>If there exist</u> $A' \in \mathfrak{F}'$ <u>and</u> $A'' \in \mathfrak{F}''$ <u>such that</u> $A' \neq_a A''$ <u>for some</u> a, <u>then</u> $B \neq_{a-(\varepsilon_1+\varepsilon_2)} B''$ <u>for all</u> $B' \in \mathfrak{F}'$ <u>and</u> $B'' \in \mathfrak{F}''$.

<u>Proof</u>. If $A' \neq_a A''$, then $d_i(A',A'') > 1-a$. Since $d_i(A',B') \leq 1-\varepsilon_1$ and $d_i(A'',B'') \leq 1-\varepsilon_2$ for all $B' \in \mathfrak{F}'$ and $B'' \in \mathfrak{F}''$, we must have $d_i(B',B'') > 1 - a - (1 - \varepsilon_1) - (1 - \varepsilon_2) = 1 - (a - (\varepsilon_1 + \varepsilon_2))$, which is equivalent to the assertion.

We have also

THEOREM. Let $\mathfrak{F}', \mathfrak{F}'' \subset \mathfrak{F}$ be, respectively, an ε_1-bundle and ε_2-bundle in $\mathfrak{F}$. If there exist $A' \in \mathfrak{F}'$ and $A'' \in \mathfrak{F}''$ such that $d_i(A',A'') \leq \varepsilon_3$, then $\mathfrak{F}' \cup \mathfrak{F}''$ forms an $(\varepsilon_1+\varepsilon_2+\varepsilon_3)$-bundle.

Indeed, for any $B' \in \mathfrak{F}'$ and $B'' \in \mathfrak{F}''$ we may write, using the triangle inequality, $d_i(B',B'') \leq d_i(B',A') + d_i(A',A'') + d_i(A'',B'') \leq \varepsilon_1+\varepsilon_2+\varepsilon_3$, which proves the assertion. We may now formulate

COROLLARY. If the ε_1- and ε_2-bundles $\mathfrak{F}'$ and $\mathfrak{F}''$ are not disjoint, then their union forms an $(\varepsilon_1+\varepsilon_2)$-bundle.

For the proof it suffices to take in the preceding theorem $A' = A'' \in \mathfrak{F}' \cap \mathfrak{F}''$ and put $\varepsilon_3 = 0$.

STRUCTURE OF CONCEPTS

We shall now try to explore the structural aspects of concepts and attempt to relate them with the properties of their membership functions, as well as with some methods of estimation of these functions.

The formal framework within which we shall try to explicate the structural aspects of concepts will be the system

$$\langle X, \mathcal{G}, L, \oplus, u \rangle$$

where the interpretation of elements is as follows. As before, X is the basic set of objects to which the considered concepts apply, e.g. persons in case of most psychological constructs.

Next, $\mathcal{G} = \{G_1, G_2, \ldots\}$ is a class of fuzzy subsets of X, to be interpreted as elementary properties of elements in X. In practical applications, e.g. in constructing measurement tools in psychology (tests for measurements of constructs), the elements of will correspond to the properties used to build questionnaire items. For instance, $G_1 \in \mathcal{G}$ may be the fuzzy set of persons who "usually wake up fresh and optimistic", $G_2 \in \mathcal{G}$ may be the set of persons who "think that the modern society is too permissive", etc.

The subsequent considerations are related to the class $\mathcal{G}$. Naturally, in any practical construction this set ought to be complete in the appopriate sense. This requirement was well understood by Cattell (see [2],[9]), who stressed its importance in his work on personality theory and measurement. His starting point was the set of adjectives, as complete as possible (without synonyms) describing simple traits. He then extracted the structure of personality by means of factor analysis.

Next, L is the set of labels. The elements of L will be the names of concepts under consideration, e.g. L may consist of all concepts from one or more theories. In the case of psychology, L would contain terms such as motivation, personality, schizothymia, superego strength, shyness, etc.

The symbol $\oplus$ will signify empty label, corresponding to any concept bearing no name.

To describe the interpretation of u, let $\mathcal{G}'$ denote the class of all sets which may be formed from elements of $\mathcal{G}$ by means of a finite number of operations of unions, intersections, and complementations. Thus, $\mathcal{G}'$ will contain fuzzy sets such as $(G_1 \cap G_2) \cup (G_3 \cap \bar{G}_5)$, $\sim(G_2 \cap G_4 \cap G_6)$, and so on. In other words, $\mathcal{G}'$ is the smallest class containing $\mathcal{G}$ and closed under unions, intersections and complementations.

Roughly, the class $\mathcal{G}'$ will consist of possible candidates for composite concepts built out of elementary concepts from $\mathcal{G}$.

Accordingly, u will be a function which maps $\mathcal{G}'$ into $L \cup \{\oplus\}$. If $G \in \mathcal{G}'$ is some fuzzy set built of elements of $\mathcal{G}$, and $u(G) = s \in L$, it means that G is a concept labelled s, or that s represents the meaning of G, or that s is a linguistic representation of G. If $u(G) = \oplus$, then the fuzzy concept G bears no name.

Thus, $u^{-1}(L) = \{G \in \mathcal{G}' : u(G) \neq \oplus\}$ is the class of all concepts which bear labels from L. If $s \in L$, the set

$$u^{-1}(s) = \{G \in \mathcal{G}' : u(G) = s\}$$

is the class of all concepts labelled s expressed in terms of elements of $\mathcal{G}$.

Take now a fixed $s \in L$ and assume that $u^{-1}(s)$ is not empty. Let G be one of the elements of this set, i.e. $u(G) = s$.

Suppose that G can be represented as an expression of the form

$$G = G^{(1)} \cup \ldots \cup G^{(k)} \quad \text{or} \quad G = G^{(1)} \cap \ldots \cap G^{(k)}$$

such that all $G^{(i)}$ are in $\mathcal{G}'$, and

(1) each $G^{(i)}$ belongs to $u^{-1}(L)$, i.e. constitutes a concept bearing a label;

(2) no intersection or union of j $(1 < j < k)$ distinct sets $G^{(i)}$ is in $u^{-1}(L)$, i.e. the concepts $G^{(1)}, \ldots, G^{(k)}$ cannot be combined in such a way as to form a concept bearing a label, except by taking one of them at a time, or all of them.

Then $(u(G^{(1)}), \ldots, u(G^{(k)}))$ will be called the first level decomposition of the concept s, or equivalently, of the concept G.

In a similar way we define the notion of higher level decomposition by decomposing the concepts appearing in the first level decomposition of G.

In this manner we obtain one of the possible behavioral correspondents of s, in form of a hierarchy of concepts, each of them eventually reducible to the elementary concepts from $\mathcal{G}$.

We may now define the relation $\prec$ holding between labels of concepts, representing the domination in the level of generality: $s' \prec s''$ if s'' appears as a component at some lower level in a decomposition of s'.

Clearly, the decomposition of a concept s need not be unique, as there may, in general, be several sets G with $u(G) = s$. Any component s which appears at some level of <u>any</u> decomposition of s may be called essential for s. A component which appears in some, but not all, decompositions, may be called dispensable.

In other words, the essential components are those which are invariant under all possible choices of behavioral correspondents of the concept s.

To use an example, any decomposition of the concept such as motivation should have, as one of its components, say, sexual motivation, since without considering the latter the concept of motivation would not be complete.

Consider now again the set $u^{-1}(s)$, and define

$$K(s) = \bigcap_{G \in u^{-1}(s)} G .$$

The set K(s) may be called the kernel of s, and each G in $u^{-1}(s)$ is an explication of s.

Similarly,

$$S(s) = \bigcup_{G \in u^{-1}(s)} G$$

will be called the support of s.

Moreover, any $x \in X$ such that $f_{K(s)}(x) = 1$ will be called an ideal type of s, while any $x \in X$ such that $f_{S(s)}(x) = 1$ will be called an exemplar of s.

We have then

THEOREM. <u>If</u> x <u>is an ideal type of</u> s, <u>then</u>

$$u(G) = s \implies f_G(x) = 1.$$

Indeed, if $f_{K(s)}(x) = 1$, then $f_G(x) = 1$ for any $G \in u^{-1}(s)$, since $f_{K(s)}(x) = \inf_{G \in u^{-1}(s)} f_G(x)$, which proves the assertion.

We shall say that s leads to ambiguity of the first kind, if there exist exemplars which are not ideal types. We have

THEOREM. <u>If</u> s <u>leads to ambiguity of the first kind, then there exist explications</u> G_1 <u>and</u> G_2 <u>of</u> s, <u>and an element</u> x <u>such that</u> $f_{G_1}(x) = 1$ <u>and</u> $f_{G_2}(x) < 1$.

Concepts which do not lead to ambiguity of the first kind, but for which $K(s) \neq S(s)$ will be called leading to ambiguity of the second kind: though there exist differences in explications of the concept, there is no disagreement as to the ideal types.

RELATIONSHIP TO MEMBERSHIP FUNCTION AND MEASUREMENT PROCEDURES

Let us consider a fixed label s, and one of the sets G with u(G) = s. As already stated, G is ultimately representable in the form of a fuzzy set obtained by set-theoretical operations from elements of $\mathcal{G}$. This suggests a way of constructing a measurement tool for estimating the values of membership function $f_G(x)$.

The value of $f_G(x)$ is obtained from the values $f_{G_1}(x)$, $f_{G_2}(x)$,... by the operations of taking maxima, minima and subtraction from 1.

These values are, however, as a rule unknown, and cannot be estimated easily. For constructs which are of sufficiently high level of generality, such that the set G, if written explicity, would be a long and involved expression, the standard procedure is based on the following assumption.

ASSUMPTION 1. If a subject is asked about his membership in any of the sets G_1, G_2,... from $\mathcal{G}$, the probability of a positive response is an increasing function of the value $f_{G_i}(x)$, that is

$$P\left\{\begin{array}{l}\text{the subject } x \text{ replies "yes" to the question:}\\ \text{"Do you belong to the set } G_i\text{?"}\end{array}\right\} = v(f_{G_i}(x)),$$

where v is an increasing function mapping $[0,1]$ into $[0,1]$.

Needless to say, the question in quotation marks is usually formulated in some natural way, e.g. "Do you often feel tired in the morning?" (= "Do you belong to the fuzzy set of persons who are often tired in the morning?), etc.

ASSUMPTION 2. The total number of responses "yes" to all questions about sets G_1, G_2,... which enter into decomposition of G is monotonically related to the value of the membership function $f_G(x)$.

This, essentially, is the way in which the psychological tools of measurement are constructed. They are simply lists of items, not necessarily having the form of questions, equipped with a scoring key for computing the test score.

Naturally, some items are formulated in such a way that the positive response indicates the membership in the complement of the corresponding fuzzy set. This amounts to reversing the score for a given item. Also, Assumption 2 can be reasonably expected to be satisfied only for the concepts s which are coherent in the sense of the following definition.

DEFINITION. A concept s is coherent, if any G such that $u(G) = s$ may be represented as a formula involving some of the fuzzy sets G_1, G_2,... from $\mathcal{G}$, satisfying the following property: either G_i or its complement, but not both, appear in the formula.

The intuitive justification of Assumption 2 is that if a concept is coherent and involves a large number of elements from $\mathcal{G}$, some sort of averaging out, or large number effect, takes place.

From this description of the construction of psychological measurement tools, one can make the following recommendations (see [7]):

(A) Generally, the higher the level of generality of the construct, the longer the test has to be in order to be adequate.

(B) Moreover, as such constructs would usually have a number of essential components, logically and psychologically unrelated, or only mildly related, it is advisable to split the test score into the appropriate components, corresponding to the components of the construct in question. In traditional terms, such constructs are referred to as multidimensional, and the test score must also be vector valued.

To use an obvious example, it would make little sense to represent a person's personality in terms of a single number. Instead, the usual way is to distinguish components of this construct, traits, and give the test score in form of the so-called personality profile, i.e. a vector of values representing the estimates of values of different traits. Naturally, there may exist constructs which are multidimensional, and on which one can impose unidimensionality artificially. Such cases, however, occur as a rule on lower levels of generality.

These recommendations are, to some extent, similar to those given in [10] for modelling purposes, by Sutherland. He distinguished some basic fuzzy dimensions of structure of phenomena. These correspond to the level of generality of the constructs, such as their degree of stochasticity, and level of generality of the phenomenon. He made recommendations as to the type of model, corresponding in this case to the tool of measurement, proper for a given type of phenomenon.

Let now G and G' satisfy the relation $u(G) = u(G') = s$, and let $G \dot{-} G'$ be the symmetric difference of the fuzzy sets G and G', i.e. $G \dot{-} G' = (G \setminus G') \cup (G' \setminus G)$. It is obvious that even when this difference is small, the concepts G and G' need not be close in the sense of metric d_1. To see this, it suffices to take two nonfuzzy sets differing by one element. However, the tools of measurement of G and G' might well yield results close to one another, if the number of subjects who have high membership in one of these fuzzy sets and low in another is relatively small. Thus, the construction of measurement tools for the membership functions leads in a natural way to the definition of closeness, or approximate equality, of the concepts, based on d_2. The random variables representing the respective test scores for G and G' must be close to one another in an appropriate sense.

To make this concept precise, let us examine in some detail the basic concepts of test theory.

Let G be a fixed concept, and let $T(x)$ be the test score for the subject x. In test theory, $T(x)$ is treated as a random variable, to be distinguished from its value when the test is applied in any particular instance.

The relation between $T(x)$ and $f_G(x)$ is, of course, unknown. Hence, the best one can do is to simply rescale the values $T(x)$ so that they fall between 0 and 1, and impose some assumptions under which the test scores may serve as estimates of $f_G(x)$.

Without loss of generality, assume that the test scores are already rescaled, and $0 \leq T(x) \leq 1$. The reliability of the test, defined as the ratio of the variance, averaged over all subjects, of the expectations of test scores (the so-called true scores), to the variance of the test in population, may serve as a measure of how exact is the assessment of the true values.

To be more precise, assume that the expectation of $T(x)$ for any person x, equals to the value of the membership function in the set G, i.e. that $ET(x) = f_G(x)$, and write

$$T(x) = f_G(x) + e(x),$$

where e(x) is a random variable, the error score. By construction, we have $Ee(x) = 0$. Let $\mathrm{Var}\, e(x) = \sigma^2(x)$.

If the person x is sampled randomly, then $f_G(x)$ may be regarded as a value of the random variable, say f_G, with the variance $\sigma^2 f_G$. One can then write, for the test score of a randomly selected person

$$T = f_G + e,$$

and one can check easily that e and f_G are uncorrelated. For any x, we have indeed $E(f_G(x)e(x)) = f_G(x)Ee(x) = 0$. Thus,

$$\sigma^2(T) = \sigma^2 f_G + \sigma^2 e.$$

The reliability of the test is defined as the ratio

$$r_T = \sigma^2 f_G/\sigma^2(T) = \sigma^2 f_G/(\sigma^2 f_G + \sigma^2 e).$$

One can show (see [4]) that $r_T = \rho(T,T')$, where T and T' are two parallel applications of the test. Parallel applications have the means and variances of test scores which are the same for every person, and for which the errors are uncorrelated. Thus, the correlation $\rho(T,T')$ may be estimated experimentally, and so may the variance $\sigma^2(T)$, of the observed test scores in the population.

Consequently, one obtains an estimate of the error variance from the formula

$$\sigma^2(e) = \sigma^2(T)(1 - r_T) = \sigma^2(T)(1 - \rho(T,T')).$$

In a similar way, one can show that if T_1 and T_2 are two tests measuring the concepts G_1 and G_2, and the reliabilities of these tests are r_1 and r_2, then the correlation between their true values, i.e. in this case, the correlation between membership values $f_{G_1}(x)$ and $f_{G_2}(x)$ when the subject is chosen randomly, is

$$\rho(G_1,G_2) = \rho(T_1,T_2)/(r_1 r_2)^{1/2} ,$$

where $\rho(T_1,T_2)$ is the correlation between observed scores in tests T_1 and T_2. This is the so-called attenuation formula (see [4]). Thus, as the quantities at the right hand side are, at least in principle, estimable from experimental data, one can estimate the correlation between the values of membership functions.

These formulas provide us with an empirical access to the degree of fuzzy identity of the concepts G_1 and G_2 measured, respectively, by tests T_1 and T_2. Indeed, $d_2(G_1,G_2) = \int (f_{G_1}(x) - f_{G_2}(x))^2 p(x)dx$ where p(x) is the sampling scheme under consideration. Hence $d_2(G_1,G_2) = E(f_{G_1} - f_{G_2})^2$.

Opening the parentheses, adding and subtracting the terms $E^2 f_{G_i}$

for i = 1,2, and $2Ef_{G_1}f_{G_2}$, the last expression is easily transformed to

$$d_2(G_1,G_2) = \sigma^2(f_{G_1}) + \sigma^2(f_{G_2}) - 2\sigma(f_{G_1})\sigma(f_{G_2})\rho(G_1,G_2)$$

$$+ (Ef_{G_1} - Ef_{G_2})^2.$$

Using the formulas $\sigma^2(f_{G_i}) = r_{T_i}\sigma^2(T_i) = \rho(T_i,T_i)\sigma^2(T_i)$, $Ef_{G_i} = E(T_i)$, and attenuation formula, one gets

$$d_2(G_1,G_2) = \sigma^2(T_1)\rho(T_1,T_1) + \sigma^2(T_2)\rho(T_2,T_2)$$

$$- 2\sigma(T_1)\sigma(T_2)\rho(T_1,T_2) + (E(T_1) - E(T_2))^2,$$

and all quantities on the right hand side are empirically accessible.

RELATIONS WITH THEORIES OF MEASUREMENT AND SCALING

The concept of closeness, or ε-identity, based on the metric d_2 is weaker than that based on the metric d_1. Instead of requiring that the values of membership functions be close to one another for all x, they may now differ considerably, provided such x occur seldom, i.e. constitute a sufficiently small fraction of the set X in question.

Admittedly, the weakening here is considerable, as compared with the theory of fuzzy sets. One would treat as approximately equal some pairs of concepts which would be considered as very much different within the framework of the theory of fuzzy sets. However, at least in psychology, this approach may be regarded as acceptable. The concepts may be treated as sufficiently close, if big differences occur seldom. The gain from such a weakening of the requirement is that one gets at least a partial empirical access to the values of membership functions.

Also, the consideration of the measure of imprecision of concepts, as outlined in the preceding sections, suggests that - at least in the social sciences - there exists a threshold of discriminability, which usually cannot be crossed.

In connection with the suggested scheme of measurement of the distance d_2 between membership functions by means of tests, it is worth while to consider briefly the relation of this scheme with the theories of measurement and scaling.

We begin with some basic definitions. A set, and a number of relations over this set (unary, binary, etc.) is called a relational system. Let $\mathcal{A} = \langle A,R_1,R_2,\ldots,R_n\rangle$ and $\mathcal{B} = \langle B,S_1,S_2,\ldots,S_n\rangle$ be two relational systems, such that the corresponding relations (R_i and S_i) have the same number of arguments. A function f mapping A into B is a homomorphism between $\mathcal{A}$ and $\mathcal{B}$, if it preserves the structure imposed by the relations (e.g. whenever $x,y \in A$ are in relation R_i, then $f(x),f(y) \in B$ are in relation S_i).

Measurement theory is a domain of mathematical research which studies problems of existence and properties of homomorphisms between the idealized forms of various empirical relational systems $\mathcal{A}$ and appropriate numerical systems $\mathcal{B}$.

The empirical relational system $\mathcal{A}$ is a representation of a set of objects of some kind and the results of empirical procedures performed on these objects, e.g. their comparisons in pairs, according to some property specified in the instruction, comparison of perceived distances between pairs, etc. The types of relational systems studied are, of course, inspired by practical situations, as well as the choice of properties, called usually axioms, from which one tries to deduce the existence of a homomorphism. Beyond this, measurement theory is a domain of pure mathematics. It is, of course, up to the experimenter to verify that the particular set of axioms is satisfied in the situation under analysis.

When successful, and when the homomorphism is in the numerical domain in which the set B is the real line, one can say that the existence of a trait, measured on the scale of a given type, is proved. The type of scale is determined by the group of transformations which preserve the homomorphism. The main types of scales are ratio scale, interval scale, ordinal scale, and nominal scale. The first is invariant under change of unit of measurement, i.e. under transformations $y = ax$ $(a > 0)$. The second is invariant under transformations of the form $y = ax + b$ $(a > 0)$, i.e. change of unit and zero of scale. The third is invariant under all monotonically increasing transformations, and the fourth - under all one-to-one transformations.

The set B need not be one-dimensional. For example, in multidimensional scaling, one tries to map the empirical system into a Euclidean space with more than one dimension.

The theory of measurement is by no means a closed domain. It will not be as long as empiricists investigate representations of new types of empirical relational systems connected with new types of experiments. At present, it appears that the main demand will be to establish appropriate representations for relational systems in which the relations are fuzzy. For some results in this direction, although without specific reference to fuzzy sets, see [5].

The theory of psychophysical scaling is a much simplified version of the above scheme, but with one essential complication added. One studies stimuli arranged on some physical continuum. Hence there is no problem of existence of a scale. The object of the study is, roughly speaking, the perception of this scale by the subjects. The complication, as compared with the theory of measurement, lies in the fact that the relations are not observable. Instead, one has only the subject's reports about his perception, and these are generally inconsistent. In a typical case, a person is asked to tell the brighter of the two lights, when their brightness is very close one to another. This creates fuzziness, and causes, among other effects, intransitivity of reports from different displayed pairs. Thus, if the data were taken directly as empirical relations, representation would almost never be possible. One assumes a perception-and-choice model of some kind (see [11],[6]) and constructs the psychophysical scale on the basis of statistical regularities of responses.

The theory of psychological tests has been developing outside the conceptual framework of measurement and scaling theory, mostly because it was not possible to indicate which underlying relational system was being represented by test scores. The development of test theory borrowed heavily from statistics, assuming that the test scores represent measurement on at least an interval scale, so that one can legitimately compute means, variances and correlations.

If one looks closely at the Assumptions 1 and 2 from the preceding sections, one can see that they are in effect very close to scaling theory and measurement theory assertions. Firstly, Assumption 1 postulates the connection between the response probability and the underlying value of the membership function. Hence it is of the type considered in scaling theory. That an assumption of the kind postulated in Assumption 1 can be expected to hold, seems beyond question. However, the form of the function v connecting the values in question is probably more elusive. It is hard to visualize any experiment which could shed some light on it. In this case, it is probably best to identify the probability of positive response simply with the value of membership function.

Finally, Assumption 2 asserts in effect that the expectation of the test score for a given person (not the expectation of test scores in population of persons) constitutes a numerical representation of the value of membership function for this person in a composite fuzzy set. Thus, a characterization of the class of fuzzy sets for which this assertion holds would be a proper task for the measurement theory specialists. Coherence of the measured concept is probably a necessary condition for Assumption 2.

If such a question could be satisfactorily resolved, one would have an intrinsic characterization of those concepts which allow adequate measurement by tests. This question is, at present, difficult even to formulate in sufficiently precise and unambiguous terms, let alone solve it. However, the treatment of complex fuzzy relational systems and their representations might ultimately lead to incorporation of test theory into the framework of measurement theory.

CLASSIFICATION SCHEMES

This section will be devoted to a somewhat different approach to the problems of empirical access to measures of fuzziness. The starting point will now be a classification scheme.

The central role will be played by the system consisting of two primitives

$$\langle U, \mathcal{C} \rangle;$$

the remaining concepts will vary depending on the type of the considered situation.

The above notions are: the universe of classified objects U, and the set $\mathcal{C}$ of categories, where

$$\mathcal{C} = \{C_1, C_2, \ldots, C_k\}.$$

By classification of the object $x \in U$ we shall understand assigninto to it an element of the set $\mathcal{C}$. Thus, a classification leads to a function (perhaps partial)

$$t: U \longrightarrow \mathcal{C}.$$

If the function t is defined for all elements of the set U, we speak of full classification; sometimes this function is defined only for a certain subset $U' \subset U$: this means that only the objects from U' were classified.

The function t is defined empirically: the relation $t(x) = C_j$ means that the person under consideration classified object x to category C_j.

Of course, with each person one may associate a different function t; also, the function t may change for the same person (this means that the same person may classify a given object to one category on one occasion, and to another category on a different occasion.)

We shall now introduce two basic types of classifications.

Type 1 comprises those classifications in which it is possible to define the "true category" of an object. Type 2 comprises classifications in which the notion of "true category" makes no sense. The examples of classifications of Type 1 are obvious. Examples of classifications of Type 2 may be answers to questionnaire items, where the subject is asked to classify his "internal state", as specified by the item, to one of the categories, usually described linguistically, e.g. "often", "average", "seldom", etc.

For classifications of Type 1 it is natural to introduce the function

$$w: U \longrightarrow \mathcal{C},$$

where the value w(x) is the "true" category, to which the object x should be assigned. This, in turn, allows us to define the notion of the error of classification: we say that object x was classified erroneously under classification t, if $t(x) \neq w(x)$, i.e. if the category to which the object x was assigned is not the one to which it ought to be assigned.

The error consisting of classifying the object from category C_j (i.e. the object x such that $w(x) = C_j$) to category C_i, i.e. $t(x) = C_i$, may have different consequences, depending on i and j. To describe this fact, one should introduce the matrix $B = (b_{ij})$, $i,j = 1,\dots,k$, where b_{ij} is a numerical expression of the seriousness of errors; $b_{ii} = 0$, and the matrix need not be symmetric, of course.

Naturally, the choice of this matrix depends on the aim of classification.

One of the central problems of classification of Type 1 defined above is to find methods of minimization of probability of error. This may be achieved by means of construction of a new classification function t, based on other functions, $t_1, t_2, \dots$ obtained from different persons who classify independently the same set of objects.

To use an example, one of such functions may be defined as follows. Let us introduce an additional category C', corresponding to "object left unclassified". Given three functions, t_1, t_2 and t_3, one can define function t using the "majority principle", i.e.

$$t(x) = C_j \text{ if } t_1(x) = t_2(x) = C_j \text{ or } t_1(x) = t_3(x) = C_j \text{ or } t_2(x) = t_3(x) = C_j$$

and

$$t(x) = C' \quad \text{in the remaining cases.}$$

Thus, according to the rule t, the object x is classified to the category C_j, if at least two out of three persons were consistent in their classifications, and put this object to C_j. If all classifications are different, the object is left unclassified.

It is intuitively obvious that the function t should lead to less errors than each of the functions t_1, t_2, t_3 separately. One can then pose the problem of finding the optimal rule built out of three (or more) functions t_i. Optimality is defined here as the requirement of the minimization of the expected error given by the matrix B. Naturally, when this matrix is changed, one obtains a different criterion of optimality.

The solution of optimization problem depends here on the classification probabilities, i.e. on the values $p_j(x)$ = P(object x will be classified to category C_j). These probabilities may depend on the classifying subject.

In case of classifications of Type 2, there is no true category, hence it is meaningless to speak of errors in classification. Thus, it is not possible to define the function w describing the "true" category of objects, nor the matrix B.

As a consequence, one may use here only the functions t_1, t_2,... A measure of the quality of classification in this case may be the probability than an object will be classified in the same way under repetition of classification, i.e.

$$a(x) = P(t_1(x) = t_2(x)),$$

where t_1 and t_2 are two independently obtained classification functions.

If $P(t_i(x) = C_j) = p_j(x)$ does not depend on the index i of the classifying subject and/or occasion, then

$$a(x) = P(t_1(x) = t_2(x) = C_1 \text{ or } \dots \text{ or } t_1(x) = t_2(x) = C_k) =$$

$$= \sum_{j=1}^{k} p_j^2(x).$$

In the general case, if i_1 and i_2 signify two subjects or two occasions for the same subject, and the probability $P(t_i(x) = C_j)$ depends on i, we have

$$a(x;i_1,i_2) = P(t_{i_1}(x) = t_{i_2}(x)) = \sum_{j=1}^{k} p_j(x;i_1)p_j(x;i_2).$$

Now, the quantity 1 - a(x), which is empirically accessible from observations, may be regarded as a value of membership function of a certain fuzzy set, namely the fuzzy set of imprecise classications.

Indeed, a(x) = 1 if and only if the distribution $\{p_j(x), j = 1,..,k\}$ is concentrated entirely on one category, so that there is no confusion in classification. On the other hand, the more uniform is this distribution, the smaller value a(x), and when the number of categories increases, and the distribution remains uniform, the value a(x) tends to zero.

When one considers a family of all classification schemes, some of them are more imprecise than others. To give examples, consider, on the one hand, classifications of Type 1, such as classifying persons into males and females, etc., and on the other hand, classifications which are common in everyday practice, expressed in the natural language by attaching adjectives to nouns representing objects. Thus, one book may be "interesting", and the other "dull" (a classification scheme, with categories such as "dull", "interesting", and possibly others); persons may be "tall", "short", "of medium height", paintings may be "dynamic" or "static", cities may be "very large", "large", "small", and so on. Each of these classification schemes of the kind described above belongs to the family of classifications of Type 2, and one may distinguish among such schemes a fuzzy subset, consisting of those classifications which are "imprecise". The value 1 - a(x), or rather, the average value of this quantity,

$$\frac{1}{n}\sum_{m=1}^{n}(1 - a(x_m))$$

taken over a suitably defined set $x_1,\ldots,x_n$ of objects, is the value of membership function of the classification scheme in question in the class of imprecise classification schemes.

In connection with this approach, it is of considerable empirical interest to analyse the relations between imprecision of the classification scheme, as measured by the value of membership function 1 - a(x) described above, and the richness of linguistic representation of categories.

Consider namely a linguistic scale, as defined by two antonyms, a and A, such as "large-small", "ugly-beautiful", "wide-narrow", etc. Let m stand for a moderator, such as "very". One can then form a potentially infinite number of classification schemes, with categories (a,A), (ma,a,A,mA), (mma,ma,a,A,mA,mmA), etc. where the iteration mm of moderators may or may not have its own linguistic representation (e.g. "very very large" may be represented as "huge", etc.)

As the number of categories of classification scheme increases, the classification becomes, in theory, more rich, by providing better means of distinguishing the objects. On the other hand, it seems intuitively obvious that the average value of 1 - a(x) will increase, i.e. the classifications will become more and more imprecise.

This negative relation between richness and imprecision accounts for the tendency towards numerical representations, accompanied by reliable measurement procedures, since such representations (it is hoped) provide less imprecise classification schemes. In the case of the social sciences, this tendency results in abundance of various indices for measurement of such or other phenomena. However, without the proper measurement-theoretical foundations, such indices provide only spurious precision: in fact, the classification remains only as precise as is allowed by the natural language, and discriminating ability of persons.

LINGUISTIC MEASUREMENT

Let us now consider some linguistic problems connected with classification and measurement. First of all, it is clear that we deal here not with one, but with many classification schemes at the same time; the characteristic features here are that (1) many schemes form "scales", i.e. categories are linearly ordered in a natural way, (2) the categories may be modified by "stretching" or "enriching" the scales, by applying modifiers, such as mentioned in the preceding section, (3) there is a possibility of forming categories "ad hoc", and (4) it is possible to introduce operations on categories, such as alternative or conjunction, leading to composite categories.

The sentences expressing classifications may be joined by the operations of conjunction, alternative, implication, etc. They may also be preceded by quantifiers (e.g. "Almost all x are ...").

An important special case concerns classifications of events, or sentences, according to categories expressing possibility or truth.

Formally, we deal here with a system of the form

$$\langle U, W, G, G' \rangle$$

where U is the universe of classified objects. The remaining three elements constitute a linguistic space, with the structure as described below.

Depending on the context, the elements of U may be objects, real or imagined, events, psychological states, etc. Also, the objects x in U may themselves form complex relational structures.

The linguistic space (W,G,G′) is interpreted as follows. First of all, W is a set, whose elements are admissible names of categories. The set W contains simple expressions, such as "black", "long", etc., as well as composite expressions, such as "white as snow", etc. It is important to observe that W may also contain verbs, as classifiers of actions.

Formally, W is a certain subset of V^*, where V^* is the class of all finite strings formed out of vocabulary V.

Next, G is a family of classification systems formed out of elements of W. Thus,

$$G = \{\mathcal{C}^{(1)}, \mathcal{C}^{(2)}, \ldots\}$$

where each $\mathcal{C}^{(i)}$ is of the form

$$\mathcal{C}^{(i)} = \{w_1^{(i)}, w_2^{(i)}, \ldots\}$$

with $w_j^{(i)} \in W$. The categories need not be ordered in any way.

The interpretation here is such that $w_j^{(i)}$ is the name of j-th category in the i-th classification. The examples here may be the names of colors, names of family relationships, classes of verbs of motion, etc.

Finally, G′ is a subfamily of G, consisting of those classification schemes in which the categories are ordered in a natural way.

The examples here may be systems of categories {small, medium, large}, or {very small, small, medium, large, very large}, etc.

The elements of G may be related to one another; the most important relations here are stretching and refinement of scales.

Formally, these relations may be described as follows. Let $\mathcal{C} = \{w_1, \ldots, w_n\}$ and $\mathcal{C}' = \{w'_1, \ldots, w'_m\}$ be two elements of G', written in their natural order.

DEFINITION. $\mathcal{C}$ is <u>embedded</u> in $\mathcal{C}'$, if $n < m$, and there exists a subsequence $w'_{i_1}, w'_{i_2}, \ldots, w'_{i_n}$ of $\mathcal{C}'$ such that $w'_{i_1} = w_1, \ldots, w'_{i_n} = w_n$.

DEFINITION. $\mathcal{C}'$ is an <u>extension</u> of $\mathcal{C}$, if
(a) $\mathcal{C}$ is embedded in $\mathcal{C}'$,
(b) $i_1, \ldots, i_n$ is a sequence of consecutive natural numbers.

DEFINITION. $\mathcal{C}'$ is a <u>simple refinement</u> of $\mathcal{C}$, if
(a) $\mathcal{C}$ is embedded in $\mathcal{C}'$,
(b) $i_1 = 1$, $i_n = n$.

The intuitive content of these definitions is as follows. Embedding of one system in another means simply that the system embedded is linguistically poorer, in the sense that it contains only some of the categories of the richer system. The latter is an extension, if the additional categories (not appearing in $\mathcal{C}$) occur only at the ends of the scale. If the ends remain the same, while new categories appear in between the old ones, we have simple refinement.

The most frequently occurring refinement of a linguistic scale is connected with the use of modifiers, e.g. "very", mentioned briefly in the preceding section.

Clearly, we have

THEOREM. <u>The relation of embedding is transitive, i.e. if $\mathcal{C}$ is embedded in $\mathcal{C}'$ and $\mathcal{C}'$ is embedded in $\mathcal{C}''$, then $\mathcal{C}$ is embedded in $\mathcal{C}''$.</u>

Besides the above mentioned, there exist other operations on scales. For a single scale, one can, of course, group its categories, forming unions of successive (or even not successive, e.g. extreme) classes, This corresponds to joining the names of the categories by the functor of alternative.

The most common operation consists of crossing two classifications. If $\mathcal{C} = \{w_1, \ldots, w_n\}$ and $\mathcal{C}' = \{w'_1, \ldots, w'_m\}$ are two elements of G, then the cross-classification $\mathcal{C} \times \mathcal{C}'$ is obtained as the Cartesian product of the sets of categories, i.e.

$$\mathcal{C} \times \mathcal{C}' = \{w_1 \,\&\, w'_1, w_1 \,\&\, w'_2, \ldots, w_n \,\&\, w'_m\}.$$

Such an operation leads, in general, to an element of G, even if both $\mathcal{C}$ and $\mathcal{C}'$ are elements of G', i.e. there exists no natural ordering of elements of $\mathcal{C} \times \mathcal{C}'$.

Finally, another operation is a construction of a scale which is hierarchically higher, i.e. assigning a name of a more general category to one or more categories (from the same, or from different scales). More precisely, the problem here is in expressing the conjunction

(x) $$x \in C_i \text{ and } y \in C'_j$$

as a single term of the form $z \in D_k$. One can distinguish here several cases.

(a) $x = y$. The conjunction (x) is then of the form $x \in C_i \,\&\, x \in C'_j$. Such sentence is inconsistent, if C_i, C'_j are different classes from the same scale, or it reduces to $x \in C_i$ if $C_i = C'_j$.

If C_i and C'_j are from different scales, or - more generally - different classification schemes, then the product class $C_i \cap C'_j$ may have its own name, or may be a part of a class with its own name.

It may also happen that there is no "ready" name in the language, in which case the names are sometimes created "ad hoc". If adequate, such names are often accepted in the language (e.g. in Polish, "nastolatek" means a teenage boy, while "nastolatka" means a teenage girl).

(b) $x \neq y$, $C_i = C'_j$. In this case, there is no need for any new category, and the conjunction (x) is expressible by means of the word such as "both".

(c) $x \neq y$, $C_i \neq C'_j$. In this case, the search for a description of the form $z \in D_k$ consists, firstly, in usage of some expression for the pair (x,y), and secondly, as D_k one can take a class in a hierachically higher classification, which comprises both C_i and C'_j i.e. $C_i \subset D_k$ and $C'_j \subset D_k$. Usually, D_k is the smallest class with this property. However, if no such class exists, e.g. if C_i and C'_j belong to opposite extremes of a scale, one uses a descriptive term such as "various", "different", etc.

Let us now consider a very important special case of a linguistic classification scheme, in which the classified objects are the classification propositions, and categories refer to various psychological continua, such as subjective probability, deontic continua, etc. Thus, we have here two families of linguistic classification schemes of the form described above. Denoting the first of them as before by $\langle U, \mathcal{C} \rangle$, where U is the set of objects, and $\mathcal{C} = \{C_1, \ldots, C_k\}$ is the set of categories, we introduce a second scheme, in which the objects are propositions of the form $x \in C_j$, and categories are various modal frames $M_1, M_2, \ldots$. The general form of a classification is then "$x \in C_j'' \in M_i$", which is more conveniently written as $M_i(x \in C_j)$.

The analysis of formal structure of such classification schemes, in which the categories $M_1, M_2, \ldots$ are special types of modal frames, namely the so-called motivational functors, leads to a certain logical calculus, called motivational calculus (see [8]). The functors considered there were obtained by abstraction from the content of sentences used in explaining, justifying, evaluating, etc the past or future decisions or actions. These functors may be partitioned into the following main categories: (1) epistemic functors (e.g. "I know that", "I am certain that", "I doubt that"), (2) emotional functors (e.g. "I am glad that"), (3) motivational functors (e.g. "I want", "I prefer"), and (4) deontic functors (e.g. "I must", "I ought to"). The main problem was to construct a set of rules of inference from sentences containing these functors.

In construction of motivational calculus, the basic concept was that of semantic implication: A implies semantically B, if the sentence "A and not B" is not admissible. The notion of admissibility is, of course, fuzzy and as a consequence, the same is true for the notion of implication. In fact, one can define the strength of semantic implication, by putting the strength of implication $A \rightarrow B$ as 1 - admissibility of "A and not B".

The calculus contains some 40 implications; to give some examples, we have $Dp \longrightarrow T(-p) \;\&\; -Cr(-p)$, which reads "I doubt that p" implies "I think that not-p, but I am not certain that not-p".

By construction of a set-theoretical model, it was possible to prove the consistency of the constructed calculus.

These 40 implications (or sometimes equivalences) could be regarded as rules of transformation of sentences which preserve their meaning. This, in fact, was the first work on the foundation of formalization of approximate reasoning in the natural language. In this analysis, for the first time the meaning of a sentence was not analysed through the concept of truth, but through the notion of semantic implication, being an extension of material implication.

It is important to remark that motivational calculus allows for iteration of some modal frames, and provides rules of transformations of some frames or their combinations into other frames. Also, various groups of modal frames are elements of G', i.e. they form certain scales, which are linguistic representations of different psychological continua. Taken jointly, these continua constitute a multidimensional cognitive space; the motivational space is a subspace of it.

Generally, the classification schemes from G' correspond to certain perceptual scales. In most cases they consist of four groups of elements: (1) a basis, formed out of a pair of antonyms, such as "large-small", etc, (2) modification of the base, obtained by applying modifiers (intensifiers, such as "very", or moderators, such as "rather"), (3) expressions built up of elements in 2, by means of negation and conjunction, and (4) additional expressions, such as analogies or metaphors, which are formed in ways other than described in (3).

Symbolically, denoting by $<$ the relation which orders the scale, and by a and A the pair of antonyms, we have $a < A$. The intensifiers, say M, M',... and moderators, say m, m',... act as follows.

An intensifier applied to an expression built out of a given basic expression, gives an expression which is more distant from the other antonym than the original antonym itself. Thus, we have $a < A < MA < MMA < \ldots$ while $MMa < Ma < a < A$. Conversely, the moderators yield expressions which are closer to the other antonym, i.e. $a < mmA < mA < A$ and $a < ma < mma < A$. It is interesting to observe the operation of the functor of negation, and connected with it effects on conjunction.

Thus, while with respect to the elements of G (not ordered classes) the negation is simply the usual complementation to the whole set of categories, i.e. $\sim w_i = \{w_1,\ldots,w_{i-1},w_{i+1},\ldots,w_n\}$, in case of systems from G' (ordered) the negation acts in a different way: a negation of an expression containing a given element of the basis yields a class which comprises the part of the scale extending from the negated expression towards its antonym. Thus, for instance, the negation of Ma comprises a, ma, mma,... mA, A, MA, MMA,..., but does not comprise, for example, MMa.

Schematically, this may be expressed as

The special way of building negations enables one to form "intervals" by negations and conjunctions. For instance, "neither large nor small", hence the expression of the form $\sim a \ \& \sim A$, comprises the intersection of the sets corresponding to these negations, i.e. ma $<$ ... $<$ mA.

As opposed to this, in the class G of unordered systems, the conjunction of negations yields nothing beyond what is given by the de Morgan rule: $\sim(p \vee q) = \sim p \ \& \sim q$.

CONSTRUCTION OF A LINGUISTIC SCALE: ANALOGIES AND METAPHORS

The elements of linguistic scales discussed so far, despite the theoretical possibility of building an infinite number of expressions, are nevertheless rather poor, since the iterations of modifiers do not always give new meanings. For instance, MMMMA is the same, in perception of most persons, as MMMA.

For further extension of the scale, in the sense of increasing the number of categories, and increase of the discriminativeness, one uses analogies, and their special form, namely metaphors. The class obtained by use of analogies is of the form "such as N", where N is the name of some object, which is to be compared with respect to some characteristic with the object evaluated (classified). Formally, this may be described by introducing into the classification $\mathcal{C}$ (not necessarily forming a scale, i.e. not necessarily being an element of G') additional categories C^N, where N is the name of an object. Inclusion of x into the class C^N will be written as $x =_{an,q} N$, which means that x is "such as N" with respect to the feature q.

The equality which appears in the above formula is a fuzzy relation; its membership function may be described as follows. If q(x) and q(N) are the values of the trait q of objects x and N, then the degree of membership in the fuzzy set of good analogies depends on the arguments q(x) and q(N), say F(q(x),q(N)); naturally, F has values between 0 and 1. Clearly, if q(x) = q(N), then F = 1. It is worth to observe, however, that F need not be a symmetric function of its arguments, which corresponds to the asymmetry of an analogy: if $x =_{an,q} N$, then it is not necessary that $N =_{an,q} x$.

Introduction of an analogy to a given scale, in form of a set N_1, N_2, ... of objects, enriches the scale, by increasing its discriminativeness. However, as in the case of intensifiers, also here there exists a limit beyond which the scale does not become enriched. One can formulate a hypothesis that the richness of linguistic representation of a given scale is closely related to the possibilities of perception, and more precisely - possibilities of discriminating stimuli with respect to a given characteristic. For example, for visual stimuli, classification or scales are usually rich (e.g. the classification into colors). The representation is also rich for such scales as subjective probability (see for instance [8]).

On the other hand, there exist scales whose linguistic representation is very poor, e.g. deontic scale, which contains only functors "I must" and I ought to". According to the above hypothesis, this is connected with the fact that the deontic scale is, in principle, only binary and its categories cannot be refined into "layers" - at least not to the extent which would enable one to differentiate the degree to which one "ought to do something", and communicate these degrees to others. However, deontic scale may be refined by expressions referring to certain situations, in which a given action is necessary, or important. For a thorough discussion of this and other scales, see [8].

FORMAL SCHEME OF LINGUISTIC MEASUREMENT

The full system of linguistic measurement is of the form

$$\langle U, S, O, W, G, G', F, K \rangle$$

where U is the set of objects, S is the set of persons, O is the set of occasions, (W,G,G') is the linguistic space, as described in the preceding section. Finally, F is a family of classification mappings, which to every person s, occasion o, object x, and the classification scheme $\mathcal{C}$ assigns a category $f(x,s,o) \in \mathcal{C}$ (where, for simplicity, the arguments s and o will be omitted). Thus, $f(x) = C_i$ means that object x was classified to category C_i, or equivalently, $f(x) = C_i$ describes the sentence "x is C_i" (in opinion of person s, on occasion o).

The last element, K, signifies the knowledge, which to every person s, occasion o, object x, classification system $\mathcal{C}$, and assignment f(x) (or equivalently, to every sentence "x is C_i") assigns an admissibility distribution $h_{x,i}(s)$ on the set of all possible states of object x, denoted by M_x. Thus, K is a structure, which has the form of a mapping M, which to each object assigns the set M_x of its possible states, and the mapping h, being the admissibility function.

In other words, each object is characterized by the set of distributions of its possible states for various scales (features, attributes, etc.).

Given two objects, x and y, and their classifications C_i and C_j (with respect to the same, or different classification schemes), one can consider operations on admissibility distributions, which correspond to various operations on sentences "x is C_i" and "y is C_j".

For an illustration, let us consider the case of conjunction. Thus, to the sentence "x is C_i and y is C_j" there corresponds an admissibility distribution concentrated on the Cartesian product of the sets, on which the components of the sentence are defined. Thus, if $h_{x,i}(s)$ is defined on A_x, and $h_{y,j}(s)$ is defined on B_y, then the admissibility distribution for the conjunction is defined on $A_x \times B_y$. Denoting this distribution by $h_{x,y}(s,s')$, we have

$$h_{x,y}(s,s') = \min (h_{x,i}(s), h_{y,j}(s')).$$

In a similar way, for the alternative "x is C_i or y is C_j", the admissibility distribution is $\max (h_{x,i}(s), h_{y,j}(s'))$. The negation of the sentence "x is C_i" has the admissibility distribution $1 - h_{x,i}(s)$, and the implication "if x is C_i, then y is C_j" is treated in the usual way, by first transforming it to the form of an alternative. Thus, for the pair of states s,s', the value of the admissibility function will be here $\max ((1 - h_{x,i}(s)),h_{y,j}(s'))$.

Using the concept of admissibility distribution, one can define the notion of semantic equivalence; this definition, in a somewhat different form, was used for the construction of motivational calculus, discussed in the preceding section.

Consider two expressions, w and w', to which there correspond categories C_i and C_j (of the same, or different scales). Consider next the sentences "x is C_i" and "y is C_j", and the corresponding admissibility distributions $h_{x,i}(s)$ and $h_{y,j}(s)$.

DEFINITION. The expressions w and w' are <u>semantically equivalent,</u> if

$$(\forall x \in U)(\forall s \in M_x): h_{x,i}(s) = h_{x,j}(s).$$

This definition may be weakened by introducing fuzziness, i.e. by requiring that the above equality holds only approximately for majority of states s, and for majority of objects x.

It is clear that semantic implication may be defined in an analogous way, by requiring the inequality in the above definition.

Moreover, one can define the strength of semantic equivalence, for instance as

$$1 - \iint (h_{x,i}(s) - h_{x,j}(s'))^2 p(s)ds\ q(x)dx$$

where p(s) and q(x) are probability distributions on the set M_x of states and set of objects U. The ideas of measuring the strength of semantic implication and semantic equivalence were outlined in the preceding sections.

The semantic equivalence partitions the set W into equivalence classes. With each $w \in W$ one can therefore associate the class Q(w) of its semantic equivalents, where $w \sim w'$, for all $w,w' \in Q(w)$.

Thus, one can form semantic equivalents of scales, as well as replace classification propositions by equivalent ones, perhaps more complex, or simpler. Of course, if one allows for fuzziness of semantic equivalence, the equivalence classes become fuzzy, where fuzziness is reflected linguistically by the modifiers of the strength of semantic implication and equivalence.

MEASUREMENT BY ANALOGY

The considerations above concerned a single act of measurement, i.e. a classification of an object with respect to a fixed scale $\mathcal{C}$, either of an ordinal character (from G'), or of a nominal character (from G). Each such classification is of the form "x is C_i"; in particular cases, C_i is a category obtained by analogy, i.e. the expression corresponding to category C_i is of the form "$=_{an,q}$" (is such as N, with respect to the characteristic q).

These classes may, of course, be modified by expressions such as "almost as such as", "quite such as", "a little such as", etc.

A linguistic measurement is in this case reducible to a conjunction of propositions $f(x) = C_j$ for various scales. If these scales are $\mathcal{C}^{(1)}, \mathcal{C}^{(2)}, \ldots$, then the result of measurement is a conjunction of the form

$$(*) \qquad x \in C^{(1)}_{i_1} \;\&\; x \in C^{(2)}_{i_2} \;\&\; \ldots .$$

It is clear that "adequate" description of the form (*) must provide independent information by each term of the conjunction, i.e. it should not contain repetitions, nor should it contain propositions which follow logically from some other propositions in (*).

Moreover, an adequate description must be "exact", in the sense that it should allow differentiation between x and those elements y,z,... from which it ought to be distinguished in the intention of the speaker. This requirement may be expressed by requirement that (*) should be false for y,z,... .

From the point of view of the person who receives the information contained in (*), the situation looks as follows. The successive terms of (x) modify the admissibility distribution, according to the principles outlined in the preceding section. Thus, the first term of (*) gives the admissibility distribution $h_1(s)$ (omitting for simplicity the index x). This is then modified to the term $h_{1,2}(s)$ when the second factor in the conjunction (x) appears, and so on, until the final admissibility distribution is obtained, say $h_{1,2,\ldots,n}(s)$. The object x is determined - to some extent - by the maximum of this function, in linguistic space.

The logical scheme of inference from a conjunction of such type, expressed in form of an admissibility distribution, is based on a weakened Modus Ponens Rule. The conclusion has the form of a family of sets

$$\{ M^{inf}_c \subset M_x,\ 0 \le c \le 1 \},$$

where

$$M^{inf}_c = \{ s:\ h_{1,2,\ldots,n}(s) \geq c \}$$

is the set of all those states which have admissibility at least c. Here the parameter c expresses the level of certainty of inference; for c = 1 the inference is most certain. Since, by definition, $h_{1,2,\ldots,n}(s) = \min_k h_{x,i_k}(s)$, we have

$$s \in M^{inf}_c \quad \text{iff} \quad (\forall k)\ h_{x,i_k}(s) \geq c.$$

The inference may proceed here in two main directions. Firstly, it may help in recognition and discrimination of the object x, i.e. distinguishing it from a set of other objects. Secondly, it may lead to distinguishing a new category of objects, with some common features arising from composition of other features.

It is worth to point out that the latter type of inference may be called inference by analogy. The characteristic feature here is the supposition that if many traits of the objects x and N are common, the subsequent traits may also be common. The same concerns the laws connecting the considered object with other objects. This may be expressed formally as follows: if $x =_{an,q_1} N$, $x =_{an,q_2} N, \dots,$ $x =_{an,q_n} N$, then one may expect that $x =_{an,q_{n+1}} N$. The latter conclusion is usually formulated as a hypothesis. If this hypothesis is not based on premises with sufficiently high level of admissibility, one often proceeds by means of selection of premises, e.g. propositions which make the conclusion more probable, or more certain subjectively. If such premises cannot be found, one usually rejects the conclusion.

At the end it is worth to stress that the suggested system of linguistic measurement is wider and richer than the topics studied recently by Zadeh ([17]). In the latter case, the situation is as follows. Given some classification to fuzzy categories, one tries to determine the numerical representation of fuzziness. In the cited paper, Zadeh constructs a semantic language PRUF (Possibilistic Relational Universal Fuzzy), which allows calculations of numerical values for some linguistic variables, such as sentences, expressions, or words, and pass then to composite linguistic variables again.

As opposed to that, in this section, the topic is rather cognitive than algorithmic, and concerns a wider class of linguistic variables. Also, it allows for an empirical access to classifications, through the function a(x), introduced in the preceding sections.

EXTENSION OF SIMPLE CLASSIFICATORY SENTENCES: SEMANTIC LANGUAGE

The linguistic measurement was identified with classification with respect to one or more categories. The result is a family of sentences of the form "x is C_i". The problem is how to extend the classification, and correspondingly, the sentences, to compound sentences. Some of such extensions were already discussed: use of propositional functors, use of modifiers, use of modal frames, with possible iterations, etc. What was omitted, is the use of quantifiers, which lwad to sentences of the form "All x are C_i" or "Some x are C_i".

With the use of admissibility distribution, the notion of semantic implication and semantic equivalence were introduced. This construction, as it was mentioned, was based on the idea of transformation rules preserving meaning, introduced in motivational calculus (see [8]).

The next step consists of considering dynamic aspects of classifications, when - due possibly to some actions of the subject, or of other objects - the object classified changes its category. The examples of such situations were already given in [8]. The general form is a sentence-frame M(a,P,t) or Mz(a,P,t), where M is a modal frame, (a,P,t) stands for "a has property P at time t", and z(a,P,t) is "(to) make a to have property P at time t". The basic operations of conjunction and alternative may concern each of the arguments, leading to composite sentences, such as M(a & b,P,t), M(a,P $\vee$ Q,t), M_1Mz(a,P,t), etc.

Naturally, such sentences may be modified by modifiers, and also preceded by quantifiers. The examples may be sentences such as "I think that it would be very good if a had the property P", "I believe that both a and b will cease to have the property P by time t", and so on.

The calculus of such expressions constitutes an abstract tool for the description of the changes of the state of the object, and/or relations between objects. This gives the possibility of defining the degree of change, e.g. by simple distance between categories (if applicable).

One can now pose a very crucial and interesting problem of invariance of certain characteristics of objects under the change of those objects. Alternatively, one can ask for the conditions of stability of classifications under change, so that for an object, its category would be invariant under classifications.

As stressed already several times, the linguistic measurement is based on semantic calculus. Despite the fact that there is no full semantic theory at present for the natural languages (for recent developments see [17]), one can visualize the following construction. Suppose for a moment that a semantic theory exists; such a theory contains a list of symbols for semantic markers, as well as special symbols, such as parentheses, and so on. To each sentence there corresponds a string of such symbols, or a set of strings (in case of multiple meanings).

Let B denote the "alphabet" of the considered semantic theory, and let B* be the monoid over B. Thus, to each sentence or string of sentences (in a dialogue, or text), there corresponds a subset of B*, which establishes a map from the class of all strings of sentences, to 2^{B^*}, the power set of the monoid B*. If Q is a set of strings of sentences under consideration, one can study the set (denoting the map mentioned above by h):

$$(*) \qquad h(Q) = \bigcup_{u \in Q} h(u) .$$

Thus, we have defined a "semantic language" h(Q), corresponding to Q, and one can study the syntax of this language, by the usual methods developed in formal linguistics. Needless to say, it is not at all necessary to restrict the summation in (*) to a subset of the language; one can take the whole natural language as well.

One can expect that the grammar of such "semantic language" (despite the fact that it is defined in a somewhat arbitrary way, the arbitrariness stemming from the fact that the choice of "alphabet" B and composition rules depend on the chosen semantic theory) will reveal some structural properties of the reality which is described by the language under consideration. In short, one may expect a formal similarity of structure of what one speaks about, and its formal representation.

REFERENCES

[1] Bellman, R.E. and Zadeh, L.A., Decision making in fuzzy environment, Management Science 17 (1970) 141-164.

[2] Cattell, R.B., Personality and Motivation: Structure and Measurement (World Book Co., New York, 1957).

[3] Gottinger, H.W., Towards fuzzy reasoning in the behavioral sciences, in: Leinfellner, W. and Köhler, E. (eds.) Developments in the Methodology of Social Sciences (Reidel, Dordrecht, 1974), 287-308.

[4] Lord, F.M. and Novick, M.R. (with contribution of Birnbaum, A.) Statistical Theories of Mental Test Scores (Addison-Wesley, Readings, Mass., 1968).

[5] Luce, R.D., Semiorders and a theory of utility discrimination, Econometrica 24 (1956) 178-191.

[6] Luce, R.D., Individual Choice Behavior (Wiley, New York, 1959).

[7] Nowakowska, M., Some psychological problems of psychometry, General Systems XII (1967) 96-102.

[8] Nowakowska, M., Language of Motivation and Language of Actions (Mouton, The Hague, 1973).

[9] Nowakowska, M., The limitations of the factor analytical approach to psychology, with special application to Cattell's research strategy, Theory and Decision 3 (1973) 109-139.

[10] Sutherland, J.W., A General Systems Philosophy for the Social and Behavioral Sciences (G. Braziller, New York, 1973).

[11] Thurstone, L.L., A law of comparative judgment, Psychological Review 34 (1927) 273-286.

[12] Zadeh, L.A., Fuzzy sets, Information and Control 8 (1965) 338-353.

[13] Zadeh, L.A., Fuzzy algorithms, Information and Control 12 (1968) 94-102.

[14] Zadeh, L.A., Similarity relations and fuzzy orderings, Information Sciences 3 (1971) 177-200.

[15] Zadeh, L.A., Fuzzy languages and their relations to human and machine intelligence, Electronics Res. Lab. Univ. of Calif. Berkeley Reports (1971).

[16] Zadeh, L.A., Towards a theory of fuzzy systems, in: Kalman, R.E. (ed.) Aspects of Network and Systems Theory (Rinehart and Winston, New York, 1971).

[17] Zadeh, L.A., PRUF - a meaning representation language for natural languages, International Journal of Man-Machine Studies 10 (1978) 395-460.

NOTE: This paper is a substantial modification of the paper entitled "Methodological Problems of Measurement of Fuzzy Concepts in the Social Sciences" which appeared in Behavioral Sciences, Vol.22, 1977. It is being published with the Editor's permission.

ADVANCES IN FUZZY SET THEORY AND APPLICATIONS
M.M. Gupta, R.K. Ragade, R.R. Yager (editors)

EFFECTS OF CONTEXT ON FUZZY MEMBERSHIP FUNCTIONS[1]

Harry M. Hersh
Pattern Analysis and Recognition Corporation
Rome, New York, USA

Alfonso Caramazza
Department of Psychology
The Johns Hopkins University
Baltimore, Maryland, USA

Hiram H. Brownell
Department of Psychology
University of Southern California
Los Angeles, California, USA

Abstract

Research on vagueness in natural language has demonstrated that fuzzy set-theoretic techniques can be used to describe the interpretation of imprecise terms such as large and small. Most of these studies have been performed under laboratory conditions with the contexts carefully controlled. Yet, it is clear that the meanings of such vague terms can vary as the context changes.

In this study, context was experimentally manipulated in order to determine the nature of context effects upon the interpretation of a set of natural language terms. However, in order to adequately evaluate the results, a technique was needed to compare two fuzzy membership functions. Borrowing concepts from probit analysis and multiple regression, a weighted least squares procedure was developed which could be used to determine whether two empirical membership functions are significantly different from one another. Using this procedure, it was possible to show that the distribution of unique elements did influence the membership functions, while variations in element frequency did not. The results are discussed in terms of their implications for theories of meaning representation and in terms of the utility of the techniques.

Introduction

As one looks at the literature concerning fuzzy set theory, it is evident that most of the work has been of a theoretical nature. Significant portions of the literature are concerned with the generalization of various mathematical

1. Portions of this research were presented at the Operations Research Society of America/The Institute of Management Science meeting, New York, 1978. Harry M. Hersh is now at Digital Equipment Corporation, 146 Main Street, Maynard, MA 01754.

concepts and with similar issues of theoretical import [13]. Although one motivational factor in the development of fuzzy set theory was to provide a formal system for describing human reasoning, decision making, and other vaguely specified processes and phenomena, very little of the fuzzy set literature reflects applications in these areas.[2] Noting this lack of empirical work relating to fuzzy set theory, Watanabe [32] has recently stated that it is not possible to determine empirical fuzzy membership functions. He then used this statement to motivate a generalized fuzzy set theory.

Watanabe's objections notwithstanding, there has been some significant empirical research in fuzzy set theory, although the amount of published research has been relatively small. In psychology, fuzzy set theory has been used as a formal system for characterizing the vagueness inherent in natural language processing and in reasoning. Hersh and Caramazza [18] showed that the interpretation of vague terms such as small and large can be described by empirically derived membership functions. These authors also demonstrated that for unidimensional concepts negation was interpreted as the complement of the fuzzy subset, and that the intensifier very as in very large was interpreted as a translation operation on the "base" membership function.

Other research has replicated and extended these findings. McCloskey and Glucksberg [26] have shown that the vagueness in verbal concepts is an integral part of the concept rather than a result of the summing of variable responses over individuals. Fuzzy set theoretic techniques have been used to study age terms [16, 19], height [24], multidimensional concepts [25], and the development of verbal concepts in children [2]. Empirical membership functions have been used to argue against a prototype theory of meaning representation [6], and to describe the integration of vague information [17, 20]. The operations of union and intersection of fuzzy subsets have also been empirically investigated [27, 28].

Most of these studies have been performed under laboratory conditions, and thus the exact experimental contexts have been carefully controlled. That is, the labels for the fuzzy subsets (e.g., young, old) and the candidate elements have been defined a priori. One might say that the contexts (i.e., the entire relevant environments) have been held constant. One of the findings across these

2. The dearth of empirical work can be traced to two significant practical problems. The first problem is the determination of fuzzy membership functions. The second concerns the evaluation of the empirical functions: e.g., objectively determining whether two sample membership functions are random variants of the same underlying population membership function. Both problems will be addressed in the next section.

studies has been that when the range of the contexts have been equated the interpretation of modifiers (operations) appears to be independent of the base term modified. For example, the effect of very on small and large for a set of ordered squares is the same as its effect on young and old for a set of ages, when the age range and the square size range are equated [19].

More generally, it has been found that the membership functions for unidimensional concepts vary with the range of the particular context under consideration. Thus, for example, the set of small objects can be considered to be a fuzzy subset of the universe of objects. But contrast a small mouse with a small elephant. Clearly in each instance it is appropriate to talk about a particular item as being a member of the fuzzy subset labelled small, but the range of items that has at one time or another been described as small is so varied that it is unreasonable to assume that a particular item is small in any absolute sense.

Rather, it appears as if the fuzzy subset is defined in terms of the immediate context. Halff et al. [15] have shown that the interpretation of red varies systematically as the reference context changes from red apple to red wine. Generally, however, the research on comprehension of vague terms [17, 18, 19] has shown that people perform some type of normalization of the scale underlying the base term, with the normalized membership function being mapped onto the range in question. When the contextual range is specified, as in laboratory experiments, people use this explicitly stated context to map the membership functions. When the range is implicit, people supply their own default context based on their experiences. Hersh [19] has found, for example, that if not constrained by the situation, people will define vague concepts such as young and old in terms of their own culturally defined age range.

The relations among the empirical membership functions obtained in the various studies lead to the hypothesis that the exact membership function for a particular vague concept is related only to the range of the context. However, it soon became apparent that for many concepts of interest, the range of the context was as vague as the concepts themselves! In addition, other research on context has suggested that the context is provided by the entire distribution. Halff et al. [15] hypothesized that people use stored distributional information rather than plausibility bounds. Similarly, Walker [31] argued that people determine membership in a category by retrieving the subjective distribution (i.e., membership function) for the category. Thus to determine whether a turkey can weigh 30 pounds, Walker speculated that one relates the interpretation of 30 pound turkey to the distribution of turkeys-I-have-known, i.e., to an experientially supplied context. Tversky and Kahneman [30] refer to this process as appealing to availability: people estimate frequencies and probabilities by examining the set of items that are easiest to recall from memory.

Much of the discussion on context has been in the form of speculation on the parameters that affect the outcome. The research we report directly examines the effects of context by experimental manipulation, in order to determine what aspects of context affect the membership functions for verbal concepts.

Methodological and Analytical Problems

As pointed out in the introduction, there has not been much empirical research in fuzzy set theory to date. A reason for the imbalance between the amount of theoretical and empirical work may be two problems that plague vague experimentation. One is the difficulty in obtaining comprehensive, reliable data; the other is the difficulty in evaluating results such as, for example, the difference between two empirically derived fuzzy membership functions. Both difficulties will be discussed in this section, along with a statistical solution to the problem of comparing two membership functions.

For substantive areas within the behavioral sciences, the problem of obtaining data, the fuzzy membership functions, falls within the realm of experimental design and (psychophysical and psychological) scaling.[3] The problem of obtaining an empirical membership function, or even of determining the grade of membership for a single element, is essentially a variant of classical problems in psychological scaling techniques and can be treated accordingly. For verbal concepts, if the grade of membership of an element in a fuzzy subset is defined as the proportion of judgements agreeing that the element is a member of the set (following Black [3]), then the grade of membership can be determined using the same paradigms by which sensory thresholds and subjective probability distributions are generated [9, 14, 29].

Using classical scaling techniques to obtain empirical membership functions provides a firm basis for evaluating the reliability of obtained data. For example, Hersh and Caramazza [18] were able to show that two different scaling techniques used with two different sets of subjects produced the same membership functions for several verbal concepts. When scaling procedures are not taken

3. Factors concerning experimental design (such as eliminating the effects of potentially confounding variables by counterbalancing, by randomizing, or by holding these variables constant) are beyond the scope of this paper and can be found in a range of basic references [7, 8, 9, 10, 21].

into consideration, very different results can follow. For example, although fuzzy set theoretic techniques can be used to describe how people process vague, natural language terms, it is not at all clear that people can accurately articulate the numeric grade of membership of a particular element in a particular fuzzy subset. It especially cannot be assumed a priori. The awareness of a grade of membership is an empirical question just as is the awareness of a subjective probability distribution. Yet several researchers [24, 27] have simply assumed that people have the ability to state reliably grades of membership for elements. Such tacit assumptions lead to results that are, at best, difficult to interpret. Thus, for example, Oden [27], while claiming to examine how fuzzy set theory can be used to describe subjects' processing of natural language concepts, actually examined subjects' ability to use fuzzy logic.

In a paper by MacVicar-Whelan [24], on the other hand, one result was that membership functions were linear, although it is very unlikely that the grades of membership in the subsets short and tall increase linearly with physical height. The linearity was a result of the techniques used. With his linear categorization method, subjects were literally forced to produce linear membership functions. MacVicar-Whelan also used a method of monotonically increasing height to obtain membership functions. The technique is one form of the classical method of limits or serial exploration for obtaining sensory thresholds [14, 29]. To avoid bias, this technique requires a series of both increasing and decreasing sequences of stimuli. However, because MacVicar-Whelan only used increasing sequences, there is the possibility that the obtained responses were systematically biased. The result of recognizing that scaling techniques are needed to obtain fuzzy membership functions implies that substantial efforts are required to produce sufficient amounts of data, but such is the price for obtaining reliable and valid membership functions.

Once membership functions have been obtained, there remains a problem of statistical analysis - how to determine if two membership functions are significantly different from one another. The problem of comparison was not addressed in the earlier empirical work where the emphasis was on demonstrating that fuzzy set theory was a viable approach to describing the vagueness in natural language. In contrast, the present study is concerned with whether variations in context modify the membership functions. It therefore becomes important to determine whether a particular treatment significantly changes a membership function more than chance alone would allow.

In general, it is not clear how one would specify a particular membership function for a fuzzy subset (e.g., the set of important novels). However, for a unidimensional continuum onto which is mapped a monotonically increasing or decreasing membership function (as in the set of small animals or bald men), the problem is tractable. In fact, dosage-mortality curves have been quantitatively analyzed in toxicological research for almost half a century using probit analysis [4, 5, 12]. Given the similarity between dosage-mortality curves and unidimensional fuzzy membership functions (e.g., one could interpret the height of a dosage-mortality curve as the grade of membership in the fuzzy set of lethal substances), it appeared that probit analysis was a reasonable starting point for developing a quantitative approach to membership function comparison.

In toxicology research in order to statistically compare the dosage-mortality curves for two substances, one must first approximate the empirical functions formally. With dosage-mortality functions (as with the class of fuzzy membership functions under consideration here), the curves are sigmoid and, for purposes of probit analysis, have typically been assumed to approximate cumulative normal curves [4, 5, 12] or, occasionally, logistic functions [1]. The techniques for probit analysis are quite similar to those used for successive intervals scaling of psychophysical data [11] in that the data are transformed from proportions to corresponding normal deviate values. A probit line (weighted regression line) is then fit to the transformed data using a maximum likelihood criterion. The weighting is usually the Müller-Urban weights, x^2/pq, where x is the ordinate of the normal distribution corresponding to a proportion p, and $q = 1-p$. (This weighting scheme tends to normalize the variances of the proportions.) For bioassay applications, the probit line is then used to estimate the dosage corresponding to 50% killed. One can then determine whether the 50% dosages for two substances are significantly different using a χ^2 test [12].

As stated above, since the grade of membership of an element in a fuzzy subset can be defined as the proportion of positive responses, fuzzy membership functions can be considered equivalent to dosage-mortality curves. Thus membership functions can likewise be transformed and regression lines fit to them. However, advances in statistical methodology allow a more direct evaluation than χ^2 tests that have been used in the past. More specifically, if transformed fuzzy membership functions can be described by regression lines (using a least squares criterion rather than maximum likelihood) then well known techniques (e.g., analysis of variance) can be used for determining:

1. Whether two regression lines have different slopes; and

2. Whether two regression lines have different intercepts, given that their slopes are not different [22].

Thus the evaluative techniques to be employed here involve combining two previously established techniques: finding a weighted regression line for transformed membership functions (using a least squares criterion) and determining whether the slopes and intercepts of the regression lines are significantly different.

To arrive at these techniques, we first define the following quantities:

X_i = element of the universe under consideration,

$\mu_A(X_i)$ = grade of membership of X_i in fuzzy subset $\underline{A}$,

Y_i = normal deviate value corresponding to a proportion equivalent to $\mu_A(X_i)$,

w_i = Müller-Urban weight corresponding to a proportion equivalent to $\mu_A(X_i)$,

n_i = number of judgements concerning element X_i,

and N = number of elements being examined for which $0.0 < \mu_A(X_i) < 1.0$.

Using a weighted least squares criterion, the following statistics can be calculated (Note, all summations are from i = 1 to i = N):

Sample means: $$\overline{X} = \frac{\Sigma n_i w_i X_i}{\Sigma n_i w_i} \qquad \overline{Y} = \frac{\Sigma n_i w_i Y_i}{\Sigma n_i w_i} \tag{1}$$

Sample variances: $$s^2_x = \frac{\Sigma n_i w_i (X_i - \overline{X})^2}{\Sigma n_i w_i} \qquad s^2_y = \frac{\Sigma n_i w_i (Y_i - \overline{Y})^2}{\Sigma n_i w_i} \tag{2}$$

(weighted) produce moment correlation: $$r = \frac{\Sigma n_i w_i (X_i - \overline{X})(Y_i - \overline{Y})}{(\Sigma n_i w_i) s_x s_y} \tag{3}$$

The weighted regression equation is then

$$Y_i = bX_i + a, \tag{4}$$

where $$b = r\frac{s_y}{s_x}, \tag{5}$$

$$a = \overline{Y} - r\frac{s_y}{s_x}\overline{X}, \tag{6}$$

and $\hat{Y}_i$ = estimate of Y_i .

A similar regression equation can be calculated for each fuzzy membership function under consideration. Once these regression equations are determined, significant differences between them can be evaluated using multiple regression techniques. For example, if using a common slope for two regression equations substantially increases the error variance of the estimated $\underline{Y_i}$ values, then it can be concluded that the slopes of the separate regression equations (and thus the corresponding membership functions) are significantly different.

If two empirically derived membership functions, $\underline{A}$ and $\underline{B}$, containing N and M elements $(0.0 < \mu_A(X_i), \mu_B(X_i) < 1.0)$, respectively, are arranged according to Table 1, then the methods developed by Kerlinger and Pedhazur [22, pp. 233-238] can be used to compare the two functions. Actually, it is not necessary to compute the regression equations directly; all that is needed is the intercorrelation matrix for the five variables from Table 1 calculated according to Equation (3). From the correlation matrix, the multiple correlations $R_{Y.123}$ and $R_{Y.14}$ can be computed. The analysis of variance test for significantly different slopes is then

$$F = \frac{(R^2_{Y.123} - R^2_{Y.14})}{(1-R^2_{Y.123})/(N+M-4)}, \qquad (7)$$

with 1 and (N+M-4) degrees of freedom.

Table 1
Data Organization for Membership
Function Analysis

	Variables			
Y	1	2	3	4
Y_{1A}	1	X_{1A}	0	X_{1A}
Y_{2A}	1	X_{2A}	0	X_{2A}
.	.	.	.	.
.	.	.	.	.
.	.	.	.	.
Y_{NA}	1	X_{NA}	0	X_{NA}
Y_{1B}	-1	0	X_{1B}	X_{1B}
Y_{2B}	-1	0	X_{2B}	X_{2B}
.	.	.	.	.
.	.	.	.	.
.	.	.	.	.
Y_{MB}	-1	0	X_{MB}	X_{MB}

Once it is shown that the slopes are not significantly different from one another, the difference between the intercepts can be tested using a similar F-ratio:

$$F = \frac{(R^2_{Y.14} - R^2_{Y.4})}{(1-R^2_{Y.14})/(N+M-3)} \tag{8}$$

with 1 and (N+M-3) degrees of freedom.

Equations (7) and (8) allow one to evaluate whether two empirical membership functions, obtained under different conditions, are significantly different from one another. For example, we can test whether the effect of the intensifier very on a fuzzy subset is to increase the slope of the membership function or to translate the membership function towards the extreme without changing the slope [18, 23, 33]. The two tests we have proposed can be used to decide among the alternative hypotheses. In the present study these techniques will be used to evaluate the effects of varying context on unidimensional membership functions.

Summary of Experiments

This study was concerned with how the distribution of elements within a given range influences the membership functions. It has been hypothesized [15, 31] that the subjective distribution of elements is an important aspect of the context. The hypotheses is certainly plausible, for although concepts such as height, weight, age, and length can be considered continuous over the population, in fact, no person can ever perceive a continuous distribution. Rather, knowledge concerning the distribution of a real-world concept builds up gradually as the result of the observation of discrete samples. Thus discrete approximations to continuous distributions will be used in these experiments with the realization that even continuous distributions are learned in discrete steps by subjects.

The domain of line lengths was chosen for these experiments. The fuzzy subsets to be mapped onto the line lengths in each experiment were:

short	long
very short	very long
sort of short	sort of long

There were four experiments, and in each the procedure was identical. Slides, each depicting a black line on a white background, were used as the physical stimuli. A screen was placed one meter in front of a slide projector and the subject sat 1/2 meter behind the lens of the projector. Subjects were first shown a group of slides depicting the range of line lengths to give them a general idea of the continuum. They were then told the six phrases to be used. At the start of each trial, the experimenter advanced the slide tray and orally presented a phrase. The subject then pushed an initiate button that caused a line length to appear on the screen for two seconds. The subject responded yes or no whether he thought the phrase accurately described the line length and also gave a confidence rating for his response using the integers 1 to 5, with 5 indicating complete confidence and 1 a pure guess. Subjects were told to use their intuitions about the meanings of the phrases to make their decisions. In each experiment every phrase was paired with each line length at least once. The specific conditions of each experiment, shown in Table 2, are discussed below. The sequence of line length and phrase pairs was completely randomized in each condition with the constraint that a line length-phrase pair did not occur twice in immediate succession. The subjects, who served in only one experiment each, were undergraduates at The Johns Hopkins University. Each subject was a native speaker of English.

Experiment 1. The first experiment was the control condition for the study. Twelve line lengths that varied along an exponential scale were used. The distribution of line lengths was rectangular, so that each of the six phrases was paired with each of the twelve line lengths exactly three times. There were 13 subjects in this experiment.

Experiment 2. This experiment examined whether, within the same range as in Experiment 1, the set of unique line lengths would influence the application of the six phrases. That is, the only change from Experiment 1 was to replace the exponentially increasing line lengths with linearly increasing lengths (see Table 2). Thirteen subjects also served in this experiment.

Experiment 3. Here the line lengths were the same as in Experiment 1. The difference was the frequency with which a subject saw each line length. In Experiment 1 the line lengths were uniformly distributed. In Experiment 3 they were normally distributed. Thus the 20 subjects saw the same line lengths as in Experiment 1; however, some lengths were seen only six times (once for each phrase) while others were observed as many as 48 times.

Experiment 4. The final condition was to determine whether the number of unique elements made any difference. For the set of 12 line lengths used in Experiment 1, 11 additional lengths were interpolated along the exponential scale. As shown in Table 2, the mean and standard deviation of the line lengths for Experiments 1 and 4 are virtually the same: only the number of unique elements has changed.

Table 2

Relative Line Lengths (The number in parentheses is the number of times a subject was shown that line length).

	Experiment Number			
	1	2	3	4
	1.00 (18)	1.00 (18)	1.00 (6)	1.00 (12)
				1.10 (12)
	1.22 (18)	1.73 (18)	1.22 (6)	1.22 (12)
				1.35 (12)
	1.49 (18)	2.46 (18)	1.49 (12)	1.49 (12)
				1.65 (12)
	1.82 (18)	3.19 (18)	1.82 (30)	1.82 (12)
				2.01 (12)
	2.22 (18)	3.92 (18)	2.22 (42)	2.22 (12)
				2.46 (12)
	2.72 (18)	4.65 (18)	2.72 (48)	2.72 (12)
				3.00 (12)
	3.32 (18)	5.38 (18)	3.32 (48)	3.32 (12)
				3.67 (12)
	4.06 (18)	6.11 (18)	4.06 (42)	4.06 (12)
				4.48 (12)
	4.95 (18)	6.84 (18)	4.95 (30)	4.95 (12)
				5.47 (12)
	6.05 (18)	7.57 (18)	6.05 (12)	6.05 (12)
				6.68 (12)
	7.39 (18)	8.30 (18)	7.39 (6)	7.39 (12)
				8.17 (12)
	9.03 (18)	9.03 (18)	9.03 (6)	9.03 (12)
Mean Length	3.71	5.02	3.33	3.77
Sd	2.36	2.52	1.58	2.49
Number of Trials	216	216	288	276

Experimental Results

As in previous studies [6, 18], subjects could be classified into two groups on the basis of how they interpreted the meanings of the experimental phrases. Most subjects felt that the term long also applied to lines that were very long and that the category short included very short. These individuals were termed "logically" responding subjects since, for them, a logical entailment relationship existed; a line's length being very long entailed its being long. The other group of subjects, termed "linguistically" responding, included those who felt that long or short excluded the extreme values on the continuum. For these individuals, a given line length could not be both long and very long or short and very short. In each of the experiments there were only a few linguistically responding subjects (2, 4, 4, and 2 in experiments 1-4, respectively). Although these subjects are of interest in their own right, they were excluded from further analysis in this study (see [6]).

In order to obtain empirical membership functions from the binary responses and the confidence intervals, the responses were converted to grades of membership by the following relation:

$$\mu(X) = 0.5 + d\left(\frac{r}{10}\right),$$

$$\text{where } d = \begin{Bmatrix} 1 \text{ for } \underline{\text{yes}} \text{ response} \\ -1 \text{ for } \underline{\text{no}} \text{ response} \end{Bmatrix}$$

and r = confidence value.

(Hersh and Caramazza [18] have shown that grades of membership obtained in this way are comparable to grades of membership based only on binary judgements.) These grades of membership were then averaged over replications and subjects for each phrase in each experiment to obtain sets of empirical membership functions. The resulting membership functions for Experiment 1 are shown in Figures 1 and 2. Similar membership functions were obtained for the other experimental conditions as well. (As expected, the functions for sort of short and sort of long are not sigmoid, but unimodal. They were included in the experiment mainly to increase the task complexity and thus to discourage subjects from adopting peculiar response strategies. Data obtained with the modifier sort of will not be considered further here.)

Each membership function was approximated by a weighted regression equation as in Equation (4). For Experiments 1, 3, and 4, a log transformation was taken of the exponentially related line lengths to obtain an effective linear scale. Table 3 gives the slopes and intercepts for each transformed function. Figure 3 shows the empirical membership function, and the membership function reconstructed from the regression equation, for short and long from Experiment 2. As can be seen in the figure, the fit is quite close. To more objectively evaluate the goodness of fit, χ^2 tests were computed for each membership function. The fit of

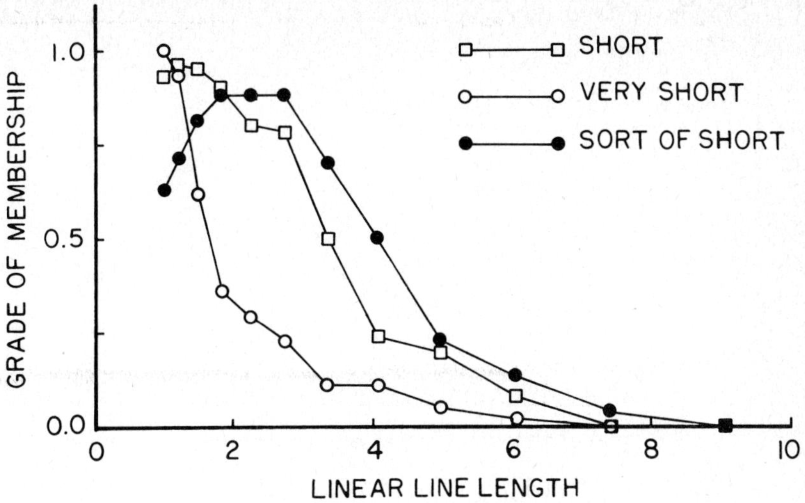

Figure 1. Membership functions from Expt. 1 for short, very short, and sort of short.

the regression equations to the membership functions approached significance for only one of the 16 membership functions: very short in Experiment 4 ($\chi^2(7) = 24.07$, $p < .01$). Even here the lack of adequate fit was not from any systematic error, but from the increased variability in the grades of membership, perhaps due to the larger number of elements and the fewer judgements per element.

Given that, overall, the weighted regression equations adequately fit the transformed membership functions, it is then possible to test the effects of the change in context. Table 4 gives the F-ratios comparing Experiments 1 and 3; i.e., comparing the differences in the frequency distributions of the elements with the same unique elements in both conditions. Clearly there were no significant differences for either short or long from Experiment 3.

Table 5 contains the F-ratios for comparisons based on Experiments 1 and 4. Here the only difference between the two distributions was the number of elements. The analyses of variance indicate that there is no difference between the slopes for either short or long. In fact, an examination of these slopes in Table 3 shows them to be remarkably similar (-2.14 vs. -2.10 and 2.61 vs. 2.61). The

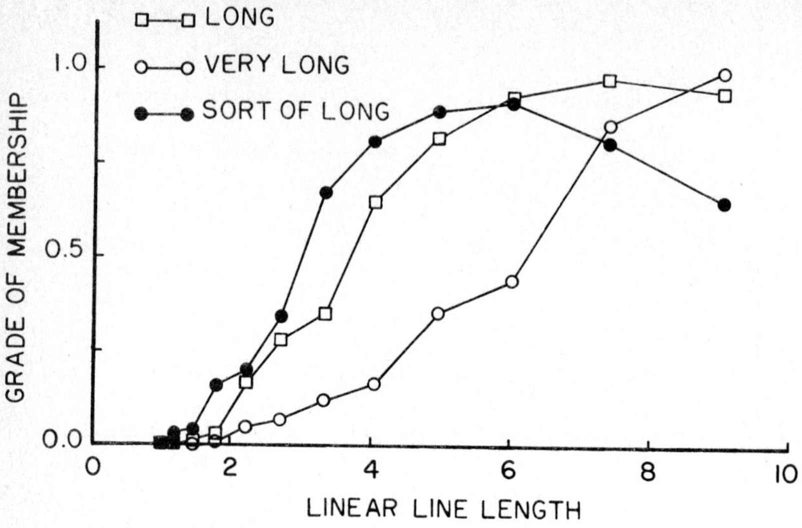

Figure 2. Membership functions from Expt. 1 for long, very long, and sort of long.

Table 3

Slopes and intercepts from regression equations of transformed membership functions.*

Experiment	Phrases			
	Short	Very short	Long	Very long
1 Slope	-2.14	-1.79	2.61	2.10
1 Intercept	7.54	6.05	1.72	1.33
2 Slope	-4.58(-1.14)	-2.76(-0.96)	3.15(0.64)	4.37(0.75)
2 Intercept	11.22(9.54)	7.28(7.31)	0.06(1.78)	-3.31(-0.12)
3 Slope	-2.88	-2.87	3.07	2.36
3 Intercept	8.12	6.89	1.24	0.86
4 Slope	-2.10	-2.27	2.61	2.41
4 Intercept	6.50	5.70	2.24	0.89

* All equations refer to log line lengths. For Experiment 2, the coefficients for the equations referring to raw line lengths are included in parentheses.

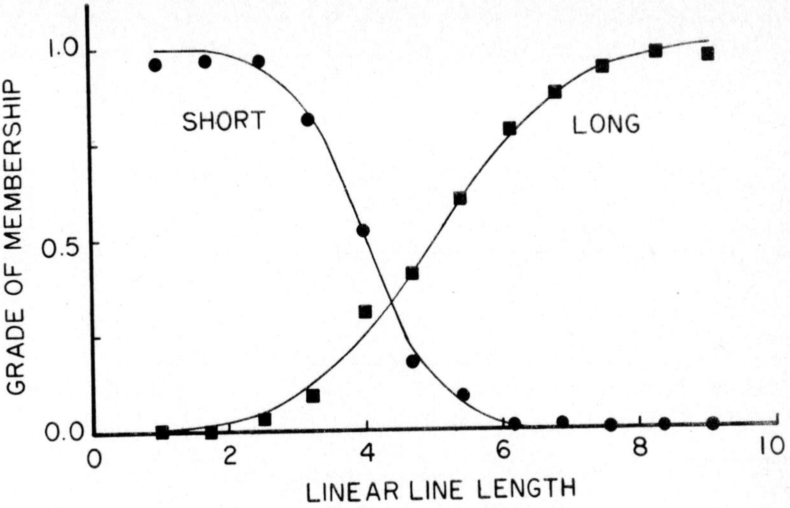

Figure 3. Empirical membership functions for short and long from Experiment 2, and fitted functions reconstructed from regression solutions.

Table 4. F-ratios for Experiment 1 vs. Experiment 3. (Numbers in parentheses are degrees of freedom.)*

	Short	Long
Slope	3.78 (1,15)	3.61 (1,14)
Intercept	1.16 (1,16)	0.79 (1,15)

* None of the F-ratios reach significance.

intercepts for both short and long, however, are significantly different. The differences (from Table 3) indicate that for both phrases, the membership functions have been translated toward the shorter end of the continuum.

The results of comparing the membership functions from Experiments 1 and 2 are contained in Table 6 using both raw line lengths and log lengths. Similar patterns of results are obtained in both cases. The slope for short is significantly steeper for the linearly related elements than for the exponentially related elements. However, for long, the slopes are essentially the same, but the intercepts are different. The membership function for the linear condition is translated toward the lower end of the continuum. Possible causes for these differences and for the differences detected in Experiment 4 will be discussed in the next section.

Table 5. F-ratios for Experiment 1 vs. Experiment 4.

	Short	Long
Slope	0.03 (1,26)	0.00 (1,22)
Intercept	102.55 (1,27)*	22.92 (1,23)*

*$p < .001$

Table 6. F-ratios for Experiment 1 vs. Experiment 2. (Numbers in parentheses are degrees of freedom.)

Raw line lengths

	Short	Long
Slope	36.69 (1,11)*	6.10 (1,14)**
Intercept	2.74 (1,12)	85.10 (1,15)*

Log line lengths

	Short	Long
Slope	15.04 (1,11)†	0.51 (1,14)
Intercept	2.68 (1,12)	63.60 (1,15)*

*$p < .001$

**$p < .05$

†$p < .005$

One final note of interest. To determine the effects of the intensifier very in each of the four experiments, short was compared against very short and long against very long. The results are presented in Table 7. For each of the eight comparisons, no pair of slopes was significantly different. However, differences for each pair of intercepts were highly significant. These results suggest that the effect of very on both short and long across the four context conditions was the same: Very acts to translate the 'base' membership function in the extreme direction. The slopes of the membership functions do not change.

Table 7. F-rations for evaluating the effects of very.
(Numbers in parentheses are degrees of freedom.)

Comparison	Experiment 1	2	3	4
Short-very short				
Slope	1.43(1,14)	1.23(1,9)	0.00(1,13)	0.35(1,35)
Intercept	74.06(1,15)*	67.72(1,10)*	86.46(1,14)*	48.81(1,36)*
Long-very long				
Slope	2.55(1,13)	2.69(1,14)	4.80(1,12)	0.40(1,31)
Intercept	67.30(1,14)*	124.23(1,15)*	121.69(1,13)*	136.55(1,32)*

*$p < .001$

Conclusions

The result of the comparisons among context conditions suggests that the frequency of occurrence of the elements does not influence the location or form of the fuzzy membership functions. Perhaps this is only reasonable. If one were to establish the fuzzy subset of young people, only unique individuals would be considered: no one would be judged multiple times. And most (fuzzy) sets one encounters every day are of this type. Sets where there are multiple identical elements are certainly in the minority. Thus it is reasonable that subjects would attend to the unique elements under consideration and discount the importance of frequency information.

By contrast, the number and value of the unique elements does appear to make a difference. Simply adding interpolated elements, as was done in Experiment 4, was sufficient to translate the membership functions toward the shorter end of the continuum. Relating the elements by a (physically) linear scale rather than an exponential scale as in Experiment 2 also affected the form of the membership functions, but in a different manner. Some reflection on the relationships in Table 2 can help understand these changes. If subjects did learn the context for a particular experimental condition on the basis of only the unique elements, then it is clear why there was no difference between Experiments 1 and 3, for each has exactly the same unique elements. Now, examining the set of elements in Experiment 4, there are a larger number of unique elements within the same range as in Experiment 1. The result is that there is a relatively larger proportion of elements near the short end of the continuum. The subjects' responses reflected this fact in that the acceptable range for short was restricted somewhat, and the range for long was extended toward the shorter end of the continuum. Perhaps the slope remained constant because the distribution of the unique elements remained in a constant exponential relationship to one another.

The results of Experiment 2, where the linear scale was used, is more complex, yet the same arguments hold. A physically linear scale can look quite different from an exponential scale across the range. From Table 2 it can be seen that for the line lengths in Experiment 2, the ones at the shorter end of the range are few and increase in length rapidly compared to the exponentially distributed lengths. The result is a membership function for short that reflects this distribution in that it falls off much more rapidly than does the comparable function for the exponential distribution. At the long end of the continuum, the elements of the linear scale are more closely grouped than are the elements from the exponential scale. Thus the membership function for long reflects this fact: subjects in Experiment 2 responded in such a manner that the membership function for long was translated somewhat toward the long end of the continuum, restricting the range of applicability.

In general, it appears as if the subjects were sensitive to the individual elements in the distributions. Where the distribution was subjectively linear throughout, the relations among the membership functions reflected this fact. When the distribution was not subjectively linear (as in Experiment 2), then it appeared as if subjects were attending to relevant subsets of the distributions, and were using this local context to influence the mapping of the membership function.

The results discussed above were made possible by the analytic techniques developed in previous sections. The pairing of probit analysis and multiple regression techniques, and the use of these techniques to compare empirical membership functions, can be a valuable tool for applications of fuzzy set theory. For example, using these techniques in this study, it was possible to show that the effect of very is to translate the membership function being modified: very does not change the shape of the membership function.

Unfortunately, problems still remain which make empirical investigations difficult. At this time psychological scaling procedures are still needed to obtain unbiased, reliable data, and large amounts of data are needed for many types of investigations. Perhaps the recognition of these problems will motivate the development of more efficient paradigms and scaling procedures which will, in turn, stimulate additional research into vagueness and imprecision in human reasoning and in other related areas.

References

[1] J. Berkson, (1953), A statistically precise and relatively simple method of estimating the bioassay with quantal reponse, based on the logistic function, J. of the Amer. Stat. Assoc., 48, 565-599.

[2] R.S. Berndt, (1977), The acquisition of basic and modified size terms: An interaction of cognitive and linguistic operations, Unpublished doctoral dissertation, The Johns Hopkins University.

[3] M. Black, (1937), Vagueness, Philosophy of Science, 4, 427-455.

[4] C.I. Bliss, (1935), The calculation of the dosage-mortality curve, Annals of Applied Biology, 22, 134-167.

[5] C.I. Bliss, (1935), The comparison of dosage-mortality data, Annals of Applied Biology, 22, 307-333.

[6] H.H. Brownell & A. Caramazza, (1978), Categorizing with overlapping categories, Memory & Cognition, 6, 481-490.

[7] D.T. Cambell & J.C. Stanley, (1963), Experimental and Quasi-experimental Design for Research, Rand McNally, Chicago.

[8] C.H. Coombs, (1964), A theory of Data, Wiley, New York.

[9] C.H. Coombs, R.M. Dawes, and A. Tversky, (1970), Mathematical Psychology, Prentice-Hall, Englewood Cliffs, New Jersey.

[10] D.R. Cox, (1958), Planning of Experiments, Wiley, New York.

[11] G.W. Diederich, S.J. Messick, and L.R. Tucker, (1957), A general least squares solution for successive intervals, Psychometrika, 2, 159-173.

[12] D.J. Finney, (1964), Probit Analysis, 2nd Edition, Cambridge University Press.

[13] B.R. Gaines & L.J. Kohout, (1977), The fuzzy decade: a bibliography of fuzzy systems and closely related topics, Int. J. of Man-Machine Studies, 9, 1-68.

[14] J.P. Guilford, (1936), Psychometric Methods, McGraw-Hill, New York.

[15] H.M. Halff, A. Ortony, and R.C. Anderson, (1976), A context-sensitive representation of word meanings, Memory & Cognition, 4, 378-383.

[16] H.M. Hersh and J. Spiering, (1976), How old is old? Paper presented at the meeting of the Eastern Psychological Association, Philadelphia.

[17] H.M. Hersh, (1976), Fuzzy reasoning: The integration of vague information, Unpublished doctoral dissertation, The Johns Hopkins University.

[18] H.M. Hersh and A. Caramazza, (1976), A fuzzy set approach to modifiers and vagueness in natural language, J. of Exp. Psych.: General, 105, 254-276.

[19] H.M. Hersh, (1977), A fuzzy set-theoretic analysis of age terms, unpublished manuscript.

[20] H.M. Hersh, (1977), A fuzzy model of human reasoning, Paper presented at the ORSA/TIMS meeting, Atlanta.

[21] B.H. Kantowitz and H.L. Roediger, (1978), Experimental Psychology, Rand McNally, Chicago.

[22] F.N. Kerlinger and E.G. Pedhazur, (1973), Multiple Regression in Behavioral Research, Holt, Rinehard and Winston, New York.

[23] G. Lakoff, (1973), Hedges: A study in meaning criteria and the logic of fuzzy concepts, J. of Philosophical Logic, 2, 458-508.

[24] P.J. MacVicar-Whelan, (1978), Fuzzy sets, the concept of height, and the hedge very, IEEE Trans. on Systems, Man, and Cybernetics, SMC-8, 507-511.

[25] R.C. Martin, (1978), The effects of modifiers on adjectives with more than one dimension of meaning, Paper presented at the Eastern Psychological Association, Washington, D.C.

[26] M.E. McCloskey & S. Glucksberg, (1978), Natural categories: Well defined or fuzzy sets? Memory & Cognition, 6, 462-472.

[27] G.C. Oden, (1977), Integration of fuzzy logical information, J. of Exp. Psych.: Human Percep. and Perf., 3, 565-575.

[28] U. Thole, H.J. Zimmermann, and P. Zysno, (1978), The connective "and" in fuzzy decision making, an empirical study, unpublished manuscript.

[29] W.S. Torgerson, (1958), Theory and Methods of Scaling, Wiley, New York.

[30] A. Tversky and D. Kahneman, (1973), Availability: A heuristic for judging frequency and probability, Cognitive Psych., 5, 207-231.

[31] J.H. Walker, (1975), Real-world variability, reasonableness judgements, and memory representations for concepts, J. of Verbal Learning and Verbal Behavior, 14, 241-252.

[32] S. Watanabe, (1978), A generalized fuzzy-set theory, IEEE Trans. on Systems, Man, and Cybernetics, SMC-8, 756-760.

[33] L.A. Zadeh, (1972), A fuzzy-set-theoretic interpretation of linguistic hedges, J. of Cybernetics, 2, 4-34.

ADVANCES IN FUZZY SET THEORY AND APPLICATIONS
M.M. Gupta, R.K. Ragade, R.R. Yager (editors)

FUZZY PROPOSITIONAL APPROACH TO PSYCHOLINGUISTIC PROBLEMS: AN APPLICATION OF FUZZY SET THEORY IN COGNITIVE SCIENCE

Gregg C. ODEN

Department of Psychology
University of Wisconsin
Madison, Wisconsin

There is a long history of fuzzy thinking in psychology. Actually, to be more precise, the notions of fuzziness have only recently been recognized to be important for psychological models of cognition but there is a long history of realizing that most of our subjective experience is based on continuous underlying dimensions and that our perception and understanding of things and events is seldom clearcut and exact. Since the earliest days of the field of psychology, it has been recognized that qualities such as good versus bad or tall versus short are subjectively continuous. It has furthermore been recognized that concepts such as "good meal", "tall person" and even "chair" are applied by people to quite broad and varied ranges of objects but that there are also "normal" instances of these concepts and also other instances which are seen to be more extreme or atypical.

Despite both these realizations, traditional psychological models of the representation of concepts have either abandoned the continuousness of subjective experience in order to concentrate on the structure and interrelations of concepts or else have treated concepts as if they corresponded to a distinct point along some dimension or in some multidimensional space. Until recently, even the few hydrid models in which continuous qualities were included in propositional representations only allowed a given concept to have a particular exact value along any given dimension. At most, some models which included continuous dimensions allowed for random fluctuation about the precise, normal value to try to account for variation in typicality. However, this approach produces such anomalies as imagining that a person would think an object was definitely a chair on some occasions and definitely not a chair on other occasions but never to think that it is anything in between such as sort of a chair and sort of a bench.

Because of the shortcomings of these traditional views, the development of the theories of fuzzy sets and fuzzy logic (e.g. Zadeh, 1965, 1975a) has been received with considerable enthusiasm by a number of psychologists who are interested in formulating models of the semantic structures and processes in humans.

My own efforts have been directed at developing fuzzy propositional models of human semantic information processing. Fuzzy propositional models are an elaboration of the currently popular propositional or network models of semantic memory (e.g., Anderson, 1976; Kintsch, 1974; Norman & Rumelhart, 1975). The elaboration is that the primitive semantic relations and properties that are the foundation of a propositional model are now considered to be fuzzy predicates (Zadeh, 1975b) which may be true to intermediate degrees. The important characteristic of this approach is that it allows for the fuzziness of subjective concepts while still emphasizing and capturing their rich structural interrelatedness. Furthermore, with this approach, the representation of the structure of complex concepts remains discrete and propositional but the concept so defined is itself fuzzy because the underlying primitives that it is built upon are fuzzy predicates. The rest of this paper describes some of my research on the psychological processes which deal with fuzzy semantic information.

Human Fuzziness Competancy

Several recent experiments (e.g. Rips, Shoben & Smith, 1973; Rosch, 1973, 1975a, 1975b) have shown that subjective degree of class membership is a psychologically real variable which affects many cognitive processes in measurable ways. In particular, Rosch (1973, 1975a) has demonstrated that people are competent at dealing with fuzzy information in that their ratings of the degree to which various objects belong to a particular subjective category are consistent both from one person to another and within a given person from one time to another. This result is elementary but important since the use of ratings provides fairly direct information about the cognitive processes used by humans to handle fuzzy information. This contrasts with the less direct approach of examining these processes by means of their influence on other processes. Of course, we generally cannot monitor the internal psychological processes themselves directly, but judgmental tasks appear to minimize the amount of intervening, extraneous processes.

The first experiment to be reported here (Oden, 1977a) extended this line of research by using a direct judgment task to investigate whether people can reason with fuzzy information in a consistent and systematic way. The particular experimental task was derived from questions such as "Which is more of a bird: an eagle or a pelican?" Such questions are natural in style and appearance as well as being intrinsically concerned with fuzzy class membership. Furthermore, since such questions involve a comparison of two class membership relations, they require at least some simple reasoning using fuzzy information. In addition, using two membership relations allows factorial stimulus designs to be constructed by independently varying each relation. This greatly increases the information which the experiment can provide about subjects' reasoning processes.

In the form presented so far, the question is merely that of deciding which item is the better exemplar for the category "birds". However, whichever item the person chooses, additional information can be obtained about his fuzzy information processes by following the first question with "How much more of a bird is [the chosen item] than [the alternative]?" In fact the actual experiment involved rephrased questions which were more flexible. The subjects were asked:

"Which of the following statements is truer and how much more true is it?

'An eagle is a bird.'

'A pelican is a bird.'"

In this form the task could be extended to include statements comparing degree of set membership in two different classes. For example,

"Which of the following statements is truer and how much more true is it?

'A sparrow is a bird.'

'A table is furniture.'"

A simple model was proposed to describe how people perform this task which assumes that people do in fact deal directly with fuzzy information in making the judgment. According to the model, the degree to which each statement is true is first evaluated and then the truthfulness of one statement is compared to that of the other to determine the relative truthfulness:

$$R_{ij} = \frac{t(a_i)}{t(a_i) + t(b_j)} \qquad (1$$

where a_i stands for the first statement, b_j stands for the second statement, t(x) is the degree of truth of x and R_{ij} is the subject's response to stimulus (i,j). For example, for the second question posed above, the rule becomes:

$$R = \frac{t(BIRD(sparrow))}{t(BIRD(sparrow)) + t(FURNITURE(table))} \quad (2$$

That is, the truth of the first statement is compared to the total truth of both of the statements to determine its relative truthfulness. If the two statements are varied independently in a factorial design, this relative judgment model can be tested using the procedures of information integration theory (Anderson, 1974).

The results of the experiment supported the proposed model and, therefore, the hypothesis that humans are competent processors of fuzzy information. The subjects in this study used their knowledge about the degree of membership in the class of birds, furniture and so on to perform the judgment task consistently and systematically. We know that the subjects used the fuzzy information consistently simply because of the success of the proposed judgment model since the model used each of its parameters (the truth values for each individual statement) to make predictions about several different responses. For example, the model was able to account for the responses of each subject to 10 different stimuli which included the statement "A SPARROW IS A BIRD" while using only a fixed value for the parameter corresponding to the subjective degree to which a sparrow belongs to the class of birds. The model's success, therefore, implies that the subjects also had a consistent, uniform conception of the "birdiness" of sparrows that they used in solving each of these judgmental problems.

Given that category membership is a matter of degrees, it is of interest to consider how categories may be internally structured; that is, how the degree of membership is determined for each exemplar. One common hypothesis (e.g. Rosch, 1973) is that there is a prototype for each category and that the degree of membership for an item is directly related to the similarity of the item to the prototype. With this hypothesis in mind, the pattern of the obtained parameter values for this first experiment is quite interesting. Specifically, for both birds and furniture there seems to be a fairly highly varied collection of items which are considered to be quite good members. For example, while tables and chairs do not seem to be highly similar to each other, both were treated as very good exemplars of furniture. Similarly, eagles and robins differ in size, habitat, shape, diet, beauty of song, ferocity and a number of other properties but both are considered to be good birds. Furthermore, even though eagles and buzzards appear to share as many qualities as do eagles and robins, buzzards were considered to be much less good exemplars of birds.

While not totally incompatible with the general prototype hypothesis, the obtained pattern of class membership does restrict it somewhat. Three possible restricted prototype hypotheses which would be consistent with the present results come to mind: (1) Degree of class membership is not proportional to degree of similarity to the prototype but is related by a roughly ogival function. (2) Degree of class membership is proportional to similarity. However, the prototype is somewhere between robins and hawks for birds and between chairs and tables for furniture, etc. (3) Degree of class membership is proportional to degree of similarity to prototypes but categories are allowed to have more than one prototype.

The first of these more restricted hypotheses seems natural for other fuzzy concepts, such as "tall" or "close to 5" and is what is often implicitly assumed (e.g. Lakoff, 1972; Zadeh, 1975). The second hypothesis has received some measure of empirical support from a multidimensional scaling study (Rips, Shoben and Smith, 1973) in which the point for "bird" fell between "robin" and "eagle".

However, a shortcoming of both of these hypotheses is that they make the strongly non-intuitive prediction that taking a perfect table and changing it to make it more similar to a chair (e.g. by adding a back and padding) should cause it to become an even better exemplar of furniture. In contrast, the third hypothesis avoids making this prediction by allowing separate chair-furniture and table-furniture prototypes. Similarly, for the category of birds, this hypothesis allows both a song-bird prototype and a bird-of-prey prototype. Especially in this latter example, the multiple prototypes approach seems to be intuitively attractive.

Operations of Fuzzy Psycho-logic

Conjunction and disjunction are fundamental operations in most logical systems and, in the case of fuzzy logics, may be defined in a number of ways. Goguen (1969) has suggested two rules of fuzzy conjunction each of which are intuitively attractive as models of subjective conjunction. One rule:

$$t(A \wedge B) = \min(t(A), t(B)) \quad (3$$

is that the degree to which statements A and B are both true is equal to the truth value of whichever statement is least true. This rule is directly analogous to Zadeh's (1965) definition of intersection on fuzzy sets. Goguen's alternative rule:

$$t(A \wedge B) = t(A)*t(B) \quad (4$$

is that the truth of a conjunction is equal to the product of the truth values of the component statements. With the conventional mapping in which zero represents absolutely false and one absolutely true, both of these rules reduce to the conjunction of standard logic when only absolute truth and falsity are involved.

There are some times when the Minimum rule of Equation 3 seems to be the way we use conjunction in natural reasoning. For example, consider a room in which there are a number of people all of whom are having their nineteenth birthday today. How true is it that all of these people are young: that is, how true is it that this person and this person and ... are young? Here the conjunction seems to be true to the degree to which any single person is young. This agrees with the Minimum rule for which the truth of the conjunction remains the same when statements which are equally true are added or deleted.

In contrast, in other situations the Minimum rule leads to nonintuitive results and the Multiplying rule of Equation 4 seems much more appropriate. Goguen (1969) describes such a case which is based on a paradox devised by the Greek philosopher Eubulides. For example, most of us would agree that a one day old infant is young and that anyone who was young yesterday is still young today. Nevertheless, through the mere passage of days everyone somehow ends up being old! Obviously this paradox is a result of considering youth to be an all-or-none property. However, as Goguen shows, a fuzzy logic with a Minimum rule definition of conjunction does not resolve the problem. The paradox involves a conjunctive expression corresponding roughly to "A one day old child is young and living another day does not destroy one's youth and living still another day does not destroy one's youth and ..." The Multiplying rule is able to resolve the paradox because the conjunction becomes less true as we add more statements which are not absolutely true. Thus, with this rule, the conjunctive expression will grow less true as the child grows old. This resolution of the paradox seems natural and therefore gives intuitive support to the Multiplying rule.

For each rule for conjunction there is a corresponding rule for disjunction. Each of these disjunction rules is also intuitively appealing. To arrive at these rules we define negation and then assume that De Morgan's Law holds for fuzzy logic. With the zero-to-one truth scale, the most obvious definition for negation is:

$$t(\neg A) = 1 - t(A). \quad (5$$

Thus, starting from Equation 3 we arrive at the following definition of disjunction corresponding to the Minimum rule for conjunction:

$$t(A \vee B) = 1 - \min(1-t(A),\ 1-t(B)) = \max(t(A),\ t(B)). \quad (6$$

This formula, the Maximum rule, states that the truthfulness of a disjunction of two statements is equal to the truthfulness of whichever statement is most true. On the other hand, starting with the Multiplying rule for conjunction results in the following definition for disjunction:

$$t(A \vee B) = 1 - (1-t(A))*(1-t(B)) = t(A) + t(B) - t(A)*t(B). \quad (7$$

that is, the degree to which a disjunction is true is equal to one minus the product of the degrees to which the component statements are false, or in other words, the falsity of a disjunction is equal to the falsity of one statement times the falsity of the other. I call this rule the Inverted Multiplying rule in that it involves a product of terms which is then inverted about the neutral point on the truth scale.

To test between these alternative rules for fuzzy disjunction and conjunction, I (Oden, 1977b) used another set of judgment tasks. One task was based on questions such as "How true is it that both a sparrow is a bird and a penguin is a bird?" In addition to questions involving conjunction, like the preceding example, and questions involving disjunction, such as "How true is it that either a sparrow is a bird or a penguin is a bird?", subjects were also asked questions about the average truthfulness of statements. These averaging questions were included since the results of the "average" task could be depended upon to provide a validation of the subjects' use of the response scale as has been shown in numerous judgment studies (Anderson, 1974). This task also emphasized to the subjects that the "and" task was concerned with the logical sense of the word "and" rather than its aggregational sense.

As in the first experiment, the relative validity of the alternative rules was assessed using the procedures of information integration theory (Anderson, 1974). This required the construction of factorial stimulus designs by independently varying the truth values of the statements composing a stimulus. Each rule then makes a distinct prediction about the pattern of the data for such stimulus matrices.

The results of this experiment indicated that the psychological rules for both the conjunction and disjunction of fuzzy logical information are multiplicative in form at least for the kind of natural reasoning tasks used in this study. While the Minimum and Maximum rules were reasonable alternatives and were actually fairly successful at fitting the data, the Multiplying and Inverted Multiplying rules provided a substantially better fit to the data for every matrix and for the great majority of the subjects.

Since the evidence strongly supported the Multiplying rule as the psychological process of conjunction used by subjects in this experiment, it is reasonable to ask why the Minimum rule seemed so natural for the example concerning the roomful of 19 year olds. One explanation is that for problems such as this one for which all of the items are identical to each other it may be that prior to considering truth values the subjects reduced the problem to a simpler form. For example, people may interpret a question such as "How true is it that <u>both</u> a penguin is a bird <u>and</u> a penguin is a bird?" as being semantically equivalent to "How true is it that a penguin is a bird?" If this kind of semantic "preprocessing" occurs before the logical evaluation of such questions then they would actually not involve logical conjunction at all.

On the other hand, it is not unreasonable to expect that different rules may be used for conjunction under different situations. The experiment reported here only examined the logical operations involved in making judgments about non-identical statements concerning class membership. The Minimum rule or perhaps even other integration rules might be used in other situations. This plurality of rules is a reasonable possibility because the different rules have different properties which may make each of them more or less appropriate for different problems. Still, since the problems used here do seem to be representative of natural questions concerning fuzzy logical operations, the multiplicative rules for conjunction and disjunction will be assumed in the rest of the work to be reported.

Implicit and Explicit Negation on Semantic Continua

It is not uncommon for people to hedge a statement by using the negation of a term on a semantic continuum such as "It's not hot today" or "It's not unreasonable to assume that ..." Unlike dichotomous semantic oppositions where, for example, "not absent" necessarily means "present", with semantic continua the explicit negation of a term is less extreme than the opposite term. Intuitively, the explicit negation includes the neutral middle ground; for example, "not cold" includes "chilly", "luke warm", and so on, which are not included in "hot." However, this notion of including more of the continuum is incompatible with the traditional featural and dimensional models of semantic continua for which terms like "hot", "warm" and "cold" are represented as discrete points on the underlying dimension.

To explain this phenomenon, Oden and Hogan (1977) proposed a propositional semantic representation based on fuzzy logic. Thus, words like "hot" are not treated as semantic constraints but rather are considered to be fuzzy predicates and adverbs such as "not" and "very" are considered to operate upon the truth value given by the base predicate according to the usual rules:

$$t(NOT(P(x))) = 1 - t(P(x)) \tag{8}$$

$$t(VERY(P(x))) = P(x)^V \tag{9}$$

where $\underline{v}$ is presumably greater than 1.0 for the adverb "very." In addition, it was proposed, following Langendoen and Bever (1973), that a marked adjective for a continuum is actually an implicit <u>intensified</u> negative of its corresponding positive term. Thus, for example, "cold" is the intensified negative of "hot." Consequently, within the fuzzy propositional model, the fuzzy predicate for "cold" is:

$$t(COLD(x)) = t(VERY(NOT(HOT(x))))$$
$$= (1 - t(HOT(x)))^q \tag{10}$$

where $\underline{q}$ represents the degree of intensitification for this term, which may not be identical to that for the explicit "very." This formulation directly expresses the fact that "cold" is more extreme than "not hot" and also that, for any given level of truthfulness, the latter includes more of the middle temperatures.

To test these ideas, two experiments were run in which subjects read statements like "If ice melts quickly, then it is probably true that the weather is not hot." In the first experiment, subjects made a speeded true/false forced choice and time to respond was measured. In the second experiment, the same subjects judged the truth value of each statement using a continuous line mark rating scale.

Four 5x4x2 stimulus matrices based on different semantic continua were constructed. For each matrix, the first factor consisted of five phrases such as "ice melts quickly" which specified a range of subjective values along the underlying continuum. The second factor consisted of four terms or phrases, such as "hot", "chilly", "extremely cold", etc., which specified concepts defined over the semantic continuum. Of these four phrases, two were implicit negatives. The third factor had two levels which were whether or not the explicit negative "not" was included in the second clause of the statement.

For both the judgmental data and the verification times, the data were fit with models based on the hypothesized fuzzy propositional representations of the different terms. Chandler's (1969) STEPIT routine was used to obtain the best fitting predictions for each set of data.

For the judgments, the proposed definitions of the fuzzy predicates were applied directly to the data. The model provides a good account of the data and, therefore, supports the fuzzy propositional representation.

For the verification times, three additional important assumptions were made: (1) it takes different amounts of time to apply the base predicate to the subjective value depending on the resulting degree of truthfulness. In particular, it was expected that this process would take longer for the cases when the base predicate is less true. (2) The application of each operator takes an additional fixed amount of time. (3) It was assumed that the final degree of truthfulness is compared to some criterion in order to determine the forced choice response. It was expected that the farther the truth value is from this criterion, the more rapidly this final choice would be made. As a gross approximation to this final decision selection time, we used a function which dropped off linearly from a maximum time at .5 true. The predicted truth values from the judgmental experiment were used in estimating the times for this response stage. While there are a few large discrepancies, the model provides a reasonably good first approximation to the data and, therefore, gives additional converging support for the proposed fuzzy propositional representations.

Identification of Letters and Speech Sounds

Pattern identification is often considered to be a matter of extracting features or properties from stimuli and then matching the obtained features with descriptions of the alternative candidate patterns stored in long term memory. Here again, however, the traditional models of this sort have made the simplifying and undoubtably distorting assumption that each relevant feature is perceived to be either definitely present or definitely absent in a given stimulus. To overcome this shortcoming, a fuzzy logical model of pattern identification has been developed. The model consists of three conceptually distinct operations: the feature evaluation operation determines the degree to which each feature is present in a given stimulus, the prototype matching operation determines how close each candidate pattern comes to providing an absolute match to the stimulus and the pattern classification operation compares how well each pattern matches the stimulus relative to the goodness of match for the other patterns being considered.

Feature evaluation. For letter identification, feature evaluation must involve two processes. First, the optimal mapping must be found between the stimulus properties and the features of each pattern (Hayes-Roth, 1976). The optimal mapping is presumably that which maximizes the match of the pattern to the stimulus. Second, the degree to which each feature is present in the stimulus must be determined. Since features are considered to be fuzzy sets, this process corresponds to referring to the set membership function for each feature to determine the degree to which the relevant property of the stimulus is in fact an exemplar of the feature. Equivalently (Goguen, 1969; Zadeh, 1975b), we can think of the membership function of each feature as a fuzzy predicate and of the feature evaluation process as being the application of the predicate to the stimulus with the result being the degree to which it is true that the feature is present in the stimulus.

For example, consider a pattern prototype for which one of the features is the relation PARALLEL(*1,*2). According to the proposed model, there will be a fuzzy predicate for this feature, call it P12, that will be true to the degree to which the stimulus components that the mapping process associates with *1 and *2 are perceived to be parallel to each other. Thus, for a given stimulus, S, if the relevant lines are perceived to be quite parallel, the result of the feature evaluation process might be that it is .8 true that this feature is found in the stimulus. This can be expressed as:

$$P12 = t(P12(S)) = .8 \qquad (11$$

where again the notation t(A) represents the degree of truth of proposition A and P12 is used as a shorthand equivalent of t(P12(S)).

Prototype matching. Each letter pattern is defined by a prototype in long-term memory. These prototypes are propositions consisting of relational and componential features, as expressed by primitive semantic predicates, connected together by logical conjunction, disjunction and negation. To determine how well any given prototype matches the stimulus, the information about the degrees to which each of the features of the prototype is present in the stimulus must be integrated into a single value representing the goodness of match. According to the fuzzy logical model, the fuzzy truth values obtained by feature evaluation are combined through the use of fuzzy conjunction, disjunction and negation following the recipe specified by the prototype. In other words, the logical expression that is the prototype for the pattern is translated directly into a fuzzy logical expression by replacing all of the predicates by their associated fuzzy predicates and replacing all logical connectives by analogous fuzzy logical connectives. The resulting fuzzy logical expression is the matching function for the pattern and will be true to the degree that the prototype matches the stimulus.

As an example, consider the somewhat simplified description of an equal sign given by the following proposition:

$$\text{EQUAL-SIGN:}\quad \text{LINE}(*1) \wedge \text{LINE}(*2) \wedge \text{PARALLEL}(*1, *2) \wedge \text{NOTHING-ELSE}. \qquad (12$$

Based on the previous work (Oden, 1977b; Oden & Massaro, 1978) let

$$t(\neg A) = 1 - t(A) \qquad (13$$

$$t(A \wedge B) = t(A)*t(B) \qquad (14$$

and $$t(A \vee B) = 1 - (1 - t(A))*(1 - t(B)) \qquad (15$$

respectively. Then the matching function, Φ_{EQ}, corresponding to the pattern prototype of Equation 12 would be

$$\Phi_{EQ}(S) = L1*L2*P12*N \qquad (16$$

where L1, L2, P12 and N are the truth values of the fuzzy predicates associated with the four respective features of Equation 12.

Pattern classification. In the final operation, the goodness of match of each candidate pattern h is compared to that of all of the other alternative patterns to determine its relative goodness of match to the stimulus:

$$r = \frac{\Phi_h(S)}{\sum_{g=1}^{n} \Phi_g(S)} \qquad (17$$

In a forced choice situation, the person would presumably choose the pattern with the highest relative goodness of match value. In other situations, the relative goodness of match may be more directly reflected in each response.

In general, the selection among the patterns under consideration will also be influenced by external constraints such as that exerted by the orthographic, syntactic and semantic structure of the context in which the stimulus occurs. Contextual constraints may themselves be continuous and integrated along with the featural information to determine the overall goodness of a pattern as the identification of the stimulus (e.g. see Massaro, 1977; Oden, 1978b,c).

This model has received empirical support from several psychological experiments, both with speech stimuli (e.g. Oden, 1978a; Oden & Massaro, 1978) and with letter stimuli (e.g. Oden, in press). In addition, it has proven useful in accounting for certain psychological phenomena which would otherwise be difficult to explain (see Massaro & Oden, 1978; Oden & Massaro, 1978).

Continuous Semantic Constraints and Language Processing

One high level semantic process which is especially interesting and important is the role that semantic constraints play in language processing to determine whether or not a sentence is sensible. Semantic constraints are restrictions on what various parts of sentences may be if the entire sentence is to be sensible given the meaning of other parts. However, since it has been argued that meaning is based on fuzzy propositions, we will obviously need to consider the degree to which two sentence parts are semantically compatible. In other words, the degree to which a sentence is sensible will be determined by the degree to which its semantic constraints are satisfied. Supporting this notion is the fact that Oden and Anderson (1974) found that subjects judge the sensibleness of sentences by integrating the degree to which the various semantic constraints governing the sentence are satisfied.

Semantic constraints are known to provide important information which can be used by other stages of language processing, e.g. in speech perception (Reddy, Erman & Neely, 1973; Woods & Makhoul, 1973) and in syntactic analysis (Schank, 1972; Winograd, 1973). Language comprehension is a complex problem and appears to require the thorough use of the information provided by all of the many kinds of language structure. In particular, we should expect the language comprehension processes to make full use of the continuous information derived from semantic constraints. In fact, Oden (1978b) has shown that certain ambiguous sentences cannot be resolved to obtain the most often intended meaning without using the information provided by the fuzziness of semantic constraints. Furthermore, experiments have been performed which show that the degree to which people prefer one interpretation versus the alternative is a direct function of their relative sensibleness (Oden, 1978c) and also that the interpretation which is actually obtained, even when people do not necessarily know that the sentence is ambiguous, depends on which interpretation is most sensible as determined by continuous semantic constraints (Oden & Ekberg, 1978).

Building upon these results, I have proposed a model for how these continuous semantic constraints based upon fuzzy propositional semantic knowledge may be used to assist in the syntactic analysis of sentences. For the details of this model see Oden (1978b).

SUMMARY

The representation of knowledge is the foundation upon which many cognitive models must be based. There have been two general traditional approaches to modeling semantic memory. One approach, that of multidimensional spatial models, has stressed the continuousness of natural concepts but is inadequate at representing the interrelationships between the component features of concepts. The other approach, that of semantic network models, is patterned after formal logic and, consequently, is rich in structure but at the cost of not easily handling semantic continua. The fuzzy propositional approach described in the present paper appears to combine the major advantages of these two traditional approaches. In this formulation, which is based directly on fuzzy set theory and fuzzy logic, the logical propositions of semantic network models are adopted but the semantic primitives from which the propositions are constructed are assumed to be fuzzy predicates. In this way a minimum of additional cognitive apparatus is required beyond that which is necessary for conventional semantic network models: Once the primitives are made to be fuzzy, the complex concepts which are composed from these primitives also become fuzzy as a natural consequence. This approach has had considerable success in modeling psycholinguistic phenomena such as ambiguity resolution, quantifier understanding, implicit negation on continua, and letter and phoneme identification. Thus, the competancy for processing fuzzy semantic information appears to be a fundamental characteristic of human cognition. While still in its infancy, the psychological study of this fuzziness competancy promises to lead to more powerful and thereby also more accurate models of human thought processes.

References

Anderson, J. R. (1976) Language, memory and thought. Hillsdale, N.J.: Lawrence Erlbaum.

Anderson, N. H. (1974) Information integration theory: A brief survey. In D.H. Krantz, R.C. Atkinson, R.D. Luce, & P. Suppes (Ed.), Contemporary developments in mathematical psychology. Volume 2. San Francisco: Freeman.

Chandler, J. P. (1969) Subroutine STEPIT--Finds local minima of a smooth function of several parameters. Behavioral Science, 14, 81-82.

Goguen, J. A. (1969) The logic of inexact concepts. Synthese, 19, 325-373.

Hayes-Roth, F. (1976) Representation of structured events and efficient procedures for their recognition. Pattern Recognition, 8, 141-150.

Kintsch, W. (1974) The representation of meaning in memory. New York: Halsted.

Lakoff, G. (1972) HEDGES: A study in meaning criteria and the logic of fuzzy concepts. Papers from the eighth regional meeting of the Chicago Linguistic Society. Chicago: University of Chicago Linguistics Department.

Langendoen, D. T., & Bever, T. G. (1973) Can a not unhappy person be called a not sad one? In S.R. Anderson & P. Kiparsky (Eds.) A festschrift for Morris Halle. New York: Holt, Rinehart & Winston.

Massaro, D. W. (1977) Reading and listening. (Technical Report No. 423). Madison, Wisconsin: Wisconsin Research and Development Center for Cognitive Learning.

Massaro, D. W. & Oden, G. C. (1978) Evaluation and integration of acoustic features in speech perception. (WHIPP Report No. 9). Madison, Wisconsin: Wisconsin Human Information Processing Program.

Norman, D. A., & Rumelhart, D. E. (1975) Explorations in cognition. San Francisco: W. H. Freeman.

Oden, G. C. (1977a) Fuzziness in semantic memory: Choosing exemplars of subjective categories. Memory & Cognition, 5, 198-204.

Oden, G. C. (1977b) Integration of fuzzy logical information. Journal of Experimental Psychology: Human Perception and Performance, 3, 565-575.

Oden, G. C. (1978a) Integration of place and voicing information in the identification of synthetic stop consonants. Journal of Phonetics, 6, 83-93.

Oden, G. C. (1978b) On the use of semantic constraints in guiding syntactic analysis. (WHIPP Report No. 3). Madison, Wisconsin: Wisconsin Human Information Processing Program.

Oden, G. C. (1978c) Semantic constraints and judged preference for interpretations of ambiguous sentences. Memory & Cognition, 6, 26-37.

Oden, G. C. (in press) A fuzzy logical model of letter identification. Journal of Experimental Psychology: Human Perception and Performance, 5.

Oden, G. C., & Anderson, N. H. (1974) Integration of semantic constraints. Journal of Verbal Learning and Verbal Behavior, 13, 138-148.

Oden, G. C. & Ekberg, K. (1978) Semantic constraints determine the relative sensibleness of interpretations of ambiguous sentences. Paper presented at the meeting of the Midwestern Psychological Association, Chicago.

Oden, G. C., & Hogan, M. E. (1977) A fuzzy propositional model of negation on semantic continua. Paper presented at the meeting of the Midwestern Psychological Association, Chicago.

Oden, G. C. & Massaro, D. W. (1978) Integration of featural information in speech perception. Psychological Review, 85, 172-191.

Reddy, D. R., Erman, L. D., & Neely, R. B. (1973) A model and a system for machine recognition of speech. IEEE Transaction on Audio and Electroacoustics, AU-21, 229-238.

Rips, L. J., Shoben, E. J., & Smith, E. E. (1973) Semantic distance and the verification of semantic relations. Journal of Verbal Learning and Verbal Behavior, 12, 1-20.

Rosch, E. H. (1973) On the internal structure of semantic and perceptual categories. In T.E. Moore (Ed.), Cognitive development and the acquisition of language. New York: Academic Press.

Rosch, E. H. (1975a) Cognitive representations of semantic categories. Journal of Experimental Psychology: General, 104, 192-233.

Rosch, E. H. (1975b) The nature of mental codes for color categories. Journal of Experimental Psychology: Human Perception and Performance, 1, 303-322.

Schank, R. C. (1972) Conceptual dependency: A theory of natural language understanding. Cognitive Psychology, 3, 552-631.

Winograd, T. (1973) A procedural model of language understanding. In R.C. Shank & R.M. Colby (Eds.) Computer models of thought and language. San Francisco: W. H. Freeman.

Woods, W. A., & Makhoul, J. (1973) Mechanical inference problems in continuous speech understanding. Proceedings of the Third International Joint Conference on Artificial Intelligence, 200-207.

Zadeh, L. (1965) Fuzzy sets. Information and Control, 8, 338-353.

Zadeh, L. A. (1975a) Calculus of fuzzy restrictions. In L.A. Zadeh, K.S. Fu, K. Tanaka, & M. Shimura (Eds.), Fuzzy sets and their applications to cognitive and decision processes. New York: Academic Press.

Zadeh, L. A. (1975b) The concept of a linguistic variable and its application to approximate reasoning-II. Information Sciences, 8, 301-357.

Acknowledgements

The research reported in this paper was supported in part by grant BNS77-15820 from the National Science Foundation and by grants from the Wisconsin Alumni Research Foundation. This work has benefitted greatly from the suggestions and criticisms of many people including especially Norman Anderson, Lola Lopes and Dominic Massaro.

ADVANCES IN FUZZY SET THEORY AND APPLICATIONS
M.M. Gupta, R.K. Ragade, R.R. Yager (editor)

COMPOSITIONS OF FUZZY RELATIONS

E. SANCHEZ

Laboratoire de Biomathématiques, Statistiques et Informatique Médicale
Faculté de Médecine, Marseille - France.

ABSTRACT

In this paper we recall some results related to sup-min (or max-min) compositions of fuzzy relations equations. We describe a variety of problems for the existence and determination of solutions by means of maximizing and minimizing operators or compositions. Moreover, this paper serves as a basis to a companion paper "Medical Diagnosis and composite fuzzy relations" in which we follow the various presentations of the present one, but in terms of medical diagnosis applications.

KEY-WORDS

Fuzzy Relations, Compositions, Inference, Fuzzy Inequations, Brouwerian logic, Inverses, Eigen fuzzy sets, Mapping rule.

I - INTRODUCTION

The max-min composition of a fuzzy relation with a fuzzy set is closely related to the extension principle which extends the notion of a function acting on subsets of a given set, in fuzzy logic it is at the basis of rules of inference and it generalizes, of course, the Boolean sum-product from which many applications are derived.

In the following sections, we propose a kind of survey of our previous works on the max-min composite fuzzy relation equations and related, or derived, concepts such as (in)equations in Brouwerian logic, extensions of multi-valued mappings with their different kinds of inverses, eigen fuzzy sets, mapping rules in pattern classification.

2. FUZZY RELATIONS. SUP-MIN COMPOSITIONS.

In this section, we synthesize some relevant definitions and properties of binary fuzzy relations and sup-min compositions. Binary fuzzy relations are characterized by compatibility, or membership, functions defined over the cartesian product of two, possibly different, non fuzzy sets. So that a fuzzy relation R from X to Y is a fuzzy subset of XxY, characterized by its membership function

$$\mu_R : XxY \to [0,1]. \tag{1}$$

When A is a fuzzy subset of X, the sup-min composition of R with A, yielding B = RoA which is a fuzzy subset of Y, is defined by

$$\mu_{RoA}(y) = \sup_{x \in X} [\min (\mu_A(x), \mu_R (x, y))], \text{ for all } y \text{ in } Y. \tag{2}$$

Equation (2) is usually rewritten with the $\vee$ (sup) and $\wedge$ (min) operators, deleting "$\in X$" when no ambiguity arises, yielding

$$\mu_{RoA} (y) = \bigvee_x [\mu_A (x) \wedge \mu_R (x, y)] , \text{ for all } y \text{ in } Y. \tag{3}$$

When X and Y are finite sets, one may make use of a matrix representation to compute values derived from equations (2) or (3). Moreover, "sup" may be replaced by "max" and B = RoA is refered to as a max-min composition.

In the particular case with R being a (non fuzzy) function from X to Y, and where as usual the notation $y = R(x)$ replaces $\mu_R (x,y) = 1$,equation (3) yields, for all y in Y,

$$\mu_{RoA}(y) = \bigvee_{x \in R^{-1}(y)} \mu_A(x) \quad \text{for } R^{-1} (y) \neq \emptyset \tag{4}$$
$$= 0 \quad \text{otherwise}$$

The right-hand member of (4) is a formulation of the extension principle, see Zadeh [17] , defining $\mu_{R(A)}$ (y) which allows the domain of the function R to be extended from points in X to fuzzy subsets of X. So that, when R is a (non fuzzy) function from X to Y and A is a fuzzy subset of X, we have

$$RoA = R (A). \tag{5}$$

Sup-min composition of fuzzy relations. Let Q be a fuzzy relation from X to Y and R be a fuzzy relation from Y to Z, the sup-min composition of R with Q, yielding T = RoQ which is a fuzzy relation from X to Z, is defined by

$$\mu_{RoQ}(x,z) = \bigvee_y [\mu_Q(x,y) \wedge \mu_R (y,z)], \text{ where } y \in Y, \text{ for all } (x,z) \text{ in } XxZ \tag{6}$$

Remarks. When R and Q are non fuzzy relations, i.e. when their membership functions are {0, 1}-valued, the following property is derived from (6)

(x,z) ∈ RoQ
iff (7)
the exists y in Y such that (x,y)∈ Q
and (y,z) ∈ R,

which corresponds to a usual notation and definition of the composition of two (non fuzzy) relations.

Moreover, when R and Q are functions,

$$(x,z) \in RoQ \tag{8}$$

is denoted

$$z = (RoQ)\ (x) \quad \text{or} \quad z = R\ (Q\ (x)).$$

Such remarks, together with formula (5), were introduced to explain the notations RoA or RoQ, but one must be aware that the notation AoR, resp. QoR, is used in many papers for RoA, resp. RoQ, which are defined in (3), resp. (6).

3. COMPOSITE FUZZY RELATIONS EQUATIONS.

The most general problem of (sup-min) composite fuzzy relations equations consists in finding the solutions of T = RoQ, see (6), where T and Q (problem I) or T and R (problem II) are given fuzzy relations.

In fact, problem I and problem II are dual problems, for defining for any fuzzy relation S from X to Y, its inverse (or converse) S^{-1} which is a fuzzy relation from Y to X, such that

$$\mu_{S^{-1}}\ (y,\ x) = \mu_S\ (x,y) \quad ,\ \text{for all } (y,\ x) \text{ in } YxX, \tag{9}$$

one easily checks that

$$(RoQ)^{-1} \ = \ Q^{-1} o\ R^{-1}, \tag{10}$$

so that RoQ = T is equivalent to $Q^{-1} o R^{-1} = T^{-1}$ and a simple transformation allows one to pass from solutions in problem I to solutions in problem II, or vice-versa.

Because of the nature of the sup (or max) and min operators, one surmises that, in general and in both problems, when a solution exists it is not unique ; that was already the case with {0,1}-valued membership functions (Boolean case), [8].

For the existence and the determination of solutions in our dual problems, we need now to introduce the following operators.

For any elements a and b in [0,1], we define

$$\begin{aligned} a\ \alpha\ b &= 1 \quad \text{for } a \leqslant b \\ &= b \quad \text{for } a > b. \end{aligned} \tag{11}$$

Let Q be a fuzzy relation from X to Y and R be a fuzzy relation from Y to Z, we define T = Q ⓐ R as a fuzzy relation from X to Z by

$$\mu_{Q \circledast R}(x,z) = \bigwedge_{y \in Y} (\mu_Q\ (x,y)\ \alpha\ \ \mu_R\ (y,z)\ ),\ \text{for all } (x,z) \text{ in } XxZ, \tag{12}$$

where $\wedge$ is the "inf" operator, and α is the operator defined in (11).

Let us now denote by $\mathcal{R}$, resp. $\mathcal{Q}$, the set of solutions (when they exist) of problem I, resp. problem II, i.e.

$$R \in \mathcal{R} \text{ iff } T = R \circ Q \quad \text{(T and Q are given)} \tag{13}$$

$$Q \in \mathcal{Q} \text{ iff } T = R \circ Q \quad \text{(T and R are given)}. \tag{14}$$

We are now able to recall the following basic result, see [9 - 10].

$\mathcal{R}$ is not void if, and only if, the fuzzy relation $\check{R} = Q^{-1} \textcircled{\alpha} T$ is an element of $\mathcal{R}$, moreover $\check{R}$ is the greatest element in $\mathcal{R}$. (15)

In other words, given the fuzzy relations T, from X to Z, and Q, from X to Y, to know whether there exists a fuzzy relation R, from Y to Z, such that T = R o Q, it is sufficient to check that

$$\check{R} \circ Q = T, \text{ where } \check{R} = Q^{-1} \textcircled{\alpha} T. \tag{16}$$

If $\check{R} \circ Q \neq T$, there exists no R such that $T = R \circ Q$.

When (16) is verified, then $\check{R}$ is an element of $\mathcal{R}$ and for all R in $\mathcal{R}$, $R \subset \check{R}$, where "$\subset$" is the symbol of containment :

$$R \subset \check{R} \text{ iff } \mu_R(y,z) \leqslant \mu_{\check{R}}(y,z), \text{ for all } (y,z) \text{ in } Y \times Z \tag{17}$$

There is, of course, a dual result.

$\mathcal{Q}$ is not void if, and only if, the fuzzy relation $\check{Q} = (R \textcircled{\alpha} T^{-1})^{-1}$ is an element of $\mathcal{Q}$, moreover $\check{Q}$ is the greatest element in $\mathcal{Q}$. (18)

Remarks

Theorems (15) and (18) are still valid with more general fuzzy sets in the sense of L-fuzzy sets, see [5], where membership functions take values in a lattice L. For our purposes, L is allowed to be a complete Brouwerian lattice :

- a lattice L is said complete when each of its subsets M has a least upper bound and a greatest lower bound, in L.

- a Brouwerian lattice is a lattice L in which, for any given elements a and b, the set of all x in L such that $a \wedge x \leqslant b$, where $a \wedge x$ denotes the greatest lower bound of a and b, contains a greatest element denoted $a \,\alpha\, b$.

The sup-min composition of fuzzy relations is distributive on unions — but not on intersections — of fuzzy relations, hence the set of solutions of problem I or problem II, i.e. the elements of $\mathcal{R}$ or the elements of $\mathcal{Q}$ form a sup-semilattice.

4. RESOLUTION OF B = R o A

The equation under study, B = R o A defined in (3), is in fact a particular case of T = R o Q, see (6). The problems consist in finding the solutions of B = R o A, where B and A (problem III) or B and R (problem IV) are given fuzzy relations.

For practical purposes, we shall now assume that membership functions are defined over finite sets, so that solutions have a better description.

Let us denote by $\mathcal{S}$, resp. $\mathcal{A}$, the set of solutions (when they exist) of problem III, resp. problem IV, i.e.

$$R \in \mathcal{S} \text{ iff } B = R \circ A \quad (B \text{ and } A \text{ given}) \tag{19}$$

$$A \in \mathcal{A} \text{ iff } B = R \circ A \quad (B \text{ and } R \text{ given}). \tag{20}$$

STUDY OF $\mathcal{S}$.

Problem III is a particular case of problem I, so that from (15) one derives

$\mathcal{S}$ is not void if, and only if, the fuzzy relation

$A^{-1} \circledα B$ is an element of $\mathcal{S}$, moreover $A^{-1} \circledα B$ (21)

is the greatest element in $\mathcal{S}$,

where

$$\mu_{A^{-1} \circledα B}(x,y) = \mu_A(x) \;\alpha\; \mu_B(y) \;, \text{ for all } (x,y) \text{ in } X \times Y. \tag{22}$$

In order to study minimal solutions in $\mathcal{S}$, we need now to introduce a minimization operator σ, see [12].

For any elements a and b in [0,1], we define

$$a \;\sigma\; b = 0 \qquad \text{for } a < b \tag{23}$$

$$ = b \qquad \text{for } a \geqslant b$$

In the Boolean case, the σ operator is the Boolean product, otherwise for any elements a and b in [0,1], one has

$$a \;\sigma\; b \;\leqslant\; a \wedge b \;\leqslant\; a \;\alpha\; b \tag{24}$$

Let us now define the $\circledσ$ product of two fuzzy sets.

Let A be a fuzzy subset of X and let B be a fuzzy subset of Y, we define the fuzzy relation $A \circledσ B$ from X to Y by

$$\mu_{A \circledσ B}(x,y) = \mu_A(x) \;\sigma\; \mu_B(y), \text{ for all } (x,y) \text{ in } X \times Y \tag{25}$$

Dnoting by

$$R^{*} = A \circledcirc B \tag{26}$$

it is shown that

$$\mathcal{S} \text{ is not void iff } R^{*} \in \mathcal{S}. \tag{27}$$

Let us nox recall that a fuzzy relation R_m from X to Y is a minimal element of $\mathcal{S}$ if, and only if, R_m is an element of $\mathcal{S}$, and, if R is an element of $\mathcal{S}$ such that $R \subset R_m$, then $R = R_m$.

It is shown [12] that :

If $\mathcal{S}$ is not void, then $\mathcal{S}$ has minimal elements and the union of all minimal elements in $\mathcal{S}$ is equal to the fuzzy relation R^{*}. (28)

The determination of a minimal element R_m is as follows :

For all y in Y, the only x in X such that $\mu_{R_m}(x,y) \neq 0$ may be only one element that verifies $\mu_A(x) \geqslant \mu_B(y)$ and, moreover, $\mu_{R_m}(x,y) = \mu_B(y)$.

Let R_1 and R_2 be two fuzzy relations from X to Y, and bet A be a fuzzy subset of X. From the property

$$R_1 \subset R_2 \implies R_1 \circ A \subset R_2 \circ A, \tag{29}$$

it follows that, if R_1 and R_2 are two elements of $\mathcal{S}$ such that $R_1 \subset R_2$, then any fuzzy relation R from X to Y, such that $R_1 \subset R \subset R_2$, is an element of $\mathcal{S}$.

STUDY OF $\mathcal{A}$.

The determination of solutions in problem IV bas been investigated by C.P.PAPPIS and M.SUGENO [7], A.KAUFMANN [6], Y TSUKAMOTO [18], D.DUBOIS and H.PRADE [4] with not very different presentations.

The authors deal with interval-valued fuzzy sets, and properties related to the α or σ operators. They give practical algorithms for the determination of solutions in the finite case.

5.COMPOSITE FUZZY RELATIONS INEQUATIONS

GREATEST SOLUTION

The general problem is stated here in terms of fuzzy relational inequations of the type

$$R \circ Q \subset T , \tag{30}$$

where R is a fuzzy relation from Y to Z, Q is a fuzzy relation from X to Y and T is a fuzzy relation from X to Z.

With the preceding notations, the following properties were shown [12].

If b and d are two elements in [0,1] and if $b \leqslant d$, them $a \alpha b \leqslant a \alpha d$ for all element a in [0,1]. (31)

If R_1 and R_2 are two fuzzy relations from Y to Z, such that $R_1 \subset R_2$, them $Q \textcircled{\alpha} R_1 \subset Q \textcircled{\alpha} R_2$ for all fuzzy relation Q from X to Y. (32)

We now have two dual results.

For every pair of fuzzy relations Q, from X to Y, and T, from X to Z, the set of all fuzzy relations R, from Y to Z, such that $R \circ Q \subset T$, contains a greatest element $Q^{-1} \textcircled{\alpha} T$. (33)

For every pair of fuzzy relations R, from Y to Z, and T, from X to Z, the set of all fuzzy relations Q, from X to Y, such that $R \circ Q \subset T$ (34) contains a greatest element $(R \textcircled{\alpha} T^{-1})^{-1}$.

In both cases, the null relation is, of course, the least element of the set of solutions.

INEQUATIONS AND BROUWERIAN LOGIC

The reason why we speak of Brouwerian logics in this paper is that a Brouwerian Logic is associated with a Brouwerian lattice and as special cases, membership functions associated with fuzzy sets are Brouwerian lattices, see DE LUCA and TERMINI [3]. So, with our notations, let us recall a definition and a theorem, see BIRKHOFF [2].

A Brouwerian logic is a propositional calculus that is a lattice with 0 and I, in which

$$\begin{cases} (P \rightarrow Q) = I \quad \text{iff } P \leqslant Q & (35) \\ P \rightarrow (Q \rightarrow R) = (P \wedge Q) \rightarrow R \quad \text{for all } P, Q, R. & (36) \end{cases}$$

A Brouwerian logic is a Brouwerian lattice with $P \alpha Q$ relabelled as $P \rightarrow Q$.

In [12] we showed that a Brouwerian lattice, with $P \rightarrow Q$ relabelled as $P \alpha Q$, is associated with a Brouwerian logic. In fact, assertion (35) is evident in a Brouwerian lattice with I, so that we needed only to prove the property related to (36), i.e.

If L is a Brouwerian lattice, for a, b and c in L, we have $(a \wedge b)\ \alpha\ c = a\ \alpha\ (b\ \alpha\ c)$. (37)

Returning now to fuzzy relations, considered in a Brouwerian logic, one may show (see our second paper related to applications) that implications of the type

$$R \longrightarrow (Q \longrightarrow T) \qquad (38)$$

imply the fuzzy inequation

$$R \circ Q \subset T \,. \qquad (39)$$

Equation (38) generalizes a classical Boolean model of medical diagnosis assistance.

6. INVERSES OF 'MULTI-VALUED" FUZZY RELATIONS

The fuzzy relations under study here, are said "multi-valued" for they extend multi-valued mappings, so that their membership functions deal with subsets of given sets.

With multi-valued mappings, two kinds of inverses are usually defined, a lower inverse and an upper inverse, see C.BERGE [1], so that one surmises that two kinds of inverses will be defined in the fuzzy case. Moreover, for practical purposes we have introduced a ponderation of the upper inverse by means of a fuzzy relation of inclusion, see [15].

We first assume that a "multi-valued" fuzzy relation R from X to Y is characterized by its membership function

$$\mu_R : X \times 2^Y \longrightarrow [0,1] \qquad (40)$$

where 2^Y stands for the power set of Y.

μ_R is not a multi-valued mapping, but the fuzzy relation R is said "multi-valued" for its membership function associates a point in X and a subset of Y with a degree of compatibility lying in the unit internal.

Then, we extend the domain of μ_R to the product of two power sets by defining for each subset A of X, and for all B included in Y,

$$\mu_R\,(A,B) = \operatorname*{SUP}_{x \in A}\ \mu_R\,(x,B) \qquad (41)$$

When R is a "multi-valued" fuzzy relation from X to Y and if A is a (fuzzy) subset of X, one may define R o A, a fuzzy subset of 2^Y, by

$$\mu_{RoA}\,(B) = \operatorname*{SUP}_{x \in X}\,[\mu_A\,(x) \wedge \mu_R(x,B)]\,,\ \text{forall } B \subset Y \qquad (42)$$

So that, when A is not fuzzy one derives

$$\mu_R\,(A,B) = \mu_{R\,o\,A}\,(B) \qquad (43)$$

Moreover, (41) implies that

$$\text{for all } x \text{ in } X \text{ and for all } B \subset Y, \quad \mu_R(\{x\}, B) = \mu_R(x,B) \tag{44}$$

so that we need only deal , in the sequel, with fuzzy relations from 2^X to 2^Y and when writing that R is a fuzzy relation from X to Y, it is understood that its membership function is defined over $2^X \times 2^Y$, and we denote it by $R \subset 2^X \times 2^Y$.

So, let $R \subset 2^X \times 2^Y$, we define its lower inverse as $R_* \subset 2^Y \times 2^X$ by

$$\forall B \subset Y, \forall A \subset X, \ \mu_{R_*}(B,A) = \sup_{\substack{C \subset Y, \\ C \neq \emptyset, \ C \subset B}} \mu_R(A,C) \tag{45}$$

and we define its upper inverse as $R^* \subset 2^Y \times 2^X$ by

$$\forall B \subset Y, \forall A \subset X, \ \mu_{R^*}(B,A) = \sup_{\substack{C \subset Y \\ C \cap B \neq \emptyset}} \mu_R(A,C) \tag{46}$$

Definitions (45) and (46) are extensions of the definitions of lower and upper inverses of multi-valued mappings, and we have

$$R_* \subset R^* \quad . \tag{47}$$

Let us now introduce a type of fuzzy relation S expressing the degree of inclusion of two (non-fuzzy) subsets of a given set Y

$S \subset 2^Y \times 2^Y$, and for all subsets C and B of Y, $\mu_S(C, B)$ expresses the degree to which $C \subset B$. (48)

Such a fuzzy relation S is called a fuzzy relation of inclusion in Y. For example, if Y is a finite set and if $|D|$ denotes the cardinality of a (non-fuzzy) subset D of Y, the fuzzy relation S from Y to Y defined by fo all subsets C and B of Y,

$$\mu_S(C,B) = \begin{cases} \dfrac{|C \cap B|}{|C|} & \text{if } C \neq \emptyset \\ = 0 & \text{if } C = \emptyset \end{cases} \tag{49}$$

may serve as a fuzzy relation of inclusion.

For each subset B of Y, we define

$$f(B) = \{C \subset Y \mid C \cap B \neq \emptyset \text{ and } C \not\subset B\} \quad , \tag{50}$$

so that from (45) and (46) we derive, for all $B \subset Y$ and for all $A \subset X$,

$$\mu_{R^*}(B,A) = \mu_{R_*}(B,A) \vee \left[\sup_{C \in f(B)} \mu_R(A,C) \right] \tag{51}$$

In the right-band number of (51), in f (B) we take into account all the subsets C of Y which have a common part with B and which are not included in B, whatever might be the cardinality of the intersection compared with the cardinality of the involved sets. Hence it seems natural, for applications, to introduce the following ponderation by means of a fuzzy relation of inclusion S.

Let $R \subset 2^X \times 2^Y$, we define its ponderated upper inverse as $R^{*'} \subset 2^Y \times 2^X$ by $\forall B \subset Y, \forall A \subset X$,

$$\mu_{R^{*'}}(B,A) = \mu_{R_*}(B,A) \vee \left[\sup_{C \in f(B)} (\mu_R(A,C) \wedge \mu_S(C,B)) \right] \tag{52}$$

and we derive

$$R_* \subset R^{*'} \subset R^* \quad . \tag{53}$$

One may mow define the following composition.

Let $Q \subset 2^X \times 2^Y$ and $R \subset 2^Y \times 2^Z$; $T = R \circ Q$, $T \subset 2^X \times 2^Z$ is characterized by

$\forall A \subset X, \forall C \subset Z$,

$$\mu_{R o Q}(A,C) = \sup_{B \subset Y} [\mu_Q(A,B) \wedge \mu_R(B,C)] \tag{54}$$

so that we can rewrite $R^{*'}$ as follows.

$$R^{*'} = (S \circ R)_* \tag{55}$$

or, equivalently,

$$\forall B \subset Y, \forall A \subset X \ , \quad \mu_{R^{*'}}(B,A) = \mu_{S o R}(A,B) \ . \tag{56}$$

7. EIGEN FUZZY SETS

Returning now to the equation $B = R \circ A$ in the finite case, see (3) and section 4, we assume here that $X = Y$ and we study $A = R \circ A$. We call eigen fuzzy set, associated with a given fuzzy relation R from X to X, any fuzzy subset A of X such that : $R \circ A = A$ (57)

In the family of solutions, a major role is played by the greatest of them (the smaller one being of course the null set); in order to determine it,one needs to consider the sequence $(A_n)_n$ of fuzzy subsets of X defined by

$$\mu_{A_1}(x') = \underset{x}{\mathrm{MAX}}\ \mu_R(x,x') \quad (58)$$

$$\mu_{A_o}(x) = \underset{x'}{\mathrm{MIN}}\ \mu_{A_1}(x') \quad (59)$$

and

$$A_2 = R \circ A_1$$

$$A_3 = R\circ A_2 = R^2 \circ A_1 \quad (60)$$

$$A_{n+1} = R \circ A_n = R^n \circ A_1$$

A first property is that the sequence $(A_n)_n$ is decreasing and bounded by A_o and A_1.

$$A_o \subset \ldots \subset A_{n+1} \subset A_n \subset \ldots \subset A_3 \subset A_2 \subset A_1. \quad (61)$$

A basic result is that there exists an interger n less or equal to the cardinality of X and such that A_n is the greatest eigen fuzzy set association with R.

In [14] we proposed three methods, illustrated by examples, for the determination of A_n.

8. THE MAPPING RULE IN PATTERN CLASSIFICATION.

In the framework of the linguistic approach in pattern classification [16] , the mapping rule described in terms of a max-min composition provides a natural setting for the formulation of a linguistic interpolation.

Informally, the originality of Zadeh's presentation consists in converting an opaque recognition algorithm, which specifies some grades of membership of a fuzzy set F over natural objects, into a transparent fuzzy recognition algorithm defined on an associated space of mathematical objects.

For a natural object i,

$$A_{opaque}(i) = \mu_F(i) \quad (62)$$

$$A_{transparent}(M(i)) = A_{opaque}(i), \quad (63)$$

where the mathematical objects M(i) represent explicit or implicit attributes associated with i.

The two major problems are :

1) determination of measurement procedures for M(i)

2) choice and description of a "well-behaving" transparent recognition algorithm.

In general, the correspondence from M(i) to μ_F (i) is relational, so that F is assumed to be characterized by a relationnal tableau R with linguistic entries. The entries are allowed to be linguistic in order to set translation rules of fuzzy propositions related to ill-defined, not precise, coarse dependencies.

The relational tableau R is then translated into a fuzzy relation $\tilde{R}$. Finally, given a fuzzy set G related to attributes, the computation of the max-min composition $\tilde{R}$ o G corresponds to a linguistic interpolation yielding new grades of membership for F.

REFERENCES

[1] C. Berge, Topological Spaces, Including a Treatment of Multi-valued Mappings, The Mac Millan Company, New-York (1963).

[2] G. Birkhoff, "Lattice Theory", 3rd ed., American Mathematical Society Colloquium Publications, Vol. XXV, Providence, RI (1967).

[3] A.De.Luca and S. Termini, "Algebraic Properties of Fuzzy Sets", J. Math. Anal. Appl., 40, 373-386 (1972).

[4] D. Dubois and H. Prade, Fuzzy Sets and Systems : Theory and Applications, Academic Press, to appear.

[5] J.A. Goguen, "L - Fuzzy sets", J. Math. Anal. Appl. 18, 145-174 (1967).

[6] A. Kaufmann and E. Sanchez (Coll.), Compléments sur les concepts flous, Recherches et Applications, Vol. III, To appear.

[7] C.P. Pappis and M. Sugeno, "Fuzzy Relational Equations and the Inverse Problem", Internal Report, Queen Mary College, London (1976).

[8] E. Sanchez, "Matrices et Fonctions en Logique Symbolique", Thèse Mathématiques, Marseille (1972).

[9] E. Sanchez, "Equations de Relations Floues", Thèse Biologie Humaine, Marseille (1974).

[10] E. Sanchez, "Resolution of Composite Fuzzy Relation Equations", Information and Control, 30, 38-48 (1976).

[11] E. Sanchez, "Eigen Fuzzy Sets", Approximate Reasoning and Approximate Algorithms in Computer Science, NCC, New-York (1976).

[12] E. Sanchez, "Solutions in Composite Fuzzy Relation Equations. Application to Medical Diagnosis in Brouwerian Logic" in Fuzzy Automata and Decision Processes, M.M. Gupta, G.N. Saridis and B.R. Gaines eds, Elsevier North-Holland, New-York (1977).

[13] E. Sanchez, "Eigen Fuzzy Sets and Fuzzy relations". Memo No UCB/ERL M77/20, University of California, Berkeley, CA (1977).

[14] E. Sanchez, "Resolution of Eigen Fuzzy Sets Equations", Fuzzy Sets and Systems, 1, 69-74 (1978).

[15] E. Sanchez, "Inverses of Fuzzy Relations. Application to Possibility Distributions and Medical Diagnosis", Fuzzy Sets and Systems, 2, 75-86, (1979).

[16] L.A. Zadeh, "Fuzzy Sets and their Application to Pattern Classification and Cluster Analysis" Memo No ERL-M607, University of California, Berkeley, CA (1976).

[17] L.A. Zadeh, "Theory of Fuzzy Sets" in Encyclopedia of Computer Science and Technology, J. Belzer, A. Holzman and A. Kent (eds), Marcel Dekker, New-York (1977).

[18] Y. Tsukamoto, "Fuzzy logic based on Lukasiewicz logic and its applications to diagnosis and control". Ph. D. Dissertation, Tokyo Institute of Technology (1979).

****

PART THREE

APPLICATIONS

ADVANCES IN FUZZY SET THEORY AND APPLICATIONS
M.M. Gupta, R.K. Ragade, R.R. Yager (editors)

MEDICAL DIAGNOSIS AND COMPOSITE FUZZY RELATIONS

E. SANCHEZ

Laboratoire de Biomathématiques, Statistiques et Informatique Médicale
Faculté de Médecine, Marseille - FRANCE.

ABSTRACT

This paper illustrates problems of medical diagnosis based on max - min composite fuzzy relation equations which are described in a companion paper "compositions of fuzzy relations". They correspond to three stages : determination of symptoms, of a "medical knowledge", of diagnosis, all in the sense of degrees of membership of fuzzy sets or fuzzy relations.

KEY-WORDS : Medical diagnosis, Inference, Composite fuzzy relations, Brouwerian logic, Inverses, Mapping rule.

I - INTRODUCTION

In many cases, medical diagnosis involves processes that are susceptible to approximate rather than precise analysis. In this paper, the suggested basis, from which practical procedures can effectively be developed, is the use of methods based on the max - min composition of fuzzy relations. Such methods are a rich generalization of well-known medical diagnosis procedures derived from the study of Boolean sum-product equations.

The problems that arise in the resolution of composite fuzzy relation equations are presented in the companion paper "compositions of fuzzy relations" which will be refered to as CFR. The proposed methods were designed for applications and we illustrate in simple words how they can be translated in terms of medical diagnosis assistance in the framework of fuzzy set theory. The sections of this paper are in correspondence with the ones of CFR.

In a given pathology, we denote by $\mathcal{S}$ a set of symptoms, $\mathcal{D}$ a set a diagnosis and $\mathcal{P}$ a set of patients. What we call "medical knowledge" is a fuzzy relation, generally denoted by R, from $\mathcal{S}$ to $\mathcal{D}$ expressing associations between symptoms, or syndroms , and diagnosis, or groups of diagnosis.

Our aim is to describe or determine three stages related to medical diagnosis in which fuzzy concepts are inherent : the symptoms that are observed or supposed to be presented by patients, the "medical knowlededge", the diagnosis assigned to patients.

2 - ON THE INFERENCE OF DIAGNOSIS

Let A be a fuzzy subset of $\mathcal{S}$ related to a patient and let R be a fuzzy relation from $\mathcal{S}$ to $\mathcal{D}$, then the computation of the max - min composition B = R o A is assumed to describe the state of the patient in terms of diagnosis as a fuzzy subset B of $\mathcal{D}$, characterized by its membership function

$$\mu_B (d) = \underset{s \in \mathcal{S}}{\mathrm{MAX}} [\mu_A(s) \wedge \mu_R(s,d)] \ , \ d \in \mathcal{D} . \qquad (1)$$

In fuzzy logic, equation (1) expresses a <u>compositional rule of inference</u> (ZADEH), or a <u>fuzzy meta - implication</u> (KAUFMANN).
It generalizes a well - known procedure. If we have a curve represented by y = f(x), from a given assignment x = a, we can infer an assignment of y denoted by b = f(a).

So that we merely interpret (1) as follows.

If the state of a given patient p is described in terms of a fuzzy subset A of symptoms in $\mathcal{S}$, then p is assumed to be assigned diagnosis in terms of a fuzzy subset B of $\mathcal{D}$, through a fuzzy relation R of "medical knowledge", from $\mathcal{S}$ to $\mathcal{D}$.

Such a fuzzy relation R is assumed to be given by a physician who can translate his/her own perception of the fuzziness involved in the degrees of associations between symptoms and diagnosis.

What we propose in the next section are some methods designed to help the physician in the judgments that sometimes translate a great part of subjectivity.

3 - ANALYSIS FROM PATIENTS

R is still a fuzzy relation from $\mathcal{S}$ to $\mathcal{D}$, but we consider now several patients which are members of a set called $\mathcal{P}$. For each patient we have a corresponding A, i.e. a fuzzy subset of $\mathcal{S}$, and a fuzzy subset of $\mathcal{D}$ for diagnosis. In other words, we consider Q, a fuzzy relation from $\mathcal{P}$ to $\mathcal{S}$, and T, a fuzzy relation from $\mathcal{P}$ to $\mathcal{D}$ such that T = R o Q, i.e.

$$\mu_T\ (p,d) = \underset{s\,\in\,\mathcal{S}}{\mathrm{MAX}}\ [\mu_Q(p,s) \wedge \mu_R(s,d)] \quad , (p,d) \in \mathcal{P} \times \mathcal{D} \tag{2}$$

If $\mathcal{P}$ is reduced to a single element, equation (2) reduces to equation (1) of course. In fact, as indicated before, T = R o Q may be viewed as $B_p = R \circ A_p$ taking all p's in $\mathcal{P}$, where

$$\mu_{A_p}\ (s) = \mu_Q\ (p,s) \tag{3}$$

$$\mu_{B_p}\ (d) = \mu_T\ (p,d)\ . \tag{4}$$

We focus our attention here on the patients, for it is a convenient way of computing the degrees of compatibility of the relations Q and T, i.e. row after row. For the study of a given symptom s observed on the patients p in $\mathcal{P}$, we should have a fuzzy subset P_s of the patients and

$$\mu_{P_s}\ (p) = \mu_Q\ (p,s) \tag{5}$$

which constitutes an other way of computing the degres of compatibility of Q. Analogous remarks hold for T.

From the knowledge of Q and T, one may compute, when it exists, the greatest R such that T = R o Q, see (15) in CFR, i.e. the most significant fuzzy relation translating the higher degrees of associations of symptoms with diagnosis, an approach to the "medical knowledge".

Let us note the following result which gives a necessary condition for the existence of F from the knowledge of T and Q. It is easily checked from (2) :

for any patient p in $\mathcal{P}$

$$\mu_T\ (p,d) \leqslant \underset{s\,\in\,\mathcal{S}}{\mathrm{MAX}}\ \mu_Q\ (p,s), \text{ for all d in } \mathcal{D}\ . \tag{6}$$

Such inequalities may be useful for applications.

From the computed R, one may then infer diagnosis from symptoms in the sense of section 2. If results are not satisfactory for the physician, he/she must modify the relation R, making use of new information.

4 - RETURNING TO B = R o A

A patient p in $\mathcal{P}$ is fixed here and from the knowledge of A and B one may know the structure of "$\mathcal{S}$", i.e all the R's such that B = R o A.

Another way of using the results of CFR, section 4 (study of "$\mathcal{S}$") is the following.

Assume that a fuzzy relation of "medical knowledge" R is known so that from a fuzzy subset A of $\mathcal{S}$, one may infer a fuzzy subset B of $\mathcal{D}$ (it was the object of section 2). Now, with these given A and B we know, of course, that "$\mathcal{S}$" is not void, but it may be interesting to know how much we can increase or reduce the degrees μ_R (s,d), keeping unchanged A and B.

For a given ideal patient, from the knowledge of a fuzzy subset B of diagnosis and R, the study of "$\mathcal{A}$" allows one to derive the different fuzzy subsets of the symptoms, that is all A's such that B = R o A.

5. ON INEQUATIONS

The notations of R, Q and T have here the same meaning than in section 3 and we return to the determination of R, from the knowledge of T and Q. It may happen that the problem has no solution, in fact it is a very common case, and in such a case, on may always compute a "better" approximation of R, i.e. the greatest R such that R o Q $\subset$ T, see (33) in CFR.

What happens now in terms of Brouwerian logics? Let us consider (36) in CFR, with our notations :

$$\mu_R\ (s,d) \rightarrow (\mu_Q\ (p,s) \rightarrow \mu_T\ (p,d)) \qquad (7)$$

or, equivalently,

$$(\mu_R\ (s,d) \wedge \mu_Q\ (p,s)) \rightarrow \mu_T\ (p,d) \qquad (8)$$

Let us remark that in Boolean propositional calculus, in terms of {0,1}-valued membership functions, (7) or (8) are easily interpreted and they correspond to classical models of medical diagnosis.

From (35) in CFR, we have (8) = I if, and only if,

$$\mu_R (s,d) \wedge \mu_Q (p,s) \leqslant \mu_T (p,d) \quad (9)$$

which implies

$$\underset{s}{\mathrm{MAX}} (\mu_R (s,d) \wedge \mu_Q (p,s) \leqslant \mu_T (p,d) \quad (10)$$

or, equivalently,

$$\mu_{R \circ Q} (p,d) \leqslant \mu_T (p,d) . \quad (11)$$

If we assume that (11) is valid for all p in $\mathcal{P}$ and for all d in $\mathcal{D}$, we derive

$$R \circ Q \subset T , \quad (12)$$

which is a composite fuzzy relational inequation .

6 - ON "MULTI-VALUED" FUZZY RELATIONS

We deal here with power sets of $\mathcal{P}$, $\mathcal{S}$ and $\mathcal{D}$.

Let R be a fuzzy relation from $\mathcal{S}$ to $\mathcal{D}$, in the sense of a generalization of multi-valued mappings, see section 6 in CFR, which is denoted by $R \subset 2^{\mathcal{S}} \times 2^{\mathcal{D}}$.

The fuzzy relations R_*, R^* and $R^{*'}$ (from $\mathcal{D}$ to $\mathcal{S}$) enable a complete study of inverse problems which consist in the determination of degrees of combinationof diseaseswith groups of symptoms.

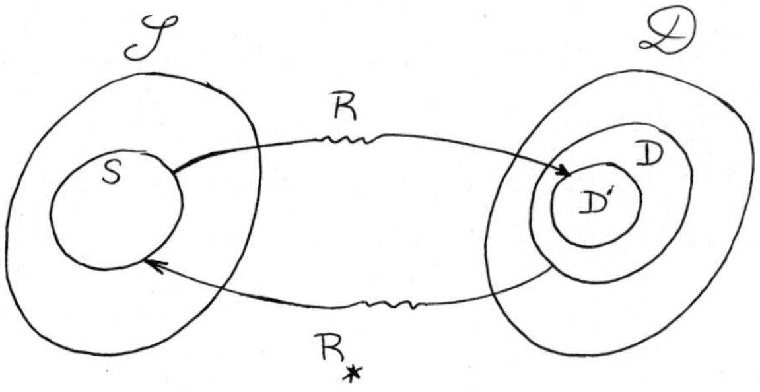

For all subsets D of $\mathcal{D}$ and S of $\mathcal{S}$,

$$\mu_{R_*}(D,S) = \underset{\substack{D' \subset D \\ D' \neq \emptyset}}{\mathrm{SUP}} \ \mu_R(S,D') \qquad (13)$$

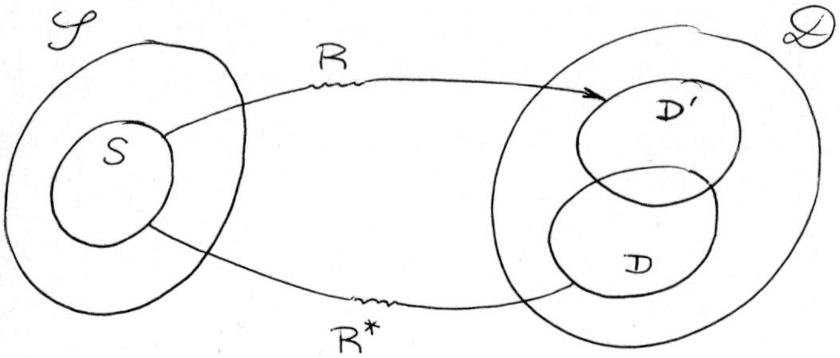

For all subsets D of $\mathcal{D}$ and S of $\mathcal{S}$,

$$\mu_{R^*}(D,S) = \underset{\substack{D' \\ D' \cap D \neq \emptyset}}{\mathrm{SUP}} \ \mu_R(S,D') \qquad (14)$$

Moreover, the ponderated upper inverse is

$$\mu_{R^{*'}}(D,S) = \mu_{R_*}(D,S) \vee \underset{\substack{D' \\ D' \cap D \neq \emptyset \\ D' \not\subset D}}{\mathrm{SUP}} \left[\mu_R(S,D') \wedge \mu_\Sigma(D',D)\right] \qquad (15)$$

where $\Sigma \subset 2^{\mathcal{D}} \times 2^{\mathcal{D}}$ is a fuzzy relation of inclusion, $\mu_\Sigma(D',D)$ expressing the degree to which D' is included in D.

Let us note that when D consists of a single diagnosis, e.g. D = {d}, the following property holds

$$\mu_{R_*}(d,S) = \mu_R(S,d) \qquad (16)$$

Moreover, we always have for all $D \subset \mathcal{D}$,

$$\mu_R(S,d) \leqslant \mu_{R_*}(D,S) \ , \ \forall\, d \in D\,. \qquad (17)$$

Finally, the use of $R^{*'}$ should be prefered to R^* in which the supremum involved in its definition takes into account some intersections which may not be of paramount significance.

Let us now indicate a formulation which shows some analogies with already proposed models.

For all subsets S of $\mathcal{S}$ and D of $\mathcal{D}$, let us assume that

$$S \longrightarrow D \qquad (18)$$

merely means "all patients, with observed symptoms in S, must present the diagnosis in D", and that

$$S \overset{\theta}{\longrightarrow} D \qquad (19)$$

means $S \longrightarrow D$ or, there exists a subset S' of S such that $S' \longrightarrow D$.

Defining now $\Sigma_{\#} \subset 2^{\mathcal{S}} \times 2^{\mathcal{D}}$ and $\Sigma^{\#} \subset 2^{\mathcal{S}} \times 2^{\mathcal{D}}$ by

$$\mu_{\Sigma_{\#}}(S,D) = 1 \text{ for } S \longrightarrow D, \quad = 0 \text{ otherwise} \qquad (20)$$

$$\mu_{\Sigma^{\#}}(S,D) = 1 \text{ for } S \overset{\theta}{\longrightarrow} D, \quad = 0 \text{ otherwise,} \qquad (21)$$

for $R \subset 2^{\mathcal{P}} \times 2^{\mathcal{S}}$, we derive

$$\mu_{R_*}(D,P) = \sup_{S \to D} \mu_R(P,S) = \sup_{S} [\mu_R(P,S) \wedge \mu_{\Sigma_{\#}}(S,D)] \qquad (22)$$

and

$$\mu_{R^*}(D,P) = \sup_{S \overset{\theta}{\to} D} \mu_R(P,S) = \sup_{S} [\mu_R(P,S) \wedge \mu_{\Sigma^{\#}}(S,D)] \qquad (23)$$

which allows one to infer associations of patients with diagnosis from observed symptoms.

7 - ON EIGEN FUZZY SETS

The application of eigen fuzzy sets equations, of the type R o A = A, see section 7 in CFR, is yet a marginal one and we suggest the following direction.

Given a fuzzy relation R, between symptoms, expressing the action of a drug on patients in a given therapy, what is the greatest intensity (in the sense of degrees of membership) of each symptom on which R produces no effect ?

8 - ON THE MAPPING RULE

In [1] we proposed an application at the thyroid pathology of Zadeh's linguistic approach in pattern classification.

Our purpose was to characterize different aspects of hyperthyroidism and euthyroidism, that played the role of the fuzzy sets approached by opaque algorithms given by a physician.

The attributes of the objects (patients) where biological tests : T4 test, I.T.L., T3 RIA, with linguistic values such as increased, very low, not so high, etc....

In order to improve the relational tableau R characterizing the fuzzy subsets,we performed the max-min composition of the mapping rule with the original data contained in R. The effect was smoother membership characterizations of the involved fuzzy sets after very few iterations of the algorithm.

R E F E R E N C E

[1] J.L. San Marco, E. Sanchez, G. Soula, R. Sambuc, J. Gouvernet. "Classification de formes floues. Application au diagnostic médical". Int. Coll. of Fuzzy Set Theory and Applications, Marseille (1978).

A more complete and general list of references is at the end of the companion paper "Compositions of fuzzy relations".

ADVANCES IN FUZZY SET THEORY AND APPLICATIONS
M.M. Gupta, R.K. Ragade, R.R. Yager (editors)

THE APPLICATION OF FUZZY SET THEORY TO MEDICAL DIAGNOSIS*

Wilfred A. FORDON and James C. BEZDEK

Abstract

Modern medicine must deal with the problems of organizing and analyzing large amounts of data. The digital computer can be one of the doctor's most useful assistants to help solve these problems.

Among the first attempts to utilize the computer was as an aid in medical diagnosis, applying Bayes' rule to a set of patient's symptoms in order to determine the disease. [1]

When it became evident that improved nosology was not the answer for clinical medicine, later investigators used a sequential approach using statistical decision theory and the computer to present the data in a form suitable for the physician to make decisions at each step in the diagnostic process.

It has also been demonstrated that even large amounts of data can not always be processed successfully for diagnosis if they are organized around "parametric" statistics. The remedy is a recourse to clustering techniques.

However, it appears that there is a class of problems for which even the robustness of clustering is insufficient. This occurs where the structures of the data are such as to cause a large amount of overlap among the various classes. In this case, the fundamental shortcoming often lies in the attempt to formulate the approach on the basis of binary set theory. An alternative is to supplant binary ("hard") set theory with fuzzy set theory.

This chapter exemplifies these concepts by applying them to the problem of performing a differential diagnosis of a number of surgically curable forms of hypertension. Here, the inadequacy of the parametric approach is demonstrated. Next, the successful application of a hard nonparametric technique to the form of hypertension known as primary aldosteronism is described. Finally, the application of hard nonparametric statistics to renovascular hypertension, and its potentially more effective analysis is expounded via a fuzzy clustering technique.

---

*This research partially supported by NSF Grant MCS77-00855.

1. Introduction

It is often believed in clinical medicine that computers and computer algorithms have had an insignificant impact on medical diagnosis.[2] Some of the major reasons are outlined by Feinstein,[3] as summarized below.

Historically, nosology has undergone many changes. For example, at the time of Hippocrates the names of diseases were identified directly by observing the patient at his bedside. This observation gave rise to symptoms which were given names such as, 'fever', 'cyanosis', and 'consumption'. In order to determine these symptoms, therefore, no chain of diagnostic reasoning was required.

In time, the concept of disease changed, so that by the 19th century many of the names of diseases had changed. The use of necropsy and microscopy created new forms of medical evidence, and the concept of disease was transformed from a clinical into an anatomic entity. By then, new methods of examing the patient had been developed (e.g. percussion, auscultation, thermometry, sphygmomanometry). By the end of the 19th century the concepts of 'physical diagnosis' had developed in much the same form as we know them today.

In the late 19th and early 20th centuries, there were further technological advances which resulted in an expansion of the morphologic concept of disease to include additional entities, such as bacteriology and virology, functional abnormalities as revealed by physiology and bio-chemistry, and abnormalities as demonstrated by immunology, electron microscopy, and biophysics.

Today, many names of diseases reflect abnormalities of morphology (e.g. carcinoma, atherosclerosis, cholelithiasis, appendicitis). Other disease names are taken from microbiology (e.g., amebiasis, streptococcal infection); other names are taken from biochemistry (e.g., porphyria, phenylketonuria, hyperlipemia); and others still from physiology (e.g., achalasia, atrial fibrillation).

Thus, history has had the effect of emphasizing the "name" of the disease rather than the diagnostic techniques involved. The contrast here is between symptomatology (e.g., hypertension) and pathophysiology (e.g., primary aldosteronism).

Because of the major influence of chronology on nosology, there are at least four major logical barriers to the depiction of the diagnostic process in a simplistic scheme, based on a single sequence of logic, or a single probabilistic formula. These barriers are discussed below.

Altered Disease Spectra

No matter what algorithm is developed for diagnosis, the scheme will lose its validity if the name of the disease is altered from a clinical to a morphologic, or from a morphologic to a bio-chemical term. This will serve to change the number of patients who have the disease. Thus, when disease A is converted to disease B, statistical data based on one form (e.g. morphologic) is not likely to be valid for another form (e.g. bio-chemical).

The Choice of Pathognomonic Evidence

When the names of diseases change, it is also possible to make more complete use of technology to develop evidence that is more specific, objective, preservable, and repeatable than clinical symptoms and signs. This can sometimes obviate the ambiguity arising when a number of diseases display the same symptoms and signs. Thus, modern clinical decisions are based on an appropriate test, rather than diagnostic reasoning alone.

The Choice of Diagnostic End Point

The many changes of disease nomenclature have sometimes resulted in difficulty in deciding when diagnosis has terminated.

For example, is 'myocardial infarction' a satisfactory explanation for a certain type of chest pain, or should the array of terms be extended sequentially to include 'coronary thrombosis', coronary atherosclerosis', 'hyperlipemia,' 'nutritional imbalance', and 'neurotic personality'? Here, the resolution depends heavily on the clinician's judgment and the constraints imposed by the individual patient.

Multiple Disease Concurrence

Frequently, it has been found that multiple diseases may co-exist in a single patient. If the diseases have no symptomatic or anatomic overlap, their different clinical contributions can be readily distinguished. However, in many practical cases, large overlap can make such a separation impossible.

To summarize, then, the process of diagnostic reasoning contains a complex series of logical branches, with the process ending not with a diagnostic name, but a decision to perform adjunctive tests or to administer therapy. Because of this complex process, techniques based only upon Bayes' Theorem, in which the input is a set of signs, symptoms, and tests, and the output is the most likely disease, (or diseases), are not likely to find much favor with clinicians who are primarily concerned with treating the patient.

In order to obviate these difficulties, there have been developments in the past ten years in which the diagnostic process has been modeled by means of sequential techniques.[4] In many complex diagnostic problems, such techniques are effective in improving the diagnosis--since in addition to coming closer to the reasoning employed by the modern clinician, they also specifically include quantification of loss of life and disability in the decision process. [5]

Despite the great potential improvement afforded to clinical decision making by sequential techniques, there are still difficulties. One of the greatest of these lies in the nature of the data base.

Differential diagnosis, by its very nature, involves applying what is generally known about signs, symptoms, and tests to the determination of the status of a specific individual--with the objective of improving it, if necessary. We therefore rely heavily on the size and specificity of the retrospective data base and the statistics that may be generated from it about the patient.

If the retrospective data base is large, including a sufficient number of patients with diseases similar to the patient's, a determination of the precise a priori probability distributions and their parameters is feasible. However, even large amounts of data cannot always be organized successfully to facilitate effective clinical decisions via parametric techniques.

The primary reason for this is that very often accurate diagnosis depends upon many factors; and unless a sufficiently large number of samples are available in the data base concerning patients with similar profiles (e.g. age, sex, geographic location) a precise determination of the parameters of the probability distribution is impractical--not to speak of determining the underlying distribution itself.

A secondary problem, but often almost as vexing, is that if one bases a diagnosis on parametric techniques in which an underlying distribution is assumed, large errors result when that assumption is false.

A possible remedy to these problems is the use of cluster analysis--a technique which seeks to separate data into constituent groups. This technique lies in the realm of nonparametric statistics.

In general, clustering procedures yield a description of the data base in terms of clusters--groups of data points which possess strong internal similarities. More formal procedures use a criterion function, such as the sum of the squared distances from the cluster centers, seeking the grouping that extremizes the criterion function.

The robustness of clustering (i.e., the lack of sensitivity of the technique to assumptions about underlying distributions) makes it applicable to many difficult medical diagnostic problems.[6] However, despite its efficacy, clustering analysis based upon conventional set theory can yield an uncomfortably large number of false positives in many situations.

A possible remedy to this shortcoming is to generalize the concept of memberships as a binary function, (either a sample is a member of the set or it is not), with the use of fuzzy set theory.

The following sections include discussion of some of these problems as exemplified in the differential diagnosis of two forms of hypertension--primary aldosteronism and renovascular hypertension--for which surgical correction is potentially feasible.

2. Bayes' Rule and the Decision-Theoretic Approach

Bayes decision theory is a fundamental statistical approach to the problem of pattern classification. This approach is based on the assumption that the decision problem is posed in probabilistic terms, and that all of the relevant probability values are known. [7]

In the words of decision theory, as applied to medical diagnosis, we say that the patient is in one of a number of possible states. These states can correspond to a number of discrete diseases, (or lack of disease). The decision in which state the patient belongs is determined by various features--as described by a set of signs, symptoms, and tests. If the a priori probabilities of each disease is known, as well as the disease-conditional probability densities of each of the features, it is possible to compute the posterior probability that a given set of features belong to a specific disease.

An example of a successful use of this technique is the diagnosis of polycythemic states using hematologic findings on patients. [8] The data were fitted to an assumed distribution, (e.g. normal, log-normal). The program was able to diagnose 95 cases out of 100 cases of polycythemia rubra vera or normal correctly where experienced hematologists correctly identified 76, and general practitioners 65.

As a less than successful example, one of the authors, (WAF), utilized this technique to attempt to obtain a differential diagnosis of three classes of hypertension: essential, (or idiopathic), primary aldosteronism, and renovascular hypertension. Lack of a sufficient data base and errors arising from false assumptions of the underlying distributions in this retrospective study contributed to the poor results. [9]

3. Clustering Techniques as applied to the Differential Diagnosis of Primary Aldosteronism

As an example of the successful application of clustering techniques, one of the writers, (WAF), has applied the 'k-nearest-neighbor rule' to the determination of hypertensive patients whose hypertension was the result of an excess production of aldosterone by the adrenal glands caused by adenoma or adrenal hyperplasia.

The k-nearest-neighbor rule involves calculating the 'distance' between each of the features of all new patients to the features of a retrospective patient data base in which diagnoses had already been determined. The distances were ordered, and the diagnoses of the k retrospective patients whose distances were closest to the new patient were examined. We assign to the new patient the diagnosis corresponding to the majority of the k-nearest retrospective patients, (diagnosis known). [7]*

The technique was applied to data obtained from patients subjected to a four-or-five-day protocol at the Specialized Center of Research (SCOR) in Hypertension of the Indiana University Medical School. [10]**

The parameters of critical importance in the analysis were determined to be Post-Saline Plasma Aldosterone (PAS) and Post-Lasix Plasma Renin Activity (PRAL), determined by radioimmunoassay. A total of 562 patients were examined for which these measurements had been obtained. The breakdown of the diagnoses as determined by the protocol is shown in Table 3.1.

* p. 233

** These studies were supported, in part, by USPHS grants HL14159, Specialized Center of Research (SCOR) in Hypertension and RR00750 (General Clinical Research Center).

Table 3.1 - Hypertensive Patients Classified as to Diagnosis - Special Center for Research in Hypertension, Indiana University School of Medicine

Diagnosis	Number of Patients
1. Normotensive	158
2. Essential Hypertension	276
3. Renovascular Hypertension	28
4. Uncertain	66
5. Miscellaneous*	12
6. Primary Aldosteronism	22
	562

Table 3.2 depicts a more detailed breakdown of the diagnoses and data for Primary Aldosteronism patients.

Using the two critical parameters (PAS and PRAL), digital computer runs were obtained for the k-nearest-neighbor algorithm for k up to 6 (i.e., at least 6 nearest neighbors of each patient were examined). The results are also shown in Table 3.2.

* Includes Page Kidney, Estrogen-Induced Hypertension, Pheochromocytoma, Renal Parenchymal disease, Pyelonephritis, and Coarctation of the Aorta.

Table 3.2 - Detailed Breakdown of the Data and Diagnoses for Primary Aldosteronism Patients at the Special Center for Research in Hypertension of the Indiana University School of Medicine

Clinical Diagnosis	Patient #	PAS (ng/100 mℓ)	PRAL (ng/mℓ/3h)	AGE (Years)	KNN* Diagnosis
Cause Uncertain	371	15.4	4.2	55.6	PA
Cause Uncertain	375	19.8	5.0	59.2	PA
Cause Uncertain	389	12.2	2.1	51.8	PA
Cause Uncertain	495	22.0	4.9	50.2	PA
Cause Uncertain	813	20.7	1.6	58.2	PA
Cause Uncertain	770	15.1	1.6	46.3	PA
Adenoma	772	67.1	1.4	52.8	PA
Adenoma	775	27.0	0.6	50.9	PA
Adenoma	780	33.7	0.4	52.0	PA
Adenoma	783	77.5	0.2	42.3	PA
Adrenal Hyperplasia	771	40.4	1.3	54.2	PA
Adrenal Hyperplasia	782	11.8	2.0	56.7	PA
Adrenal Hyperplasia	789	30.0	0.6	47.0	PA
Adrenal Hyperplasia	792	13.8	0.5	59.3	PA
Adrenal Hyperplasia	793	12.5	2.1	47.7	PA
Adrenal Hyperplasia	794	63.0	3.1	40.6	PA
Adrenal Hyperplasia	795	30.0	0.7	51.9	PA
Uncertain	800	21.3	1.5	55.0	PA
Uncertain	801	12.4	2.3	68.1	PA
Uncertain	803	13.0	0.7	42.7	PA
Uncertain	805	14.6	0.4	42.4	PA
Uncertain	799	10.7	4.1	62.8	E
Number		22	22	22	
Mean		26.55	1.88	52.17	
Sample Deviation		19.25	1.50	7.08	

* PA: Primary Aldosteronism
E : Essential Hypertension

In addition to the patients whose clinical diagnosis was Primary Aldosteronism, there were 4 other patients selected by the kNN algorithm, as shown in Table 3.3.

Table 3.3 - Additional Patients Diagnosed by the kNN Algorithm as PA

Clinical Diagnosis	Patient #	PAS (ng/100 mℓ)	PRAL (ng/mℓ/3h)	AGE (Years)	KNN Diagnosis
Essential Hypertension	526	10.7	4.2	43.1	PA
" "	556	14.8	3.0	40.8	PA
" "	606	11.3	0.2	55.1	PA
" "	668	10.8	2.0	52.2	PA
Number		4	4	4	
Mean		11.90	2.35	47.80	
Sample Deviation		1.95	1.69	6.92	

Table 3.4 is a summary of the results.

Table 3.4 - Predictive Value, Sensitivity, Specificity, and Efficiency Matrix for Primary Aldosteronism Patients at the Special Center for Research in Hypertension

		KNN Predicted Diagnosis: PA	NPA	Totals		
Clinical Diagnosis	PA	21 (TP)	1 (FN)	22	0.955	SE
	NPA	4 (FP)	536 (TN)	540	0.993	SP
		25	537	562		
		0.840	0.998		0.991	
		P+	P-		Eff	

TP = True Positive, P+ = Positive Predictive Value,
TN = True Negative, P- = Negative Predictive Value,
FP = False Positive, SE = Sensitivity,
FN = False Negative, SP = Specificity,
NPA = Non-Primary Aldosteronism, Eff = Efficiency, where

$$
\begin{aligned}
P+ &= TP/(TP + FP), \\
P- &= TN/(TN + FN), \\
SE &= TP/(TP + FN), \\
SP &= TN(TN + FP), \text{ and} \\
Eff &= (TN + TP)/(TN + TP + FN + FP).
\end{aligned}
$$

It is apparent that PAS and PRAL are sensitive and reliable indicators of Primary Aldosteronism. As Table 3.2 shows, only one false negative was produced, and the clinical diagnosis was uncertain. All confirmed cases were selected by the algorithm. In addition, there were four patients diagnosed by the algorithm as falling into the PA class which were classified as False Positives since the clinical diagnosis was Essential Hypertension. These patients are being more carefully evaluated by the clinic.*

The reason for the algorithm's effectiveness can be clarified by examining the means and standard deviations of the parameters for the PA and NPA classes. These are shown in Table 3.5, which shows a significant separation between the means of the two parameters for each class.

Table 3.5 - Comparison of Statistics of PA and NPA Classes

	Category			
	NPA		PA	
Parameter	Mean	Standard Deviation	Mean	Standard Deviation
Post-Saline Plasma Aldosterone (PAS)	5.02	3.28	26.55	19.25
Post-Lasix Plasma Renin Activity (PRAL)	22.59	20.28	1.88	1.50

4. The Application of Fuzzy Set Theory to the Diagnosis of Renovascular Hypertension

The previous example, in which conventional cluster analysis was successful in the determination of primary aldosterone suggests the application of the technique to other diseases. One of the writers, (WAF), has applied it to the diagnosis of renovascular hypertension. [9]

Initial attempts to utilize the kNN algorithm for renovascular hypertension led to indifferent success. Table 4.1 shows the results.

Table 4.1 - 3NN Algorithm-Renovascular Hypertension-PRAL & PAS as Parameters

		Predicted Diagnosis RV	E		
Retrospective Diagnosis	RV	6	22	28	0.214 = Sensitivity
	E	5	196	20	0.972 = Specificity
		11	218	229	
		P+ = 0.545	P- = 0.899		0.882 = Efficiency

A sequential technique was developed, based upon the kNN algorithm previously described--with some heuristic extensions. The features employed were post-saline plasma aldosterone, (PAS), post-lasix plasma renin activity, (PRAL), and pre-lasix plasma renin activity, (PRAPL); together with consideration of patients with abnormal intravenous pyelograms and abdominal bruits. Table 4.2 shows a summary of the results.

*One of these patients has subsequently been re-evaluated, found to have laboratory results consistent with the diagnosis of primary aldosteronism and operated on for the removal of an adenoma, curing the hypertension.

Table 4.2 - Sequential Heuristic Diagnostic Technique for Renovascular Hypertension

		Predicted Diagnosis RV	E	Total	
Retrospective	RV	28		28	1.000 = Sensitivity
Diagnoses	E	64	136	200	0.680 = Specificity
		92	136	228	0.719 = Efficiency

P+ = 0.304 P- = 1.000

It is apparent from Table 4.2 that the sensitivity of the test is excellent, although because of its ad hoc nature it would be difficult to automate. The biggest problem appears to lie in its relatively low specificity due to the large number of false positives (i.e., 64). The nature of this difficulty lies in the large overlap between the two classes, (E and RV), as seen from Figure 4.1, which is the scatter diagram of the two most sensitive parameters employed, (PRAL and PAS).

In order to improve the separation, a Fuzzy k-means algorithm [11] was applied to the Indiana University data.

The parameters selected for analysis included both binary and continuous types as described in the protocol. They were selected on the basis of previous heuristic and statistical procedures. [9] Table 4.3 lists the parameters. Table 4.4 is a breakdown of the diagnoses of the 285 patients.

Table 4.3 - Patient Parameters Included in the Analysis

	Parameter	Abbreviation	Feature #
Binary Parameters	Systolic Abdominal Bruit	BA2	1
	Continuous Abdominal Bruit	BA3	2
	Abnormal Intra-venous Pyelogram	IVP	3
	Serum Potassium $\leq$ 3.8 m Eq/ℓ	K	4
	Blood Urea Nitrogen > 20 mg/100 m	BUN	5
Continuous Parameters	Post-Lasix Plasma Renin Activity (ng/mℓ/3 hr)	PRAL	6
	Post-Saline Plasma Aldosterone (ng/100 mℓ)	PAS	7
	Pre-Lasix Plasma Renin Activity (ng/mℓ/3 hr)	PRAPL	8
	Pre-Saline Plasma Renin Activity (ng/mℓ/3 hr)	PRAPS	9

The binary and continuous parameters were selected primarily because they appeared to be useful in screening patients for Renovascular Hypertension.

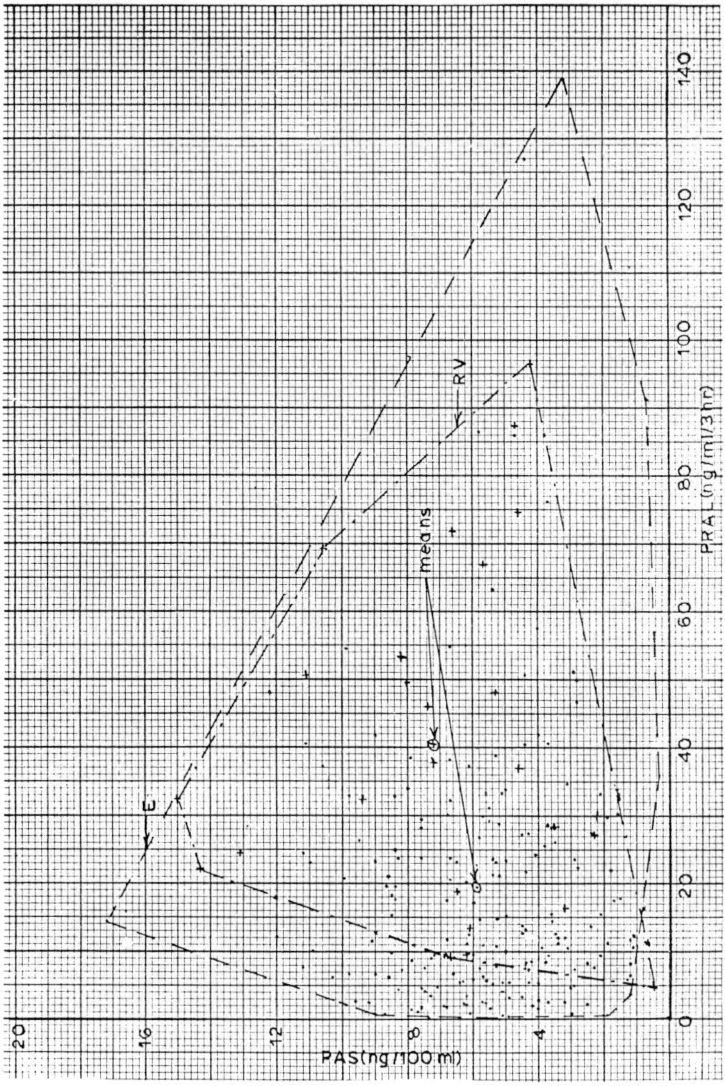

Figure 4.1 - Scatter diagram - PRAL/PAS - E, RV patients

Table 4.4 - Diagnoses of 285 Hypertensive Patients [10]

Diagnosis	Number of Patients
Essential (E)	194
Renovascular (RV)	24
Primary Aldosteronism (PA)	26
Uncertain (U)	34
Renal (R)	3
Pyelonephritis (PY)	3
Coarctation of the Aorta (CA)	1

This algorithm is described at length elsewhere ([11]-[13]). A brief summary follows. For $X = \{\underline{x}_1, \ldots \underline{x}_n\} \subset R^d$ any finite data set, a (cxn), $2 \leq c < n$ matrix $U = [u_{ik}]$ is called a fuzzy c-partition of X in case

$$\text{(1a)} \quad 0 \leq u_{ik} \leq 1 \quad \forall \; i,k \quad ;$$

$$\text{(1b)} \quad \sum_{i=1}^{c} u_{ik} = 1 \quad \forall \quad k \quad ;$$

$$\text{(1c)} \quad \sum_{k=1}^{n} u_{ik} > 0 \quad \forall \; i \quad .$$

We denote by M_{fc} the set of all such matrices in real (cxn) matrix space. The name partition alludes to our interpretation of each row of U as (values of) a membership function $u_i : X \rightarrow [0,1]$, whose values $u_{ik} = u_i(\underline{x}_k)$ denote the grades of a membership of each $\underline{x}_k$ in fuzzy set or fuzzy cluster u_i. If at (1a) every $u_{ik} = 0$ or 1, U is a conventional (or "hard") c-partition of X. Fuzzy k-means generates fuzzy c-partitions of X by iteratively optimizing the fuzzy least squared error functional:

$$\text{(2)} \quad J_m(U,\underline{v}) = \sum_{k=1}^{n} \sum_{i=1}^{c} (u_{ik})^m \| \underline{x}_k - \underline{v}_i \|^2 \quad ; \quad 1 \leq m < \infty \; ,$$

where $U \in M_{fc}$; $\underline{v} = (\underline{v}_1, \underline{v}_2, \ldots \underline{v}_c)$; $\underline{v}_i \in R^d$ is a fuzzy "cluster center" or prototype for fuzzy cluster u_i; and $\| \cdot \|$ is any differentiable norm on feature space R^d.

It was shown in (11) that $(\hat{U}, \hat{\underline{v}})$ might minimize J_m only if

$$\text{(3a)} \quad \hat{u}_{ik} = \frac{1}{\left[\sum_{j=1}^{c} \left(\frac{\| \underline{x}_i - \underline{v}_i \|}{\| \underline{x}_j - \underline{v}_j \|} \right)^{\frac{2}{m-1}} \right]}$$

(3b)
$$\hat{\underline{v}}_i = \frac{\sum_{k=1}^{n} (\hat{u}_{ik})^m \underline{x}_k}{\sum_{k=1}^{n} (u_{ik})^m} \quad \forall i$$

Equations (3) provide a loop for iterative minimization of J_m.

(A1) Fix m, $1 < m < \infty$

(A2) Guess $U_o \varepsilon M_{fc}$.

(A3) Calculate $\{\hat{v}_i\}$ with (3b).

(A4) Update $U_o \to \hat{U}$ with (3a).

(A5) Compare U_o to $\hat{U}$ with a cutoff criterion ε ; if

$||U_o - \hat{U}|| \leq \varepsilon$, stop. Otherwise, put $\hat{U} \to U_o$ and go to (A3).

Parameter c, the number of clusters in X, may or may not be prespecified. If it isn't, several scalar measures of partition quality described in reference [13] may be useful in addressing the question of cluser validity. In the example to follow, c is assumed known and fixed for loop (A1)-(A5) at c=2.

One standard approach to classification via decision functions is to utilize prototypes for each of the c clusters to define the decision boundaries. To this end, we define

(4) $$\hat{d}_j(\underline{x}) = \langle \underline{x}, \hat{\underline{v}}_j \rangle - 1/2 ||\hat{\underline{v}}_j||^2 ; 1 \leq j \leq c.$$

In (4) $\langle \cdot , \cdot \rangle$, $||\cdot||$ are the Euclidean inner product and norm, respectively. Using the decision strategy,

(5) $$\text{decide } \underline{x} \; \varepsilon \; i \iff \hat{d}_i(\underline{x}) = \max_{1 \leq j \leq c} \{\hat{d}_j(\underline{x})\},$$

with ties resolved arbitrarily, results in a piece wise linear classifier whose decision boundaries are (pieces of) hyperplanes in R^d occurring where $\hat{d}_i(x) = \hat{d}_j(x)$, i.e., by the hyperplanes $H_{ij} = \{(\underline{x} \, \varepsilon \, R^d | \langle \underline{x}, \hat{\underline{v}}_i - \hat{\underline{v}}_j \rangle = 1/2 \, (||\hat{\underline{v}}_i||^2 - ||\hat{\underline{v}}_j||^2)\}$. Different error rates accrue by using prototypes $\{\hat{\underline{v}}_j\}$ obtained by various training algorithms. In any case, we call $\{\hat{d}_j(x)\}$ a one-nearest prototype (1-NP) linear classifier; in particular, a fuzzy 1-NP linear classifier in case the $\hat{\underline{v}}_j$'s derive from equation (3b) above. It will be impossible for us to attain the error rate of Bayes' risk (P_b) achieved by using $\hat{d}_j(x) = \Pr$ (class $j/x) = p_j g(\underline{x}/j) / \sum_{i=1}^{c} p_i g(\underline{x}/i)$, which is optimal with respect to the Neyman-Pearson probability of error. [7] It is our supposition, however, that the softening of partitioning boundaries by fuzzy partitions of X will provide a classifier with less sensitivity to mixed samples with noise and errors than most hard-training algorithms. The numerical example below supports this conjecture.

Numerical Example

In the discussion to follow, algorithmic parameters for fuzzy k-means were fixed as (chosen by preprocessing with (A1-A5):

Number of Subjects n=285
Number of Classes c=2
Weighting Exponent m=1.33
Norm $||\cdot||$ for J_m=Euclidean : Cutoff Criterion, $\varepsilon = .01$
Cutoff norm $||U - V|| = \max_{i,k} \{|U_{ik} - V_{ik}|\}$

Runs were made on 3 feature subsets, viz., features {1, 2, 3} ; {6, 8, 9}; and {1, 2, 3, 6, 8, 9}. We shall be interested in observing whether--as we suspect--the values of continuous features {6, 8, 9} are sufficient in magnitude to dominate the binary features {1, 2, 3} in the Euclidean norm.

A total of 20 runs were made, using a modification of Toussaint's rotation method [14] for training and testing. Specifically: X was sampled at random 35 times; this set was processed with fuzzy ISODATA (A1)-(A5) to secure $\{\hat{\underline{v}}_j\}$'s, and hence $\{\hat{d}_j(x)\}$ via (4); the remaining 250 points in X were then attached to $\hat{\underline{v}}_1$ or $\hat{\underline{v}}_2$ via minimum distance (equivalently, maximum $\hat{\underline{d}}_j(x)$) to fuzzy prototype; mistakes on the labelled test points were counted. The data was then restored to 285 points, and randomly sampled again. This procedure was repeated 10 times for each of the feature subsets {1, 2, 3} , {6, 8, 9} , and {1, 2, 3, 6, 8, 9} . We should emphasize exactly how errors were recorded. At the end of a training session, $\hat{d}_1$ and $\hat{d}_2$ are labelled numerically but not physically. Each labelled test vector was then adhered to either $\hat{\underline{v}}_1$ or $\hat{\underline{v}}_2$ via the expedient of minimum distance. At the end of testing, the number of labelled members of each of the seven classes attached to each $\hat{\underline{v}}_i (i=1, 2)$ was counted, and $\hat{\underline{v}}_i$ received the physical label corresponding to the maximum number from a single class. If, for example, $\hat{\underline{v}}_2$ had 101 E's, 19 RV's, 17 PA's, 6 U's, and 1 PY at the end of testing, $\hat{\underline{v}}_2$ is declared a prototype for E, and we ascribe 43 errors in 144 tries to these assignments. It may happen, of course, that one class actually ends up with more than one prototype using this strategy, while others are not represented at all!

After 10 runs of this type were made, the average empirical error rate was calculated. Note that we have not used convex weighting of the rotation and resubstitution methods as advocated by Toussaint, so the error rates listed for these runs is somewhat biased towards the pessimistic side. A summary of the observed performance of fuzzy 1-NP classification using these protocols is found in Table 4.5. These figures indicate that features {1, 2, 3} may contain classification information of significant value since the average error rate is about 2% less in every instance than when continuous features are used. Indeed, the 10 runs on {6, 8, 9} and {1, 2, 3, 6, 8, 9} (same random seed for each of two feature subsets) yielded identical classifiers, neither of which was as accurate as the one designed on features {1, 2, 3} . This verifies that one must use a feature weighting transformation if the information carried by the binary features is to have any influence in classifier performance.

Since 34 of the 285 patients are "uncertain," these were automatically mistakes. In every case, the larger cluster was essential (E) hypertension, as one might expect from the predominance of this subclass in the data; and the second cluster in every case was renovascular (RV) although (PA) came within one patient of a tie on several runs. Table 4.5 also exhibits the empirical error rate of the $\{\hat{d}_j\}$'s when viewed as a "strainer" for class (E) patients only: error rates drop to about 29% from 36% using any of the three feature subsets. This is probably because this discounts the 34 certain errors of the "U" class patients. Again, the binary features seem to possess an advantage, albeit a slight one, and are dominated in norm by the continuous ones.

Table 4.5 - Empirical Estimates for Probability of Error Using a Fuzzy 1-NP Linear Machine, Percent

	Features {1, 2, 3}		Features{ 6, 8, 9} or {1, 2, 3, 6, 8, 9}	
Run	All Classes	Class E Only	All Classes	Class E Only
1	38	31	52	30.8
2	38	30	35.6	30.8
3	35.2	28.2	35.2	30.7
4	34.8	27.2	32	27.5
5	36.8	29.2	34.8	30
6	38.8	31.9	48	28
7	36	27.6	44	31.4
8	34	28	36	26.7
9	35.2	29.5	32.8	31.5
10	36.5	28.3	31.2	29.5
Average	36.3	29.1	38.2	29.7

Excluding uncertain diagnoses, and a few patients diagnoses as having hypertension resulting from renal disease, pyelonephritis and coarctation of the Aorta, there are 244 patients in three principal classes: E(194), RV(24), and PA(26).

It has already been demonstrated that Primary Aldosteronism can be reliably detected by the application of a kNN "hard" algorithm. The screening of patients for Renovascular hypertension is less effective by hard non-parametric techniques, as was discussed previously.

If we concentrate on RV and E patients, a total of 218 patients' data were examined via the Fuzzy k-means algorithm.

A simple classifier was assumed. The membership functions of each of the patients resulting from the Fuzzy ISODATA algorithm using only the continuous parameters PRAL, PAS, PRAPL, and PRAPS were examined. The case for c=4 and m=1.75 was examined. Class 4 was assumed to be characteristic of Essential hypertension, and classes 1, 2, and 3 were assumed to contain Renovascular patients. Membership in the resulting two class problem was assumed to be a membership function >0.5. In addition, all E or RV patients having an abnormal IVP were classified as RV. The resulting combined error matrix is shown in Table 4.6.

Table 4.6 - Combined Error Matrix - Fuzzy k-means and Abnormal IVP for 218 Essential and Renovascular Hypertensive Patients

		RV	E	Totals		
Actual	RV	21 (TP)	3* (FN)	24	0.875	Sensitivity
Diagnoses	E	39 (FP)**	155 (TN)	194	0.799	Specificity
		60	158	218		
		P+ = 0.350	P- = 0.981	Efficiency = 0.807		

* Includes 2 patients for which an IVP was not administered.
** Includes 35 False positives from Fuzzy ISODATA and 4 False positives as a result of Abnormal IVP.

As seen in Table 4.6 the overall efficiency of the algorithm is 80%. This is better than a sequential algorithm based upon a 3-Nearest-Neighbor, (3NN), technique[9] as shown in Table 4.2. The superiority of the Fuzzy k-means algorithm over the sequential technique based upon a 3NN algorithm lies in the significantly fewer False positives resulting from the Fuzzy ISODATA algorithm (39 as opposed to 64).

Summary

In view of the difficulties outlined in the introduction, total reliance by clinicians upon computerized medical diagnosis seems unrealistic. Considering the risks involved, the medical community is entitled to demand proven success before it can be asked to rely on the diagnostic suggestions of a computer. It is our conviction, however, that the pessimism displayed in [15] concerning this enterprise is due at least in part to failures which were caused not so much by the impossibility of finding useful algorithms, but rather, by the restrictivness imparted to the underlying model by the axioms of conventional set theory. The numerical example presented above seems to corroborate this supposition. Our contention is that computerized diagnostic advice based on sufficiently large data bases can supply the clinician with valuable insight and direction. References [16-18] suggest that fuzzy sets may assume an important place in the evolution of this new application of data processing.

References

[1] Ledley, R. S., L. B. Lusted, "Reasoning Foundations of Medical Diagnosis," Science, 130: 9-21, 1959.

[2] Conn, H. F., R. B. Conn Jr., (Eds.), "Current Diagnosis-4", W. B. Saunders Co., 1974.

[3] Feinstein, A. R., "An Analysis of Diagnostic Reasoning," Yale Jour. Biol. Med. 46, 1973, I. The Domains and Disorders of Clinical Macrobiology, pp. 212-232, II. The Strategy of Intermediate Decisions, pp. 264-283; Vol. 1, 1974, III. The Construction of Clinical Algorithms, pp. 5-32.

[4] Gorry, G. A., G. O. Barnett, "Experience with a Model of Sequential Diagnosis," Comp. Biomed. Res. 1. 1968, pp. 490-507.

[5] Barnoon, S., H. Wolfe, "Measuring the Effectiveness of Medical Decisions-An Operations Research Approach," Charles C. Thomas, 1972.

[6] Jacquez, J. A., (Ed.), "Computer Diagnosis and Diagnostic Methods," Charles C. Thomas, 1972.

[7] Duda, R. O., P. E. Hart, "Pattern Classification and Scene Analysis," John Wiley and Sons, 1973.

[8] Bishop, C. R., H. R. Warner, "A Mathematical Approach of Medical Diagnosis: Application of Polycythemic States utilizing clinical findings with values continuously distributed," Comp. Biomed. Res., 2: Oct. 1969, pp. 486-493.

[9] Fordon, W. A., "Computer-Aided Differential Diagnosis of Hypertension," Ph.D. Thesis, Purdue University, W. Lafayette, IN., 1976.

[10] Grim, C. E., M. H. Weinberger, et. al., "Diagnosis of Secondary Forms of Hypertension-A Comprehensive Protocol," JAMA 237: 1331-1335, 1977.

[11] Bezdek, J. C., "Fuzzy Mathematics in Pattern Classification," Ph.D. Thesis, Cornell University, Ithaca, New York, 1973.

[12] Bezdek, J. C., "Numerical Taxonomy with Fuzzy Sets," J. Math. Biol., 1-1, 57-71, 1974.

[13] Bezdek, J. C., "Some Current Trends in Fuzzy Clustering," in The General Systems Paradigm, Proc. SGSR, J. White, Ed., SGSR Press, Washington, D. C., 347-353, 1977.

[14] Toussaint, G., P. Sharpe, "An Efficient Method for Estimating the Probability of Misclassification Applied to a Problem in Medical Diagnosis," Comp. Bio. Med., 4, 269-278, 1975.

[15] Croft, D., "Mathematical Methods in Medical Diagnosis, " Annals Bio. Engr., 2, 68-89, 1974.

[16] Bezdek, J., "Feature Selection for Binary Data-Medical Diagnosis with Fuzzy Sets," Proc. NCC, J. White, ed., 1057-1068, 1976.

[17] Bezdek, J., "Prototype Classification and Feature Selection with Fuzzy Sets," IEEE Trans. SMC, SMC-7, 2, 87-92, 1977.

[18] Sanchez, E. "Fuzzy Relations, Possibility Distributions, and Medical Diagnosis," Proc. 1978 IEEE-CDC, San Diego, 1979.

ADVANCES IN FUZZY SET THEORY AND APPLICATIONS
M.M. Gupta, R.K. Ragade, R.R. Yager (editor)

FUZZINESS AND CATASTROPHE IN ESTIMATION AND DECISION PROCESSES

Masasumi KOKAWA
Systems Development Lab.*
Hitachi Ltd.
1099, Ohzenji, Tama-ku
Kawasaki, 215 Japan

Kahei NAKAMURA and Moriya ODA
Automatic Control Lab.
Faculty of Engineering
Nagoya University
Nagoya, 464 Japan

Abstract

This paper describes the results of inference and hint-effect experiments, and analyses the human decision-making process from fuzzy-theoretic and concept-formational viewpoints. To begin with, we conduct a fuzzy-theoretic experiment on inference and show an interrelation among memory, newly-acquired information, and inference. Next, we discuss the problem of the decision-making process subject to a specific and newly-given information. In other words, a fuzzy-theoretic experiment with some hints in the estimation and decision processes of the card arrangement is conducted. Then, the concept-formational process is analysed clearly by means of the introduction of both the cusp-curved surface in catastrophe theory and fuzziness concept in fuzzy set theory. Finally, we show the hint effect in the estimation and decision processes.

1. Introduction

The studies on the human thinking process can roughly be divided into two approaches. One is the microscopic approach, e.g., physiological-and-molecular-biological approach [1], [2], and the other is the macroscopic approach, e.g., psychological approach [3], [4]. The human thinking process is usually accompanied by the property of "fuzziness." The concept of fuzzy set, introduced by L. A. Zadeh, does seem appropriate for the study of the human thinking process with fuzziness, as the fuzzy set theory can deal effectively with the subjective fuzziness of human [25]. In order to formulate the human thinking process, we have reported several works on a human decision-making model [5], [6], a fuzzy process of memory behavior [7], a fuzzy-theoretic and concept-formational approaches to inference [8], fuzzy-theoretic dimensionality reduction method of multidimensional quantity [9], [10], a handling method of fuzzy data [11], and

---

* This study was made while the author was working at Nagoya University.

so on, by means of the macroscopic approach.

As a continuation of the authors' serial studies, this paper describes the results of inference and hint-effect experiments, and analyses the decision-making process from fuzzy-theoretic and concept-formational viewpoints. Moreover, the concept-formational process including a jump of logic, which is not explained uniformly and generally in conventional studies of psychology and psychological engineering, is explained uniformly and generally by means of the introduction of both the cusp-curved surface in catastrophe theory* and fuzzy set theory. In Section 2, we conduct a fuzzy-theoretic experiment on inference, that is, the experiment to estimate and decide the hidden regularity of the arranged playing cards, and show a quantitative interrelation among memory, newly-acquired information, and inference. In Section 3, we conduct a fuzzy-theoretic experiment on the effect of external hint, that is, an experiment to estimate and decide the hidden regularity of a card arrangement under some hints. We thus analyse (1) the concept-formational process and fuzzy catastrophe and (2) the hint effect to a subject in his estimation and decision processes. Section 4 is the conclusion of the paper.

2. Fuzzy-Theoretic Experiment on Inference and Analysis of Results

2.1 Fuzzy-theoretic experiment on inference

The experimental material was a set of playing cards. Five men and five women aged 19 to 29 served as subjects for each test arrangement of the cards. The experiment was carried out according to the following procedure:

(1) The subject was given one of seven kinds of the 4 × 13 two-dimensional array of 52 playing cards (each kind of array had its own regularity).

(2) All cards were initially arranged with the reverse side up.

(3) At each trial, the subject was instructed to guess the position of a particular card specified by the experimenter.

---

* The catastrophe theory is a geometric theory of the construction of a model. R. Thom emphasizes in his book [18], that "Although the goal is to construct the quantitative global model, this may be difficult or even impossible. However, the local dynamical interpretation of the singularities of morphogenesis is possible and useful and is an indispensable preliminary to defining the kinematics of the model." Agreeing with his opinion, we construct the model of concept-formational process.

(4) The subject declared a degree of confidence in his guess as a number in [0, 1] and was allowed to see if his guess was correct by turning over the card in the guessed position, either permanently (Case I) or temporarily (Case II).

(5) Steps (3) and (4) were repeated until the subject could guess correctly the specified card with the degree of confidence 1.0, that is, the regularity of the card specification was completely known to the subject.

Two cases of the experiment (Case I and Case II) were carried out for each subject two weeks or more apart. In Case I, the card in the position of his guess was left turned over so that the regularity of the card arrangement could be inferred from seeing the turned-up cards. In Case II, the subject was allowed to see the card and check that his guess was correct, but the card remained with the reverse side up during the whole period of experiment. So he had to memorize the outcomes of the previous trials to infer the hidden regularity of the card arrangement. A total of seventy experiments were made on subjects. From the results of the two kinds of experiments, we can make a comparative investigation of inference with memorizing effort (Case II) and without memorizing effort (Case I).

2.2 Concept-formational analysis of the result

The results of the experiment are summarized in Figs. 1 and 2 for Case I and Fig. 3 for Case II.

A concept-formational process in a subject is a series of repetitions of formation, confirmation, and reformation (or modification) of hypothesis, which is

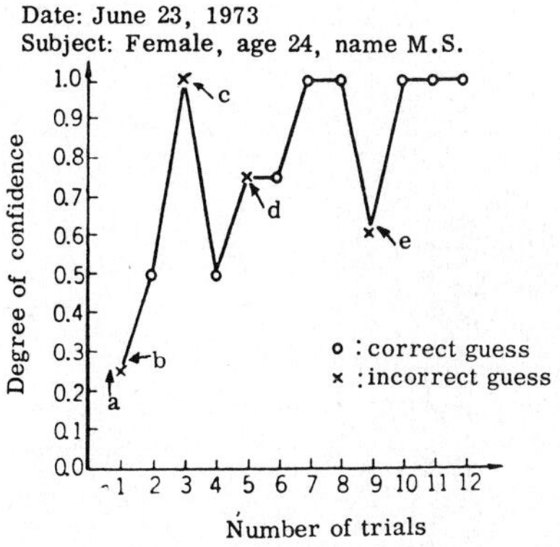

a: Formation of the first hypothesis.

b: Modification of the first hypothesis
Formation of the second hypothesis

c: Modification of the second hypothesis
Formation of the third hypothesis

d: Modification of the third hypothesis
Formation of the fourth hypothesis

e: Modification of the fourth hypothesis
Formation of the fifth hypothesis

Fig. 1. Result of Case I-6-8 (6 is No. of card arrangement, 8 is No. of subject).

manifested by ups and downs of the curves shown in Fig. 1.

The subject guesses a position of a specified card according to his own hypothesis regarding the card arrangement. If his guess is found correct, his confidence in his hypothesis rises and so does his confidence in his next guess (the curve ascends). If the guess is found incorrect, he must modify his hypothesis, and his confidence in his next guess naturally decreases (the curve descends).* The complexity of the card arrangement has a tendency to increase the number of ups and downs on the curve.

2.3 Interrelation among memory, new information, and inference

The influence of memory on the inference may be measured by the difference q between confidence responses for Case I and Case II depicted in Figs. 2 and 3, respectively: In Case II the subject must memorize the cards turned-up in previous trials for future inference, while in Case I he must not. Hence, the degree of confidence of the turned-up card in Case II is smaller by the quantity of fuzziness caused by memorizing than that in Case I.

The degree of memory effect q_i at the ith trial is defined as

$$q_i = \begin{cases} f_i^I - f_i^{II}, & f_i^I - f_i^{II} \geqq 0, \quad i = 1, 2, \ldots, N^{II} \\ 0 & , \ f_i^I - f_i^{II} > 0 \end{cases} \tag{1}$$

where N^{II} is the number of trials needed for a subject to find the card arrangement completely. f_i^I and f_i^{II} are the degrees of confidence for the ith trial in Case I and Case II, respectively. The quantity of memory effect q on the inference is defined by

$$q = \sum_{j=1}^{N^{II}} q_i \, \Delta_i, \quad i = 1, 2, \ldots, N^{II} \tag{2}$$

where Δ_i is the interval between the ith and (i+1)th trials (see Fig. 4).

Next, we define the amount of new information (simply called "information") p of the turned-up cards as non-probabilistic entropy which is introduced in terms of

---

* Sometimes the curve ascends even though the guess is found to be incorrect. This phenomenon can be interpreted as follows: There are several hypothesis and only one of them is confirmed or modified or replaced as the result in incorrect guess. The overall confidence may increase in this situation.

the membership function $f(x_j)$ of the event x_j [12].

$$p = -K\left(\sum_{j=1}^{N^I} f(x_j)\,\ln f(x_j) + \sum_{j=1}^{N^I} \bar{f}(x_j)\,\ln \bar{f}(x_j)\right),$$

$$j = 1, 2, \ldots, N^I \qquad (3)$$

where $f(x_j)$ is the membership function of the card x_j specified at the jth trial, $\bar{f} = 1 - f$, K is a positive constant and N^I is the total number of trials in Case I needed for a subject to find the card arrangement completely.

The reason for the choice of this definition, Equation (3), is that the amount of information p should be maximized when the event x_j is most fuzzy, i.e., $f(x_j) = 0.5$, as manifested by variations in confidence responses summarized in Figs. 2 and 3. A vertical bar at each trial indicates a range of the variation in confidence responses of ten subjects. Its length is greatest when $f(x_j)$ is about 0.6 to 0.8. In other words, the data plotted on Fig. 3 at each trial are the mean, the maximum, and the minimum of the confidence responses obtained from ten subjects. The mean value may be regarded as the confidence response of the average person and the interval between the maximum and the minimum may represent fuzziness in his response.

Using f^I of Fig. 2 and Equation (3) with $K = 1/N^I$, we can calculate the amount of information p with respect to the number of trials as shown in Fig. 5. f^I of Case I may be regarded as the sum of inference r and information p, that is, the degree of inference r_j of the jth trial may be defined by

$$r_j = f_j^I - p_{j-1}, \quad j = 1, 2, \ldots, N^I \qquad (4)$$

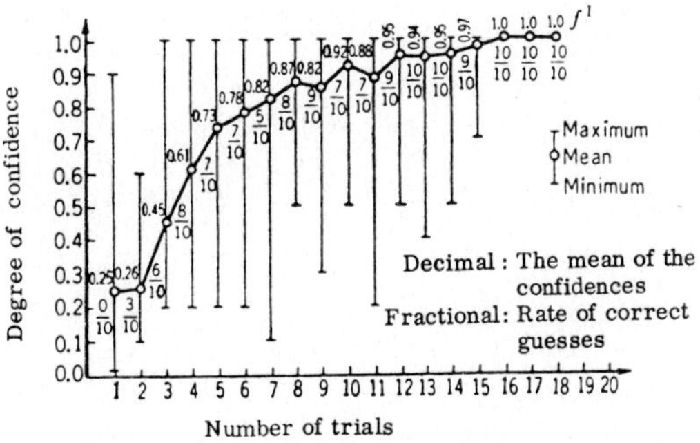

Fig. 2. Mean value of ten subjects in Case I-4.

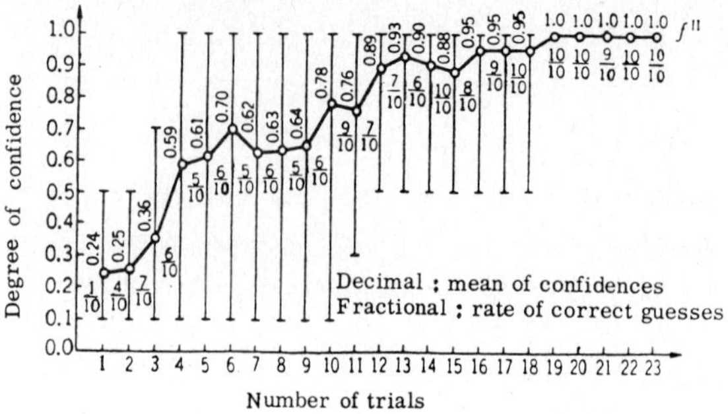

Fig. 3. Mean value of ten subjects in Case II-4.

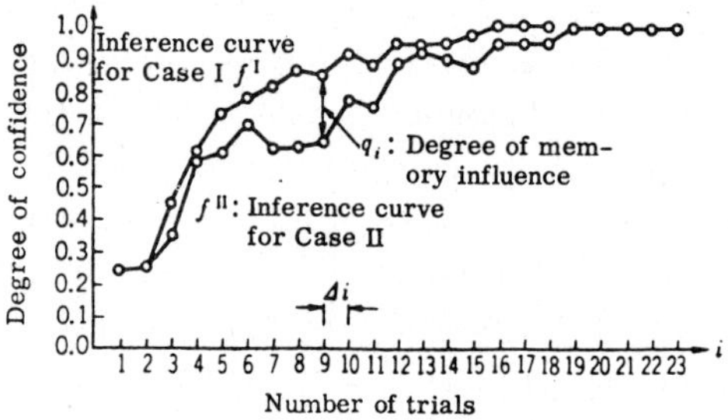

Fig. 4. Degree and quantity of memory effect (mean values in Cases I-4 and II-4).

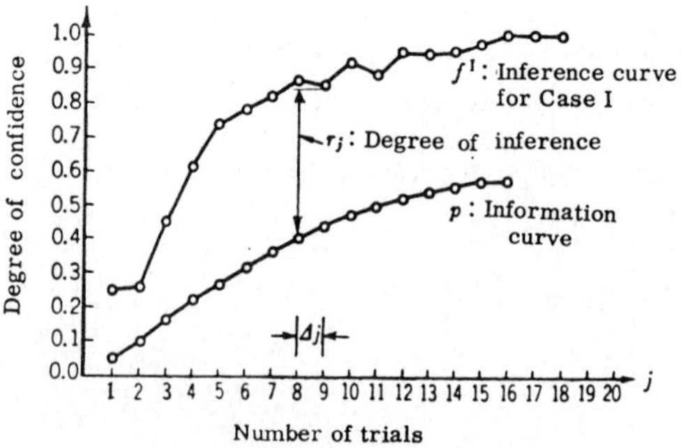

Fig. 5. Quantity of information, degree of inference and quantity of inference.

where p_{j-1} is the amount of information of the cards turned-up until the (j-1)th trial. The overall quantity of inference is defined by

$$r = \sum_{j=1}^{N^I} r_j \Delta_j, \quad j = 1, 2, \ldots, N^I \tag{5}$$

3. Fuzzy-Theoretic Experiment on Hint Effect and Analysis of Results

3.1 Fuzzy-theoretic experiment on hint effect

3.1.1 Card arrangement, subject, and hint

The material was again a set of playing cards. Fifty-two cards were arranged in the 4 × 13 two-dimensional array of six kinds (each kind had its own regularity, see Fig. 6). A total of sixty experiments were made on five men and five women aged 19 to 26, who were graduate students and members of staff of Nagoya University. The hints which were given to subjects were as follows:

Hint A: "The cards are not always arrayed from left side to right side in the order of the number." This hint was given just before the experiment.

Hint B: "The cards of each row are not arrayed by all in the same suit." This hint was given when the degree of confidence in guess reached 0.5, which is the most ambiguous point (cf. section 2.3).

Hint C: "Three cards (one card in Arrangement 6) of the same color are arrayed mutually in each row." This hint was given when the subject requested. But, all subjects requested to give this hint just before the experiment. Then,

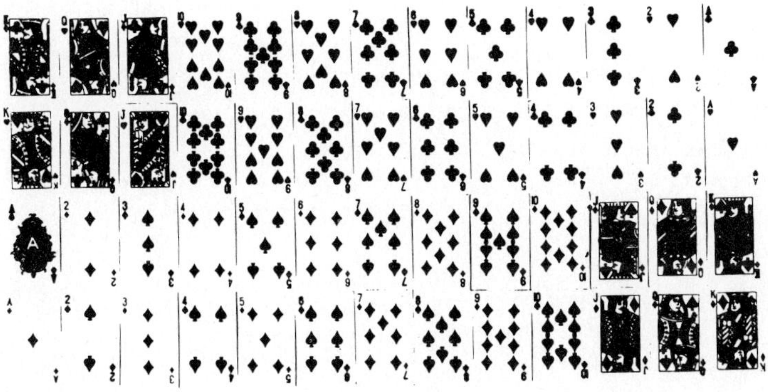

Fig. 6. An example of the card arrangement (arrangement 3).

the time to give this hint was the same as the time of Hint A. Hint A was given for the estimation problem on the card arrangements 1 and 2, and Hint B for 3 and 4, and Hint C for 5 and 6 (cf. Fig. 6).

3.1.2 Procedure of the experiment

The experiments was carried out as follows:

(1) The subject was given one of six kinds of the 4 × 13 two-dimensional array of 52 playing cards.

(2) All cards were arranged with the reverse side up.

(3) At each trial, the subject was instructed to guess the specification of a card pointed to the position by the experimenter.

(4) The subject declared a degree of confidence in his guess as a number in [0, 1] and was allowed to see if his guess was correct by turning over the card in the pointed position. The card was left turned over during the whole period of experiment.

(5) Steps (3) and (4) were repeated until the hidden regularity of the card arrangement was completely found by the subject, that is, his degree of confidence reached 1.0 and his answer was correct.

(6) After the experiment, the subject was requested to answer what kinds of concept for the card arrangement were formed or modified or confirmed in the estimation and decision processes of the regularity of the card arrangement.

By the result of the procedure (6), we can analyse the concept-formational process in the estimation and decision processes of the hidden regularity of the card arrangement.

Two cases of the experiment (Case I and Case II) were also carried out for each subject two weeks or more apart. In Case I, the hint on the card arrangement was not given to the subject, on the other hand, in Case II, the hint was given. The aim of the experiment was to compare the result of Case I with that of Case II, and to analyse the hint effect in the estimation and decision processes of the hidden regularity of the card arrangement.

3.2 Analysis of results

3.2.1 Concept-formational process

The results of the experiment are summarized in Table 1 and Fig. 7 for Case I of the card arrangements 1 and 3 (Test 1-I and Test 3-I). Table 1 shows that, for example, the subject N.S. formed the first hypothesis before the first trial:

Table 1

Concept-Formational Process on Estimation and Decision of the Hidden Regularity of Card Arrangement

test	subject \ No. of trials	0	1	2	3	4	5	6	7	8	...
1 - I	N.S.	(f) No.:left to right		(c)	No.:also right to left (c)	No. of 1st and 2nd row: right to left (c)					
			3rd row: (c)	4th row: (c)	1st row: (c)	Su.: find			Ar.: find		
	K.H.*	(f) No.:left to right		(c)	No.:also right to left (c)						
		(f) Co. of column: R, B, R, B		(m)	B, R, B, R (c)	Su.: find	Ar.: find				
3 - I	T.D.	(f) No.:left to right		(c)	No.: (m)		(f) Su.:oblique			row: 2 suits are arrayed mutually (c)	
			3rd row: (c)	4th row: (c)							Ar: find
	S.T.*	(f) No.:left to right		(c)	No.: (m)		row:4 kinds of suits are not arrayed mutually (c)			row: 2 suits are arrayed mutually (c)	
			3rd row: (c)	4th row: (c)	1st row:not (c)						Ar: find

Note: *, female; (f), formation of hypothesis; (c), confirmation of hypothesis; (m), modification of hypothesis; No., number; Su., suit; Co., Color; Ar, arrangement of the cards; R, red; B, black

"The cards were arrayed from left side to right side in the order of the number," and after the second trial the subject confirmed the first hypothesis. From Table 1 and other results, we can state that the hypothetic concept is formed, modified, and confirmed in the estimation and decision processes. The hypothetic concept is also formed roughly and wholly for the arrangement of the dimensions [13] of the card such as number, suit, color, and so on, at the first step. Then, the hypothetic concept for the card arrangement is partially modified and confirmed repeatedly until the final hypothesis is confirmed. This process is similar to the process of the heuristic search of the multimodal optimum point [14].

Figure 7 depicts the relation between the number of trials and the degree of coincidence in guess of the card, and shows that (1) concept-formational curve for each subjects has the similar form for each card arrangement, (2) depending on the complexity of the regularity of the card arrangement, the concept-formational curve extends to the right side. Basing on the above-mentioned results, the concept-formational curve can be modeled as the curves in Fig. 8.

The curves in Fig. 7 or Fig. 8 show the degree of confidence in guess of the specified card. We can consider that the degree of confidence in guess of the specified card is dependent on the confidence of the hypothetic concept for the card arrangement or the arrangement of the dimension (number, suit, or color) of the card. Then, the curves in Fig. 7 or Fig. 8 depict indirectly the degree of confidence of the hypothetic concept. Figure 9 depicts the mean, the maximum,

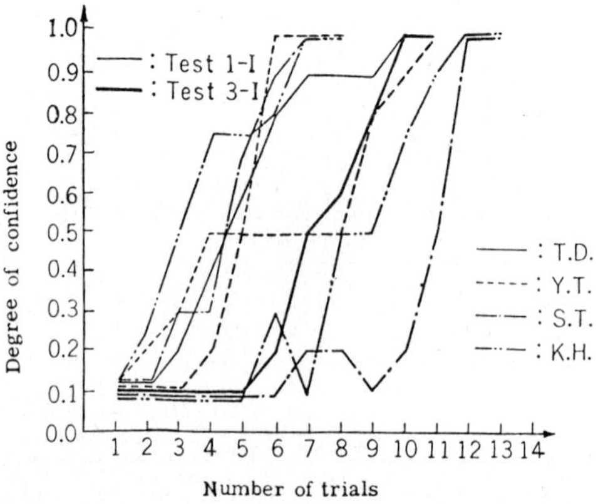

Fig. 7. Individual process of concept formation in Tests 1-I and 3-I.

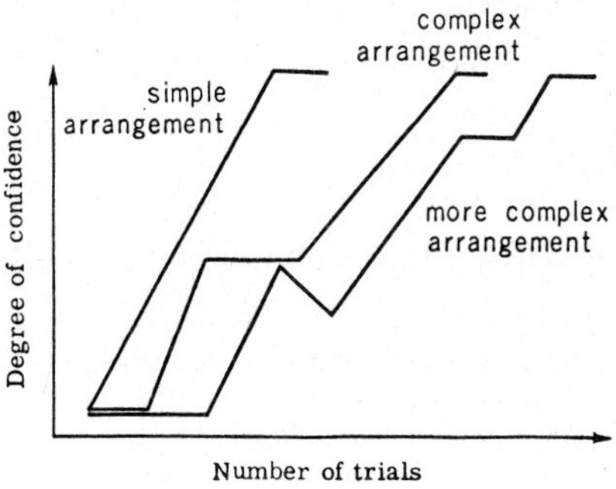

Fig. 8. A model of concept-formational process due to complexity of card arrangement.

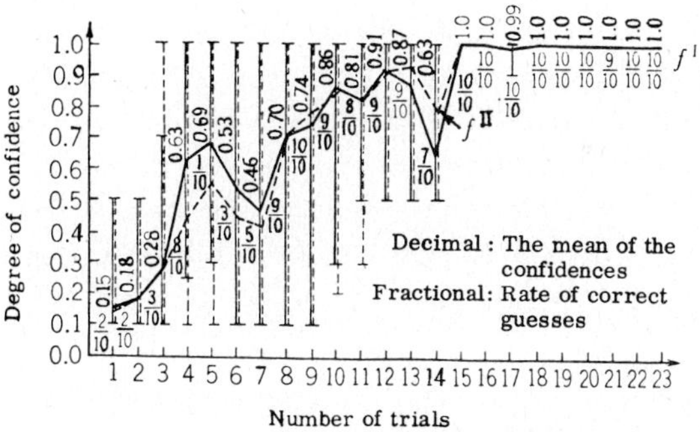

Fig. 9. Averaged process and its deviation (fuzziness) of concept formation in Tests 2-I and 2-IIA (Decimal and fraction are omitted in Test 2-IIA).

and the minimum of the degree of confidence of the hypothetic concept for ten subjects. The difference between the maximum and the minimum can be considered the fuzziness of the mean of confidences. From Fig. 9, we can read that the degree of confidence of the hypothetic concept is most fuzzy when the degree of confidence is about 0.5.

3.2.2 Concept-formational process and fuzzy catastrophe

In this section, we introduce the concepts of catastrophe and fuzziness in order to explain uniformly and generally the concept-formational proces, which has many different forms with a jump of logic [15], [16], [17].

(1) Introduction of catastrophe

Catastrophe theory is a geometric theory of the construction of a model [18], and is currently applied to many fields [19], [20], [21]. We introduce the catastrophe theory in order to explain the model of the concept-formational process uniformly and generally.

We have sometimes experienced to give up to solve a very hard problem, against to our many efforts. But, we also experience that the problem is solved unexpectedly on one day. This phenomenon is arised by the jump of logic (catastrophe). The jump of logic can be shown on the concept-formational curve. In Fig. 7, the catastrophe is found at the jumping point of the confidence degree, where the jump of logic is considered to be generated. This phenomenon can be explain as follows by using the cusp-curve surface which is one of the seven types of catastrophe. The curve A in Fig. 10 (this curve is vaguely depicted for the purpose of the analysis in the next paragraph, while in this paragraph, we deal with the curve without any vagueness) is depicted as the locus A on the cusp-curved surface, and shows a model of concept-formational curve in the case of simple regularity of the card arrangement. When the hidden regularity of the card arrangement becomes complex, the concept-formational curve is shown as the locus B on the cusp-curved surface, where the jumps of logic appear at the points 3 and 7 in Fig. 10.

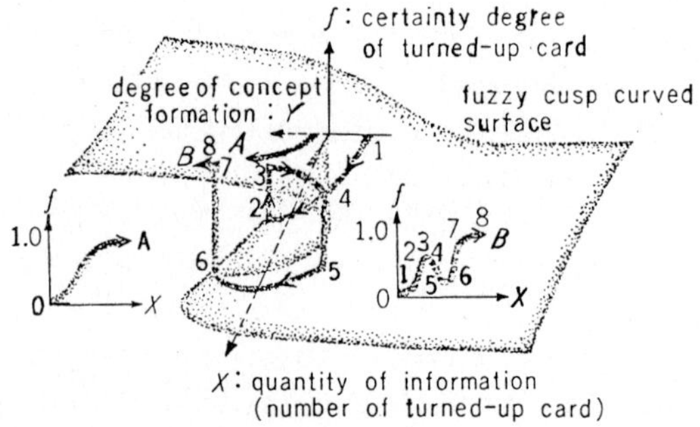

Fig. 10. Fuzzy concept-formational process on fuzzy cusp-curved surface.

(2) Introduction of fuzziness

We introduce the concept of "fuzziness" for the practical explanation of the concept-formational curve. The degree of confidence of guessing a concept has its own fuzziness of, for example, "about 0.5," as discusses in Section 3.2.1. Then, the concept-formational curve is depicted as the curves A and B being corresponding to the loci A and B on the fuzzy cusp-curved surface in Fig. 10.

In this section, the concept-formational process, which is not explained clearly in the conventional studies of psychology, is explained more clearly by means of the introduction of both the cusp-curved surface in catastrophe theory and fuzziness in fuzzy set theory.

3.2.3 Hint effect in estimation and decision processes

A number of psychological papers on the hint effect have been reported and the following results are derived [22]:

(a) The hint has a good effect when it is given to the subject who is in the confused situation.

(b) The hint of specifying an object has a better effect than the one which does not specify the object.

(c) The concretely specified hint has a better effect than the abstractly specified hint.

(d) The hint which conforms to the thinking pattern of the subject and is close to the development of his thinking process has a good effect.

The result of the experiments on Tests 2-I and 2-IIA (Card Arrangement 2, Case II, and Hint A) is summarized in Fig. 9. Figure 11 illustrates the variation of average confidence of the ten subjects for Tests 2-I, 2-IIA, 5-I, and 5-IIC. Figure 12 illustrates the hint effect, the difference between the Case II and Case I, for each card arrangement. The numeral in the circular form in Fig. 12 is the mean of the degrees of confidence at the peak point of each hint-effect curve. We summarize the results of the experiments from these figures as follows:

(1) The hint has sometimes the negative or adverse effect (see Arrangement 1 in Fig. 12): This is because of the confusion of thinking caused by the given hint. In other words, the concept-formational process is confused sometimes by the given hint when the given problem is rather easy. The hint effect in Test 1-IIA (The cards are not always arrayed from left side to right side in order of the number) is not generated in the concept-formational process of Test 1-I. Accordingly, we can consider that the development of the concept formation is sometimes disturbed by the given hint A. This is the adverse phenomenon compared

with the above phenomenon (d) which is specified in the psychological study. In other words,

(1') The hint, which is not conformed to the thinking pattern of the subject and is different from the development of the thinking process, has a negative effect.

(2) The positive effect of the hint increases as the difficulty of the problem increases (see Arrangements 1 and 2, Arrangements 3 and 4, and Arrangements 5 and 6 in Fig. 12).

The phenomenon (2) can be proved by the following facts. At first, from the results of Case I, we can see that the regularity of the Arrangements 2, 4, and 6 is more difficult than that of the Arrangements 1, 3, and 5 respectively. Because, the subject needed more trials until he/she found the regularity completely in the experiment of the card arrangements 2, 4, and 6 than that of the card arrangements 1, 3, and 5. Next, although the hint A is given for Tests 1-IIA (Arrangement 1, Case II, and Hint A) and 2-IIA, hint B for Tests 3-IIB and 4-IIB, and hint C for Tests 5-IIC and 6-IIC, the results of Tests 2-IIA, 4-IIB, and 6-IIC show more positive effect of the hints than those of Tests 1-IIA, 3-IIB, and 5-IIC, respectively. Then, we can summarize the results as the above item (2).

(3) The hint produces the maximum effect independent of both the polarity of effect and the time to give the hint, if it is given at the time when the subject is in the most confusing state in the thinking process. This phenomenon is based on the fact that, in Fig. 12, the hint has the maximum effect when the degree of confidence in guess is nearly equal to 0.5 (from 0.33 to 0.76). This phenomenon is more general and detailed than the above phenomenon (a) which is specified in the psychology text book [22].

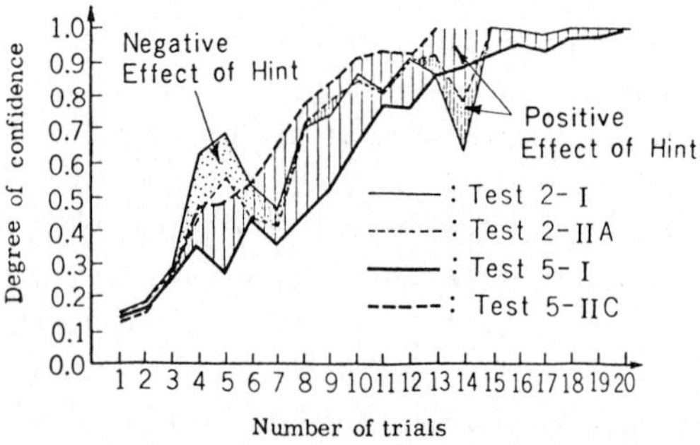

Fig. 11. Averaged processes of concept formation with or without hint (in Tests 2-I, 2-IIA, 5-I, and 5-IIC).

The above-mentioned three items are the results of the hint effect obtained through our experiments

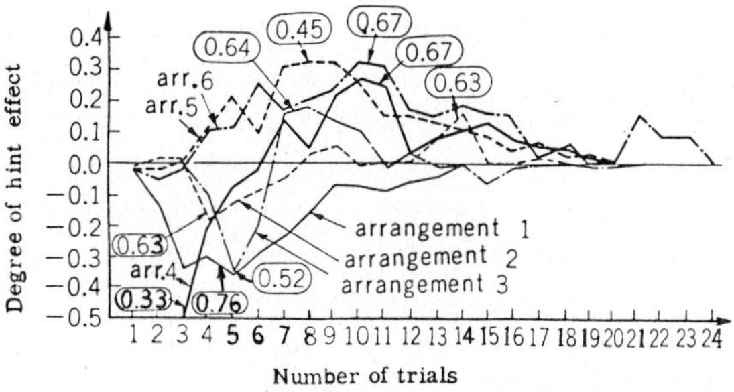

Fig. 12. Positive and negative effects of hint in estimation process of card arrangement.

The model of the hint effect is depicted as the curve in Fig. 13.

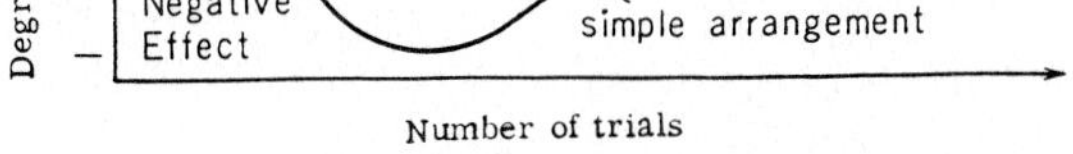

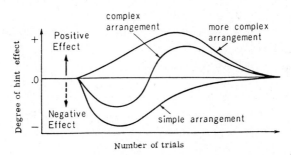

Fig. 13. Model of hint effect due to complexity of card arrangement.

4. Conclusion

In this paper, we conducted the experiments concerning with the inference and the hint effect in the estimation and decision processes of the hidden regularity of the card arrangement, and also analysed the concept-formational process with fuzziness and the quantitative interrelation among memory, new information, inference, and hint effect from viewpoints of fuzzy set theory, information theory, concept formation theory, and catastrophe theory.

In Section 2, we examined the inference function which is a key factor in the decision-making process, by performing fuzzy-theoretic experiments and introduced the concepts (definitions) of quantities of inference, new information and memory effect which play an important role in the study of artificial intelligence and formulation of human decision-making process.

In Section 3, we examined the hint effect which is also a key factor in human decision-making process, by performing fuzzy-theoretic experiments. From the results of our study, the following items are summarized:
(1) Under the same conditions (the same hint and the same arrangement of the cards), the subject forms or modifies or replaces almost the same hypothesis (or subconcepts) in the concept-formational process.
(2) The concept is always accompanied by the property of fuzziness in the concept-formational process.
(3) The fuzziness of the confidence degree of guessing a concept (or subconcept) becomes most ambiguous when the confidence degree of turned-up card is around 0.5.
(4) The concept-formational process which is not explained clearly in the conventional studies of psychology, becomes clear by means of the introductions of both the cusp-curved surface in catastrophe theory and fuzziness concept in fuzzy theory.
(5) Three new facts on the hint effect to the subject in the estimation and decision processes are found.
(6) The hint effect is defined and measured.

Future research problems include the generalization of f^{I} and f^{II} and the effects of the subject's knowledge upon the amounts of new information [23], [24].

Acknowledgement

The authors gratefully acknowledge the kind and helpful suggestions and criticisms of the members of our laboratory.

References

[1] N.M. Amosov: Modeling of thinking and the mind (Spartan Books, New York, 1967).

[2] H. Takahashi, et al.: Thinking process and information science (Sangyo-Tosho Publisher, Tokyo, 1972).

[3] B. Berelson and G. A. Steiner: Human behavior (Harcourt, Brace and World Inc., New York, 1964).

[4] M. Yamauchi, et al.: What is thinking? (Diamond Publisher, Tokyo, 1970).

[5] M. Kokawa, K. Nakamura, and M. Oda: A formulation of human decision-making process, Res. Rept. of Auto. Cont. Lab., Nagoya Univ., 19 (1972) 3-10.

[6] M. Kokawa, K. Nakamura, and M. Oda: Experimental approach to fuzzy simulation of memorizing, forgetting and inference process, Fuzzy Sets and Their Applications to Cognitive and Decision Processes, edited by L.A. Zadeh, et al. (Academic Press, New York, 1975) 409-428.

[7] M. Kokawa, K. Nakamura, and M. Oda: Fuzzy process of memory behavior, Trans. Society of Instrument and Control Engineers, 10, 3 (1974) 385-386.

[8] M. Kokawa, K. Nakamura, and M. Oda: Fuzzy theoretical and concept formational approaches to memory and inference experiments, Trans. Institute of Electronics and Communication Engineers of Japan, 57-D, 8 (1974) 487-493.

[9] M. Kokawa, M. Oda, and K. Nakamura: Fuzzy theoretical one-dimensionalizing method of multidimensional quantity, Trans. Society of Instrument and Control Engineers, 11, 5 (1975) 8-14.

[10] M. Kokawa, M. Oda, and K. Nakamura: Fuzzy-theoretical dimensionality reduction method of multidimensional quantity, Fuzzy Automata and Decision Process, edited by M.M. Gupta, et al. (North-Holland, Amsterdam, 1977) 235-249.

[11] M. Kokawa, M. Oda, and K. Nakamura: A study of handling method of fuzzy data, Trans. Institute of Electronics and Communication Engineers of Japan, 62-A, 1 (1979) 97-102.

[12] A. Deluca, et al.: A definition of a non-probabilistic entropy in the setting of fuzzy set theory, Information and Control, 20 (1972) 301-312.

[13] S. Ono: Experiments on Learning (Kyoritsu Publisher, Tokyo, 1971).

[14] M. Oda, K. Nakamura, and B.F. Womack: A survey of heuristic search method of multimodal optimum point, Learning Systems and Intelligent Robots, edited by K.S. Fu and J.T. Tou (Plenum Press, New York, 1974) 145-169.

[15] M. Kokawa, K. Nakamura, and M. Oda: Hint effect and a jump of logic in a decision process, Trans. Institute of Electronics and Communication Engineers of Japan, 58-D, 5 (1975) 256-263.

[16] M. Kokawa, K. Nakamura, and M. Oda: Fuzzy-theoretic and concept-formational approaches to inference and hint-effect experiments in human decision processes, Proc. of the IEEE Conference on Decision and Control (1977) 1330-1337.

[17] M. Kokawa, K. Nakamura, and M. Oda: Fuzzy-theoretic and concept-formational approaches to hint-effect experiments in human decision processes, Fuzzy Sets and Systems, 2, 1 (1979) 25-36.

[18] R. Thom: Structural Stability and Morphogenesis (W.A. Benjamin, Inc., Massachusetts, 1975).

[19] T. Poston and I. Stewart: Nonlinear modeling of multistable perception, Behavioral Science, 23, 5 (1978) 318-334.

[20] H. Noguchi: Topics of topology (Nippon-Hyoronsha Publisher, Tokyo, 1973).

[21] H. Noguchi: Catastrophe theory (Kodansha Publisher, Tokyo, 1973).

[22] C. Tatsuno: Psychology of problem solving (Kaneko-Shobo Publisher, Tokyo, 1970).

[23] V.A. Trapeznikov: Man in the control system, Automation and Remote Control, 33, 2 (1972) 171-179.

[24] H. Murata: Information, system and behavior, bit, 5, 11 (1973) 51-55.

[25] H.M. Hersh and A. Caramazza: A fuzzy set approach to modifiers and vagueness in natural language, Journal of Experimental Psychology, 105, 3 (1976) 254-276.

ADVANCES IN FUZZY SET THEORY AND APPLICATIONS
M.M. Gupta, R.K. Ragade, R.R. Yager (editors)

CONTRIBUTION OF THE FUZZY SETS THEORY TO MAN-MACHINE SYSTEM

by D. WILLAEYS, N. MALVACHE
LABORATOIRE D'AUTOMATIQUE INDUSTRIELLE ET HUMAINE
UNIVERSITE DE VALENCIENNES
59326 - VALENCIENNES (FRANCE)
TEL. (20) 46.66.08

ABSTRACT.

The authors present briefly, in the first part, their work on human perception of visual and vestibular information of a human operator making a watch-and-decision task, in order to point out the imprecise nature of the human problem solving strategy. In the second part, the authors dwell on the problem of description of physical values by a human operator. The third part develops a mathematical tool which permits accounting, at the formulation level, subjective data from the human observer. The next part presents a fuzzy tool for diminishing the number of data to be stored in a computer, for the representation of fuzzy relations. The last part is devoted to an example of fuzzy calculus consisting in the elaboration of a fuzzy regulator from a fuzzy model.

INTRODUCTION.

One of the goals of automation is to design systems that improve human life, by avoiding muscular and psycho-sensory over-work for human beings. The former is avoided more and more ; certain systems, in the field of robotics greatly help Man and replace him in accomplishing certain functions. Technical progress, however, is not always accompanied by similar progress in human science, and Man sometimes finds himself the victim of his own technicity. The role of communication is more and more important, and the functions of decision and data treatment become more and more complex. Psycho-sensory overwork is still a reality in many cases for it is often forgotten that systems remain in communication with Man. The rules of "comfort" in this Man-machine communication are based not only on technical criteria but also, especially, on objecto-subjective concepts. The subjective element is inherent in all systems where Man is still present, whatever his function is. It is from this point of view that certain automators have attempted to bring to the theories of automation a contribution from the theory of fuzzy sets, which seems to us a suitable tool for seizing complex systems where the subjective data constitute part of the information for decision-making and control-evolving [4,5].

However, there are problems that arise at the interface level between objective and subjective methods. We believe that herein lie possibilities of pure contributions of applying fuzzy sets theory. Let us point out that this orientation is based on two observations made in Man-machine systems :

- the "shaded" strategy of human perceptions, which even appears imprecise from an objective view-point ;
- the difficulty of dominating and exploring the parameters or variables, whether in the hard-to-predict environment or in Man. Let us cite only fatigue, attention, motivation, which can be predominant factors for the analysis of Man-machine systems.

The subjective character of the individual and the impossibility, in certain cases of putting a phenomenon into probabilities are revealed quite simply on examining the subject's verbal responses, in fuzzy or shaded terms, when he is asked questions about his task and his knowledge of the actual system.

We expose our contribution, in this field of research, to process control, in the case where human intervention still plays an important part.

The human operator, in accomplishing a task of supervision and manual action calling on the properties of the visio-manual loop, uses visual information (such as position, speed, distance...) to which he applies neither measurements nor statistical calculations, but which he "evaluates". It is on the basis of this subjective evaluation of values that he decides to take such and such action. Moreover, when he has several tasks to do, his decisions are weighted, and depend on their respective difficulties. For instance, if one task requires all his attention in the central visual field, he may be led to put aside the treatment of the information relating to another task, while keeping an eye on it in peripheral vision.
Let us take for example a job in a telephone information center and present it as a 3-level structure :
- standard automatic operations,
- operations requiring predictable treatment,
- non-predicted operations requiring the intervention of a (human) subject on the system.

These may be analogous to the 3 principal levels, in Man, of the treatment and action process :

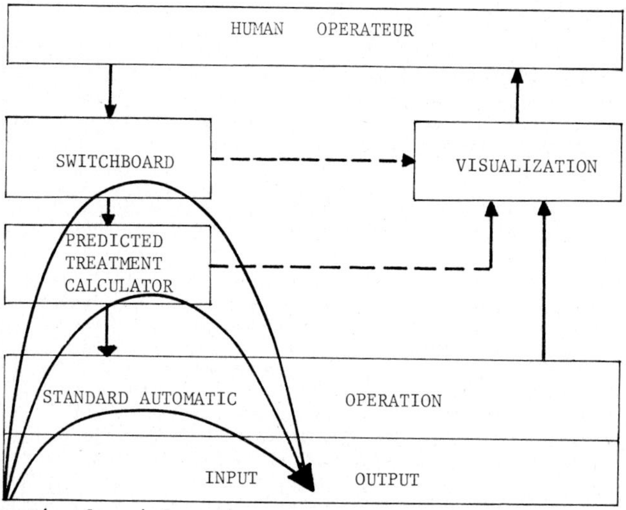

Figure 1. Synoptic of an information center seen as a 3-level structure.

- reflex level operations,
- cerebellum level operations,
- brain cortex level operations.

These echelons are according to the required task or to a measure of the complexity of the information treatment.
Upon analysis of this system, it is obvious that the subjective parameters, such as :-the human operator's initiative before the machine, by a "ready-to-receive" acknowledgement, his over-all knowledge of the system, which enables him to be informed, with no delay, of the results of his action (effort feedback) are the more important parameters in good system operation with minimum fatigue.
These observations lead us to state that the system's operation should be adapted to that of the human being, who can always act on the semi-automatic subsystems.
The object of this expose is not particularly to analyze Man-Machine systems, therefore, let us merely show a few results which may explain certain contradictory measurements, if the subjective factors are not taken into account.

The experimental protocol [18] consists, globally, in studying a subject's behavior during a work period, the daily chronogram of which is shown in Fig.2, and testing his reactions after each work period.

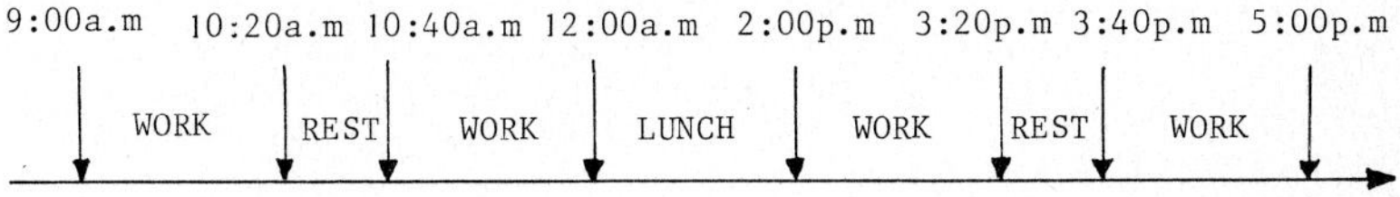

Figure 2. Job chronogram.

The job is to supply answers, through a keyboard, to visual messages of variable frequency of apposition (Tasks T_1, T_2, T_3) (Fig. 3).

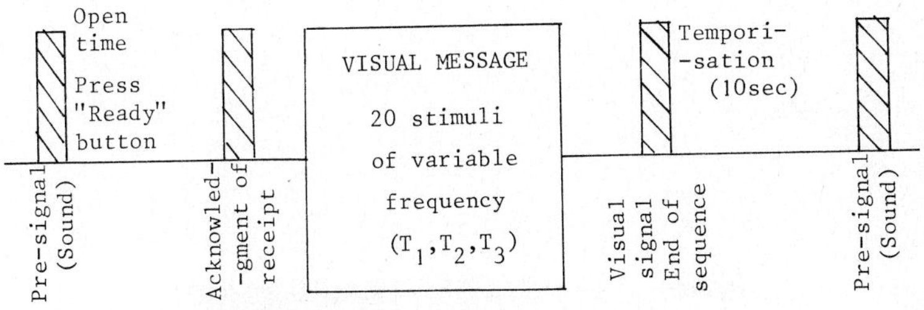

Figure 3. Development of a work sequence.
Time between stimuli $T_1[2s,5s]$, $T_2[3s,6s]$, $T_3[4s,6s]$

The manipulation is run by a computer ; and the fatigue evaluation tests at the end of the day show that the subject's fatigue, measured objectively, may sometimes be less in the evening than in the morning, whereas the human operator says he is "tired" in qualified, fuzzy terms.

Attempts to automate this process should bear in mind, in dividing tasks between Man and machine, Man's desire to retain his initiative ; and that determining the overall comfort indicators must be based on objecto-subjective criteria, in order to avoid informational type overloads.

It is from this point of view that the Laboratory of Industrial and Human Automation decided to approach the study of fuzzy systems ; but, having very quickly stumbled over this theory's lack of methods and tools for its use (due assentially to its youth), was obliged to try to develop a technique for using the theory of fuzzy sets in the field of automation.

This article represents the status of the Laboratory's work in this field, putting the accent less on formal presentation, which the reader can find in the bibliographical references, than on the significance and the operational mode of the "fuzzy" calculations used. The first part of the article presents the beginning of current work, relating to the identification of fuzzy sub-sets corresponding to adjectives used by a human operator for defining physical values.

The second part is devoted to the technique of developping a fuzzy system model, and the third describes the notion of fuzzy discretisation, which allows reducing the size of the tables representing a fuzzy model. The last part presents a technique of control calculation from a fuzzy model.

I - GRASPING AND DESCRIBING PHYSICAL VALUES ON A HUMAN OPERATOR'S PART.

One of the problems encountered in using the theory of fuzzy sets is the definition of the subjective membership functions of a fuzzy set corresponding to a value perceived by a human subject. Let us take a simple example of value evaluation in an industrial process of continuous pouring of steel rounds. More complex examples could have be chosen, in other fields such as:(1) a diagnosis based on a rapid, overall observation of a lot of of E.E.G tracings :(2) a sample of the tracing seemingly "identical" with other samples, can reveal significant clues to the doctor, if the sample is observed comprehensively, with an eye, particularly, to the"tendencies" of signal evolution.

The dialogue between two human beings is easy, when based on their mutual knowledge of a subject. With this in mind, we built an operator's console for process observation and control (Fig. 4). In fact, the dialogue between Man and a machine rests on two very different structures ; and in order to facilitate this communication, a special keyboard (Fig.5) wad evolved for qualifying the deviations or evolution of a value in the process. This keyboard, though arbitrary, has been determined by preliminary tests based on observations of the language of a human operator acquainted with process control. He transmits his "knowledge" to another operator in "fuzzy" terms, like :

"If the temperature in zone 1, near the ingot mold, is very high, then slow down a little the extraction speed on the round, to avoid a breakthrough"; or else :

"If the temperature in zone 3 is kind of low, then speed up the extraction rate of the round, so as to step up the output and keep the round from being too hard at the cutting point".

The console is designed to represent an image of the language referential between "master" and apprentice, who is in this case the computer associated with the continuous pouring process.

Figure 4. An example of an operator's console for process control and observation.

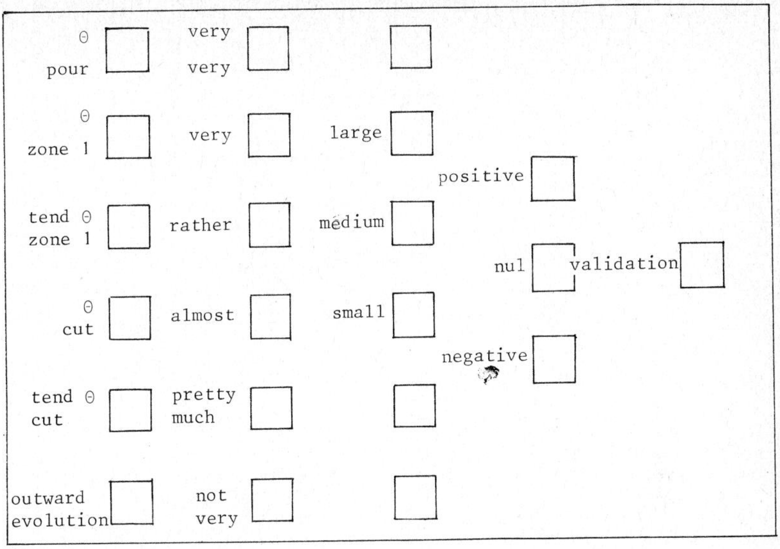

Figure 5. A special keyboard for describing values in fuzzy language.

The following illustrations show the results obtained when a human operator is placed in front of a pointer dial indicating the evolution of one of the values of the above process. The adjectives used are : small, medium, large ; and the adverbs : very-very, very, rather, almost, pretty much, not very.
These illustrations show the number of times each adjective occurs as a function of the pointer's position ; they are the result of a pre-treatment of the data and have to be refined by a treatment aiming at identifying the fuzzy sub-sets corresponding to the adjectives and adverbs used.

		MOYEN		(CODE 30)	APPARAIT	51 FOIS
	ASSEZ	GRAND		(CODE 25)	APPARAIT	5 FOIS
	PLUTOT	GRAND		(CODE 23)	APPARAIT	4 FOIS
	PLUTOT	PETIT		(CODE 43)	APPARAIT	8 FOIS
	ASSEZ	PETIT		(CODE 45)	APPARAIT	7 FOIS
TRES	TRES	PETIT		(CODE 41)	APPARAIT	13 FOIS
TRES	TRES	GRAND		(CODE 21)	APPARAIT	21 FOIS
	PLUTOT	MOYEN		(CODE 33)	APPARAIT	25 FOIS
	TRES	GRAND		(CODE 22)	APPARAIT	28 FOIS
	PRESQUE	MOYEN		(CODE 34)	APPARAIT	25 FOIS
		GRAND		(CODE 20)	APPARAIT	20 FOIS
		PETIT		(CODE 40)	APPARAIT	16 FOIS
	PRESQUE	PETIT		(CODE 44)	APPARAIT	18 FOIS
	ASSEZ	MOYEN		(CODE 35)	APPARAIT	5 FOIS
	PRESQUE	GRAND		(CODE 24)	APPARAIT	25 FOIS
	PRESQUE		NUL	(CODE 204)	APPARAIT	7 FOIS
	TRES	PETIT		(CODE 42)	APPARAIT	19 FOIS
			NUL	(CODE 200)	APPARAIT	3 FOIS

Figure 6. An example of the number of occurences of the various responses.

RANG	PHRASE			CODE	VI	VF	DT
1			MOYEN	30	4.05	4.02	3.62
2		ASSEZ	GRAND	25	5.31	5.46	5.98
3		PLUTOT	GRAND	23	7.34	7.34	4.04
4		PLUTOT	PETIT	43	2.65	2.76	5.30
5		ASSEZ	GRAND	25	6.89	6.89	12.06
6		ASSEZ	PETIT	45	2.28	2.33	4.52
7	TRES	TRES	PETIT	41	0.01	0.01	3.08
8	TRES	TRES	GRAND	21	9.31	9.31	2.98
9		PLUTOT	MOYEN	33	3.08	2.94	3.54
10		TRES	GRAND	22	8.12	8.12	3.28
11			MOYEN	30	4.05	4.04	3.68
12	TRES	TRES	PETIT	41	0.50	0.46	3.82
13		TRES	GRAND	22	8.08	8.06	5.64
14			MOYEN	30	5.80	5.78	4.04
15			MOYEN	30	5.58	3.57	2.76
16		PLUTOT	PETIT	43	1.56	1.56	3.72
17		PRESQUE	MOYEN	34	2.75	2.76	3.78
18			GRAND	20	7.02	7.18	4.86
19	TRES	TRES	PETIT	41	0.29	0.30	2.88
20			GRAND	20	7.46	7.36	3.16
21			PETIT	40	1.50	1.40	3.20
22			MOYEN	30	4.84	4.98	2.08
23		PRESQUE	PETIT	44	2.50	2.47	4.84
24		TRES	GRAND	22	8.74	8.76	3.40
25		PLUTOT	PETIT	43	1.21	1.24	3.70

Figure 7. An example of operator's responses for different pointer positions located between 0 and 10 Volts.
IV - Initial value of the position
FV - Final value at the instant of the operator's response
DT - Decision time (response).

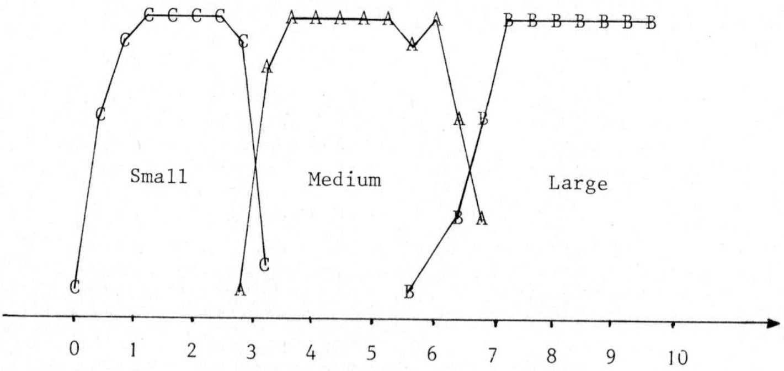

Figure 8. An example of the histogram of 3 adjectives, obtained by combining the adjectives qualified by an adverb with the basic adjectives. i.e. "Very large" combined with "large".

This sort of manipulation enables us to better define the "zones of influence" of the different adjectives used, and also, their stability in time and with different operators. It should also allow us to visualize, at the level of meaning, the influence of an adverb on the adjective it is attached to.

In resume, the use of fuzzy sets enables a computer to utilize the subjective information on values gathered by a human operator. We shall now see that the theory of fuzzy sets allows us to go even further, by formalizing not only an operator's perception, but also the knowledge he has acquired through his experience in controlling or observing a system.

II - DEVELOPMENT OF A FUZZY SYSTEM MODEL.

When a system is considered in view of developing a model, the first knowledge available is that belonging to the humans belonging to this system, or who observe it or control it. It is most infrequent to have at one's disposal a sufficient quantity of objective, statistical measurements to elaborate directly a model of a system. Therefore, the essential problem, it is found, consists in extracting from a human operator's grasp of the process, information which can be integrated at the system control level, or which can be used to improve the "comfort" of the Man-machine pair.
It is here that the theory of fuzzy sets, thanks to the notions of fuzzy languages and fuzzy relationships, allows the formulation of the subjective information supplied by a human operator on the perception of a process values. In fact, it can be supposed that this operator, having defined "touching" membership classes, applies a subjective treatment leading to an internal, subjective model. This internal model, which is the result of knowledge about and training on the system, is what the operator can describe, and what will be formalized as a fuzzy model. This type of fuzzy model, obtained from a human operator's subjective knowledge of the process, is complementary to those obtained by deterministic or stochastic

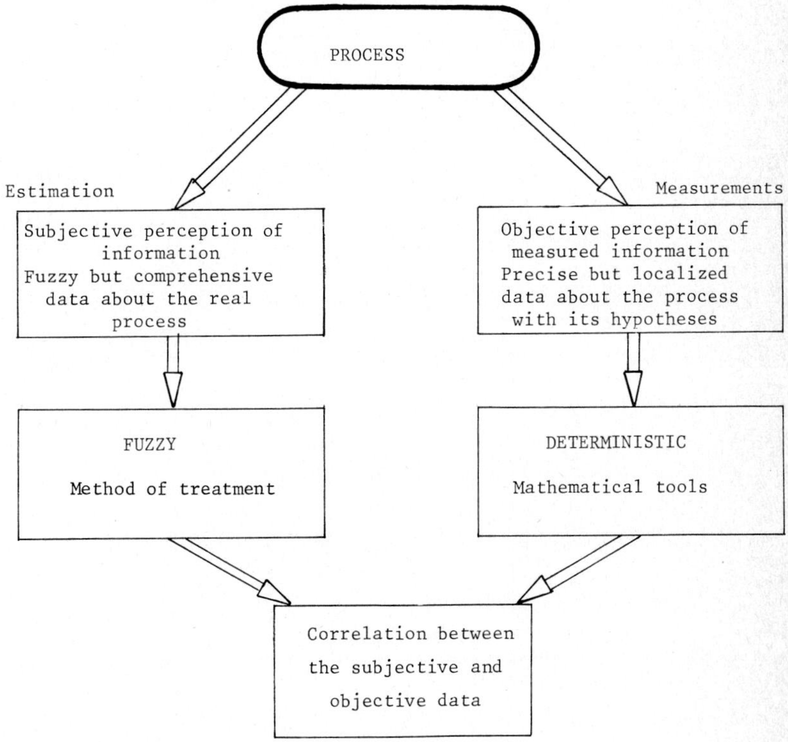

Figure 9. Synoptic diagram of two possible pathways for system analysis.

methods. This subjective method can even be considered as a first approach in modelizing a complex system in order to arrive at a model sub-set and can contribute to helping the objective decision.
We can represent these two parallel path-ways, each of which has advantages and disadvantages, but which in association can contribute to furthering our understanding of systems.
The rest of the expose is devoted to presenting our contribution in this field of research, which strives to adapt the theory of fuzzy sets for the analysis of Man-machine systems. Thanks to the notions of fuzzy language, fuzzy algorithms, and fuzzy relationhips, the concepts defined in the theory of fuzzy sets allow the description or interpretation of the environment perceived by the human operator [1,2,8,15,16]. Three possibilities are open : to describe the operational tactic worked out (Fig.10) by the human operator on the process ; or to characterize a process'operation, using the subjective information supplied by the human operator ; or else, to estimate the characteristic parameters of the process.

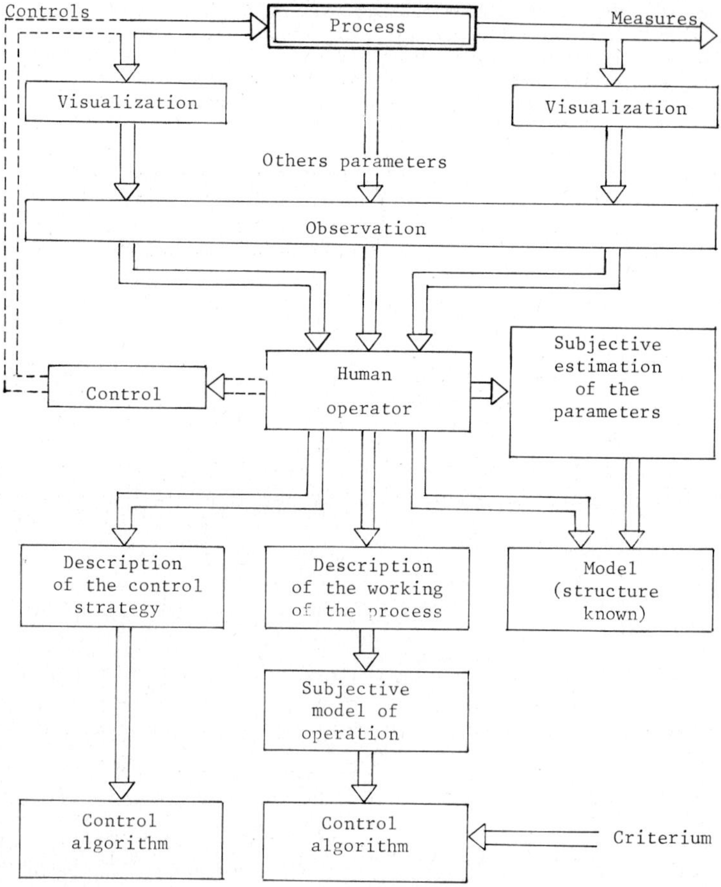

Figure 10. Synoptic diagram of the principle of using the information supplied by a human operator observing a process.

For the moment, we are more interested in the first two approaches ; in our expose we put particular emphasis on the second one, which consists in modelizing the evolution of a value in a process. Let us first state that the fuzzy regulator, for control, is extracted directly from the fuzzy model, and not from a fuzzy algorithm illustrating the tactic used by an operator for conducting his process. We feel that this approach can lead to better results, for it rests on the fact that the human operator treats and filters a certain amount of information, himself. Moreover, the observer retains the initiative of choosing the evaluation instant and the sequence which to him seems characteristic of the process' evolution. Examination of the samples clearly shows that the subject chooses essential sequences, giving them a maximum of nuances, and that his decision strategy always leaves a certain degree of initiative. In fact, he rarely adjusts to the "maximum", but rather to a little less than the maximum, which gives him an impression of comfort, safety and initiative.
Let us now consider the space of a variable x and, as we have seen previously, let us attempt to characterize the "shaded" sub-sets by membership functions $\mu_{\chi_j}(x)$.

Let X be a real space. We define in X a series χ_j of fuzzy sub-sets, such that all the x are in at least one fuzzy sub-set.

$$\mu_{\chi_j}(x) \neq 0 \quad \forall x \in X$$

This operation consists in dividing a space X, not into disconnected sub-sets, but into non-disconnected, fuzzy sub-sets.

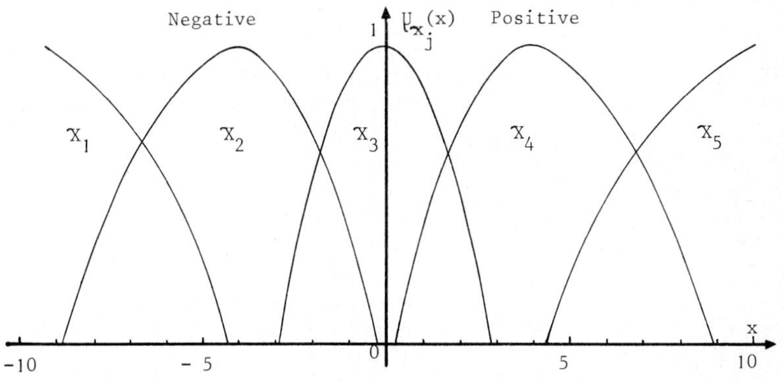

Figure 11. An example of fuzzy sets used for describing values.

The human operator's description may then be composed of associations between different variables of the process, and can be formalized by using the notion of a fuzzy algorithm.
Let us consider a simple example of a process where an input, U, and an output S, are visualized. The observer is sentisized to the tendency of variations in the desired goal, S_o, or with the desired evolution. Suppose that S is to be maintained constant at a certain value S_o. The subject then estimates the value $e = S - S_o$, generally in 2 dimensions, position and speed;and the operator describes his perception of the process, for instance, in the statement : "If at instant nT , u_n is 4.7 volts, e_n is medium positive and Δe_n large positive then e_{n+1} will be null and Δe_{n+1} medium positive at instant (n+1)T". This statement represents the subject's subjective perception of the process'evolution at two instants of

observation, nT and (n+1)T ; it contains both fuzzy and non-fuzzy descriptions [12]. Noting $u_n \in U$, $e \in E$, $\Delta e \in E$, we now have to interpret this statement ; the method consists in taking the direct product of the fuzzy variables in the space UxEx ΔEx E x ΔE , i.e., in calculating :

$$(1) \quad \mu_{f_j}(e_n, \Delta e_n, u_n, e_{n+1}, \Delta e_{n+1}) = \underset{\substack{\text{medium}\\ \text{positive}}}{\mu(e_n)} \wedge \underset{\substack{\text{large}\\ \text{positive}}}{\mu(\Delta e_n)} \wedge \mu_{4,7}(u_n) \wedge \underset{\text{null}}{\mu(e_{n+1})} \wedge \underset{\substack{\text{medium}\\ \text{positive}}}{\mu(\Delta e_{n+1})}$$

$f_j \in E \times \Delta E \times U \times E \times \Delta E$ being the interpretation of the jth statement. In other words, this statement, j, will be represented by the direct product of the fuzzy sets $(\text{null})_E$, $(\text{medium positive})_{\Delta E}$, $(4.7 \text{ volts})_U$, $(\text{medium positive})_E$, and $(\text{large positive})_{\Delta E}$.

For a set of j statements, the algorithm will then be interpreted by the union of these direct products, and will give us the fuzzy model M of the system

$$M = \bigcup_j f_J$$

$$(2) \quad \mu_M(e_n, \Delta e_n, a_n, e_{n+1}, \Delta e_{n+1}) = \bigvee_j \mu_{f_j}(e_n, \Delta e_n, u_n, e_{n+1}, \Delta e_{n+1})$$

This sort of fuzzy relationship, M can correspond to the subjective description of the process'variables ; it represents the subjective model M in the 5-dimensional space $E_n \times \Delta E_n \times U \times E_{n+1} \times \Delta E_{n+1}$.

This type of model is difficult to represent graphically ; however, we have a representation of it in the case of a very simple example in the fourth part. Representing a fuzzy model consequently, requires the use of a computer, which in turn imperatively requires the chopping of the variables'real space. Now, for a 5-dimensional model like the one we just examined, the computer must be able to store k^5 values, if k is the number of segments of one of its dimensions.

In orden to reduce the memory capacity necessary, and taking into consideration the fact that this model uses fuzzy sub-sets, we have developped a technique of fuzzy discretisation, which we feel is also close to the human division corresponding to a task of classifying values x in fuzzy subjective classes.

III - THE USE OF FUZZY DIVISION OF A REAL SPACE.

This operation consists in breaking down a space, not into ordinary, disconnected sub-sets, but into non-disconnected fuzzy sets [10,17] .

Let X be real space.

We define on X a series χ_j of fuzzy sets, Fig. 11 or 12 . The membership function of a sub-set χ_j , written $\mu_{\chi_j}(x)$ is such that there exists at least one χ_j such that :

$$\mu_{\chi_j}(x) \neq 0 \ \forall x \in X .$$

i.e. that every point of X is "known" in at least one χ_j .

Now let A be any sub-set in X with the membership function $\mu_A(x)$.

We can define this sub-set in terms of the series χ_j , used as the "subjective referential χ" , as follows :

$$(3) \quad \mu_A(\chi_j) = \bigvee_x (\mu_A(x) \wedge \mu_{\chi_j}(x))$$

By this definition, a point x in X is defined by

(4) $\mu_x(\chi_j) = \mu_{\chi_j}(x)$

and in this referential, this point x will be known only by :

$$\bigvee_j \mu_{\chi_j}(x) .$$

Conversely, we can define the inverse passage from χ to $X_R \subset X$, X_R being the best knowledge of X in χ by :

$$\mu_{A_R}(x) = \bigvee_j (\mu_A(\chi_j) \wedge \mu_x(\chi_j))$$

(4)

$$= \bigvee_j (\mu_A(\chi_j) \wedge \mu_{\chi_j}(x))$$

X is entirely reconstituted ; that is, $X_R = X$ if

$$\max_j \mu_{\chi_j}(x) = 1 \ \forall x \text{ i.e } \bigcup_j \chi_j = X$$

Note that definition (3) has the same equation as the definition of the measurement of possibility of a fuzzy set established by L.A. ZADEH, $\mu_{\chi_j}(x)$ being considered the distribution of possibility of a variable χ_j taking its values in X and $\mu_A(\chi_j)$ being the measure of possibility that χ_j is A [14].

If we compare the operations of union and intersection of two fuzzy sets A and B carried out in X and χ, by noting $\mu_A(x)$ the membership function of A in X and $\mu_A(\chi_j)$ the discrete membership function of A in χ , we obtain the following two relationships :

(5) $\mu_A(\chi_j) \wedge \mu_B(\chi_j) = \mu_{A\cup B}(\chi_j)$

(6) $\mu_A(\chi_j) \wedge \mu_B(\chi_j) \geqslant \mu_{A\cap B}(\chi_j)$

in which $\mu_{A\cup B}(\chi_j)$ represents the discret membership function in χ of the fuzzy sub-set "union of A and B" and $\mu_{A\cap B}(\chi_j)$, the intersection.

Figure 12, below, represents a series χ of 6 fuzzy sets : large negative, medium negative, null negative, null positive, medium positive, large positive.

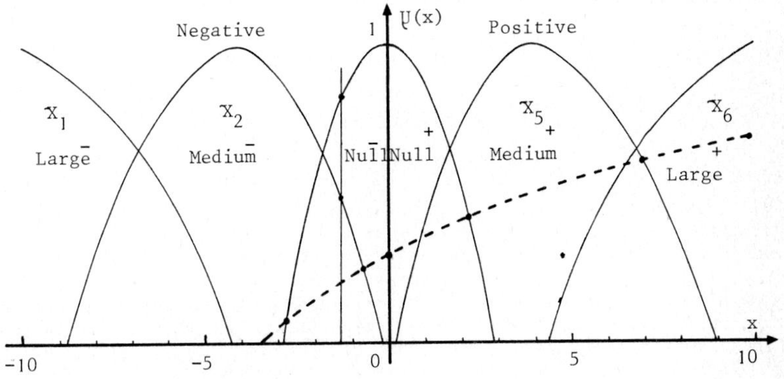

Figure 12. An example of fuzzy chopping.

The fuzzy set A is then divided by

$$\mu_A(\chi_j) = \{0 \; ; \; 0.24 \; ; \; 0.28 \; ; \; 0.41 \; ; \; 0.60 \; ; \; 0.78\}$$

and point x = -1.5 by

$$\mu_{-1.5}(\chi_j) = \{0 \; ; \; 0.58 \; ; \; 0.79 \; ; \; 0 \; ; \; 0 \; ; \; 0\}$$

As for the reconstitution of a fuzzy set, the property

$$(7) \qquad \mu_{A_R}(x) \geq \mu_A(x) \wedge \left(\bigvee_{\chi_j} \mu_{\chi_j}(x) \right)$$

indicates that in the case where X is entirely known in χ, i.e

$$\bigvee_j \mu_{\chi_j}(x) = 1 \quad \forall \; x$$

then $A_R \supseteq A$.

Figures 13a and 13b illustrate the reconstitution of the initial fuzzy set A , and that of any fuzzy set G having the membership function

$$\mu_G(\chi_j) = \{0.8 \; ; \; 1 \; ; \; 0.6 \; ; \; 0 \; ; \; 0 \; ; \; 0\}$$

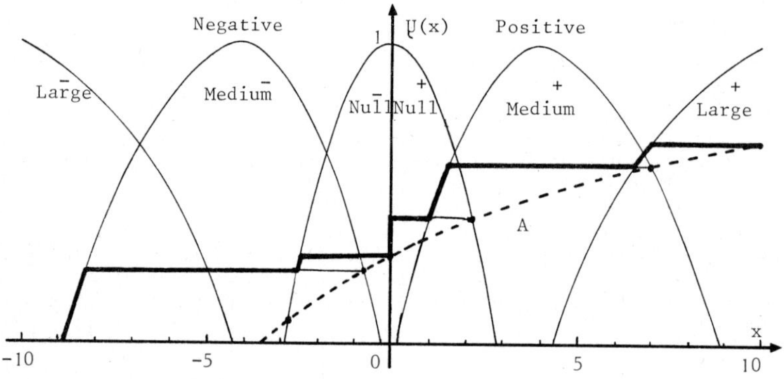

Figure 13a. The reconstitution A_R of a fuzzy set A in X , with the membership function in the series χ, of

$$\mu_A(\chi_j) = \{0 \; ; \; 0.24 \; ; \; 0.28 \; ; \; 0.41 \; ; \; 0.60 \; ; \; 0.78\}$$

The application of this technique to the development of a fuzzy model consists using as the series χ a certain number of basic adjectives in the human description and in referencing the description phrases according to this series χ .The space thus saved expedites the implementation of the calculations in small computers, while maintaining a performance level comparable to ordinary chopping [10]. Thus, when we consider the example of a 5-dimensional fuzzy model, if we break each dimension down into 10 segments, the model will comprise $(10)^5$ values, whereas chopping into fuzzy sets like those in fig.12 leads to the memorization of $(6)^5 = 7\,776$ values.

This fuzzy discretisation, moreover, allows us to use, indiscriminately, membership functions defined either by points, or by explicit equations, as is the case

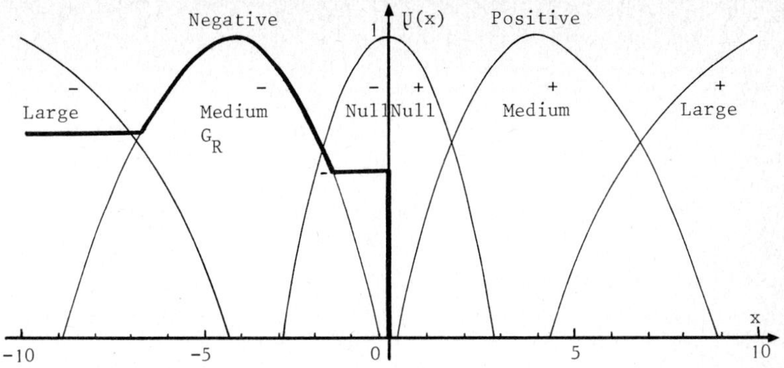

Figure 13b. The reconstitution G_R of a fuzzy set G with the membership function $\mu_G(x_j) = \{0.8 ; 1 ; 0.60 ; 0 ; 0 ; 0\}$

in the experiment recounted in [12], in which a 5-dimensional fuzzy relationship referenced in five fuzzy sets is constructed.

In resume, we have seen now, by means of the notions of fuzzy language and fuzzy algorithm, it is possible to formalize the subjective information supplied by a human operator concerning a system that he is acquainted with ; this is done as a fuzzy, n-dimensional relationship, representative of a fuzzy model. We have next seen a technique for diminishing the number of data to be stored in a computer ; this technique is not limited to the development of fuzzy models, but can also be profitably used for treating fuzzy algorithms, or other calculations bases on the theory of fuzzy sets.

The subjective information contained in a fuzzy model can be used in different ways to aid in the decision-making or control development of a system. In principle, the calculation of a control algorithm now calls on the human operator only to establish the target, or the possible constraints on the control mechanisms [3,8,9]; and it allows a solution, no matter how complex the model used. The example presented in the last part of this expose is that of calculating a regulation algorithm for a system, using a fuzzy model. This last part corresponds to one of the many possibilities of a subjective model's use, and aims at showing how the calculation are used.

IV - USING A FUZZY MODEL FOR CALCULATING A CONTROL ALGORITHM.

The general principle for calculating a control algorithm consists in,using a model of the system, finding a relationship between the state of the system and the control action to be applied to it in order to bring the system to a state within the field that constitutes the target zone.

Let us again consider the model, M (paragraph II) with the membership function :

$$\mu_M(e_n, \Delta e_n, u_n , e_{n+1} , \Delta e_{n+1})$$

and let us use the notation :

$x_n = \{e_n , \Delta e_n\}$ the state of the system at instant nT

$x_{n+1} = \{e_{n+1}, \Delta e_{n+1}\}$ the state of the system at instant (n+1)T .

The membership function of M then becomes :

$$\mu_M(x_n, u_n, x_{n+1}) ;$$

and M is an application of $X_n \times U$ in X_{n+1}

with $x_n \in X_n$, $u_n \in U$, $x_{n+1} \in X_{n+1}$.

The development of a fuzzy process control relationship now presupposes the definition of a target G , which is another fuzzy sub-set of space X representing subjectively a zone where we wish to bring the state of the system, between x_n and x_{n+1} , and to maintain it .

Our method consists in developing by computer a fuzzy relationship R , using the proceeding fuzzy relationship M, insuring the recall and holding of the process' state in G [12,13].

This fuzzy relationship R has the membership function $\mu_R(x_n,u_n)$ and is obtained by seeking, successively the relations between X and U in M which bring the system state x back into G , in one, two,., ℓ-1,ℓ control periods.

Consider, for example, an order 1 regulator. We seek the control u_n that must be applied to the model in order to bring it from some state, x_n ,to another state x_{n+1} included in the fuzzy target G. The membership function is then given by the expression :

$$(8) \qquad \mu_{r_1}(x_n , u_n) = \bigvee_{x_{n+1}} \left[\mu_M(x_n, u_n, x_{n+1}) \wedge \mu_G(x_{n+1})\right]$$

or, with the usual notation for the composition of fuzzy relationships :

$$r_1 = G o M$$

In other words, the value of the membership function $\mu_{r_1}(x_n,u_n)$ connecting state x_n with control action u_n is given by the maximum value of the fuzzy set intersection, attained by the model at instant (n+1)T when the control action is applied and it (the model) is in state x_n , with the fuzzy target G.

Is is not always possible to reach the target in just one control period : this fact results in insufficient or null values in the control relationship, i.e, when the system is in state x_n , the fuzzy control set is either null or too weak to allow a control decision.

Therefore, the calculation must be re-effected, with the goal of reaching the target in 2, then 3, then...ℓ control periods.

But, since the control can be effected only at instant nT, and the controls applied to the system later are not known, we must now do this calculation while examining all the possibilities of later controls, i.e, while seeking to reach the target regardless of what these later controls may be. This amounts to using, not relationship M, but its projection M' on u :

$$(9) \qquad M' = \operatorname{Proj}_u M : \mu_{M'}(x_n, x_{n+1}) = \bigvee_{u_n} \mu_M(x_n, u_n, x_{n+1})$$

which is the relationship enabling us to find the zone of X_{n+1} attainable from X_n , no regardless of the control applied. This the relationship r_2 will be calculated on the following principle :

$$\{x_n, u_n\} \overset{M}{\rightsquigarrow} x_{n+1} \overset{M'}{\rightsquigarrow} x_{n+2} \subset G$$

by the expression

$$(10) \qquad \mu_{r_2}(x_n,u_n) = \bigvee_{x_{n+1}} \left[\bigvee_{x_{n+1}} (\mu_M(x_n, u_n, x_{n+1}) \wedge \mu_{M'}(x_{n+1}, x_{n+2})) \wedge \mu_G(x_{n+2})\right]$$

which can also be written

$$r_2 = G o M' o M$$

The relationship r_i enabling us to reach the target in i control periods will be obtained by

$$(11) \qquad r_i = G o (M'^{\,i-1} o M)$$

with the convention

$A^k = A \circ A \circ \ldots \circ A$ k times

et $A^o = I$ identify

which allows us to find $r_1 = G \circ M'^o \circ M = G \circ I \circ M = G \circ M$

The complete regulator of order ℓ, R_ℓ, is then the union of the various r_i, that is

(12) $R_\ell = \bigcup_{i=1}^{\ell} r_i$

In practice, the number of times the calculation is done will be determined by the quantity of information contained in the relationship; that is, when each state x_n is associated with a control sub-set with a membership function sufficient for control decision-making. However, it has been shown in [19] that there exists a maximum value ℓ, which is given by the expression

(13) $R_{MAX} = G \circ (I \cup \hat{M}') \circ M$.

in which $\hat{A}$ designates the transitive closure of relationship A

$\hat{A} = A \cup A^2 \cup \ldots \cup A^k$,

k being, at most, equal to the cardinality of the breakdown of the variables'space X . In order, to elucidate these calculations, let us present them in a simple example, which comprises a fuzzy model having an input, u, and an output, x . The model is the fuzzy representation of a system with an integration, has advantage of still being usable manually, and allows a graphic visualization of the calculations evolution.

In order to represent the model by tables, the space of the system's variable x is quantified into 5 fuzzy sets, A_1, A_2, A_3, A_4, A_5, and that of its control u , into 3 fuzzy sets ;

$U_n = B_1$; x_n \ x_{n+1}	A_1	A_2	A_3	A_4	A_5
A_1	1	0	0	0	0
A_2	.9	.3	0	0	0
A_3	.4	.8	0	0	0
A_4	0	.2	.9	.3	0
A_5	0	0	.2	.8	.5

$U_n = B_2$; x_n \ x_{n+1}	A_1	A_2	A_3	A_4	A_5
A_1	1	0	0	0	0
A_2	0	1	0	0	0
A_3	0	0	1	0	0
A_4	0	0	0	1	0
A_5	0	0	0	0	1

$U_n = B_3$; x_n \ x_{n+1}	A_1	A_2	A_3	A_4	A_5
A_1	.5	.8	.2	0	0
A_2	0	.3	.9	.2	0
A_3	0	0	0	.8	.4
A_4	0	0	0	.3	.9
A_5	0	0	0	0	1

Figure 14. The membership function of model M.

$X = \{A_1, A_2, A_3, A_4, A_5\}$

$U = \{B_1, B_2, B_3\}$.

The A_i and B_i can correspond to the fuzzy sets "large negative, medium negative, null, etc...."; or, in the case of ordinary chopping, to segments of the space. Let M, then, be the model which can be represented by the table (3x5x5) in Fig.14, which shows the integration at the control $u = B_2$, for which the system does not change state.

Let us now consider the target $G \subset X$:

$$\mu_G(x) = \{0 ; 0.2 ; 1 ; 0.2 ; 0\}.$$

using expression (8), let us calculate $\mu_{r_1}(x_n, u_n)$; and the coefficient $\mu_{r_1}(A_2,B_1)$ is :

$$\mu_{r_1}(A_2,B_1) = \bigvee_{x_{n+1}} (\mu_M(A_2,B_1, x_{n+1}) \wedge \mu_G(x_{n+1})) ,$$

with $\mu_M(A_2,B_1,x_{n+1}) = \{0.9 ; 0.3 ; 0 ; 0 ; 0\}$.

$$\mu_{r_1}(A_2,B_1) = \vee\{0.9 \wedge 0 ; 0.3 \wedge 0.2 ; 0 \wedge 1 ; 0 \wedge 0.2 ; 0 \wedge 0\}$$
$$= \vee\{0 ; 0.2 ; 0 ; 0 ; 0\} = 0.2$$

Carrying out this calculation for each pair $\{A_i,B_j\}$, we obtain R_1 (fig. 15) .

μ_{R_1} — U_n / x_n	B_1	B_2	B_3
A_1	0	0	.2
A_2	.2	.2	.9
A_3	.2	1	.2
A_4	.9	.2	.2
A_5	.2	0	0

Figure 15. The membership function of R_1.

To calculate R_2, we must calculate M', using expression (9) : take, for example :

$$\mu_{M'}(A_3,A_2) = \bigvee_{u_n} \mu_M(A_3,u_n,A_2)$$
$$= \vee\{8 ; 0 ; 0\} = .8$$

The repetition of this calculation leads to the table in Figure 16, which shows that the largest fuzzy zone attainable from, for instance, state A_4 , is $\{0 ; 0.2 ; 0.9 ; 1 ; 0.9\}$.

The calculation of r_2 is then carried out by means of expression (10),

$$r_2 = G \circ M' \circ M$$

Let us for example $\mu_{r_2}(A_2,B_1)$ first carying out the operation $M' \circ M$.

$$\mu_{M'oM}(A_2,B_1,x_{n+2}) = \bigvee_{x_{n+1}} (\mu_M(A_2,B_1,x_{n+1}) \wedge \mu_{M'}(x_{n+1},x_{n+2}))$$

which gives, for $x_{n+2} = A_3$, for instance ,

$\mu_{M'}$ \ x_{n+1} / x_n	A_1	A_2	A_3	A_4	A_5
A_1	1	.8	.2	0	0
A_2	.9	1	.9	.2	0
A_3	.4	.8	1	.8	.4
A_4	0	.2	.9	1	.9
A_5	0	0	.2	.8	1

Figure 16. The membership function of $M' = \mathrm{Proj}_u\, M$

$$\mu_{M'oM}(A_2,B_1,A_3) = \vee\{0.9\wedge 0.2 \;;\; 0.3\wedge 0.9 \;;\; 0\wedge 1 \;;\; 0\wedge 0.9 \;;\; 0\wedge 0.2\}$$
$$= \vee\{0.2 \;;\; 0.3 \;;\; 0 \;;\; 0 \;;\; 0\}$$
$$= 0.3 \;;$$

and $\mu_{M'oM}(A_2,B_1,x_{n+2}) = \{0.9 \;;\; 0.8 \;;\; 0.3 \;;\; 0.2 \;;\; 0\}$.
Next we carry out Go(M'oM) by means of the expression

$$\mu_{r_2}(x_n,\, u_n) = \bigvee_{x_{n+2}} (\mu_{M'oM}(x_n,\, u_n,\, x_{n+2})\wedge \mu_G(x_{n+2}))$$
$$= \vee\{0.9\wedge 0 \;;\; 0.8\wedge 0.2 \;;\; 0.3\wedge 1 \;;\; 0.2\wedge 0.2 \;;\; 0\wedge 0\}$$
$$= \vee\{0 \;;\; 0.2 \;;\; 0.3 \;;\; 0.2 \;;\; 0\} = 0.3$$

R_2 is calculated by $R_2 = r_1 \cup r_2$, which gives the table in Fig.17
The property of the fuzzy relation M' ,

$$I \subset M' \subset M'^2 \subset M'^3 \subset M'^4 = M'^5$$

μ_{R_2} \ U_n / x_n	B_1	B_2	B_3
A_1	.2	.2	.8
A_2	.3	.9	.9
A_3	.8	1	.8
A_4	.9	.9	.3
A_5	.8	.2	.2

R_2

$\mu_{R_{MAX}}$ \ U_n / x_n	B_1	B_2	B_3
A_1	.8	.8	.8
A_2	.8	.9	.9
A_3	.8	1	.8
A_4	.9	.9	.8
A_5	.8	.8	.8

$R_3 = R_4 = R_5 = R_{MAX}$

Figure 17. The membership function of R_2 and R_{MAX}.

allows the deduction that $I \cup \hat{M}' = M'^4$, and thus that R_{MAX} can be calculated by means of the expression derived from (13).

$$R_{MAX} = G \circ M'^4 \circ M ;$$

the membership function is given by the table in Fig.17. With knowledge of a regulator R, it is then possible to calculate and visualize the looped system's fuzzy trajectories by calculating the fuzzy relation of the model with the regulator, by means of the expression :

$$B = M \circ R$$

$$(14) \qquad \mu_B(x_n, x_{n+1}) = \bigvee_{u_n} \mu_R(x_n, u_n) \wedge \mu_M(x_n, u_n, x_{n+1})$$

In resume, the method is as follows :
Starting point : a human operator's subjective description of the evolution of the variables in a process he is well acquainted with. The subject perceives these variables through a control panel, or through recorded data, while also taking into account subjective, non-measured variables, such as noise, the look of a product (the steel round in continuous pouring, for example), associating its auditory, visual, olfactive, and even vestibulary and taste properties.
The subjective data are entered in the computer, where the membership functions of each descriptive adjective are programmed by a subjectively defined fuzzy set bearing on a value zone of the value corresponding to the adjective ; these zones are not disconnected. The membership function can be easily programmed by using straight-line, parabolic or hyperbolic arc segments. The adverbs may be characterized by functions like $[\mu(x_i)]^2$, $\sqrt{\mu}(x_i)$, $[1 - (\mu(x_i)^2]$.
These data, entered through a specialized terminal or not, are the object of treatment based on a union of the direct products corresponding to the subjective terms of the sentences. This treatment results in a fuzzy relation representative of the subjective model of the process'variables'evolution. From this model, an algorithm for process control is extracted, in the form of a fuzzy relation, not by interpreting a description of the observer's manoeuvres or control tactic [6,7,11] but by calculating the fuzzy control relation directly from the discrete fuzzy model of the process. This is done by considering a fuzzy sub-set G, of X, representing the "target". The calculation of the regulator r_i consists in seeking the fuzzy control set or sets likely to bring the system from a state x_n to a state within the target G, in i periods then corresponding to the state x_{n+i}.

CONCLUSION.

This type of approach appears to us in no way contradictory with the deterministic or stochastic approaches to automation ; it can only be complementary.
The formulation of the human operator's subjective information allows orientation of the results of an analysis of objective measurements, and even the completion of efforts to modelize the system.
The advantage of the theory of fuzzy sets lies in the fact that it can grasp human subjectivity and make an interesting contribution, in the case of complex systems. Moreover, this tool permits the development of Man-machine interfaces better adapted to Man's subjectivity, presenting him with the fuzzy control set.
The different notions are applied... by the human operator for attempting of characterize the work charge. The theory of fuzzy sets then acts like a doctor questioning a patient in order to direct his diagnosis, trying to take into consideration parameters such as vigilance, concentration, fatigue, in relationship with the variation ranges of speed position, acceleration, and intensity of the input stimuli, and attempting to interpret phenomena which are difficult to interpret through objective measurements.

REFERENCES.

[1] L.A. ZADEH, Fuzzy sets, inf. and control, 8, (1965).

[2] L.A. ZADEH, Fuzzy algorithms, inf. and control, (1969).

[3] R.E. BELLMAN, L.A. ZADEH, decision-making in a fuzzy environment, Management science, vol. 17, n°4, (1970).

[4] S.S.L. CHANG, L.A. ZADEH, On fuzzy mapping and control, I.E.E.E. TSMC 2, (1972).

[5] C.V. NEGOITA, D.A. RALESCU, Fuzzy systems and artificial intelligence, Kybernetes, 3, (1974).

[6] E.H. MANDANI, S. ASSILIAN, An experiment in linguistic synthesis with fuzzy logic controller, Int. Journ. Man-Machine Studies, 7, (1975).

[7] P.J. KING, E.H. MANDANI, The application of fuzzy control system to industrial processes, Proc. 6Th Trienn IFAC World Congress, Boston, (1975).

[8] A. KAUFMANN, Introduction à la théorie des sous-ensembles flous, Tomes 1, 2,3,4, Masson Edit. (1975).

[9] C.V. NEGOITA, D.A. RALESCU, Applications of fuzzy sets to systems Analysis, Birkhäuser Verlag, Basel und Stuttgart, (1975).

[10] D. WILLAEYS, N. MALVACHE, Using referentiel of fuzzy sets : application to fuzzy algorithm. Int. Conf. on System Science, Wroclaw, Pologne (1976).

[11] J. VAN AME RONGEN, R. VAN NAUTA LEMKE, I. VAN DER VEEN, An autopilot for ships designed with fuzzy sets. Proc. 5ème conf. Int.IFAC/IFIP, La Haye (1977).

[12] D. WILLAEYS, P. MANGIN, N. MALVACHE, Use of fuzzy sets for systems modelizing and control : application to the speed regulation of a strongly perturbed motor, Proc. 5ème conf. Int. IFAC/IFIP, La Haye (1977).

[13] D. WILLAEYS, N. MALVACHE, Utilisation of fuzzy sets for system modelling and control. Proc. Congrès Int. I.E.E.E. New Orleans, (1977).

[14] L.A. ZADEH, Fuzzy sets as a basis for a theory of possibility, ERL, Berkeley, M 77/12 (1977).

[15] L.A. ZADEH, A theory of approximate reasoning, ERL, Berkeley, M 77/58 (1977).

[16] L.A. ZADEH, Pruf : a meaning representation language for natural languages. ERL, Berkeley, M 77:61, (1977).

[17] D. WILLAEYS, N. MALVACHE, Utilisation de la discrétisation floue pour le traitement d'informations floues par calculateur. Proc. Colloque Int. sur la théorie et les applications des sous-ensembles flous, Marseille (1978).

[18] N. MALVACHE, Etude du temps de réaction de l'homme à une information alphanumérique simple en fonction de son environnement. Contrat CNET-PTT, n°78-9B-075-BCI/DZS (1978).

[19] D. WILLAEYS, N. MALVACHE, Use of fuzzy model for process control, Proc. Int Conf. on Cybernetics and Society, Tokyo, (1978).

ADVANCES IN FUZZY SET THEORY AND APPLICATIONS
M.M. Gupta, R.K. Ragade, R.R. Yager (editors)

EXPLORING LINGUISTIC CONSEQUENCES OF ASSERTIONS IN SOCIAL SCIENCES

Fred WENSTØP

Bedriftsøkonomisk Institutt
P. Box 69
1341 Bekkestua
NORWAY

Abstract

Social Sciences, notably social psychology, are concerned with human behavior on the individual level. The prime source of knowledge about human behavior stems from psychological insight. As a consequence, most theories in this field are formulated verbally, making possible general and approximate statements about loosely defined phenomena. Although rich in content, verbal theories can not be systematically analyzed since no method for this is known. However, if theories are formulated structurally analogous to conventional simulation models where linguistic values replace numbers, Zadeh's concepts of linguistic variables and approximate reasoning can be applied to deduce consequences. We present a language called ZL-2 which makes this approach operational. Since ZL-2 is written in the computer language APL which has a syntax similar to that of English, it becomes easy to implement and natural to use. It can also be conviently expanded or improved. We show several illustrations of how ZL-2 performs. At its present state, ZL-2 can accomodate variables with linguistic values exclusively. If variables are operationally defined, a higher level of precision than can presently be offered by ZL-2 is most often required. The use of ZL-2 implies a general level of precision somewhere between that of conventional verbal theories and that of mathematical models but considerably closer to the former. Since few have experience in working in this intermediate area, one should expect difficulties in adjusting to it. Since costs are low, however, and benefits obvious, we think the approach is worthwhile pursuing.

INTRODUCTION

In this paper, our purpose is to demonstrate that it is possible to build models of the same general structure as conventional simulation models but where *values* are given verbally rather than numerically. We shall do this by presenting a particular language named ZL-2 in honour of Zadeh [1] and Lakoff [2] who originated the idea of using fuzzy sets to model the meaning of linguistic values. We are not arguing that ours is a final solution that immediately can be put to significant applications. On the contrary, the path towards a more general acceptance of the principles advocated here, is obviously still full of stumbling blocks. We are arguing, however, that verbal modelling *can* be done and very conveniently too. It also *may* be useful. This paper shows one way of doing it and that it is not difficult - neither conceptually nor technically. We also provide some illustrative examples which should make it possible for the reader to get a feeling for the general merits of the approach.

VERBAL MODELLING IN THE SOCIAL SCIENCES

We shall focus on how to represent knowledge about human behavior in the social sciences. This is of course a large and complex problem, and rather than debating the whole issue we shall here pursue a narrow train of thought. We start out with the following premisses:

* We want to represent knowledge of human behavior because we want to be able to make statements about expected behavior under different circumstances.

* The best form of representation is at a level of precision and in a language as closely as possible corresponding to that of the available knowledge.

For obvious reasons, knowledge about human behavior stems predominantly from intuitive insight and personal experience and as such finds its expression in natural languange - a medium excellent for summarizing experience in general - and therefore fuzzy -statements. We are thus led to believe that we should choose natural language - or at least an auxiliary language with similar properties - as our medium for representation. This does not automatically solve the requirement of our first premiss that statements about behavior should lend themselves to deductions. An example will serve to clarify this point? Consider the following statement:

> "The level of bureaucratization is more than proportional to the size of the organization."

This is a very general proposition about how organizational size affects the amount of red tape. But supposing that we knew the size of organization A, it is not wholly impossible to infer *something*, however vague, about the level of bureaucratization in A. For instance, assuming that

"Organization A is of medium size",

few people would object to a conclusion like

"the level of bureaucratization in A is below medium but not low"

or any similar conclusion with approximately the same content. The problem is that there is no *systematic*, *consistent* and *automatic* way to reach a conclusion in cases like this. It is of course possible to substitute the verbal formulations with mathematical formulas thus making the question of inference procedure trivial. According to our premisses, however, this must be done while paying due attention to the *meaning* of the words used so that we do not lose or excessively distort the original information. An important feature of words like "more than proportional" and "medium" is that their meaning is fuzzy, the very property that makes such statements possible.

One of the principal virtues of Zadeh's *fuzzy set theory* [3] is that it provides a systematic means for meaning representation of linguistic values - like "medium" - and causal relations - like "more than proportional" - together with a principle for fuzzy inference.

In the sequel, we shall present the APL-implemented auxiliary language ZL-2. Before we proceed, however, we have to discuss a very pertinent question.

ARE VERBALLY FORMULATED CONCLUSIONS USEFUL?

Let us for the moment assume that ZL-2 contains a generally acceptable meaning representation of a limited number of English words. Suppose further that our knowledge about organization A and bureaucracy is encoded in ZL-2. It reads:

```
SIZE ← MEDIUM
BUREAUCRATIZATION ← INCREASINGLY GROWING WITH SIZE
LABEL BUREAUCRATIZATION
```

ZL-2 now returns the implied level of *BUREAUCRATIZATION:*

```
FUZZILY BELOW MEDIUM BUT NOT LOW
```

Did we learn anything? The problem is, that from the answer it is impossible to draw numerical conclusions, like for instance a probability or possibility distribution [4] around a bureaucracy index number. This situation is rather uncomfortable for a quantitatively trained person who is used to deal in numbers. It is fair to say that he might be inclined to dismiss the whole approach as nonsense. But in ordinary discourse, people are in fact frequently using linguistic values like "high", "above medium" etc. -without having anything like a numerically valued reference scale in the back of the mind.

In the social sciences one is less concerned with *predicting* behavior than *understanding* it and for the latter purpose the current approach may prove useful. Understanding is a phenomenon finding its expression at the linguistic level whereas prediction is more precise and quantitative in nature.

In this context, it becomes important to distinguish between two cases:

A: The variable in question is operationally defined so that actual measurements of its numerical value is possible. In this case, the universe of discourse is the range of possible values which constitutes a ratio- or interval scale. The meaning of linguistic values would then be defined as fuzzy subsets of the scale.

B: The variable in question is not operationally defined. The universe of discourse is a psychological continuum [5] constituting an interval scale. The meaning of linguistic values are defined as fuzzy subsets of this continuum.

The variable *level of bureaucratization* surely belongs to category B. One really cannot expect an answer to the question "What do you actually mean by a "rather high" level?" On the other hand, *organization size* is readily made operational by using the number of members. Thus, one can be expected to answer the question "What do you mean by a "medium" sized organization?" The answer might run like "50 people to degree .5, 100 people to degree 1.0 and 200 people to degree .5".- However, it is also possible that one would refuse to answer the question, arguing that it would depend on the kind of organization considered. "Medium size" obviously does not mean the same whether one is talking about high schools or universities. Hence, we see that in case A where we are able to define rather more precisely the meaning of linguistic values, we immediately run into the problem of context dependency.

In ZL-2, we assume that all variables belong to category B above. This means that we lose the power of being rather more precise in cases where this would have been possible and desirable. All our statements and conclusions are hence bound to be rather general and vague. At this price, however, we achieve simplicity and efficiency.

THE SALIENT VIRTUES OF APL

Consider the following statements:

"C's value is more or less medium"
"B's value is rather high"
"A will be somewhat higher than B if C is below upper medium but not low"

Although peculiar, these are legitimate English sentences about assignments of values to the variables A, B and C. Our concern is how these value assignments are to be effectuated. We assume that we have decided on the principle that fuzzy-set theoretic meaning of the individual words shall be fed into a computer which then shall perform the necessary tasks. But how is this best done? Let us, in the spirit of Lakoff and Zadeh, consider the syntax of the last assignment statement. Since Germanic languages are typically right-generative [6], meaning that words tend to affect the meaning of words to come rather than words already uttered, we are probably best advised to start from the right. We observe: "Low" is a value. "Not" modifies this value. "But" connects the values "below upper medium" and "not low". "Medium" is a value which is modified by "upper", the resulting value again being modified by "below". "Is" compares the value of C with all that is to the right of "is", the result being informally a truth value. "If" connects the truth value with the value "somewhat higher than B", "higher" being a relation from the value of B and modified by "somewhat". "Than" connects the relation "somewhat higher" with the value of B. Finally, "will be" assigns the complete value to A.

In this process, we have distinguished four syntactical categories:

1. *Constants* which do not affect the meaning of other parts of the statement. These are: higher, medium low.

2. *Monadic functions* which affect the meaning to their right. These are: somewhat, below, upper, not.

3. *Dyadic functions* which connect words on both sides. These are: than, if, is, but.

4. *Value assigners* which transport a value and assign it to an attribute. These are: will be, 's value is.

If a computer is to read and process sentences like the ones above, one has to pay due attention to their syntax. Now it turns out that the computer language *APL* [7] operates with just the same four syntactic categories. In APL, the three statements would read:

```
C ← MOREORLESS MEDIUM
B ← RATHER HIGH
A ← ((SOMEWHAT HIGHER) THAN B) IF C IS (BELOW UPPER MEDIUM)  BUT NOT
LOW
```

The parentheses are here used more or less in the same way as commas, e.g. to make sure that *SOMEWHAT* affects nothing else than *HIGHER*. If the meaning of *MOREORLESS, RATHER, SOMEWHAT, THAN* etc, is represented as predefined APL-functions, and *MEDIUM, HIGH, LOW,* and *HIGHER* are predefined APL constants, the three statements will be *automatically* executed yielding the value of *A*, which would be labelled as:

POSSIBLY MOREORLESS HIGH

Syntactically speaking, it is an interesting coincidence that APL, which originally was designed not as a computer language, but as a language for mathematical notation corresponds so very closely to natural language when viewed from a fuzzy set theoretic reference frame. Moreover, the special ability of APL in handling vectors and matrices, makes also the implementation of the semantic content of the statements straightforward.

DESCRIPTION OF ZL-2

By choosing APL as the computer language in which to implement ZL-2, we derive the additional benefit of convenient and simple documentation. Natural language words are defined as self contained mathematical functions or constants carrying the same name. A simple grammar specifies the set of allowed statements so that we don't run the risk of meaningless constellations.

SEMANTICS

The semantics of ZL-2 is completely defined by a set of APL constants and functions. When a grammatically correct statement is written on an APL terminal, the semantical system automatically comes into play and carries out the intended operations.
Example:
We type,
LABEL HIGH OR RATHER HIGH
and get the response,
MOREORLESS HIGH

What happens is that APL first computes the meaning of the compound value *HIGH OR RATHER HIGH*, then it finds a label which appropriately covers this meaning.

To achieve this, the system has to use a method for *internal representation* of meaning. The meaning of a linguistic value is encoded as a number array:

HIGH ← 0 0 0 0 0 0 0 0 .1 .65 1

The numbers can be thought of as occuping positions in the psychological interval of possible values of the variable in question. The leftmost number occupies the lowest possible value, etc. The numbers themselves indicate the degree to which the position is included by the linguistic value. In other words, the numbers define a fuzzy set which is the meaning of the linguistic value. Thus, on an 11-point scale, *HIGH* is assumed to include the highest value to degree 1.0, the next-to-the-highest value to degree 0.65, etc.
Similarly:

MEDIUM ← 0 0 0 .03 .43 1 .43 .03 0 0 0

In ZL-2, the number of points on the scale can be chosen arbitrarily to provide a tradeoff possibility between computing time and resolution.

The operations of words which modify or connect meaning are defined as mathematical functions using such number arrays as arguments. Below are shown some examples of function definitions in ZL-2 and how they operate. The symbols ∇....∇ enclose an APL function definition. Notice that when ZL-2 receives an input from the terminal, it will be indented. The response from ZL-2, however, is flushed left. The direction of the communication is therefore always clear.

* *VERY* squares the numbers of the array:

```
  ∇X ← VERY Y
   X ← Y×Y ∇
```

Example:

```
  VERY HIGH
0    0    0    0    0    0    0    0    .01  .42  1
```

* *MOREORLESS* takes the roots of the numbers in the array:

```
  ∇X ← MOREORLESS Y
   X ← Y*0.5∇
```

Example:

```
  MOREORLESS HIGH
0    0    0    0    0    0    0    0    .32  .81  1
```

* *NOT* inverts the number array:

```
  ∇X ← NOT Y
   X ← 1 - Y∇
```

Example:

```
  NOT MEDIUM
1    1    1    .97  .57  0    .57  .97  1    1    1
```

* *BELOW* is a one-sided NOT:

```
  ∇X ← BELOW Y
   X ← 1 - ⌈\Y
   X [1] ← 1∇
```

Example:

```
  BELOW MEDIUM
1    1    1    .97  .57  0    0    0    0    0    0
```

* *AND* takes the minimum of the two adjacent arrays:

```
∇Z ← X AND Y
 Z ← X⌊Y∇
```

Example:

```
  (MOREORLESS HIGH) AND NOT VERY HIGH
0    0    0    0    0    0    0    0    .32  .58  0
```

* *OR* takes the maximum of the two adjacent arrays:

```
∇Z ← X OR Y
 Z ← X⌈Y∇
```

Example:

```
  MEDIUM OR HIGH
0    0    0    .03  .43  1    .43  .03  .1   .65  1
```

IS computes the max-min inner product of the two adjacent arrays. The result is a scalar representing a truthvalue:

```
∇Z ← X IS Y
 Z ← X⌈.⌊Y∇
```

Example:

```
  (MOREORLESS HIGH) IS NOT VERY HIGH
0.58
```

IF takes the minimum of the truthvalue to the right and the components of the array to the left:

```
∇Z ← X IF Y
 Z ← X⌊Y∇
```

Example:

```
  MEDIUM IF (MOREORLESS HIGH) IS NOT VERY HIGH
0    0    0    .03  .43  .58  .43  .03  0    0    0
```

It should be clear from these examples that little programming effort is needed to define the meaning of words. Technically, it is easy to change the meaning of words, or to add new ones. The vocabulary as a whole, however, should form an internally consistent system. In ZL-2, the definitions of *IF* and *IS* are consequences of Zadeh's principle of compositional inference [8].
There is in this context an ongoing debate among fuzzy set theorists regarding the best definiton of *AND* and *OR* - whether they should be min-max-operators or product-sum as in statistics [9] or something in between [10]. ZL-2 is built up according to Zadehs original belief that min-max- are the most suitable operators. We are not going to take up this discussion here, only point out that ZL-2 in a matter of minutes can be changed so as to accomodate any one of the proposed alternatives.

Below, we present some more examples of ZL-2's meaning representation of composite linguistic values. Note again that the number arrays are the automatic results of the system's evaluations.

```
     INDEED HIGH
0    0    0    0    0    0    0    0    .02  .76  1

     FUZZILY HIGH
0    0    0    0    0    0    0    0    .22  .58  1

     AROUND HIGH
0    0    0    0    0    0    0    0    .18  .81  1

     RATHER HIGH
0    0    0    0    0    0    0    0    .17  1    .26

     MEDIUM TO RATHER HIGH
0    0    0    .03  .43  1    1    1    1    1    .26

     LOWER HIGH
0    0    0    0    0    0    0    .12  .7   1    .46
```

UPPER MEDIUM

0	0	0	0	.03	.4	1	.46	.03	0	0

(BELOW MEDIUM) BUT NOT LOW

0	.35	.9	.97	.57	0	0	0	0	0	0

LINGUISTIC APPROXIMATION

So far, we have been concerned with the transformation of composite linguistic values into ZL-2's *internal* meaning representation scheme. The user of ZL-2, however, does not have to be concerned with how this is done. He may exclusively work on the level of natural language in communication with the system. For this purpose, it is neccessary to have a function which transforms any number array back to natural language. This is called *linguistic approximation* [3] and is performed by the function *LABEL*, to be discussed presently.

LABEL is a function from the set of n-ary arrays to a set of approximately 10^{10} different linguistic values. This is achieved by merely 22 APL-statements utilizing a branching structure combined with recursive function calls.

In principle, *LABEL* operates in two steps. When applied to a given fuzzy set X , X is first decomposed into constituent fuzzy sets which are *unimodal* and *reasonably steepsided*. Next, each set is assigned one of the 37 prespecified labels shown in figure 1. The 37 labels are arranged in a coordinate system with *LOCATION* -or mean value - on the horizontal axis and *IMPRECISION* - or sum of membership values - on the vertical axis. The two parameters are calculated for the set to be labelled, the resulting point entered into the *LOCATION* - *IMPRECISION* diagram of fig.1 and the closest one of the 37 prespecified labels selected. If a fuzzy set is subnormal in the sense of Zadeh [8], it is normalized and its label preceeded by *POSSIBLY*. In figure 2 is illustrated how *LABEL* handles a complicated fuzzy set.

The advantage of our approach is that the resulting linguistic value is concocted piece by piece requiring limited computation and little storage.

Below, we show some illustrations of how ZL-2 operates. When ZL-2 receives its input (which is indented), it first computes the internal meaning representation of the linguistic value. Then it applies *LABEL* to this and responds at the left margin with a linguistic value with approximately the same meaning. Although these examples do not show how ZL-2 can be used in complicated system modeling, they provide a general idea of how it works.

```
      LABEL HIGH OR RATHER HIGH
MOREORLESS HIGH

      LABEL MEDIUM OR ABOVE MEDIUM
NOT BELOW MEDIUM

      LABEL (ABOVE MEDIUM) OR MEDIUM OR BELOW MEDIUM
(RATHER LOW) TO RATHER HIGH

      LABEL NOT MEDIUM
UNKNOWN EXCEPT MEDIUM
```

```
     LABEL HIGH OR LOW
UNKNOWN EXCEPT NEITHER HIGH NOR LOW

     LABEL (NOT HIGH) AND NOT LOW
NEITHER HIGH NOR LOW

     LABEL (ABOVE RATHER LOW) BUT NOT RATHER HIGH
(NOT LOW) EXCEPT RATHER HIGH

     LABEL ABOVE LOW
NOT LOW

     LABEL RATHER NOT HIGH
(RATHER HIGH) OR POSSIBLY NOT HIGH

     LABEL NOT RATHER HIGH
UNKNOWN EXCEPT RATHER HIGH

     LABEL MEDIUM OR POSSIBLY (BELOW MEDIUM) OR ABOVE MEDIUM
MEDIUM OR POSSIBLY UNKNOWN
```

The following examples show how *LABEL* handles more general problems where *X* not neccessarily is the meaning of relatively simple linguistic values. This is often the case in practical simulation.

```
     LABEL    .1   .5   1   1   1   1   .5   .1   0   0   0
(RATHER LOW) TO MEDIUM

   LABEL    0    0    0    0    0   .1   .2   .3   .5   .7   1
ABOVE UPPER MEDIUM

   LABEL    1   .2    0    0   .1    1   .1    0    0   .2   1
UNKNOWN EXCEPT ((RATHER LOW) TO RATHER HIGH)
EXCEPT MEDIUM)

   LABEL    1   .5    1   .5    1   .5    1   .5    1   .5   1
UNKNOWN EXCEPT POSSIBLY (((RATHER LOW) TO RATHER HIGH)
EXCEPT ((NEITHER (MOREORLESS HIGH) NOR MOREORLESS LOW)
EXCEPT (((LOWER MEDIUM) TO UPPER MEDIUM) EXCEPT (( MOREORLESS
 AROUND MEDIUM) EXCEPT MEDIUM)))
```

Even remembering that the parentheses are non-ambigious commas, the last response is admittedly somewhat far out. But try to do it better!

It is our experience that *LABEL* operates effectively if not always efficiently. The procedure is somewhat *ad hoc* and can once in a while give objectionable results, mainly because the number of predifened labels currently is too sparse. However, the perhaps more straightforward method of using the entire metric of *X* to pick out the closest linguistic value demands large storage capacity with regard to linguistic values.

ZL-2 also contains a provision for addition and subtraction of linguistic values, operations which can be performed on an interval scale. The introduction of multiplication and division requires ratio scale and is not implemented here [11]. Addition and subtraction are defined by max-min convolution which follows from Zadeh's principle of fuzzy composition [12].

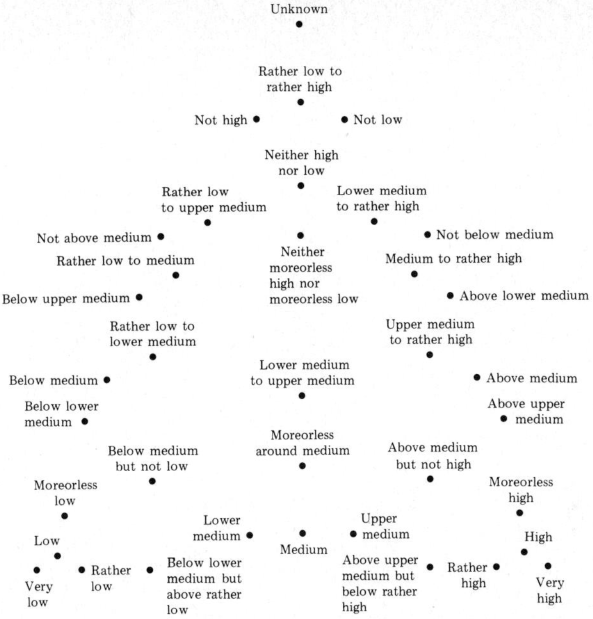

Figure 1. The 37 pre-specified labels of the linguistic approximation function *LABEL* in ZL-2 were selected to spread out evenly in the coordinate system above. The *meaning* of a label is a fuzzy set. It is measured by the parameters *location* (horizontal axis) and *imprecision* (vertical axis) which are defined as the mean value and the area under the set, respectively. Linguistic approximation of a given *regular* fuzzy set *X* - meaning that it is normal, unimodal and rather steepsided - is performed by plotting its position in the diagram above and selecting the closest label. If *X* is irregular, it is first broken down into regular constituents as shown in figure 2.

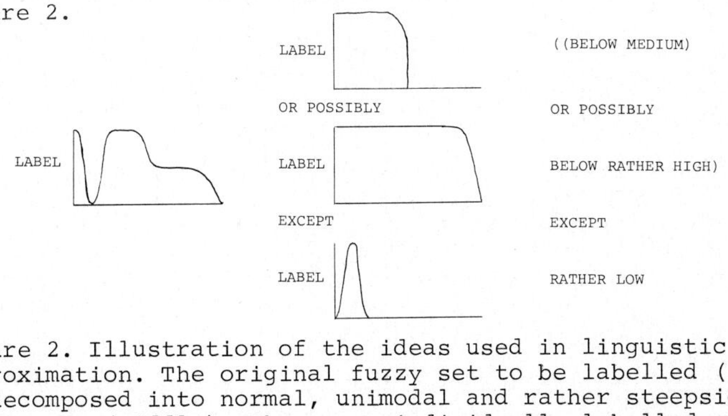

Figure 2. Illustration of the ideas used in linguistic approximation. The original fuzzy set to be labelled (left) is decomposed into normal, unimodal and rather steepsided fuzzy sets (middle). These are individually labelled using the prespecified labels in figure 1 yielding the final result (right).

Some illustrations of how it works are shown below.

LABEL HIGH PLUS HIGH
HIGH

LABEL HIGH PLUS UNKNOWN
NOT BELOW MEDIUM

LABEL HIGH PLUS RATHER HIGH
RATHER HIGH

LABEL HIGH MINUS HIGH
MEDIUM

LABEL HIGH MINUS MEDIUM
(ABOVE UPPER MEDIUM) BUT BELOW RATHER HIGH

LABEL HIGH PLUS LOW
MEDIUM

LABEL (NOT LOW) PLUS NOT HIGH
(RATHER LOW) TO RATHER HIGH

SYNTAX

For the computer to properly understand verbal assignment statements, it is neccessary to adhere to a strict set of syntactical rules when they are formulated. In ZL-2, the syntax is defined by a vocabulary and a generative grammar [6]. Owing to the similarities of the syntax of English and of APL, the generative grammar becomes simple. As a matter of practical experience, the rules soon appear so natural to the user that he does not have to refer to the formal grammar.

THE VOCABULARY OF ZL-2

CATEGORY	*SYMBOL*	*CATEGORY MEMBERS*
Primary terms	*T*	*HIGH, LOW, MEDIUM, UNDEFINED, UNKNOWN*
Hedges	*H*	*ABOVE, BELOW, AROUND, UPPER, LOWER, RATHER, MOREORLESS, VERY, NOT, NEITHER, POSSIBLY, TRULY, INDEED, FUZZILY*
Connectives	*C*	*AND, OR, BUT, NOR, PLUS, MINUS, TO, EXCEPT*
Trend Mode	*M*	*INCREASINGLY, DECREASINGLY, LINEARLY*
Trend direction	*D*	*FALLING, CLIMBING, GROWING*

Relation connective	*RC*	*THEN*
Pointer	*W*	*WITH*
Truth evaluator	*IS*	*IS*
Conditionalizer	*IF*	*IF*
System variable	*X*	*no restriction*

THE GENERATIVE GRAMMAR OF ZL-2

In addition to the category symbols listed above, we use the following non-terminal symbols:

S is an assignment statement
V is a linguistic value
R is a linguistic relation
N is a truthvalue

Five rewriting rules defines the grammar:

1) $S \rightarrow X \leftarrow V$
2) $V \rightarrow X, T, (H\ V),\ (V\ C\ V),\ (V\ IF\ N),\ (R\ W\ X)$
3) $N \rightarrow (N\ C\ N),\ (X\ IS\ V)$
4) $R \rightarrow (R\ RC\ R),\ D$
5) $D \rightarrow (H\ D),\ (M\ D)$

Together with the vocabulary, this grammar formally defines the infinite set of all admissible assignment statements.

Example:
When we want to write an assignment statement *S*, this must according to rule 1) be rewritten as

$X \leftarrow V$

Where *X* is the name of the system variable to be assigned the linguistic value *V*. We now must select one of the alternative rules under 2) according to the known deep structure of the final product. For instance,

$X \leftarrow (V\ IF\ N)$

V and *N* must again be rewritten using appropriate rules under 2) and 3), respectively. For instance,

$X \leftarrow ((R\ W\ X)\ IF\ (X\ IS\ V))$

Continuing in this fashion along a path which we know to be appropriate with regard to the intended final product, we rewrite successively:

$X \leftarrow (((R\ RC\ R)\ W\ X)\ IF\ (X\ IS\ (H\ V)))$
$X \leftarrow (((D\ RC\ D)\ W\ X)\ IF\ (X\ IS\ (H(H\ V))))$
$X \leftarrow ((((H\ D)\ RC\ (M\ D))\ W\ X)\ IF\ X\ IS\ (H(H\ T))))$

Finally, we rewrite the non-terminal category symbols with appropriate terminal words and remove redundant parentheses:

Y ← (((FUZZILY CLIMBING) THEN INCREASINGLY FALLING) WITH Z)
IF W IS NOT ABOVE MEDIUM

This assignment statement is now grammatically correct and will be accepted by a ZL-2 active terminal, returning *Y*'s consequential value provided the values of *Z* and *W* are prespecified.

THE USE OF VERBAL RELATIONS IN ZL-2.

In natural language it is customary to specify a functional relationship from one quantity to another by a sequence of conditional statements, e.g.:

A is below lower medium if *B* is very low,
A is around medium if *B* is below medium but not very low
A is rather high if *B* is medium,
A is more or less high if *B* is above medium

This can of course be translated directly into ZL-2. However, besides being cumbersome, the method has an operational drawback. If, during simulation, the actual value of *B* happens to fall in the boundary between two prespecified values let us say that *B* were around "upper medium",- then the resulting value of *A* would be either disproportionally fuzzy or poorly defined.

In fact, in ZL-2 it becomes *POSSIBLY ABOVE UPPER MEDIUM*.
This can be seen from the matrix below which is ZL-2's meaning representation of the implicit relation. It can be intuitively viewed as a fuzzy graph [3] and we notice that as such it has (unintended) overlaps and holes:

0	0	0	0	0	0	.57	.97	1	1	1
0	0	0	0	0	0	.57	.81	.81	.81	.81
0	0	0	0	0	0	.32	.32	.32	.32	.32
0	.07	.07	.07	.07	0	0	0	0	0	0
0	.53	.53	.53	.53	0	0	0	0	0	0
0	.58	.99	.97	.57	0	0	0	0	0	0
0	.53	.53	.53	.53	0	0	0	0	0	0
.57	.42	.07	.07	.07	0	0	0	0	0	0
.97	.42	.01	0	0	0	0	0	0	0	0
1	.42	.01	0	0	0	0	0	0	0	0
1	.42	.01	0	0	0	0	0	0	0	0

We therefore should provide alternative ways of verbally expressing fuzzy relations between variables.

At the outset, one should think that a person's mental picture of a fuzzy relation would be of a continuous kind. The reason that he uses a discontinous scheme to express it, is that natural language offers no good alternatives. The language of mathematics does, however, and in ZL-2 we have chosen to allow mathematically inspired formulations like:

A ← DECREASINGLY GROWING WITH B

which pretty much covers the meaning expressed with more effort above and renders it in a more continuous fashion.

The *meaning* of the relation *DECREASINGLY GROWING* is shown below. Here the numbers form a fuzzy graph which is ⎛-shaped:

0	0	0	.01	.03	.01	.03	.1	1	1	1
0	0	0	.03	.08	.07	1	1	.56	.06	0
0	0	.01	.09	.79	1	.07	.04	.08	.01	0
0	.01	.03	1	1	.04	.02	.01	.04	.01	0
0	.02	.2	.77	.12	.01	.01	.01	.02	0	0
0	.07	1	.11	.05	.03	0	0	0	0	0
0	1	.07	.05	.03	0	0	0	0	0	0
0	.64	.03	.03	.02	0	0	0	0	0	0
.01	.11	.02	.02	.02	0	0	0	0	0	0
.06	.06	.01	.02	.01	0	0	0	0	0	0
1	.04	.01	.02	.01	0	0	0	0	0	0

Similarly, the meanings of the following relations are in principle shaped like / /\ V and ν, respectively:

GROWING
INCREASINGLY GROWING
DECREASINGLY FALLING
FALLING THEN GROWING
INCREASINGLY FALLING THEN INCREASINGLY CLIMBING

We admit that these verbally formulated relations are neither parts of natural English nor easily comprehensible. They do fill a purpose, however. -Besides, a user is free as to whether he wishes to use them.
Below, we show some illustrations of how ZL-2 evaluate relations. The dyadic function *WITH* performs max-min composition. It is important to bear in mind that both the dependent and independent variables are supposed to range over the corresponding psychological intervals.

X ← MEDIUM

LABEL Y ← GROWING WITH X
MOREORLESS AROUND MEDIUM

LABEL Y ← (INCREASINGLY GROWING) WITH X
(BELOW MEDIUM) BUT NOT LOW

LABEL Y ← (DECREASINGLY GROWING) WITH X
(ABOVE MEDIUM) BUT NOT HIGH

LABEL Y ← (FALLING THEN CLIMBING) WITH X
MOREORLESS LOW

SUMMARY AND CONCLUSION

ZL-2 is a program package consisting of a total of 220 APL-statements defining an approximate model of the context independent meaning of 32 English words. These words can be combined to statements describing behavior of systems in the disguised form of a computer program. The description will be more or less approximate in nature and quantites will be given verbally.- Hence it can be called a *verbal model*. ZL-2 can simulate the behavior of verbal models and state the approximate results verbally, using an internal representation of linguistic values. -An important question in this respect is whether the internal representation adequately renders the common meaning of the words. In figure 3 we show the ZL-2 definitions of the primary terms *HIGH*, *MEDIUM* and *LOW* together with a rather different set of alternative definitions.

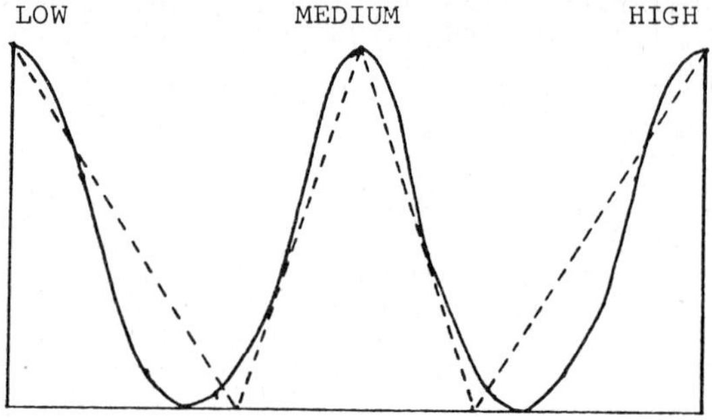

Figure 3. The ZL-2 definitions of the primary terms *HIGH* , *MEDIUM* and *LOW* (unbroken line) shown together with an alternative set of definitions (broken line). The difference made no appreciable change in the input - output properties of ZL-2.

We have run ZL-2 on numerous tests where we *LABEL*led complicated linguistic values. It turned out that we without exceptions got identical results or results with insignificant differences in meaning. This means that on the purely linguistic level - where ZL-2 is intended to operate - the exact form of the internal definitions of primary terms matters little, as long as they do not violate obvious common sense principles. If we were to advance a step in precision and take into account objectively measurable base variables [13] and context dependence, this would probably not be right anymore.

The big question is to which degree our special approach holds prospects. To be sure, there has so far been a limited number of applications where established social psychological theories has been reviewed [14],[15] or man-machine systems analyzed [17], [18], [19]. One of the premisses of the current approach has been that theorists in the social sciences in fact use linguistic values like high, medium, not high etc. They do so informally without having an objective measure scale in mind. It is a wide gap from this to quantitative models with numerical variables - and somewhere in the middle of this gap lies our approach with deductive verbal models. The problem with using verbal models is that it is neccessary to use linguistic values more formally and with greater diligence than conventionally. This requires a new attitude which it takes time to adapt to. Before, it was no point arguing whether a variable is below medium or more or less low, but using the analytic power of ZL-2 or similar languages, it is - because consequenses can be derived.

If a researcher in the social sciences wants to build causal models of human behavior and is willing to think harder than usual about linguistic values, we believe that a language like ZL-2 can be of useful assistance in helping him draw conclusions when he is faced with inexact premisses. He then has the power of finding out to which extent the degree of imprecision in his information about the system affects his possibilites of drawing conclusions about its behavior. If the input is sufficiently approximate, the linguistic output value will be *UNKNOWN* - a warning that the available information does not permit conclusions. This is an important analytical principle which makes it possible to study the trade-off between precision and significance in model formulation [18].- And since a good interactive language can be defined by approximately more or less around 250 APL statements, there are small costs and great fun experimenting with it.

REFERENCES

[1] L.A. Zadeh, "A fuzzy set interpretation of linguistic hedges", *J. of Cybernetics*, vol. 8, pp. 4-34, 1972.

[2] G. Lakoff, "Hedges: A study in meaning criteria and the logic of fuzzy concepts", in *Proc. 8th Regional Meeting of Chicago Linguistic Society*, pp. 183-228, 1972.

[3] L.A.Zadeh, "The concept of a linguistic variable and its application to approximate reasoning -II", *Information Science*, Vol. 8, pp. 301-357, 1975.

[4] L.A.Zadeh, "Fuzzy sets as a basis of possibility", *Fuzzy Sets and Systems*, Vol. 1, pp. 3-28, 1978.

[5] W.S.Torgerson, *Theory and Methods of Scaling*, John Wiley, New York, 1958.

[6] N.Chomsky, *Aspects of the Theory of Syntax*, MIT press, Cambridge 1965.

[7] K.E.Iverson, *A Programming Language*, John Wiley, New York 1962.

[8] L.A.Zadeh, "A theory of approximate reasoning", Memorandum No. UCB/ERL M77/58, Univ. of Calif., Berkeley 1977.

[9] P.Albert, "The algebra of fuzzy logic", *Fuzzy Sets and Systems*, Vol. 1, pp. 203-230, 1978.

[10] R.R.Yager, "On a general class of fuzzy connectives", Technical report ‡ RRY 78-18, Iona College, New York, 1978.

[11] D. Dubois and H.Prade, "Operations on fuzzy numbers", *Int.J. Systems Sci.* Vol. 9, pp. 613-626, 1978.

[12] F.E.Wenstøp, "Application of linguistic variables in the analysis of organizations", Dissertation, School of Bus. Ad., Univ. of Cal., Berkeley, 1975.

[13] M. Zeleny, "Membershipfunctions and their assessment", in J.Rose, ed., *Current topics in cybernetics and systems*, pp. 391-392, Springer Verlag, Berlin, 1978.

[14] F.E.Wenstøp, "Deductive verbal models of organizations", *Int. J. Man-Machine Studies*, Vol. 8, pp. 293-311, 1976.

[15] H.Koppelaar et al., "Verbale modelvorming", *Sociologische Gids*, pp. 201-211, 1978/3.

[16] W.J.M.Kickert and E.H.Mamdani, "Analysis of a fuzzy logic controller", *Fuzzy Sets and Systems*, Vol. 1, 1977.

[17] J.M.Adamo and M.Karsky, "Application of fuzzy logic to the design of a behavioral model in an industrial environment", in J.Rose, ed., *Current topics in cybernetics and systems*, pp. 361-365, Springer Verlag, Berlin, 1978.

[18] F.E.Wenstøp, "Fuzzy set simulation models in a systems dynamics perspective", *Kybernetes*, Vol. 6, pp. 209-218,1977.

[19] W.J.M.Kickert, "An example of linguistic modeling: The case of Mulder's theory of power", *This volume*.

ADVANCES IN FUZZY SET THEORY AND APPLICATIONS
M.M. Gupta, R.K. Ragade, R.R. Yager (editors)

AN EXAMPLE OF LINGUISTIC MODELLING:

the case of Mulder's theory of power

Walter J.M. KICKERT

Department of Industrial Engineering
Eindhoven University of Technology
P.O. Box 513, 5600 MB Eindhoven/Netherlands

Summary

A new kind of simulation models is introduced, namely a linguistic model. This linguistic modelling approach is based on the theory of fuzzy sets. Besides the general rationale and description of the approach, a factual linguistic model of Mulder's theory of power is presented. This linguistic model is compared to a similar but numerical simulation model of Mulder's power theory.

1. Introduction

This paper deals about a relatively new kind of modelling, namely linguistic modelling. This method makes use of linguistic variables and linguistic causal relationships in stead of the numerical variables and relations which are usual in systems modelling. The whole approach is based on the theory of fuzzy sets.
Up till now the method of linguistical modelling has mainly been applied to technical systems, namely in the form of fuzzy-logic-control. Promising results have been obtained in this field of application [14,18], and research is still going on there [17]. However we feel that this intrinsically vague and imprecise approach had better be applied to the so-called "soft sciences" in stead of a "hard science" like control theory. It is only very recently that an application study appeared where the method was used to model organizational behaviour [24].

In order to show the differences between the linguistic approach and the numerical approach to simulation models, we have chosen to model the power theory of M. Mulder [19]. This choice was mainly made because a simulation study of this theory by means of numerical simulation techniques was recently performed [8]. The second reason for choosing this social-psychological theory is that the theory has been presented in a quite unambiguous manner, namely in the form of fourteen clear theses. This avoids a lot of subjective interpretation of the theory, which would otherwise have to be performed before being able to model anything at all.

This paper is ment to introduce the linguistic modelling approach but mainly intends to show its usefulness in modelling social scientific processes and theories about those phenomena. Contrary to the lot of fuzzy set research we do not want to show how fuzzy sets can be applied but we want to solve a practical problem. No attention will be paid to the basic theory of fuzzy sets, which is supposed to be known. For an extensive treatment of the the theory of fuzzy sets the reader is referred to Kaufman [9]. A clear introduction is given in Zadeh [25].

2. Mulder's power theory

The Dutch social-psychologist M. Mulder has developed a theory about power which has some interesting properties [19]. The outstanding novelty in this theory is that he states that the exercise of power per se leads to satisfaction. This is contrary to the usual ideas about "rational man" as usual in social sciences. According to these theories man pursues a goal, generally speaking the maximizing of some kind of profit. In order to attain this goal man can use power. Power gives him an advantageous bargaining position and therefore a better chance to obtain his goal. In those theories about human behaviour, power is only a means to

arrive at some desired state [4].
Mulder states that this is a false starting-point. Motivation for power does not have to be derived from other motives; the exercise of power per se can lead to satisfaction (thesis 1 of Mulder's theory). Man strives for power, for more power than he has. Based on this fundamental thesis, Mulder proposes a theory about the reduction of the power-distance. His theory has the advantage of being dynamic: it predicts changes in power (distance) levels. The power theory essentially is a theory about power processes in small groups and it wants to give an explanation of the increases and decreases in power of the persons of the group. Mulder has conducted fairly extensive laboratory and field experiments to validate his theory. The theory is laid down in fourteen clear theses which can roughly be divided into a group of theses about the primary tendencies of people behaving in power situations (theses 1 to 5) and into a group of theses which describe secundary effects, such as personality factors and crisis situations (theses 6 to 14). In table 1 all theses are presented.

Table 1: Mulder's fourteen theses [19]

thesis 1	the mere exercise of power will give satisfaction
thesis 2	the more powerful person will strive to maintain or to increase the power-distance to the less powerful
thesis 3	the greater this distance from the less powerful person, the stronger the striving to increase it
thesis 4	individuals will strive to reduce the power distance between themselves and more powerful persons
thesis 5	the smaller this distance from the more powerful person, the stronger the tendency to reduce it
thesis 6	the power-distance reduction tendency will occur regardless of a recent upward movement to a more powerful position or a recent well-earned promotion
thesis 7	the expected costs increase more sharply than the profits with reduction in power distance in reality
thesis 8	more participation in decision-making will not reduce but increase a great power distance
thesis 9	the quantity of power, i.e. power distance, is a more decisive factor than the quality of power (its proper or improper use)
thesis 10	in crises a social system requires leadership which shows great self-confidence and is capable of strong exercise of power
thesis 11	if leaders exercise their power forcefully, people will attribute great self-confidence to them
thesis 12	people with great self-confidence and strong power motives show a stronger power distance reduction tendency
thesis 13	when less powerful individuals find that they have more self-confidence than the powerful person, they will show a stronger tendency to reduce the distance to the powerful person
thesis 14	when an individual builds up an inverted Y-structure in which he imagines he is halfway between the powerless and the powerful, he will also manifest a power-distance reduction tendency

As already mentioned before, the foundation of the whole theory is laid down in the first thesis: power is aspired for its own sake. The basic dynamical principles of the theory are presented in the theses 2 until 6, where theses 2 and 3 deal with the tendency of the more powerful while theses 4 and 5 deal with the less po-

werful. The tendency of the more powerful to increase the power distance is positively reinforced by a larger power distance while on the contrary the tendency of the less powerful to reduce the power distance increases as the power distance diminishes.
Thesis 7 represents the following situation. When a powerless person feels he would like to take over the position of the man in power, this aspiration is characterized by a low level of reality. Should he be faced with the real situation of actually being able to take over the more powerful position, he will realise that the exercise of power has a lot of disadvantages: a loss of personal contact, a risk of failing, the tension, the responsibility, etc. The barriers to becoming more powerful are harder to overcome in reality than in imagination. According to thesis 7 the cost factor will resist the tendency of power-distance-reduction to increase as expressed in thesis 5. In this sense thesis 7 represents an additional dynamic element in the theory. Because it does not lie in the scope of this paper to give an exhaustive description and analysis of Mulder's power theory, if only because of the author is no expert on the field of social psychology, the interpretation of the other theses will be left to the reader (see [19]). Roughly speaking they add a theory of personality to the power distance reduction theory. For practical reasons this extension will not be dealt with in this paper.

3. A systems model of the power theory

Although Mulder has presented his theory about power distance reduction in a series of clear theses, some interpretation still has to be done before the theory can be modelled as a consistent formal system.
In this section the main theses will be reviewed and translated into systems language.

The fundamental thesis of the theory, that power per se leads to satisfaction, will be interpreted to mean that the system is closed. This means that the model of power behaviour is not a part of a larger system, such as a decision process in which power only plays a part as a means to obtain more preferred decisions, but that this model is self-containing; power behaviour can be considered for its own sake irrespective of the possible surroundings, which is the very definition of a closed system.

Secondly, we can identify several feedback loops in the system. The first loop is contained in theses 2 and 3. The tendency of the powerful to increase the power distance is itself influenced by the power distance; the larger the power distance the larger this tendency will be. Obviously we have to deal here with a positive feedback loop.
In a similar way the power distance influences the tendency of the less powerful to reduce the power distance (theses 4 and 5). According to the theory this reduction tendency increases as the power distance decreases. Hence in this case we have to deal with a negative feedback loop.
The interpretation of thesis 7 causes somewhat more trouble, mainly because of its shortness. Mulder has explicitely stated that the power distance has an influence on the costs and benefits. Litterally nothing more is said. However, this thesis was ment to represent a resisting factor in the tendency of power-distance-reduction. Interpreted in that sense, this thesis adds an extra feedback loop to the model: the power distance influences the costs and benefits which on their turn influence the power-distance-reduction tendency.
This tendency obviously affects the power distance.
A diagram of the theory as formulated up till now, is presented in figure 1.

Note that a symmetrical cost/benefit subsystem is added to the behaviour at the side of the powerful; although Mulder does not explicitely mention that this mechanism also holds for these people, there is no reason why it should not exist or at least be tried out in the simulation study. Eventually it can always be discarded.

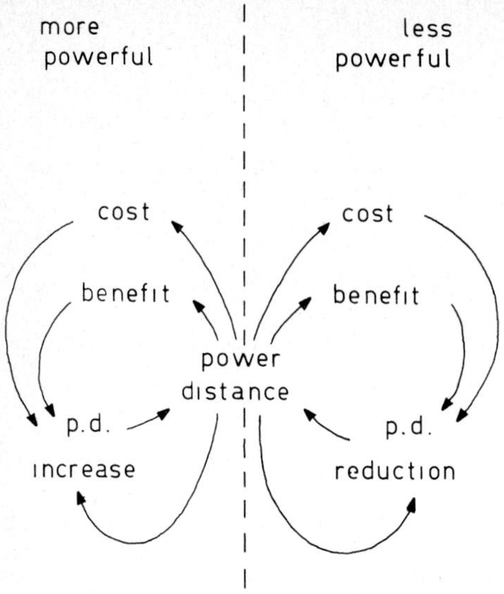

FIGURE 1 STRUCTURAL MODEL OF THE POWER THEORY

At least as important as stating what is included in the system is to mention what is not incorporated. As stated before we restrict ourselves to the modelling of the primary power behaviour as expressed in the seven first theses. This implies that no features of personality, such as self-confidence are included (see theses 11, 12 and 13) Neither does the model account for the so-called "crisis" situation (thesis 10). Moreover there is one factor which is explicitely mentioned in thesis 7 which is deliberately not incorporated, namely the so-called level of reality (also referred to in thesis 14). It is simply assumed that the whole process of power behaviour does take place at a high level of reality, namely in reality. Apart from the reason that the model will remain simpler, the main reason to exclude this extra factor is that a lot of interpretation would otherwise have to be done. Again, like in the cost/benefit explanation, it is not very clear how the factor reality should exactly be dealt with. Therefore it is simply abandoned. As an illustration of what system might result if all these additional factors were included, we present the diagram of the system used in [8] (see figure 2). This sytem will be discussed in more detail in section 6 where we present our factual model.

We have now arrived at the stage where the model is given in the most elementary form of a system, namely as a set of elements and a set of relations between those elements. The model, as visualized in figure 1, only describes the structure of the system. In many cases this might be a satisfactory result and one might proceed to analyse this structural model in the usual way, e.g. by means of graph theory [6]. This might reveal the connectedness of subsystems, critical paths, hierarchies, etc. Apart from the fact that this analysis does not seem to open much new realms in this particular case, our aim is to go beyond this essentially static structure analysis; we want to model the theory dynamically to be able to predict future behaviour. For note that though figure 1 represents a directed graph of the system and gives the causal relations, dynamics are not yet incorporated.

Because it goes beyond the scope of this report to discuss the general sense of dynamic modelling, we will proceed directly to discuss some possible methods of dynamic simulation models.

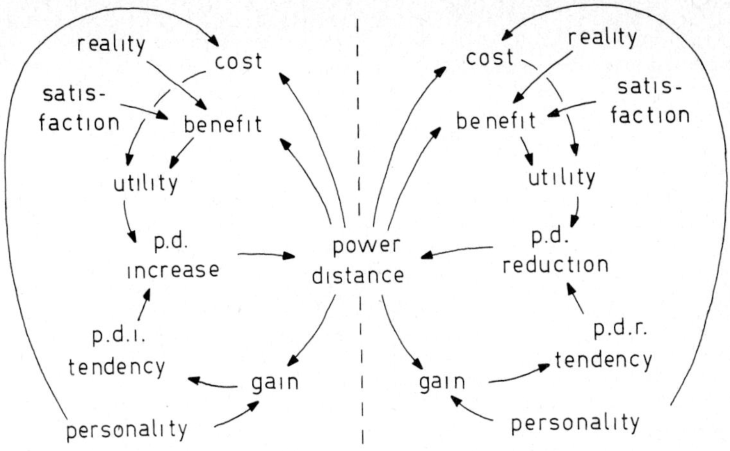

FIGURE 2 EXTENDED STRUCTURAL MODEL

4. Numerical versus linguistic models

At first it should be pointed out that the concept of model and modelling as used in this report, does not refer to the general notion of building theories, but refers to system models and that the method used is that of computer simulation. This kind of modelling can as well be used to simulate empirical situations as to simulate theories - like in our case. Computer simulation studies in social sciences are not new. Taking, for example, organization theory, several simulation models have been proposed to explain and predict organizational behaviour (for example [2,3]). The information processing analysis of human problem solving is well-known to be based on computer simulations [20].
For simulation models in social science we refer e.g. to the journal "Behavioral Science". An introduction to the problem of computer simulation in the social sciences is given in [7], which gives a summary of the advantages of computer simulation models, of which we mention a few:

- the use of a formalized language forces a theorist to express himself clearly and precisely
- the logical structure between the concepts and the propositions has to made clear
- they enable us to discover gaps in our knowledge
- the system of propositions can be tested empirically without the use of re-interpretations and ad hoc explanations to save the model or theory from falsification
- they permit fast and correct deductions from complex systems of propositions that are not disturbed by or adapted to wishful thinking
- they show how processes progress in time, they make a theory dynamic
- they make it easy to effectively use large amounts of data and are easily adapted to new data.

However, there are disadvantages as well such as:

- the danger of "model overstraining"; the danger of too rigorously reducing the complex reality in order to fit it in a simulation model

- the danger of not using the right empirical data
- the danger of adapting a theory to a computer language, to the possibilities of a computer. This danger is considerably reduced by the availability of a large amount of computer languages.

In general it can be stated that simulation models offer a relatively clear, easy, fast and cheap method of investigating theories. So far the general rationale for this kind of modelling.

There exist numerous simulation techniques, such as analog computers or digital computer languages like SIMULA, CSMP or DYNAMO. All techniques have in common that they are numerical; the variables assume numerical values.

Here we arrive at the crucial argument of this paper, which after all attempts to introduce a qualitatively different kind of modelling, namely non-numerical linguistic modelling. It is argued that this numerical character of models constitutes a major disadvantage of the usual simulation method.

The history of science is characterized by an ever increasing use of formal mathematical tools. This is surely true of the natural sciences but it might also be stated of the social sciences. No one will deny that this development was useful, looking for example at the massive results in natural sciences and the indespensability of statistics in the social sciences. However one can not circumvent the problems posed by the introduction of mathematical methods in fields like social sciences. The very existence of a theory of measurement might highlight this fact. Many problems arise when trying to use mathematics; we will concentrate on one of those problems, possibly the most serious one, namely that of the required precision. In order to be able to use precise and exact techniques like mathematics, the quantities have to be measurable in that same precise and exact way. Variables have to be exactly defined and they have to be numerically measurable. The same holds for any relationships used. Every practising scientist knows from experience that this often raises difficulties. Take for example the problems of validity, reliability and accuracy.
We believe that one of the main difficulties is caused by the requirement of numerical precision. The more precisely and exactly one wants to work, the more simplifications and approximations one has to introduce, and hence the greater the gap between the reality and the derived theory. One might state that - often - precision is complementary to reliability.
Specifying these general remarks for the case of simulation models, we would like to add a few extra disadvantages of such numerical simulation models:

- the danger of "overstraining" the empirical data to meet the requirement of numerical precision
- the danger of "over interpreting" the numerical results of the model
- the danger of "overstraining" all kinds of actually vague relationships into exact relations, usually by means of simplification, complexity reduction and approximations.

One possible way to diminish the required amount of precision is to use linguistic variables in stead of numerical values. Examples of linguistic values are: "high", "low", "very low", "rather low", etc. Similarly one might use linguistic relations between variables in stead of numerical relations, such as:
"A is similar to B", "A becomes much higher than B if B is rather high", etc.
Hence the two constituting parts of any system - its elements and its relationships - have become linguistic. We will call such a model a linguistic model. We hope that such models will be more reliable and significant because of their implicit inexactness and vagueness.

Remember that our aim still is to simulate these models on a computer. Although on first sight this might seem a contradiction we will try to show how this can be done by means of the theory of fuzzy sets.

5. Linguistic models

In this section we will firstly elaborate the idea of linguistic variables. A theory will be proposed which defines linguistic variables in a syntactic and in a semantic way (section 5.1). The semantic meaning of a linguistic value will be defined as a fuzzy set. The theory of fuzzy sets is assumed to be known. Secondly the framework of linguistic systems will be presented (section 5.2).

5.1. Linguistic variables

Let us begin to give an illustrative example of a linguistic variable, which at the same time clarifies the parallels with the more usual notion of a numerical variable, namely that a variables assumes values. For example, the numerical variable "age" might assume the values: 15, 20, 47, 65, etc., each of which is a numerical value of the variable. In a parallel way the linguistic variable "age" might assume the values: "young", "old", "rather old", "very young", etc. each of which is a linguistic value of the variable.
In the same sense as the numerical values that a variable can assume, are bounded - e.g. they have to belong to the set of integer numbers, fractions, real numbers or irrational numbers - we want to put a restriction on the linguistic values that a linguistic variable can assume. We want to define a set of linguistic values where any possible value should belong to in order to be an admissable value of a variable. This set will be called the term-set. This term-set is defined by syntactic rules which generate the possible values; in other words, this term-set is defined as the language of a generative grammar [1]. A simple illustration of such a syntactical definition of the term-set is the following:

Take a context-free grammar $G = \{V_N, V_T, P, S\}$ where the non-terminal symbols V_N are denoted by capital letters, the set of terminal symbols is V_T = {young, old, very, not, and, or} and S is the starting symbol. The production rules P are given by:

$S \to A$	$B \to$ not C
$S \to S$ or A	$C \to D$
$A \to B$	$C \to$ very C
$A \to A$ and B	$C \to E$
$B \to C$	$D \to$ young
	$E \to$ old

A term-set T which can be generated by this grammar is T(age) = {young, old, young or old, young and old, not young, very young, ..., young or (not very young and not very old), ...} .
As in the case of numerical values, the set of possible or admissible values has so been defined in a structural way and not by simple enumeration.

On the other hand we want to define the meaning - the semantics - of the linguistic values of the term-set. That is where fuzzy set theory enters the scene, for each linguistic value is defined as a fuzzy set. A fuzzy set is a function which assigns grades of membership of elements to vague concept. E.g. the fuzzy set "young" might be defined as

$\mu_{young}(20) = 1.0$; $\mu_{young}(25) = 0.9$

$\mu_{young}(30) = 0.8$; $\mu_{young}(35) = 0.6$, and so on,

which denotes that we adhere to the (numerical) age of 20 a grade of membership of the fuzzy set "young" of 1.0, that means: 20 completely belongs to "young". The age of 25 belongs with a grade of 0.9 to "young", etc.
Returning to our original example of the linguistic variable "age", the several

relationships between a variable, the values and the semantics can be illustrated by the following figure [26].

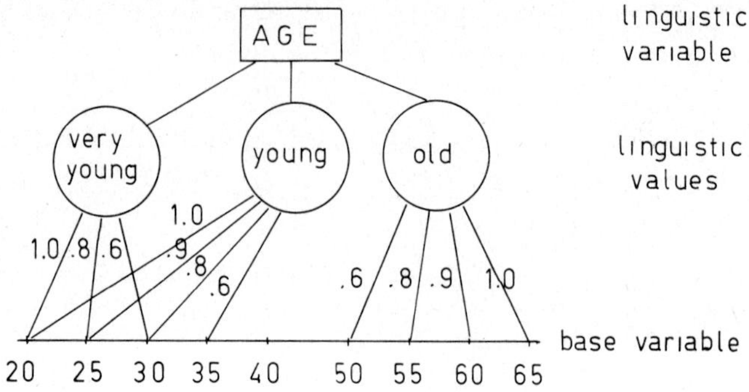

FIGURE 3 EXAMPLE OF LINGUISTIC VARIABLE

We now present the formal definition of the concept of a linguistic variable. A linguistic variable will be defined by a quintuple {A, T(A), U, G, M} in which A is the name of the linguistic variable, T(A) is the term-set of A, that is, the set of names of linguistic values that A can assume, where each linguistic value of A, denoted by X, is a fuzzy set over the universe of discourse U. G is a syntactic rule (usually a generative grammar) for generating the names of the values of A, that is, for generating the term-set T(A). M is a semantic rule for assigning to each X from T(A) its meaning M(X), which is a fuzzy subset over U. A particular name of a linguistic value, X, is called a term [26].

The semantic rule M requires somewhat more explanation. This rule essentially serves the following purpose: given the meanings of the basic linguistic values and connectives "young" and "old", (defined as fuzzy sets), one would like to be able to derive the meaning of a composite term like X = "young or (not very young and not very old)", in other words, to derive the membership function of X. This is possible by taking the following semantic rules for the four connectives:

$$M(A \text{ and } B) = M(A) \wedge M(B)$$

$$M(A \text{ or } B) = M(A) \vee M(B)$$

$$M(\text{not } A) = 1 - M(A)$$

$$M(\text{very } A) = (M(A))^2$$

The computation of the meaning of a composite term is performed by first constructing the syntactic tree of the term, then filling in the meaning of the terminal symbols and working up the tree to the composite term at the top. The meaning of the example will thus become

$$M(X) = M(\text{young}) \vee ((1-(M(\text{young}))^2)) \wedge (1-(M(\text{old}))^2)))$$

5.2. Linguistic systems

Having defined how to handle linguistic variables as the possible elements of a system, we automatically arrive at the second constituent part of a system: its relationships.

We will discuss the fuzzification of the notion of a systems relation in two phases. Firstly we will try to show on a general level how step for step a fuzzy system mapping can be generated, beginning with an ordinary mapping, via an "ordinary mapping on fuzzy sets" and a "fuzzy mapping on ordinary sets" up to a "fuzzy mapping on fuzzy sets". In the second phase we will describe a specific kind of fuzzy systems relations, namely the linguistic causal relations used in our final model.

Fuzzy relation

Let X and Y be ordinary sets. The cartesian product X x Y is the collection of ordered pairs (x,y) with $x \in X$, $y \in Y$. A fuzzy relation R between a set X and a set Y is defined as a fuzzy subset of X x Y, characterized by a bivariate membership function $\mu_R(x,y)$.

Fuzzy mapping

A mapping F in ordinary set theory is defined as a specific kind of relation, namely a relation $F \subset X x Y$ where to each $x \in X$, one $y \in Y$ is assigned with $(x,y) \in F$, written as: $F : X \to Y$ or $F(x) = y$ (see figure 4).

FIGURE 4 ORDINARY MAPPING

The definition of a mapping cannot be extended directly in a fuzzy sense; it is not possible to assign to an $x \in X$ exactly one $y \in Y$ in the case of fuzzy sets. This is inherent to their nature.

One form of this extenstion could be to define a kind of "ordinary mapping on fuzzy sets", where the mapping itself remains classical:

Let F be an ordinary mapping from set X to set Y, written as $F(x) = y$, $x \in X$ and $y \in Y$. Let $\mu_A(x)$ be the membership function of a fuzzy set A in X. Then the mapping F assigns to a fuzzy set A a fuzzy set B in Y in the following way:

$$\mu_B(y) = \max_{x=F^{-1}(y)} \mu_A(x) \qquad \text{(see figure 5)}$$

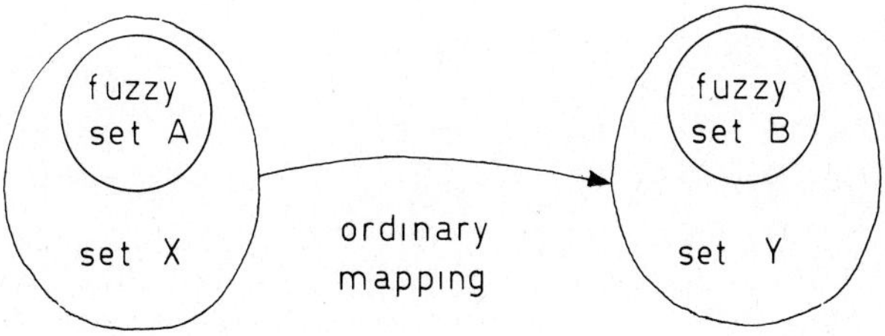

FIGURE 5 ORDINARY MAPPING ON FUZZY SETS

Clearly this is not a completely fuzzy mapping definition.
Another form of the extension to a fuzzy mapping is to define a "fuzzy mapping on ordinary sets" as being a fuzzy subset F on the Cartesian product XxY with bivariate membership function $\mu_F(x,y)$. This is identical to the definition of fuzzy relation (see figure 6).

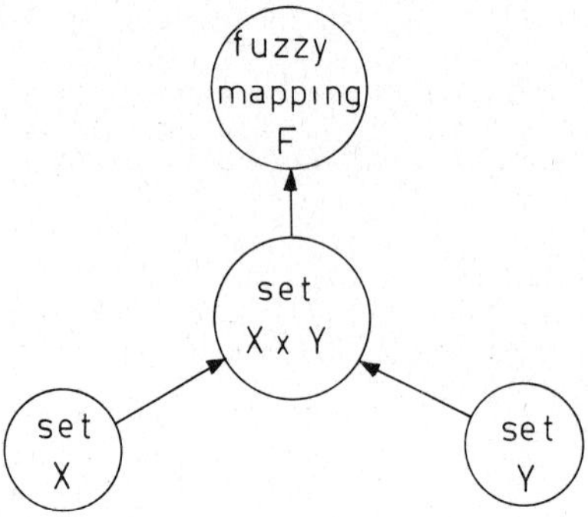

FIGURE 6 FUZZY MAPPING
ON ORDINARY SETS

The next step should then be to define a "fuzzy mapping on fuzzy sets":
Let the fuzzy set A on X induce a fuzzy set B on Y. So the fuzzy set B on Y is the fuzzy mapping of the fuzzy set A on X; the membership function $\mu_B(y)$ is defined by:

$$\mu_B(y) = \mu_{F(A)}(y) = \max_{x\in X} \min \{\mu_A(x);\ \mu_F(x,y)\} \qquad \text{(see figure 7)}$$

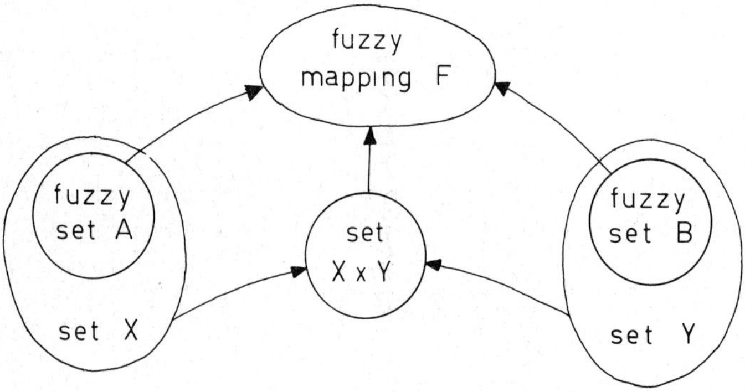

FIGURE 7 FUZZY MAPPING
ON FUZZY SETS

This equation can be interpreted as the fuzzy system response definition: while the fuzzy relation F describes the fuzzy system transformation, this last formula defines which fuzzy output B results from a particular fuzzy input A. The formula is known as the compositional rule of inference [25].

Linguistic causal relations

The kind of relationships we used to model the systems relations are causal relations of the form:

if A is high then B is low.

Clearly this is an implication between two fuzzy sets. A definition of a fuzzy implication together with the compositional rule of inference enables us to construct linguistic strings of inference like:

if A is high then B is low

A is rather high

---

thus B is rather low

We then have the framework to handle linguistic systems where the constituent elements are linguistic variables and where the systems relations consist of linguistic cause-effect relationships. This seems to unlock an area of systems where neither the constituent elements nor the coupling relationships could be made precise but where those concepts could at most be described in words and sentences.

The definition of a fuzzy implication S : if A then B where A is a fuzzy set on X and B is a fuzzy set on Y is given by its membership function as

$$\mu_S(y,x) = \min \{\mu_A(x);\ \mu_B(y)\}.$$

This is the semantical rule for the meaning of a fuzzy implication. Given a fuzzy implication S of the form: if A then B and a fuzzy implicand A' on X, then the implied fuzzy set B' on Y is defined by its membership function as

$$\mu_{B'}(y) = \max_x \min \{\mu_{A'}(x);\ \mu_S(y,x)\}.$$

This is the compositional rule of inference. In terms of linguistic variables this rule constitues the semantic meaning of the fuzzified "modus ponens".
Of course, the system cannot be described by only one relationship. The system is described by a set of fuzzy implications. The final system is considered to behave as the union of all these causal relationships

S : if A_1 then B_1 or, if A_2 then B_2, ..., if A_n then B_n

is defined by

$$\mu_S(y,x) = \max_i [\min \{\mu_{A_i}(x);\ \mu_{B_i}(y)\}]i = 1, 2, \ldots, n.$$

The system thus defined will result in a fuzzy output set. This fuzzy set will have to be transformed back into a linguistic value. This is done by generating the linguistic values of the term-set (by means of the semantic grammar) and successively fitting between those values and the fuzzy set. As the fitting criterium the least sum of squares or the least sum of absolute difference can be taken. This last operation is called the linguistic approximation [26].

6. The linguistic power-distance-reduction model

As the title already suggests this whole alternative simulation attempt is up to some extent a reaction on the simulation attempt in [8]. The main argument against their simulation model, namely the numerical character of it, has already been extensively elaborated. However there are more differences between this model and that in [8]. In section 3 both the structure of the present model and the structure of the model in [8] are given (figure 1 and 2 respectively). Apart from the already mentioned omissions (personality factors and level of reality index) the main difference lies in the interpretation of Mulder's thesis 7. Hezewijk et al derive from this thesis four distinct system equations. Several remarks can be made about these equations (see [8], p. 56). Firstly it is not clear why they let the costs depend on different factors from those on which the benefits depend. Secondly the language they used, namely DYNAMO, caused them to split the power distance reduction into an observable tendency and a non-observable tendency corresponding to Forrester's difference between levels and rates. None of these refinements is mentioned in [19]. To make a long story short, our criticism against their actual model comes down to the already mentioned "danger of adapting a theory to a computer language" and "the danger of model overstraining", that is, putting more in the model of the theory than the theory itself actually says.

The simulation model of Hezewijk et al was further elaborated by Koppelaar [15]. Koppelaar reformulated their DYNAMO-model into a system of linear first-order differential equations. This enabled him to analyse the stability of the model by means of the phase plane method. The analysis resulted in a set of conditions for stability: depending on the sign of the parameter

$$- \text{PERVAE} + \text{PERVAI} + (\text{PERVAE} - \text{GENE})\ \text{REALE} + (\text{PERVAI} - \text{GENI})\ \text{REALI}$$

the power distance will oscillate or explode. According to Hezewijk et al [8] the symbols used have the following meaning:

PERVA. = personality variable, a linear combination of power motivation, perception constant, self confidence and abilities
GEN. = the satisfaction the person derives from power
REAL. = constant for the level of reality the person operates on
The . refers to an I or E for respectively the less or the more powerful person. We sincerely have our doubts about the psychological interpretation of this condition. We question whether it will ever be possible to measure as well a personality as a satisfaction as a reality factor accurately enough to be able to calculate this composite parameter and determine its sign. In our view the analysis in [15] gives an excellent example of the "danger of over interpreting the numerical results of the model".

Last but not least the time has come to present our own simulation model, the structure of which has been illustrated in figure 1 (section 3). Keeping this structure in mind one can place the actual dynamic relationships between the variables, which are presented in table 2.

Table 2. The rules of the power-distance-reduction model

INFLUENCE OF POWER DISTANCE (PD) ON POWER DISTANCE INCREASE (PDI)

PDI_t becomes ((high if PD_t is high) or (rather high if PD_t is rather high) or (rather low if PD_t is rather low) or (low if PD_t is low))

INFLUENCE OF POWER DISTANCE (PD) ON POWER DISTANCE REDUCTION (PDR)

PDR_t becomes ((low if PD_t is high) or (rather low if PD_t is rather high) or (rather high if PD_t is rather low) or (high if PD_t is low)).

INFLUENCE OF POWER DISTANCE (PD) ON COSTS FOR THE MORE-POWERFUL (COSM)

$COSM_t$ becomes ((low if PD_t is rather low) or (high if PD_t is rather high) or (very high if PD_t is high)).

INFLUENCE OF POWER DISTANCE (PD) ON BENEFITS FOR THE MORE-POWERFUL (BENEM)

$BENEM_t$ becomes ((low if PD_t is rather low) or (rather high if PD_t is rather high) or (high if PD_t is high)).

INFLUENCE OF POWER DISTANCE (PD) ON COSTS FOR THE LESS-POWERFUL (COSL)

$COSL_t$ becomes ((low if PD_t is rather high) or (high if PD_t is rather low) or (very high if PD_t is low)).

INFLUENCE OF POWER DISTANCE (PD) ON BENEFITS FOR THE LESS-POWERFUL (BENEL)

$BENEL_t$ becomes ((low if PD_t is rather high) or (rather high if PD_t is rather low) or (rather high if PD_t is low)).

INFLUENCE OF COSTS (COSM) AND BENEFITS (BENEM) ON THE POWER DISTANCE INCREASE (PDI)

PDI_{t+1} becomes ((lower than PDI_t if $COSM_t$ is higher than $BENEM_t$) or (somewhat lower than PDI_t if $COSM_t$ is somewhat higher than $BENEM_t$) or (slightly lower than PDI_t if $COSM_t$ is slightly higher than $BENEM_t$)).

INFLUENCE OF COSTS (COSL) AND BENEFITS (BENEL) ON THE POWER DISTANCE REDUCTION (PDR)

PDR_{t+1} becomes ((lower than PDR_t if $COSL_t$ is higher than $BENEL_t$) or (somewhat lower than PDR_t if $COSL_t$ is somewhat higher than $BENEL_t$) or (slightly lower than PDR_t if $COSL_t$ is slightly higher than $BENEL_t$)).

INFLUENCE OF POWER DISTANCE INCREASE (PDI) AND POWER DISTANCE REDUCTION (PDR) ON POWER DISTANCE (PD)

PD_{t+1} becomes ((lower than PD_t if PDR_{t+1} is higher than PDI_{t+1}) or (somewhat lower than PD_t if PDR_{t+1} is somewhat higher than PDI_{t+1}) or (similar to PD_t if PDR_{t+1} is similar to PDI_{t+1}) or (somewhat higher than PD_t if PDR_{t+1} is somewhat lower than PDI_{t+1}) or (higher than PD_t if PDR_{t+1} is lower than PDI_{t+1})).

These rules require somewhat more explanation.
As will be clear from the inspection of table 2, we have used two rule structures in this model, namely causal relationships of the forms:
a) if A is high then B is low
b) if A is higher than B then C is lower than D.

Obviously two more alternative rule structures could be:

c) if A is high then B is lower than C
d) if A is higher than B then C is low.

Another possible rule structure could be:

e) A is higher than B,

but this last kind of rule does not seem to be a causal relationship any longer. The difficulty with these four different rule structures is that statements like "the higher A the lower B" can not unquestionably be translated into one of those four rules. All four rules

- if A_t is high then B_t is low
- if A_t is high then B_{t+1} is lower than B_t
- if A_{t+1} is higher than A_t then B_{t+1} is lower than B_t
- if A_{t+1} is higher than A_t then B_{t+1} is low

could be appropriate descriptions of the statement. The morale of this remark is that evidently there remains a danger of interpretation with linguistics too. (Note that e.g. theses 3 and 5 of Mulder's power theory have this ambiguous form.)

Some examples of the semantics of the linguistic values used, are given in table 3.

Table 3. SEMANTICS OF SOME LINGUISTIC VALUES

high :	0	0	0	0	0	0	0.2	0.7	1.0	1.0
very high :	0	0	0	0	0	0	0	0.2	0.7	1.0
rather high :	0	0	0	0	0	0.1	0.5	1.0	0.5	0.1
sortof high :	0	0	0	0	0.1	0.5	1.0	0.5	0.1	0
medium :	0	0	0	0.2	0.7	1.0	0.7	0.2	0	0
sortof low :	0	0	0.1	0.5	1.0	0.5	0.1	0	0	0
rather low :	0.1	0.5	1.0	0.5	0.1	0	0	0	0	0
very low :	1.0	0.7	0.2	0	0	0	0	0	0	0
low :	1.0	1.0	0.7	0.2	0	0	0	0	0	0
undefined :	0	0	0	0	0	0	0	0	0	0
somewhat higher than :	0	0.1	0.5	1.0	0.5	0.1	0	0	0	0
	0	0	0.1	0.5	1.0	0.5	0.1	0	0	0
	0	0	0	0.1	0.5	1.0	0.5	0.1	0	0
	0	0	0	0	0.1	0.5	1.0	0.5	0.1	0
	0	0	0	0	0	0.1	0.5	1.0	0.5	0.1
	0	0	0	0	0	0	0.1	0.5	1.0	0.5
	0	0	0	0	0	0	0	0.1	0.5	1.0
	0	0	0	0	0	0	0	0	0.1	0.5
	0	0	0	0	0	0	0	0	0	0.1
	0	0	0	0	0	0	0	0	0	0

There are two remarks to be made about these data. Clearly all fuzzy sets have a ten-point support set. This implies that all universes of discourse are computationally identical. Note that this does not imply that the linguistic values like "low", "rather low", etc. have exactly the same meaning independent of the variables; the support sets could easily be transformated while keeping all arithmetics the same.
Secondly it should be noted that the theory about linguistic hedges, such as "very", "rather" and "sortof" is not applied; this is mainly due to the fact that a strict application of that theory does not seem practical here.

In figure 8 the diagram of the simulation model is given.

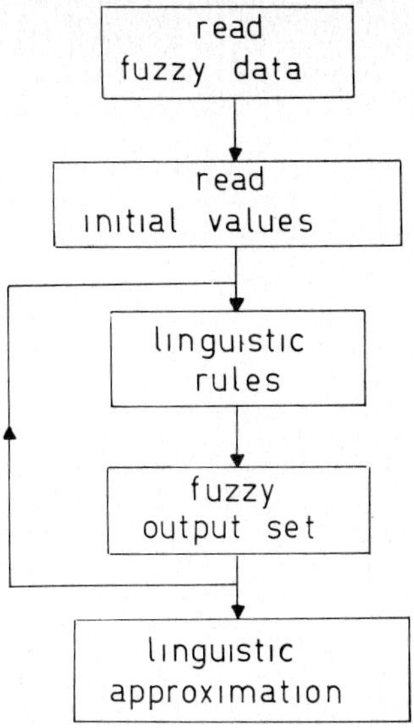

FIGURE 8

LINGUISTIC MODEL

The actual program has been written in FORTRAN IV and implemented on a PDP-11/40. The reason that an essentially numerical language as FORTRAN was used instead of a language like APL, which is much better suited for this kind of linguistic data handling [24], is that most computers have a FORTRAN compiler and that most scientists know FORTRAN. This is not always the case with a language as APL.

7. Simulation results

Note that the actual rule configuration as presented in table 2 implies that only one initial state had to be determined to start the model, namely the initial power distance. Below some results are shown where all variables are reported linguistically.

Simulation 1

Initial power distance:	low
Output:	
period 1	
power distance is	rather low or medium
power distance increase is	very low
power distance reduction is	medium or rather low or sortof high
period 2	
power distance is	rather high or rather low or sortof low
p.d. increase is	sortof high or rather low
p.d. reduction is	sortof high or rather low
period 3	
power distance is	rather high or rather low or sortof low
p.d. increase is	sortof high or rather low
p.d. reduction is	sortof high or rather low

Simulation 2

Initial power distance:	medium
Output:	
period 1	
power distance is	rather high or sortof low
period 2	
power distance is	rather high or rather low or sortof low
period 3	
power distance is	rather high or rather low or sortof low

Simulation 3

Initial power distance: rather high
Output:

period 1
power distance is rather high

period 2
power distance is rather high or sortof low

period 3
power distance is rather high or sortof low

period 4
power distance is rather high or rather low or sortof low

These results are quite remarkable in two senses. Firstly there is an important difference between these results and the results reported in [8] which almost always showed an ever-increasing power distance. In fact the power distance exponentially increased, that is, the process was non-stable. Koppelaar [15] analytically proved under what conditions the process was stable or unstable. In our case the model displays a definite stable character. The output tends to strive at some "golden mean" for nearly all situations.
Secondly the model results show a tendency to become more and more fuzzy as time increases, up to a level where significance becomes doubtful (extra information about the results can be obtained by displaying the fuzzy sets themselves besides their linguistic label). On close inspection this fact is not so surprising; feeding a fuzzy input into a fuzzy relation will evidently result in a still more fuzzy output. Because of the iterative character of the model a steady increase in fuzziness will occur. Of course this kind of intuitive explanation does not prove anything. A general mathematical analysis of this sort of processes would be very useful, but is still missing except for some incidental attempts [11,13,21]. In most numerical simulation models this steady increase of fuzziness is suppressed by the deterministic precision requirements. Non-deterministic models such as statistical regression however have the same characteristic: when extrapolating beyond the region of available data, the variance rapidly increases.

7.1. Simulation of other rule structures

As stated in section 6 the sort of rules used - see table 2 - are arbitrary. This consideration together with the stable simulation results which are quite contrary to the numerical simulation results, induced us to try a different kind of rules as well. We changed the rule structure of the first six blocks of rules of table 2 into rules of the form: "if X_t high, then Y_t higher than Y_{t-1}" e.g. The very first rule of table 2 now becomes:

PDI_t becomes ((higher than PDI_{t-1} if PD_t is high) or (somewhat higher than PDI_{t-1} if PD_t is rather high) or (somewhat lower than PDI_{t-1} if PD_t is rather low) or (lower than PDI_{t-1} if PD_t is low)).

Although the choice of this rule structure might look rather arbitrary there was a good reason for choosing this one. One might see an intuitive similarity between the four kind of rules and some kinds of differential, integral or algebraic equations. For instance a rule like "if X_t is high, then Y_t is higher than Y_{t-1}" might be considered as the linguistic counterpart of the numerical equation $Y_t - Y_{t-1} = K \cdot X_t$. This latter equation is the discrete counterpart of a differential equation

$$\frac{d}{dt} Y(t) = \lim_{\Delta t \downarrow 0} \frac{Y(t) - Y(t-1)}{\Delta t} = K' \cdot X(t)$$

Hence the above-mentioned linguistic rule can be viewed as a linguistic differential equation of the form $\frac{d}{dt} Y(t) = K'X(t)$. In this same intuitive way one might argue that a rule of the form "if X_t is higher than X_{t-1} then Y_t is high" represents a linguistic integral equation of the form $Y(t) = K' \frac{d}{dt} X(t)$ and that a rule like "if X_t is high, then Y_t is high" represents a linguistic algebraic equation of the form $Y(t) = K \cdot X(t)$. However, we should be aware of the fact that there is no rigid mathematical basis for these analogies. This of course does not prevent that there might be a convincing practical basis for the analogies. Previous research on fuzzy logic controllers has indeed established some qualitative comparability between linguistic rules and differential, integral and proportional equations [10].
If indeed this similarity actually held, this change in rule structure of the first six blocks of rules of table 2 should result in a linguistic model which is almost the linguistic analogon of the system of differential equations used in [8].

Some results are shown below.

(Three cases were tested where the power distance should respectively grow, remain the same and diminish.)

Simulation 1

Initial values: power distance is rather high, p.d. increase is sortof high, p.d. reduction is sortof low

Output:

period 1

power distance is	rather high
power distance increase is	rather high or sortof low
power distance reduction is	low or sortof low

period 2

power distance is	undefined
p.d. increase is	undefined
p.d. reduction is	undefined

Simulation 2

Initial values: p.d. is sortof low, p.d. increase is sortof high, p.d. reduction is sortof high

Output:

period 1

power distance is	rather high or rather low or sortof low
p.d. increase is	rather low or medium
p.d. reduction is	rather high or rather low or sortof low

period 2

power distance is	rather high or rather low or sortof low
p.d. increase is	rather high or rather low or sortof low
p.d. reduction is	rather high or rather low or sortof low

Simulation 3

Initial values: p.d. is very low, p.d. increase is very low, p.d. reduction is very high

Output:

period 1

power distance is	undefined
p.d. increase is	undefined
p.d. reduction is	undefined

These results are still less informative than the previous runs. The tendency of the model outputs to become more fuzzy with each iteration, has even increased in

this case. As a matter of fact the above-mentioned intuitive explanation of the fuzzification tendency would imply this result: we have inserted an extra set of vague relations between the vague variables, hence the final vagueness will further increase.
Although to some extent this might be a reasonable and intuitively logical result, one can not deny that it is rather annoying: it makes long-term predictions impossible.
As a matter of fact one might be more interested to know what happens in the long run than to know what happens in the very near future; it might be more interesting to predict that the power-distance will eventually become infinite than to predict that this power-distance decreases during the first few steps. One could state that by avoiding "the danger of over-interpreting numerical results" we now ended up at the complementary "danger of insignificance of linguistic results". Therefore some possible ways of solving this problem will be discussed here.

7.2. Reducing the fuzziness

The most obvious way of reducing fuzziness would to be sharpen the definitions of the constituent fuzzy sets and fuzzy relations: by diminishing the spread of the fuzzy sets their fuzziness will decrease. However, this would come down to the arbitrariness of the meaning of words, that is, the linguistic values. We think it not sensible to shift and change those meanings at pleasure. Moreover the model seems to be quite insensitive to changes in the basic fuzzy sets.

The second possibility for decreasing the fuzziness in this linguistic system might be to adopt a different set of definitions for fuzzy logic. Although we have chosen a particular definition for fuzzy implication and fuzzy modus ponens (compositional rule of inference) many other definitions are possible.
The only condition which these definitions should satisfy is the following argument:
Given an implication $A \rightarrow B$, an implicant A and the infered consequence $A*(A \rightarrow B)$. Suppose that the corresponding truth values are μ_A and $\mu_{A \rightarrow B}$.

Now we do not want the truth value of the consequence $\mu_A * \mu_{A \rightarrow B}$ to exceed the truth value of B, that is, μ_B. On the other hand we would like to have it as large as possible.
This argument in fact is a sort of intuitively appealing description of a fuzzy modus ponens.
Remark that the definitions which we have adopted indeed satisfy the condition. However the following combinations of definitions satisfy the condition as well

$$\mu_{A \rightarrow B} = \max\,[1 - \mu_A;\ \mu_B]$$

with the minimum operator for * [16], or

$$\mu_{A \rightarrow B} = \begin{cases} \mu_B/\mu_A & \text{if } \mu_A \geq \mu_B \\ 1 & \text{otherwise} \end{cases}$$

with multiplication for * [5].
Preliminary simulation tests however indicate that none of these alternative definitions leads to a decrease in fuzziness of the simulation results.

A third possible way of reducing the fuzziness is to insert a transformation between the successive model iterations. In stead of feeding the linguistic output value, that is, the fuzzy output set, directly back into the next model iteration, the vague output data are first transformed into exact data before being fed into the model. This results in a model input with a "membership function" which is equal to zero except at one point x_o where its value is one:

$$\mu_{A'}(x) = 1 \text{ at } x = x_o$$
$$0 \text{ elsewhere}$$

This degenerated fuzzy set of course represents a non-fuzzy exact value. With this input the compositional rule of inference reduces to

$$\mu_{B'}(y) = \max_x \min \{\mu_{A'}(x); \mu_S(y,x)\} = \mu_S(y,x_o)$$

The idea behind the fuzzy-exact transformation is that one represents the linguistic value by a non-fuzzy exact value. This exact value should be a good substitute for the fuzzy set. A possible way of doing this is to take that value at which the membership function is maximal:

$$y_o \text{ at which } \mu_{B'}(y_o) = \max_y \mu_B(y)$$

or to take the value

$$y_o = \sum_i y_i\, \mu(y_i)/\sum_i \mu(y_i)$$

The latter representation was chosen, firstly because it gives an unique value, secondly because it takes into account the whole shape of the fuzzy set. We are surely aware of the fact that in fact this way of reducing the fuzziness in the model results is no solution at all, but merely a rather artificial way of bypassing the problem. In every step vagueness is just removed. This intermediate transformation of linguistic values into numerical values in fact touches the very bases of the linguistic approach in modelling.
It suffices to give one counter argument: it works.
Not only does it work with this simulation model, it also is exactly the way the succesful fuzzy logic controllers work [14,17,18]. Some results are shown below.

Simulation 1

Initial values: power distance is medium, p.d. increase is sortof high, p.d. reduction is medium.

Output:
period 1
power distance is sortof high
p.d. increase is sortof high or rather high
p.d. reduction is rather low or sortof low

period 2
power distance is very high
p.d. increase is very high
p.d. reduction is low

Evidently the power distance explodes in this case.

Simulation 2

Initial values: power distance high, p.d. increase sortof high, p.d. reduction very high

Output:
period 1
power distance is very high
p.d. increase is high
p.d. reduction is medium

period 2
power distance is very high
p.d. increase is very high
p.d. reduction is rather low

Although initially the reduction tendency prevailed, the power distance still exploded.

Simulation 3

Initial values: power distance is rather low, p.d. increase is medium, p.d. reduction is medium
Output:
period 1
power distance is rather low
p.d. increase is sortof low
p.d. reduction is medium or sortof high

period 2
power distance is very high
p.d. increase is rather low or sortof low
p.d. reduction is very high

An initially low power distance will tend to further decrease, even with an initially prevailing p.d. increase tendency, like the next simulation shows.

Simulation 4

Initial values: power distance is rather low, p.d. increase is rather high, p.d. reduction is rather low
Output:
period 1
power distance is sortof low
p.d. increase is sortof high
p.d. reduction is rather low or sortof low

period 2
power distance is very low
p.d. increase is sortof low
p.d. reduction is very high

8. Conclusions and discussion

Conclusions about the linguistic model can be made from two points of view. The first point of view is the field of application: what was the sense of this computer simulation for social sciences and more in particular, for Mulder's theory of power? The second point of view is that of the theory of fuzzy sets: is fuzzy set theory the right basis for the semantic interpretation of linguistics? We will start with the latter question.

Like in most practical applications of fuzzy set theory, one of the main problems is how to obtain the particular fuzzy sets and how to be sure that they do represent the meaning of the linguistic terms. Wenstøp [23] reported a method of finding acceptable meanings for the primary terms by means of questionnaires. He also noted that people easily adapt to the slightly different use of natural language. On the other hand there are indications that the usual interpretation of meanings by fuzzy sets is not the one actually used [22]. Obviously the question is not settled yet.
An argument in favor of this fuzzy sets semantics is that the functioning of the model is quite insensitive to changes in the definitions of the primary linguistic terms. This result was also reported by Wenstøp [24] and might turn out to be one of the major advantages of the fuzzy-logic-controller type of application of the method [17].

We now arrive at the other type of question: what sense do linguistic models have for social sciences in general and Mulder's theory of power in particular? We still believe that the use of linguistic variables and relationships in the modelling of human or social processes is to be preferred to the use of numerical models. The problem of the required precision and exactness often shows to be huge and sometimes seems to be insurmountable.
Like Hezewijk et al [8] frequently report there is a great danger in "overstrain-

ing" and "overinterpreting" the numerical data. The much more approximate, vague and unpretentious linguistic data do not have these disadvantages. Remark that the linguistic model should also be compared to other numerical but non-deterministic models such as e.g. statistical regression models, in order to clearly state the differences and advantages. Such a comparison would certainly lead us to the relationships between fuzziness and probability, a subject which unfortunately appears to be quite underestimated in fuzzy literature [12: pg. 22, 50]. Actually it turns out that the linguistic model every period steadily increases the fuzziness of the results. On one hand this seems an evident and right phenomenon, on the other hand this blocks the possibility of long-term predictions. We have proposed a way to circumvent this problem (in a recent mathematical analysis of linguistic models [11] some alternative methods have been presented to analyse the long-term dynamic behaviour of the model, particularly the stability of the model. A stability analysis of a particular kind of linguistic model - the fuzzy logic controller type - can be found in [13]).

The obtained results seem to differ from those found in [8] in the sense that the stable state where the power distance either tends to zero or to a medium position, seems to be of frequent occurrence. It has been suggested that this kind of structural behaviour might be dependent on the sort of linguistic rules used, for there seems to exist an intuitive similarity between several linguistic causal relations and the conventional integral, differential or algebraic equations.

Two general remarks are left to be made.
Firstly it should be emphasized that the practical usefulness of the linguistic model approach can only be proved by actually applying the method. A lot more application studies will have to be performed to study this question.
Secondly it seems to be very useful if one could develop a mathematical tool for the analysis of the linguistic model behaviour. Up till now only the fuzzy-logic-controller type of linguistic systems have been studied in this sense [13,21]. Like Wenstøp [24] remarked this would certainly add to the powerfulness of linguistic models (since the completion of the study reported in this paper, an attempt towards the analysis of linguistic models has been made in [11]).

Acknowledgement

I would like to thank H. Koppelaar and R. van Hezewijk for the helpful and interesting discussions about this paper.

9. References

[1] Chomsky, N., Aspects of the Theory of Syntax, MIT press, Cambridge, Mass., 1965.
[2] Cohen, K.J. and Cyert, R.M., Simulation of organizational behaviour in: J.G. March (ed.): Handbook of Organizations, Chicago, 1965.
[3] Cohen, M.D., March J.G. and Olsen, J.P., A garbage can model of organizational choice, Administrative Science Quarterly 17, 1972, 1-25.
[4] French, J.R.P. and Raven, B.H., The bases of social power, in: D. Cartwright (ed.): Studies in Social Power, Ann Arbor, Michigan, 1959.
[5] Goguen, J.A., The logic of inexact concepts, Synthese 19, 1968, 325-373.
[6] Harary, F., Norman, R.Z. and Cartwright, D., Structural Models, New York, 1965
[7] Harbordt, S., Computersimulation in den Sozialwissenschaften, Reinbek, 1974.
[8] Hezewijk, R. van, Kanters, H. and Melief, A., Playing the game of power, Annals of Systems Research 4, 1974, 39-60.
[9] Kaufmann, A., Introduction to the theory of Fuzzy Subsets, New York, 1975.
[10] Kickert, W.J.M., Further analysis and application of fuzzy logic control, int. report, Queen Mary College, Dept. of Electrical Engineering, London, 1975.
[11] Kickert, W.J.M., Towards an analysis of linguistic modelling, paper presented at the fuzzy system session of the 4th Int. Congress on Cybern. Syst., Amsterdam, August 1978. Accepted for publication in Fuzzy Sets and Systems.
[12] Kickert, W.J.M., Fuzzy theories on decision-making, Nijhoff, Leiden/Boston, 1978.

13 Kickert, W.J.M. and Mamdani, E.H., Analysis of a fuzzy logic controller, Fuzzy Sets and Systems, 1(1978), 29-44.
14 Kickert, W.J.M. and Nauta Lemke, H.R. van, Application of a fuzzy controller in a warm water plant, Automatica 12 (1976), 301-308.
15 Koppelaar, H., Predictive power theory, Annals of Systems Research, 3 (1976), 1-5.
16 Lee, R.C.T., Fuzzy logic and the resolution principle, Journal of the ACM, 19 (1972), 109-119.
17 Mamdani, E.H., Advances in linguistic synthesis of fuzzy controllers, International Journal of Man-Machine Studies, 8 (1976), 669-679.
18 Mamdani, E.H. and Assilian, S., An experiment in Linguistic Synthesis with a fuzzy logic Controller, International Journal of Man-Machine Studies, 7 (1975), 1-13.
19 Mulder, M., The daily power game, Leiden, 1977.
20 Newell, A. and Simon, H.A., Human Problem Solving, Englewood Cliffs, 1972.
21 Pappis, C.P. and Sugeno, M., Fuzzy relational equations and the inverse problem, int. rep. Queen Mary College, Dept. of Electrical Engineering, London, 1976.
22 Rödder, W., On "and" and "or" connectives in fuzzy set theory, Working Paper 75/07, RWTH Aachen, 1975.
23 Wenstøp F.E., Application of linguistic variables in the analysis of organizations, PhD thesis, Univ. of California, Berkeley, 1975.
24 Wenstøp F.E., Deductive verbal models of organizations, International Journal of Man-Machine Studies, 8 (1976), 293-311.
25 Zadeh, L.A., Outline of a new approach to the analysis of complex systems and decision processes, IEEE Transactions on Systems, Man and Cybernetics, vol. SMC-3, 1 (1973), 28-44.
26 Zadeh, L.A., The concept of a linguistic variable and its application to approximate reasoning. I, Information Sciences, 8 (1975), 199-249; II, Information Sciences, 8 (1975), 301-357; III, Information Sciences, 9 (1975), 43-80.

ADVANCES IN FUZZY SET THEORY AND APPLICATIONS
M.M. Gupta, R.K. Ragade, R.R. Yager (editors)

TIME-OF-USE PRICING OF ELECTRICITY: A POLICY ASSESSMENT METHODOLOGY

Joseph Fiksel, John Diffenbach, Alan Renda

Arthur D. Little, Inc., Cambridge, Massachusetts
and Boston College, Chestnut Hill, Massachusetts

Implementation of time-of-use pricing of electricity in the United States could potentially lead to significant societal and political impacts. Policymakers have a need to identify the plausible policy choices and to assess the resulting first-order and higher-order impacts in both the near-term and the long-term. However, such a policy assessment is difficult due to the complexity of the overall problem, and the imprecision associated with the projection of socio-economic and political attributes. A methodology was developed, using possibility theory, which systematically incorporates subjective evaluations into an assessment of possible future scenarios.

1. INTRODUCTION

Possibility theory offers an innovative approach for coping with the fuzziness inherent in policy assessment studies. Policy analysis is characterized by uncertainty which stems from the vagueness or imprecision of relevant societal factors. In the past, these factors have been treated in a qualitative manner, but for complex studies this approach makes it difficult to describe all the intricacies of the potential outcomes of a policy choice. This paper presents a proposed application of possibility theory to a policy assessment of time-of-use electricity rates. The first part of the paper is devoted to explaining the policy issues involved and the research strategy by which possibility theory is applied. The latter part of the paper then describes in detail the mathematical formulation of the possibility approach.

The assessment of future impacts of time-of-day electricity rate structures is a formidable task when considered as a whole. Depending upon the existing economic or political conditions at the state of local levels, there are a multitude of different policies which could be adopted to govern time-of-use pricing. The immediate impacts may be felt by utility companies as well as electricity consumers, in the form of reduced peak loads, changing use patterns, etc. As time passes, these "first-order impacts" may vanish, stabilize, or give rise to more subtle higher-order impacts, such as changes in living schedules. Assessment of these indirect impacts must also account for fluctuations and trends in the socio-economic setting, or context, for which a policy is being evaluated. There is a large amount of uncertainty inherent in such an assessment, due to vagueness in our understanding of the underlying principles of human behavior, as well as vagueness in our prediction of the tides of future events. Accordingly, a methodology was developed to methodically examine all of the above phenomena, while expressing this inherent uncertainty in an explicit manner.

Although the methodological construct using possibility analysis is presented here in the context of a specific policy assessment, namely time-of-use electricity rates, the potential usefulness of the proposed approach can extend to policy assessments in general. It provides a framework whereby the various features and relationships perceived by knowledgeable parties can be integrated in a systematic fashion and used to describe potential future scenarios.

2. POLICY CONTEXT OF TIME-OF-USE PRICING

2.1 TIME-OF-USE PRICING

Time-of-use (TOU) pricing of electricity, as the name implies, entails the pricing of electricity service on a time-differentiated basis. That is, the price per kilowatt-hour (kWh) varies according to when the customer uses (or consumes) the electricity. Prices are set higher during peak usage periods and lower for off-peak times. TOU pricing can be based on time-of-day (TOD), day-of-week, season-of-year, or a combination of these.

The economic justification for TOU pricing is that a customer's electricity needs during an electric utility company's peak period impose a greater capacity demand (i.e., demand for capacity in the form of power plants) upon the company's system, and thus should bear a correspondingly higher portion of the fixed capacity charges.[1] This is so, it is argued, because the power plant capacity level that must be made available by the company is dictated not by total electricity consumption but rather by the level of highest demand at any single time. Even if this peak occurs only once a year, for example, at 4:30 p.m. on the hottest day of the year, there must be capacity available to meet this demand. Inasmuch as it will be the incremental additions to demand at the peak time that will force the company to invest in additional capacity, the logic is that peak period demand incurs a greater liability for capacity costs and thus should be priced higher. It follows from this marginal cost argument, therefore, that peak rates will be higher than offpeak rates.

TOU rates are intended to at least make those who are most responsible for the utility company's peak demand pay more per unit of electricity and hopefully as a result of price elasticity lead to a shift in some of the consumption from peak periods to less expensive offpeak periods. If such a shift can be achieved, the rate of growth of required generating capacity can be limited. Since capacity costs constitute a major portion of electricity prices, the ultimate effect will be to slow the rise in electricity prices to customers.

Despite the marginal cost justification for TOU pricing, the application of TOU rate structures will be guided by achievement of desired practical objectives and not by theoretical integrity. This means that the choice of peak/offpeak price differentials and the designation of peak and offpeak time period will be geared to practical considerations and will not rigidly follow marginal costing. Therefore, it must be understood that TOU pricing is a pricing mechanism and not a costing mechanism. Although it will more closely reflect marginal costs than do the traditional declining block rate structures, it nonetheless is, at most, a "second best" approach to marginal cost theory.

2.2 THE POLICY ISSUE

The policy issue surrounding TOU pricing is quite simple: what form of TOU pricing will give the best results? No one really knows, yet pressure is mounting rapidly to introduce TOU pricing on a widespread basis in the United States. At best, there is only a hazy and uncertain picture of what TOU pricing could lead to. One thing, however, is clear - the widespread adoption of TOU pricing would represent a radical departure in design from existing declining block rate structures that have been employed for decades.[2] Thus, there are legitimate concerns that new TOU rate structures could ultimately lead to significant societal rearrangements and disruptions. Such impacts, of course, could be either adverse or beneficial. Policymakers want to choose the TOU pricing policy that will minimize the potential for bad impacts and maximize the likelihood of good impacts.

What makes the policymaker's task so difficult is the high variety of design and implementation structures and the many different potential impacts resulting from the various policy choices. Moreover, the immediate direct impacts can trigger second, third and higher order impacts stretching over many years. Finally, many of the potential societal impacts suffer from vague definitions, do not lend themselves to very precise measurement, and are highly uncertain. The policy choice challenge really comes down to this - how can we best marshall whatever information or expertise we have about the potential for these impacts into a form usable by policymakers? Thus the criterion for such an assessment of impacts is not whether full knowledge is achieved, but rather the extent to which existing knowledge and expertise can be best put to use.

2.3 POLICY CHOICES

The choice of a policy for a TOU rate structure involves two fundamental sets of decisions: the design of the rate structure and the mode of implementation. There are both design and implementation variables, each possessing a set of values that constitute choices for the policymaker. A policy choice, therefore, consists of a unique set of values for the design and implementation variables. As discussed below, there is a wide range of TOU policy choices. The introduction of this policy choice variety into the policy assessment will complicate the task considerably. The design and implementation of TOU rate structures in practice, therefore, will not be as simple and straightforward as the simplified description above might imply.

There are several important design variables. The first is the price differential variable; for example, the peak/offpeak price differential could be 3:2, 2:1, 3:1, 4:1, etc. Second, is the scheduling variable for peak and offpeak periods. For example, the offpeak prices could be effective for 6 hours at night, for 8 hours at night, or for 10 hours at night in addition to weekends. To further complicate matters, the pricing and scheduling choices could change for summer and winter. Another variable reflects whether the rates are applied on an individual customer basis, or for groups of customers exhibiting similar consumption patterns. Still another variable relates to the choice of a scheme for the electric utility companies to distribute revenue surpluses and deficits resulting from TOU pricing.

The implementation options constitute another set of policy choice variables. All of the approximately 214 large electric utility companies may introduce TOU rates, or only certain ones might choose to do so. Clearly, the potential impacts of TOU pricing in this country would no doubt depend on how widespread these new rate structures became. It also could turn out that TOU pricing is applied only to selected customer classes, e.g., large users, or only industrial customers, and not to all customers. Another implementation variable concerns the use of load controls, devices for automatically cutting off power to appliances of customers when a prearranged demand limit is reached.[3] Finally, another implementation variable has to do with whether TOU pricing is implemented rapidly or in a gradual, incremental fashion. Although this variable is hardly amenable to precise measurements, it should be considered in an assessment of impacts.

2.4 POTENTIAL SOCIETAL IMPACTS

The potential for significant societal consequences - favorable and unfavorable - exists because of (1) the pervasive consumption of electricity in our economy and society and (2) the temporal quality of TOU rate technology. Any TOU-induced changes in electricity consumption will take the route of time shifts and changes in the level of residential and commercial-industrial electricity consumption. The greatest kilowatt-hour potential for residential shifts or changes in consumption level is related to the usage of electric heating and cooling devices, electric water heaters, refrigerators and freezers, electric ranges, clothes dryers and color televisions.

It is conceivable that higher order impacts of such changes will be adjustments to eating, sleeping and socializing schedules which could in turn lead to subtle but significant alterations in the family structure and community socialization. In addition, attempts to reduce the peak premium rates could well spur the development of solar heating and cooling technologies as well residential energy storage technology. Installation of such devices could in turn add to the investment required for home ownership and thus delay or preclude home ownership for many families.

Industrial patterns of electricity consumption would probably have the greatest economic and social impact through an increase in evening or graveyard workshifts. It has been found for instance that synchronization of diurnal rhythms of a night shift employee will occur after a period of time, but not if the employee is switched back and forth between day and night shifts.[1] Other late workshift impacts might be significant increases in clerical and management salaries as well as in labor wages.

In addition to the higher-order impacts of workshifts, other potential impacts of TOU rate technology in the industrial sector have to do with structural economic dislocations. For example, there could be realignments in the relative economic competitiveness of electricity-intensive products and services. This could lead to a migration of electricity-intensive businesses, especially if the application of TOU rate technology is not uniform. In addition there could be disruptions in regional economic development and structural (geographic) unemployment. Moreover, there could be an even greater dependency on foreign sources for certain materials such as aluminum and paper, which involve energy-intensive manufacturing processes. Finally, TOU rate technology could possibly accelerate the trend for basic industries to be transferred from the United States to developing nations.

3. RESEARCH STRATEGY

3.1 POLICY ASSESSMENT RESEARCH QUESTIONS

A policy assessment is a systematic, comprehensive study of the potential societal impacts associated with alternative policy choices for introducing a technology, an institutional arrangement, or a process. In this sense, TOU pricing is considered a process. Societal impacts include consequences of an economic, social, political, technological, ecological or institutional nature that have a significant effect on society. To identify not only the direct impacts, but also the indirect second, third and higher order impacts, requires assessing causal chains of impacts extending from the short to the long term. A policy assessment does not attempt to predict; rather, the objective is to provide information about the possible state of the world. While it would be desirable for a policy assessment to result in "solutions" for the policymaker, in reality it can only be expected to lead to better-informed policymakers, who will develop "better" policies not only at the introduction stage, but also at later stages when corrective policy interventions are necessary.

The research questions to be addressed in a policy assessment of TOU pricing are as follows:

1. What are the more plausible policy choices for TOU pricing?
2. What are the more plausible impacts of these policy choices?

These are the questions which are dealt with in the possibility analysis to be described below.

3.2 QUANTIFICATION OF VAGUE PHENOMENA

Because of the inexact nature of the predictive process, and the complexity of the societal and institutional patterns being considered, a strict quantitative modelling approach is not justified. However, a logical, systematic framework is needed to cope with the profusion of qualitative and subjective information that will be encountered. The approach described here has been designed specifically for TOU pricing but it is applicable to any investigation of the projected implications of alternative policies in the face of an uncertain future.

It was Lotfi Zadeh [2] who first drew a clear distinction between uncertainty due to randomness and uncertainty due to vagueness. In the former case, the events of interest are chance outcomes of precisely-defined situations. Multiple observations of these situations (e.g., the rolling of a die) can lead to improved knowledge about the chance events, expressed in the form of frequency distributions or probabilities. On the other hand, in the case of purposeful systems, such as human beings or societies, Zadeh's "principle of incompatibility" states that the complexity of these systems precludes precise descriptions. Furthermore, repeated observations are not possible, due to the unique, irreproducible nature of any purposeful system. In short, complex organisms defy generalization or reduction to mechanistic descriptions. Consequently, it is inappropriate to attempt to use a probability measure as a descriptor of these systems. Zadeh [3] suggests as an alternative the concept of possibility, which is a measure of the relative potential for occurrence of various events, based upon a holistic perception of the system in question.

In order to permit the manipulation of vague concepts, we will make use of "linguistic variables" - that is, variables which take on linguistic rather than numerical values. For example, JOHN's AGE may be considered a linguistic variable with possible values YOUNG, MIDDLE-AGED, and OLD. Of course, any selection of values for linguistic variables is necessarily vague, or <u>fuzzy</u>, since semantic concepts (such as COLD) do not have clearly-defined boundaries. This is dealt with by assigning degrees of membership, or possibility ratings, to each possible value. Even in the case of numerical variables such as "rate differential" a possibility distribution may be used to express uncertainty about their values. The variables to be studied are defined in the next section.

A considerable literature already exists on methods for dealing with "fuzzy" cognitive frameworks, including the achievement of group consensus concerning the values of linguistic variables [4]. The assignment of possibility distributions to variables will require informed judgments on the part of experts from several disciplines, including engineering, management, economics and sociology. The beliefs of these experts can be quantified in the form of possibility ratings for specific combinations of events. Where possible, microeconomic or behavior models may be used to supplement or guide the expert judgments. However, the scope of the policy assessment encompasses so much complexity that for higher-order impacts one is forced to rely upon the integrative power of the human intellect. By assembling the experts in working groups and interrogating them about specific portions of the overall study, possibility distributions may be derived for the different variables. This process may require several Delphi-style iterations before consensus is reached.

3.3 RESEARCH VARIABLES

A research strategy must consider, at least, <u>policy choice variables</u> and <u>impact variables</u>. In addition, because one can not assume static exogenous influences, it is necessary to incorporate <u>contextual variables</u> which might affect policy choices. Brief descriptions and examples of these three sets of variables follow.

Policy choice variables identify the alternative design and implementation strategies for applying TOU pricing. A policy choice is defined as a unique combination of a design and an implementation strategy. Below are examples of policy choice variables.

POLICY CHOICE VARIABLES

Design of TOU Rate Structures

C_1 = Schedule of peak/offpeak hours
C_2 = Peak/offpeak rate differential
C_3 = Distribution of revenue surpluses or deficits
C_4 = Rates applied to individual customers or to groups of customers

Implementation of TOU Rate Structures

I_1 = All or selected electric utilities
I_2 = Residential or industrial customers, or both
I_3 = With or without load controls
I_4 = Rapid or gradual implementation

Contextual variables represent potentially significant exogenous forces emanating from the prevailing economic, social, political and technological environment. Examples are cited below:

EXAMPLES OF CONTEXTUAL VARIABLES

F_1 = fuel costs
F_2 = energy consumption
F_3 = population
F_4 = household size
F_5 = electric heat saturation
F_6 = public image of electric utilities
F_7 = regulatory climate
F_8 = economics of solar energy
F_9 = commercial readiness of breeder technology

Impact variables describe the potential first, second and higher order societal consequences that could result from the policy choices. First-order impacts are defined as those which directly involve users or producers of electricity. Second-order impacts are those which relate to the energy sector of the economy; for example, decreased oil consumption. Third-and higher-order impacts will include any indirect effects, such as lifestyle variations. Examples of some of the possible impact variables are as listed below.

EXAMPLES OF IMPACT VARIABLES

S_1 = Change in cost per kilowatt-hour of electricity

S_2 = Change in level of consumption of electricity

S_3 = Change in financial stability of electric utilities

S_4 = Change in activity patterns of individuals and families

S_5 = Change in the amount of shift work and its effects on:

- S_{51} = Productivity
- S_{52} = Change in number of two-income families
- S_{53} = Structural unemployment

S_6 = Change in relative costs of electricity-intensive goods and services and its effects on:

- S_{61} = Industry migration
- S_{62} = Process changes for producing goods and services

S_7 = Change in the rate of commercial diffusion of solar heating and cooling technology.

Impact occurrences will be specified according to generation and level. Generation refers to chronological timing. Impacts will be identified as either first generation (0 to 2 years from implementation), second generation (2 to 10 years in the future) or third generation impacts (10 or more years in the future). The level of impact on the other hand describes its position in a causal hierarchy of impacts; i.e., first order, second order and higher order impacts. The relationship of impact generations and levels is shown in the diagram below.

		(1-2 yrs.) Near Term	(2-10 yrs.) Medium Term	(10-100 yrs.) Long Term
(Direct)	First Order	•	•	•
(Energy-Use Related)	Second Order	•	•	•
(Indirect)	Higher Order	•	•	•

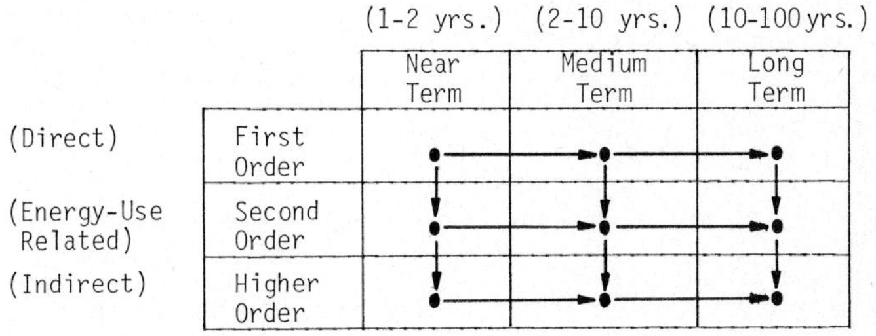

3.4 MODULAR STRUCTURE

The research strategy was designed to take into account the following factors: (1) the extreme complexity of the subject, (2) the inherent fuzziness of the variables, and (3) the need to rely heavily on subjective, inter-disciplinary judgments about impact and contextual variables. Consequently, the approach incorporates four different features:

- A disaggregated, sequential modular approach for structuring the study in order to cope with the complexity issue, and to introduce validation checks.

- A "reference set" incorporating relevant existing statistical data and baseline extrapolations, as well as empirical findings of other studies.

- Inter-disciplinary panels for integrating relevant expertise and for collecting subjective judgments.

- Possibility analysis to systematically provide structure and logical consistency in the collection and manipulation of subjective data, and to cope with the fuzziness of the variables.

The modular structure is designed to cope with the complexity of the problem by initially disaggregating the field of interest and subsequently synthesizing the findings of the disaggregated analysis in an iterative fashion. The six modules are:

1. Determination of plausible policy choices
2. Assessment of first generation
3. Assessment of second generation
4. Assessment of third generation
5. Description of possible impact-chains arising from policy choices
6. Sensitivity analysis to determine important policy parameters

Each of these modules is described briefly below; the detailed analytic approach is presented in the next section.

Module 1

In this module, both TOU design configuration strategies and implementation strategies are considered in arriving at possibility estimates of TOU policy choices, which represent specific design-implementation combinations. Based on these estimates, a representative subset of more plausible TOU policy choices will be identified. The influence of contextual variables must also be accounted for.

Module 2

Module 2 focuses on the first generation impacts (1G impacts) of the individual policy choices selected in Module 1. The assessment of the 1G impacts will include a categorization of the impacts, an estimate of their plausibility, an analysis of the causal network linking the impacts, and estimates as to how the impacts will affect various actors. The final step will be to identify a representative subset of more plausible 1G impacts.

Module 3

Module 3, with a time dimension of 2 to 10 years following the implementation of TOU policy choices, will focus on the more plausible second-generation impacts (2G impacts) that could result from the 1G impacts. The procedure will be similar to that for Module 2 except that the potential for uncertain changes in the relevant contextual environment of TOU pricing will be introduced by constructing a reasonable number of representative alternative scenarios.

Possibility estimates will be developed, again using panel groups. It is conceivable that there will be 2G impacts directly related to the original policy choice and not strongly linked to any intermediate 1G impacts, i.e., first-order, 2G impacts. The final step will be to identify a representative subset of more plausible 2G impacts.

Module 4

Module 4, with a time dimension of beyond 10 years after implementation of TOU policy choices, will focus on the more plausible third generation impacts (3G impacts) that could result from those 2G impacts assessed as more plausible in Module 3. As with Module 3, contextual uncertainty will be entered into the assessment by means of alternate contextual scenarios for the third generation time horizon. The end product will be the identification of a representative subset of 3G impacts.

Module 5

Module 5 represents a synthesis of the 1G, 2G and 3G impacts of Modules 2-4 with the TOU policy choices identified as more plausible in Module 1. Two valuable by-products that can be expected from this synthesis are (1) a validation of the impacts found in Modules 2-4 and (2) the possibility of gaining additional insights and perhaps revealing subtle impacts that are more apparent in a holistic perspective then in the narrower "link-by-link" perspectives offered in Modules 2-4.

The synthesis will be accomplished largely by testing various policy-impact chains with expert panel groups and with experts on an individual basis.

Module 6

Module 6 is intended to offer information about the sensitivity of categories of impacts to the various design and implementation variables constituting the original TOU policy choices. For example, sociological impacts stemming from graveyard shifts would be highly sensitive to the implementation option: "application to all or selected customer classes". There are likely to be less obvious sensitivities, and if so, it will be important for the policymaker to be aware of them not only for formulating initial TOU applications but also for subsequent interventions.

4. POSSIBILITY ANALYSIS METHODOLOGY

4.1 DETERMINATION OF PLAUSIBLE POLICY CHOICES

A policy choice will consist of two distinct aspects: the choice of an implementation strategy, and the choice of a design strategy. Implementation pertains to the type of rate structures imposed, whereas design pertains to the selection of parameters within a given structure. We will designate the vector of possible implementation variables as:

$$\tilde{I} = \left(I_1, I_2 \ldots\right)$$

and the vector of possible design variables as:

$$\tilde{C} = \left(C_1, C_2 \ldots\right)$$

A policy choice can then be defined as a selection of specific values for each variable in the above two vectors. We will designate the set of all policies as:

$$\bar{P} = \left\{P_1, P_2 \ldots\right\}$$

where each P_i is a realization of $(\tilde{I},\tilde{C})$; i.e., a particular value.

Let the domain of each fuzzy variable W be denoted by D(W), and assume that it is discrete and finite. Then the set of all possible policies can be written as:

$$\bar{P} = D\ (\tilde{I})\ \ \times\ \ D\ (\tilde{C})$$

The set $\bar{P}$ may be thought of as a set of discrete points lying in a multi-dimensional policy space, whose dimensions correspond to the policy variables.

For a given variable W, a possibility distribution is an assignment of numerical ratings to each element of the domain D(W). These ratings must lie between zero and one, and are denoted by

$$\pi\ (W=t)\ \text{for}\ t \in D(W)$$

Henceforth the symbol $\wedge$ will denote the minimum operator, and V the maximum operator. Two variables W and Z are said to be non-interactive if their joint possibility distribution is the minimum of their individual possibility distributions; that is:

$$\pi\ (W=t,Z=u) = \pi\ (W=t) \wedge \pi\ (Z=u)$$

In the case of policy variables, the non-interactive property can clearly be violated, since two highly possible policy components may be mutually incompatible. Hence, it is necessary to estimate the joint possibility distributions. The following method is proposed: For two interactive variables I_j and I_k, $j \neq k$ we have the relation:

$$\pi\ (I_j=t,I_k=u) \le \pi(I_j=t) \wedge \pi(I_k=u)$$

In general, $\pi(P_i) \le \bigwedge_Z \pi(Z)$

where Z ranges over all events

$\left\{I_j=t\right\}, \left\{C_k=u\right\}$ in policy $P_i \epsilon \bar{P}$

A good upper bound on $\pi(P_i)$ is:

$$\lambda(P_i) = \bigwedge_{j,k} \pi(Z_j=t,\ Z_k=u)$$

$$\text{where } Z \text{ is I or C, } t\epsilon D(Z_j),\ u\epsilon D(Z_k)$$

The majority of interactions among policy variables can probably be accounted for through these pairwise comparisons, although in some instances three-way interactions may have to be considered.

We can then obtain as an estimate of the overall possibility of policy P_i the expression $\lambda(P_i)$. The function λ induces a ranking on the policy space $\bar{P}$, indicating the relative plausibility of different policies. It may be interpreted as the prior possibility of policy P_i being adopted, without regard to specific exogenous conditions. The operation of finding a joint possibility by minimizing over the possibilities of all constituent events, may be thought of as a "weakest link" approach. In essence, a policy can be no more possible than its least possible component.

4.2 INFLUENCE OF CONTEXTUAL VARIABLES

A necessary step in Module 1 is to identify the exogenous forces or conditions that will influence the choice of a policy. These conditions may be divided into categories such as geographic, climatic, political, or industrial forces. Corresponding to each category, we will define a linguistic variable F_i with several different values which represent different situations within that category. For example, if category 2 is climatic conditions, then the variable F_2 might take on the values WARM and HUMID, TEMPERATE, COLD. The vector of exogenous variables may then be written:

$$\tilde{F} = (F_1,\ F_2\ \ldots)$$

Henceforth, the symbol F_i will be used to denote both the linguistic variable (e.g., CLIMATE) and its particular realization (e.g., COLD) in a specific instance. There should be no ambiguity, since the context in which F_i is discussed will make clear which meaning is intended. Care must be exercised in restricting these categories to conditions which are expected to have an important impact upon TOU rate policies. Otherwise, the amount of effort required to "model the world" may become unmanageable.

If we assume that Z is held at a given value u, then we can speak of the conditional possibility of W given Z, analogous to the notion of conditional probability.[5] This is written:

$$\pi(W=t|Z=u) \text{ for } t\epsilon D(W) \text{ or more concisely: } \pi(W|Z)$$

Thus, given any realization of $\tilde{F}$, the possibility may be assessed of having a particular choice of implementation variable I_k. This possibility will be written as:

$$\pi(I_k|\tilde{F})$$

which symbolizes the conditionality of the policy selection with regard to exogenous forces. It would be preferable to estimate $\pi(\tilde{I}|\tilde{F})$ for each $\tilde{I}$ as a whole, but the effort involved might become prohibitive. For design variables, we will assume that the policy choice is also contingent upon $\tilde{F}$, and this will be expressed by the possibility distribution:

$$\pi(C_K|\tilde{F})$$

The above possibility expressions can again be combined in a pairwise manner to obtain an estimate for the conditional possibility of a policy choice $P_i = (\tilde{I},\tilde{C})$ given $\tilde{F}$.

$$\pi(\tilde{I},\tilde{C}|\tilde{F})$$

Now, in order to find the overall possibility of a policy $P_i = (\tilde{I},\tilde{C})$, we may compute the maximum of its possibilities under various exogenous conditions. In other words,

$$\pi(\tilde{I},\tilde{C}) = \bigvee_{\tilde{f}} \pi(\tilde{I},\tilde{C}|\tilde{F}=\tilde{f})$$

where the maximization is done over all realizations of $\tilde{F}$. Again, V is the <u>maximum</u> operator. This calculation may be used to validate the unconditional estimate obtained above.

Let $\emptyset(P_i)$ be defined as that $\tilde{F}$ which achieves the maximum in the above equation. Then the set $\bar{P}$ of policies can be ranked according to their possibility of occurrence, and associated with each P_i is an exogenous situation $\tilde{F} = \emptyset(P_i)$ under which that policy has the greatest potential for occurrence.

The outcome of the calculations described above will be a possibility distribution on the set of policies, denoted by $\pi(\tilde{I},\tilde{C})$. Each policy will be associated with exogenous conditions $\tilde{F}$, under which it is expected to occur. For the balance of the study, a specific subset of <u>representative</u> policies, namely:

$$\bar{P}_0 \subseteq \bar{P}$$

will be selected for impact assessment. Since many of the policies will differ only slightly in terms of the policy variables, it would be fruitless and time-consuming to analyze them exhaustively. The purpose of choosing a representative set $\bar{P}_0$ is to identify policies whose possibility is significant, and to provide a thorough coverage of the policy space. Indeed, the high-possibility policies may be relatively clustered, so that it will be advisable to select policies which lie at extremities of the policy space, even though they may have low possibility. It is important to examine the full range of alternative time-of-use rate policies, rather than merely focusing on the most plausible ones.

4.3 ASSESSMENT OF FIRST GENERATION IMPACTS

The impacts of any policy must be considered at several different levels, and within several different time frames. The immediate impacts may be short-felt, whereas longer-term impacts may be of a substantially different nature. Moreover, the impacts at any given time may be in the form of direct consequences upon utility companies and their users, or indirect consequences upon social and economic patterns. In order to effectively study these various phenomena, we will employ a classification scheme for impacts, as described earlier. If one considers all possible chains of events, the picture can become extremely complex. For this reason, we will restrict ourselves to identifying the important effects within each compartment of the classification scheme.

In the first generation, it is anticipated that only first-and second-order impacts will emerge. Effects on residential, commercial, and industrial users of electricity will be assessed, and the implications for other forms of energy supply will be investigated. These impacts will be suitably categorized and reviewed by panels of experts until consensus has been reached. In the course of this analysis, models of economic behavior may be used to obtain approximate quantitative results.

With regard to the use of models of economic behavior, numerous microeconomic models currently exist in the literature which address the key first-order impacts of TOU pricing. For example, single period demand models have been generated which analyze the probable consumer response to TOD pricing. Further, numerous econometric studies have been carried out which attempt to forecast consumer response by estimating the elasticity of demand for electricity. In addition, rate of return models have been used to predict the probable behavior of utility companies if and when TOD pricing is imposed. These models will be carefully examined and distilled. The essentials and important components of each model will be used as a boundary check to our expert estimates of possible effects. Most of the models that exist address the first-order, first generation effects. Thus, it will be more difficult and less meaningful to use such models as consistency checks on expert estimates of higher-order or later generation effects.

To facilitate the above investigation, the impact will be further categorized into types or spheres of impact. It is difficult to predict at this point exactly what categories would be appropriate, but in the course of identifying the impact this should become clear. One example of impact category is SHIFT TO NIGHT USE OF ELECTRICITY. Again, we may define a linguistic variable corresponding to this category, whose values would simply describe the extent of the shift in usage.

Let $\tilde{S} = (S_1, S_2 \ldots)$ be the vector of impact variables defined in this way. For a given variable S_k, we are actually dealing with an array of impacts, corresponding to the classification scheme discussed earlier. To denote the individual compartments of this array, we will use the following notation. Both T and L may take on the values 1,2, or 3.

$$S_k\ (T,L) = \text{Lth - order impact in category k, during time frame T}$$

The results of impact assessment in the first generation can then be expressed as a possibility distribution over the values of $\tilde{S}$, conditioned on the initial policy choice and contextual factors. Notationally, this would be written as:

$$\pi(S_k\ (1,L) = s \mid P = (\tilde{I},\tilde{C});\ \tilde{F}=\tilde{f})$$

where s is in the domain $D(S_k)$ of the impact variable.

For ease of presentation, we will occasionally suppress the contextual variable, assuming that P has been identified with a specific scenario $\tilde{F} = \emptyset\ (P)$. Incidentally, a consolidated result can be obtained if we wish to know the unconditional possibility of experiencing impact S_k at some value. This is obtained by looking at all policy choices, as follows:

$$\pi(S_k\ (T,L) = s) = \bigvee_{P \in \bar{P}_0} \pi(S_k\ (T,L) = s \mid P)$$

4.4 ASSESSMENT OF SECOND AND THIRD GENERATION IMPACTS

Having constructed a framework for estimating possibilities, we can describe the balance of the work relatively concisely. The possibility of a later-generation impact can be evaluated by determining all earlier impacts which could have given rise to it. Hence, it is sufficient to determine all such "coupling effects" between the different categories of impacts. For example, the possibility of a second-order second generation effect in category S_1 might be expressed as the conjunction of two possible causal chains, as shown in the diagram below:

Policy $\tilde{I},\tilde{C}$	1G	2G
First-Order	S_3 →	S_2 ↓
Second-Order	S_1 →	S_1

Some caution must be exercised in the application of this methodology for two reasons:

1) Causal links may exist other than the horizontal and vertical ones postulated here.

2) The use of coupling may ignore subtle dependencies of a later-generation impact upon the initial policy choice.

Thus, adjustments to the possibility estimates may be called for in certain cases. These adjustments should again make use of experts' judgments, after the implications of their initial judgments have been computed.

In general, a coupling effect may be postulated between time frames and between orders of impacts. This would imply that the possibility of impacts could often be evaluated with the aid of simpler expressions of the form:

$$\pi(S_k\ (T,2)|S_j\ (T,1))$$
or
$$\pi(S_k\ (2,L)|S_j\ (1,L))$$

Note that an effect in one category may result from previous effects in different categories. Although the notation can become unwieldy, this coupling is simply a formalization of the causal links shown by arrows in the impact classification array. For the 1G impacts, of course, T = 1 would be fixed, and the only possible coupling would be in the form of second-level impacts being generated from first-level ones.

An additional complication must be introduced at this point, since later-generation impacts may be influenced by changing economic, demographic, or political conditions. Our uncertainty about exogenous forces in the future must be accounted for by constructing alternative scenarios. These scenarios may be thought of as an extension of the exogenous variables $\tilde{F}$ defined in Module 1. Some of these variables may fluctuate over time, and this will be denoted by the index T = 1, 2, or 3. F_k (1) simply denotes the initial value of the variable F_k; if this variable is time-invariant (e.g., CLIMATE) then for each T,

$$F_k\ (T) = F_k\ (1)$$

Thus we may define a scenario as a vector of exogenous conditions $\tilde{F}$ (T). While $\tilde{F}$ (1) is known and fixed, $\tilde{F}$ (2) and $\tilde{F}$ (3) are uncertain, and must be described by possibility distributions, contingent upon the initial exogenous conditions, namely $\pi(\tilde{F}\ (T)|\tilde{F}\ (1))$. In the preceding discussion, it was assumed that the initial conditions $\tilde{f} = \emptyset\ (P_i)$ were fixed for each policy P_i. If exogenous conditions are permitted to fluctuate, then the possibility distributions defined above must also be conditioned upon $\tilde{F}$ (T), where T is the time frame being considered.

Assignment of possibility ratings would proceed just as outlined above for policy choices, and a representative set of scenarios will be selected for the balance of the study. Finally, returning to the assessment of impacts, any possibility estimate must be conditioned not only on the initial policy choice, but also on the assumed contextual scenario. Thus, when future scenarios are included, the expression for the overall possibility of experiencing impact S_k becomes:

$$\pi(S_k\ (T,L)) = \bigvee_{\substack{P \in \bar{P}_0 \\ \tilde{f} \in D(\tilde{F})}} \left[\pi(S_k\ (T,L)|P;\tilde{F}) \wedge \pi(\tilde{F}\ (T)=\tilde{f}|\emptyset\ (P))\right]$$

In other words, we must investigate all possible scenarios and all possible policy choices which could have led to that impact. The above expression constitutes a "law of total possibility" entirely analogous to the probabilistic case.

4.5 DESCRIPTION OF POSSIBLE IMPACT-CHAINS

Once all major impacts have been identified, and their possibilities assessed, the results may be combined to predict the collective impacts of specific policy choices. The chains of effects which were observed during the analysis may be grouped together to derive the total impact in any compartment of the impact array. Much of this work will already have been accomplished if the coupling effects in steps 2 and 3 were fully investigated. However, there is certainly a chance that more subtle coupling effects were not accounted for, and the impact assessments may need to be adjusted accordingly. It is here that one must draw back from the detailed analysis and look for insights or interpretations that fall beyond the scope of the methodology. In order to focus upon important impacts, it may be desirable to restrict attention to those chains whose joint possibility exceeds some lower bound.

The notion of an impact chain is predicated upon the assumption of a quasi-Markovian property of possibility distributions. This property may be phrased as follows:

Let the state of the system at stage r be denoted by S(r). Then the next state depends only upon S(r), and not upon the previous history of the system. In notation,

$$\pi\left[S(r+1)|S(o), \ldots, S(r)\right] = \pi\left[S(r+1)|S(r)\right]$$

The state of the system in the present context would be the values of $\tilde{S}(T,L)$ and $\tilde{F}(T)$. The "transition possibilities" are simply the coupling effects postulated earlier. However, the resulting process is comparable to a <u>Markov field</u> rather than a Markov chain, since transitions may occur not only in time but also within the impact hierarchy.

Once the possibility analysis has been completed, an <u>impact chain</u> may be defined as a sequence of impacts linked by coupling transitions, leading from an initial policy choice to a final impact. Associated with the impact chain would be a sequence of scenarios under which the impacts occurred.

The results of this module would have the following form.

- Identification of the representative policy choices in the set $\bar{P}_0$.
- For each of these policies, the exogenous conditions under which it has the highest possibility of occurrence.
- For each set of initial conditions and policy choice, the major impacts that are expected to result from them in the first generation.
- For each set of initial conditions and policy choice, and for each possible future scenario, the second-and higher-order effects that are anticipated, along with the corresponding chains of impacts.
- Possibility ratings for all of the impacts chains cited above.

4.6 SENSITIVITY ANALYSIS

Having completed the impact assessment and the identification of impact chains, we can next inquire about the relative importance of our initial assumptions regarding policy choices and exogenous factors. By varying the possibility estimates for

$\tilde{I}$,$\tilde{C}$ and $\tilde{F}$, we can observe how strongly these variations affect the impact possibilities. It is likely that certain variables will critically determine the ultimate outcomes, while others have only slight effects. This analysis should be performed using a computer model of the impact hierarchy, preferably written in APL, which has a unique capacity for matrix manipulation and max-min operations. Such a model would be extremely useful in controlling the maze of computations described in the previous steps, although they could be performed manually under a suitably systematic approach.

The results of the sensitivity analysis, apart from being useful in actual policy selection, will point out those variables which are dominant in determining future impacts. Additional attention will then be focused upon these variables to ensure that the estimates obtained concerning their values are as accurate as possible. Furthermore, the analysis could be extended to include more detail for the dominant variables, while eliminating those variables which had relatively little impact.

5. ILLUSTRATION

An example of how this methodology would be applied is shown below in tabular form. To simplify this illustration we assume that contextual conditions are fixed and consider only two possible implementation variables and one possible design variable. The results of Module 1 are shown in the first table, in which the policy framework is identified and possibilities of various policies are evaluated. Some entries have been circled indicating that an interaction between policy variables was taken into account. The universe of discourse for each variable is assumed to have two values and the possibility distributions for individual variables have been used to estimate joint possibilities for the policies.

The resulting policy space is shown next with the particular realization corresponding to each of the eight different policies and the possibility values assigned to them. In this case, a representative set of policies was chosen by selecting those either with high possibility or representing extremes of a policy space. For simplicity, only three impact categories were considered, two of them first-order and one of them (S_3) a second-order impact.

The last two tables illustrate the assessment of possible impacts for a given policy, in this case, P_3. The possibilities for the 1G impacts may be directly evaluated through use of expert panels. To evaluate possibilities of 2G impacts, one must combine the 1G impact possibility distributions with the coupling effects between different categories of impact. For this illustration coupling effects are shown only within a given impact category. In general, we would have to evaluate the possibility of S_1 having value u in the second generation given that S_2 has value v in the first generation.

POLICY FRAMEWORK - ILLUSTRATION

Assume exogenous conditions $\tilde{F}$ are given.

Let $\tilde{I} = (I_1, I_2)$, where I_1 = Fraction of utilities implementing TOD

I_2 = Average time until full implementation

$\tilde{C}$ = (C_1) where C_1 = Peak/offpeak rate differential

The results of policy evaluation are expressed as follows:

Policy Variable		I_1		I_2		C_1	
Possible Values v		>50%	>80%	>3 yr.	>10 yr.	>2:1	>5:1
$\pi(v\mid\tilde{F})$		.7	.3	0.2	0.9	1.0	.2
I_2	>3 yr.	0.2	0.1			Joint possibilities	
	>10 yr.	0.7	0.3			$\pi(I_1, I_2)$ etc.	
C_1	>2:1	0.7	0.3	0.2	0.9	Circled entries are	
	>5:1	0.2	0.1	0.1	0.2	interactive pairs.	

Policies $P_i = (\tilde{I}, \tilde{C})$

	I_1	I_2	C_1	From above table $\pi(\tilde{I}, \tilde{C}\mid\tilde{F})$
P_1	>50	>3	>2:1	0.2
P_2	>50	>3	>5:1	0.1
P_3	>50	>10	>2:1	0.7
P_4	>50	>10	>5:1	0.2
P_5	>80	>3	>2:1	0.1
P_6	>80	>3	>5:1	0.1
P_7	>80	>10	>2:1	0.3
P_8	>80	>10	>5:1	0.1

As a representative set of policies, we choose $\bar{P}_0 = (P_1, P_3, P_7)$

To examine 1G impacts, we assume $\tilde{F}$ remains constant. For 2G impacts, we assume that a scenario $\tilde{F}(2)$ has been specified.

IMPACT ASSESSMENT - ILLUSTRATION

Let $\tilde{S} = (S_1, S_2, S_3)$ where S_1 = Load factor impact

S_2 = Electricity consumption/capita

S_3 = Fraction of people in shiftwork

The following coupling effects are assumed:

S_k	v	u	$\pi[S_k(2,L) = u \mid S_k(1,L) = v]$		
S_1	0	0,1	0.1	0.5	
	1	0,1	0.0	0.8	
S_2	1	1,0,-1	0.4	0.6	0.2
	0	1,0,-1	0.8	0.8	0.6
	-1	1,0,-1	0.2	0.3	0.4
S_3	0	0,1	0.6	0.7	
	1	0,1	0.0	0.9	

Impacts of Policy P_3			Conditional on Policy P_3		Conditional on P_3 and $\tilde{F}(2)$	
Variable		Possible Values	$\pi[S_k(1,1)]$	$\pi[S_k(1,2)]$	$\pi[S_k(2,1)]$	$\pi[S_k(2,2)]$
S_1	0	No significant change	0.5		0.1	
	1	Increase significantly	0.6		0.6	
S_2	1	Significant annual increase	0.7		0.5	
	0	No significant change	0.5		0.6	
	-1	Significant annual decrease	0.1		0.5	
S_3	0	Remain near present level		0.8		0.6
	1	Increase significantly		0.2		0.7

The above table is obtained by:

1. Assessing 1G impacts $\pi[S_k(1,L)|P_3 = (\tilde{I},\tilde{C})]$
2. Using the coupling effects to assess 2G impacts.

6. CONCLUSIONS

We have shown that a policy assessment must consist of a systematic examination of the economic, social, environmental and institutional impacts of alternative policy choices. The domain of interest chosen here was time-of-use electricity rate policies. However, the methodology that was developed is sufficiently general to to be applied to any policy assessment or any exercise which seeks to explore the potential impacts of present decisions upon entities or organizations in the future. The methodology is particularly useful in situations where quantitative modelling is difficult due to unavailability of information and uncertainty in the variables of interest.

For a given set of conditions, the set of policy choices can be rated in terms of which ones are more or less possible for adoption in the future. This rating may be performed on a scale from 0 to 1, where a possibility of zero indicates a completely implausible choice, whereas a possibility of one indicates a choice which is among the most plausible compared to all conceivable choices. In a similar manner, the possibility of various impacts arising from these policy choices may be evaluated. It must be emphasized that this possibility rating is not a measure of the chance, likelihood, or probability of a particular event. Rather it is an approximate estimation of the potential for this event, given the totality of surrounding circumstances. More concisely, it addresses the question of whether the event <u>can</u> occur, rather than whether it <u>will</u> occur.

The approach described above is currently being pursued under the support of the National Science Foundation on a pilot basis. The initial exercise will address a subset of the total policy space and a limited set of categories of impacts. If, in fact, the methods prove practical to implement, then the policy assessment will be extended to the full scope of TOU rate structures. The intent of such a study is to provide findings which are useful to policymakers, both for initial policy formulation and also for continuing policy intervention.

FOOTNOTES

[1] Fixed capacity charges are referred to as "demand" costs as opposed to "energy" costs consisting of the variable running costs for fuel, and operating and maintenance. A utility company's cost of capacity - passed through to its customers - can be approximated by the cost of capital for the investment tied up, plus property taxes and insurance.

[2] Time-of-use pricing of electricity has been used to a considerable extent in England and France for years, but only in the past two years has it been seriously considered in the United States.

[3] One can argue that the objective of reducing the growth rate of peak demand by having customers shift consumption from peak to offpeak periods is not likely to be achieved with TOU pricing alone. This is because TOU prices are simply price signals inducing voluntary shifts in consumption. By combining TOU pricing with load controls, the electric utility companies can be more sure of limiting peak demand, especially, for example, on the third afternoon of a heat spell when customers might well be willing to pay peak load prices to gain relief from their electricity-intensive air conditioners.

REFERENCES

[1] Shift Working: the Arrangement of Hours on Night Work, Nature, December 11, 1965, 1127-1128.

[2] Ragade, R.K., (1976) Fuzzy Interpretive Structural Modelling, Journal of Cybernetics, 6; 189-211.

[3] Zadeh, L.A., (1965) Fuzzy Sets, Information and Control, 8; 338-353.

[4] Zadeh, L.A., (1977) Probability Theory vs. Possibility Theory in Decision-Making, IEEE Symposium on Fuzzy Set Theory and Applications, New Orleans.

[5] Nguyen, H.T., (1978) On Conditional Possibility Distributions, Journal of Fuzzy Sets and Systems, Vol. 1, No. 4

ADVANCES IN FUZZY SET THEORY AND APPLICATIONS
M.M. Gupta, R.K. Ragade, R.R. Yager (editors)

THE CONSTRUCTION AND EVALUATION OF FUZZY MODELS

Richard M. Tong

Dept. Electrical Engineering and Computer Sciences
University of California
Berkeley CA94720, USA.

1. Introduction

This paper discusses the idea of a fuzzy model, ways in which such models may be constructed and ways in which their performance may be evaluated.

The first part of the paper is concerned with the relevance of fuzziness as a concept in systems modelling. It is argued that, because fuzzy set theory has the ability to handle linguistic information, the most natural definition of a fuzzy model is in terms of finite discrete relations.

The second part of the paper reviews some techniques for model construction. These fall into three broad categories, namely verbalisation, fuzzification and identification. Each is considered in turn and its advantages and disadvantages examined.

The next section considers ways in which the performance of a fuzzy model may be assessed. The idea of model quality is introduced and is seen to be a multi-faceted concept which includes complexity, uncertainty and accuracy.

The main body of the paper concludes with two examples. The first of these is a well studied problem in the system identification field. The second is a problem in water quality modelling and shows quite clearly the main features of the fuzzy modelling methodology.

2. The Fuzzy Model

There are many situations in which our understanding of process behaviour is imprecise. In the steelmaking industry, for example, dynamic models of the basic oxygen furnace rely on a representation of very complex chemical and physical phenomena in terms of non-linear ordinary differential equations. This may be adequate in a given context, but no modeller of this process would argue that it is a true understanding. Similarly, in economics it is possible to construct time series models for the highly inter-connected variables which characterise the system. However, such models are often very poor.

The limited performance of the models in these examples, and in other cases as well, may be a result of the "Principle of Incompatibility". This is an intuitive idea introduced by Zadeh(1) which states that as systems become more complex, it becomes increaseingly difficult to make statements about them which are both meaningful and precise.

Fuzzy set theory, originated by Zadeh(2) and developed by many others, is a tool for handling imprecision. What it allows the modeller to do is make statements about process behaviour which, although inaccurate in the usual sense, convey an understanding of the basic characteristics.

A distinction should be made though between 'subjective' and 'objective' fuzziness. That is, between fuzziness which is due to the modellers perception of the process and that which is an actual property of the process. The former is typified by the steelmaking example. Here the imprecision lies in our inability to gather sufficient information from the process rather than in any fuzziness in the laws governing its behaviour. In the economic system, though, it is possible to argue that the overall behaviour is dependent on the decisions of each individual operating within the economy and, as such, is an example of the objective type of fuzziness.

There are many forms which a fuzzy model could take. It could simply be a static description of relationships between variables. Or it could be a forecasting model for use in economic planning. In this paper the emphasis is on dynamic systems models for use in the control engineering type of application.

One particularly interesting form is derived from the study of fuzzy numbers (Dubois and Prade(3)). Let $\mathbb{R}$ be the set of real numbers and $\underset{\sim}{\mathbb{R}}$ be the power set of strongly convex and normal fuzzy sets in $\mathbb{R}$. A fuzzy system may be defined as a function

$$f: \underset{\sim}{\mathbb{R}}^n \times \underset{\sim}{\mathbb{R}}^m \times \underset{\sim}{\mathbb{R}}^d \rightarrow \underset{\sim}{\mathbb{R}}^n$$

such that

$$X(t+1) = f(X(t),V(t),W)$$

where

$$X^T(t) = (x_1(t), \ldots ,x_n(t))$$
$$V^T(t) = (v_1(t), \ldots ,v_m(t))$$

are the state and input vectors of the fuzzy system at time 't' and

$$W^T(t) = (w_1(t), \ldots ,w_d(t))$$

is the parameter vector. The vector components $x_i(t)$, $v_i(t)$ and $w_i(t)$ being elements in $\underset{\sim}{\mathbb{R}}$. Thus the function f and the parameter vector W define a model of the system.

Whilst there are few results using this formulation, it is hoped that concepts in conventional systems theory can be generalised to what mught be called a fuzzy systems theory (Zadeh(4)).

However, this approach does not explicitly utilise the linguistic properties of fuzzy set theory. One of the reasons for the success of fuzzy set theory in tackling control problems is exactly this parallel with natural language (Tong(5)). It is unfortunate, therefore, that this is lost when pusuing the algebraic notions outlined above.

An alternative is to model the system in terms of linguistic relations. In doing so, the model becomes the dual of a fuzzy logic controller and fits naturally into the theoretical framework of fuzzy control systems described in Tong(6).

Thus, in the remainder of the paper, a fuzzy model is defined as a finite set of linguistic relations, $A = \{r_i ; i=1, \ldots ,n\}$, which together form an algorithm for determining the outputs of the process from some finite number of past inputs and past outputs. For implementation on a digital computer the output space and the input space of the system have to be discretised. These will be denoted S and U respectively. The set of numerical definitions of the primary fuzzy sets used in specifying A will be denoted F, so that the fuzzy model is now defined in terms of a quadruple, (S,U,F,A). Then using the calculus of fuzzy sets it can be rewritten in terms of a finite discrete relation, R, and a composition operator, $\circ$, such that $(S,U,F,A) \rightarrow (R, \circ)$.

3. Model Construction

There appear to be three ways in which information about a systems behaviour may be obtained. They are not mutually exclusive and, indeed, it is reasonable to suppose that some combination of all three would be the most effective way of constructing a 'good' model.

3.1 Verbalisation

This is the process of recording verbal descriptions and formalising them as fuzzy relations. The descriptions may be given by process operators, plant managers or process technologists. In fact, anyone who has some 'feel' for the process could be considered a possible source of information.

The idea of asking operators to describe their actions is not new and has been well studied by workers in the field of industrial psychology (Bainbridge(7)). There are, however, some severe problems with this approach, principally methodological. It is, for example, difficult to ensure that in observing the operator all significant responses have been recorded (Umbers(8)). Similarly, there is no guarantee that the operators verbal responses are accurate descriptions of his mental model.

Nonetheless, this approach has been used successfully in the study of the activated sludge sewage treatment process (Beck, Latten, Tong(9)). In this work the plant manager was given a carefully constructed questionnaire in which he was asked to describe the behaviour of the process under certain hypothetical operating conditions. From his answers it was possible to construct a fuzzy control algorithm which embodied several important features of the plant's dynamic behaviour.

Much work needs to be done to perfect this technique. At present, there is no systematic way of approaching a given verbal modelling problem although some obvious guidelines may be drawn up. Difficulties, apart from those already mentioned, include selecting the primary fuzzy sets and the appropriate discretisations.

3.2 Fuzzification

Zadeh's "Extension Principle" (10) can be used to extend the meaning of ordinary mathematical expressions from points to fuzzy sets. In this way any exact relationship known to hold for the process can be transformed into a fuzzy relation.

Thus if 'f' is a scalar function of the n-vector $\underline{x}=[x_1,\ldots,x_n]$ such that $y=f(\underline{x})$ and if $\mu(x_i)$ is the membership function associated with the 'ith' element, then

$$\mu(y)=\max_{y=f(\underline{x})}\left[\min_i\left(\mu(x_i)\ :\ i=1,\ldots,n\right)\right]$$

In practice it is necessary to restrict the values that $\underline{x}$ and y can take to be both finite and discrete. This presents some implementation problems which are basically concerned with the discrete representation of fuzzy sets. Suppose a fuzzy set of temperature is defined on the continuous range [0,100], then a finite discrete representation corresponds either to a set of representative points in [0,100] or a set of intervals on [0,100]. The first of these may be termed a 'point-set' and the second an 'interval-set'.

Obviously, there are four ways of combining representations of $\underline{x}$ and y. However, only one of these leads to a straightforward computational problem and that is to choose $\underline{x}$ to be a point-set and y an interval-set.

Of course the different representations will give different fuzzifications of the original function, although they may be similar linguistically. Very little practical work has been done on the evaluation of the Extension Principle and it is not at all clear if the choice of fuzzy set representation is of real significance.

3.3 Identification

In many cases, industrial processes for example, it is possible to perform data logging experiments. These produce various amounts of non-fuzzy data relating to input variables and output variables. The generation of fuzzy relational descriptions of this data, and hence the process, is called 'identification'.

A technique for doing this has been reported by Tong(11). It is essentially a method for testing fuzzy propositions about the process against the data to see if they are 'true'. Those which are then constitute valid fuzzy descriptions for incorporation in the model.

The technique has deficiencies, principally because it does not eliminate the need to do some kind of correlation analysis on the input and output variables. However, it remains the only published method for utilising input-output data.

What is required in this area is a theoretical investigation of the relationship between models and data, analagous to that done in conventional system identification (see Eykhoff(12) for example).

4. Model Evaluation

After the model has been constructed it must be tested and evaluated. In the conventional modelling excercise this rarely consists of more than the calculation of some accuracy measure between the model and the data. In the fuzzy case this is inadequate and a more detailed inspection of the model is required.

Usually, several models have been obtained and the modeller wishes to choose the best amongst them. The first requirement is that the models all come from the same 'class'. In the context of the model definition given in section two, this means that S,U and F must be the same. Thus the only variable is A, the set of fuzzy relations.

Of primary importance is model complexity, denoted p1. Clearly, if the models are the same in all other respects then the one with least complexity is preferred. It is natural to choose as a measure for p1 the number of linguistic relations which make up A.

Related to complexity is the notion of accuracy. This is problematic in fuzzy models since they generate fuzzy sets as outputs, in contrast to the measured data which is non-fuzzy. Although non-fuzzy values can be generalised to fuzzy singletons, the distance measures usually proposed for fuzzy sets (Kaufmann(13) for example) are not really appropriate. Furthermore, the model operates with discretised variables and consequently there are bounds on the accuracy that can be achieved.

One way of overcoming these difficulties is to de-fuzzify the outputs of the fuzzy model. Note that whilst this seems appropriate in the current context it is not necessarily a general solution. De-fuzzification may be achieved in several ways and there is no published evidence that any method is superior. One obvious way is simply to select the value corresponding to the peak in the fuzzy set, averaging when there are several such values. Another is to select the value which divides the area under the membership function curve in half.

Having fixed on a method, two accuracy measures seem appropriate.

The first of these is simply the squared error, such that

$$p2(i) = (\hat{y}(i) - y(i))^2$$

where $\hat{y}(i)$ is the de-fuzzified model output at the 'ith' data point and y(i) is the 'ith' non-fuzzy observation. This accuracy measure is useful because it is the most commonly employed measure in non-fuzzy modelling. The second accuracy measure is the absolute difference between the discretised de-fuzzified model output and the discretised measurement. That is

$$p3(i) = \left| D[\hat{y}(i)] - D[y(i)] \right|$$

where D[.] denotes a discretisation mapping and is such that

$$D: y \rightarrow \{j \ ; \ j=1,\ldots,L\}$$

where L is the number of discretisation points (or levels) for the output space. The value of this accuracy measure is that it does not discriminate against the inherent inaccuracy introduced by using discretised fuzzy sets.

A third, and final, aspect of model evaluation is the notion of uncertainty. It is obviously the case that output sets with different membership functions can generate the same de-fuzzified output but, intuitively, the 'sharper' the output set the better. The concept of non-probalistic entropy developed by DeLuca and Termini(14) is of some use in this context. A simpler and more direct alternative, which is not strictly a measure of sharpness, is to set

$$p4(i) = 1 - \mu[\hat{y}(i)]$$

where $\mu[.]$ is the membership function of the model output set.

Measures p2(i), p3(i), and p4(i) are for single data points only and thus for a given data set of 'N' points they become

$$p2 = \frac{1}{N}\sum_{i=1}^{N} p2(i) \qquad p3 = \frac{1}{N}\sum_{i=1}^{N} p3(i) \qquad p4 = \frac{1}{N}\sum_{i=1}^{N} p4(i)$$

So any model can be characterised in relation to any set of data by means of complexity, p1, accuracy, p2 and p3, and uncertainty, p4.

In general, the more complex the model the higher the accuracy and the smaller the uncertainty. However, it may not be a trivial task to impose an ordering on the models on the basis of such measures. The trade-offs between complexity and accuracy and between complexity and uncertainty may require the modeller to use some external criteria in order to select a 'best' model.

5. Application Studies

The following two examples illustrate some of the characteristics of fuzzy models and emphasise the points made above. The first example is a model of Box and Jenkins(15) gas furnace data and shows how effective the model construction techniques can be. The second example consists of two models of Beck's(16) river water quality data. The first is simple and appears satisfactory, but has some unexpected consequences. The second is much more complex but overcomes the deficiencies of the first.

5.1 The gas furnace data of Box and Jenkins

The data of Box and Jenkins is extremely well known and is often used as a standard test for identification techniques. It provides, therefore, a useful starting point for assessment of the methods proposed in section three. The data consists of 296 pairs of input-output measurements. The input is gas flow rate into the furnace, the output is the CO_2 concentration in the outlet gases and the sampling interval is nine seconds.

Using a combination of the techniques described earlier, the fuzzy model shown in Figure 1 was obtained. The table, which in this case is 'complete' since all the cells are filled, is often called a 'transition table' and should be interpreted as follows.

The model is a relation between gas flow four sampling intervals ago, CO_2 concentration one sampling interval ago and current CO_2 concentration. That is

$$y(t) = f(y(t-1), u(t-4))$$

The row indices are mnemonics for primary fuzzy sets of gas flow. The column indices are mnemonics for primary fuzzy sets of CO_2 concentration, as are the entries in the table.

The primary sets were chosen after an analysis of the data for range and distribution of values. The mnemonics are ordered to reflect increasing values of the corresponding variables. Thus C6 represents a 'higher' concentration of CO_2 than C1 and PQ3 represents a 'higher' gas flow rate than NQ3.

Each table entry then forms an 'elementary' relation of the form

$$\text{IF } \{u(t-4) \text{ is NQ1 } \& \ y(t-1) \text{ is C3}\} \text{ THEN } \{y(t) \text{ is C4}\}$$

where & corresponds to disjunction.

Interpretation of implication is, of course, a subject of debate but in this example, and in the following one, a Cartesian product form is used (ie. $A \Rightarrow B \triangleq A \times B$). The reason for this is simply that it is unreasonable for the modeller to infer anything at all if proposition A is not true. Thus the transition table can be thought of as a tabulation of a fuzzy function.

y(t-1)

u(t-4)	C1	C2	C3	C4	C5	C6
NQ3	C4	C4	C5	C5	C6	C6
NQ2	C3	C3	C4	C4	C5	C6
NQ1	C3	C3	C4	C4	C5	C5
ZEQ	C2	C3	C4	C4	C4	C6*
PQ1	C2	C2	C3	C4	C5	C6*
PQ2	C1	C2	C2	C2	C4*	C4*
PQ3	C1	C1	C2	C2	C3	C3

y(t)

Figure 1. Complete fuzzy model of the gasfurnace

To test the model it is necessary to generate one-step-ahead, OSA, predictions from the data. This is straightforward. The procedure is to take the non-fuzzy measurements corresponding to y(t-1) and u(t-4), construct the fuzzy singleton $\underset{\sim}{y(t-1)\&u(t-4)}$, calculate $\underset{\sim}{\hat{y}(t)}=\underset{\sim}{y(t-1)\&u(t-4)} \circ R$ and then de-fuzzify to give $\hat{y}(t)$.

The gas furnace model performs very well indeed. The OSA predictions are shown in Figure 2 (continuous line) together with the measured output data (solid circles). The complexity is equal to the number of compound relations (ie. relations which fill more than one cell in the table) and here is equal to 19. The other measures are

$$p2 = 0.469 \qquad p3 = 0.558 \qquad p4 = 0.220$$

so that the discretised de-fuzzified model output coincides with the discretised measured output about half the time. The average value of the membership function at the output value is 0.780.

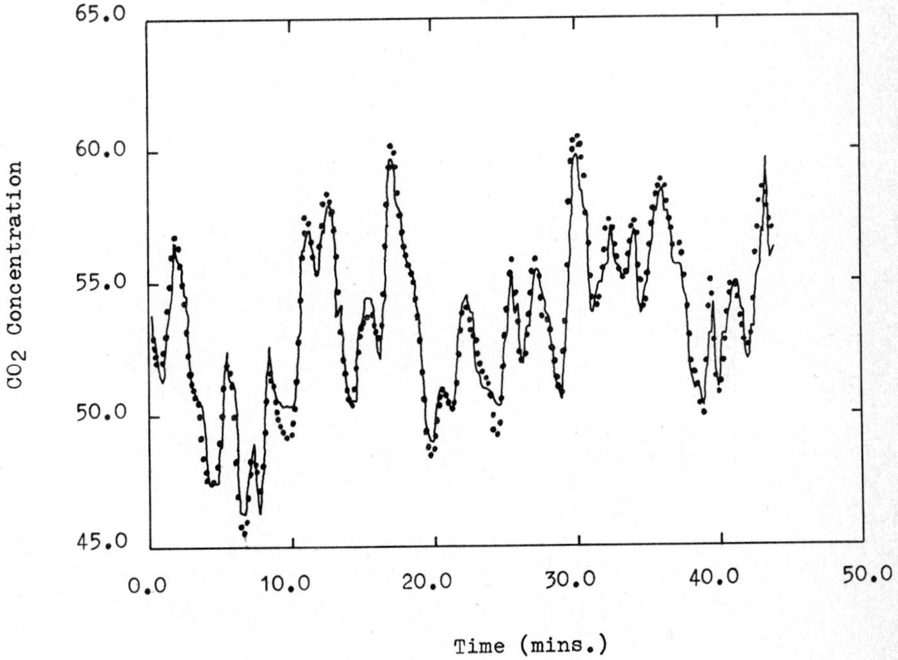

Figure 2. OSA Predictions for the complete model

For comparison, the OSA predictions of Box and Jenkins deterministic model are shown in Figure 3. Their model has the form

$$y(t) = - \frac{0.53 + 0.37B + 0.51B^2}{1.00 - 0.57B - 0.01B^2} \cdot u(t-3)$$

where B is the backward time shift operator, and gives an equivalent p2 measure of 0.202. Note that for a non-fuzzy model, p1,p3 and p4 have no real meaning.

The most interesting thing about the fuzzy model is that it fits the last section of the data better than Box and Jenkins model. The data is known to exhibit some non-stationarity over the last forty samples or so (Young et al(17)) and it is clear that the inherent non-linearity of the fuzzy model gives it the ability to compensate for this. Inspection of Figure 1 shows that the transition table has a broadly 'monotonic' form except for a few cells. Analysis of the model shows that those cells marked with '*' are only in operation during the last portion of the data and thus account for the improved performance.

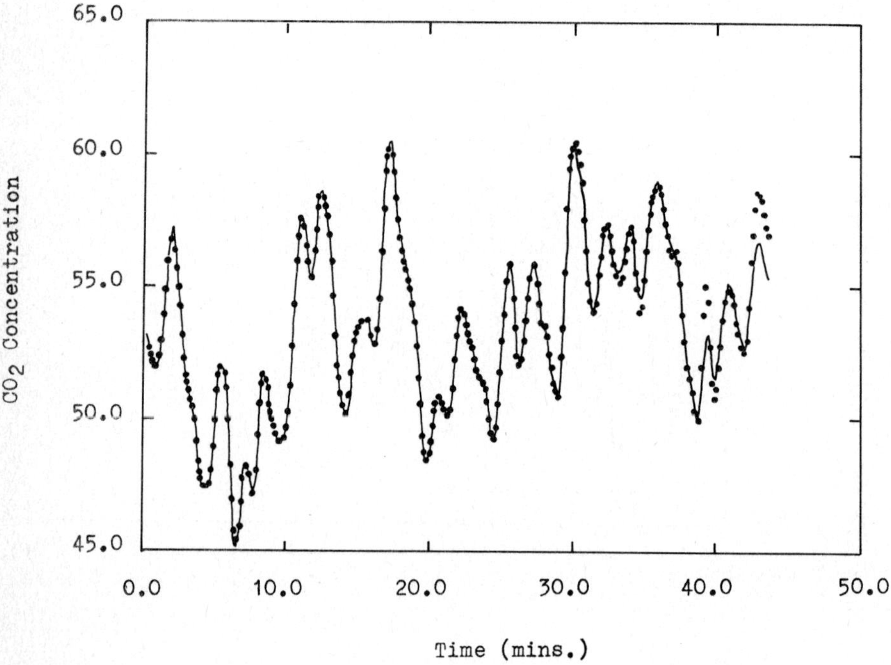

Figure 3. OSA Predictions for Box and Jenkins' model

5.2 The river water quality data of Beck

The river can be thought of as a five-input two-output process, see Figure 4, with biochemical oxygen demand, BOD, and dissolved oxygen, DO, used as measures of water quality. The data consists of 81 consecutive daily sampled values of upstream BOD, upstream DO, volumetric flow rate, river water temperature, hours of sunlight incident on the river during each day, the downstream BOD and the downstream DO.

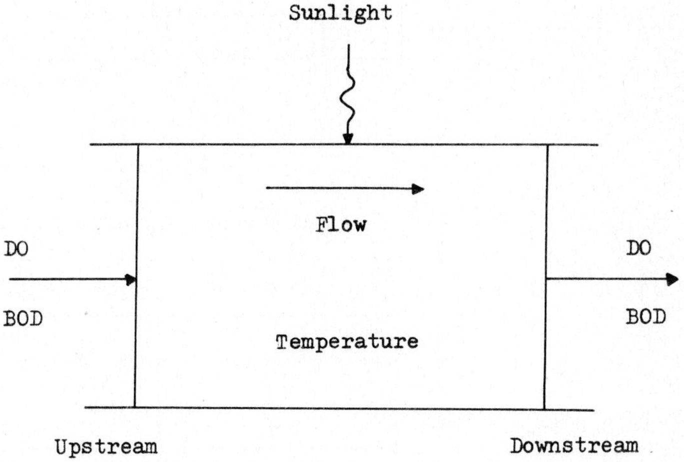

Figure 4. Schematic of the river system

A preliminary analysis of the data using only identification suggests that downstream DO is a function only of itself and downstream BOD. It also suggests that downstream BOD is a function of itself and upstream BOD. The model developed is shown in Figure 5 and is essentially two single-input single-output processes in series. As in the gasfurnace example, the mnemonics for DO and BOD reflect the underlying values of these variables.

The OSA predictions are shown in Figure 6 and the quality measures are

DO:	p1 = 9	p2 = 0.4204	p3 = 1.2468	p4 = 0.2727
BOD:	p1 = 8	p2 = 0.5127	p3 = 1.1606	p4 = 0.3182

Downstream DO(t-1)

Downstream BOD(t-3)

Downstream DO(t)	DO1	DO2	DO3	DO4	DO5
BOD1		DO2	DO3		
BOD2	DO2	DO2	DO3	DO5	DO4
BOD3	DO1	DO2	DO3	DO5	DO5
BOD4	DO2	DO2	DO3	DO4	
BOD5	DO1	DO2	DO2		

Downstream BOD(t-1)

Upstream BOD(t-3)

Downstream BOD(t)	BOD1	BOD2	BOD3	BOD4	BOD5
BOD1	BOD1	BOD2			
BOD2	BOD2	BOD2	BOD3		
BOD3		BOD3	BOD3	BOD4	BOD5
BOD4			BOD3	BOD4	BOD4
BOD5		BOD2	BOD3	BOD4	BOD4

Figure 5. First fuzzy model of the river

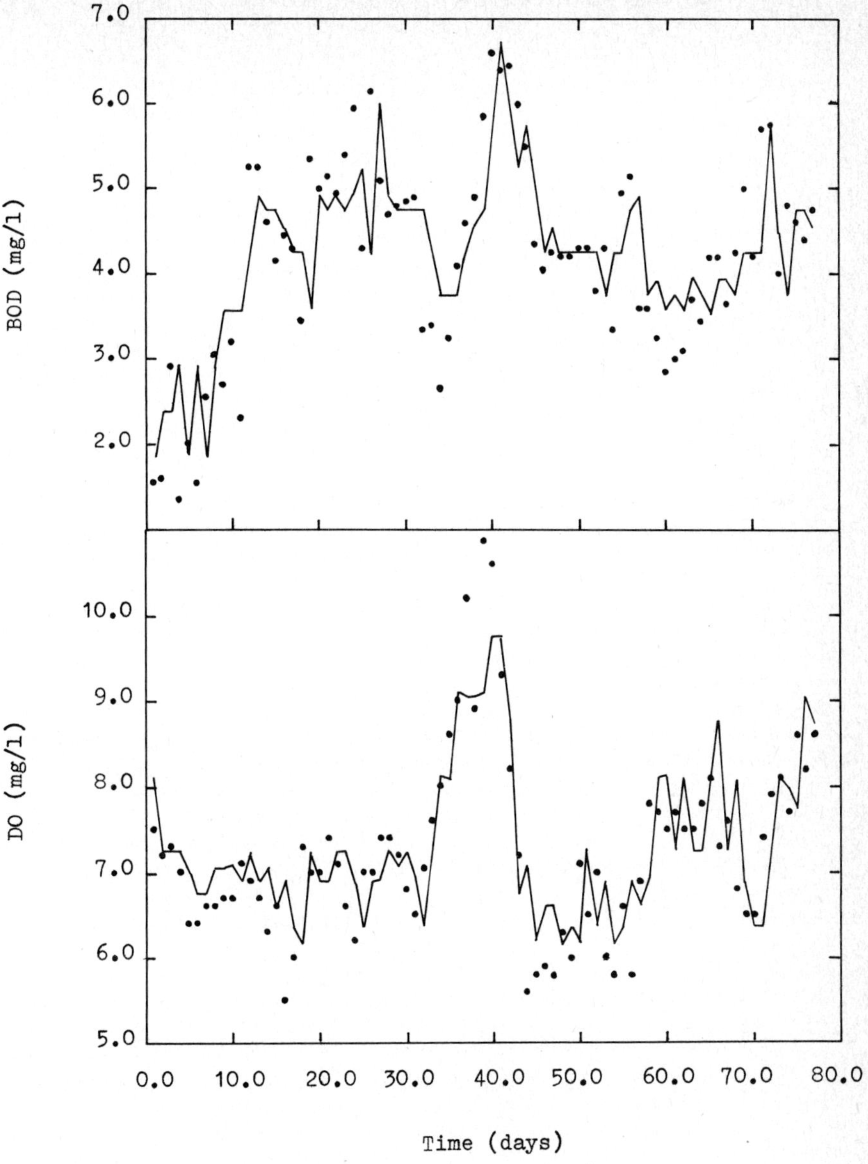

Figure 6. OSA Predictions for the first model

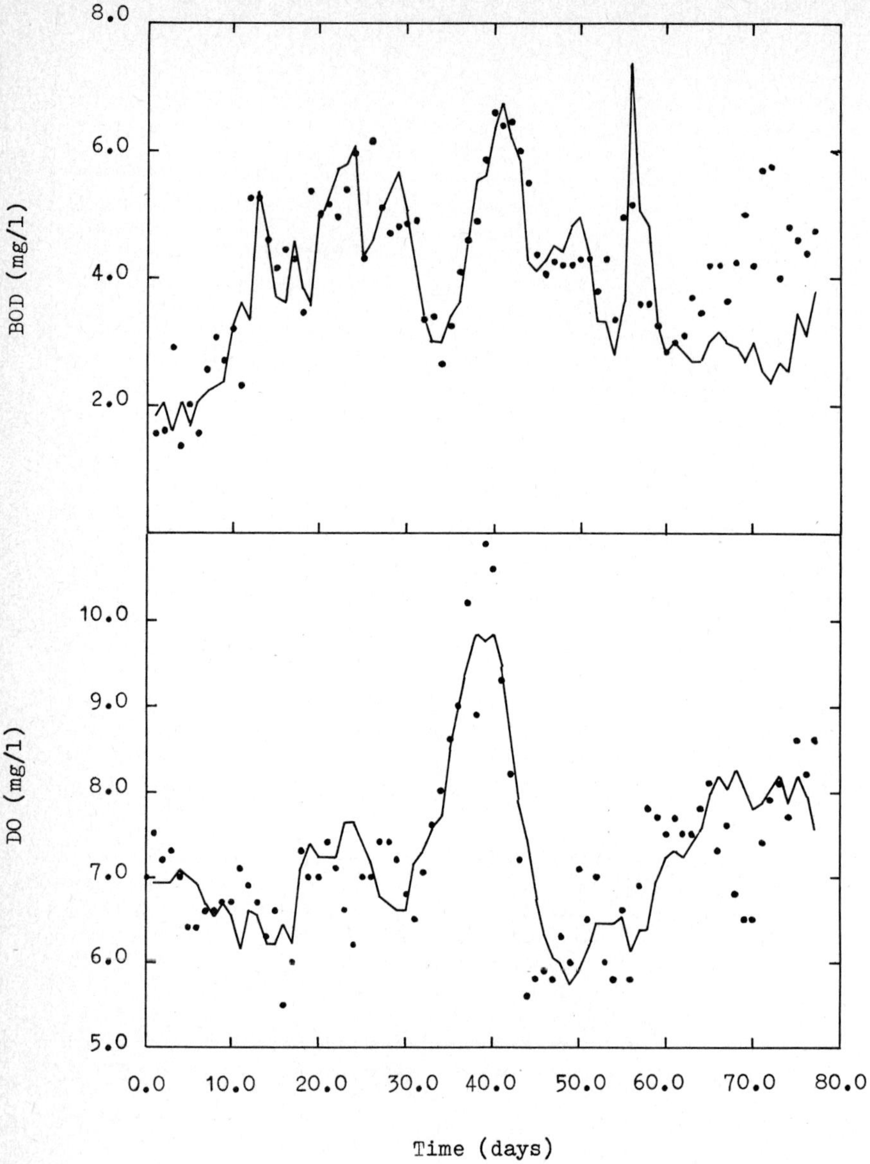

Figure 7. Response of Beck's model

For comparison, Beck's model gives the output shown in Figure 7. His differential equation model has equivalent accuracy measure, p2, of DO: 0.4313 and BOD: 1.0364.

On the whole, the fuzzy model does rather well. But inspection of the transition tables in Figure 5 shows that the model is essentially "next output equals current output". Thus if the model is run as a pure predictor (ie. using de-fuzzified model outputs instead of measured outputs) the results are as given in Figure 8. The quality measures are now

DO:	p1 = 9	p2 = 1.1648	p3 = 1.8276	p4 = 0.3039
BOD:	p1 = 8	p2 = 0.9782	p3 = 1.6646	p4 = 0.3169

and it is obvious that the model is really a crude approximation to the river's dynamic behaviour.

The reasons for this become clear after a study of Beck's model. This shows that sunlight, or rather a weighted moving average of sunlight, plays an important role. In particular, it is largely responsible for the peaks which occur in both the downstream DO and downstream BOD at about forty days.

A second model was developed accordingly. It is much more complicated, having a structure such that downstream DO is a function of itself, downstream BOD, upstream BOD and sunlight, and such that downstream BOD is a function of itself, upstream BOD and sunlight. The model may not be conveniently described by a transition table, but consists of 20 relations describing downstream DO behaviour and 24 relations describing downstream BOD behaviour.

The OSA predictions for this second model are shown in Figure 9. The quality measures are

DO:	p1 = 20	p2 = 0.6155	p3 = 1.2756	p4 = 0.5455
BOD:	p1 = 24	p2 = 0.5912	p3 = 1.2306	p4 = 0.4727

Thus despite the large increase in complexity, the accuracy is worse. However, using the model in a purely predictive mode gives the output shown in Figure 10 and the quality measures

DO:	p1 = 20	p2 = 0.9306	p3 = 1.7148	p4 = 0.5922
BOD:	p1 = 24	p2 = 0.9420	p3 = 1.5526	p4 = 0.5182

So, whilst these are still poor, the second model is a better predictor than the first.

It is clear that neither of the fuzzy models is as good as Beck's model. However, it would be surprising if they were. What they do show, though, is an ability to capture the underlying behaviour of the river. It is also the case that neither model is 'best' in any absolute sense. It depends on the proposed application. Thus the first model might be useful in controller design whilst the second might serve as a qualitative description for non-specialist personnel.

6. Conclusions

Fuzzy models can obviously be made to work, and work very well indeed. However it is important to define those situations in which a fuzzy model is appropriate.

The big advantage of a fuzzy model is that it is relatively simple to construct and is in itself quite a simple structure. It does not require the modeller to have a deep mathematical insight but relies more on intuition and experience of the process. Its greatest value must be, therefore, in those areas where such qualitative process knowledge is predominant and essential for understanding.

It seems likely that in 'hard' technological areas, where precision is often an over-riding consideration, fuzzy models would be

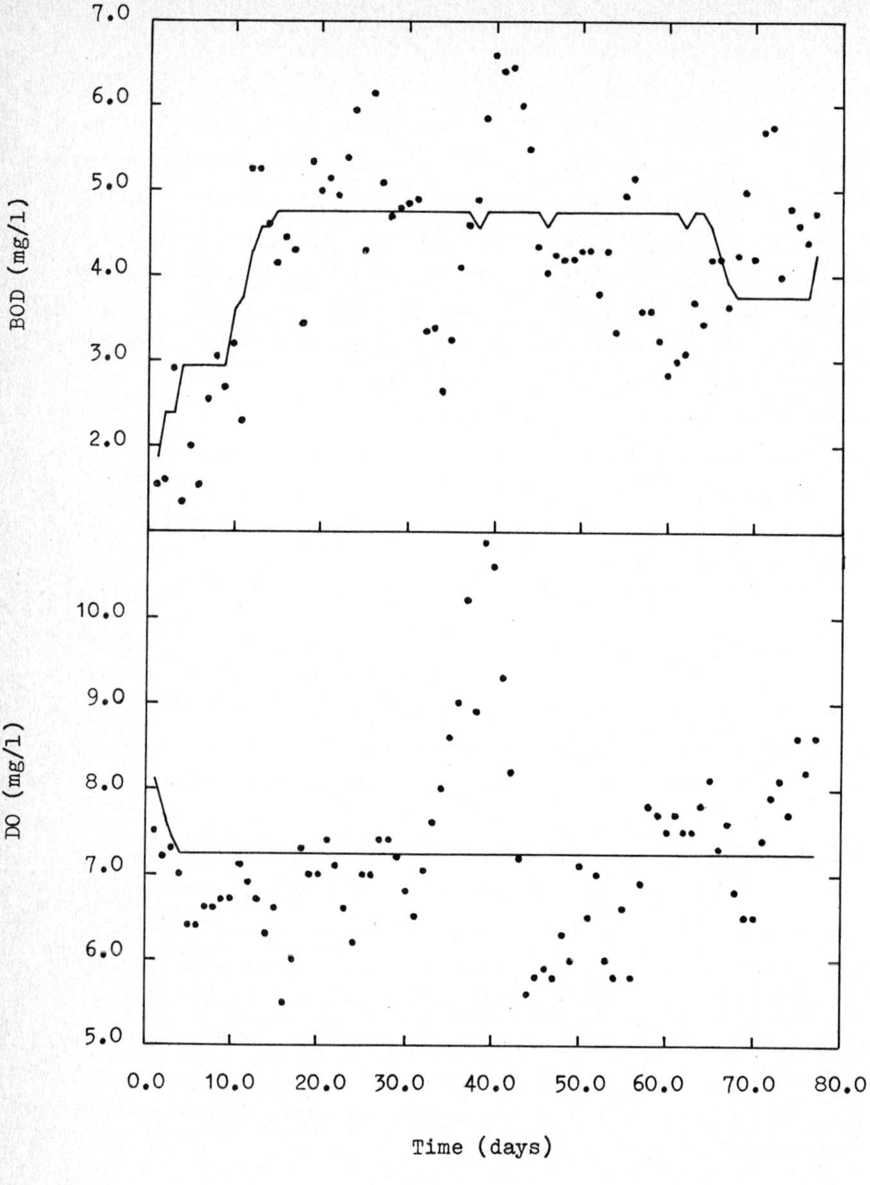

Figure 8. Pure predictions for the first model

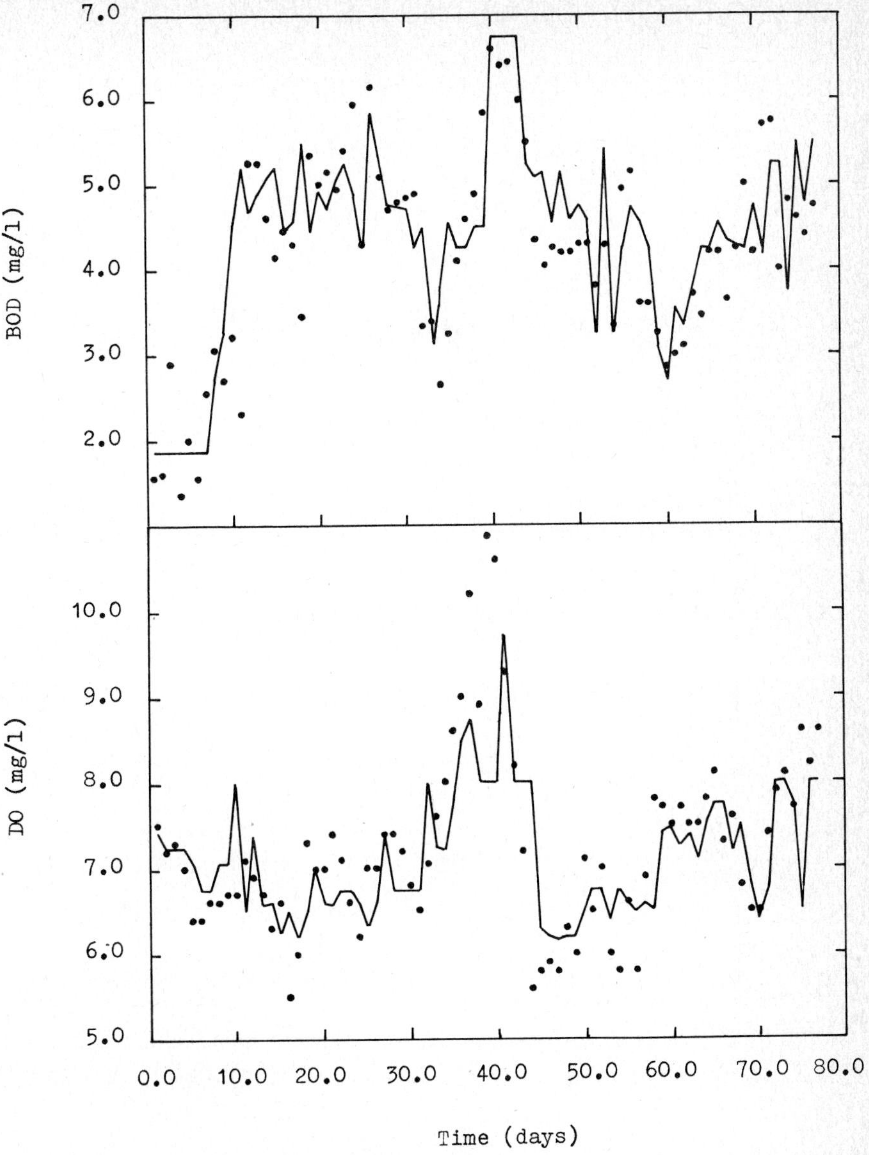

Figure 9. OSA Predictions for the second model

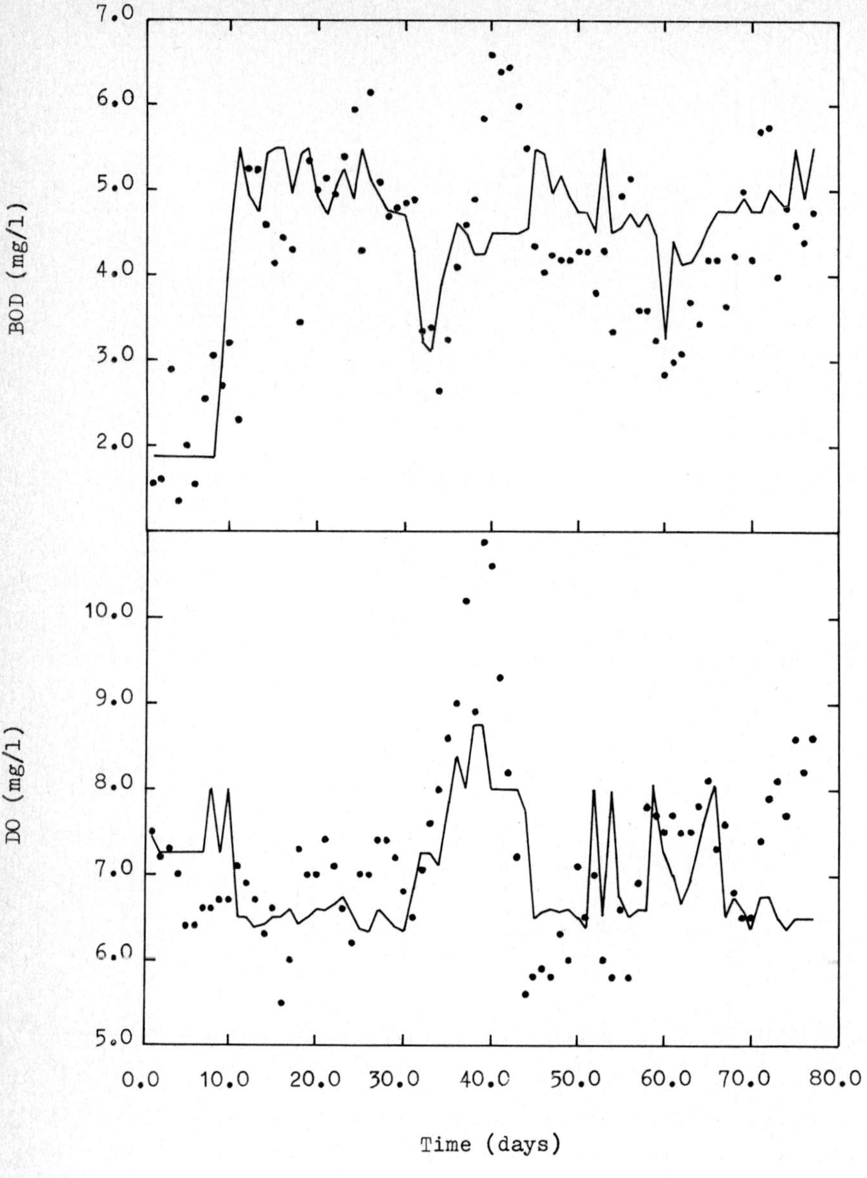

Figure 10. Pure predictions for the second model

of most value as devices for assessing approximate behaviour rather than as tools for detailed engineering design. However, in 'softer' areas, such as water quality control, where the goals are usually less clearly defined, fuzzy models may be useful in a wider range of tasks.

This paper has not been concerned with the use of fuzzy models in socio-economic systems, although the work of Wenstop(18) and Kickert(19) gives indications of the difficulties. It does seem that for the ideas discussed in the preceding sections to be generalised to such systems a considerable amount of work needs to be done. In particular, the relationship between data set size and model complexity needs to be examined, as does the concept of 'model purpose'. Perhaps a combination of fuzzy and non-fuzzy models would allow broad generalisations to be made about systems without sacrificing accuracy where this is required?

The main conclusion must be that, whilst fuzzy models can be successfully constructed, the overall concept needs a more detailed investigation before its true worth can be evaluated.

Acknowledgement

The author would like to acknowledge the financial support provided by a NATO/SRC Postdoctoral Research Fellowship.

References

(1) L.A.Zadeh, Outline of a new approach to analysis of complex systems and decision processes, IEEE Trans. Systems Man Cyber., SMC-3 (1973) 28-44.
(2) L.A.Zadeh, Fuzzy sets, Inform. Control, 8 (1965) 338-353.
(3) D.Dubois and H.Prade, Operations on fuzzy numbers, Int. J. Sys. Science, 9 (1978) 613-626.
(4) L.A.Zadeh, Towards a theory of fuzzy systems, in: R.E.Kalman and N.DeClaris (eds.), Aspects of Network and Systems Theory (Holt, Rinehart and Winston, 1971).
(5) R.M.Tong, A control engineering review of fuzzy systems, Automatica, 13 (1977) 559-569.
(6) R.M.Tong, Analysis and control of fuzzy systems using finite discrete relations, Int. J. Control, 27 (1978) 431-440.
(7) L.Bainbridge, The process controller, in: W.T.Singleton (ed.), The Study of Real Skills (Academic Press, 1977).
(8) I.G.Umbers, A Study of Cognitive Skills in Complex Systems, Ph.D. Thesis, Univ. of Aston, England (1976).
(9) R.M.Tong, M.B.Beck and A.Latten, Modelling and Operational Control of the Activated Sludge Process in Wastewater Treatment, IIASA Professional Paper PP-78-10 (Nov. 1978).
(10) L.A.Zadeh, The concept of a linguistic variable and its application to approximate reasoning; part 1, Inf. Sci., 8 (1975) 199-249.
(11) R.M.Tong, Synthesis of fuzzy models for industrial processes, Int. J. General Sys., 4 (1978) 143-162.
(12) P.Eykhoff, System Identification (Wiley Interscience, 1974).
(13) A.Kaufmann, Introduction to the Theory of Fuzzy Subsets (Academic Press, 1975).
(14) A.DeLuca and S.Termini, A definition of non-probalistic entropy in the setting of fuzzy sets theory, Inform. Control, 20 (1972) 301-312.
(15) G.E.P.Box and G.M.Jenkins, Time Series Analysis Forecasting and Control (Holden Day, 1970).

(16) M.B.Beck and P.C.Young, A dynamic model for DO-BOD relationships in a non-tidal stream, Water Research, 9 (1975) 769-776.
(17) P.C.Young, S.H.Shellswell and C.G.Neethling, A Recursive Approach to Time Series Analysis, Cambridge Univ. Eng. Dept. report CUED/B-Control/TR16 (1971).
(18) F.Wenstop, Fuzzy set simulation models in a systems dynamic perspective, Kybernetes, 6 (1977) 209-218.
(19) W.J.M.Kickert, Towards an Analysis of Linguistic Modelling, Proc. 4th Int. Congress of Cybernetics and Systems, Amsterdam (1978).

ADVANCES IN FUZZY SET THEORY AND APPLICATIONS
M.M. Gupta, R.K. Ragade, R.R. Yager (editors)

APPLICATION OF FUZZY SETS FOR THE ANALYSIS OF COMPLEX SCENES

Ramesh Jain

Department of Computer Science
Wayne State University
Detroit Mi 48202

A scene analysis system should be able to work with imprecise data and should integrate pieces of evidence collected from different knowledge sources. Many efforts are being made to develop scene analysis systems which can exploit knowledge from various sources to interpret scenes of sufficiently general nature. It is shown in this paper that fuzzy set theory can be used for considering the imprecise data and for employing approximate reasoning. In this system fuzzy properties, such as monotonicity or fillness, of a sequence are used to analyze the motion of objects in complex real world scenes.

INTRODUCTION

About two decades of research in scene analysis have made it clear that it is a complex task. Though good understanding of a very limited class of scenes (scenes containing only polyhedral blocks) has been achieved; no system exists which can analyze scenes containing assorted objects in a realistic situation {2}. The complexity in the task of analysis of a scene arises because in order to analyze a scene, the system should be able to work with imprecise data and should be able to integrate knowledge from various sources. Conventional mathematical tools are usually not effective because of the impreciseness of the data and influence of many factors in the formation of an image of a scene. The heuristic techniques developed for many applications are tailored to fit a particular domain. These heuristic techniques do not perform well when applied to other domains.

Obviously, the problem of complexity and impreciseness of the data is not limited to scene analysis. Researchers in various fields, particularly those working with humanistic systems, have been facing these problems. To solve such problems Zadeh advanced the concept of fuzzy sets {9}. We believe that fuzzy sets can be used in many situations in scene analysis also. In this paper first we identify some areas in scene analysis, where fuzzy sets may be useful, and then describe a scene sequence analysis system in which fuzzy sets have been applied.

SCENE ANALYSIS

In scene analysis usually cues from many sources, such as image intensities, position and number of light sources, the type of objects, stereo, motion etc. are integrated to obtain consistent interpretation of the scene {1}. The interaction of these sources is not well defined and is not well understood, and hence conventional logic can not be applied to integrate in a meaningful way the information available from various cues. Approximate reasoning based on fuzzy set theory {10} may be useful in integrating cues from various sources. Fuzzy logic controllers {11} have been successfully applied in situations where approximate rather than exact reasoning is used.

In the fields of pattern recognition and scene analysis, newer and newer methods are being developed for making features and properties more flexible {7,8}.

Fuzzy features have flexibility as is given by rubber mask features or spring loaded templates. For computing fuzzy feature values one may use a model for the feature, and using a suitable membership function the similarity of a given feature and the model may be determined {7}. This may result in insensitivity to noise and may offer flexibility in feature detection processes by noting even weak evidence of the presence of a feature.

It is a common practice in model based vision systems to direct the algorithm on the basis of presence or absence of a specific property. There are many situations where it is not pertinent to compute the presence or absence of such a property. In such situations one may compute all possible supports for the presence and the absence of such a feature and then combine them to assign a fuzzy value for the presence of that feature. The monotonicity of a sequence will be computed using such an idea in this paper.

In scene analysis systems, redundancy should be intentionally allowed as it may help in keeping track of that weak evidence alive which may later become strong and influence interpretation significantly. Using fuzzy sets we may keep all those processes alive which have even a weak possibility of becoming significant. At the nth level (or at the nth time instant) the strength of an evidence may be considered on the basis of a) strength of the evidence at (n-1)th level, and b) strength of the evidence ath the nth level. This prevents the interpretation from getting too much influenced by a single level. This process may be modeled using fuzzy set theory.

A SCENE SEQUENCE ANALYSIS SYSTEM

In this section we discuss a scene sequence analysis system. This system has been discussed in {3-6} with emphasis on scene analysis. Here we consider some aspects of this system dealing with the applications of the fuzzy set theory. First we discuss this system very briefly. The reader interested in details of this system may see references {5,6}.

The input to this system is a digitized TV-frame sequence having 572 rows and 512 columns at 8 bit greylevel resolution. Each of these frames is condensed into 96X128 pixels by representing each pixel of the condensed frame by the mean and variances for each group of four consecutive columns in six consecutive rows. For every TV frame in the sequence, each pixel data is compared to the data of the condensed frame at the same pixel position using a criterion based on second order statistics. An accumulative difference picture is prepared using this criterion. Each position in the First Order Difference Picture (FODP) contains the number of times from the second frame to the current frame for which the data were found to be different. Starting with the third frame, the Second Order Difference Picture (SODP) is also prepared for each frame. The SODP for the nth frame is a binary picture having an entry '1' in only those pixel positions where the difference pictures for (n-1)th and nth frame have different entries. A group of at least N connected nonzero entries in an FODP or SODP is called a region.

It has been shown that if a reasonably large homogeneous object is moving against a contrasting background with constant velocity, then, before the object has moved over the distance D equal to its projection in the reference frame, there are two regions in FODP each having a corresponding region in SODP. The entries along the rows and columns of the FODP region form monotonic sequences having decreasing values along the direction of motion. After the object has moved over the distance D, the two FODP regions join and form a single FODP region. This FODP region contains two SODP regions at those pixel positions where the object was in the first frame and where the object is in the current frame.

In real world scene sequences, a moving object may be of arbitrary shape, may have many greylevel components, may move in any possible direction, may be occluding and/or getting occluded, etc. In other words, the situation in real world scenes is much more complex than indicated by the above example. It was found, however, that even in real world scenes some information may be obtained by carefully selecting fuzzy features and properties and then applying fuzzy inference rules. These features and properties are discussed in what follows.

SODP REGION PROPERTIES

Assuming that the lighting in the scene and the camera position remain unchanged during the sequence, an entry in a SODP is always the result of either covering or uncovering of a pixel which contains the greylevel of the background by the pixel which contains the greylevel of some moving object image. The region due to covering of the background by the object image is termed O region; due to uncovering of the background is termed B region; the region where the object was in the reference frame but has now moved away completely is termed S region; and the region where the object is in the current frame after crossing its D is termed as M region. It has been shown that the ratio

CURREF = current frame edge length/reference frame edge length

may be used for the classification of an SODP region {5,6}. The value of this ratio for O region is always MORE THAN 1; for B region LESS THAN 1; for M region VERY HIGH; and for S region VERY LOW. Note the clear overlap of values for O and M, and B and S type regions. Based on the value of CURREF we used following functions for assigning a suitable grade of membership in various classes to a given region:

FO = ((CURREF-1)/T1)**k1

FB = ((1.0-CURREF))**k2

FS = (1.0-CURREF)**k3

FM = ((CURREF-1.0)/T1)**k4

where k1, k2, k3, k4 and T1 are constants and k4>k1. The negative values of these functions are converted to zero, and values above 1 are converted to 1.

The final values assigned to a region in nth frame are defined as

$$P(n) = 0.75*P(n-1) \oplus P$$

where P(i) denotes final grade of membership in a type P region in the ith frame, and P denotes grade of membership in the type P region based on the CURREF for the current frame only. The operator $\oplus$ is defined as

$$A \oplus B = A + B - A * B$$

The class of a region is usually fuzzy. Because of noise in images and coincidences in real world scenes, it is difficult to distinguish clearly O from M, and B from S. If CURREF is close to 1, which may happen in case of one moving object image occluding another moving object image or for a moving object whose image has many greylevel regions, then also it is not clear to which class the region should be assigned. By considering the class of a region to be fuzzy, and by accumulating evidence for the class over many frames, we do not tend to take decision arbitrarily. It is found that this method gives in most cases correct type of the region. Though a region may have membership in more than one of the classes, its membership, as given by this method, in the correct class is usually higher than its membership in other classes.

FODP REGION PROPERTIES

The two most important properties of an FODP region are MONOTONICITY and FILLNESS. For computation of these properties of a region each row(column) is considered a sequence and the following functions are used:

$$F1 = \{(abs(asc-desc))/(asc+desc)\}**L1$$

$$F2 = \{1.0+gap\}**(-L2)$$

$$F3 = \{\sum_i (lasc(i)+ldesc(i))/length\}**L3$$

where asc, desc, lasc(i), and ldesc(i) denote number of ascending subsequences, number of descending subsequences, length of ith ascending subsequence, and length of ith descending subsequence, respectively; length and gap are the length of the sequence and the number of subsequences of zeros in the sequence; and L1, L2, and L3 are positive constants.

MONOTONICITY

We observe each sequence and although that sequence may not be strictly monotonic, we assign monotonicity to the sequence on the basis of the numbers of ascending and descending subsequences using F1. The monotonicity of each sequence of the region is determined and then these sequence monotonicities are used in assigning monotonicity to the region. If F1 for a sequence is positive then the sequence is an ascending sequence, otherwise the sequence is a descending sequence. Now suppose that there are Nr rows in a region of which Mr are ascending. The horizontal component of the monotonicity of this region is

$$HM(R) = \{abs(Mr-(Nr-Mr))/Nr\}**L1$$

If the region has Nc columns of which Mc columns are ascending then the vertical component of the monotonicity is

$$VM(R) = \{abs(Mc-(Nc-Mc))/Nc\}**L1$$

The positive value of a component denotes ascending nature. The monotonicity of a region gives the direction of the displacement of the image. The region has descending monotonicity in the direction of the displacement. The horizontal and vertical monotonicities of a region give the direction of the displacement of an image component along rows and columns in the image.

High value of the monotonicity is a strong indication of the displacement of an image component from frame to frame. Low value of the monotonicity indicates that either there is high noise in the picture, the object image being nonhomogeneous, or there are two overlapping object images moving in different directions.

FILLNESS

The fillness of a sequence is

$$SF = F2 \text{ AND } F3 = Min(F2, F3)$$

The fillness of the region is the average fillness of the sequences of the region. The horizontal and vertical components of fillness of a region are

$$HF(R) = 1.0/Nr \sum_{i=1}^{Nr} SFri$$

$$VF(R) = 1.0/Nc \sum_{i=1}^{Nc} SFci$$

where SFri and SFci are the fillnesses of ith row and ith column of the region, respectively.

Fillness is high if the number and lengths of the gaps in the region are small. A high value of fillness indicates that the region is either one of the two regions at the rear and front ends of the moving object, or that the object image is already displaced by its D. For a nonhomogeneous moving object there will be several regions, which may join much before the object image is displaced by its D. In such a situation the FILLNESS of the region is low.

RESULTS

Using these fuzzy features and the inference rules given in Table 1, an algorithm has been developed. This algorithm has been implemented in a Pascal program on a KI-10 processor. The input to this program is a sequence of digitized TV frames. The program initially finds the displacement, if any, in the scene, next it gives estimates of the displacement rate and of the size of the object image. After the object image is displaced from its position of the reference frame, the algorithm can focus its attention on the object image. When the algorithm has entered the attention focussing phase, it modifies the reference frame by substituting the greyvalue characteristics of the background from the current frame for those pixels which were occupied by the image of the object in the reference frame. When the greyvalue characteristics of all nonstationary components have been substituted in the reference frame, the reference frame contains only stationary image components.

We present here only those results which are related to the fuzzy properties. In Figures 1 and 2 two frames from a sequence are shown. Note the white car at the intersection and the pedestrian on the left side of the road. The car is heading to the right and the pedestrian is walking to the left. In Figures 3, 4, and 5 the FODPs and SODPs after 6, 13, and 19 frames are shown. These figures actually give only parts of FODPs and SODPs covering the car and the pedestrian. In Table 2 the values of horizontal monotonicity, horizontal fillness, and the direction of motion inferred from the monotonicity of the regions of FODP are given. It was found that even if no sequence of a region was a true monotonic sequence, the direction of motion given by fuzzy MONOTONICITY was correct. The grade of memberships of regions of the SODP in various classes and CURREF for the current frame are given in Table 3. It is found that though a SODP region may have grade of memberships in various classes, its correct memberships is almost always stronger than other membership values. Note that before the pedestrian has moved over its D, the SODP region has membership in both O and B classes. After it has crossed its D, the situation becomes clear.

Our algorithm predicted after 5th frame that the car may cross its D in 15th frame which came out to be true. After 19th frame sufficient confidence has been gained in the images of the car and the pedestrian. The reference frame was modified by substituting the greylevel characteristics of the background from the current frame in the reference frame for those pixels which were covered by the moving object images in the reference frame. Figure 6 shows the reference frame after the modification. Note that, though, no frame of the sequence is without the images of the moving objects, the reference frame does not contain these objects. After this modification the SODPs for these two objects give their images in the current frame.

CONCLUSION

Our aim in this paper has been to show that fuzzy sets can be applied in scene analysis. The system discussed in this paper is in its evolutionary phase. The functions used for computing various features and properties are at best heuristic. We intend to study these functions carefully. The results obtained so far, however, strengthen our belief that fuzzy sets can be useful in modeling some situations in scene analysis.

ACKNOWLEDGEMENTS

The scenê analysis system described in this paper was developed with H.H. Nagel while the author was with Fachbereich Informatik, Universitaet Hamburg. It is a pleasure to acknowledge the help of Prof. Nagel and other members of picture processing group during the course of development of this system.

REFERENCES

|1| Barrow, H.G. and Tenenbaum, J.M., Recovering intrinsic scene characteristics from images, Technical report, SRI, April 1978.
|2| Jain, R. and Aggarwal, J.K., Computer analysis of curved objects, to appear in Proc. IEEE, April 1979.
|3| Jain, R., Militzer, D., and Nagel, H.-H., Separating non-stationary from stationary scene components in a sequence of real world TV-images, Proc. IJCAI-77, pp 612-618, Cambridge, Mass., Aug. 22-25, 1977.
|4| Jain, R. and Nagel, H.-H., Analyzing a scene sequence using fuzziness, Proc 1977 IEEE Decision and Control, Vol 2, pp 1367-1372, New Orleans, Dec 7-9, 1977.
|5| Jain, R. and Nagel, H.-H., On a motion analysis process for image sequences of real world scenes, Technical Report B-48, Fachbereich Informatik, Universitat Hamburg, May 1978.
|6| Jain, R. and Nagel, H.-H., On the analysis of accumulative difference pictures from image sequences of real world scenes, to appear in IEEE Trans. on Pattern Analysis and Machine Intelligence, April 1979.
|7| Lee, E.T., The shape-oriented dissimilarity of polygons and its application to the classification of chromosome images, Pattern Recognition, 6, pp 47-60, 1974.
|8| Widrow, B., The rubber mask technique I. Pattern measurement and analysis, Pattern Recognition, 5, pp 175-194, 1973.
|9| Zadeh, L.A., Fuzzy sets, Information and Control, 8, pp 338-353, 1965.
|10| Zadeh, L.A., The concept of linguistic variable and its application to approximate reasoning, Part I, Information Sciences, 8, pp 199-249, 1975; Part II, Information Sciences, 8, pp 307-357, 1975; Part III, Information Sciences, 9, pp 43-80, 1975.
|11| Pappis, C. and Mamdani, E.H., A Fuzzy Logic Controller for a Traffic Junction, IEEE Trans, SMC-7, pp 707-717, 1977.

Table 1

Fuzzy Inference Rules

1. If an FODP region has HIGH H-FILLNESS AND HIGH V-FILLNESS and if there are two regions in SODP for this FODP region, then the nonstationary image component has been displaced by more than its distance D.

2. If the FODP region has HIGH H-MONOTONICITY OR HIGH V-MONOTONICITY, then a search is made to locate other FODP region which may be due to the same object.

3. FIMO=(H-FILLNESS OR V-FILLNESS) AND (H-MONOTONICITY OR V-MONOTONICITY)

4. If FIMO is HIGH, then it is assumed that the object is small.

5. If FIMO is MEDIUM, then it is assumed that the object is nonhomogeneous.

6. If FIMO is LOW, then nothing may be predicted about this FODP region.

Table 2a

Number of Frames Observed = 6

Region No.	MONOTONICITY	FILLNESS	Direction	Strength of Evidence
1	0.675	1.00	Right	Good
2	1.00	0.958	Right	Certain
3	0.047	1.00	Left	Very Weak

Table 2b

Number of Frames Observed = 13

Region No.	MONOTONICITY	FILLNESS	Direction	Strength of Evidence
1	0.384	0.778	Right	Weak
2	0.769	0.902	Left	Good

Table 2c

Number of Frames Observed = 19

Region No.	MONOTONICITY	FILLNESS	Direction	Strength of Evidence
1	0.031	0.820	Right	Very Weak
2	0.75	0.93	Left	Good

Table 3a

Number of Frames Observed = 6

Region	STATIC	MOBILE	O-GROW	B-GROW	CURREF
1	0.00	0.00	0.00	0.89	0.235
2	0.00	0.00	0.782	0.00	3.00
3	0.00	0.00	0.00	0.495	0.8

Table 3b

Number of Frames Observed = 13

Region	STATIC	MOBILE	O-GROW	B-GROW	CURREF
1	0.489	0.00	0.00	0.912	0.392
2	0.00	0.755	0.915	0.00	7.33
3	0.00	0.00	0.265	0.607	0.900

Table 3c

Number of Frames Observed = 19

Region	STATIC	MOBILE	O-GROW	B-GROW	CURREF
1	0.59	0.00	0.00	0.866	0.575
2	0.00	0.765	0.894	0.00	5.00
3	0.886	0.00	0.063	0.974	0.083
4	0.00	1.00	1.00	0.00	100.0

Figures 1 and 2. Two frames from a TV-frame sequence representing a street intersection with traffic. In this paper, our discussion is centered around the white car at the center moving to the right and the pedestrian on the left hand side of road (for the observer) moving to the left.

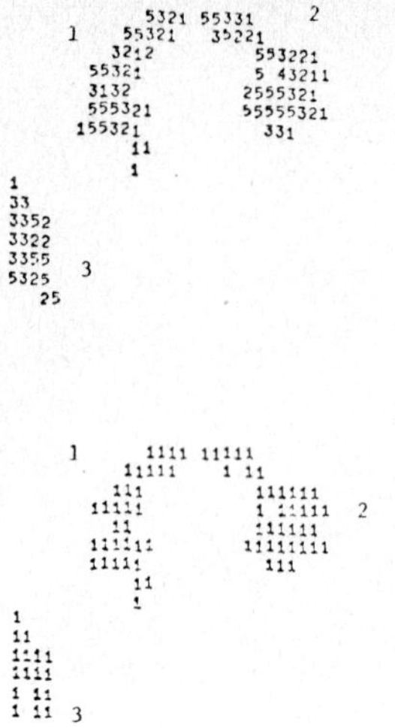

Figure 3. FODP and SODP for the region of the scene having maximum activity--a moving car and a pedestrian--after 6 frames. For car there are two FODPs but for pedestrian there is one. Each FODP has a corresponding SODP.

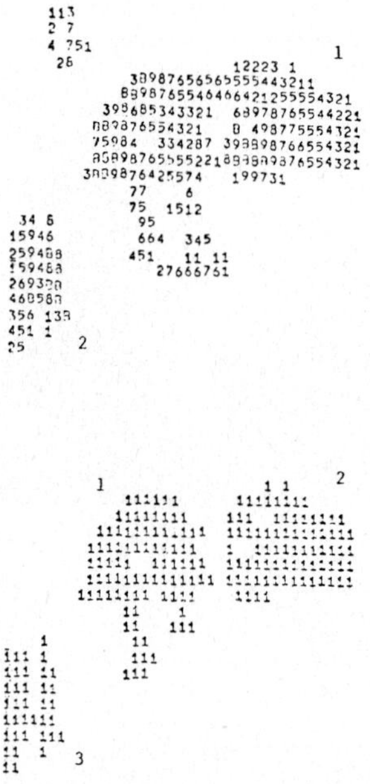

Figure 4. After 13 frames the two FODPs of the car have merged together. The car has not moved over its D -- this is shown by gaps in FODP. Pedestrian is also about to cross over its D, as shown by SODP.

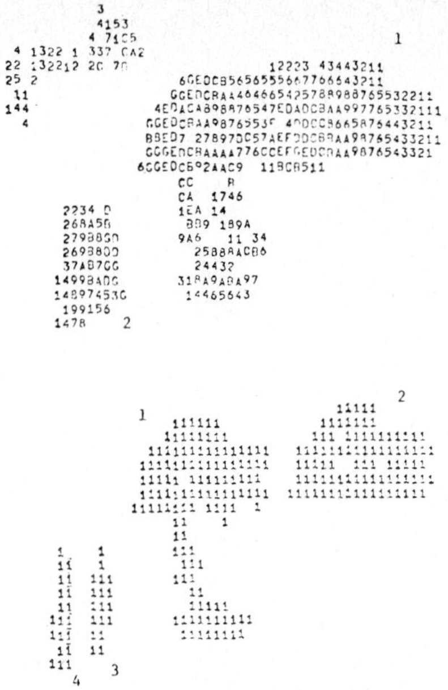

Figure 5. After 19 frames both objects have moved over their D's. For each FODP there are two SODPs.

Figure 6. The reference frame after the greylevels for the car and the pedestrian have been replaced by the greylevels of the background.

ADVANCES IN FUZZY SET THEORY AND APPLICATIONS
M.M. Gupta, R.K. Ragade, R.R. Yager (editors)

A GENERAL PURPOSE POLICY CAPTURING DEVICE USING FUZZY PRODUCTION RULES

Jonathan J. WEISS and Michael L. DONNELL

Decisions and Designs, Inc.
Suite 600, 8400 Westpark Drive, P. O. Box 907
McLean, Virginia 22101, U. S. A.

Through the use of fuzzy policy capturing, it is possible to represent and simulate the performance of human experts at complex tasks. A general-purpose (as opposed to task-specific) system is described which uses linguistic production rules to guide actions. A fuzzy production rule system consists of four basic structures: A set of rules; a set of input data; a method for evaluating any proposed action; a method for generating actions and for determining when to stop searching for better ones. A sample fuzzy production rule system is described.

1.0 INTRODUCTION

A number of difficult problems in public policy, corporate planning, and military command and control are further complicated by the following conditions:

- they require speedy real-time decisions;
- they recur with slight variations, so that the proper actions depend on information which cannot be assessed in advance;
- the proper actions depend on complex relationships involving a large number of potentially relevant observations;
- precise readings may be impossible or prohibitively expensive to obtain--often a linguistic description is the best information available;
- although some experts exist who can make effective decisions, their scarcity makes it unlikely that they would be present when their advice is required.

Because of the many slow and costly assessments entailed in performing a complete decision analysis from scratch, it would be more efficient to simulate the human expert's information-processing strategies. The approach we shall refer to as "fuzzy policy capturing" capitalizes on the expert's knowledge by having him describe in approximate terms the set of available actions, and which configurations of input data would indicate each of the

This work was performed for the Naval Electronics Systems Command Under Contract No. N00039-77-C-0252 options. The various input patterns so described would partition the space of possible inputs into regions so that whenever a new situation is encountered it can be compared with the given configurations and assigned to the correct action category.

Using approximate descriptions of these input configurations affords the decision maker a number of advantages. In general, fewer assessments will be necessary to determine the proper action than would be needed in a conventional decision analysis. Furthermore, such assessments could tolerate a certain amount of imprecision or inaccuracy, allowing faster and simpler assessments without jeopardizing the correctness of the prescribed action. And, because the logic of the evaluation process corresponds closely to human reasoning, the procedures and recommendations of a fuzzy policy capturing device could use inputs which were linguistic rather than numerical.

Since the original work of Mamdani (1976), who applied fuzzy algorithms to steam engine control, a number of applications have been developed to deal with other problems, for example, problems related to a heat exchanger, a sinter plant in steelmaking, water baths, a limekiln, and a traffic controller. In most cases, the fuzzy algorithms used were both practical and efficient at performing complex control tasks in which a limited amount of information was utilized.

We are currently developing a procedure which extends the general ideas of these systems to decision-making situations in which the information load is much larger, and the rules for processing inputs correspondingly more numerous and complex. The advances which have been made, and those upon which work is continuing, include establishing a firm theoretical basis for a fuzzy production rule version of policy capturing, implementing a working system for dealing with fuzzy production rules on a portable computer, developing a methodology for assessing the expert's rules and the decision maker's inputs, and investigating potentially useful applications in decision-making problems.

2.0 THEORETICAL FOUNDATIONS

A fuzzy production rule system consists of four basic structures:

- o a set of rules which represents the policies and heuristic strategies of the expert decision maker;
- o a set of input data assessed immediately prior to the actual decision;
- o a method for evaluating any proposed action in terms of its conformity to the expressed rules, given the available input data;
- o a method for generating promising actions and for determining when to stop searching for better ones.

The following principles govern the operation of such a system.

2.1 Rule Structure

Every rule consists of some restriction on the range of acceptable actions, which may be conditional upon the degree to which the inputs match a particular configuration. These restrictions need not be absolute: ordinarily, more than one action will be acceptable in a given situation (which is certainly true if you consider arbitrarily fine distinctions between similar actions). Furthermore, the fuzzy approach to such restrictions allows us to consider some actions as partially acceptable, if they are not so good as the best possible course, but nonetheless superior to certain other possibilities.

2.2 Membership Functions

Every restriction on the range of possible actions is associated with a membership function, which specifies the degree to which any action satisfies the given restriction. This function takes on values between zero and one, such that a value of one indicates the best possible performance with respect to that restriction, and a value of zero indicates the worst performance (for intermediate values, the higher the value the better the performance). As an example, in response to the command "bring me about a dozen nails," the performance (corresponding to a membership function of one) would be to bring exactly twelve nails to the proper place, within a reasonable time. It would not be disastrously worse in most circumstances to bring ten or fifteen nails, or to deposit them near the desired place, or to comply a bit slowly; on the other hand, to bring no nails at all, or to be extremely slow in complying, might well be considered completely unsatisfactory, and therefore assigned a membership function of zero.

The same sorts of membership functions also apply to the conditions which govern certain restrictions, except that in this case the membership functions are defined as the degree to which the input situation satisfies various preconditions. For example, the precondition "visibility is good" might be satisfied perfectly in the case of bright sunlight, somewhat less perfectly in the case of a cloudy day, still less on a moonlit night, and not at all on a very dark night.

In fuzzy production rule systems, the actual membership functions will be assessed from the expert at the time he describes his rules. Ordinarily, there will be general agreement across a community of decision makers concerning the meaning of the terms associated with these membership functions; that is, the decision maker's inputs will be based upon the same definitions as the experts's rules. If such is not the case, it is always possible to specify unusual or special definitions by explicitly describing one's own membership functions.

2.3 Evaluating Actions with Respect to a Single Rule

2.3.1 Unconditional rules - An unconditional rule is simply the membership function associated with a restriction that applies regardless of any input information; examples might be overriding goals that supersede any of the rules described. The "performance score" for an action with respect to a single unconditional rule is simply the value of the membership function upon that action.

Figure 2-1 illustrates the membership function corresponding to the unconditional restriction "drive quickly."

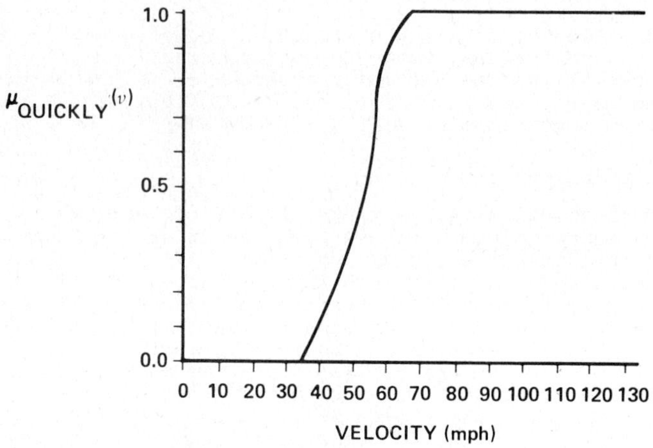

Figure 2-1

A MEMBERSHIP FUNCTION FOR THE RESTRICTION "DRIVE QUICKLY"

Note that for any particular velocity v, we can find the value of the function $\mu_{Quickly}(v)$ ("the degree to which v corresponds to the restriction 'drive quickly'"); this value is the performance score for that simple rule. For example, if the actual velocity chosen were 60 mph, then the performance score would be about 0.85.

2.3.2 <u>Conditional rules</u> - Because the majority of actions in decision problems do depend upon input information (otherwise there would be no difficulty!), most of the rules actually encountered will be conditional ones. Such rules will take the form, "if the input data satisfy condition C, then the action must conform to restriction R." Here, the performance score will depend not only on the degree to which R is satisfied, but also on the degree to which the inputs satisfy the precondition C. The following rules, which constitute a generalization of the logical formula defining implication, define the performance score for a single conditional rule.

<u>If the condition C is perfectly satisfied</u> (i.e., if the membership function corresponding to C has the value of exactly one for the given inputs), then the restriction R applies just as if the rule were unconditional. The performance score is therefore equal to the membership function for R, evaluated for the particular action under consideration. For example, the command "if 2+2=4 then drive quickly" is logically the same as "drive quickly."

If the condition C is not at all satisfied (i.e., if the membership function corresponding to the precondition C is exactly zero), then the restriction R need not apply at all. In this case, any action whatever will receive a performance score of one (perfect compliance), since the rule in question imposes no restriction at all. An example of such reasoning in everyday life involves a driver who was accused of disobeying the rule "if the traffic light is red, then stop": it would be logically sufficient for the defendant to prove that the light was not in fact red at the time of the alleged violation, in order to establish his innocence.

If the condition C is only partially satisfied (let the value of the membership function corresponding to C be some number x between zero and one), then the restriction R applies, but only up to a certain degree. In particular, the performance score will equal the value of the membership function for the restriction R, so long as that value exceeds 1-x; otherwise, the performance score for the rule is equal to 1-x. In other words, if we were to start off with a score of one and then assess a penalty for non-compliance with the restriction R, the maximum penalty assessable would be x.

Another way of looking at this portion of the definition is in the context of classical two-valued logic. There, the conditional formula "if P then Q" is considered true whenever Q is true or P is false (i.e., the conditional is false only when P is true but Q is false). Therefore, the formula for "if P then Q" is identical to that for "(not P) or Q," as shown by the truth table below:

P	Q	~P	(~P) v Q	P→Q
T	T	F	T	T
T	F	F	F	F
F	T	T	T	T
F	F	T	T	T

According to the calculus of fuzzy logic, as developed by Lotfi Zadeh and his associates, the following extensions charactize fuzzy logic:

- truth values correspond to the values of a membership function in which false statements get values of zero, true statements get values of one, and partially true statements take on intermediate values;
- the negation of a proposition P has a truth value equal to one minus the truth value of P;
- the disjunction of two statements P and Q has a truth value equal to the truth value of P or the truth value of Q, whichever is greater.

Applying either mode of reasoning to the definition, we obtain the following general formula for the conditional rule "if the inputs satisfy condition C, then the actions must conform to restriction R":

$$\text{performance score} = \max[1-\mu_C(\text{inputs})), \mu_R(\text{action})].$$

It should be noted that although the formula defined here is logically consistent and agrees with the intuitive properties and the practical requirements for such a definition, other possible definitions which are equally consistent may arrive at very different formulas. In fact, there is an infinite number of different formulas which might be "correct," among which the choice will depend on a combination of empirical usefulness, analytic expediency, and psychological validity.

One warning is, however, in order: the formula used by Mamdani in his various fuzzy control papers is not among the category just mentioned, although it performs satisfactorily in the applications he reports. The problem is that Mamdani (1976) treats "if P then Q" and "both P and Q" as synonymous (and is saved because his other rules always approximate the missing term "or else not P"). For practical purposes, Mamdani's algorithm is as good as any other in the limited applications he describes; but to apply such formulas to more complex input structures would eventually lead to difficulties.

2.4 Evaluating Actions with Respect to a System of Rules

2.4.1 Conjunctions of rules When two or more rules must be jointly satisfied, the proposed action is evaluated with respect to each component rule separately, and then assigned a performance score equal to the lowest of the scores on the individual rules. For example, suppose a system of two rules consists of the unconditional restriction, "turn on the heat," and the conditional rule, "if it is raining hard, then close the window." The input information here would consist of whether or not it is raining hard; hypothetically, let us assume that it is raining fairly hard, which might correspond to a membership value of 0.8 for the precondition "raining hard." Now, to evaluate an action which consists of turning the heat on all the way and closing the window halfway, the following steps must be taken:

- evaluate the first (unconditional) rule--since the action turns the heat on all the way, let us assume the membership function for the first restriction is equal to 1.0;

- evaluate the second (conditional) rule--as we mentioned, the precondition is satisfied at the level of 0.8, and since the window has been closed only halfway, let us assign a membership score of 0.5 to the corresponding restriction, so that the performance score for the whole conditional rule is max((1-.0.8),0.5), which equals 0.5 (according to the formula at the end of Section 2.3.2);

- assign an overall performance score equal to the minimum of the individual performance scores--in this case, min(1.0,0.5) is equal to 0.5.

It should be noted that the performance score for a conjunctive system of rules, once it has been obtained, can function exactly as if it were obtained from a single rule. Thus, for example, if Rule A_1, and Rule A_2, and so on are members of a system of rules which must be jointly satisfied, then there is a single Rule A which consists of the restriction, "obey Rules A_1, A_2, etc."

2.4.2 Nested conditions - It is possible that some input configuration or precondition indicates not a simple restriction on the action to be chosen, but instead a more complex rule or system of rules. For example, consider the rule "if you are in Massachusetts, obey the Massachusetts state laws." This has the same form as a simple conditional rule, except for the fact that obeying the Massachusetts state laws is not a simple restriction, but rather a complex system of subordinate rules (many of which are themselves conditional. Nonetheless, as long as it is possible to obtain a performance score for the rule or system of rules that might apply depending on the precondition, the same formula may be adapted from Section 2.3.2:

$$\text{performance score} = \max[(1-\mu_C(\text{inputs})), \mu_{\text{rule}}(\text{action})].$$

A few observations on the properties of this definition will prove enlightening. First, the nested rule which takes the form, "if A, then 'if B then C'" is logically equivalent to the simpler rule, "if both A and B, then C." Referring to the formulas defined in Section 2.3.2, the former rule will have a performance score equal to

$$\max[(1-\mu_A), \max[(1-\mu_B), \mu_C]],$$

or in other words, equal to the greatest of the three quantities, $(1-\mu_A)$, $(1-\mu_B)$, and μ_C. Similarly, the latter version has a performance score equal to

$$\max[1 - \min(\mu_A,\mu_B)), \mu_C].$$

Simply by checking the six possible ways of ordering the three quantities $(1-\mu_A)$, $(1-\mu_B)$, and μ_C, it is possible to prove that the two formulas always yield the same values for the performance score, and are therefore equivalent for the purpose of evaluation. An everyday illustration of this principle is the equivalence of the two statements, "if it is raining, then if you are going outside carry an umbrella," and "if it is raining and you are going outside, carry an umbrella."

A straightforward extension of this principle is mulitple nesting: any time a rule can be expressed in the form, "if A_1, then (if A_2 then (...(if A_n then B)...))," it may be equivalently worded, "if (A_1 and A_2 and...A_n), then B," and vice versa.

Still another useful property is known as a distributive law. This asserts the equivalence of the two forms, "(if A then B) and (if A then C)," and "if A then both B and C." Again, it is possible to extend this concept to systems of more than two rules, so that if a system of rules includes a number of restrictions dependent upon the same precondition, it may be re-formulated as a single rule with the precondition, whose restriction consists of the conjunction of the several restrictions in the original system. Thus, the system,

(if A then B_1)
and (if A then B_2)
and (if A then B_3)

is equivalent to the single statement, "if A, then (B_1 and B_2 and B_3)," and vice versa.

2.5 Canonical Structure for Fuzzy Rule Systems

By using the properties developed in the previous section, plus a few other easily derived ones, we can transform any set of rules (as we have been describing them here) into a format which admits a clear, simple representation and a straightforward method of evaluation.

We shall define a canonical rule structure as follows: a canonical rule structure is a conjunction of rules R_1, R_2,...R_n, such that any unconditional restrictions appear before any conditional rules. Here, we define an unconditional restriction as any membership function which applies regardless of the value of any input.

Conditional rules consist of two parts: a precondition, which is a membership function based only upon the inputs; and a restriction, which may be any canonical rule structure. Note that in particular, the restriction may be a membership function applied to the proposed action. It will be convenient to represent conditional rules in the following format (although this is merely for notational simplicity, not for any theoretical advantage):

IF condition C
THEN restriction R.

Note that "restriction R" might in fact be either a single restriction or a set of rules all of which are contingent upon condition C (recall the distributive property discussed in Section 2.3.2).

The following is an example of a more complex canonical rule structure:

- restriction R_1
- restriction R_2
- restriction R_3
- IF condition C_1
  - THEN restriction R_4
  - restriction R_5
- IF condition C_2
  - THEN
    - restriction R_6
    - IF condition C_3
      - THEN restriction R_7
    - IF condition C_4
      - THEN restriction R_8
- IF condition C_5
  - THEN
    - IF condition C_6
      - THEN restriction R_9.

For simplicity, all canonical structures longer than a single restriction have been marked by vertical brackets. The whole system consists of six rules--three unconditional and three conditional ones. Each of the three conditional rules has as its restriction another canonical structure; note that the condition under whose heading the structure appears applies to the entire structure (again, recall the distributive property).

Another way of structuring the same set of rules (and one which is logically equivalent) might be the following:

restriction R_1
restriction R_2
restriction R_3
<u>IF</u> (C_1) <u>THEN</u> R_4
<u>IF</u> (C_1) <u>THEN</u> R_5
<u>IF</u> (C_2) <u>THEN</u> R_6
<u>IF</u> (C_2 <u>and</u> C_3) <u>THEN</u> R_7
<u>IF</u> (C_2 <u>and</u> C_4) <u>THEN</u> R_8
<u>IF</u> (C_5 <u>and</u> C_6) <u>THEN</u> R_9.

The two formats are both canonical structures for the same set of rules; furthermore, there are a number of alternate equivalent formulations which are also canonical structures. The latter format has the desirable property that there is exactly one rule for each restriction, although there is a certain amount of inefficiency involved in repeating the conditions C_1 and C_2 on several lines; nonetheless, once the various conditions have been evaluated (i.e., once the various membership functions have been applied to the input data), this format is the most compact representation possible. A further advantage is that all the rules have been reduced either to unconditional restrictions or to simple (rather than nested) conditional rules. Therefore, the compact form is quite simple to evaluate numerically. In fact, the procedure involves evaluating the various simple rules (according to the methods described in Section 2.3), and then taking the minimum performance score over the entire set of rules (according to the procedure in Section 2.4.1).

Since the compact canonical form is so simple, one might well wonder about the purpose of ever bothering with canonical forms which do include nested conditions (we shall call such forms "open" forms). The answer is that the psychology of human reasoning sacrifices the compactness of representation in favor of a simpler, more natural mode of thinking. Thus, for example, in the example we have been studying, the entire canonical structure headed by <u>"IF</u> condition C_2" represents a coherent pattern to the human expert; such a pattern might well be represented on a macroscopic level by a name or number, and its specific structure spelled out only for the benefit of a non-expert.

Such "chunking" of information is a vital part of the human ability to encompass far more information than a more naive (but compact) strategy would allow, and to communicate a great deal of information effectively. It would be ridiculous, for example, if a quarterback had to describe the entire conditional structure of each play in a football game, rather than refer to it by number. Nonetheless, it would be theoretically possible to transform that knowledge into a set of compact rules, if for some reason a novice with a super-human information-processing capacity but no experience had to replace the quarterback.

It also turns out that in some situations it might be important to limit the number of input data assessments to a minimum. Referring back to the prototype rule structure, if condition C_2 turned out to be completely false, we could skip over the entire canonical structure it dominates, thereby saving us the effort of evaluating restrictions R_6, R_7, and R_8, and conditions C_3 and C_4.

There is one final important reason for allowing open-form canonical structures. A number of instances arise in which, after a number of specific conditions and their corresponding restrictions have been enumerated, the expert would like to use the phrase "or else" to indicate the restrictions which would apply should none of the specified conditions be satisfied. Such a "default option" would prove particularly useful in the event that there were a large number of special situations which demanded particular actions, but also an overall set of background rules or "standard operating procedures" to be followed unless one of the special conditions was in fact observed. The formula for interpreting the condition "ELSE" is simple the degree to which none of the "IF" conditions preceding it (within the same canonical structure) is satisfied, or

$$1 - \max_i [\mu_{C_i}].$$

It is sufficient to treat the word "else" in this context as an abbreviation for the phrase "if none of the above conditions is satisfied," rather than as a new concept. Therefore, we need no new theoretical analysis to extend our definition of canonical forms to include "ELSE" clauses; there is, however, the restriction that within each canonical structure, an "ELSE" clause may appear only when there is at least one "IF" clause, and then only once, as the last clause in the structure.

To illustrate, again using our prototype canonical model, there would be three places at which one could insert an "ELSE" clause. The following represents the open-form canonical model augmented to include all three:

```
restriction R1
restriction R2
restriction R3
IF  condition C1
    THEN  restriction R4
          restriction R5
IF  condition C2
    THEN  restriction R6
          IF  condition C3
              THEN  restriction R7
```

```
                 IF  condition C4
                          THEN  restriction R8
                 ELSE  restriction R10
IF  condition C5
     THEN  IF  condition C6
                    THEN  restriction R9
                 ELSE  restriction R11
ELSE        restriction R12.
```

Note that each occurrence of "ELSE" is the last rule of a canonical structure (although not necessarily of the complete system). Of course, just like any other restriction, restrictions R_{10}, R_{11}, and R_{12} could be replaced by more complicated canonical structures.

In summary, canonical structures are operationally convenient methods of representing a system of rules in such a manner that they can easily be elicited from experts and applied to actual situations in order to evaluate proposed actions according to a well-defined set of algorithms or formulas. A given set of rules may have a number of different equivalent canonical representations, including some which are compact and easy to evaluate quickly, and others which are more extensive, but closer to the natural style of human thought. Since it is easy to transpose one canonical structure into another, it will be advantageous to shift back and forth, using both kinds of structure according to their convenience.

3.0 A SAMPLE FUZZY PRODUCTION RULE SYSTEM

The following example illustrates how the principles developed in Section 2.0 can be applied in a simple control situation. Of course, the situation is hypothetical and the rules a bit stylized, but the overall process is a fair representation of the kind of modelling involved in fuzzy policy capturing.

The situation we shall describe involves controlling the temperature of a room, by any combination of three methods:

- opening or closing the window;
- adjusting the heating unit; and
- turning the air conditioning on or off.

The information which might potentially affect such controls could concern the current temperature of the room, how heavily it is raining (if at all), and the present settings of the three controls (see Figure 3-1). In the set of fuzzy rules described below, the preconditions will mention only these input data; the restrictions will regulate the three controls which together specify the action taken, depending on the various input values.

3.1 Formal Description of the Temperature Control Problem

3.1.1 Action space - Each action shall be characterized by a vector containing three quantities:

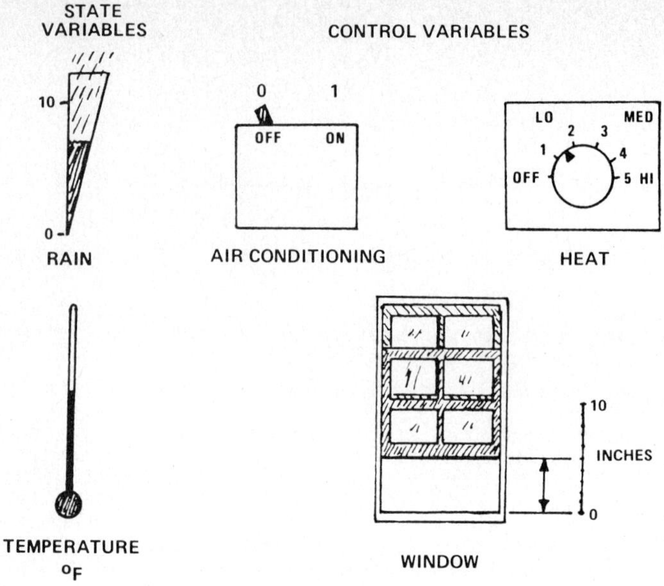

Figure 3-1
TEMPERATURE CONTROL PROBLEM

- the desired window opening in inches;
- the heat setting, which may range from 0 ("off") to 5 ("high"), continuously; and
- the air conditioning state desired, which may be either 0 ("off") of 1 ("on").

3.1.2 Inputs - Five quantities will need to be assessed or measured:

- the room temperature in degrees (Fahrenheit);
- the rate of rainfall, in millimeters per hour;
- the current window opening in inches;
- the current heat setting (between 0 and 5); and
- the current air conditioning state (0 or 1).

3.1.3 Generic rule system (in open canonical form) - Figure 3-2 contains the hypothetical set of rules which might be assessed from a temperature control "expert." Each of the conditions and restrictions mentioned will be defined as a fuzzy membership function, as shown in Figure 3-3. When specific values are determined for the inputs and the action dimensions, each of those expressions can be evaluated, and then the performance score for

```
IF   it is raining hard
     THEN    close the window
IF   the room temperature is hot
     THEN
             IF    the heat is on
                   THEN    turn the heat lower
             IF    (the window is closed) and (the air conditioning is off)
                   and (it is not raining heavily)
                   THEN    open the window
             IF    (the window is open) and (the air conditioning is off)
                   THEN    close the window
                           turn the air conditioning on
IF   the room temperature is cold
     THEN
             IF    the air conditioning is on
                   THEN    turn the air conditioning off
                           IF    the room temperature is very cold
                                 THEN    open the window
             ELSE
                           IF    the window is open
                                 THEN    close the window
                           ELSE
                                 turn the heat higher
ELSE
     no change in window, heat, or air conditioning
```

Figure 3-2

A SET OF RULES FOR REGULATING ROOM TEMPERATURE

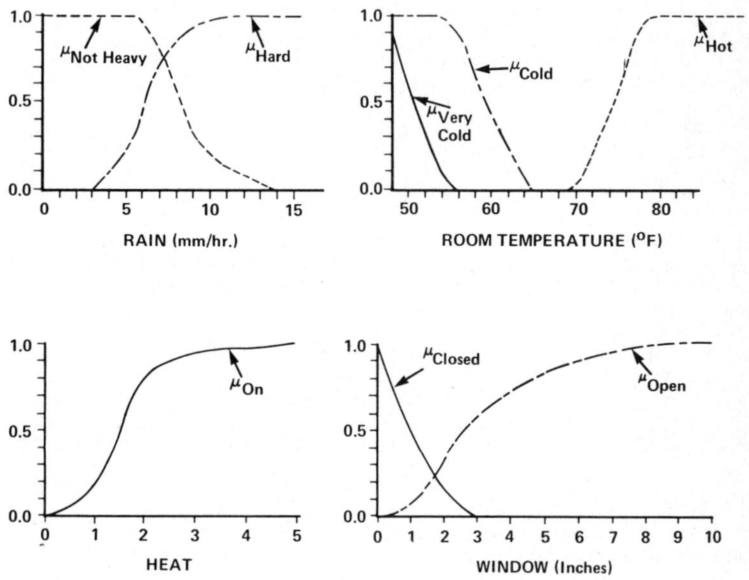

Figure 3-3

MEMBERSHIP FUNCTIONS FOR TEMPERATURE CONTROL PROBLEM

for the entire system can be calculated, according to the principles developed in Section 2.0.

3.2 A Sample Instance

Here, we shall specify some hypothetical input values, and a possible action to be evaluated, and then proceed through the evaluation.

3.2.1 Input values - Let us assume that

- the room temperature is 57°,
- it is raining at the rate of 5 mm/hr,
- the window is currently open 1 inch,
- the heat is presently off (i.e., zero), and
- the air conditioning is off (i.e., zero).

3.2.2 Action to be evaluated - Suppose the proposed action is to close the window completely and turn the heat to "low"; in particular, the action values chosen will be as follows:

- the new window opening will be 0 inches,
- the new heat setting will be 1.4, and
- the air conditioning will remain at 0.

3.2.3 Evaluation of conditions and restrictions - By looking up the values of the membership functions shown in Figure 3-3, where the input and action values correspond to the ones given in the two preceding sections, we can assign a numerical value to each of the conditions and restrictions in the model.

Thus, for example, the first condition ("the rain is hard") is satisfied "at the 0.2 level" by the actual rainfall specified (assuming that 5 mm/hr represents a light drizzle), since the membership function evaluated at 5 mm/hr equals 0.2. Similarly, the restriction "close the window" is completely satisifed (i.e., at the 1.0 level) by the given action; the value of the corresponding membership function at 0 inches is 1.0. In a like manner, all the other conditions and restrictions can be assigned numerical values.

3.2.4 Calculating an overall performance score - There are a number of equivalent methods for arriving at the performance score; the one we shall illustrate is perhaps the simplest. Our overall strategy will be to transform the open-form model in Figure 3-2 into a compact-form model, substituting numerical values for the various conditions and restrictions, according to the evaluation performed in Section 3.2.3. Figure 3-4 shows the open-form model with evaluations for the various formulas in parentheses, and Figure 3-5 contains the compact canonical form arrived at by applying the distributive law (see Section 2.4.2) to the model (the verbal conditions are omitted for clarity).

Now, for each compact rule, we can calculate the level at which the preconditions are jointly satisfied (by taking the

```
IF   it is raining hard (0.2)
     THEN    close the window (1.0)
IF   the room temperature is hot (0.0)
     THEN
             IF   the heat is on (0.0)
                  THEN    turn the heat lower (0.0)
             IF   (the window is closed) and (the air conditioning is off)
                  and (it is not raining heavily) (0.5)
                  THEN    open the window (0.0)
             IF   (the window is open) and (the air conditioning is off) (0.1)
                  THEN    close the window (1.0)
                          turn the air conditioning on (0.0)
IF   the room temperature is cold (0.8)
     THEN
             IF   the air conditioning is on (0.0)
                  THEN    turn the air conditioning off (1.0)
                          IF   the room temperature is very cold (0.0)
                               THEN   open the window (0.0)
             ELSE (1.0)
                          IF   the window is open (0.1)
                               THEN   close the window (1.0)
                          ELSE (0.9)
                               turn the heat higher (0.9)
ELSE (0.2)
     no change in window, heat, or air conditioning (0.6).
```

Figure 3-4

TEMPERATURE CONTROL RULES WITH NUMERICAL MEMBERSHIP VALUES

	Condition		*Restriction*	*Score*
IF	(0.2)	THEN	(1.0)	1.0
IF	(0.0 and 0.0)	THEN	(0.0)	1.0
IF	(0.0 and 0.5)	THEN	(0.0)	1.0
IF	(0.0 and 0.1)	THEN	(1.0)	1.0
IF	(0.0 and 0.1)	THEN	(0.0)	1.0
IF	(0.8 and 0.0)	THEN	(1.0)	1.0
IF	(0.8 and 0.0 and 0.0)	THEN	(0.0)	1.0
IF	(0.8 and 1.0 and 0.1)	THEN	(1.0)	1.0
IF	(0.8 and 1.0 and 0.9)	THEN	(0.9)	0.9
IF	(0.2)	THEN	(0.6)	0.8*
			Total performance score:	0.8

*The critical rule (i.e., the one which was least satisfied) was "else, no change in window, heat, or air conditioning"; to improve overall score, it is necessary to conform better to the restriction it imposes.

Figure 3-5

COMPACT-FORM MODEL WITH NUMERICAL SCORES

minimum over the values for the individual conditions); then, treating that minimum value as the precondition, we can obtain a performance score as if the rule were a simple conditional rule (see Section 2.3.2). These values appear in the right-hand column of Figure 3-5.

Finally, the overall performance score is easily calculated by taking the minimum score over the various individual compact rules, as shown at the bottom of Figure 3-5. The model's usefulness, however, extends well beyond this simple performance score: it can also be clearly seen, by referring back to the model, just which rule received this minimum value; furthermore, one can determine exactly which aspect or aspects of the action would have to be modified in order to improve the overall score (inputs, of course, cannot be changed as they represent data observed prior to the decision).

In the example shown here, the rule with the lowest score is the very last one ("ELSE no change in settings"). To improve overall performance, we would have to improve performance on this rule; therefore, if there is any better action, it is in the direction of a less extreme change in the settings (i.e., the proposed action constitutues an overrreaction to the given situation).

The procedures illustrated here will prove sufficient to evaluate actions for any well-structured fuzzy production rule model (i.e., any one which can be phrased in a canonical form), however large. It is not necessarily possible with every set of inputs in every model for a single action to achieve a perfect score on all the rules simultaneously; instead, that action or group of actions must be located for which no further improvement on the overall performance is possible.

4.0 SUMMARY AND CONCLUSION

A general-purpose fuzzy production rule system has been described. We have gone on to develop and test a prototype computerized fuzzy policy capturing system. A critical component of this system is the optimizing routine that controls the automatic generation of candidate action and terminates when it has determined the action or actions resulting in the highest possible overall performance scores. With the necessary support, we hope to continue our theoretical and applied efforts in the area of fuzzy policy capturing.

Reference

E. H. Mamdani: Advances in the linguistic synthesis of fuzzy controllers, International Journal of Man-Machine Studies 8 (1976) 669-678.

ADVANCES IN FUZZY SET THEORY AND APPLICATIONS
M.M. Gupta, R.K. Ragade, R.R. Yager (editors)

FUZZY CLUSTERING WITH A FUZZY COVARIANCE MATRIX

Donald E. GUSTAFSON and William C. KESSEL

Scientific Systems, Inc.
186 Alewife Brook Parkway
Cambridge, MA 02138

A class of fuzzy ISODATA clustering algorithms has been developed previously which includes fuzzy means. This class of algorithms is generalized to include fuzzy covariances. The resulting algorithm closely resembles maximum likelihood estimation of mixture densities. It is argued that use of fuzzy covariances is a natural approach to fuzzy clustering. Experimental results are presented which indicate that more accurate clustering may be obtained by using fuzzy covariances.

1. INTRODUCTION

The notion of fuzzy sets, first put forth by Zadeh [1], is an attempt to modify the basic conception of a space--that is, the set on which the given problem is defined. By introducing the concept of a fuzzy, i.e., an unsharply defined set, a different perspective is provided for certain problems in systems analysis, including pattern recognition.

One of the significant difficulties in development of a systematic approach to pattern recognition is that the phenomena of interest are modeled by equations which contain functions and operators which may appear simple and natural, but which yield some solutions which could be regarded as pathological. The difficulty stems from our desire to differentiate between classes in a manner which is simple and easy to visualize. In doing so, we restrict the solutions in an unknown way. The use of fuzzy sets is an attempt to ameliorate this problem.

Pattern classification problems have provided impetus for the development of fuzzy set theory. Recently, fuzzy sets have provided a theoretical basis for cluster analysis with the introduction of fuzzy clustering. The use of fuzzy sets in clustering was first proposed in [2] and several classification schemes were developed [3]. The first fuzzy clustering algorithm was developed in 1969 by Ruspini [4], and used by several workers [5]. Following this, Dunn [6] developed the first fuzzy extension of the least-squares approach to clustering and this was generalized by Bezdek [7] to an infinite family of algorithms.

Several problems in medical diagnosis have been attacked using fuzzy clustering algorithms. Adey [8] achieved promising results in interpreting EEG patterns in cerebral systems. Bezdek [9] has studied its use in differentiating hiatal hernia and gallstones. It appears that medical diagnosis may be an especially fruitful area of application for fuzzy clustering, since biological systems are extremely complex and the boundaries between "distinct" medical diagnostic classes are not sharply defined. This has been suggested for cardiovascular investigations [10].

In a "hard" clustering algorithm, each pattern vector must be assigned to a single cluster. This "all or none" membership restriction is not a realistic one, since many pattern vectors may have the characteristics of several classes. It is more natural to assign to each pattern vector a set of probabilities, one for each class.

The implication of this is that the class boundaries are not "hard" but rather are "fuzzy." Another problem is that the set of all partitions resulting from a "hard" clustering algorithm is extremely large, making an exhaustive search extremely complicated and expensive. Fuzzy clustering will generally lead to more computational tractability [11]. Another advantage of fuzzy clustering is that troublesome or outlying members of the data set are more readily recognized than with hard clustering, since the degree of membership is continuous rather than "all-or-none." Bezdek and Dunn [12] have noted the relationship of fuzzy clustering to estimating mixture distributions, but retained the Euclidean metric. Here, a generalization to a metric which appears more natural is made through the use of a fuzzy covariance matrix.

2. PROBLEM FORMULATION

The definition of a fuzzy partition used here agrees with that of Ruspini [4], Dunn [6] and Bezdek [13] and is a natural extension of the conventional partitioning definition. An ordinary, or "hard" partition is a k-tuple of Boolean functions $w(\cdot) = \{w_1, w_2, \ldots, w_k\}$ on the feature space $\Gamma \subset R^n$ which satisfy

$$w_j(x) = 0 \text{ or } 1, \quad \forall\, x \in \Gamma,\ 1 \leqq j \leqq k \tag{1}$$

$$\sum_{j=1}^{k} w_j(x) = 1 \quad \forall\, x \in \Gamma \tag{2}$$

If Γ_j represents the j-th class, with $\Gamma_i \cap \Gamma_j = \phi\ \forall\, i \neq j$ and $\cup_{j=1}^{k} \Gamma_j = \Gamma$, then $w_m(x) = 1$ means that $x \in \Gamma_m$ and (2) insures that x is a member of precisely one class. It is possible to pass from this definition to a corresponding fuzzy partition by retaining (2) but replacing (1) with the relaxed condition $0 \leqq w_j \leqq 1$. Thus, a fuzzy partition is a k-tuple of membership functions

$$w(\cdot) = \{w_1(x), w_2(x), \ldots, w_k(x)\}$$

which satisfy

$$0 \leqq w_j(x) \leqq 1, \quad \forall\, x \in \Gamma,\ 1 \leqq j \leqq k \tag{3}$$

$$\sum_{j=1}^{k} w_j(x) = 1, \quad \forall\, x \in \Gamma \tag{4}$$

Equation (3) suggests a probabilistic interpretation for the membership functions, as discussed by Ruspini [4]. However, this may or may not be a correct interpretation.

In devising a conventional clustering algorithm, one typically looks for a scalar performance index which attains its minimum for a partition which maximally separates the naturally-occurring clusters. There should exist a feasible algorithm for minimizing the performance index. The weighted within-class squared error is a useful performance measure.

Denote the distance from a point x to the j-th class by

$$d_j(x) = d(x, \theta_j); \quad d_j(x) > 0, \tag{5}$$

where the j-th class is parameterized by θ_j. For an indexed set of samples $x_1, x_2, x_3, \ldots, x_N$ we denote the distance measure and membership function by

$$d_j(x_i) = d_{ij}, \quad w_j(x_i) = w_{ij} \tag{6}$$

We are interested in minimizing the following cost:

$$J(w,\theta) = \sum_{i=1}^{N} \sum_{j=1}^{k} w_{ij}{}^{\alpha} d_{ij} \;; \quad \alpha \geqq 1 \tag{7}$$

where $\theta = \{\theta_j\}$, $w = \{w_{ij}\}$, k is the number of classes, and α is a smoothing parameter which controls the "fuzziness" of the clusters. For $\alpha = 1$, the clusters are separated by hard partitions and

$$w_{ij} = 0 \text{ or } 1 \tag{8}$$

As α increases, the partitions become more fuzzy.

3. DETERMINATION OF FUZZY CLUSTERS

3.1 Determination of Optimal Membership Functions

Now consider the problem of minimizing J with respect to (fuzzy) w, subject to $\alpha>1$ and the constraints (3) and (4). We defer for later the determination of the optimal parameters by minimizing J over θ. Constraint (3) may be eliminated by setting

$$w_{ij} = S_{ij}{}^2 \tag{9}$$

with S_{ij} real. Using (9), we adjoin the constraints (3) and (4) to J with a set of Lagrange multipliers $\{\lambda_i\}$ to give

$$\bar{J}(S,\theta,\lambda) = \sum_{i=1}^{N} \sum_{j=1}^{k} S_{ij}{}^{2\alpha} d_{ij} + \sum_{i=1}^{N} \lambda_i \left(\sum_{j=1}^{k} S_{ij}{}^2 - 1 \right) \tag{10}$$

The first-order necessary conditions for optimality are found by setting the gradients of $\bar{J}$ with respect to S to zero. Now,

$$\frac{\partial \bar{J}}{\partial S_{ij}} = 2\alpha S_{ij}{}^{2\alpha-1} d_{ij} + 2S_{ij}\lambda_i \tag{11}$$

By setting $\frac{\partial \bar{J}}{\partial S_{ij}}$ to zero we obtain the following first-order necessary conditions:

$$S_{ij}^{*}(\alpha S_{ij}^{*}{}^{2(\alpha-1)} d_{ij} + \lambda_i^{*}) = 0; \quad \forall\ i,j \tag{12}$$

$$\sum_{j=1}^{k} S_{ij}^{*}{}^2 = 1; \quad \forall\ i \tag{13}$$

where the asterisk denotes association with optimality.

Equations (12)-(13) comprise a set of $Nk + N$ equations which can be solved for the $Nk + N$ unknowns $w^{*} = \{w_{ij}{}^{*}\}$, and $\lambda^{*} = \{\lambda_i^{*}\}$. We proceed by assuming that $S_{ij}^{*} \neq 0\ \forall\ i,j$. This is consistent with the assumption that $\alpha > 1$. With this assumption we have

$$w_{ij}^{*} = (-\lambda_i^{*}/\alpha d_{ij})^{1/(\alpha-1)} \tag{14}$$

By summing over j and using (4)

$$(-\lambda_i^{*})^{1/(\alpha-1)} = \frac{1}{\sum_{j=1}^{k} \left(\frac{1}{\alpha d_{ij}}\right)^{1/(\alpha-1)}} \tag{15}$$

and (14) becomes

$$w^*_{ij} = \frac{1}{\sum_{\ell=1}^{k} (d_{ij}/d_{i\ell})^{1/(\alpha-1)}} \tag{16}$$

Then, from (7), <u>for any θ</u>, the associated extremum of $J(w,\theta)$ is

$$J^*(\theta) = \min_{w} J(w,\theta)$$

$$= \sum_{i=1}^{N} \left[\sum_{j=1}^{k} (d_{ij})^{1/(1-\alpha)} \right]^{1-\alpha} \tag{17}$$

Note that (17) holds for any values of the parameters θ.

<u>Limiting Case When $\alpha \to 1$</u>

If $\alpha \to 1$,

$$J \to \sum_{i=1}^{M} \sum_{j=1}^{k} w_{ij} d_{ij} \tag{18}$$

and the argument given by Dunn [6] will establish that $\forall$ i,k

$$w^*_{ik} \to \begin{cases} 1; & d_{ik} = \min_j (d_{jk}) \\ 0; & \text{otherwise} \end{cases} \tag{19}$$

<u>Addition of Risk Factors</u>

Let

$$J = \sum_{i=1}^{N} \sum_{j=1}^{k} \gamma_i w_{ij}{}^{\alpha} d_{ij} \tag{20}$$

with γ_i a risk factor for x_i.
Then by defining $\tilde{d}_{ij} = \gamma_i d_{ij}$ we have

$$J = \sum_{i=1}^{N} \sum_{j=1}^{k} w_{ij}{}^{\alpha} d_{ij} \tag{21}$$

which is of the form of the original cost function (7). Thus a patient indexed risk factor can be included naturally within the framework of the original problem by simply altering the distance measures. A high value of γ assigned to several records will tend to force their pattern vectors closer to the cluster centroids by changing the cluster shape and size.

3.2 <u>Determination of Optimal Parameters</u>

We now turn to the problem of finding the optimal parameter set $\theta^* = \{\theta^*_1, \theta^*_2, \ldots, \theta^*_k\}$. From (7) we have

$$\frac{\partial}{\partial \theta_j} \bar{J}(w,\theta,\lambda) = \sum_{i=1}^{N} w_{ij}{}^{\alpha} \frac{\partial}{\partial \theta_j} d_{ij} \tag{22}$$

The first-order necessary conditions for a local minimum of J are (12), (13) and

$$\sum_{i=1}^{N} w^*_{ij}{}^{\alpha} \frac{\partial}{\partial \theta_j} d_{ij} \Big|_{*} = 0 \ \forall \ j \tag{23}$$

To proceed we need to specify the parameterization of d_{ij}.

Fuzzy ISODATA

Let

$$d_{ij} = (x_i - \theta_j)^T A(x_i - \theta_j); \quad A > 0 \tag{24}$$

Then (23) gives

$$\sum_{i=1}^{N} w^{*\,\alpha}_{ij}(x_i - \theta^*_j) = 0 \;\forall\; j \tag{25}$$

This is equivalent to

$$\theta^*_j = \frac{\sum_{i=1}^{N} w^{*\,\alpha}_{ij} x_i}{\sum_{i=1}^{N} w^{*\,\alpha}_{ij}}; \quad j = 1,\ldots,k \tag{26}$$

$$\triangleq m_{fj}$$

We will call m_{fj} the fuzzy mean of class j in recognition of its limiting property under hard partitioning. This case comprises fuzzy ISODATA [14].

Hard ISODATA

As $\alpha \to 1$ and the partitioning becomes hard

$$w^{*\,\alpha}_{ij} \to \begin{cases} 1; & j = m \\ 0; & j \neq m \end{cases} \tag{27}$$

where

$$d_{im} = \min_j d_{ij} \tag{28}$$

That is, under the one-nearest-neighbor rule, $w^{*\,\alpha}_{ij}\Big|_{\alpha=1} = 1$ for all pattern vectors x_i assigned to class j and is zero otherwise. Thus, for hard partitioning

$$\sum_{i=1}^{N} w^*_{ij} = N_j \tag{29}$$

where N_j is the number of pattern vectors assigned to Γ_j and

$$\theta^*_j\Big|_{\alpha \to 1} \to \frac{1}{N_j} \sum_{x_i \in \Gamma_j} x_i \tag{30}$$

$$= \hat{m}_j$$

where $\hat{m}_j$ is the sample mean of Γ_j. This is the hard k-means algorithm; it constitutes the basic idea underlying hard ISODATA [15].

3.3 Generalization to Include Fuzzy Covariance

Now consider replacing (24) by an inner product induced norm metric of the form

$$d_{ij}(\theta_j) = (x_i - v_j)^T M_j(x_i - v_j); \quad 1 \leq j \leq k \tag{31}$$

with M_j symmetric and positive-definite. If $\theta_j = v_j$, equation (26) for θ^*_j still holds [14]. If, however, we take $\theta_j = \{v_j, M_j\}$, a class of algorithms more general than ISODATA will ensue. Note that J is now linear in M_j, giving a singular problem. The cost J may be made as small as desired by simply making M_j less positive-definite. To get a feasible solution, we must constrain M_j in some manner. Ideally we would like the metric to handle different scalings along each direction in feature space. That is, we would like to allow variations in the shape of each class induced by the metric but not let the metric grow without bound. A way of accomplishing this by using only one parameter is to constrain the determinant $|M_j|$ of the matrix M_j. This induces a volume constraint.

Consider the set of constraints

$$|M_j| = \rho_j, \quad \rho_j > 0 \tag{32}$$

with ρ_j fixed for each j. The augmented cost is now

$$J(w,\theta,\lambda,\beta) = \sum_{i=1}^{N} \sum_{j=1}^{k} w_{ij}{}^{\alpha} d_{ij}(\theta_j) + \sum_{i=1}^{N} \lambda_i \left(\sum_{j=1}^{k} w_{ij} - 1 \right) + \sum_{j=1}^{k} \beta_j \left(|M_j| - \rho_j \right) \tag{33}$$

where $\{\beta_j\}$ is a set of Lagrange multipliers.

The partial derivatives with respect to w_{ij} are the same as before. However, the partial derivatives with respect to θ_j now change.

From (33), the necessary conditions are

$$\left.\frac{\partial \bar{J}}{\partial v_j}\right|_{*} = -2 \sum_{i=1}^{N} w_{ij}{}^{\alpha} M_j (x_i - v_j^*) = 0; \quad j = 1,2,\ldots,k \tag{34}$$

which is equivalent to (25) and

$$\left.\frac{\partial \bar{J}}{\partial M_j}\right|_{*} = 0 = \sum_{i=1}^{N} w_{ij}{}^{\alpha} (x_i - v_j)(x_i - v_j)^T + \beta_j |M_j^*| M_j^{*-1} \tag{35}$$

To get (35), we have used the identities

$$\frac{\partial}{\partial A}(x^T A x) = x x^T, \qquad \frac{\partial}{\partial A}|A| = |A| A^{-1}$$

which hold for a non-singular matrix A and any compatible vector x. Equation (34) gives (26) again:

$$v_j^* = \frac{\sum_{i=1}^{N} w_{ij}{}^{\alpha} x_i}{\sum_{i=1}^{N} w_{ij}} \tag{36}$$

For the optimal membership functions $(w_{ij} = w_{ij}^*)$, v_j^* is the fuzzy mean of Γ_j. Equation (35) gives, for $v_j = v_j^*$,

$$M_j^{*-1} = \frac{1}{\beta_j |M_j^*|} \sum_{i=1}^{N} w_{ij}{}^{\alpha} (x_i - v_j^*)(x_i - v_j^*)^T \tag{37}$$

Now define the <u>fuzzy covariance</u> matrix for Γ_j by

$$P_{fj} = \frac{\sum_{i=1}^{N} w_{ij}{}^{\alpha}(x_i - m_{fj})(x_i - m_{fj})^T}{\sum_{i=1}^{N} w_{ij}{}^{\alpha}}; \quad \alpha > 1 \tag{38}$$

Then, using (38) in (37) gives

$$M_j^{*-1} = \frac{\sum_{i=1}^{N} w_{ij}{}^{\alpha}}{\beta_j |M_j^*|} |P_{fj}| \tag{39}$$

The β_j are found by using (38), which yields

$$M_j^{*-1} = \left[\frac{1}{\rho_j |P_{fj}|}\right]^{1/n} P_{fj} \tag{40}$$

where n is the feature space dimension. In the sequel, a hard covariance matrix refers to P_{fj} of (38), evaluated at $\alpha = 1$. In view of (27), a hard covariance matrix is simply the sample class covariance matrix under the cluster assignment rule (28).

The previous discussion suggests the following iterative algorithm for finding stationary points of $J(w,\theta)$. Given data $\{x_i\}$ and an initial partition $\theta_j^{(0)} = \{m_{fj}^{(0)}, P_{fj}^{(0)}\}$, we proceed as follows:

(i) compute $\{d_i(\theta_j{}^{(k)})\}$ using (31).

(ii) compute $\{w_{ij}^{(k)}\}$ using (16). If $d_{ik} = 0$ for some k, set $w_{ik} = 1$, $w_{i\ell} = 0 \ \forall \ell \neq k$.

(iii) compute new estimates $\theta_j^{(k+1)}$ using (36), (38) and (40). Recycle to (i) until a specified convergence criterion is satisfied.

4. RELATION TO MAXIMUM LIKELIHOOD ESTIMATION

There is an intimate relationship between fuzzy ISODATA algorithms and algorithms designed to estimate mixture density parameters under the Gaussian assumption. The objective of this discussion is to demonstrate this relationship explicitly and to discuss the implications which it has to problems of unsupervised learning.

Maximum likelihood estimation of parameters has been studied for a long time (see, e.g., Rao [16]), and the theory is quite well understood. The problem in applications is developing numerical techniques which can efficiently solve, or approximately solve, the problem. The development here follows the work of Wolfe [17].

Let $p(x|\Gamma_j)$ be the probability density for the random vector $x \in R^n$, conditioned on x being a member of the j-th class $(x \in \Gamma_j)$, and let P_j be the a priori probability associated with Γ_j. We assume that Γ_j is parameterized by a set of parameters $\theta_j \in R^s$ and that $p(x|\Gamma_j)$ is a twice differentiable function of θ_j. Since x can be associated with more than one class, it has a mixture density function which is, for k classes,

$$p(x) = \sum_{j=1}^{k} P_j p(x,\theta_j) \tag{41}$$

where

$$p(x,\theta_j) = p(x|\Gamma_j) \tag{42}$$

$$\sum^k P_j = 1 \tag{43}$$

The "probability of membership" of x in class j can be found by using Bayes' Rule:

$$p(\Gamma_j|x) = \frac{P_j p(x,\theta_j)}{p(x)} \tag{44}$$

Now suppose a sample of N random vectors is drawn from the mixture and denote these by $x_1,x_2,x_3,\ldots,x_N$. Then

$$p(x_1,x_2,\ldots,x_N) = \prod_{i=1}^{N} p(x_i) \tag{45}$$

and the log probability is

$$\log p(x_1,x_2,\ldots,x_N) = \sum_{i=1}^{N} \log p(x_i) \tag{46}$$

The maximum likelihood estimate of the parameters $\theta = \theta_1,\theta_2,\ldots,\theta_k$ is found by solving

$$\max_{\theta}[\log p(x_1,x_2,\ldots,x_N)] \tag{47}$$

subject to the constraint (43).

Thus, forming the augmented cost

$$L = \sum_{i=1}^{N} \log p(x_i) + \lambda(\sum_{j=1}^{k} P_j - 1) \tag{48}$$

We wish to find the parameters $\theta = \theta^*$ to maximize L. By setting $\frac{\partial L}{\partial P_j} = 0$, we obtain

$$\sum_{i=1}^{N} p(\Gamma_j|x_i) + P_j = 0 \tag{49}$$

Summing (49) over j gives $\lambda = -N$. Thus, a first order necessary condition for maximum likelihood estimation is

$$P_j^* = \frac{1}{N} \sum_{i=1}^{N} p^*(\Gamma_j|x_i) \tag{50}$$

Likewise, we can obtain, by setting $\frac{\partial L}{\partial \theta} = 0$, the other first order necessary condition

$$\sum_{i=1}^{N} p^*(\Gamma_j|x_i) \frac{\partial}{\partial \theta_j} \log p^*(x_i,\theta_j{}^*) = 0 \tag{51}$$

Now consider the special case where x is conditionally Gaussian distributed. Then

$$\log p(x,\theta_j) = -\frac{n}{2} \log 2\pi + p \log |E_j{}^{-1}| - \tfrac{1}{2}(x-m_j)^T E_j{}^{-1}(x-m_j) \tag{52}$$

where $\theta_j = \{m_j, E_j\}$ and E_j is assumed nonsingular.

Taking the indicated partial derivatives in (51) and using the fact that $\lambda = -N$, we obtain the following three equations which describe the necessary conditions to be satisfied for the maximum likelihood estimates:

$$m_j^* = \frac{1}{NP_j^*} \sum_{i=1}^{N} p(x_i, m_j^*, E_j^*) x_i \tag{53}$$

$$E_j^* = \frac{1}{NP_j^*} \sum_{i=1}^{N} p(x_i, m_j^*, E_j^*)(x_i - m_j^*)(x_i - m_j^*)^T \tag{54}$$

$$P_j^* = \frac{1}{N} \sum_{i=1}^{N} p(x_i, m_j^*, E_j^*) \tag{55}$$

The first order necessary conditions for fuzzy clustering and maximum likelihood estimation possess similarities which can be studied by imbedding both solutions in a larger class of solutions.

Consider the following set of algebraic relations:

$$Q_j = \frac{1}{N} \sum_{i=1}^{N} q_{ij}; \quad 0 \leqq q_{ij} \leqq 1$$

$$\eta_j = \frac{1}{NQ_j} \sum_{i=1}^{N} q_{ij} x_i; \quad x_i \in R^n, \ \eta_j \in R^n$$

$$M_j = \frac{\gamma_j}{NQ_j} \sum_{i=1}^{N} q_{ij} r_{ij} r_{ij}^T$$

$$r_{ij} = x_i - \eta_j$$

where N is a positive integer and γ_j is a positive scalar.

The parameter q_{ij} is the membership function of x_i relative to class j and Q_j is the average membership for class j. Thus, q_{ij} increases as x_i comes closer to class j and relatively large values of Q_j are associated with the largest or most dense classes. The parameter η_j can be regarded as the nucleus point of class j and M_j is a matrix which describes the shape and size of the class. The parameter r_{ij} is the distance from x_i to the class j nucleus. The parameters r_{ij}, M_j are combined into a measure d_{ij} which is used to evaluate the distance of x_i to the class j:

$$d_{ij} = r_{ij}^T M_j^{-1} r_{ij}$$

The values of q_{ij} and the associated constraints for fuzzy clustering and maximum likelihood estimation are summarized in Table 1. The parameter D_i is a normalization constant for x_i and C_j is a normalization constant for Γ_j. Note that q_{ij} decreases monotonically with increasing d_{ij} for both cases. It is also interesting to note that membership functions are normalized differently. With fuzzy clustering, normalization is done over the classes to get D_i, whereas normalization under maximum likelihood estimation is done over the whole space R^n to obtain c_j. Thus, q_{ij} is given a slightly different interpretation in the two methods. The constraints are quite different: a class volume constraint is used under fuzzy clustering whereas a total probability constraint is used under maximum likelihood estimation.

Even with these differences, there is a striking similarity between the two methods. Note in particular that the fuzzy covariance matrix appears naturally in the problem and appears to be more appropriate than a hard covariance matrix.

We now consider the structure of a classifier using the q_{ij}'s from either maximum-likelihood or fuzzy ISODATA. The decision rule by which x_i is assigned to a class is as follows:

> Assign x_i to class m if $q_{im} \geqq q_{ij}$; $j = 1,2,\ldots,k$. In case of ties, assign x_i to the least-numbered class.

Table 1

Comparison of Fuzzy Clustering and Maximum Likelihood Solutions

Fuzzy Clustering	parameter or condition	Maximum Likelihood
w_{ij}^{α} $(\alpha \geqq 1)$	$q_{ij} = q_j(x_i)$	p_{ij}
$w_{ij} = d_{ij}^{1/(1-\alpha)}/D_i$		$p_{ij} = C_j \exp[-d_{ij}/2]$
$\sum_{j=1}^{k} w_{ij} = 1 \Rightarrow D_i \; \forall \; i$	normalization	$\int_{X \in R^n} p(x)dx = 1 \Rightarrow C_j \; \forall \; j$
$\lvert M_j \rvert = \rho_i \Rightarrow \gamma_j$	constraints	$\sum_{j=1}^{k} Q_j = 1$

5. FUZZY CLUSTERING EXPERIMENTS

A fuzzy clustering algorithm has been developed, implemented, and tested on two types of data. First, experiments were run on labeled vectorcardiogram data to assess the differences between cardiologist-determined classes and classes determined via fuzzy clustering. The second set of experiments was made using two stylized classes which had some degree of overlap.

5.1 Experiments with Vectorcardiograms

A set of 1000 labeled vectorcardiograms (VCG) was obtained from the U.S. Air Force School of Aerospace Medicine. The pattern vector for each record was comprised of the coefficients of a 60th order Karhunen-Loeve expansion (20/axis) [18,19].

5.1.1 Performance Measure

In order to assess the performance of fuzzy clustering relative to labeled data, the following measure was used. Let $w_{ij} = w_j(x_i)$ and assume the label is $x_i \in \Gamma_i$. Let $w_{i\ell} = \max_j w_{ij}$ (i.e., $x_j \rightarrow$ cluster ℓ) and denote

$$e_{1i} = \begin{cases} 0; & \ell = k \\ w_{i\ell} - w_{ik}; & \ell \neq k \end{cases}$$

The selected basis of comparison is the error probability:

$$e_i = I(e_{1i})/N_k$$

where $I(\gamma)$ is the indicator function for γ ($I(\gamma) = 1$ if $\gamma > 0$, $I(\gamma) = 0$ if $\gamma = 0$) and N_k is the number of points in Γ_k. The overall performance measure used was the mean value of e_i (averaged over the whole data base).

5.1.2 Numerical Experiments on Actual VCG Data

The initial covariance matrix for each class was $P_0 = \beta I$, where I is the identity matrix and β was a positive scalar. Updating of the covariance matrices was done using (38) and then scaling P_{fj} to keep a constant determinant ($|P_0|$). The clusters were seeded by using the first record in each class as the seed. This was equivalent to random seeding. A total of 29 diagnostic classes were used. A summary of the results of several trials is given in Table 2. The 13 classes shown had the highest number of members and were thus the most representative statistically. The number in each class is shown in parentheses. The average error probability is shown for both hard and soft partitions and dimensions of 4 and 10. Note that the results are mixed: no single set of parameters is uniformly best for all cases. However, the objective of studying unsuperivsed learning in problems including labeled data is not for purposes of classification per se, but rather to study the natural tendencies of the labeled classes to cluster in feature space. The partition for separating two small widely-separated classes is relatively independent of the parameter α. Thus, e_2 is small and does not vary much with α for easily separated classes. Examples are LBBB, RBBB and LVH. However, if the classes are overlapping e_2 will be high and can change significantly with α. Examples are RBBB w/RAD, increased anterior forces and increased initial forces. The effect of feature space dimension can also be evaluated. Some classes have smaller overlap in 4-space (increased initial and decreased anterior forces) while others have larger overlap in 4-space (RBBB, LBBB, LVH) and some are relatively unaffected. These results give some idea as to how complex the classes are in feature space.

5.2 Two-Class Test Case

In order to study the inherent differences in the various clustering methods under study and the effect of varying the clustering parameters, a simple two-class test case was set up. The two classes are intersecting and consist of two long and narrow regions at right angles to one another in a cross pattern. The two cluster centroids coincide exactly so that the discrimination must be based on cluster shape information. In order to test the algorithms, a total of ten points in each class were chosen randomly, using a uniform distribution over each class. These points are depicted in Figure 1, with points numbered 1 to 10 selected from Class 1 and points numbered 11 to 20 selected from Class 2. All tests were run assuming two classes a priori.

The first test was made using hard ISODATA with the cluster seeds coinciding at (0,0). The resulting cluster assignments are shown in Figure 2. As expected the performance was quite bad, due to the fact that class shape is not accounted for explicitly. The algorithm converged after three iterations.

The next test was run using hard ISODATA seeded with the class means. The resulting assignments are shown in Figure 3 and are again poor for the same reason. The algorithm converged after only two passes.

Table 2

Fuzzy Clustering Performance with VCG Data

$P_0 = .001\ I$, full updating

Numerical values are average error probability

CLASS (No.)	DESCRIPTION	DIM = 10			α = 2	
		α = 1	α = 2	α = 4	DIM = 4	DIM = 10
1 (46)	normal	0.85	0.91	0.87	0.89	0.91
4 (30)	RBBB	0.37	0.20	0.13	0.70	0.20
6 (5)	RBBB w/RAD	0.60	0.40	1.00	0	0.40
7 (15)	LBBB	0.20	0.27	0.13	0.53	0.27
8 (13)	terminal IVCD	0.71	0.71	0.71	1.00	0.71
12 (14)	T wave changes	0.64	0.57	0.57	0.79	0.57
13 (17)	ST depression or straightening	0.83	0.83	0.94	0.89	0.83
14 (4)	early repolarization	0.40	0.40	0.60	0.40	0.40
15 (15)	LVH	0.18	0.18	0.29	0.59	0.18
23 (32)	increased anterior forces	0.84	0.94	0.56	0.94	0.94
24 (5)	decreased anterior forces	0.80	1.00	1.00	0.40	1.00
25 (9)	increased initial forces	0.67	1.00	0.78	0.56	1.00
29 (31)	ST + T (Class 12 + Class 13)	0.28	0.66	0.19	0.84	0.66

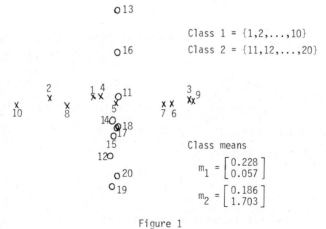

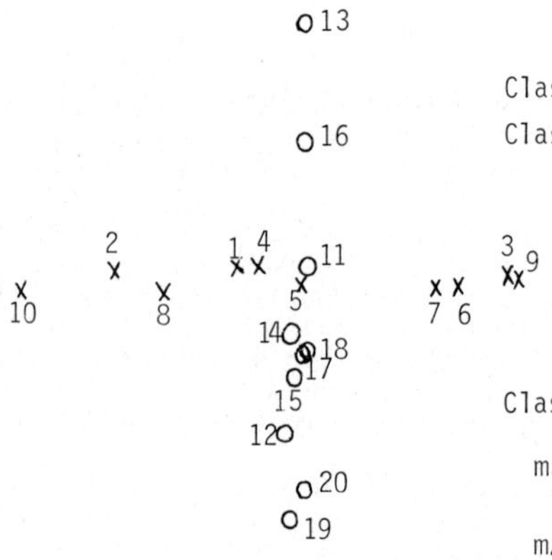

Class 1 = {1,2,...,10}

Class 2 = {11,12,...,20}

Class means

$$m_1 = \begin{bmatrix} 0.228 \\ 0.057 \end{bmatrix}$$

$$m_2 = \begin{bmatrix} 0.186 \\ 1.703 \end{bmatrix}$$

Figure 1

Two-Class Configuration

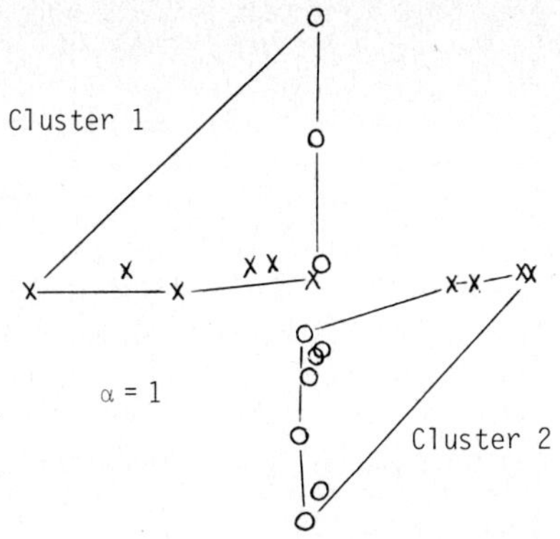

Figure 2

Cluster Assignments Using Hard ISODATA with Seeds: $S_1 = (0,0)$, $S_2 = (0,0)$

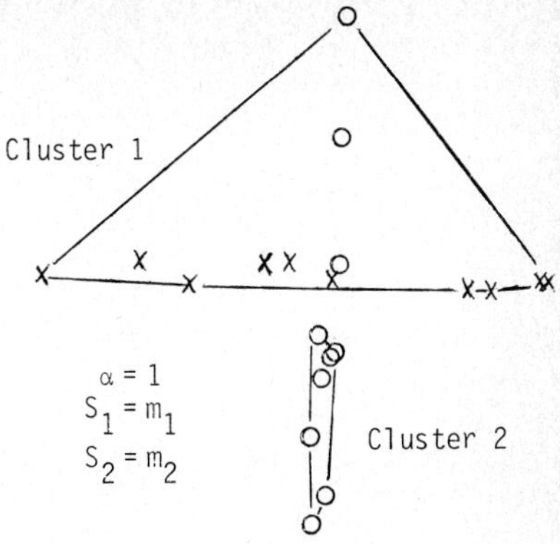

Figure 3

Cluster Assignments Using Hard ISODATA Seeded with Class Means

The next test used fuzzy ISODATA, in which the means were fuzzy but the covariance matrices all set to identity at each pass. The resulting clusters are shown in Figure 4 and are considerably different from the desired result. Cluster 1 is very large and Cluster 2 is very small, encompassing only three peripheral points of Class 1. Convergence was obtained in 4 passes.

A test was next run using fuzzy clustering with $\alpha = 2$ and using the fuzzy covariance matrix. The clusters were seeded at the actual class means. The class assignments are shown in Figure 5 and are seen to be correct for all points, although the results for #5 and #11 would appear fortuitous. The difficulty in classifying these two points is apparent from the values of their membership functions. Thirteen passes were required to meet the convergence criterion that the maximum change in any membership function be less than 0.001 in magnitude.

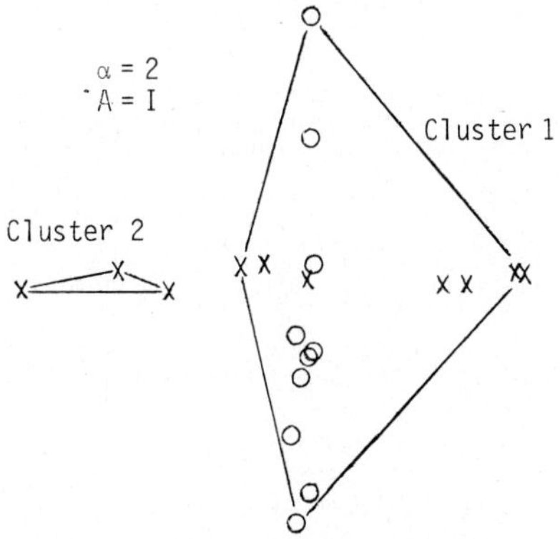

Figure 4

Cluster Assignments Using Fuzzy ISODATA with Seeds: $S_1=(0.001,0)$, $S_2=(0,0)$

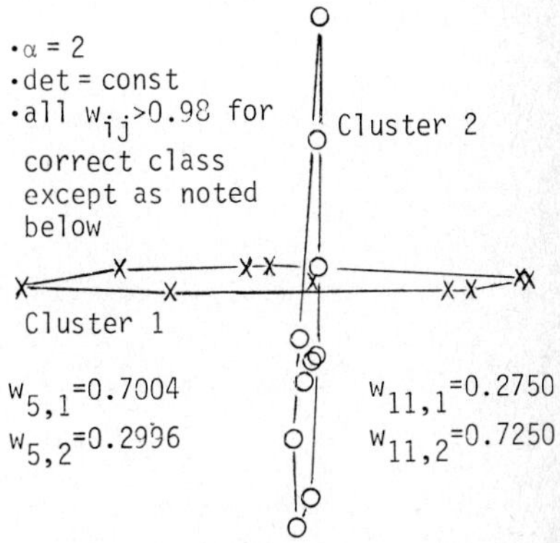

Figure 5

Cluster Assignments Using Fuzzy Clusters Seeded at Class Means

The next run was similar to the previous run except that the cluster seeds were set at $S_1 = (0.0001,0)$ and $S_2 = (0,0)$, which were used in the fuzzy ISODATA run, in order to make the discrimination more difficult. The discrimination was, in fact, more difficult. From the 9th to the 13th pass, the assignments were as shown in Figure 6(a). However, after 20 passes, the algorithm did converge to the configuration of Figure 6(b). As before, all of the assignments were correct. However, the way in which the clusters were formed was quite interesting. The histories of the membership functions for several critical points are given in Table 3 and demonstrate the nature of the iterative process. Note that $w_{3,1}$, $w_{4,1}$, $w_{10,1}$, $w_{13,2}$ and $w_{19,2}$ generally increase and approach a value of unity. This is the desired behavior and is expected for points which lie much closer to one class than the other. Note that the response of $w_{13,2}$ is relatively slow, staying close to 0.5 until the 15th pass and then increasing monotonically. Thus, for the first 14 passes point #13 is about equally distant from both clusters. Point #11 is strongly associated with Cluster 1 on the 11th through 15th passes. However, once point #10 is correctly assigned to Cluster 1, $w_{11,2}$ increases monotonically to its final value. Note that point #11 is strongly associated with Cluster 1 at pass 15 and 17, indicating that Cluster 2 does not start to form correctly until the 18th pass.

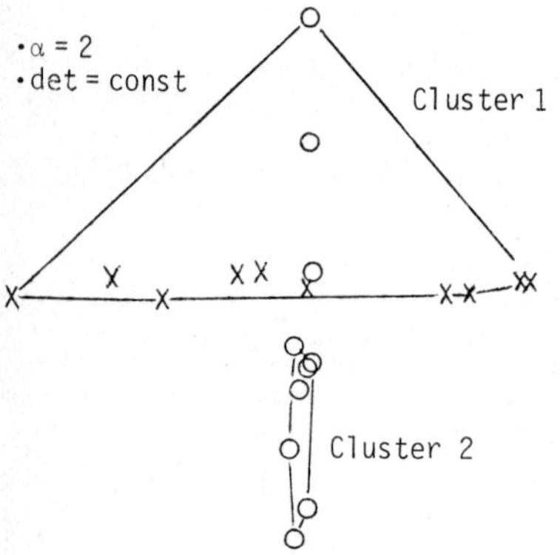

Figure 6(a)
Cluster Assignments Using Fuzzy Clusters with Seeds: $S_1=(0.0001,0)$, $S_2=(0,0)$, 9th - 13th pass

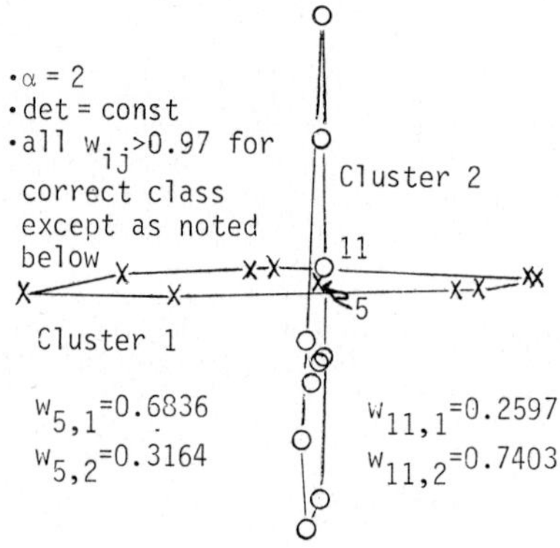

Figure 6(b)
Cluster Assignments Using Fuzzy Clusters with Seeds: $S_1=(0.0001,0)$, $S_2=(0,0)$ 20th pass (converged)

Table 3

Membership Function Histories for Case Shown in Figure 6

Pass	$w_{3,1}$	$w_{4,1}$	$w_{5,1}$	$w_{10,1}$	$w_{11,2}$	$w_{13,2}$	$w_{14,2}$	$w_{19,2}$
1	.5001	.5004	.5015	.5001	.4993	.5000	.5000	.5000
7	.5029	.5001	.5153	.5002	.4907	.4996	.5119	.5022
10	.5218	.5495	.7047	.5085	.3676	.4934	.6599	.5341
15	.8690	.9574	.8664	.8757	.0104	.6992	.9353	.9373
17	.9905	.9521	.7921	.9899	.2424	.9680	.9808	.9905
18	.9972	.9509	.6850	.9968	.6268	.9911	.9794	.9965
20	.9988	.9606	.6836	.9985	.7403	.9949	.9715	.9975

The effect of using a fuzzy covariance matrix was studied by running a case differing from the previous one only in the way the covariance matrix was calculated. A hard covariance matrix was used instead of a fuzzy one. The solution is shown in Figure 7 and was obtained after eight passes. Note that points #4 and #11 are incorrectly classified. The failure to correctly assign #11 is hardly surprising but the misassignment of #4 is judged to be a clustering error. This result suggests that the use of fuzzy covariances can enhance clustering performance. Further numerical testing is required to verify this behavior in general.

It is interesting to note that the configuration of Figure 7 is relatively insensitive to the distance measure used. A run was made in which the distance measure

$$1 - \exp(-d_{ij}/2)$$

was used rather than d_{ij} and the same cluster assignments were obtained. It should also be noted that no problems of convergence were encountered in any runs.

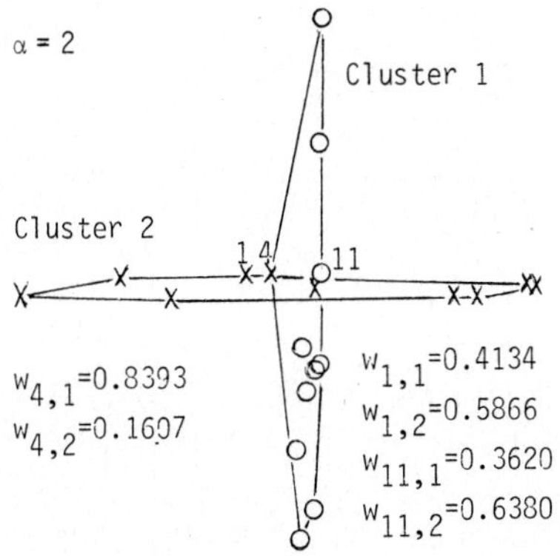

Figure 7

Cluster Assignments Using Fuzzy Clustering with Hard Covariances, Seeded at S_1=(0.001,0), S_2=(0,0)

REFERENCES

[1] Zadeh, L. A., Fuzzy Sets, Information and Control, 8 (1965) 338-353.

[2] Bellman, R. E., R. A. Kalaba and L. A. Zadeh, Abstraction and Pattern Classification, Jrnl. Math. Anal. Appl., 13 (1966) 1-7.

[3] Gitman, I. and M. Levine, An Algorithm for Detecting Unimodal Fuzzy Sets and its Application as a Clustering Technique, IEEE Trans. Computers, C-19 (1970) 917-923.

[4] Ruspini, E. H., A New Approach to Clustering, Information and Control, 15 (1969) 22-32.

[5] Larsen, L., E. Ruspini, J. McDew, D. Walter and W. Adey, A Test of Sleep Staging Systems in the Unrestrained Chimpanzee, Brain Research, 40 (1972) 319-343.

[6] Dunn, J., A Fuzzy Relative of the ISODATA Process and its Use in Detecting Compact Well-Separated Clusters, Jrnl. Cybernetics, 3 (1974) 32-57.

[7] Bezdek, J., Fuzzy Mathematics in Pattern Classification, PhD Thesis, Cornell University (1973).

[8] Adey, W., Organization of Brain Tissue: Is the Brain a Noisy Processor? Int. Jrnl. Neuroscience, 3 (1972) 271-284.

[9] Bezdek, J., Feature Selection for Binary Data - Medical Diagnosis with Fuzzy Sets, National Computer Conference (1976).

[10] Kalmanson, D. and H. F. Stegall, Cardiovascular Investigations and Fuzzy Sets Theory, Amer. Jrnl. of Cardiology, 35 (1975) 30-34.

[11] Bezdek, J., Cluster Validity with Fuzzy Sets, Jrnl. Cybernetics, 3 (1974) 58-73.

[12] Bezdek, J. and J. C. Dunn, Optimal Fuzzy Partitions: A Heuristic for Estimating the Parameters in a Mixture of Normal Distributions, IEEE Trans. Computers, C-24 (1975) 835-838.

[13] Bezdek, J. C., Numerical Taxonomy with Fuzzy Sets, Jrnl. Math. Biology, 1 (1974) 57-71.

[14] Bezdek, J. C. and P. F. Castelaz, Prototype Classification and Feature Selection with Fuzzy Sets, IEEE Trans. Systems, Man and Cybernetics, SMC-7 (1971) 87-92.

[15] Ball, G. H., Classification Analysis, Stanford Research Institute report AD-716-482 (Nov. 1970).

[16] Rao, C. R., Advanced Statistical Methods in Biometric Research (Wiley and Sons, New York, 1952).

[17] Wolfe, J. H., Pattern Clustering by Multivariate Mixture Analysis, Multivariable Behavioral Research, (1970) 329-350.

[18] Halliday, J. S., The Characterization of Vectorcardiograms for Pattern Recognition, M.S. Thesis, MIT (1973).

[19] Womble, M. E., J. S. Halliday, S. K. Mitter, M. C. Lancaster and J. H. Triebwasser, Data Compression for Storing and Transmitting ECGs/VCGs, Proc. IEEE, 65 (1977) 702-706.

A 1979 BIBLIOGRAPHY ON FUZZY SETS, THEIR APPLICATIONS, AND RELATED TOPICS

Abraham KANDEL
Florida State University
Tallahassee, FL, U.S.A.

and

Ronald R. YAGER
Iona College
New Rochelle, NY, U.S.A.

This bibliography represents our attempt to publish a complete and comprehensive list of studies in fuzzy sets, their applications, and related topics. This Volume of items, following Zadeh's original in 1965, illustrates, in our opinion both the significance and the rate of growth of this field.

The bibliography originated from a 1977 bibliography by Gaines and Kohout and is maintained and formated on a CDC Cyber 74 such that updates and printouts are simple and quick. This final product represents such a printout.

We are grateful to the many people involved in this project. Special thanks are due to Robert Coleman and Don Bucley of Iona College, and Douglas Schlak and Paul Hanna of Florida State University for their help and interest in this work.

Since the bibliography can very easily be appended and corrected we request the help of researchers in this field in correcting faulty entries that appear in our list, or updating the list by providing information on missing items.

EDITOR'S NOTE:

The first extended bibliography on Fuzzy Systems from 1965 to the middle of 1976 containing 763 entries was compiled by Gains and Kohout and it appeared in the book "Fuzzy Automata and Decision Processes," North-Holland, 1977. This bibliography contains 1,799 entries and represents a further attempt to compile the references in the field from its inception (1965) until the end of 1978. This was compiled at a very short notice; naturally the readers will find that some articles are missing whereas some other articles have appeared more than once. In spite of all this, Kandel and Yager undoubtedly deserve an appreciation for their efforts. Entries from 489 to 500 are missing; this is due to an attempt made to correct some entries.

1 ACKERMANN, R. (1967) INTRODUCTION TO MANY VALUED LOGICS, ROUTLEDGE AND KEGAN PAUL, LONDON.

2 ACZEL, J., AND PFANZAGL, J. (1966) REMARKS ON THE MEASUREMENT OF SUBJECTIVE PROBABILITY AND INFORMATION, METRIKA,5,91-105.

3 ACZEL, M.J. (1948) SUR LES OPERATIONS DEFINIES POUR LES NOMBRES REELS, BULL. SOC. MATH. FRANCAISE, 76, 59-64.

4 ADAMEK, J., AND WECHLER, W. (1976) MINIMIZATION OF R-FUZZY AUTOMATA, IN STUDIEN ZUR ALGEBRA UND IHRE ANWENDUNGEN, AKAD.-VERL., BERLIN.

5 ADAMO, J.-M. (1978) UNE IMPLEMENTATION DE LA THEORIE DES SOUS-ENSEMBLES FLOUS - APPLICATION A L'ANALYSE DE PROCESSUS DE DECISION, THESE D'ETAT ES SCIENCES, PRESENTED AT THE UNIV. CLAUDE BERNARD, LYON, MARCH.

6 ADAMO, J.M., AND KARSKY, M. (1978) APPLICATION OF FUZZY LOGIC TO THE DESIGN OF A BEHAVIORAL MODEL IN AN INDUSTRIAL ENVIRONMENT, UNIV. CLAUDE-BERNARD LYON I, MAITRISE MIAG, 43, BOULEVARD DU 11 NOV. 1918, 69621, VILLEURBANNE, FRANCE.

7 ADAMO, J.M., AND KARSKY, M. (1978) APPLICATION OF FUZZY LOGIC TO THE DESIGN OF AN INDUSTRIAL ENVIRONMENT, UNIV. CLAUDE-BERNARD LYON I, FRANCE, 4TH INT. CONGRESS ON CYBERNETICS AND SYST., AMSTERDAM.

8 ADAMS, E.W. (1965) ELEMENTS OF A THEORY OF INEXACT MEASUREMENT, PHILOS. SCI.,32, 205-228.

9 ADAMS, E.W. (1966) PROBABILITY AND THE LOGIC OF CONDITIONALS, IN HINTIKKA, J., AND SUPPES, P. (EDS.), ASPECTS OF INDUCTIVE LOGIC, NORTH-HOLLAND, AMSTERDAM, 265-316.

10 ADAMS, E.W. (1974) THE LOGIC OF " ALMOST ALL ", J. PHILOS. LOGIC,3,3-17.

11 ADAMS, E.W., AND LEVINE, H.P. (1975) ON THE UNCERTAINTIES TRANSMITTED FROM PREMISES TO CONCLUSIONS IN DEDUCTIVE INFERENCES, SYNTHESE, 30, 429-460.

12 ADAVIC, P.N., BORISOV, A.N., AND GOLENDER, V.E. (1968) AN ADAPTIVE ALGORITHM FOR RECOGNITION OF FUZZY PATTERNS, IN D.S. KRISTINKOV,J.J.OSIS, L.A. RASTRIGIN (EDS.), KIBERNETIKA I DIAGNOSTIKA,2,13-18,(IN RUSSIAN) ZINATNE, RIGA, U.S.S.R. .

13 ADEY,W.R. (1972) ORGANIZATION OF BRAIN TISSUE: IS THE BRAIN A NOISY PROCESSOR?, INT. J. NEUROLOGY,3,271-284.

14 AIDA, S. (1975) INFORMATICS IN " ECO - TECHNOLOGY ", IN SUMMARY OF PAPERS ON GENERAL FUZZY PROBLEMS, THE WORKING GROUP ON FUZZY SYSTEMS, TOKYO, JAPAN, NOV.,1-4.

15 AIZERMANN, M.A. (1975) FUZZY SETS, FUZZY PROOFS AND SOME UNSOLVED PROBLEMS IN THE THEORY OF AUTOMATIC CONTROL, SPECIAL INTEREST DISCUSSION SESSION ON FUZZY AUTOMATA AND DECISION PROCESSES, 6TH IFAC WORLD CONGRESS, BOSTON, MASS., USA, AUG. .

16 AIZERMANN, M.A. (1976) FUZZY SETS, PROOFS AND CERTAIN UNSOLVED PROBLEMS IN THE THEORY OF AUTOMATIC CONTROL, AUTOMATIKA I TELENEHANIKA, P. 171-177.

17 AIZERMANN, M.A. (1977) SOME UNSOLVED PROBLEMS IN THE THEORY OF AUTOMATIC CONTROL AND FUZZY PRCOFS, IEEE TRANS. ON AUTO. CONT., P.116-118.

18 AIZERMANN, M.A., AND SMIRNOVA, I.N. (1978) FIRST MONOGRAPH ON THE THEORY OF FUZZY SETS, AUTOMATIC CONTROL 10, P. 1581-1583.

19 ALBERT, P. (1977) THE ALGEBRA OF FUZZY LOGIC, FUZZY SETS AND SYST. 1, 203-230, NORTH-HOLLAND.

20 ALBIN, M. (1975) FUZZY SETS AND THEIR APPLICATION TO MEDICAL DIAGNOSIS, PHD THESIS, DEPARTMENT OF MATHEMATICS, UNIVERSITY OF CALIFORNIA, BERKELEY, CALIFORNIA.

21 ALLEN, A.D. (1973) A METHOD OF EVALUATING TECHNICAL JOURNALS ON THE BASIS OF PUBLISHED COMMENTS THROUGH FUZZY IMPLICATIONS: A SURVEY OF THE MAJOR IEEE TRANSACTIONS, IEEE TRANS. SYST. MAN CYBERN., SMC-3, 422-425.

22 ALLEN, A.D. (1974) MEASURING THE EMPIRICAL PROPERTIES OF SETS, IEEE TRANS. SYST. MAN CYBERN., SMC-4, 66-73.

23 ALLEY, H., BACINELLO, C.P., HIPIL, K.W. (1978) FUZZY SET APPROACHES TO PLANNING IN THE GRAND RIVER BASIN, TO APPEAR ADVANCES IN WATER RESOURCES.

24 ALTHAM, J.E. (1971) THE LOGIC OF PLURALITY, METHUEN, LONDON.

25 AMANO, T., AND KUNII, T.L. (1974) A FUZZY ACCESS TO COLOR AND TEXTURE DATA IN A GRAPHIC APPLICATION SYSTEM, CONFERENCE ON COMPUTER GRAPHICS AND INTERACTIVE TECHNIQUES, COLORADO.

26 AMERONGEN, V., LEMKE, V.N., AND VENN, V. (1977) AN AUTOPILOT FOR SHIPS DESIGNED WITH FUZZY SETS, 5EME CONF. INT. IFAC/IFIP, LA HAYE.

27 ANDERSON, A.R., AND BELNAP, N.D. (1962) THE PURE CALCULUS OF ENTAILMENT, J. SYMBOLIC LOGIC, 27, 19-25.

28 ANDERSON, A.R., AND BELNAP, N.D. (1975) ENTAILMENT, PRINCETON UNIVERSITY PRESS, NEW JERSEY.

29 AOKI, Y. (1976) OPTIMIZATION OF ENVIRONMENTAL PLANNING UNDER THE RISK-AVERSION OF NON-REPAIRABLE DAMAGE, 1-13,

SUMMARY OF PAPERS ON GENERAL FUZZY PROBLEMS, THE WORKING GROUP ON FUZZY SYSTEMS, TOYKO, JAPAN.

30 ARBIB, M. (1977) BOOK REVIEW OF APPLICATIONS OF FUZZY SETS TO SYSTEMS ANALYSIS, SIAM 19, P.753.

31 ARBIB, M.A. (1967) TOLERANCE AUTOMATA, KYBERNETIKA (PRAGUE), 3, 223-233.

32 ARBIB, M.A. (1970) SEMIRING LANGUAGES, ELECTRICAL ENGINEERING DEPARTMENT, STANFORD UNIVERSITY, CALIFORNIA, USA.

33 ARBIB, M.A. (1975) FROM AUTOMATA THEORY TO BRAIN THEORY, INT. J. MAN-MACHINE STUDIES,7, 279-295.

34 ARBIB, M.A., AND MANES, E.G. (1974) FUZZY MORPHISMS IN AUTOMATA THEORY, PROC. FIRST INTERNATIONAL SYMPOSIUM ON CATEGORY THEORY APPLIED TO COMPUTATION AND CONTROL, 98-105.

35 ARBIB, M.A., AND MANES, E.G. (1975A) A CATEGORY-THEORETIC APPROACH TO SYSTEMS IN A FUZZY WORLD, SYNTHESE, 30, 381-406.

36 ARBIB, M.A., AND MANES, E.G. (1975B) FUZZY MACHINES IN A CATEGORY, BULLETIN AUSTRALIAN MATH. SOC., 13, 169-210.

37 ARIGONA, A.O. (1978) SEMANTIC IMPLICATION AND ITS UTILIZATION IN FUZZY SET THEORY, PRESENT AT 4TH CONF. ON CYBERN. ON SYS. RES., LINZ.

38 ARIGONI, A.O. (1976) MEMBERSHIP CHARACTERISTIC FUNCTION OF FUZZY ELEMENTS FUNDAMENTAL THEORICAL BASIS, 3RD EUR. MEETING CYBERN. SYST. RES., VIENNA.

39 ASAI, K., AND KITAJIMA, S. (1971) A METHOD FOR OPTIMIZING CONTROL OF MULTIMODAL SYSTEMS USING FUZZY AUTOMATA, INFORM. SCI., 3, 343-353.

40 ASAI, K., AND KITAJIMA, S. (1971) LEARNING CONTROL OF MULTIMODAL SYSTEMS BY FUZZY AUTOMATA, IN FU, K.S. (ED.), PATTERN RECOGNITION AND MACHINE LEARNING PLENUM PRESS, NEW YORK, 195-203.

41 ASAI, K., AND KITAJIMA, S. (1972) OPTIMIZING CONTROL USING FUZZY AUTOMATA, AUTOMATICA, 8, 101-104.

42 ASAI, K., AND TANAKA, H. (1973) ON THE FUZZY MATHEMATICAL PROGRAMMING, PROCEEDINGS OF THE 3RD IFAC SYMP. ON IDENTIFICATION AND SYST. PARAMETER ESTIMATION, PT. II, NORTH-HOLLAND, AMSTERDAM.

43 ASAI, K., AND TANAKA, H. (1975) APPLICATIONS OF FUZZY SETS THEORY TO DECISION-MAKING AND CONTROL, J. JAACE, 19, 235-242.

44 ASAI, K., TANAKA, H., AND OKUDA, T. (1975) DECISION MAKING

AND ITS GOAL IN A FUZZY ENVIRONMENT, IN ZADEH, L.A., FU, K.S., TANAKA, K., AND SHIMURA, M. (EDS), FUZZY SETS AND THEIR APPLICATIONS TO COGNITIVE AND DECISION PROCESSES, ACADEMIC PRESS, NEW YORK, 257-277.

45 ASAI, K., TANAKA, K., AND OKUDA, T. (1977) ON THE DISCRIMINATION OF FUZZY STATES IN PROBABILITY SPACE, KYBERNETES 6, 185-192.

46 ASAI, K.,TANAKA, H., NEGOITA, C.V., AND RALESCU, D.A. (1978) INTRODUCTION TO FUZZY SYSTEMS THEORY, OHM-SHA CO. LTD., TOKYO, (IN JAPANESE).

47 ASENJO, F.G. (1966) A CALCULUS FOR ANTINOMIES, NOTRE DAME J. FORMAL LOGIC, 7, 103-105.

48 ASSILIAN, S. (1974) ARTIFICIAL INTELLIGENCE IN THE CONTROL OF REAL DYNAMIC SYSTEMS, PHD THESIS, QUEEN MARY COLLEGE, UNIVERSITY OF LONDON.

49 ATKIN, R.H. (1974) MATHEMATICAL STRUCTURE IN HUMAN AFFAIRS, HEINEMANN, LONDON.

50 AUBIN, J.P. (1974B) THEORIE DE JEAUX, COMP. ROND. ACAD. SCI. (PARIS), 279, A-963.

51 AUBIN, J.P. (1974C) FUZZY GAMES, MRC TECHNICAL SUMMARY REPORT 1480, MATHEMATICAL RESEARCH CENTER, UNIVERSITY OF WISCONSIN-MADISON, MADISON, USA.

52 AUBIN, J.P. (1976) FUZZY CORE AND EQUILIBRIA OF GAMES DEFINED IN STRATEGIC FORM, IN HO, Y.C., AND MITTER, S.K. (EDS.), DIRECTIONS IN LARGE-SCALE SYSTEMS, PLENUM PRESS, NEW YORK, 371-388.

53 AURAY, J.P., DURU, G. (1976) INTRODUCTION OF THE THEORY OF MULTIFUZZY SPACES INT. OF MATH. ECON., UNIV. OF DYON, FRANCE.

54 AXINN,A., AND AXINN, D. (1976) NOTES ON THE LOGIC OF IGNORANCE RELATIONS, AMER. PHILOS. QUART., 13 135-143.

55 BAAS, S.M., AND KWAKERNAAK, H. (1977) RATING AND RANKING OF MULTIPLE-ASPECT ALTERNATIVES USING FUZZY SETS, AUTOMATICA 13, 1, 47-58.

56 BALLMER, T.T. (1976) FUZZY PUNCTUATION OR THE THEORY OF CONTINUING OF GRAMMATICALITY, ERL UNIV. OF CALIF., BERKELEY, MEMO NO. ERL-M590.

57 BANASCHEWSKA, B. (1968) INJECTIVE HULLS IN THE CATEGORY OF DISTRIBUTIVE LATTICES, JOURNAL FUR DIE REINE UND ANGEWANDTE MATHEMATIK, 102-109.

58 BANASCHEWSKA, B., AND BRUNS, G. (1967) CATEGORICAL CHARACTERIZATION OF THE MAC-NEVILLE COMPLETION, ARCHIV DER MATHEMATIK, 369-377.

59 BANG, S.Y., AND YEH, R.T. (1974) TOWARD A THEORY OF RELATIONAL DATA STRUCTURE, SELTR-1, UNIVERSITY OF AUSTIN,TEXAS, USA.

60 BAR-HILLEL, Y. (1964) LANGUAGE AND INFORMATION, ADDISON-WESLEY, READING, MASS., USA.

61 BARCAN, R.C. (1946) A FUNCTIONAL CALCULUS OF FIRST ORDER BASED ON STRICT IMPLICATION, J. SYMBOLIC LOGIC, 11, 1-16.

62 BARNES, G.R. (1976) FUZZY SETS AND CLUSTER ANALYSIS, PROCEEDINGS AT THE 3RD INT. JOINT CONF. ON PATTERN RECOGNITION, IEEE CAT. NO. 76CH1140-3C.

63 BARNEV, P., DIMITROV, V., AND STANCHEV, P. (1974) FUZZY SYSTEM APPROACH TO DECISION-MAKING BASED ON PUBLIC OPINION INVESTIGATION THROUGH QUESTIONAIRES, IFAC SYMPOSIUM ON STOCHASTIC CONTROL, BUDAPEST, SEP. .

64 BARTHELMY, J.P., BORCHUT, D. (1976) NOTES OF THE UNIV. SEMINAR ON FUZZY SETS, ED. BY LUONG AND MASSONITE, UNIV. OF BEZANCON, FRANCE.

65 BECKER, J.M. (1973) A STRUCTURAL DESIGN PROCESS, PHD THESIS, DEPARTMENT OF CIVIL ENGINEERING, UNIVERSITY OF CALIFORNIA, BERKELEY, CALIFORNIA, USA.

66 BELLMAN, R. (1974) LARGE SYSTEMS, IEEE TRANS. ON AUTOMATIC CONTROL, AC-19, P. 465.

67 BELLMAN, R.E. (1970) HUMOR AND PARADOX, IN MENDEL, W.M. (ED.) A CELEBRATION OF LAUGHTER, MARA BOOKS, LOS ANGELES, CALIFORNIA, 35-45.

68 BELLMAN, R.E. (1971) LAW AND MATHEMATICS, TECHNICAL REPORT 71-34, UNIVERSITY OF SOUTHERN CALIFORNIA, LOS ANGELES, USA, SEP. .

69 BELLMAN, R.E. (1973) MATHEMATICS AND THE HUMAN SCIENCES, IN WILKINSON, J., BELLMAN, R.E. AND GARAUDY, R. (EDS.), THE DYNAMIC PROGRAMMING OF HUMAN SYSTEMS, MSS INFORMATION CORP., NEW YORK, USA, 11-18.

70 BELLMAN, R.E. (1973) RETROSPECTIVE FUTUROLOGY: SOME INTROSPECTIVE COMMENTS, IN WILKINSON, J., BELLMAN, R.E., AND GARAUDY, R. (EDS.), THE DYNAMIC PROGRAMMING OF HUMAN SYSTEMS, MSS INFORMATION CORP., NEW YORK, USA,35-37.

71 BELLMAN, R.E. (1974) LOCAL LOGICS, TECHNICAL REPORT NO. USC EE RB 74-9, UNIVERSITY OF SOUTHERN CALIFORNIA, LOS ANGELES, USA.

72 BELLMAN, R.E. (1975) COMMUNICATION, AMBIGUITY AND UNDERSTANDING, MATH. BIOSCIENCES, 26, 347-356.

73 BELLMAN, R.E. AND MARCHI, E. (1973) GAMES OF PROTOCOL: THE CITY AS A DYNAMIC COMPETETIVE PROCESS, TECHNICAL REPORT RB73-36, UNIVERSITY OF SOUTHERN CALIFORNIA, LOS ANGELES, CALIFORNIA, USA.

74 BELLMAN, R.E., AND GIERTZ, M. (1973) ON THE ANALYTIC FORMALISM OF THE THEORY OF FUZZY SETS, INFORM. SCI., 5, 149-156.

75 BELLMAN, R.E., AND ZADEH, L.A. (1970) DECISION-MAKING IN A FUZZY ENVIRONMENT, MANAGEMENT SCI., 17, 141-164.

76 BELLMAN, R.E., AND ZADEH, L.A. (1977) LOCAL AND FUZZY LOGICS, IN : J. C. DUNN AND G. EPSTEIN (EDS.), MODERN USES OF MULTIPLE-VALUED LOGIC (D. REIDEL, DORDRECHT), 103-165.

77 BELLMAN, R.E., KALABA, R., AND ZADEH, L.A. (1966) ABSTRACTION AND PATTERN CLASSIFICATION, J. MATH. ANAL. AND APPLN., 13, 1-7.

78 BELLUCE, L.P. (1964) FURTHER RESULTS ON INFINITE VALUED PREDICATE LOGIC, J. SYMBOLIC LOGIC, 29, 69-78.

79 BELLUCE, L.P., AND CHANG,C.C. (1963) A WEAK COMPLETENESS THEOREM FOR INFINITE VALUED FIRST-ORDER LOGIC, J. SYMBOLIC LOGIC, 28, 43-50.

80 BELNAP, N.D. (1960) ENTAILMENT AND RELEVANCE, 'J. SYMBOLIC LOGIC, 25, 144-146.

81 BERAN, L. (1974) GRUPY A SVAZY, SNTL-TECHNICAL PUBLISHERS, PRAGUE, (IN CZECH: GROUPS AND LATTICES).

82 BERNDT, R.S., CARAMAZZA, A. (1977) THE DEVELOPMENT OF SOME ADVERBIAL MODIFIERS OF DIMENSIONAL ADJECTIVES, PROC. 2ND ANN. BOSTON UNIV. CONF. ON LANGUAGE DEVELOPMENT.

83 BERNDT, R.S., CARAMAZZA, A. (1978) THE DEVELOPMENT OF VAGUE MODIFIERS IN THE LANGUAGE OF PRE- SCHOOL CHILDREN, IN PRESS. CHILD LANG.

84 BERTOLINI, F. (1971) KRIPKE MODELS AND MANY VALUED LOGICS, SYMPOSIA MATHEMATICA, 113-131.

85 BERTONI, A. (1973) COMPLEXITY PROBLEMS RELATED TO THE APPROXIMATION OF PROBABILISTIC LANGUAGES AND EVENTS BY DETERMINISTIC MACHINES, IN NIVAT, M. (ED.), AUTOMATA, LANGUAGES AND PROGRAMMING, NORTH-HOLLAND, AMSTERDAM, 507-516.

86 BESSON-BERRANDO, M. (1977) A LOGICAL ANALYSIS OF SIMPLE AND FUZZY DECISIONS, 6TH EURO WORKING GROUP ON FUZZY SETS.

87 BEZDEK, J.C. (1973) FUZZY MATHEMATICS IN PATTERN CLASSIFICATION, PHD THESIS, CENTER FOR APPLIED MATHETICS,

CORNELL UNIVERSITY, ITHACA, NEW YORK, USA.

88 BEZDEK, J.C. (1974) CLUSTER VALIDITY WITH FUZZY SETS, 'J. CYBERNETICS, 3, 58-73.

89 BEZDEK, J.C. (1974) NUMERICAL TAXONOMY WITH FUZZY SETS, J. MATH. BIOLOGY, 1, 57-71.

90 BEZDEK, J.C. (1975) MATHEMATICAL MODELS FOR SYSTEMATICS AND TAXONOMY, IN ESTABROOK, G. (ED.), PROCEEDINGS 8TH ANNUAL INTERNATIONAL CONFERENCE ON NUMERICAL TAXONOMY, FREEMAN, SAN FRANCISCO.

91 BEZDEK, J.C. (1976) A PHYSICAL INTERPRETATION OF FUZZY ISODATA, IEEE TRANS. SYST. MAN CBBERN, 6, 387-389.

92 BEZDEK, J.C. (1976) FEATURE SELECTION FOR BINARY DATA: MEDICAL DIAGOSIS WITH FUZZY SETS, PROC. NATIONAL COMPUTER CONFERENCE, AFIPS PRESS, MONTVALE, NEW JERSEY, JUNE.

93 BEZDEK, J.C. (1977) FUZZY PREFERENCE RANKINGS, PRES. AT ORSA/TIMS MEETING, ATLANTA, GEORGIA.

94 BEZDEK, J.C. (1977) SOME CURRENT TRENDS IN FUZZY CLUSTERING, IN: J. WHITE, ED., THE GENERAL SYSTEMS PARADIGM (PROC. SGSR), 347-352.

95 BEZDEK, J.C. (1979) PARTICLE AND GRAIN SHAPE ANALYSIS WITH FUZZY SETS, IN PROC. 1ST RESIDENTIAL WORKSHOP ON PARTICULATE MORPHOLOGY, K. BEDDOW, ED., CRC PRESS.

96 BEZDEK, J.C., AND DUNN, J.C. (1975) OPTIMAL FUZZY PARTITIONS: A HEURISTIC FOR ESTIMATING THE PARAMETERS IN A MIXTURE OF NORMAL DISTRIBUTIONS, IEEE TRANS. COMP., C-24, 835-838.

97 BEZDEK, J.C., AND FORDON, W.A. (1978) ANAL. OF HYPERTENSIVE PATIENTS BY THE USE OF FUZZY ISODATA ALGORITHM, PROC. JACC, VOL. 3, 349-355.

98 BEZDEK, J.C., AND HARRIS, J. (1977) CONVEX DECOMPOSITIONS OF FUZZY PARTITIONS, INFORMATION SCI. .

99 BEZDEK, J.C., AND HARRIS, J.D. (1977) FUZZY PARTITIONS AND RELATIONS; AN AXIOMATIC BASIS FOR CLUSTERING, FUZZY SETS AND SYST. 1, 111-127, NORTH-HOLLAND.

100 BEZDEK, J.C., AND HARRIS, J.D. (1978) FUZZY RELATIONS AND PARTITIONS: AN AXIOMATIC BASIS FOR CLUSTERING, FUZZY SETS AND SYS. 1, 111-126.

101 BEZDEK, J.C., GUNDERSON, R., EHRLICH, R., AND MELOY, T. (1979) ON THE EXTENSION OF FUZZY K-MEANS ALGOR. FOR DETECTION OF LINEAR CLUSTERS, 17TH IEEE CONF. ON DEC. AND CONTROL, 1438-1443.

102 BEZDEK, J.C., GUNDERSON, R., EHRLICH, R., MELOY, T. (1978)

ON THE EXTENSION FOR DETECTION OF LINEAR CLUSTERS, 1438-1443, PROC. 1978 IEEE CONF. ON DEC. AND CONTROL, INCLUDES THE 17TH SYMP. ON ADAPTIVE PROCESSES, 78CH1392-OCS, JAN., 1979, SAN DIEGO, CALIF. .

103 BEZDEK, J.C., SPILLMAN, B., AND SPILLMAN, R. (1977) FUZZY MEASURES OF PREFERENCE AND CONSENSUS IN GROUP DECISION-MAKING, IN: K.S. FU, ED., PROC. 1977 IEEE CONF. ON DECISION AND CONTROL (IEEE PRESS, PISCATAWY, N.J.), 1303-1309.

104 BEZDEK, J.C., SPILLMAN, B., AND SPILLMAN, R., A MANIFOLD OF FUZZY RELATIONS FOR GROUP DECISION THEORY, IN REVIEW, INT. J. OF FUZZY SETS AND SYSTEMS.

105 BEZDEK, J.C., SPILLMAN, B., SPILLMAN, R. (1977) FUZZY MEASURES OF GROUP CONSENSUS AND UNCERTAINTY, PROCEEDINGS OF THE SYMP. ON FUZZY SET THEORY AND APPLN., IEEE CONF. ON DECISION AND CONTROL, NEW ORLEANS.

106 BEZDEK, J.C., SPILLMAN, B., SPILLMAN, R. (1978) A FUZZY RELATION SPACE FOR GROUP DECISION THEORY, FUZZY SETS AND SYST. 1, 255-268, NORTH-HOLLAND.

107 BEZDEK, J.C., SPILLMAN, B., SPILLMAN, R. (1979) FUZZY RELATION SPACES FOR GROUP DECISION THEORY: AN APPLICATION, FUZZY SETS AND SYS., VOL. 2, NO. 1, JAN., 5-14.

108 BEZDEK, J.C.,(TO APPEAR) FUZZY IMBEDDINGS THAT ARE CONVEX HULLS, J. CYBERNET. .

109 BEZDEK, J.C.,AND CASTELAS, P.F. (1977) PROTOTYPE CLASSIFICATION AND FEATURE SELECTION WITH FUZZY SETS, IEEE TRANS. SYST., MAN, CYBERNET. 2.

110 BIALNICKI-BIRULA, A. (1957) REMARKS ON QUASI-BOOLEAN ALGEBRAS, BULL. DE L'ACADEMIE POLONAISE DES SCIENCES, SER. MATH., ASTR. AND PHYS., 5, 615-619.

111 BIRKHOFF, G. (1948) LATTICE THEORY, AMERICAN MATHEMATICAL SOCIETY, RHODE ISLAND, USA.

112 BLACK, M. (1937) VAGUENESS: AN EXERCISE IN LOGICAL ANALYSIS, PHILOS. SCI., 4, 427-455.

113 BLACK, M. (1963) REASONING WITH LOOSE CONCEPTS, DIALOGUE,2, 325-373.

114 BLACK, M. (1968) THE LABYRINTH OF LANGUAGE, MENTOR BOOKS, NEW YORK, USA.

115 BLACK, M. (1970) MARGINS OF PRECISION, CORNELL UNIVERSITY PRESS, ITHACA, NEW YORK, USA.

116 BLACKBURN, S. (ED.) (1975) MEANING,REFERENCE AND NECESSITY, CAMBRIDGE UNIVERSITY PRESS.

117 BLIN, J.M. (1973) PATTERNS AND CONFIGURATIONS IN ECONOMIC SCIENCE, REIDEL, AMSTERDAM.

118 BLIN, J.M. (1974) ET AL, PATTERN RECOGNITION AND MACROECONOMICS, J. OF CYBERNETES.

119 BLIN, J.M. (1974) FUZZY RELATIONS IN GROUP DECISION THEORY, J. CYBERNETICS, 4, 17-22.

120 BLIN, J.M. (1975) FUZZY RELATIONS IN MULTIPLE-CRITERIA DECISION MAKING, NORTHWESTERN UNIVERSITY, JAN.

121 BLIN, J.M. (1977) FUZZY SETS IN MULTIPLE CRITERIA DECISION MAKING, EDS. SLARR, M.K. AND ZELENY, M., MULTIPLE CRITERIA DECISION MAKING, N.Y., ACAD. PRESS.

122 BLIN, J.M., AND WHINSTON, A.B. (1973) FUZZY SETS AND SOCIAL CHOICE, J. CYBERNETICS, 3, 28-33.

123 BLYTH, T.S. AND JANOWITZ, M.F. (1972) RESIDUATION THEORY, PERGAMON PRESS, OXFORD.

124 BONA, B. (1977) AN APPLICATION OF A MULTICRITERIA FUZZY SIMILARITY RELATION TO THE CLASSIFICATION OF METROPOLITAN DISTRICTS, PROCEEDINGS OF THE SYMP. ON FUZZY SET THEORY AND APPLN., IEEE CONF. ON DECISION AND CONTROL, NEW ORLEANS.

125 BORGHI, O. (1972) ON A THEORY OF FUNCTIONAL PROBABILITY, REVISTA UN. MAT. ARGENTINA, '26, 90-106.

126 BORISOV, A., AND ERENSTEIN, R.X. (1970) COMPARISON OF SOME CRISP AND FUZZY ALGORITHMS OF RECOGNITION, METODY I SREDSTVA TEXNICESKOI KIBERNETIKI, 6, RIGA, (IN RUSSIAN).

127 BORISOV, A.N., AND KOKLE, E.A. (1970) RECOGNITION OF FUZZY PATTERNS BY FEATURE ANALYSIS, CYBERNETICS AND DIAGNOSTICS, NO. 4, RIGA, U.S.S.R. .

128 BORISOV, A.N., AND KOKLE, E.A. (1970) RECOGNITION OF FUZZY PATTERNS, IN D.S. KRISTINKOV, J.J. OSIS, L.A. RASTRIGIN (EDS.), KIBERNETIKA I DIAGOSTIKA, 4, 135-147, (IN RUSSIAN) ZINATNE, RIGA, U.S.S.R. .

129 BORISOV, A.N., AND OSIS, J.J. (1970) METHODS FOR EXPERIMENTAL ESTIMATION OF MEMBERSHIP FUNCTIONS OF FUZZY SETS, IN D.S. KRISTINKOV, J.J. OSIS, L.A. RASTRIGIN (EDS.), KIBERNETIKA I DIAGNOSTIKA, 4, 125-134, (IN RUSSIAN) ZINATNE, RIGA, U.S.S.R. .

130 BORISOV, A.N., AND OSIS, J.J. (1970) SEARCH FOR THE GREATEST DIVISIBILITY OF FUZZY SETS, IN D.S. KRISTINKOV, J.J. OSIS, L.A. RASTRIGIN (EDS.), KIBERNETIKA I DIANOSTIKA, 3, 79-88, (IN RUSSIAN) ZINATNE, RIGA, U.S.S.R. .

131 BORISOV, A.N., AND VUL'F, G.N., AND OSIS,J.J. (1972) STATE PROGNOSIS OF COMPLEX SYSTEMS USING THE THEORY OF FUZZY

SETS, IN KYBERNETIKA I DIAGNOSTIKA, 5, RIGA, (IN RUSSIAN).

132 BORISOV, A.N., VULF, G.N., AND OSIS, J.J. (1972) PREDICTION OF THE STATE OF A COMPLEX SYSTEM USING THE THEORY OF FUZZY SETS, IN D.S. KRISTINKOV, J.J. OSIS, L.A. RASTRIGIN (EDS.), KIBERNETIKA I DIAGNOSTIKA, 4, 79-84, (IN RUSSIAN) ZINITNE, RIGA, U.S.S.R. .

133 BORISOV, A.N., WOOLF, G.N., AND OSIS, J.A. (1972) APPL. OF THE THEORY OF FUZZY SETS TO STATE IDENTIFICATION OF COMPLEX SYSTEMS, CYBERNETICS AND DIAGNOSTICS NO. 5.

134 BORKOWSKI, L. (ED.) (1970) JAN LUKASIEWICZ SELECTED WORKS, NORTH-HOLLAND, AMSTERDAM.

135 BORKOWSKI, L. (1958) ON PROPER QUANTIFIERS 1, STUDIA LOGICA, 8, 65-128.

136 BORKOWSKI, L., AND SLUPECKI, J. (1958) THE LOGICAL WORKS OF LUKASIEWICZ, STUDIA LOGICA, 8, 7-50.

137 BORUVKA, O. (1937) STUDIES ON MULTIPLICATIVE SYSTEMS (SEMIGROUPS), PART 1, PUBLICATIONS DE LA FACULTE DES SCIENCES DE L'UNIVERSITE MASARYK, NO. 245, (IN ENGLISH).

138 BORUVKA, O. (1938) STUDIES ON MULTIPLICATIVE SYSTEMS (SEMIGROUPS), PART 2, PUBLICATIONS DE LA FACULTE DES SCIENCES DE L'UNIVERSITE MASARYK, NO. 265, 1-24, (IN ENGLISH).

139 BORUVKA, O. (1939) THEORY OF GROUPOIDS, PUBLICATIONS DE LA FACULTE DES SCIENCES DE L'UNIVERSITE MASARYK, NO. 275, (IN CZECH).

140 BORUVKA, O. (1941) UBER KETTEN VON FAKTOROIDEN, MATHEMATISCHE ANNALEN, 188, 41-64.

141 BORUVKA, O. (1974) FOUNDATIONS OF THE THEORY OF GROUPOIDS AND GROUPS, VEB DEUTSCHER VERLAG DER WISSENSCHAFTEN, BERLIN.

142 BOSSEL, H.H., AND HUGHES, B.B. (1973) SIMULATION OF VALUE-CONTROLLED DECISION MAKING, REPORT SRC-11, SYSTEMS RESEARCH CENTER, CASE WESTERN RESERVE UNIVERSITY, CLEVELAND, OHIO, USA.

143 BOUILLE, F. (1978) FUZZY DATA PROCESSING WITH THE HYPERGRAPH-BASED DATA STRUCTURE, UNIV. PIERREDET MARIE CURIE, FRANCE, PROCEEDINGS OF THE 1978 INT. CONF. ON CYBERNETICS AND SOCIETY, VOL2, 1222-1227.

144 BOYD, J.P. (1978) TOPAI AS MODELS OF FUZZY SETS, PROC. 22ND MEETING SOC. FOR GENERAL SYSTEMS RESEARCH, WASH., 443-450.

145 BRAAE, M., AND RUTHERFORD, D.A. (1978) FUZZY RELATIONS IN

A CONTROL SETTING, KYBERNETES 7, 185-188.

146 BREMERMANN, H.J. (1971) CYBERNETIC FUNCTIONALS AND FUZZY SETS, IN IEEE SYMPOSIUM ON SYSTEMS MAN AND CYBERNETICS, 71C46SMC, 248-253.

147 BREMERMANN, H.J. (1974) COMPLEXITY OF AUTOMATA, BRAINS AND BEHAVIOUR, IN CONRAD, M., GUTTINGER AND DALCIN, M. (EDS.), PHYSICS AND MATHEMATICS OF THE NERVOUS SYSTEM, LECTURE NOTES IN BIOMATHEMATICS, 4, SPRINGER, 304-331.

148 BROWN, G.S. (1969) LAWS OF FORM, ALLEN AND UNWIN, LONDON.

149 BROWN, J.G. (1969) FUZZY SETS ON BOOLEAN LATTICES, REP. 1957, BALLISTIC RESEARCH LABORATORIES, ABERDEEN, MARYLAND, JAN. .

150 BROWN, J.G. (1971) A NOTE ON FUZZY SETS, INFORM AND CONTROL, 18, 32-39.

151 BROWNELL, H., AND CARAMAZZA, A. (1977) CATEGORIZING, REPORT JOHN HOPKINS UNIV. .

152 BRUNNER, J. (1976) UBERLICK ZUR THEORIE UND ANWENDUNG VON FUZZY-MENGEN, VORTRAGE AUS DEM PROBLEMSEMINAR AUTOMATEN UND ALGORITHMENTHEORIE, APRIL, WEISSIG,3-15.

153 BRUNNER, J., AND WECHLER, W. (1976) THE BEHAVIOUR OF R-FUZZY AUTOMATA, MAZURKIEWICZ, A. (ED.), LECTURE NOTED IN COMPUTER SCIENCE, 45, SPRINGER-VERLAG, BERLIN, 210-215.

154 BUNGE, M.C. (1966) CATEGORIES OF SETS VALUED FUNCTIONS, PHD THESIS, DEPARTMENT OF MATHEMATICS, UNIVERSITY OF CALIFORNIA.

155 BUTNARIU, D. (1975) FUZZY AUTOMATA AND FUZZY GAMES, PAPER PRESENTED AT SEMINARUL DE TEORIA SISTEMELOR AT THE DEPT. OF ECONOMIC CYBERNETICS, ACADEMY OF ECONOMIC STUDIES.

156 BUTNARIU, D. (1975) L-FUZZY AUTOMATA DESCRIPTION OF A NEURAL MODEL, PROCEEDINGS OF 3RD INTERNATIONAL CONGRESS OF CYBERNETICS AND SYSTEMS, BUCHAREST, ROMANIA, AUG.

157 BUTNARIU, D. (1975) L-FUZZY TOPOLOGIES, BULL. MATH. SOC. SCI. MATH. R.S.R. 19 (3), 227-236.

158 BUTNARIU, D. (1976) FUZZY GAMES AND THEIR MINIMAX THEOREM, (IN ROMANIAN) ST. CERC. MATH. 28 (2), 142-160.

159 BUTNARIU, D. (1977) (L',L)-FUZZY TOPOLOGICAL SPACES, ANALS. SCI. UNIV. "AL. I. CUZA" - IASI 23 SERIA I, FASC. 1 .

160 BUTNARIU, D. (1977) FUZZY GAMES: A DESCRIPTION OF THE CONCEPT, FUZZY SETS AND SYST. 1, 181-192, NORTH-HOLLAND.

161 BUTNARIU, D. (1977) THREE-PERSON FUZZY GAMES, (IN ROMANIAN) ST. CERC. MATH. (2), 1-10.

162 BUTNARIU, D.,(TO APPEAR) EQUILIBRIUM POINTS IN FUZZY GAMES, (IN ROMANIAN), ST. CERC. MATH. .

163 CAPOCELLI, R.M., AND DE LUCA, A. (1972) MEASURES OF UNCERTAINTY IN THE CONTEX OF FUZZY SETS THEORY, IN ATTI DEL ILE CONGRESSO NATIONALE DI CIBERNETICA DI CASCIANA TERME, PISA, ITALY.

164 CAPOCELLI, R.M., AND DE LUCA, A. (1973) FUZZY SETS AND DECISION THEORY, INFORM. AND CONTROL, 23, 446-473.

165 CARGILE, J. (1969) THE SORITES PARADOX, BRIT. J. PHILOS. SCI., 20, 193-202.

166 CARLSSON, C. (1978) A SYSTEM OF PROBLEMS AND HOW TO DEAL WITH IT, REPORT ABO SWEDISH UNIV. .

167 CARLSSON, CH. (1978) FUZZY AUTOMATA AS CYBERNETIC CONTROL FUNCTIONS, INSTIT. OF MANAGEMENT SCI., ABO SWEDISH UNIV. SCHOOL OF ECONOMICS, HENRIKSGATAN 7, SF-20500 ABO 50, FINLAND.

168 CARLSTROM, I.F. (1975) TRUTH AND ENTAILMENT FOR A VAGUE QUANTIFIER, SYNTHESE, 30, 461-495.

169 CARLUCCI, D., AND DONATI, F. (1975) A FUZZY CLUSTER OF THE DEMAND WITHIN A REGIONAL SERVICE SYSTEM, IN SPECIAL INTEREST DISCUSSION ON FUZZY AUTOMATA AND DECISION PROCESSES,6TH IFAC WORLD CONGRESS, BOSTON, MASS., USA, AUG. .

170 CARLUCCI, D., AND DONATI, F. (1977) FUZZY CLUSTER OF DEMAND WITHIN A REGIONAL SERVICE SYSTEM, 379-386, IN FUZZY AUTOMATA AND DECISION PROCESSES, M.M. GUPTA, G.N. SARIDIS, B.R. GAINES, EDS., NORTH-HOLLAND.

171 CARNAP, R. (1947) MEANING AND NECESSITY, UNIVERSITY OF CHICAGO PRESS.

172 CARNAP, R. (1950) LOGICAL FOUNDATIONS OF PROBABILITY, UNIVERSITY OF CHICAGO PRESS.

173 CARNAP, R. (1963) THE PHILOSOPHER REPLIES, IN SCHILPP, P.A. (ED.), THE PHILOSOPHY OF R. CARNAP, THE LIBRARY OF LIVING PHILOSOPHERS, 11, OPEN COURT, LA SALLE, ILLINOIS, USA.

174 CARNAP, R. (1964) THE LOGICAL SYNTAX OF LANGUAGE, ROUTLEDGE AND KEGAN PAUL, LONDON, (1ST ED. 1937).

175 CARTER, G.A., AND HAGUE, M.J. (1973) FUZZY CONTROL OF RAW MIX PERMEABILITY AT A SINTER PLANT, IN MAMDANI, E.H., AND GAINES, B.R. (EDS.), DISCRETE SYSTEMS AND FUZZY REASONING, EES-MMS-DSFR-73, QUEEN MARY COLLEGE, UNIVERSITY

OF LONDON, (WORKSHOP PROCEEDINGS).

176 CASTONGUAY, C. (1972) MEANING OF EXISTANCE IN MATHEMATICS, LIBRARY OF EXACT PHILOSOPHY, 9, SPRINGER-VERLAG, VIENNA.

177 CECH, E. (1937) TOPOLOGICKE PROSTORY, CASOPIS PRO PESTOVANI MATEMATIKY A FYSIKY, 66, D225-D236.

178 CECH, E. (1966) TOPOLOGICAL SPACES, ACADEMIA, PRAGUE.

179 CECH, E. (1968) TOPOLOGICAL SPACES IN TOPOLOGICAL PAPERS OF E. CECH, ACADEMIA, PRAGUE, 436-472, (TRANS. 1937 PAPER IN CZECH.).

180 CHANAS, S., AND KOKALANOW, M. (1977) AN ASSIGNMENT PROBLEM WITH FUZZY EFFECTIVENESS ESTIMATES, KOMUNIKAT 253, WROCLAW.

181 CHANG, C.C. (1958A) PROOF OF AN AXIOM OF LUKASIEWICZ, TRANS. AMER. MATH. SOC., 87, 55-56.

182 CHANG, C.C. (1958B) ALGEBRAIC ANALYSES OF MANY VALUED LOGICS, TRANS. AMER. MATH. SOC., 88, 467-490.

183 CHANG, C.C. (1959) A NEW PROOF OF THE COMPLETENESS OF THE LUKASIEWICZ AXIOMS, TRANS. AMER. MATH. SOC., 93, 74-80.

184 CHANG, C.C. (1963A) THE AXIOM OF COMPREHENSION IN INFINITE VALUED LOGIC, MATH. SCAND., 13, 9-30.

185 CHANG, C.C. (1963B) LOGIC WITH POSITIVE AND NEGATIVE TRUTH VALUES, ACTA PHILOSOPHICA FENNICA, 16, 19-39.

186 CHANG, C.C. (1964) INFINITE VALUED LOGIC AS A BASIS FOR SET THEORY, IN BAR-HILLEL, Y. (ED.), PROCEEDINGS OF 1964 INTERNATIONAL CONGRESS FOR LOGIC METHODOLOGY AND PHILOSOPHY OF SCIENCE, NORTH-HOLLAND, AMSTERDAM, 93-100.

187 CHANG, C.L. (1967) FUZZY SETS AND PATTERN RECOGNITION, PHD THESIS, UNIVERSITY OF CALIFORNIA, BERKELEY, CALIFORNIA, USA.

188 CHANG, C.L. (1968) FUZZY TOPOLOGICAL SPACES, J. MATH. ANAL. AND APPLN., 24, 182-190.

189 CHANG, C.L. (1971) FUZZY ALGEBRA, FUZZY FUNCTIONS AND THEIR APPLICATION TO FUNCTION APPROXIMATION. DIVISION OF COMPUTER RESEARCH AND TECHNOLOGY, NATIONAL INSTITUTES OF HEALTH, BETHESDA, MARYLAND, USA.

190 CHANG, C.L. (1975) INTERPRETATION AND EXECUTION OF FUZZY PROGRAMS, IN ZADEH, L.A., FU, K.S., TANAKA, K., AND SHIMURA, M. (EDS.) FUZZY SETS AND THEIR APPLICATIONS TO COGNITIVE AND DECISION PROCESSES, ACADEMIC PRESS, NEW YORK, 191-218.

191 CHANG, C.L., AND LEE, R.C.T. (1973) SYMBOLIC LOGIC AND

MECHANICAL THEOREM PROVING, ACADEMIC PRESS, NEW YORK.

192 CHANG, R.L.P. (1976) APPLICATION OF FUZZY DECISION TECHNIQUES TO PATTERN RECOGNITION AND CURVE FITTING, PHD THESIS, DEPT. OF EECS, PRINCETON UNIV. .

193 CHANG, R.L.P., AND PAVLIDIS, T. (1976) FUZZY DECISION TREES, IEEE CONF. SYS. MAN CYBERN., WASHINGTON.

194 CHANG, R.L.P., AND PAVLIDIS, T. (1977) FUZZY DECISION TREE ALGORITHMS, VOL.7, IEEE TRANS. SYST., MAN, AND CYBERN., 28-34.

195 CHANG, R.L.P., AND PAVLIDIS, T. (1979) APPLICATIONS OF FUZZY SETS IN CURVE FITTING, FUZZY SETS AND SYSTEMS, VOL. 2, NO. 1, JAN, 67-74.

196 CHANG, S.K. (1971) AUTOMATED INTERPRETATION AND EDITING OF FUZZY LINE DRAWINGS, SJCC, 38, 393-399.

197 CHANG, S.K. (1971) FUZZY PROGRAMS-THEORY AND APPLICATIONS, IN PROC. OF POLYTECHNIC INSTITUTE OF BROOKLYN SYMPOSIUM OM COMPUTERS AND AUTOMATA, 147.

198 CHANG, S.K. (1971) PICTURE PROCESSING GRAMMAR AND ITS APPLICATIONS, INFORM. SCI., 121-148.

199 CHANG, S.K. (1972) ON THE EXECUTION OF FUZZY PROGRAMS USING FINITE STATE MACHINES, IEEE TRANS. COMP., C-21, 241-253.

200 CHANG, S.K., AND KE, J.S. (1976) DATABASE SKELETON AND ITS APPLICATION TO FUZZY QUERY TRANSLATION, DEPT. OF INFORMATION ENGINEERING, UNIV. OF ILLINOIS, CHICAGO CIRCLE, CHICAGO, ILLINOIS.

201 CHANG, S.S.L. (1969) CONTROL SYSTEMS, SUNY STONY BROOK, DEPT. OF ELEC. SCI. PROJ. NOAF 9749.

202 CHANG, S.S.L. (1969) FUZZY DYNAMIC PROGRAMMING AND APPROX. OPTIMIZATION OF PARTIALLY KNOWN SYSTEMS, PROC. 2ND HAWAII INT. CONF. ON SYST. SCIENCE, HONOLULU, 123.

203 CHANG, S.S.L. (1969) FUZZY DYNAMIC PROGRAMMING AND THE DECISION MAKING PROCESS, IN PROC. 3RD PRINCETON CONFERENCE ON INFORMATION SCIENCE AND SYSTEMS, 200-203.

204 CHANG, S.S.L. (1972) FUZZY MATHEMATICS, MAN, AND HIS ENVIRONMENT, IEEE TRANS. SYST. MAN CYBERN., SMC-2, 93.

205 CHANG, S.S.L. (1975) ON RISK AND DECISION MAKING IN A FUZZY ENVIRONMENT, IN ZADEH, L.A. ET. AL. EDS FUZZY SETS AND THEIR APPLICATIONS TO COGNITIVE AND DECISION PROCESSES, NEW YORK: ACADEMIC PRESS, 219-226.

206 CHANG, S.S.L. (1977) APPLICATION OF FUZZY SET THEORY TO

ECONOMICS, KYBERNETES 6, 203-207.

207 CHANG, S.S.L. (1977) ON FUZZY ALGORITHM AND MAPPING, 191-196, IN FUZZY AUTOMATA AND DECISION PROCESSES, M.M. GUPTA, G.N. SARIDIS, B.R. GAINES, EDS., NORTH-HOLLAND.

208 CHANG, S.S.L. (1978) DECISION IMPLICATIONS OF FUZZY SET THEORY, PROC. 1978 JACC, VOL. 3, 362.

209 CHANG, S.S.L. (1978) ON A FUZZY ALGORITHM AND ITS IMPLEMENTATION, VOL.8, NO.1, 31, IEEE TRANS. ON SYST., MAN, AND CYBERN. .

210 CHANG, S.S.L., AND ZADEH, L.A. (1972) ON FUZZY MAPPING AND CONTROL, IEEE TRANS. ON SYSTEMS MAN AND CYBERNETICS VOL.2 PP 30-34.

211 CHAPIN, E.W. (1971) AN AXIOMATIZATION OF THE SET THEORY OF ZADEH, NOTICES AMERICAN MATHEMATICAL SOCIETY, 687-02-4, 753.

212 CHAPIN, E.W. (1974) SET-VALUED SET THEORY: PART 1, NOTRE DAME J. FORMAL LOGIC, 15, 619-634.

213 CHAPIN, E.W. (1975) SET-VALUED SET THEORY: PART 2, NOTRE DAME J. FORMAL LOGIC, 16, 255-267.

214 CHEN, C. (1974) REALIZABILITY OF COMMUNICATION NETS: AN APPLICATION OF THE ZADEH CRITERION, IEEE TRANS. CIRCUITS AND SYST., CAS-21, 150-151.

215 CHEN, S.C., AND SHIMURA, M. (1974) ON-LINE RECOGNITION OF HAND PRINTED CHARACTERS USING FUZZY LOGIC, TECHNICAL REPORT OF PATTERN RECOGNITION AND LEARNING OF IECE, PRL74-65.

216 CHIARA, C.S. (1973) ONTOLOGY AND THE VICIOUS CIRCLE PRINCIPLE, CORNELL UNIVERSITY PRESS, ITACA, NEW YORK.

217 CHICHINADZE, V. (1977) SOME PROBLEMS OF ORGANIZATION, PROCEEDINGS OF THE SYMP. ON FUZZY SET THEORY AND APPLN., IEEE CONF. ON DECISION AND CONTROL, NEW ORLEANS.

218 CHILAUSKY, R., JACOBSEN, B., AND MICHALSKI, R.S. (1976) AN APPLICATION OF VARIABLE- VALUED LOGIC TO INDUCTIVE LEARNING OF PLANT DISEASE DIAGNOSTIC RULES, PROC. 6TH INT. SYMP. MULTIPLE-VALUED LOGIC, IEEE 76CH1111-4C, 233-240.

219 CHITTENDEN, E.W. (1941) ON THE REDUCTION OF TOPOLOGICAL FUNCTIONS, IN WILDER, R. L., AND AYRES, W.L. (EDS.), LECTURES IN TOPOLOGY, UNIVERSITY OF MICHIGAN PRESS, ANN ARBOR, USA, 267-285.

220 CHOMSKY, N. AND HALLE, M. (1965) SOME CONTROVERSIAL QUESTIONS IN PHONOLOGICAL THEORY, J. LINGUISTICS, 1, 97-138.

221 CHRISTOPHER, F.T. (1977) QUOTIENT FUZZY TOPOLOGY AND LOCAL COMPACTNESS, J. OF MATH. ANAL. AND APPLN., VOL. 53, N3.

222 CHYTIL, M. (1969) ON CONSTITUTING OF SEMANTICAL MODELS FOR GUHA-METHODS, CESKOSLOVENSKA FYSIOLOGIE, 18, 43-147, (IN CZECH).

223 CLEAVE, J.P. (1970) THE NOTION OF VALITY IN LOGICAL SYSTEMS WITH INEXACT PREDICATES, BRIT. J. PHILOS. SCI., 21, 269-274.

224 CLEAVE, J.P. (1974) THE NOTION OF LOGICAL CONSEQUENCE IN THE LOGIC OF INEXACT PREDICATES, Z. MATH. LOGIK GRUNDLAGEN MATH., 20, 307-324.

225 CLEAVE, J.P. (1976) QUASI-BOOLEAN ALGEBRAS, EMPIRICAL CONTINUITY AND THREE-VALUED LOGIC, Z. MATH. LOGIK GRUNDLAGEN MATH. .

226 CLEMENTS, D. (1977) EVALUATION OF COMPUTER SECURITY USING A FUZZY RATING LANGUAGE, ERL REPORT M 77/41, UNIV. OF CALIF., BERKELEY.

227 COHEN, L.J. (1975) PROBABILITY-THE ONE AND THE MANY, PROC. BRIT. ACADEMY, 61, 3-28.

228 COHEN, P.J. (1967) NON-CANTONIAN SET THEORY, SCIENTIFIC AMERICAN, DEC., 104-116.

229 COLE, P., AND MORGAN, J.L. (EDS.) (1975) SYNTAX AND SEMANTICS VOL. 3, ACADEMIC PRESS, NEW YORK.

230 CONCHE, B. (1973) A METHOD OF CLASSIFICATION BASED ON THE USE OF A FUZZY AUTOMATON, UNIV. OF PARIS - DAUPHINE.

231 CONCHE, B. (1973) ELEMENTS DE UNE METHODE DE CLASSIFICATION PAR UTILISATION DE UN AUTOMATE FLOU, J.E.E.F.L.N., UNIVERSITY OF PARIS-DAUPHINE.

232 CONCHE, B., JOUAULT, J.P., AND LUAN, P.M. (1973) APPLICATION DES CONCEPTS FLOUS A LA PROGRAMMATION EN LANGUAGES QUASI-NATURELS, SEMINAIRE BERNARD ROY, UNIVERSITY OF PARIS-DAUPHINE.

233 COOLS, M., AND PETEAU, M. (1973) STIM 5: UN PROGRAMME DE STIMULATION INVENTIVE UTILISANT LA THEORIE DES SOUS-ENSEMBLES FLOUS, IMAGO DISCUSSION PAPER, UNIVERSITY CATHOLIQUE DE LOUVAIN, BELGIUM.

234 COPPO, M., AND SAITLA, L. (1976) SEMANTIC SUPPORT FOR SPEECH UNDERSTANDING BASED ON FUZZY RELATION, PROC. INT. CONT. ON CYBERN. AND SOCIETY, WASH. D.C., 520-524.

235 CRESWELL, M.J. (1973) LOGICS AND LANGUAGES, METHUEN, LONDON.

236 CURRY, H.B. (1942) THE INCONSISTENCY OF CERTAIN FORMAL LOGICS, J. SYMBOLIC LOGIC, 7, 115-117.

237 DALCIN, M. (1975A) FUZZY-STATE AUTOMATA, THEIR STABILITY AND FAULT-TOLERANCE, INT. J. COMP. INF. SCIENCES, 4, 63-80.

238 DALCIN, M. (1975B) MODIFICATION TOLERANCE OF FUZZY-STATE AUTOMATA, INT. J. COMP. INF. SCIENCES, 4, 81-93.

239 DAMERAU, F.J. (1975) ON FUZZY ADJECTIVES, RC5340, IBM RESEARCH LABORATORY, YORKTOWN HEIGHTS, NEW YORK, USA.

240 DANES, F. (1966) THE RELATION OF CENTRE AND PERPHERY AS A LANGUAGE UNIVERSAL, TRAVAUX LINGUISTIQUES DE PRAGUE, 2., 9-21.

241 DANES, F., AND VACHEK, J. (1964) PRAGUE STUDIES IN STRUCTURAL GRAMMAR TODAY, TRAVAUX LINGUISTIQUES DE PRAGUE, 1,21-31.

242 DANIELSSON, S. (1967) MODAL LOGIC BASED ON PROBABILITY THEORY, THEORIA, 33, 189-197.

243 DAVIO,M., AND THAYSE, A. (1973) REPRESENTATION OF FUZZY FUNCTIONS, PHILIPS RESEARCH REPORTS, 28, 93-106.

244 DE FINETTI, B. (1972) PROBABILITY, INDUCTION AND STATISTICS, JOHN WILEY, LONDON.

245 DE KERF, J. (1974) VAGE VERAMELINGEN, INGENIEURSTIJDINGEN 23E JAARGANG, 581-589.

246 DE KERF, J. (1974) VAGE VERZAMELINGEN, OMEGA (VERENIGING VOOR WIS- EN NATUUR- KUNDIGEN LOVANIENSES), 2, 2-18.

247 DE KERF, J. (1975) A BIBLIOGRAPHY ON FUZZY SETS, J. COMPUTATIONAL AND APPLIED MATHEMATICS, 1, 205-212.

248 DELUCA, A., AND TERMINI, S. (1971) ALGORITHMIC ASPECTS IN COMPLEX SYSTEMS ANALYSIS, SCIENTIA, 106, 659-671.

249 DELUCA, A., AND TERMINI, S. (1972) A DEFINITION OF A NONPROBABILISTIC ENTROPHY IN THE SETTING OF FUZZY SETS, INFORM. AND CONTROL, 20, 301-312.

250 DELUCA, A., AND TERMINI, S. (1972) ALGEBRAIC PROPERTIES OF FUZZY SETS, J. MATH. ANAL. AND APPLN., 40 , 373-386.

251 DELUCA, A., AND TERMINI, S. (1974) ENTROPY OF L-FUZZY SETS, INFORM AND CONTROL, 24, 55-73.

252 DELUCA, A., AND TERMINI, S. (1977) ON THE CONVERGENCE OF ENTROPY MEASURES OF A FUZZY SET, KYBERNETES 6, 219-227.

253 DEPALMA, G.F., AND YAU, S.S. (1975) FRACTIONALLY FUZZY GRAMMARS WITH APPLICATION TO PATTERN RECOGNITION, IN ZADEH, L.A., FU, K.S., TANAKA, K., AND SHIMURA, M.

(EDS.), FUZZY SETS AND THEIR APPLICATIONS TO COGNITIVE AND DECISION PROCESSES, ACADEMIC PRESS, NEW YORK, 329-351.

254 DIAMOND, P. (1975) FUZZY CHAOS, DEPARTMENT OF MATHEMATICS, UNIVERSITY OF QUEENSLAND, BRISBANE, AUSTRALIA.

255 DIARRA, N. (1975) A PROPOS DES ENSEMBLES FLOUS, PHD THESIS, CENTRE PEDAGOGIQUE SUPERIEUR DE L'ECOLE NORMALE SUPERIEUR BAMAKO, TUNISIA, OCT. .

256 DIENES, Z.P. (1949) ON AN IMPLICATION FUNCTION IN MANY-VALUED SYSTEMS OF LOGIC, J. SYMBOLIC LOGIC, 14, 95-97.

257 DIJKMAN, J.G., AND LOWEN, R. (1976) FUZZY RELATIONS ON COUNTABLE SETS, TECHNICAL HIGHSCHOOL DELFT AND VRIGE UNIVERSITEIT BRUSSEL.

258 DILL, A.M., AND EMPTOZ, H. (1979) ENTROPY AND INDETERMINATION MEASURE IN THE SETTING OF FUZZY SETS THEORY, PROCEEDINGS OF THE IEEE INT. SYMP. ON INFORMATION THEORY, GRIGNANO, ITALY, JUNE.

259 DILLMAN,I. (1973) INTRODUCTION AND DEDUCTION, BASIL BLACKWELL, OXFORD.

260 DIMITROV, V., AND CUNTCHEV, O. (1977) EFFICIENT FUZZY GOVERNING HUMANISTIC SYSTEMS BY FUZZY INSTRUCTIONS, MODERN TRENDS IN CYBERNETIC AND SYSTEMS CWOGSC, SPRINGER, BERLIN.

261 DIMITROV, V., WECHLER, W., AND BARNEV, P. (1974) OPTIMAL FUZZY CONTROL OF HUMANISTIC SYSTEMS, NST. OF MATH. AND MECH., BULGARIA ACAD. OF SCI., SOFIA.

262 DIMITROV, V.D. (1975) EFFICIENT GOVERNING HUMANISTIC SYSTEMS BY FUZZY INSTRUCTIONS, 3RD INT. CONGRESS OF GENERAL SYST. AND CYBERNETICS, BUCHAREST.

263 DIMITROV, V.D. (1977) SOCIAL CHOICE AND SELF-ORGANIZATION UNDER FUZZY MANAGEMENT, KYBERNETES 6, 153-156.

264 DIMITROV, V.D., AND DRIANKOVA, L.D. (1977) PROGRAMME SYSTEM FOR SOCIAL CHOICE UNDER FUZZY MANAGING, IN: B. GILCHRIST, ED., INFORMATION PROCESSING 77 (NORTH-HOLLAND), AMSTERDAM.

265 DIMITROV, V.D., WECHLER, W., DRJANKOV, D., AND PETROV, A. (1975) COMPUTER EXECUTION OF FUZZY ALGORITHMS, PROC. CONF. APPLNS. MATH. MODELS AND COMPUTERS IN LINGUISTICS, VARNA, BULGARIA, MAY, (IN RUSSIAN).

266 DOCKERY, J.J. (1977) A FUZZY DEFIN. OF EFFECTIVE MEASURES FROM A STUDY OF MILITARY FORCE STRUCTURE, WORKING PAPER, U.S. ARMY CONCEPT ANALYSIS AGENCY, 8120 WOODMON AVE, MD.

267 DOCKERY, J.J. (1977) THE USE OF FUZZY SETS IN THE ANALYSIS

OF MILITARY COMMANDS, PROC. OF THE 39TH MILITARY OPERATIONS RESEARCH SOCIETY(MORS).

268 DORRIS, A.L., AND SADOSKY, TH.L. (1973) A FUZZY SET THEORETIC APPROACH TO DECISION MAKING, 44TH NATIONAL MEETING OF ORSA, SAN DIEGO, CALIFORNIA, USA, NOV. .

269 DOWKER, C.H. AND PAPERT, D. (1966) QUOTIENT FRAMES AND SUBSPACES, PROC. LONDON MATH. SOC., 16,275-296.

270 DOWKER, C.H. AND PAPERT, D. (1967) ON URSOHNS LEMMA. GENERAL TOPOLOGY AND ITS RELATIONS TO MODERN ANALYSIS AND ALGEBRA 2 (PROC. OF THE SECOND PRAGUE TOPOL. SYMP. 1966), 111-114, ACADEMIA, PRAGUE AND ACADEMIC PRESS, NEW YORK.

271 DRAVECKY, J., AND RIECAN, B. (1975) MEASUREABILITY OF FUNCTIONS WITH VALUES IN PARTIALLY ORDERED SPACES, CASOPIS PRO PESTOVANI MATHEMATIKY, 100, 27-35.

272 DREYFESS, G.R., KOCHEN, M., ROBINSON, J., AND BADRE, A.N. (1975) ON THE PSYCHOLINGUISTIC REALITY OF FUZZY SETS IN GROSSMAN, R.E., SAN, L.J., AND VANCE, T.J. (EDS.), FUNCTIONALISM, UNIVERSITY OF CHICAGO PRESS, 135-149.

273 DROSSELMEYER, E., AND WONNEBERGER, R. (1975) STUDIES ON A FUZZY SYSTEM IN THE PARACHIAL FIELD, SPECIAL INTEREST DISCUSSION SESSION ON FUZZY AUTOMATA AND DECISION PROCESSES, 6TH IFAC WORLD CONGRESS, BOSTON, MASS., USA, AUG.

274 DUBOIS, D. (1978) AN APPL. OF FUZZY SETS TO BUS TRANSPORTATION NETWORK MODIFICATION, PROC. 1978 JACC, VOL. 3, 53-60.

275 DUBOIS, D., AND PRADE, H. (1978) A SUMMARY COMMENT ON THEORY OF FUZZY SETS, FUZZY SETS AS A BASIS FOR A THEORY OF POSSIBILITY, A THEORY OF APPROXIMATE REASONING, AND PRUF-A MEANING REPRESENTATION LANGUAGE FOR NATURAL LANGUAGES, REPORT NO. TR-EE 78-13, PURDUE UNIV., SCHOOL OF ELECTRICAL ENGINEERING, E1-E20.

276 DUBOIS, D., AND PRADE, H. (1978) AN ALTERNATIVE FUZZY LOGIC, REPORT NO. TR-EE 78-13, PURDUE UNIV., SCHOOL OF ELECTRICAL ENGINEERING, F1-F12.

277 DUBOIS, D., AND PRADE, H. (1978) COMMENT ON TOLERANCE ANAL. USING FUZZY SETS AND A PROC. FOR MULT. ASPECT DEC-MAKING, INT. J. SYS. SCI. 9, 357-360.

278 DUBOIS, D., AND PRADE, H. (1978) DECISION MAKING WITH FUZZINESS, 2ND LAWRENCE SYMP. ON SYSTEM AND DECISION SCIENCE, BERKELEY.

279 DUBOIS, D., AND PRADE, H. (1978) FUZZY ALGEBRA, ANALYSIS AND LOGIC, TECH. REPORT 78-13, SCHOOL OF ELECTRICAL ENGINEERING, PURDUE UNIV. .

280 DUBOIS, D., AND PRADE, H. (1978) FUZZY REAL ALGEBRA: SOME RESULTS, A1-A37, REPORT NO. TR-EE 78-13, PURDUE UNIV. SCHOOL OF ELECTRICAL ENGINEERING.

281 DUBOIS, D., AND PRADE, H. (1978) OPERATIONS IN A FUZZY-VALUED LOGIC, REPORT NO. TR-EE 78-13, PURDUE UNIV., SCHOOL OF ELECTRICAL ENGINEERING, D1-D21.

282 DUBOIS, D., AND PRADE, H. (1978) OPERATIONS ON FUZZY NUMBERS, INT. J. SYS. SCI., 9, 613-626.

283 DUBOIS, D., AND PRADE, H. (1978) SUMMARY OF TECHNICAL MEMOS OF ZADEH, REPORT STANFORD A.I. LAB.

284 DUBOIS, D., AND PRADE, H. (1978) SYSTEMS OF LINEAR FUZZY CONSTRAINTS, REPORT NO. TR-EE 78-13, PURDUE UNIV., SCHOOL OF ELECTRICAL ENGINEERING, B1-B21.

285 DUBOIS, D., AND PRADE, H. (1978) TOWARDS FUZZY ANALYSIS: INTEGRATION AND DERIVATION OF FUZZY FUNCTIONS, REPORT NO. TR-EE 78-13, PURDUE UNIV. SCHOOL OF ELECTRICAL ENGINEERING, C1-C51.

286 DUBOIS, D., AND PRADE, H., A COMMENT ON TOLERENCE ANALYSIS USING FUZZY SETS, TO BE PUB. IN INT. JR. ON SYSTEMS SCI.

287 DUBOIS, D., AND PRADE, H., FUZZY LOGICS AND FUZZY CONTROL, PROPOSED TO INT. JR. ON MAN MACHINE STUDIES.

288 DUBOIS, D., AND PRADE, H., OPERATIONS ON FUZZY NUMBERS, TO BE PUB. IN INT. JR. ON SYSTEMS SCI.

289 DUBOIS, T. (1974) LINE METHODE D'EVALUATION PAR LES SOUS-ENSEMBLES FLOUS APPLN. A LA SIMULATION, IMAGO CENTRE, IDP 13, UNIV. CATH. DE LOUVAIN.

290 DUBOIS, T. (1974) UNE METHODE D'EVALUATION PAR LES SOUS-ENSEMBLES FLOUS APPLIQUEE A LA SIMULATION, IMAGO DISCUSSION PAPER 13, UNIVERSITIE CATHOLIQUE DE LOUVAIN, BELGIUM.

291 DUBOIS, T. (1977) A TEACHING SYSTEM USING FUZZY SUBSETS AND MULTI CRITERIA ANALYSIS INT. J. MATH. EDS. SCI. TECH., VOL. 8, 2, 203-217.

292 DUBOIS, T., JONES, A., PETEAU, M., AND HUYNEN, A.N. (1977) TOWARD CONTINUOUS LEARNING MGMT. IN CAI, IND. J. MATH. EDUC. SCI. TECH., 8, 3, 335-350.

293 DUBREIL, P., AND DUBREIL-JACOTIN, L. (1973) PROPRIETES DES RELATIONS D'EQUIVALENCE, COMP. REND. ACAD. SCI. (PARIS), 205, 704-706.

294 DUGUNDJU, J. (1940) NOTE ON A PROPERTY OF MATRICES FOR LEWIS AND LANGFORDS CALCULI OF PROPOSITIONS, J. SYMBOLIC LOGIC, 5, 150-151.

295 DUKMAN, J.G., AND LOWEN, R. (1976) FUZZY RELATIONS ON COUNTABLE SETS, TECHNICAL HIGH SCHOOL DELFT AND VRIJE UNIV. BRUSSEL.

296 DUMMETT, M.A.E. (1959) A PROPOSITIONAL CALCULUS WITH DENUMERABLE MATRIX, J. SYMBOLIC LOGIC, 24, 97-106.

297 DUMMETT, M.A.E. (1973) THE JUSTIFICATION OF DEDUCTION, PROC. BRIT. ACAD., 59, 3-34.

298 DUMMETT, M.A.E. (1975) WANGS PARADOX, SYNTHESE, 30,301-324.

299 DUNN, J.C. (1973) A FUZZY RELATIVE OF THE ISODATA PROCESS AND ITS USE IN DETECTING COMPACT WELL-SEPARATED CLUSTERS, J. CYBERNETICS, 3, 32-57.

300 DUNN, J.C. (1974) A GRAPH THEORETIC ANALYSIS OF PATTERN CLASSIFICATION VIA TAMURAS FUZZY RELATION, SMC-3, 310-313.

301 DUNN, J.C. (1974) SOME RECENT INVESTIGATIONS OF A NEW FUZZY PARTITIONING ALGORITHM AND ITS APPLICATION TO PATTERN CLASSIFICATION PROBLEMS, J. CYBERNETICS, 4, 1-15.

302 DUNN, J.C. (1974) WELL- SEPARATED CLUSTERS AND OPTIMAL FUZZY PARTITIONS, J. CYBERNETICS 4, 95-104.

303 DUNN, J.C. (1975) CANONICAL FORMS OF TAMURA'S FUZZY RELATION MATRIX: A SCHEME FOR VISUALIZING CLUSTER HIERARCHIES, PROCEEDINGS OF COMPUTER GRAPHICS, PATTERN RECOGNITION AND DATA STRUCTURE CONFERENCE, BEVERLY HILLS, CALIFORNIA, USA, MAY.

304 DUNN, J.C. (1977) INDICES OF PARTITION FUZZINESS AND THE DETECTION OF CLUSTERS IN LARGE DATA SETS, 271-284, IN FUZZY AUTOMATA AND DECISION PROCESSES, M.M. GUPTA, G.N. SARIDIS, B.R. GAINES, EDS., NORTH-HOLLAND.

305 DUNST, A.J. (1971) APPLICATION OF THE FUZZY SET THEORY, JAN. .

306 EDWARDS, W. (1962) SUBJECTIVE PROBABILITIES INFERRED FROM DECISIONS, PSYCHOLOGICAL REVIEW, 69, 109-135.

307 EDWARDS, W., PHILLIPS, L.D., HAYES, W.L. AND GOODMAN, B.C. (1968) PROBABILISTIC INFORMATION PROCESSING SYSTEMS: DESIGN AND EVALUATION, IEEE TRANS. SYST. MAN CYBERN., SMC-4, 248-265.

308 EFSTATHIOU, J., AND RAJKOVIC, V. (1978) MULTI-ATTRIBUTE DECISION MAKING AND FUZZY SET THEORY, PROC. OF WORKSHOP ON FUZZY REASONING AT QUEEN MARY COLL., LONDON.

309 EHRENFEUCHT, A., AND ORLOWSKA, E. (1967) MECHANICAL PROOF PROCEDURE FOR PROPOSITIONAL CALCULUS, BULL. ACAD. POLONAISE DES SCIENCES (SERIE MATH., ET PHYS.), 15,

25-30.

310 EL-FATTAH, Y.M. (1976) CONTROL OF COMPLEX SYSTEMS BY FUZZY LEARNING AUTOMATA, IN MAMDANI, E.H., AND GAINES, B.R. (EDS.), DISCRETE SYSTEMS AND FUZZY REASONING, EES-MMS-DSFR-76, QUEEN MARY COLLEGE, UNIVERSITY OF LONDON, (WORKSHOP PROCEEDINGS).

311 ELLIOT, J.L. (1976) FUZZY KIVIAT GRAPHS, EUROPEAN COMPUTING CONGRESS (EUROCOMP 76), ONLINE, LONDON, SEP. .

312 ELLIS, C.A. (1971) PROBABILISTIC TREE AUTOMATA, INFORM AND CONTROL, 19, 401-416.

313 EMPTOZ, H. (1974) INDICES DE FLOU ET INDICES DE DISPERSION, COLLOQUE INT. SUR LA THEORIE DES SOUS-ENSEMBLES FLOUS, MARSEILLE, SEPT. .

314 ENDO, Y., AND TSUKAMOTO, Y. (1973) APPORTION MODELS OF TOURISTS BY FUZZY INTEGRALS, ANNUAL CONFERENCE RECORDS OF SICE, JAPAN .

315 ENGEL, A.B., AND BUONOMANO, V. (1973) TOWARDS A GENERAL THEORY OF FUZZY SETS 1, INSTITUTE OF MATHEMATICS, UNIVERSITY ESTADUEL DE CAMPINAS, BRAZIL.

316 ENGEL, A.B., AND BUONOMANO, V. (1973) TOWARDS A GENERAL THEORY OF FUZZY SETS 2, INSTITUTE OF MATHEMATICS, UNIVERSITY ESTADUEL DE CAMPINAS, BRAZIL.

317 ENTA, Y. (1976) FUZZY CHOICE MODELS, REPORT DEPT. OF BUSINESS HOSEA UNIV. .

318 ENTA, Y. (1978) A MEASURE FOR THE DISCRIMINATION EFFECT OF INFORMATION, PROC. 1978 JACC, VOL. 3., 69-80.

319 ENTA, Y. (1978) DISCRIMINATIVE EFFECT OF INFORMATION, THE 3RD MEASURE OF INFORMATION MEMO NO. UCB/ERL M78-62, UNIV. OF CALIF., BERKELEY.

320 EPSTEIN, G. (1972) MULTIPLE-VALUED SIGNAL PROCESSING WITH LIMITING, SYMPOSIUM ON MULTIPLE-VALUED LOGIC DESIGN, BUFFALO, NEW YORK, USA.

321 EPSTEIN, G., AND HORN, A. (1974) P-ALGEBRAS, AN ABSTRACTION FROM POST ALGEBRAS, ALGEBRA UNIVERSALIS, 4, 195-206.

322 EPSTEIN, G., AND HORN, A. (1975A) CHAIN BASED LATTICES, PACIFIC J. MATHS., 55, 65-84.

323 EPSTEIN, G., AND HORN, A. (1975B) LOGICS WHICH ARE CHARACTERIZED BY SUBRESIDUATED LATTICES, TECH. REP. 24, INDIANA UNIVERSITY COMPUTER SCIENCE DEPARTMENT, BLOOMINGTON, INDIANA, USA.

324 EPSTEIN, G., AND SHAPIRO, S.C. (1975) THE DEVELOPMENT OF LANGUAGE AND REASONING IN THE CHILD AS CONNECTED WITH

MATHEMATICAL LINGUISTICS AND LOGIC, TECH. REP., 41, OCT. .

325 EPSTEIN, G.,FRIEDER, G., AND RINE, D.C. (1974) THE DEVELOPMENT OF MULTIPLE-VALUED LOGIC AS RELATED TO COMPUTER SCIENCE, COMPUTER, 7, 20-32.

326 ESOGBUE, A.O. (1975) ON THE APPLICATION OF FUZZY ALLOCATION THEORY TO THE MODELLING OF CANCER RESEARCH APPROPRIATION PROCESS, PROCEEDINGS OF 3RD INTERNATIONAL CONGRESS OF CYBERNETICS AND SYSTEMS, BUCHAREST, AUG. .

327 ESOGBUE, A.O., AND ELDER, R.C. (1977) FUZZY SETS AND MODELLING PHYSICIAN DEC. PROC., PART 1, INITIAL INTERVIEW INF. SESS., INDUSTRIAL AND SYS. ENG. REPORT NO. J-77-6, GEORGIA INST. OF TECH., ATLANTA.

328 ESOGBUE, A.O., AND ELDER, R.C. (1977) FUZZY SETS AND THE MODELING OF PHYSICIAN DEC. PROC., PART 2, FUZZY DIAG, DEC., IND. AND SYS. ENG. REPORT NO. J-77-6, GA. INST. OF TECH., ATLANTA.

329 ESOGBUE, A.O., AND RAMESH, V. (1970) DYNAMIC PROGRAMMING AND FUZZY ALLOCATION PROCESSES, TECHNICAL MEMORANDUM 202, OPERATIONS RESEARCH DEPARTMENT, CASE WESTERN RESERVE UNIVERSITY, CLEVELAND, OHIO, USA.

330 ETO, H. (1975) MULTIVARIATE ANALYSIS OF AMBIGUOUS OPINIONS ON OPENING THE SPORTS FACILITIES OF FIRMS TO THE PUBLIC, IN SUMMARY OF PAPERS ON GENERAL FUZZY PROBLEMS, THE WORKING GROUP ON FUZZY SYSTEMS, TOKYO, JAPAN, NOV., 5-9.

331 ETO, H. (1976) FUZZY OPERATIONAL APPROACH TO ANALYSIS OF DELPHI TECHNOLOGY FORECASTING, 25-34, SUMMARY OF PAPERS ON GENERAL FUZZY PROBLEMS, THE WORKING GROUP ON FUZZY SYSTEMS, TOYKO, JAPAN.

332 ETO, H. (1977) GENERALIZED DOMINATION AND FUZZY DOMINATION, 1-10, SUMMARY OF PAPERS ON GENERAL FUZZY PROBLEMS, THE WORKING GROUP ON FUZZY SYSTEMS, TOYKO, JAPAN.

333 EVANS, G., AND MCDOWELL, J. (EDS.) (1976) TRUTH AND MEANING, CLARENDON PRESS, OXFORD.

334 EVENDEN, J. (1974) GENERALIZED LOGIC, NOTRE DAME J. FORMAL LOGIC, 15, 35-44.

335 EYTAN, M. (1978) FUZZY SETS, A TOPOS-LOGICAL POINT OF VIEW, UNIV. RENE DESCARTES, PARIS, FRANCE.

336 EZOE, T. (1975) CAUSE PICTURE METHOD INTRODUCED INTO CATEGORICAL ANALYSIS OF MULTI-VARIABLE'S DATA, IN SUMMARY OF PAPERS ON GENERAL FUZZY PROBLEMS, THE WORKING GROUP ON FUZZY SYSTEMS, TOKYO, JAPAN, NOV., 10-13.

337 FELLINGER, L. (1974) SPECIFICATION FOR A FUZZY SYSTEM

MODELING LANGUAGE, PH.D. THESIS, OREGON STATE UNIV., CORVALIS, ORE. .

338 FELLINGER, W.L. (1974) SPECIFICATIONS FOR A FUZZY SYSTEMS MODELLING LANGUAGE, PHD THESIS, OREGON STATE UNIVERSITY, CORALLIS.

339 FENSTAD, J.E. (1964) ON THE CONSISTENCY OF THE AXIOM OF COMPRENSION IN THE LUKASIEWICZ INFINITE VALUED LOGIC, MATH. SCAND., 14, 65-74.

340 FENSTAD, J.E. (1967) REPRESENTATIONS OF PROBABILITIES DEFINED ON FIRST ORDER LANGUAGES, IN CROSSLEY, J.N. (ED.), SETS MODELS AND RECURSION THEORY, NORTH HOLLAND, 156-172.

341 FERON, R. (1976) ENSEMBLES ALEATOIRES FLOUS, C.R. ACAD. SCI. PARIS, 22 (4), 903-906.

342 FERON, R. ECONOMIC D'EXCHANGE ALTEATOIRE FLOUE, C.R. ACAD. SCI. PARIS, 282 (9), 1379-1382.

343 FEVRIER, P. (1976) ON THE REPRESENTATION OF MEASUREMENTS RESULTS BY FUZZY SETS, 3RD EUR. MEETING CYBERN. SYST. RES., VIENNA.

344 FIKSEL, J., DIFFENBACH, J., AND RENDA, A. (1978) A METHODOLOGY EMPLOYING POSSIBILITY THEORY FOR A TECHN. ASS'MT. OF TIME OF DAY EL. RTS., PRES. AT ORSA/TIMS MEETING, LOS ANGELES.

345 FILLMORE, C.J., AND JANGENDOEN, D.T. (EDS.) (1971) STUDIES IN LINGUISTIC SEMANTICS, HOLT, RINEHART AND WINSTON, NEW YORK.

346 FINE, K. (1975) VAGUENESS,TRUTH AND LOGIC, SYNTHESE, 30, 265-300.

347 FINE, T.L. (1973) THEORIES OF PROBABILITY, ACADEMIC PRESS, NEW YORK.

348 FLACHS, J., AND POLLATSCHEK, M.A. (1977) SOME PROPERTIES OF OPTIMIZATION IN L(K) NORM, TECHNICAL REPORT 113, TECHNION, ISRAEL INSTITUTE OF TECH., COMPUTER SCI. DEPT., DEC. .

349 FLACHS, J., AND POLLATSCHEK, M.A. (1978) FURTHER RESULTS ON FUZZY-MATHEMATICAL PROGRAMMING, INF. AND CONTROL, VOL. 38, 241-257.

350 FLONDER, P. (1975) ON C-SETS, WORKING GROUP ON FUZZY SYSTEMS, ACAD. OF ECONOMIC STUDIES, BUCHAREST, (UNPUBLISHED).

351 FLONDER, P. (1977) : AN EXAMPLE OF A FUZZY SYSTEM, KYBERNETES, 6, 229-230.

352 FLONDOR, P. (1975) MODELS FOR PROPERTY ASSIGNMENT, SEMINAR ON FUZZY SYSTEMS, DEPARTMENT OF CYBERNETICS, ASE, BUCHAREST.

353 FORADORI, E. (1933) STETIGKAIT UND KONTINUITAT ALS TEILBARKEITSEIGENSCHAFTEN, MONATSCKEFTE FUR MATH. UND PHYSIK,40, 161-180.

354 FOSTER, M.H., AND MARTIN, M.L. (EDS.) (1966) PROBABLITY,CONFIRMATION AND SIMPLICITY, ODYSSEY PRESS, NEW YORK.

355 FRAENKEL, A.A., BAR-HILLEL, Y., AND LEVY, A. (1973) FOUNDATIONS OF SET THEORY, NORTH-HOLLAND, AMSTERDAM.

356 FRANK, M.J. (1970) PROBABILISTIC TOPOLOGICAL SPACES, ILLINOIS INSTITUTE OF TECHNOLOGY, CHICAGO, USA, JAN.

357 FRASER, B. (1975) HEDGED PERFORMATIVES, IN COLE, P., AND MORGAN, J.L. (EDS.), SYNTAX AND SEMANTICS VOL. 3 ACADEMIC PRESS, NEW YORK, 187-210.

358 FRINK, O. (1938) NEW ALGEBRAS OF LOGIC, AMER. MATH. MONTHLY, 45, 210-219.

359 FU, K.S. (1974) PATTERN RECOGNITION AND SOME SOCIO-ECONOMIC PROBLEMS, PURDUE UNIVERSITY, WEST LAFAYETTE, INDIANA 47907, USA.

360 FU, K.S., AND LI. T.J. (1969) FORMULATION OF LEARNING AUTOMATA AND GAMES, INFORM. SCI., 1, 237-256.

361 FU, K.S., AND LI, T.J. (1968) ON THE BEHAVIOR OF LEARNING AUTOMATA AND ITS APPLICATIONS, PURDUE UNIV., LAF. SCHOOL OF ELEC. ENG. REPORT NO. TR-EE68-20.

362 FU, K.S., WEE, W.G. (1967) ON GENERAL ADAPTIVE ALGORITHMS AND APPLICATIONS OF THE FUZZY SETS CONCEPT TO PATTERN CLASS., PURDUE UNIV., LAFAYETTE SCHOOL OF ELEC. ENG. REPORT TREE67-7.

363 FUJISAKE, H. (1971) FUZZINESS IN MEDICAL SCIENCES AND ITS PROCESSING, PROCEEDINGS OF SYMPOSIUM ON FUZZINESS IN SYSTEMS AND ITS PROCESSING, PROFESSIONAL GROUP OF SYSTEM ENGINEERING OF SICE.

364 FUNG, L.W., AND FU, K.S. (1973A) DECISION MAKING IN A FUZZY ENVIRONMENT, TR-EE73-22, SCHOOL OF ELECTRICAL ENGINEERING, PURDUE UNIVERSITY, LAFAYETTE, INDIANA, USA.

365 FUNG, L.W., AND FU, K.S. (1973B) AN AXIOMATIC APPROACH TO RATIONAL DECISION- MAKING BASED ON FUZZY SETS, ELECTRICAL ENGINEERING REPORT, PURDUE UNIVERSITY, LAFAYETTE, INDIANA, USA.

366 FUNG, L.W., AND FU, K.S. (1974A) THE K'TH OPTIMAL POLICY ALGORITHM FOR DECISION MAKING IN FUZZY ENVIRONMENTS, IN EYKHOFF, P. (ED.), IDENTIFICATION AND SYSTEM PARAMETER ESTIMATION, NORTH HOLLAND, 1025-1059.

367 FUNG, L.W., AND FU, K.S. (1975) AN AXIOMATIC APPROACH TO RATIONAL DECISION MAKING IN A FUZZY ENVIRONMENT, IN ZADEH, L. A., FU, K.S., TANAKA, K., AND SHIMURA, M. (EDS.), FUZZY SETS AND THEIR APPLICATIONS TO COGNITIVE AND DECISION PROCESSES, ACADEMIC PRESS, NEW YORK, 227-256.

368 FUNG, L.W., AND FU, K.S. (1977) CHARACTERIZATION OF A CLASS OF FUZZY OPTIMAL CONTROL PROBLEMS, 209-220, IN FUZZY AUTOMATA AND DECISION PROCESSES, M.M. GUPTA, G.N. SARIDIS, B.R. GAINES, EDS., NORTH-HOLLAND.

369 FURUKAWA, M., NAKAMURA, K., AND ODA, M. (1972) FUZZY MODELS OF HUMAN DECISION-MAKING PROCESS, ANNUAL CONFERENCE RECORDS OF JAACE .

370 FURUKAWA, M., NAKAMURA, K., AND ODA, M. (1973) FUZZY VARIANT PROCESS OF MEMORIES, ANNUAL CONFERENCE RECORDS OF SICE, JAPAN .

371 GAFFNEY, J.E. (1972) NAVIGATION IN SPACE, THE FUTURE AND ARTIFICIAL INTELLIGENCE HIST. OF NAVIGATION, ORLANDO, FLORDIA.

372 GAIFMAN, H. (1964) CONCERNING MEASURES IN FIRST ORDER CALCULI, ISRAEL J. MATH., 2, 1-18.

373 GAINES, B.R. (1975A) STOCHASTIC AND FUZZY LOGICS, ELECTRONICS LETT., 11 188-189.

374 GAINES, B.R. (1975B) APPROXIMATE IDENTIFICATION OF AUTOMATA, ELECTRONICS LETT., 11, 444-445.

375 GAINES, B.R. (1975C) A CALCULUS OF POSSIBILITY, EVENTUALITY AND PROBABILITY, IN EES-MMS-FUZI-75, DEPARTMENT OF ELECTRICAL ENGINEERING SCIENCE, UNIVERSITY OF ESSEX, COLCHESTER, UK.

376 GAINES, B.R. (1975D) CONTROL ENGINEERING AND ARTIFICIAL INTELLIGENCE, LECTURE NOTES OF BCS AISB SUMMER SCHOOL, CAMBRIDGE, UK, JULY, 52-60.

377 GAINES, B.R. (1975E) MULTIVALUED LOGICS AND FUZZY REASONING, LECTURE NOTES OF BCS AISB SUMMER SCHOOL, CAMBRIDGE, UK, JULY, 100-112.

378 GAINES, B.R. (1976) SURVEY OF FUZZY REASONING, WORKSHOP ON DISCRETE SYSTEMS AND REASONING, QUEEN MARY COLLEGE, DEPT. OF ELEC. ENG. .

379 GAINES, B.R. (1976) UNDERSTANDING UNCERTAINTY, PROC. WORKSHOP ON DISCRETE SYSTEMS AND FUZZY REASONING, QUEEN

MARY COLLEGE, LONDON.

380 GAINES, B.R. (1976A) WHY FUZZY REASONING?, IN MAMDANI, E.H., AND GAINES, B.R. (EDS.), DISCRETE SYSTEMS AND FUZZY REASONING, EES-MMS-DSFR-76, QUEEN MARY COLLEGE, UNIVERSITY OF LONDON, (WORKSHOP PROCEEDINGS).

381 GAINES, B.R. (1976B) RESEARCH NOTES ON FUZZY REASONING, IN MAMDANI, E.H., AND GAINES, B.R. (EDS.), DISCRETE SYSTEMS AND FUZZY REASONING, EES-MMS- DSFR-76, QUEEN MARY COLLEGE, UNIVERSITY OF LONDON, (WORKSHOP PROCEEDINGS).

382 GAINES, B.R. (1976C) GENERAL FUZZY LOGICS, 3RD EUR. MEETING CYBERN, SYST. RES., VIENNA .

383 GAINES, B.R. (1976D) FUZZY REASONING AND THE LOGICS OF UNCERTAINTY, PROC. 6TH INT. SYMP. MULTIPLE-VALUED LOGIC, IEEE 76CH1111-4C, 179-188.

384 GAINES, B.R. (1976E) BEHAVIOUR-STRUCTURE TRANSFORMATIONS UNDER UNCERTAINTY, INT. J. MAN-MACHINE STUDIES, 8, 337-365.

385 GAINES, B.R. (1976F) SYSTEM IDENTIFICATION, APPROXIMATION AND COMPLEXITY, INT. J. GENERAL SYST., (3) .

386 GAINES, B.R. (1976H) FUZZY AND STOCASTIC PROBABILITY LOGICS, EES-MMS-FUZ-76, DEPARTMENT OF ELECTRICAL ENGINEERING SCIENCE, UNIVERSITY OF ESSEX, COLCHESTER, UK.

387 GAINES, B.R. (1976I) V-FUZZY Q-ANALYSIS, EES-MMS-QFUZ-76, DEPARTMENT OF ELECTRICAL ENGINEERING SCIENCE, UNIVERSITY OF ESSEX, COLCHESTER, UK.

388 GAINES, B.R. (1977) FOUNDATIONS OF FUZZY REASONING, 19-76, IN FUZZY AUTOMATA AND DECISION PROCESSES, M.M. GUPTA, G.N. SARIDIS, B.R. GAINES, EDS., NORTH-HOLLAND.

389 GAINES, B.R. (1977) SEQUENTIAL FUZZY SYSTEM IDENTIFICATION, PROC. 1977 IEEE CONF. ON DECISION AND CONTROL, NEW ORLEANS, 1309-1314.

390 GAINES, B.R. (1978) FUZZY AND PROBABILITY UNCERTAINTY LOGICS, INFORM. AND CONTROL, 38, 154-169.

391 GAINES, B.R., AND KOHOUT, L.J. (1975A) POSSIBLE AUTOMATA, PROC. 1975 INT. SYMP. MULTIPLE-VALUED LOGIC, IEEE 75CH0959-7C, 183-196.

392 GAINES, B.R., AND KOHOUT, L.J. (1975B) THE LOGIC OF AUTOMATA, INT. J. GENERAL SYST., 2, 191-208.

393 GAINES, B.R., AND KOHOUT, L.J. (1977) THE FUZZY DECADE: A BIBLIOGRAPHY OF FUZZY SETS, APPEARED IN 26.

394 GAINES, B.R., AND KOHOUT, L.J. (1977) THE FUZZY DECADE:

A BIBLIOGRAPHY OF FUZZY SYSTEMS AND CLOSELY RELATED TOPICS, INT. J. MAN-MACHINE STUDIES, 9, 1-68.

395 GALE, S. (1972) INEXACTNESS, FUZZY SETS AND THE FOUNDATIONS OF BEHAVIORAL GEOGRAPHY, GEOGRAPHICAL ANALYSIS, 4, 337-349.

396 GALE, S. (1974) A PROLEGOMENON TO AN INTERROGATIVE THEORY OF SCIENTIFIC ENQUIRY, WP9 RES. ON METRO. CHANGE AND CONFLICT RESOLUTION, PEACE SCI. DEPT., UNIV. OF PENNSYLVANIA.

397 GALE, S. (1974A) A RESOLUTION OF THE REGIONALIZATION PROBLEM AND ITS IMPLICATIONS FOR POLITICAL GEOGRAPHY AND SOCIAL JUSTICE, WP3 RESEARCH ON METROPOLITAN CHANGE AND CONFLICT RESOLUTION, PEACE SCIENCE DEPARTMENT, UNIVERSITY OF PENNSYLVANIA.

398 GALE, S. (1975A) BOUNDARIES, TOLERANCE SPACES AND CRITERIA FOR CONFLICT RESOLUTION, JOURNAL OF PEACE SCIENCE .

399 GALE, S. (1975B) CONJECTURES ON MANY-VALUED LOGIC, REGIONS, AND CRITERIA FOR CONFLICT RESOLUTION, PROC. 1975 INT. SYMP. MULTIPLE-VALUED LOGIC, IEEE 75CH0959-7C, 212-225.

400 GALE, S., AND ATKINSON, M. (1977) FUZZY REGIONS AND SOCIAL JUSTICE, PRES. AT ORSA/TIMS MEETING, ATLANTA, GEORGIA.

401 GALLIN, D. (1975) INTENSIONAL AND HIGHER ORDER MODAL LOGIC, NORTH-HOLLAND, AMSTERDAM.

402 GANTER, T.E., STEINLAGE, R.C., AND WARREN, R.H. (1975) COMPACTNESS IN FUZZY TOPOLOGICAL SPACES, DEPT. MATHEMATICS, UNIVERSITY OF DAYTON, DAYTON, OHIO, USA.

403 GANTER, T.E., STEINLAGE, R.C., AND WARREN, R.H. (1976) COMPACTNESS IN FUZZY TOPOLOGICAL SPACES, J. MATH. ANAL. APPL., TO APPEAR.

404 GEARING, C.E. (1975) GENERALIZED BAYESIAN POSTERIOR ANALYSIS WITH AMBIGUOUS INFORMATION, 45TH ORSA/TIMS JOINT NATIONAL MEETING, BOSTON, MASS., USA, APRIL .

405 GEARING, C.E. (1976) A FUZZY-SET-THEORETIC GENERALIZATION OF BAYES THEOREM, JOINT NAT. MEETING OF ORSA/TIMS.

406 GEHRING, H., AND ZIMMERMANN, H.J. (1975) FUZZY INFORMATION PROFILES FOR INFORMATION SELECTION, AACHEN WORKING PAPER NO. 75/04, AACHEN UNIV. .

407 GELMAN, I. (1970) ORGANIZ'L DATA, A MODEL AND COMPUTATION ALGORITHM THAT USES THE NOTION OF FUZZY SYSTEMS., THESIS, MC GILL UNIV., MONTREAL, CANADA.

408 GENTILHOMME, Y. (1968) LES ENSEMBLES FLOUS EN LINGUISTIQUE, NOTES ON THEORETICAL AND APPLIED LINGUISTICS, 5, BUCHAREST, RUMANIA .

409 GEORGESCU, G. (1971A) N-VALUED COMPLETE LUKASIEWICZ ALGEBRAS, REV. ROUM. MATH. PURES ET APPL., 16, 41-50

410 GEORGESCU, G. (1971B) THE THETA-VALUED LUKASIEWICZ ALGEBRAS 1, REV. ROUM. MATH. PURES ET APPL., 16, BUCHAREST, 195-209.

411 GEORGESCU, G. (1971C) ALGEBRAS DE LUKASIEWICZ DE ORDEN THETA II, REV. ROUM. MATH. PURES ET APPL., 16, 363-369.

412 GEORGESCU, G. (1971D) THE THETA-VALUED LUKASIEWICZ ALGEBRAS III, REV. ROUM. MATH. PURES ET APPL., 16, 1365-1390.

413 GEORGESCU, G., AND VRACIU, C. (1970) ON THE CHARACTERIZATION OF CENTRAL LUKASIEWICZ ALGEBRAS, J. ALGEBRA, 16, 486-495.

414 GERHARDTS, M.D. (1965) ZUR CHARAKTERISIERUNG DISTRIBUTIVER SCHEIFVERBANDE, MATH. ANNALEN, 161, 231-240.

415 GERHARDTS, M.D. (1969) SCHRAGVERBANDE UND QUASIORDNUNGEN, MATH. ANNALEN, 181, 65-73.

416 GILES, R. (1974A) A NONCLASSICAL LOGIC FOR PHYSICS, STUDIA LOGICA, 33 .

417 GILES, R. (1974B) A PRAGMATIC APPROACH TO THE FORMALIZATION OF EMPIRICAL THEORIES, PROCEEDINGS OF CONFERENCE ON FORMAL METHODS IN THE METHODOLOGY OF EMPIRICAL SCIENCES, WARSAW, JUNE.

418 GILES, R. (1974C) FORMAL LANGUAGES AND THE FOUNDATIONS OF PHYSICS. PROC. INTERNATIONAL RESEARCH SEMINAR ON ABSTRACT REPRESENTATION IN MATHEMATICAL PHYSICS, D. REIDEL, LONDON, ONTARIO, DEC. .

419 GILES, R. (1976A) A LOGIC FOR SUBJECTIVE BELIEF, IN HARPER, W., AND HOOKER, C.A. (EDS.), FOUNDATIONS OF PROBABILITY THEORY, STATISTICAL INFERENCE, AND STATISTICAL THEORIES OF SCIENCE, 1, D. REIDEL, DORDRECHT, HOLLAND, 41-72.

420 GILES, R. (1976B) FORMAL LANGUAGES AND THE FOUNDATIONS OF PHYSICS AND QUANTUM MECHANICS, IN HOOKER, C.A. (ED.), THE LOGICO-ALGEBRAIC APPROACH TO QUANTUM MECHANICS, 2, D. REIDEL, DORDRECHT, HOLLAND.

421 GILES, R. (1976C) LUKASIEWICZ LOGIC AND FUZZY SET THEORY, INT. J. MAN-MACHINE STUDIES, 8, 313-327.

422 GITMAN, I. (1970) ORGANIZATION OF DATA: A MODEL AND COMPUTATIONAL ALGORITHM THAT USES THE NOTION OF FUZZY SETS, PHD THESIS, MCGILL UNIVERSITY, MONTREAL, CANADA.

423 GITMAN, I., AND LEVINE, M.D. (1970) AN ALGORITHM FOR DETECTING UNIMODAL FUZZY SETS AND ITS APPLICATION AS A CLUSTERING TECHNIQUE, IEEE TRANS. COMP., C-19, 583-593.

424 GLUSS, B. (1973) FUZZY MULTISTAGE DECISION MAKING, INT. J. CONTROL, 17, 177-192.

425 GODDARD, L., AND ROUTLEY, R. (1973) THE LOGIC OF SIGNIFICANCE AND CONTEXT, SCOTTISH ACADEMIC PRESS, EDINBURGH.

426 GOGUEN, J.A. (1966) FUZZY SETS, UNIV. CAL. BERKELEY, DEPT OF MACH., PROJ. NRP 49-170, NR 314-103.

427 GOGUEN, J.A. (1967) L-FUZZY SETS, J. MATH. ANAL. AND APPLN., 18, 145-174.

428 GOGUEN, J.A. (1968) CATEGORIES OF FUZZY SETS: APPLICATIONS OF NON-CANTONIAN SET THEORY, PHD THESIS, DEPARTMENT OF MATHEMATICS, UNIVERSITY OF CALIFORNIA, BERKELEY, CALIFORNIA, USA.

429 GOGUEN, J.A. (1969) FUZZY SETS, FUZZY ALGEBRA, AND FUZZY STATISTICS, BULL AMER. MATH. SOC., 75, 622-624.

430 GOGUEN, J.A. (1969A) CATEGORIES OF V-SETS, BULLETIN OF THE AMERICAN MATHEMATICAL SOCIETY, 75, 622-624.

431 GOGUEN, J.A. (1969B) THE LOGIC OF INEXACT CONCEPTS, SYNTHESE, 19, 325-373.

432 GOGUEN, J.A. (1969C) REPRESENTING INEXACT CONCEPTS, ICR QUARTERLY REPORT NO. 20, INSTITUTE FOR COMPUTER RESEARCH, UNIVERSITY OF CHICAGO.

433 GOGUEN, J.A. (1970) MATHEMATICAL REPRESENTATION OF HIERARCHICALLY ORGANIZED SYSTEM, IN ATTINGER, E.O. (ED.), GLOBAL SYSTEM DYNAMICS, S. KARGER, BERLIN, 111-129.

434 GOGUEN, J.A. (1972) HIERARCHICAL INEXACT DATA STRUCTURES IN ARTIFICIAL INTELLIGENCE PROBLEMS, PROC. 5TH HAWAII INTERNATIONAL CONFERENCE ON SYSTEMS SCIENCES, HONOLULU, 345.

435 GOGUEN, J.A. (1973) AXIOMS, EXTENSIONS AND APPLICATIONS FOR FUZZY SETS, IBM-RESEARCH REPORT.

436 GOGUEN, J.A. (1973) SYSTEMS THEORY CONCEPTS IN COMPUTER SCIENCE, PROC. 6TH HAWAII INTERNATIONAL CONFERENCE ON SYSTEMS SCIENCES, HONOLULU, 77-80.

437 GOGUEN, J.A. (1974A) THE FUZZY TYCHONOFF THEOREM, J. MATH. ANAL. AND APPLN., 43, 734-742.

438 GOGUEN, J.A. (1974B) CONCEPT REPRESENTATION IN NATURAL AND ARTIFICIAL LANGUAGES: AXIOMS EXTENSIONS AND APPLICATIONS FOR FUZZY SETS, INT. J. MAN-MACHINE STUDIES, 6, 513-561.

439 GOGUEN, J.A. (1975A) OBJECTS, INT. J. GENERAL SYST., 1, 237-243.

440 GOGUEN, J.A. (1975B) ON FUZZY ROBOT PLANNING, IN ZADEH, L.A., FU, K.S., TANAKA, K., AND SHIMURA, M. (EDS.), FUZZY SETS AND THEIR APPLICATIONS TO COGNITIVE AND DECISION PROCESSES, ACADEMIC PRESS, NEW YORK, USA, 429-447.

441 GOGUEN, J.A. (1976) ROBUST PROGRAMMING LANGUAGES AND THE PRINCIPLE OF MAXIMAL MEANINGFULNESS, MILWAUKEE SYMPOSIUM ON AUTOMATIC COMPUTATION AND CONTROL, 87-90.

442 GOOD, I.J. (1962) SUBJECTIVE PROBABILITY AS THE MEASURE OF A NON-MEASURABLE SET,IN NAGEL, E., SUPPES, P., AND TARSKI, A. (EDS.), LOGIC, METHODOLOGY AND PHILOSOPHY OF SCIENCE, STANFORD UNVERSITY PRESS, CALIFORNIA, USA, 319-329.

443 GOODMAN, I.R. (1976) SOME RELATIONS BETWEEN FUZZY SETS AND RANDOM SETS, APRIL, (UNPUBLISHED) .

444 GOODMAN, J.S. (1974) FROM MULTIPLE BALAYAGE TO FUZZY SETS, INSTITUTE OF MATHEMATICS, UNIVERSITY OF FLORENCE, ITALY.

445 GOTTINGER, H.W. (1973) TOWARDS A FUZZY REASONING IN THE BEHAVIOURAL SCIENCE, CYBERNETICA, 113-135.

446 GOTTINGER, H.W. (1975) A FUZZY ALGORITHMIC APPROACH TO THE DEFINITION OF COMPLEX OR IMPRECISE CONCEPTS, CONFERENCE ON SYSTEMS THEORY, UNIVERSITY OF BIELEFELD, APRIL.

447 GOTTINGER, H.W. (1976) SOME BASIC ISSUES CONNECTED WITH FUZZY ANALYSIS, IN BOSSEL, H., KLACZKO, S., AND MULLER, N. (EDS.), SYSTEMS THEORY IN THE SOCIAL SCIENCES, BIRKHAUSER VERLAG, BASEL, 323-325.

448 GOTTINGER, H.W. (1976) TOWARD AN ALGEBRAIC THEORY OF COMPLEXITY AND CATASTROPHE, 3RD EUR. MEETING CYBERN. SYST. RES. VIENNA .

449 GOTTINGER, H.W. (1977) COMPLEXITY AND SOCIAL DECISION RULES, PROC. 1977 IEEE CONF. ON DECISION AND CONTROL, NEW ORLEANS.

450 GOTTINGER, H.W.,HIROTA, K., AND IIYAMA, Y. (1977) COMPLEXITY AND SOCIAL DECISION, PROCEEDINGS OF THE SYMP. ON FUZZY SET THEORY AND APPLN., IEEE CONF. ON DECISION AND CONTROL, NEW ORLEANS.

451 GOTTWALD, S. (1969) KONSTRUKTION VON ZAHLBEREICHEN UND DIE GRUNDLAGEN DER INHALTSTHEORIE IN EINER MEHRWERTIGEN MENGENLEHRE, PHD THESIS, UNIVERSITY OF LEIPZIG.

452 GOTTWALD, S. (1971A) ELEMENTARE INHALS- UND MASSTHEORIE IN EINER MEHRWERTIGEN MENGENLEHRE, MATH. NACHR., 50, 27-68.

453 GOTTWALD, S. (1971B) ZAHLBEREICHSKONSTRUKTIONEN IN EINER

MEHRWERTIGEN MENGENLEHRE, Z. MATH LOGIK GRUNDLAGEN MATH., 17, 145-188.

454 GOTTWALD, S. (1973) UBER EINBETTUNGEN IN ZAHLENBEREICHE EINER MEHRWERTIGEN MENGENLEHRE, MATH. NACHR., 56, 43-46.

455 GOTTWALD, S. (1974) FUZZY TOPOLOGY, PRODUCT AND QUOTIENT THEOREMS, J. OF MATH. ANAL. APPL. 45, 512-521.

456 GOTTWALD, S. (1974) MEHRWERTIGE ANORDNUNGSRELATIONEN IN KLASSISCHEN MENGEN, MATH. NACHR., 63, 205-212.

457 GOTTWALD, S. (1975A) EIN KUMULATIVES SYSTEM MEHRWERTIGER MENGEN, HABILITATIONESCHRIFT, UNIVERSITY OF LEIPZIG.

458 GOTTWALD, S. (1975B) A CUMULATIVE SYSTEM OF FUZZY SETS, PROC. 2ND COLLOQU. SET THEORY AND HIERARCHY THEORY, BIERUTOVICE, POLAND, SEP. .

459 GOTTWALD, S. (1976A) ON THE FORMALISM OF FUZZY LOGIC.

460 GOTTWALD, S. (1976C) UNTERSUCHEN ZUR MEHRWERTIGEN MENGENLEHRE, MATH. NACHR., 72, 297-303; 74, 329-336.

461 GOTTWALD, S. (1978) UNIVERSES OF FUZZY SETS CLOSED UNDER FUZZIFICATION, SEKTION MATHEMATIK, KARL-MARX-UNIVERSITAT, DDR-701 LEIPZIG, GERMANY.

462 GOTTWALD, S., (TO APPEAR) SET THEORY FOR FUZZY SETS OF HIGHER LEVEL, SEKTION MATH., KARL-MARX-UNIV., 701 LEIPZIG, G.D.R.

463 GOTTWALD, S., (1976) FUZZY PROPOSITIONAL LOGICS, SEKTION MATH., KARL-MARX-UNIV., 701 LEIPZIG, G.D.R.

464 GOTTWALD, S., FUZZY UNIQUENESS OF FUZZY MAPPINGS, SEKTION MATHEMATIK, KARL-MARX-UNIV., 701 LEIPZIG, G.D.R., 1-41.

465 GOVIND, R. (1978) SYNTHESIS OF FUZZY CONTROLLERS FOR PROCESS PLANTS, CARNEGIE-MELLON INSTITUTE OF RESEARCH, PITTSBURGH, PA., PROCEEDINGS OF THE 1978 INT. CONF. ON CYBERNETICS AND SOCIETY, VOL2, 1228-1230.

466 GRATTEN-GUINESS (1976) FUZZY MEMBERSHIP MAPPED ONTO INTERVAL AND MANY-VALUED QUANTITIES, Z. MATH. LOGIK GRUNDLAGEN MATH., 22, 149-160.

467 GRIGOLIA, R. (1975) ON THE ALGEBRAS COORESPONDING TO THE N-VALUED LUKASIEWICZ- TARSKI LOGICAL SYSTEMS, PROC. 1975 INT. SYMP. MULTIPLE-VALUED LOGIC, IEEE 75CH0959-7C, 234-239.

468 GROFMAN, B., AND HYMAN, G. (1973) PROBABILITY AND LOGIC IN BELIEF SYSTEMS, THEORY AND DECISION, 4, 179-195.

469 GUNDERSON, R. (1978) APPLICATION OF FUZZY ISODATA ALGORITHMS TO STAR-TRACKER POINTING SYSTEMS, 7TH TRIENNIAL IFAC

WORLD CONGRESS, HELSINKI.

470 GUPTA, M.M., NIKIFORUK, P.N., AND KANAI, K. (1973) DECISION AND CONTROL IN A FUZZY ENVIRONMENT: A RATIONALE, PROC. 3RD IFAC SYMP. IDENTIFICATION AND SYSTEM PARAMETER ESTIMATION, THE HAGUE, JUNE, 1048-1049.

471 GUPTA, M.M. (1974) INTRODUCTION TO FUZZY CONTROL, PROC. COMPUTER, ELECTRONICS AND CONTROL SYMP., CALGARY, MAY, VI 3.1-3.8.

472 GUPTA, M.M. (1975) ON THE ESTIMATION AND CONTROL IN A FUZZY ENVIRONMENT - REPORT ON THE ROUND TABLE DISCUSSION, IFAC JOURNAL OF AUTOMATICA, VOL. II, MARCH 209-212. ALSO IN THE PROCEEDINGS OF THE 3RD IFAC SYMPOSIUM ON IDENTIFICATION AND PARAMETER ESTIMATION (HAGUE).

473 GUPTA, M.M. (1975) FUZZY AUTOMATA AND DECISION PROCESSES: A DECADE, 6TH TRIENNIAL IFAC WORLD CONGRESS, BOSTON, MASS., USA, AUG.

474 GUPTA, M.M. (1976) THEORY OF FUZZY SETS AND ITS APPLICATIONS TO OPERATIONS RESEARCH AND MANAGEMENT SCI., ORSA/TIMS 1976, JOINT NAT. MEETING PHILADELPHIA, MARCH 31-APRIL 2.

475 GUPTA, M.M. (1976) FUZZINESS AND HUMAN BEHAVIOR IN DECISION MAKING, JOINT NAT. MEETING OF OPERATIONS RES. SOC. OF AMERICA (ORSA) AND THE INSTITUTE OF MANAGEMENT SCI. (TIMS), MIAMI, FLA., NOV., PAPER NO. FA 10.6.

476 GUPTA, M.M., AND MAMDANI, E.H. (1975) IFAC REPORT ON THE SECOND IFAC ROUND TABLE DISCUSSION OF FUZZY AUTOMATA AND DECISION PROCESSES, HELD AT THE 6TH TRIENNIAL IFAC WORLD CONGRESS, BOSTON/CAMBRIDGE, AUGUST. AUTOMATICA, 12, 291-296, ALSO IN THE CONGRESS PROCEEDINGS.

477 GUPTA, M.M. AND RAGADE, R.K. (1977) FUZZY SET THEORY AND ITS APPLICATIONS: SURVEY, (INVITED PAPER), PROCEEDINGS OF THE IFAC SYMP. ON MULTIVARIABLE SYSTEMS (MVTS), FREDERICTON, CANADA, JULY, 247-259.

478 GUPTA, M.M. (1977) FUZZY-ISM, THE FIRST DECADE, 5-10, IN FUZZY AUTOMATA AND DECISION PROCESSES, M.M. GUPTA, G.N. SARIDIS, B.R. GAINES, EDS., NORTH-HOLLAND.

479 GUPTA, M.M. (EDITOR), SARIDIS, G.N., AND GAINES, B.R. (ASSOCIATE EDITORS) (1977) FUZZY AUTOMATA AND DECISION PROCESSES, NEW YORK: NORTH-HOLLAND, 500 PAGES.

480 GUPTA, M.M. (1977) ED. - FUZZY SET THEORY AND APPLICATIONS, PROCEEDINGS OF THE 1977 IEEE CONFERENCE ON DECISION AND CONTROL, DEC. 8, NEW ORLEANS, 1301-1450.

481 GUPTA, M.M. (1977) THE NOTION OF FUZZINESS, A PERSPECTIVE, PROC. IEEE SYMP. ON FUZZY SETS, NEW ORLEANS, 1302.

482 GUPTA, M.M. (1978) FUZZY CONCEPTS AND ITS FUTURE, ORSA/TIMS, N.Y. CITY, MAY.

483 GUPTA, M.M. (1978) THE MANIFOLD USES OF GRADED MEMBERSHIP, RT-15 VII IFAC WORLD CONG., HELSINKI, JUNE.

484 GUPTA, M.M., AND NIKIFORUK, P.N. (1978) FEEDBACK CONTROL

OF INDUSTRIAL PROCESSES VIA FUZZY SET THEORY, SOME REMARKS, PROC. 1978 JACC, VOL. 3, 37-52.

485 GUPTA, M.M. (1979) GUEST EDITORIAL TO THE SPECIAL ISSUE ON FUZZY SETS AND APPLICATIONS, FUZZY SETS AND SYS. VOL. 2 NO. 1, JAN., 1-4.

486 GUPTA, M.M., RAGADE, R.K., AND YAGER, R.R. (1979) REPORT ON THE IEEE SYMP. ON FUZZY SET THEORY AND APPLICATIONS (SHORT COMMUNICATION), FUZZY SETS AND SYSTEMS, VOL. 2, NO. 1, JAN. 105-111.

487 GUPTA, M.M. (1978) IFAC REPORT ON THE 3RD ROUND TABLE DISCUSSION SESSION (RT-15) ON FUZZY DECISION MAKING AND APPLICATIONS, PROCEEDINGS OF THE 7TH TRIENNIAL IFAC WORLD CONG., HELSINKI, JUNE, ALSO IN AUTOMATA 1979.

488 GUPTA, M.M. (1978) A SURVEY OF PROCESS CONTROL APPLICATIONS OF FUZZY SET THEORY, 1454-1461, PROC. 1978 IEEE CONF. ON DEC. AND CONTROL, INCLUDING THE 17TH SYMP. ON ADAPTIVE PROCESSES, 78CH1392-OCS, JAN., 1979, SAN DIEGO, CALIF.

501 GUSEV, L.A., AND SMIRNOVA, I.M. (1973) FUZZY SETS: THEORY AND APPLICATIONS (A SURVEY), AUTOMATION AND REMOTE CONTROL, NO. 5, MAY, 66-85.

502 GUSTAFSON, D., AND KESSEL, W. (1979) FUZZY CLUSTERING WITH COVAR. MATRIX, 17TH IEEE CONF. ON DECISION AND CONTROL, 761-766.

503 HAACK, S. (1974) DEVIANT LOGIC, CAMBRIDGE UNIVERSITY PRESS.

504 HAACK, S. (1975) "ALTERNATIVE" IN "ALTERNATIVE LOGIC", IN BLACKBURN, S. (ED.), MEANING, REFERENCE AND NECESSITY, CAMBRIDGE UNIVERSITY PRESS, 32-55.

505 HAACK, S. (1976) THE JUSTIFICATION OF DEDUCTION, MIND, 85, 112-119.

506 HAAR, R.L. (1977) A FUZZY RELATIONAL DATA BASE SYSTEM, TR-586, DEPT. OF COMPUTER SCI. ANNUAL REPORT, UNIV. OF MARYLAND.

507 HACKING, I. (1963) WHAT IS STRICT IMPLICATION?, J. SYMBOLIC LOGIC, 28, 51-71.

508 HACKING, I. (1975A) ALL KINDS OF POSSIBILITY, PHILOSOPHICAL REVIEW, 84, 319-337.

509 HACKING, I. (1975B) THE EMERGENCE OF PROBABILITY, CAMBRIDGE UNIVERSITY PRESS.

510 HACKSTAFF, H.H. (1966) SYSTEMS OF FORMAL LOGIC, D. REIDEL, DORDRECHT, HOLLAND.

511 HAGG, C. (1977) POSSIBILITY AND COST IN DECISION ANALYSIS, FUZZY SETS AND SYST. 1, 81-86, NORTH-HOLLAND.

512 HAJEK, P. (1967) SETS, SEMISETS, MODELS, IN AXIOMATIC SET THEORY, PROC. SYMP. PURE MATH., 13, AMER. MATH. SOC., RHODE ISLAND, USA, 67-81.

513 HAJEK, P. (1968) PROBLEM OBECNEHO POJETI METODY GUHA, KYBERNETIKA (PRAGUE), 6, 505-515, (IN CZECH: THE QUESTION OF THE GENERAL CONCEPT OF GUHA-METHODS).

514 HAJEK, P. (1973B) SOME LOGICAL PROBLEMS OF AUTOMATED RESEARCH, PROC. SYMP. MATH. FOUND. COMP. SCI., HIGH TATRAS CZECHOSLOVAKIA.

515 HAJEK, P. (1973C) AUTOMATIC LISTING OF IMPORTANT OBSERVATIONAL STATEMENTS I, KYBERNETIKA (PRAGUE), 9, 187-206.

516 HAJEK, P. (1973D) AUTOMATIC LISTING OF IMPORTANT OBSERVATIONAL STATEMANTS II, KYBERNETIKA (PRAGUE), 9, 251-271.

517 HAJEK, P. (1974A) GENERALIZED QUANTIFIERS AND FINITE SETS, PROC. AUTUMN SCHOOL IN SET THEORY AND HIERARCHY THEORY, WROCLAW, POLAND.

518 HAJEK, P. (1974B) AUTOMATIC LISTING OF IMPORTANT OBSERVATIONAL STATEMENTS III, KYBERNETIKA (PRAGUE), 10, 95-124.

519 HAJEK, P. (1975) ON LOGICS OF DISCOVERY, IN BECVAR, J. (ED.), MATHEMATICAL FOUNDATIONS OF COMPUTER SCIENCE 1975, LECTURE NOTES IN COMPUTER SCIENCE, 32, SPRINGER-VERLAG, BERLIN, 30-45.

520 HAJEK, P., AND HARMANCOVA, D. (1973) ON GENERALIZED CREDENCE FUNCTIONS, KYBERNETIKA (PRAGUE), 9, 343-356.

521 HAJEK, P., AND HAVRANEK, T. (1976) ON GENERATION OF INDUCTIVE HYPOTHESES.

522 HAJEK, P., BENDOVA, K., AND RENC, Z. (1971) THE GUHA METHOD AND THE THREE VALUED LOGIC, KYBERNETIKA (PRAGUE), 7, 421-435.

523 HAJEK, P., HAVEL,I., AND CHYTIL, M. (1966) THE GUHA METHOD OF AUTOMATIC HYPOTHESES DETERMINATION, COMPUTING, 1, 293-308.

524 HALMOS, P.R. (1962) ALGEBRAIC LOGIC, CHELSEA PUBL. CO., NEW YORK.

525 HALPERN, J. (1975) SET ADJACENCY MEASURES IN FUZZY GRAPHS, J. OF CYBERN., VOL. 5, NO. 4 77-87.

526 HAMACHER, H. (1975) UBER LOGISCHE VERKNUPFUNGEN UNSCHARFER AUSSAGEN UND DEHREN ZUGEHORIGE BEWERTUNGSFUNKTICNEN, REP.75/14, LEHRSTUHL FUR UNTERNEHMENFORSCHUNG, RWTH, AACHEN, WEST GERMANY.

527 HAMACHER, H. (1976) ON LOGICAL CONNECTIVES OF FUZZY STATEMENTS AND THEIR AFFILIATED TRUTH-FUNCTIONS, 3RD

EUR. MEETING CYBERN. SYST. RES., VIENNA .

528 HAMACHER, H., LEBERLING, H., ZIMMERMANN, H.J. (1978) SENSITIVITY ANALYSIS IN FUZZY LINEAR PROGRAMMING, FUZZY SETS AND SYSTEMS 1, 269-281, NORTH-HOLLAND.

529 HAMBLIN, C.L. (1959) THE MODAL "PROBABLY", MIND, 68,234-240.

530 HANAKATA, K. (1974) A METHODOLOGY FOR INTERACTIVE SYSTEMS, IN FU, K.S., AND TOU, J.T. (EDS.), LEARNING AND INTELLIGENT ROBOTS, PLENUM PRESS, NEW YORK, 317-324.

531 HARA, F. (1975) A DYNAMIC MODEL OF COLLECTIVE HUMAN FLOW FROM BIG FIRES, IN SUMMARY OF PAPERS ON GENERAL FUZZY PROBLEMS, THE WORKING GROUP ON FUZZY SYSTEMS, TOKYO, JAPAN, NOV., 14-18.

532 HARA, F., AND KITAGAWA, S. (1977) A NEW FACE GRAPH AND ITS APPLICATION TO SYSTEM FAULT DIAGNOSIS, 126-134, SUMMARY OF PAPERS ON GENERAL FUZZY PROBLEMS, THE WORKING GROUP ON FUZZY SYSTEMS, TOYKO, JAPAN.

533 HAROCHE, C. (1975) GRAMMAR, IMPLICITNESS AND AMBIGUITY--FOUNDATIONS OF INHERENT AMBIGUITY OF DISCOURSE, FOUNDATIONS OF LANGUAGE, 13, 215-236, (IN FRENCH).

534 HARRIS, J.I. (1974) FUZZY IMPLICATION--COMMENTS ON A PAPER BY ZADEH, DOAE RESEARCH WORKING PAPER, MINISTRY OF DEFENCE, BYFLEET, SURREY, UK.

535 HARRIS, J.I. (1974) FUZZY SETS: HOW TO BE IMPRECISE PRECISELY, DOAE RESEARCH WORKING PAPER, MINISTRY OF DEFENCE, BYFLEET, SURREY, UK.

536 HART, W.D. (1972) PROBABILITY AS A DEGREE OF POSSIBILITY, NOTRE DAME J. FORMAL LOGIC, 13, 286-288.

537 HATTEN, M.L., WHINSTON, A.B., AND FU, K.S. (1975) FUZZY SET AND AUTOMATA THEORY APPLIED TO ECONOMICS, REPRINT SERIES NO. 533, PURDUE UNIVERSITY H.C. KRANNERT GRADUATE SCHOOL.

538 HAVRANEK, T. (1971) THE STATISTICAL MODIFICATION AND INTERPRETATION OF THE GUHA METHOD, KYBERNETIKA (PRAGUE), 7, 13-21.

539 HAVRANEK, T. (1974) SOME ASPECTS OF AUTOMATIC SYSTEMS OF STATISTICAL INFERENCE, PROC. EUROPEAN MEETING OF STATISTICIANS, PRAGUE .

540 HAVRANEK, T. (1975A) THE APPROXIMATION PROBLEM IN COMPUTATIONAL STATISTICS, IN BECVAR, J. (ED.), MATHEMATICAL FOUNDATIONS OF COMPUTER SCIENCE 1975, LECTURE NOTES IN COMPUTER SCIENCE, 32, SPRINGER-VERLAG, BERLIN, 260-265.

541 HAVRANEK, T. (1975B) STATISTICAL QUANTIFIERS IN OBSERVATIONAL CALCULI: AN APPLICATION IN GUHA-METHODS, THEORY AND

DECISION, 6, 213-230.

542 HAY, L.S. (1963) AXIOMATIZATION OF THE INFINITE-VALUED PREDICATE CALCULUS, J. SYMBOLIC LOGIC, 28, 77-86.

543 HEJEK, P. (1973A) WHY SEMISETS, COMMENTATIONES MATH. UNIV. CAROLINAE, 14, 397-420.

544 HEMPEL, C.G. (1937) A PURELY TOPOLOGICAL FORM OF NON-ARISTOTELIAN LOGIC, J. SYMBOLIC LOGIC, 2, 97-112.

545 HENDRY, W.L. (1972) FUZZY SETS AND RUSSELL'S PARADOX, LOS ALAMOS SCIENTIFIC LABORATORY, UNIVERSITY OF CALIFORNA, LOS ALAMOS, NEW MEXICO, USA .

546 HENKIN, L. (1963) A CLASS OF NON-NORMAL MODELS FOR CLASSICAL SENTENTIAL LOGIC, J. SYMBOLIC LOGIC, 28, 300.

547 HERSH, H., AND CARAMAZZA, A. (1975) INTEGRATING VERBAL QUANTITATIVE INFORMATION, BULL. OF THE PSYCHONOMIC SOCIETY, VOL. 6(6), 589-591.

548 HERSH, H.M. (1976) FUZZY REASONING: THE INTEGRATION OF VAGUE INFORMATION, PHD THESIS, THE JOHN HOPKINS UNIVERSITY, BALTIMORE, MD, USA.

549 HERSH, H.M., AND CARAMAZZA, A. (1975) A FUZZY SET APPROACH TO MODIFIERS AND VAGUENESS IN NATURAL LANGUAGES, DEPT. OF PSYCH., THE JOHN HOPKINS UNIVERSITY, BALTIMORE, MD.

550 HERSH, H.M., AND CARAMAZZA, A. (1975) THE QUANTIFICATION OF VAGUE CONCEPTS, PSYCHOMETRIC SOCIETY MEETING, IOWA CITY, USA, APRIL.

551 HERSH, H.M., AND CARAMAZZA, A. (1976) A FUZZY SET APPROACH TO MODIFIERS AND VAGUENESS IN NATURAL LANGUAGE, J. EXPERIMENTAL PSYCHOLOGY, 105, 254-276.

552 HERSH, H.M., AND SPIERING, J. (1976) HOW OLD IS OLD ?, EASTERN PSYCHOLOGICAL ASSOCIATION MEETING, NEW YORK, APRIL.

553 HINTIKKA, J., AND SUPPES, P.(EDS.) (1970) INFORMATION AND INFERENCE, D. REIDEL, HOLLAND.

554 HIRAI, H., ASAI, K., AND KITAJIMA, S. (1968) FUZZY AUTOMATA AND ITS APPLICATION TO LEARNING CONTROL SYSTEMS, MEMOIRS OF THE FACULTY OF ENGINEERING, 10, OSAKA CITY UNIVERSITY, 67-73.

555 HIRAMATSU, K., KABASAWA, K., AND KAIBARA, S. (1974) FUZZY LOGIC APPLIED TO THE MEDICAL DIAGNOSIS, MEDICAL ELECTRONIC AND BIOSCIENCE 12, 34-41.

556 HIROTA, K. (1977) "KAKURITSU-SHUGORON" FUNDAMENTAL RESEARCH WORKS OF FUZZY SYSTEM THEORY AND ARTIFICAL INTELLIGENCE,

(S.51 MOMBUSHO KAKEN-HI HOKOKU 193-213).

557 HIROTA, K. (1977) CONCEPTS OF PROBABILISTIC SETS, PROC. 1977 IEEE CONF. ON DECISION AND CONTROL, NEW ORLEANS.

558 HIROTA, K. (1977) PROBABILISTIC SETS - EXPANSION OF FUZZY CONCEPTS BASED ON PROBABILITY THEORY, 135-155, SUMMARY OF PAPERS ON GENERAL FUZZY PROBLEMS, THE WORKING GROUP ON FUZZY SYSTEMS, TOYKO, JAPAN.

559 HIROTA, K. (1978) EXTENDED FUZZY EXPRESSION OF PROBABILISTIC SETS, SEMIMAR ON APPL. FCTN. ANALYSIS, TSUKUBA UNIV. .

560 HIROTA, K., AND IIJIMA, T. (1976) A DECISION MAKING MODEL USING THE PROBABILISTIC SET THEORY, TECHNICAL REPORT ON PATTERN RECOGNITION AND LEARNING OF IECE, PRL76-21.

561 HIROTA, K., AND IIJIMA, T. (1976) PROBABILISTIC SET THEORY, PRL76-36, TECHNICAL REPORT ON PATTERN RECOGNITION AND LEARNING.

562 HIROTA, K., AND IIJIMA, T. (1978) A DECISION MAKING MODEL-A NEW APPROACH BASED ON THE CONCEPT OF PROBAB. SETS, IEEE INT. CONF. ON CYBER. AND SOCIETY.

563 HIROTA, K., AND IIJIMA, T. (1978) THE BOUNDED VARIATION QUANTITY AND ITS APPL. TO FEATURE EXTRACTIONS, 4TH INT. JOINT CONF. ON PATTERN RECOGN., KYOTO.

564 HISDAL, E. (1978) CONDITIONAL POSSIBILITIES INDEPENDENCE AND NONINTERACTION, FUZZY SETS AND SYST. 1, 283-297, NORTH-HOLLAND.

565 HISDALE, E. (1978) COND. AND JOINT POSSIBILITY OF TYPE Z, PARTICULARIZATION, INST. OF INFORMATICS, UNIV. OF OSLO.

566 HISDALE, E. (1978) PARTICULARIZATION-THE THEORY OF FUZZY SETS VERSUS CLASSICAL THEORIES, INST. OF INFO. UNIV. OF OSLO.

567 HISDALE, E. (1978) POSSIBILITIES AND GRADES OF MEMBERSHIP CONCRETE AND MATH. SETS, INST. OF INFORMATICS, UNIV. OF OSLO.

568 HOCKNEY, D., HARPER, W., AND FREED, B. (1975) CONTEMPORARY RESEARCH IN PHILOSOPHICAL LOGIC AND LINGUISTIC SEMANTICS, REIDEL, HOLLAND.

569 HOFFMAN, L.J., AND CLEMENTS, D. (1977) FUZZY COMPUTER SECURITY METRICS, A PRELIMINARY REPORT, UNIV. OF CALIF. AT BERKELEY, ERL MEMO M77-6.

570 HOGAN, M., AND ODEN, G.C. (1978) A FUZZY PROPOSITIONAL MODEL OF LINGUISTIC MODIFIERS, 1ST JOINT PSYCHOMETRIC SOCIETY AND THE SOCIETY FOR MATH PSYCHOLOGY.

571 HOGARTH, R.M. (1975) COGNITIVE PROCESSES AND THE ASSESSMENT OF SUBJECTIVE PROBABILITY DISTRIBUTIONS, J. AMER. STATIST. ASSN., 70, 271-294.

572 HOHLE, U. (1977) PROBABILISTIC UNIFORMIZATION OF FUZZY TOPOLOGIES, FUZZY SETS AND SYST. 1, 311-332, NORTH-HOLLAND.

573 HONDA, N. (1971) FUZZY SETS, J. INST. ELECTRON COMM. ENG. (JAPAN), 54, 1359-1363.

574 HONDA, N. (1975) APPLICATIONS OF FUZZY SET TO AUTOMATA AND LANGUAGE THEORY, 249-258, SYSTEMS AND CONTROL 19..

575 HONDA, N. (1975) APPLICATIONS OF FUZZY SETS THEORY TO AUTOMATA AND LINGUISTICS, J. JAACE, 19, 249-254.

576 HONDA, N., AND AIDA, S. (1975) ENVIRONMENTAL INDEX BY FACES METHOD, 19-22, SUMMARY OF PAPERS ON GENERAL FUZZY PROBLEMS, THE WORKING GROUP ON FUZZY SYSTEMS, TOYKO, JAPAN.

577 HONDA, N., AND AIDA, S. (1975) ENVIRONMENTAL INDEX BY FACES METHOD, IN SUMMARY OF PAPERS ON GENERAL FUZZY PROBLEMS, THE WORKING GROUP ON FUZZY SYSTEMS, TOKYO, JAPAN, NOV., 19-22.

578 HONDA, N., AND AIDA, S. (1977) AN APPROACH TO THE PRODUCTION LINE BY MAN-COMPUTER SYSTEM, 156-166, SUMMARY OF PAPERS ON GENERAL FUZZY PROBLEMS, THE WORKING GROUP ON FUZZY SYSTEMS, TOYKO, JAPAN.

579 HONDA, N., AND NASU, M. (1975) F-RECOGNITION OF FUZZY LANGUAGES, IN SPECIAL INTEREST DISCUSSION SESSION ON FUZZY AUTOMATA AND DECISION PROCESSES, 6TH IFAC WORLD CONGRESS, BOSTON, MASS., USA, AUG. .

580 HONDA, N., AND NASU, M. (1975) RECOGNITION OF FUZZY LANGUAGES, IN ZADEH, L.A., FU, K.S., TANAKA, K., AND SHIMURA, M. (EDS.), FUZZY SETS AND THEIR APPLICATIONS TO COGNITIVE AND DECISION PROCESSES, ACADEMIC PRESS, NEW YORK, 279-299.

581 HONDA, N., NASU, M., AND HIROSE, S. (1976) RECOGNITION OF FUZZY LANGUAGES, 1976 JOINT CONVENTION OF FOUR INSTITUTES OF ELECTRICAL ENGINEERS OF JAPAN.

582 HONDA, N., NASU, M., AND HIROSE, S. (1977) F-RECOGNITION OF FUZZY LANGUAGES, 149-168, IN FUZZY AUTOMATA AND DECISION PROCESSES, M.M. GUPTA, G.N. SARIDIS, B.R. GAINES, EDS., NORTH-HOLLAND.

583 HOREJS, J. (1965) CLASSIFICATIONS AND THEIR RELATIONSHIP TO A MEASURE, PUBLICATIONS DE LA FACULTE DES SCIENCES DE L'UNIVERSITE J.E. PURKYNE, NO.168, BRNO, CZECH., 475-493.

584 HORMANN, A.M. (1971) MACHINE-AIDED VALUE JUDGEMENTS USING

FUZZY SET TECHNIQUES, SP-3590, SYSTEM DEVELOPMENT CORPORATION, SANTA MONICA, CALIFORNIA, USA .

585 HOSKNEY, D., HARPER, W., AND FREED, B. (1975) CONTEMP. RES, HOLLAND: D. REIDEL.

586 HUGHES, G.E., AND CRESWELL, M.J. (1968) AN INTRODUCTION TO MODAL LOGIC, METHUEN, LONDON .

587 HUGHES, P., AND BRECHT, G. (1976) VICIOUS CIRCLES AND INFINITY, JONATHON CAPE, LONDON .

588 HUNG, N.T. (1975) INFORMATION FONCTIONELLE ET ENSEMBLES FLOUS, SEMINAR ON QUESTIONAIRES, UNIVERSITY OF PARIS 6, PARIS, FRANCE .

589 HUTTON, B. (1975) NORMALITY IN FUZZY TOPOLOGICAL SPACES, J. MATH. ANAL. AND APPLN., 50, 74-79.

590 HUTTON, B. (1977) UNIFORMITIES ON FUZZY TOPOLOGICAL SPACES, J. MATH. ANAL. APPL. 58, 559-571.

591 HUTTON, B., AND REILLY, J.L. (1974) SEPARATION AXIOMS IN FUZZY TOPOLOGICAL SPACES, UNIVERSITY OF AUCKLAND, NEW ZEALAND, MARCH.

592 ICHIKAWA, A. (1976) STRUCTURE OF MULTIDIMENSIONAL CRITERIA, 14-24, SUMMARY OF PAPERS ON GENERAL FUZZY PROBLEMS, THE WORKING GROUP ON FUZZY SYSTEMS, TOYKO, JAPAN.

593 ICHIKAWA, A., NAKAO, K., AND KOBAYASHI, S. (1975) AN ANALYSIS OF SOCIAL GROUP BEHAVIOUR BY MEANS OF A THRESHOLD ELEMENT NETWORK MODEL, 23-28, SUMMARY OF PAPERS ON GENERAL FUZZY PROBLEMS, THE WORKING GROUP ON FUZZY SYSTEMS, TOYKO, JAPAN.

594 IDESAWA, M. (1975) AUTOMATIC INPUT OF LINE DRAWING AND GENERATION OF SOLID FIGURE, IN SUMMARY OF PAPERS ON GENERAL FUZZY PROBLEMS, THE WORKING GROUP ON FUZZY SYSTEMS, TOKYO, JAPAN, NOV., 29-33.

595 IIYAMA, Y., YAMAUCHI, K., AND YANAGAWA, K. (1977) ANALYTICAL STUDY OF THE COMPUTER AIDED CONTROL SYSTEM FOR THE SHINKANSEN: AN ACTUAL BEHAVIOR, PROC. 1977 IEEE CONF. ON DECISION AND CONTROL, NEW ORLEANS.

596 INAGAKI, Y. (1976) MATHEMATICAL FUNDATION OF FUZZY SETS THEORY, 1976 JOINT CONVENTION OF FOUR INSTITUTES OF ELECTRICAL ENGINEERS OF JAPAN.

597 INAGAKI, Y., AND FUKUMURA, T. (1975) ON THE DESCRIPTION OF FUZZY MEANING OF CONTEXT-FREE LANGUAGE, IN ZADEH, L.A., FU,K.S., TANAKA, K., AND SHIMURA, M. (EDS.), FUZZY SETS AND THEIR APPLICATIONS TO COGNITIVE AND DECISION PROCESSES, ACADEMIC PRESS, NEW YORK, 301-328.

598 INAGAKI, Y., TANAKA, S., AND FUKUMURA, T. (1976) SOME

CONSIDERATION ON PROBLEMATIC INFERENCE BASED ON FUZZY LOGIC, TECHNICAL REPORT ON AUTOMATON AND LANGUAGE OF IECE, AL76-3.

599 ISHIKAWA, A. (1977) FUZZY FUNCTION ANALYSIS, PROCEEDINGS OF THE SYMP. ON FUZZY SET THEORY AND APPLN., IEEE CONF. ON DECISION AND CONTROL, NEW ORLEANS.

600 ISHIKAWA, A., AND MIENO, H. (1975) DESIGN OF A VIDEO INFORMATION SYSTEM AND THE FUZZY INFORMATION THEORY, EUROCOMP 75, BRUNEL UNIVERSITY, UK.

601 ISHIKAWA, A., AND MIENO, H. (1975) DESIGN OF A VIDEO INFORMATION SYSTEM BASED UPON THE FUZZY INFORMATION THEORY, BULL., ORSA 23 SUPP. 2, B375.

602 ISHIKAWA, A., AND MIENO, H. (1977) FUZZY FUNCTIONS ANALYSIS, PROC. 1977 IEEE CONF. ON DECISION AND CONTROL, NEW ORLEANS, 1315-1317.

603 ISOMICHI, Y. (1975) SEGMENTATION-FREE RECOGNITION OF TIME SERIES PATTERNS, SEMINAR ON STUDIES ON TIME SERIES PATTERN RECOGNITION SYSTEMS, KYOTO UNIV. .

604 ITO, T., AND KIZAWA, M. (1976) ON THE SEMANTIC STRUCTURE OF NATURAL LANGUAGE, TRANS. IECE(D) 59-D, 141-148.

605 ITZINGER, D. (1977) MEASURING LOGICAL STRUCTURES OF SMALL SOCIAL SYSTEMS, PROC. 1ST INT. CONF. OF MATH MODELLING, ST. LOUIS, 2607-2616.

606 ITZINGER, O. (1974) ASPECTS OF AXIOMATIZATION OF BEHAVIOUR: TOWARDS AN APPLICATION OF RASCH'S MEASUREMENT MODEL TO FUZZY LOGIC, IN BRUCKMAN, G., FRESCHL, F., AND SCHMATTERER, L. (EDS.), COMSTAT 1974 (PROC. SYMP. COMPUTATIONAL STATISTICS, UNIVERSITY OF VIENNA), PHYSICA-VERLAG, 173-182.

607 JACOBSON, D.H. (1976) ON FUZZY GOALS AND MAXIMIZING DECISIONS IN STOCHASTIC OPTIMAL CONTROL, 'J. MATH. ANAL. AND APPLN. .

608 JAHN, K. U. (1971) AUFBAU EINER 3-WERTIGEN LINEAREN ALGEBRA UND AFFINEN GEOMETRIE AUF GRUNDLAGE DER INTERVALL-ARITHMETIK, PHD THESIS, UNIVERSITY OF LEIPZIG.

609 JAHN, K.U. (1974) EINE THEORIE DER GLEICHUNGESYSTEME MIT INTERVALL-KOEFFIZIENTEN, Z. ANGEW. MATH. MECH., 54, 405-412.

610 JAHN, K.U. (1975) EINE AUF DER INTERVALL-ZAHLEN FUSSENDE 3-WERTIGE LINEARE ALGEBRA, MATH. NACHR., 65, 105-116.

611 JAHN, K.U. (1975) INTERVALL-WERTIGE MENGEN, MATH. NACHR., 68, 115-132.

612 JAHN, K.U. (1976), ANVENDUNGEN VON FUZZY SETS, VORTRAGE

AUS DEM PROBLEMSEMINAR AUTOMATA- UND ALGORITHMEN THEORIE, APRIL, WEISSIG, 30-43.

613 JAIN, R. (1975) OUTLINE OF AN APPROACH FOR THE ANALYSIS OF FUZZY SYSTEMS, IN SPECIAL INTEREST DISCUSSION SESSION ON FUZZY AUTOMATA AND DECISION PROCESSES, 6TH IFAC WORLD CONGRESS, BOSTON, MASS., USA, AUG. .

614 JAIN, R. (1975) PATTERN CLASSIFICATION USING PROPERTY SETS, SYMPOSIUM ON CIRCUITS, SYSTEMS AND COMPUTERS, UNIVERSITY OF CALCUTTA, INDIA, FEB. .

615 JAIN, R. (1976) CONVOLUTION OF FUZZY VARIABLES, JIETE, 22

616 JAIN, R. (1976) DECISION MAKING IN THE PRESENCE OF FUZZY VARIABLES, IEEE TRANS. SYST., MAN, CYBERN., VOL. SMC-6, 698-703.

617 JAIN, R. (1976) DECISION MAKING WITH FUZZY KNOWLEDGE ABOUT THE STATE OF THE SYSTEM, NATIONAL SYSTEMS CONFERENCE, ROORKE, INDIA, FEB. .

618 JAIN, R. (1977) A PROCEDURE FOR MULTIPLE-ASPECT DECISION MAKING USING FUZZY SETS, NT. J. SYS. SCI., VOL. B, 1-7.

619 JAIN, R. (1977) ANALYSIS OF FUZZY SYSTEMS, 251-268, IN FUZZY AUTOMATA AND DECISION PROCESSES, M.M. GUPTA, G.N. SARIDIS, B.R. GAINES, EDS., NORTH-HOLLAND.

620 JAIN, R. (1977) DECISION-MAKING IN THE PRESENCE OF FUZZINESS AND UNCERTAINTY, PROC. 1977 IEEE CONF. ON DECISION AND CONTROL, NEW ORLEANS, 1318-1323.

621 JAIN, R. (1977) TOLERANCE ANALYSIS USING FUZZY SETS, INT. JR. ON SYSTEM SCI. VOL 7, NO 12, PP 1393-1401.

622 JAIN, R. (1978) APPLICATION OF FUZZY SET THEORY FOR THE ANALYSIS OF COMPLEX SCENES, UNIV. OF TEXAS, AUSTIN, 1444-1449, PROC. 1978 IEEE CONF. ON DEC. AND CONTROL, INCLUDES THE 17TH SYMP. ON ADAPTIVE PROCESSES, 78CH1392-OCS, JAN., 1979, SAN DIEGO, CALIF. .

623 JAIN, R., AND NAGEL, H.H. (1977) ANALYZING A REAL WORLD SCENE SEQUENCE USING FUZZINESS, PROC. 1977 IEEE CONF. ON DECISION AND CONTROL, NEW ORLEANS, 1367-1372.

624 JAIN, R., AND NAGEL, H.H. (1977) ANALYZING A SCENE SEQUENCE USING FUZZINESS, PROC. 1977 IEEE DECISION AND CONTROL, VOL. 2, 1367-1372, NEW ORLEANS, DEC. .

625 JAIN, R., AND STALLINGS, W. (1978) COMMENTS ON FUZZY SET THEORY VERSUS BAYESIAN STATISTICS, VOL. 8, IEEE TRANS. ON SYST., MAN, AND CYBERN., 332-333.

626 JAKUBOWSKI, R. (1977) APPLICATION OF FORMAL LANGUAGE AND FUZZY AUTOMATA IN DESIGNING, INT. CONF. ON INFORMATION

PROCESSING, IFIP-INFOPOL-76, J. MADEY ED., NORTH-HOLLAND.

627 JAKUBOWSKI, R., AND KASPRZAK, A. (1972) ALGORYTM AUTOMATYCZNEGO PROJEKTOWANIA PROCESOW TECHNOLOGICNYCH OBROKKI SKRAWANIEM, PODSTAWY STEROWANIA, TOM 2, Z.4.

628 JAKUBOWSKI, R., AND KASPRZAK, A. (1973) APPLICATION OF FUZZY PROGRAMS TO THE DESIGN OF MACHINING TECHNOLOGY, BULLETIN OF THE POLISH ACADEMY OF SCIENCE, 21(21),17-22.

629 JAKUBOWSKI, R., AND SZELC, A. (1977) APPLICATIONS OF FORMAL LANGUAGES AND FUZZY AUTOMATA IN PROBLEM SOLVING, PODSTAWY STEROWANIA, TOM 7, Z.1, 69-72.

630 JAKUBOWSKI, R., AND SZELC, A. (1977) QUASI-LINGUISTIC REPRESENTATION IN DESCRIPTION OF MICROPROGRAMS, BULLETIN DE L'ACADEMIE POLONAISE DES SCIENCES, 10, SERIE DES SCIENCES TECHNIQUES, VOL. XXV, NO. 12.

631 JAKUBOWSKI, R., AND SZELC, A. (1977) QUASI-LINGUISTIC REPRESENTATION OF SYSTEMS, BULLETIN DE L'ACADEMIE POLONAISE DES SCIENCES, 9. SERIE DES SCIENCES TECHNIQUES, VOL. XXV, NO. 12.

632 JARVIS, R.A. (1975) OPTIMIZATION STRATEGIES IN ADAPTIVE CONTROL: A SELECTIVE SURVEY, IEEE TRANS. SYST. MAN CYBERN., SMC-5, 83-94.

633 JASKOWSKI, S. (1969) PROPOSITIONAL CALCULUS FOR CONTRADICTORY DEDUCTIVE SYSTEMS, STUDIA LOGICA, 24, 143-159, (TRANS. OF 1948 POLISH PAPER).

634 JENSEN, J.H. (1976) APPLICATION FOR FUZZY LOGIC CONTROL, NO. 1, NO. 7607, ELECTRIC POWER ENGINEERING DEPT., TECHNICAL UNIVERSITY OF DENMARK, LYNGBY, JUNE .

635 JOBE, W.H. (1962) FUNCTIONAL COMPLETENESS AND CANONICAL FORMS IN MANY-VALUED LOGICS, J. SYMBOLIC LOGIC, 28, 409-421.

636 JONES, W.T. (1976) A FUZZY SET CHARACTERIZATION OF INTERACTION IN SCIENTIFIC RESEARCH, J. AM. SOC. INF. SCI. (SEPT-OCT).

637 JONES, W.T., AND RAGADE, R.K. (1978) IDENTIFICATION OF STABLE JOURNAL INTERACTION CLUSTERS, PROC. 22ND MEETING SOC. FOR GENERAL SYSTEMS RESEARCH, WASH. D.C., 431-442.

638 JORDON, P. (1952) ALGEBRAISCHE BETRACHTUNGEN ZUR THEORIE DES WIRKUNGSKVANTUM, MATH. SEM. HAMBURG, 18, 99-119.

639 JORDON, P. (1962) HALBGRUPPEN VON IDEMPOTENTEN UND NICHTKOMMUTATIVE VARBANDE, J. REINE ANGEW, MATH. 211, 136-161.

640 JOUAULT, J.P., AND LUAN, P. M. (1975) APPLICATION DES CONCEPTS FLOUS A LA PROGRAMMATION EN LANGUAGES

QUASI-NATURELS,, INSTITUT INFORMATIQUE D'ENTREPRISE, C.N.A.M., PARIS, FRANCE.

641 JUMARIE, G. (1977) SOME TECHNICAL APPLICATIONS OF RELATIVISTIC INFORMATION THEORY, SHANNON INFORMATION, FUZZY SETS, LINGUISTICS, RELATIVISTIC SETS AND COMMUNICATION; CYBERNETICA, VOL. 20, NO. 2, 91-128.

642 JUMARIE, G. (1978) RELATIVISTIC FUZZY SETS AS A MEAN TO INTRODUCE HUMAN FACTORS IN PATTERN RECOGNITION SYSTEMS, DEPT. OF MATH., UNIV. DU QUEBEC A MONTREAL, PROCEEDINGS OF THE 1978 INT. CONF. ON CYBERNETICS AND SYST., 930-935.

643 JUMARIE, G. (1979) NEW RESULTS IN RELATIVISTIC INFORMATION AND GENERAL SYSTEMS. OBSERVED PROBABILITY, RELATIVISTIC FUZZY SETS, GENERATIVE SEMANTICS, TO APPEAR, CYBERNETICA, 1979.

644 KACPRZYK, J. (1976) FUZZY SET THEORETIC APPROACH TO THE OPTIMAL ASSIGNMENT OF WORK PLACES, IFAC SYMP. ON LARGE SCALE SYSTEMS THEORY AND APPL.

645 KACPRZYK, J. (1976) LINGUISTIC VARIABLES AND FUZZY CONDITIONAL STATEMENTS FOR THE DESCRIPTION OF COMPLEX SYSTEMS, PROC. III, NAT. CONF. INFORMATION SCI., KATOWICE.

646 KACPRZYK, J. (1977) DECISION-MAKING IN A FUZZY ENVIRONMENT WITH FUZZY TERMINATION TIME, FUZZY SETS AND SYST. 1, 169-179, NORTH-HOLLAND.

647 KACPRZYK, J. (1977) FUZZY INTEGRAL AS MODEL OF OPERATOR IN COMPLEX AUTOMATION SYSTEMS, PROC. VII NAT. CONF. AUTOM. CONTR. .

648 KACPRZYK, J. (1978) BRANCH-AND-BOUND ALGORITHMS FOR THE DECISION-MAKING IN A FUZZY ENVIRONMENT, SYST. RESEARCH INSTIT., POLISH ACADEMY OF SCI., UL. NEWELSKA 6, 01-447 WARZAWA, POLAND.

649 KACPRZYK, J. (1978) FUZZY TERMINATION TIME ON DECISION-MAKING IN A FUZZY ENVIRONMENT, 4TH INT. CONGRESOCYBERNETICS AND SYSTEMS, AMSTERDAM, 368-369.

650 KACPRZYK, J., (TO APPEAR) A BRANCH-AND-BOUND ALGORITHM FOR THE MULTISTAGE CONTROL OF A NONFUZZY SYSTEM IN A FUZZY ENVIRONMENT, CONTR. CYBERNET. .

651 KAHNE, S. (1975) A PROCEDURE FOR OPTIMIZING DEVELOPMENT DECISIONS, AUTOMATICA, 11, 261-269.

652 KALMAN, J.A. (1958) LATTICES WITH INVOLUTION, TRANS. AMER. MATH. SOC., 87, 485-491.

653 KALMANSON, D., AND STEGALL, F. (1973) RECHERCHE CARDIO-VACURAIRE ET THEORIE DES ENSEMBLES FLOUS, LA NOUVELLE PRESSE MEDICALE, 41, 2757-2760.

654 KALSARAS, A.K., AND LIU, D.B. (1977) FUZZY VECTOR SPACES AND FUZZY TOPOLOGICAL VECTOR SPACES, J. OF MATH ANAL. AND APPL., 58, 135-146.

655 KAMEDA, T., AND SADEH, E. (1977) BOUNDS ON THE NUMBER OF FUZZY FUNCTIONS, INFORMATION AND CONTROL. 35, 139-145.

656 KAMPE, J. DE FERIET (1970) MESURE DE L'INFORMATION FOURIE PAR UN EVENEMENT, COLL. INTER. C.N.R.S., PARIS 186, 191-221.

657 KANDEL, A. (1972A) TOWARD SIMPLIFICATION OF FUZZY FUNCTIONS, CSR114, COMPUTER SCIENCE DEPT., NEW MEXICO INSTITUTE OF MINING AND TECHNOLOGY, SOCORRO, NEW MEXICO, USA, JUNE.

658 KANDEL, A. (1972B) ON CODED GRAMMARS AND FUZZY STRUCTURES, CSR118, COMPUTER SCIENCE DEPT., NEW MEXICO INSTITUTE OF MINING AND TECHNOLOGY, SOCORRO, NEW MEXICO, USA, SEP. .

659 KANDEL, A. (1972C) A NEW ALGORITHM FOR MINIMIZING INCOMPLETELY SPECIFIED FUZZY FUNCTIONS, CSR127, COMPUTER SCIENCE DEPT., NEW MEXICO INSTITUTE OF MINING AND TECHNOLOGY, SOCORRO, NEW MEXICO, USA, NOV. .

660 KANDEL, A. (1973) APPLICATION OF FUZZY LOGIC TO THE DETECTION OF STATIC HAZARDS IN COMBINATIONAL SWITCHING SYSTEMS, COMPUTER SCI. REPORT 122, NEW MEXICO INSTITUTE OF MINING AND TECH., SOCORRO, N.M., APRIL.

661 KANDEL, A. (1973A) A NEW METHOD FOR GENERATING FUZZY PRIME IMPLICANTS AND AN ALGORITHM FOR THE AUTOMATIC MINIMIZATION OF INEXACT STRUCTURES, CSR126, COMPUTER SCIENCE DEPT., NEW MEXICO INSTITUTE OF MINING AND TECHNOLOGY, SOCORRO, NEW MEXICO, USA, OCT. .

662 KANDEL, A. (1973B) COMMENT ON AN ALGORITHM THAT GENERATES FUZZY PRIME IMPLICANTS BY LEE AND CHANG, INFORM. AND CONTROL, 22, 279-282.

663 KANDEL, A. (1973C) FUZZY CHAINS: A NEW CONCEPT IN DECISION-MAKING UNDER UNCERTAINTY, COMPUTER SCIENCE REPORT 123, NEW MEXICO INSTITUTE OF MINING AND TECHNOLOGY, AUG.

664 KANDEL, A. (1973D) ON MINIMIZATION OF FUZZY FUNCTIONS, IEEE TRANS. COMP., C-22, 826-832.

665 KANDEL, A. (1973E) ON THE ANALYSIS OF FUZZY LOGIC, PROC. 6TH INT. CONF. SYST. SCIENCES, HONOLULU,HAWAII, JAN.

666 KANDEL, A. (1973F) COMMENTS ON "MINIMIZATION OF FUZZY FUNCTIONS", IEEE TRANS. COMP., C-22, 217.

667 KANDEL, A. (1973G) FUZZY FUNCTIONS AND THEIR APPLICATION

TO THE ANALYSIS OF SWITCHING HAZARDS, PROC. 2ND TEXAS CONF. ON COMPUTING SYSTEMS, AUSTIN, TEXAS, USA, NOV., 42:1-6.

668 KANDEL, A. (1974A) SYNTHESIS OF FUZZY LOGIC WITH ANALOG MODULES: PRELIMINARY DEVELOPMENTS, COMPUTERS IN EDUCATION TRANSACTION (ASEE DIV.), 6, 71-79.

669 KANDEL, A. (1974B) ON FUZZY MAPS: SOME INITIAL THOUGHTS, CSR131, COMPUTER SCIENCE DEPARTMENT, NEW MEXICO INSTITUTE OF MINING AND TECHNOLOGY, SOCORRO, NEW MEXICO, USA.

670 KANDEL, A. (1974C) SIMPLE DISJUNCTIVE DECOMPOSITIONS OF FUZZY FUNCTIONS, CSR132, COMPUTER SCIENCE DEPT., NEW MEXICO INSTITUTE OF MINING AND TECHNOLOGY, SOCORRO, NEW MEXICO, USA, JULY.

671 KANDEL, A. (1974D) ON THE THEORY OF FUZZY MATRICES, CSR135, COMPUTER SCIENCE DEPT., NEW MEXICO INSTITUTE OF MINING AND TECHNOLOGY, SOCORRO, NEW MEXICO, USA, OCT. .

672 KANDEL, A. (1974E) GENERATION OF THE SET REPRESENTING ALL FUZZY PRIME IMPLICANTS, CSR136, COMPUTER SCIENCE DEPT., NEW MEXICO INSTITUTE OF MINING AND TECHNOLOGY SOCORRO, NEW MEXICO, USA, OCT. .

673 KANDEL, A. (1974F) ON THE ENUMERATION OF FUZZY FUNCTIONS, 12TH HOLIDAY SYMB. "DEVELOPMENT IN OMBINATORICS", NEW MEXICO STATE UNIVERSITY, LAS CRUCES, NEW MEXICO, USA, DEC. .

674 KANDEL, A. (1974G) APPLICATION OF FUZZY LOGIC TO THE DETECTION OF STATIC HAZARDS IN COMBINATIONAL SWITCHING SYSTEMS, INT. J. COMP. INF. SCIENCES, 3, 129-139.

675 KANDEL, A. (1974H) ON THE PROPERTIES OF FUZZY SWITCHING FUNCTIONS, J. CYBERNETICS, 4, 119-126.

676 KANDEL, A. (1974I) ON THE MINIMIZATION OF INCOMPLETELY SPECIFIED FUZZY FUNCTIONS, INFORM. AND CONTROL, 26, 141-153.

677 KANDEL, A. (1974J) CODES OVER LANGUAGES, IEEE TRANS. SYST. MAN CYBERN., SMC-4, 135-138.

678 KANDEL, A. (1974K) FUZZY REPRESENTATION CNF MINIMIZATION AND THEIR APPLICATION TO FUZZY TRANSMISSION STRUCTURES, 1974 SYMPOSIUM ON MULTIPLE-VALUED LOGIC, IEEE 74CH0845-8C, 361-379.

679 KANDEL, A. (1975A) FUZZY HIERARCHICAL CLASSIFICATIONS OF DYNAMIC PATTERNS, NATO ASI PATTERN RECOGNITION AND CLASSIFICATION, FRANCE, SEP. .

680 KANDEL, A. (1975B)FUZZY MAPS AND THEIR APPLICATION IN THE SIMPLIFICATION OF FUZZY SWITCHING FUNCTION, 6TH INT. SYMP. MULTIPLE-VALUED LOGIC, IEEE 76CH1111-4C, MAY .

681 KANDEL, A. (1975C) PROPERTIES OF FUZZY MATTRICES AND THEIR APPLICATIONS TO HIERARCHICAL STRUCTURES, 9TH ASILOMAR CONF. CIRCUITS, SYSTEMS AND COMPUTERS, PACIFIC GROVE, CALIFORNIA, USA, NOV. .

682 KANDEL, A. (1975D) BLOCK DECOMPOSITION OF IMPRECISE MODELS, 9TH ASILOMAR CONF. CIRCUITS, SYSTEMS AND COMPUTERS, PACIFIC GROVE, CALIFORNIA, USA, NOV. .

683 KANDEL, A. (1976) FUZZY MAPS AND THEIR APPL. IN THE SIMPLIFICATION OF FUZZY SWITCHING SYS., PROC. 6TH INT. SYMP. ON MULTIVALUED LOGIC, UTAH ST., LOGAN, UTAH.

684 KANDEL, A. (1976A) INEXACT SWITCHING LOGIC, IEEE TRANS. SYST. MAN CYBERN. 6, 215-219.

685 KANDEL, A. (1976C) FUZZY SYSTEMS AND THEIR APPLICATIONS TO SIMULATIONS, PROC. 9TH HAWAII INT. CONF. SYST. SCI., HONOLULU, HAWAII, JAN. .

686 KANDEL, A. (1976D) ON THE DECOMPOSITION OF FUZZY FUNCTIONS, IEEE TRANS. COMP., C-25, 1124-1130.

687 KANDEL, A. (1977) A NOTE ON THE SIMPLIFICATION OF FUZZY FUNCTIONS, INFORMATION SCI. 13, 91-94.

688 KANDEL, A. (1977) COMMENTS ON COMMENTS BY LEE, INFORMATION AND CONTROL, 35, 109-113.

689 KANDEL, A. (1978) FUZZY STATISTICS AND FORECAST EVALUATION, IEEE TRANS. ON SYSTEMS, MAN, AND CYBERNETICS, SMC-8, NO. 5, 396-401.

690 KANDEL, A. (1978) IMPRECISE SWITCHING STRUCTURES AND THEIR APPLICATIONS IN MODELS USING APPROXIMATE INFORMATION, PROCEEDINGS OF THE INT. CONF. ON CYBERNETICS AND SOCIETY, 926-929, NOV., JAPAN.

691 KANDEL, A. (1978) ON THE COMPACTIFICATION AND ENUMERATION OF FUZZY SWITCHING FUNCTIONS, PROCEEDINGS OF THE 8TH INTERNATIONAL SYMP. ON MULTIPLE-VALUED LOGIC, 87-90, CHICAGO, MAY.

692 KANDEL, A., AND BYATT, W.J. (1978) FUZZY DIFFERENTIAL EQUATIONS, PROCEEDINGS OF THE INT. CONF. ON CYBERNETICS AND SOCIETY, 1213-1216, NOV., JAPAN.

693 KANDEL, A., AND BYATT, W.J. (1978) FUZZY SETS, FUZZY ALGEBRA, AND FUZZY STATISTICS, PROCEEDINGS OF THE IEEE, 66, NO. 12, 1619-1639, DEC. .

694 KANDEL, A., AND DAVIS, H.A. (1976) THE FIRST FUZZY DECADE (BIBLIOGRAPHY ON FUZZY SETS AND THEIR APPLICATIONS), CSR140, COMPUTER SCIENCE DEPT., NEW MEXICO INSTITUTE OF MINING AND TECHNOLOGY, SOROCCO, NEW MEXICO, USA, APRIL.

695 KANDEL, A., AND HUGHES, J.S. (1977) APPLICATIONS OF FUZZY ALGEBRA TO HAZARD DETECTION IN COMBINATIONAL SWITCHING CIRCUITS, INT. J. OF COMPUTER AND INFORMATION SCI., VOL. 6, NO. 1, 71-82, MARCH.

696 KANDEL, A., AND LEE, S.C. (1979) FUZZY SWITCHING AND AUTOMATA: THEORY AND APPLICATIONS, 303 PAGES, CRANE, RUSSAK AND CO., INC. N.Y.

697 KANDEL, A., AND NEFF, T.P. (1977) SIMPLIFICATION OF FUZZY SWITCHING FUNCTIONS, INT. J. OF COMPUTER AND INFORMATION SCI., VOL. 6, NO. 1. 55-70, MARCH.

698 KANDEL, A., AND OBERHAUF, T.A. (1974) ON FUZZY LATTICES, CSR 128, COMPUTER SCIENCE DEPARTMENT, NEW MEXICO INSTITUTE OF MINING AND TECH., SOCORRO, NEW MEXICO, USA.

699 KANDEL, A., AND RICKMAN, S.M. (1976) TABULAR MINIMIZATION OF FUZZY SWITCHING FUNCTIONS, IEEE TRANS. ON SYSTEMS, MAN AND CYBERNETICS, 766-769, NOV. .

700 KANDEL, A., AND RICKMAN, S.M. (1977) COLUMN TABLE APPROACH FOR THE MINIMIZATION OF FUZZY FUNCTIONS, INFORMATION SCI. 12, 118-128.

701 KANDEL, A., AND YELOWITZ, L. (1974) FUZZY CHAINS, IEEE TRANS SYST. MAN CYBERN., SMC-4, 472-475.

702 KARTTUNEN, L. (1972) POSSIBLE AND MUST, IN KIMBALL, J.P. (ED.), SYNTAX AND SEMANTICS VOL.1, SEMINAR PRESS, NEW YORK, 1-20.

703 KATZ, J.J. (1962) THE PROBLEM OF INDUCTION AND ITS SOLUTION, UNIVERSITY OF CHICAGO PRESS, CHICAGO, USA.

704 KAUFMAN, A. (1972) THEORY OF FUZZY SETS, MERSON, PARIS.

705 KAUFMAN, A. (1975) THEORY OF FUZZY SUBSETS, ACADEMIC PRESS, N.Y.

706 KAUFMAN, A., COOLS, M., AND DUBOIS, T. (1973) STIMULATION INVENTIVE DANS UN DIALOGUE HOMME-MACHINE UTILISANT LA METHODE DES MORPHOLOGIES ET LA THEORIE DES SOUS-ENSEMBLES FLOUS, IMAGO DISCUSSION PAPER 6, UNIVERSITE CATHOLIQUE DE LOUVAIN, BELGIUM.

707 KAUFMANN, A. (1973) INTRODUCTION A LA THEORIE DES SOUS-ENSEMBLES FLOUS, 1: ELEMENTS THEORETIQUES DE BASE, MASSOM ET CIE, PARIS, FRANCE.

708 KAUFMANN, A. (1975) INTRODUCTION TO A FUZZY THEORY OF THE HUMAN OPERATOR, SPECIAL INTEREST DISCUSSION SESSION ON FUZZY AUTOMATA AND DECISION PROCESSES, 6TH IFAC WORLD CONGRESS, BOSTON, MASS., USA, AUG.

709 KAUFMANN, A. (1975A) INTRODUCTION A LA THEORIE DES

SOUS-ENSEMBLES FLOUS, 2: APPLICATIONS A LA LINGUISTIQUE ET A LA SEMANTIQUE, MASSON ET CIE, PARIS, FRANCE.

710 KAUFMANN, A. (1975B) INTRODUCTION A LA THEORIE DES SOUS-ENSEMBLES FLOUS, 3: APPLICATIONS A LA CLASSIFICATION ET LA RECONNAISANCE DES FORMES, AUX AUTOMATES ET AUX SYSTEMES, AUX CHOIX DES CRITARES, MASSON ET CIE, PARIS, FRANCE.

711 KAUFMANN, A. (1975C) INTRODUCTION TO THE THEORY OF FUZZY SUBSETS VOL. 1, ACADEMIC PRESS, NEW YORK.

712 KAUFMANN, A. (1976) INTRODUCTION A LA THEORIE DES SOUS-ENSEMBLES FLOUS, TOMES IV, ED. MASSON, PARIS.

713 KAUFMANN, A. (1977) PROGRESS IN MODELING OF HUMAN REASONING OF FUZZY LOGIC, 11-18, IN FUZZY AUTOMATA AND DECISION PROCESSES, M.M. GUPTA, G.N. SARIDIS, B.R. GAINES, EDS., NORTH-HOLLAND.

714 KAUFMANN, A., COOLS, M., AND DUBOIS, T. (1975) EXERCISE AVEC SOLUTIONS SUR LA THEORIE DES SOUS-ENSEMBLES FLOUS, MASSON ET CIE, PARIS.

715 KAUFMANN, F. (1974) A SURVEY OF FUZZY SETS THEORY AND APPLICATION TO LANGUAGES AUTOMATA AND ALGORITHMS, U.S. BERKELEY SEMINAR ON FUZZY SETS AND THEIR APPLICATION BERKELEY.

716 KAY, AND MCDANIEL (1975) COLOR CATEGORIES AS FUZZY SETS, WORKING PAPER NO. 44, UNIVERSITY OF CALIFORNIA, BERKELEY, CALIFORNIA.

717 KERRIDGE, D.F. (1961) INACCURACY AND INFERENCE, J. ROY. STATIST. SOC. (SER. B), 184-194.

718 KHATCHADOURIAN, H. (1965) VAGUENESS, MEANING AND ABSURITY, AMER. PHILOS. QUART., 2, 119-129.

719 KICKERT, W.J.M. (1974) APPLICATION OF FUZZY SET THEORY TO WARM WATER CONTROL, THESIS, DELFT UNIVERSITY OF TECHNOLOGY, (IN DUTCH).

720 KICKERT, W.J.M. (1975A) ANALYSIS OF FUZZY LOGIC CONTROLLER, FUZZY LOGIC WORKING GROUP REP. F/WKI/75, QUEEN MARY COLLEGE, UNIVERSITY OF LONDON, UK, JUNE .

721 KICKERT, W.J.M. (1975B) OFF-LINE ANALYSIS OF THE FUZZY RULES, FUZZY LOGIC WORKING GROUP REP., QUEEN MARY COLLEGE, UNIVERSITY OF LONDON, UK, JULY .

722 KICKERT, W.J.M. (1975C) FURTHER ANALYSIS AND APPLICATION OF FUZZY LOGIC, FUZZY LOGIC WORKING REP. F/WK2/75, QUEEN MARY COLLEGE, UNIVERSITY OF LONDON, UK, AUG.

723 KICKERT, W.J.M. (1976) AN EXAMPLE OF LINGUISTIC MODELING, A 2ND ATTEMPT AT SIMULATING MULDER'S THEORY, REPORT

NO. 30, UNIV. OF TECH., EINDHOVE, NETERLANDS

724 KICKERT, W.J.M. (1976) FUZZY THEORIES ON DECISION MAKING: A CRITICAL SURVEY, REPORT NO. 29, TECH. UNIV. ENDHOVEN, SEPT.

725 KICKERT, W.J.M. (1978) TOWARDS AN ANALYSIS OF LINGUISTIC MODELLING, DEPT. OF INDUSTRIAL ENGINEERING, EINDHOVEN UNIV. OF TECH., 4TH INT. CONGRESS ON CYBERNETICS AND SYSTEMS, AMSTERDAM.

726 KICKERT, W.J.M., AND KOPPELAAR, H. (1976) APPLICATIONS OF FUZZY SET THEORY TO SYNTACTIC PATTERN RECOGNITION OF HANDWRITTEN CAPITALS, IEEE TRANS. SYST. MAN CYBERN., 6, 148-151.

727 KICKERT, W.J.M., AND MAMDANI, E.H. (1977) ANALYSIS OF A FUZZY LOGIC CONTROLLER, FUZZY SETS AND SYSTEMS1 (1978) 29-44., NORTH-HOLLAND PUBLISHING CO.

728 KICKERT, W.J.M., AND VAN NAUTA LEMKE, H.R. (1976) APPLICATION OF A FUZZY CONTROLLER IN A WARM WATER PLANT, AUTOMATICA, 12, 301-308.

729 KIM H.H., MIZUMOTO, M., TOYODA, J., AND TANAKA, K. (1974) AUTOMATED EDITING OF FUZZY LINE DRAWINGS FOR PICTURE DESCRIPTION, TRANS. IECE 57-A, 216-223.

730 KIM, H.H., MIZUMOTO, M., TOYODA, J., AND TANAKA, K. (1974) LATTICE GRAMMARS, SYSTEMS, COMPUTERS, CONTROLS, 5, 1-9, (ORIG. TIECE 57-D, 253-260).

731 KIM, H.H., MIZUMOTO, M., TOYODA, J., AND TANAKA, K. (1975) L-FUZZY GRAMMARS, INFORM. SCI. 8, 123-140.

732 KIMBALL, J.P. (ED.) (1972) SYNTAX AND SEMANTICS VOL. 1, SEMINAR PRESS, NEW YORK.

733 KIMBALL, J.P. (ED.) (1973) SYNTAX AND SEMANTICS VOL. 2, SEMINAR PRESS, NEW YORK.

734 KIMBALL, J.P. (ED.) (1975) SYNTAX AND SEMANTICS VOL. 4, ACADEMIC PRESS, NEW YORK.

735 KING, P.J., AND MAMDANI, E.H. (1977) THE APPLICATION OF FUZZY CONTROL SYSTEMS TO INDUSTRIAL PROCESSES, 321-330, IN FUZZY AUTOMATA AND DECISION PROCESSES, M.M. GUPTA, G.N. SARIDIS, B.R. GAINES, EDS., NORTH-HOLLAND.

736 KISE, V.A., AND OSIS, J.J. (1969) SEARCH METHODS FOR ESTABLISHING OF MAXIMAL SEPARABILITY OF FUZZY SETS; IN D.S. KRITINKOV, J.J. OSIS, L.A. RASTRIGIN, (EDS.), KIBERNETIKA I DIAGNOSTIKA, 3, 79-88, (IN RUSSIAN) ZINATNE, RIGA, U.S.S.R. .

737 KITAGAWA, T. (1973) BIOROBOTS FOR SIMULATION STUDIES OF LEARNING AND INTELLIGENT CONTROLS, IN US-JAPAN SEMINAR

ON LEARNING CONTROL AND INTELLIGENT CONTROL, GAINESVILLE, FLORIDA, USA.

738 KITAGAWA, T. (1973) THREE COORDINATE SYSTEMS FOR INFORMATION SCIENCE APPROACHES, INFORM. SCI., 15, 159-169.

739 KITAGAWA, T. (1975) FUZZINESS IN INFORMATIVE LOGICS, IN ZADEH, L.A., FU, K.S., TANAKA, K., AND SHIMURA, M. (EDS.), FUZZY SETS AND THEIR APPLICATIONS TO COGNITIVE AND DECISION PROCESSES, ACADEMIC PRESS, NEW YORK, 97-124.

740 KITAHASHI, T. (1975) A SURVEY OF STUDIES ON APPLICATIONS OF MANY-VALUED LOGIC IN JAPAN, PROC. 1975 INT. SYMP. MULTIPLE-VALUED LOGIC, IEEE 78CH0959-7C, 462-467.

741 KITAJIMA, S., AND ASAI, K. (1970) LEARNING CONTROLS BY FUZZY AUTOMATA, JOURNAL OF JAACE, 14, 551-559.

742 KITAJIMA, S., AND ASAI, K. (1972) LEARNING MODEL OF FUZZY AUTOMATON WITH STATE-DEPENDENT OUTPUT (3), ANNUAL JOINT CONFERENCE RECORDS OF JAACE .

743 KITAJIMA, S., AND ASAI, K. (1974) A METHOD OF LEARNING CONTROL VARYING SEARCH DOMAIN BY FUZZY AUTOMATA, IN FU, K.S., AND TOU, J.T. (EDS.), LEARNING SYSTEMS AND INTELLIGENT ROBOTS, PLENUM PRESS, NEW YORK, 249-262.

744 KLABBERS, J.H.G. (1975) GENERAL SYSTEM THEORY AND SOCIAL SYSTEMS: A METHOLOGY FOR THE SOCIAL SCIENCES, NEDERLANDS TIJDSCHRIFT VOOR DE PSYCHOLOGIE, 30, 493-514.

745 KLAUA, D. (1965) UBER EINEN ANSATZ ZUR MEHRWERTIGEN MENGENLEHRE, MONATSB. DEUTSCH. AKAD. WISS. (BERLIN), 7, 859-867.

746 KLAUA, D. (1966) UBER EINEN ZWEITEN ANSATZ ZUR MEHRWERTIGEN MENGENLEHRE, MONATSB. DEUTSCH. AKAD. WISS. (BERLIN), 8, 161-177.

747 KLAUA, D. (1966B) GRUNDBEGRIFFE EINER MEHRWERTIGEN MENGENLEHRE, MONATSB. DEUTSCH. AKAD. WISS. (BERLIN), 8, 782-802.

748 KLAUA, D. (1967A) EIN ANSATZ ZUR MEHRWERTIGEN MENLEHRE, MATH. NACHR., 33, 273-296.

749 KLAUA, D. (1967B) EINBETTUNG DER KLASSISCHEN MENGENLEHRE IN DIE MEHRWERTIGE, MONATSB. DEUTSCH. AKAD. WISS. (BERLIN), 9, 258-272.

750 KLAUA, D. (1968) PARTIELL AEFINLERTE MENGEN, MONATSB. DEUTSCH. AKAD. WISS. (BERLIN), 10, 571-578.

751 KLAUA, D. (1969A) PARTIELLE MENGEN MIT MEHRWERTIGEN GRUNDBEZIEHUNGER, MONATSB. DEUTSCH. AKAD. WISS. (BERLIN), 11, 585-599.

752 KLAUA, D. (1970) STETIGE GLEICHMACHTIGKEITEN KONTINUIERLICH-WERTIGER MENGEN, MONATSB. DEUTSCH. AKAD. WISS. (BERLIN), 12, 749-758.

753 KLAUA, D. (1972) ZUM KARDINALZAHLBEGRIFF IN DER MEHRWERTIGEN MENGENLEHRE, IN THEORY OF SETS AND TOPOLOGY, DEUTSCHER VERLAG DER WISSENSCHAFTEN, BERLIN, 313-325.

754 KLAUA, D. (1973) ZUR ARITHMETIK MEHRWERTIGEN ZAHLEN, MATH. NACHR., 57, 275-306.

755 KLEENE, S.C. (1952) INTRODUCTION TO METAMATHEMATICS, VAN NOSTRAND, NEW YORK.

756 KLING, R. (1973A) FUZZY PLANNER, TECH. REP. 168, COMPUTER SCIENCE DEPARTMENT, UNIVERSITY OF WISCONSIN.

757 KLING, R. (1974) FUZZY-PLANNER: REASONING WITH INEXACT CONCEPTS IN A PROCEDURAL PROBLEM-SOLVING LANGUAGE, J. CYBERNETICS, 4, 105-122.

758 KLIR, G.J. (1975A) PROCESSING OF FUZZY ACTIVITIES OF NEURAL SYSTEMS, IN TRAPPEL, R., AND PICHLER, F.R. (EDS.), PROGRESS IN CYBERNETICS AND SYSTEMS RESEARCH, 1, 21-24.

759 KLIR, G.J. (1975B) ON THE REPRESENTATION OF ACTIVITY ARRAYS, INT. J. GENERAL SYST., 2, 149-168.

760 KLIR, G.J. (1976) IDENTIFICATION OF GENERATIVE STRUCTURES IN EMPIRICAL DATA, INT. J. GENERAL SYST., 3, 89-104.

761 KLIR, G.J., AND UTTENHOVE, H.J.J. (1976A) PROCEDURE OF GENERATING HYPOTHICAL STRUCTURES IN THE STRUCTURE IDENTIFICATION PROBLEM, 3RD EUR. MEETING CYBERN. SYST. RES., VIENNA .

762 KLIR, G.J., AND UTTENHOVE, H.J.J. (1976B) COMPUTERIZED METHODOLOGY FOR STRUCTURE MODELLING, IN STENFERT, H.E. (ED.), ANNALS OF SYSTEMS RESEARCH, 4, KROESE, LEIDEN, HOLLAND.

763 KNEALE, W., AND KNEALE, M. (1962) THE DEVELOPMENT OF LOGIC, CLARENDON PRESS, OXFORD.

764 KNOPFMACHER, K. (1975) ON MEASURES OF FUZZINESS, J. MATH. ANAL. AND APPLN., 49, 529-534.

765 KNUTSON, T.J., AND HOLDREDGE, W.E. (1975) ORIENTATION BEHAVIOR, LEADERSHIP, AND CONSENSUS: A POSSIBLE FUNCTIONAL RELATIONSHIP, SPEECH MONO., 107-114.

766 KOBAYASHI, S. (1976) ON INTERACTIVE SOLUTION FOR MULTIPLE CRITERIA PROBLEMS - AN APPROACH BY PAIRWISE COMPARISON -, 42-47, SUMMARY OF PAPERS ON GENERAL FUZZY PROBLEMS, THE WORKING GROUP ON FUZZY SYSTEMS, TOYKO, JAPAN.

767 KOCHEN, M. (1975) APPLICATIONS OF FUZZY SETS IN PSYCHOLOGY,

IN ZADEH, L.A., FU, K.S., TANAKA, K., AND SHIMURA, M. (EDS.), 'FUZZY SETS AND THEIR APPLICATIONS TO COGNITIVE AND DECISION PROCESSES, ACADEMIC PRESS, NEW YORK, 395-408.

768 KOCHEN, M. (1977) IMPRECISION IN COPING AND ATTENDING PROCESSES, PROC. IEEE SYMP. ON FUZZY SETS, NEW ORLEANS, 1324-1329.

769 KOCHEN, M. (1977) ON FUNDAMENTALS, THEORY, AND APPLICATIONS OF FUZZY SETS, PROCEEDINGS OF THE SYMP. ON FUZZY SET THEORY AND APPLN., IEEE CONF. ON DECISION AND CONTROL, NEW ORLEANS.

770 KOCHEN, M. (1979) ENHANCEMENT OF COPING THROUGH BLURRING, FUZZY SETS AND SYSTEMS, VOL. 2, NO. 1, JAN., 37-52.

771 KOCHEN, M., AND BADRE, A.N. (1974) ON THE PRECISION OF ADJECTIVES WHICH DENOTE FUZZY SETS, J. CYBERNETICS, 4, 49-59.

772 KOCHEN, M., AND DREYFUSS-RAIMI,G. (1974) ON THE PSYCHOLINGUISTIC REALITY OF FUZZY SETS: EFFECT OF CONTEXT AND SET, UNIVERSITY OF MICHIGAN MENTAL HEALTH RESEARCH INSTITUTE, ANN ARBOR, USA, JUNE.

773 KOCZY, L.T. (1975) FUZZY ALGEBRAS AND SOME QUESTIONS OF THEIR TECHNICAL APPLICATIONS, DISSERTATION (IN HUNGARIAN). UNIV. OF ENGINEERING, BUDAPEST.

774 KOCZY, L.T. (1975) R-FUZZY ALGEBRA AS A GENERALIZED FORMULATION OF THE INTUITIVE LOGIC, DEPARTMENT OF PROCESS CONTROL, TECHNICAL UNIVERSITY, BUDAPEST, HUNGARY.

775 KOCZY, L.T. (1976) SOME QUESTIONS OF SIGMA-ALGEBRAS OF FUZZY OBJECTS OF TYPE N, 3RD EUR. MEETING CYBERN. SYST. RES., VIENNA .

776 KOCZY, L.T., AND HAJNAL, M. (1975) A NEW FUZZY CALCULUS AND ITS APPLICATION AS A PATTERN RECOGNITION TECHNIQUE, PROCEEDINGS OF 3RD INTERNATIONAL CONGRESS OF CYBERNETICS AND SYSTEMS, BUCHAREST, RUMANIA, AUG.

777 KOCZY, L.T., AND HAJNAL, M. (1976) A KAROMETRIC CLASSIFICATION ALGORITHM BASED ON R-FUZZY SET CALCULUS, PROC. 2ND NAT. MEETING ON BIOPHYSICS AND BIOTECHNOLOGY, ESPOO, FINLAND, FEB. .

778 KOCZY, L.T., AND HAJNAL, M. (1977) CLUSTER ANALYSIS IN KARYOMETRY APPLING A NEW FUZZY ALGEBRA, IN : W.J. PERKINS, ED., BIOMEDICAL COMPUTING (PITMAN MEDICAL PUBLISHING CO. LTD., TUNBRIDGE WELLS).

779 KOHOUT, L.J. (1974) THE PINKAVA MANY-VALUED COMPLETE LOGIC SYSTEMS AND THEIR APPLICATIONS IN THE DESIGN OF MANY-VALUED SWITCHING CIRCUITS, IEEE 74CH0845-8C, PROC. 1974 INT. SYMP. MULTIPLE-VALUED LOGIC, MAY, 261-284.

780 KOHOUT, L.J. (1975) GENERALIZED TOPOLOGIES AND THEIR RELEVANCE TO GENERAL SYSTEMS, INT. J. GENERAL SYST., 2, 25-34.

781 KOHOUT, L.J. (1976A) AUTOMATA AND TOPOLOGY, IN MAMDANI, E.H., AND GAINES, B.R. (EDS.), DISCRETE SYSTEMS AND FUZZY REASONING, EES-MMS-DSFR-76, QUEEN MARY COLLEGE, UNIVERSITY OF LONDON, (WORKSHOP PROCEEDINGS).

782 KOHOUT, L.J. (1976B) APPLICATION OF MULTI-VALUED LOGICS TO THE STUDY OF HUMAN MOVEMENT CONTROL AND OF MOVEMENT DISORDERS, PROC. 6TH INT. SYMP. MULTIPLE- VALUED LOGIC, IEEE 76CH1111-4C, 224-231.

783 KOHOUT, L.J. (1976C) REPRESENTATION OF FUNCTIONAL HIERARCHIES OF MOVEMENT IN THE BRAIN, INT. J. MAN-MACHINE STUDIES, 8, 699-709.

784 KOHOUT, L.J., AND PINKAVA, V. (1976) THE FUNCTIONAL COMPLETENESS OF PI-ALGEBRAS AND ITS RELEVANCE TO BIOLOGICAL MODELLING AND TO TECHNOLOGICAL APPLICATIONS OF MANY-VALUED LOGICS, IN MAMDANI, E.H., AND GAINES, B.R. (EDS.), DISCRETE SYSTEMS AND FUZZY REASONING, EES-MMS-DSFR-76, QUEEN MARY COLLEGE, UNIVERSITY OF LONDON, (WORKSHOP PROCEEDINGS).

785 KOKAWA, M. (1977) FUZZY THEORETIC AND CONCEPT-FORMATIONAL APPROACHES TO INFERENCE AND INFORMATION EXPERIMENTS IN HUMAN DECISION PROCESSES, PROCEEDINGS OF THE SYMP. ON FUZZY SET THEORY AND APPLN., IEEE CONF. ON DECISION AND CONTROL, NEW ORLEANS.

786 KOKAWA, M., NAKAMURA, K., AND ODA, M. (1972) A FORMULATION OF HUMAN DECISION- MAKING PROCESS, 19, AUTOMATIC CONTROL LABORATORY, NAGOYA UNIVERSITY, JAPAN, 3-10.

787 KOKAWA, M., NAKAMURA, K., AND ODA, M. (1974) FUZZY PROCESS OF MEMORY BEHAVIOUR, TRANS. SOC. INSTRUM. CONTROL ENGRS. 10, 385-386.

788 KOKAWA, M., NAKAMURA, K., AND ODA, M. (1976) FUZZY DESCRIPTION OF DECISION-MAKING PROCESS AND EXPERIMENTAL APPROACH TO FUZZY SIMULATION OF MEMORIZING, FORGETTING AND INFERENCE PROCESS, 1976 JOINT CONVENTION OF FOUR INSTITUTES OF ELECTRICAL ENGINEERING.

789 KOKAWA, M., NAKAMURA, K., AND ODA, M. (1979) FUZZY THEORETIC AND CONCEPT FORMATIONAL APPROACHES TO HINT EFFECT EXPERIMENTS IN HUMAN DECISION PROCESSES, FUZZY SETS AND SYS., VOL. 2, NO. 1, JAN., 25-36.

790 KOKAWA, M., ODA, M., AND NAKAMURA, K. (1975) FUZZY THEORETICAL ONE-DIMENSIONALIZING METHOD OF MULTIDIMENSIONAL QUANTITY, TRANS. SOCIETY OF INSTRUMENT AND CONTROL ENGINEERS, VOL. 11, NO.5, 8-14.

791 KOKAWA, M., ODA, M., AND NAKAMURA, K. (1977) FUZZY-THEORETICAL

DIMENSIONALITY REDUCTION METHOD OF MULTI-DIMENSIONAL QUANTITY, 235-250, IN FUZZY AUTOMATA AND DECISION PROCESSES, M.M. GUPTA, G.N. SARIDIS, B.R. GAINES, EDS., NORTH-HOLLAND.

792 KOKOWA, M., NAKAMURA, K., AND ODA, M. (1973) FUZZY EXPRESSION OF HUMAN EXPERIENCE-TO MEMORY PROCESS, RESEARCH REPORTS OF AUTOMATIC CONTROL LABORATORY, 20, AUTOMATIC CONTROL LABORATORY, NAGOYA UNIVERSITY, JAPAN, JUNE, 27-33.

793 KOKOWA, M., NAKAMURA, K., AND ODA, M. (1974A) FUZZY-THEORICAL APPROACHES TO FORGETTING PROCESSES AND INFERENCE, 21, AUTOMATIC CONTROL LABORATORY, NAGOYA UNIVERSITY, JAPAN, 1-10.

794 KOKOWA, M., NAKAMURA, K., AND ODA, M. (1974B) FUZZY THEORETICAL AND CONCEPT FORMATIONAL APPROACHES TO MEMORY AND INFERENCE EXPERIMENTS, TRANS. INST. ELECTRON. COMM. ENG. (JAPAN), 57-D, 487-493.

795 KOKOWA, M., NAKAMURA, K., AND ODA, M. (1975A) HINT EFFECT AND JUMP OF LOGIC IN A DECISION PROCESS, TRANS. INST. ELECTRON. COMM. ENG. (JAPAN), 58-D .

796 KOKOWA, M., NAKAMURA, K., AND ODA, M. (1975B) EXPERIMENTAL APPROACH TO FUZZY SIMULATION OF MEMORIZING, FORGETTING AND INFERENCE PROCESS, IN ZADEH, L.A., FU, K.S., TANAKA, K., AND SHIMURA, M. (EDS.), FUZZY SETS AND THEIR APPLICATIONS TO COGNITIVE AND DECISION PROCESSES, ACADEMIC PRESS, NEW YORK, 409-428.

797 KOKOWA, M., NAKAMURA, K., AND ODA, M. (1977) FUZZY-THEORETIC AND CONCEPT FORMATIONAL APPROACHES TO INFERENCE AND HINT-EFFECT EXPERIMENTS IN HUMAN DECISION PROCESSES, PROC. 1977 IEEE CONF. ON DECISION AND CONTROL, NEW ORLEANS, 1330-1337.

798 KOLIBIAR, M. (1972) DISTRIBUTIVE SUBLATTICES OF A LATTICE, PROC. AMER. MATH. SOC., 34, 359-364.

799 KONRAD, E., AND BOLLAMN, P. (1976) FUZZY DOCUMENT RETRIEVAL, 3RD EUR. MEETING CYBERN. SYST. RES. VIENNA .

800 KORNER, S. (1957) REFERENCE, VAGUENESS AND NECESSITY, PHILOS, REV., 66, JULY.

801 KORNER, S. (1959) CONCEPTUAL THINKING, NEW YORK.

802 KORNER, S. (1966) EXPERIENCE AND THEORY, ROUTLEDGE AND KEGAN PAUL, LONDON.

803 KORNER, S. (1970) CATEGORIAL FRAMEWORKS, BASIL BLACKWELL, OXFORD.

804 KORNER, S. (1971) FUNDAMENTAL QUESTIONS OF PHILOSOPHY, PENGUIN BOOKS.

805 KORNER, S. (1976A) EXPERIENCE AND CONDUCT, CAMBRIDGE UNIVERSITY PRESS.

806 KORNER, S. (1976B) PHILOSOPHY OF LOGIC, BASIL BLACKWELL, OXFORD.

807 KOTAS, J. (1963) AXIOMS FOR BIRKHOFF-V.NEUMANN QUANTUM LOGIC, BULL. DE L'ACADEMIE POLONAISE DES SCIENCES, SER. MATH., ASTR. AND PHYS., 11, 629-632.

808 KOTOH, K., HIRAMATSU, K. (1973) A REPRESENTATION OF PATTERN CLASSES USING THE FUZZY SETS, SYSTEMS, COMPUTERS, CONTROLS, 1-8, (ORIG. TIECE 56-D, 275-282).

809 KOUTSKY, K. (1947) SUR LES LATTICES TOPOLOGIES, COMPTES RENDUS (PARIS), 225, 659-661.

810 KOUTSKY, K. (1952) THEORIE DES LATTICES TOPOLOGIQUES, PUBLICATIONES DE LA FACULTE DES SCIENCES DE L'UNIVERSITE MASARYK, NO. 337, BRNO, CZECHOSLOVAKIA, 133-171.

811 KOVACS, F. (1977) DEFINING MEMBERSHIP FUNCTIONS OF FUZZY SETS, IONA COLLEGE THESIS.

812 KRAMOSIL, I. (1975) A PROBABILISTIC APPROACH TO AUTOMATON-ENVIRONMENT SYSTEMS, KYBERNETIKA (PRAGUE), 11, 173-206.

813 KRAMOSIL, I., AND MICHALEK, J. (1975) FUZZY METRICES AND STATISTICAL METRIC SPACES, KYBERNETIKA (PRAGUE), 11, 336-344.

814 KRANTZ, D.H., LUCE, R.D., SUPPES, P., AND TVERSKY, A. (1971) FOUNDATIONS OF MEASUREMENT, ACADEMIC PRESS, NEW YORK.

815 KRIVINE, J.L. (1974) LANGAGES A VALEURS REELLES ET APPLICATIONS, FUNDAMENTA MATHEMATICAE, 81, 213-253.

816 KUBINSKI, T. (1958) NAZWY NIEOSTRE (VAGUE TERMS), STUDIA LOGICA, 7, 115-179.

817 KUBINSKI, T. (1959) SYSTEMY POZORIE SPRECZNE, ZESZYTY NAUKOWE UNIWERSYTETU WROCLAWSKIEGO, SERIA B, MATEMATYKA, FIZYKA, ASTRONOMIA (1959), 53-61.

818 KUBINSKI, T. (1960) AN ATTEMPT TO BRING LOGIC NEARER TO COLLOQUIAL LANGUAGE, STUDIA LOGICA, 10, 61-75.

819 KULIKOV, V.F., AND GURIEV, E.K. (1977) MAKING COMPROMISE DECISION IN TWO-LEVEL CONTROL SYSTEM, PROCEEDINGS OF THE SYMP. ON FUZZY SET THEORY AND APPLN., IEEE CONF. ON DECISION AND CONTROL, NEW ORLEANS.

820 KULIKOV, V.F., AND GURJEV, E.K. (1977) COMPROMISE DECISION MAKING IN THE TWO-LEVEL CONTROL SYSTEM, PROC. 1977 IEEE CONF. ON DECISION AND CONTROL, NEW ORLEANS.

821 KUMAR, A. (1977) A REAL TIME SYSTEM FOR PATTERN RECOGNITION OF HUMAN SLEEP STAGES BY FUZZY SYSTEMS ANAL., PATTERN RECOG., PERGAMON PRESS, VOL. 9, 43-46.

822 KUNII, T.L. (1976) DATAPLAN: AN INTERFACE GENERATOR FOR DATABASE SEMANTICS, INFORM. SCI. 10, 279-298.

823 KUROSU, K., MURAYAMA, Y., KOBAYASHI, M., AND INASAKA, F. (1976) DISPLAY METHOD USING TREE PATTERN, 35-41, SUMMARY OF PAPERS ON GENERAL FUZZY PROBLEMS, THE WORKING GROUP ON FUZZY SYSTEMS, TOYKO, JAPAN.

824 KWAKERAAK, H. (1978) FUZZY RANDOM VARIABLES, PART1, DEFINITIONS AND THEOREMS, MEMO. NR. 193, TECHNISCHE HOGESCHOOL TWENTE, 1-33.

825 KWAKERNAAK, H. (1978) FUZZY RANDOM VARIABLES, PART2, ALGORITHMS AND EXAMPLES FOR THE DISCRETE CASE, MEMO. NR.212, TECHNISCHE HOGESCHOOL TWENTE, 1-25.

826 KYBERG, H.E. (1970) PROBABILITY AND INDUCTIVE LOGIC, MACMILLAN, LONDON.

827 LABOV, W. (1973) THE BOUNDARIES OF WORDS AND THEIR MEANINGS, IN BAILEY, AND SHUY (EDS.), NEW WAYS OF ANALYSING VARIATIONS IN ENGLISH, WASHINGTON, GEORGETOWN UNIVERSITY PRESS.

828 LAKE, J. (1974A) SETS,FUZZY SETS, MULTI-SETS AND FUNCTIONS, DEPARTMENT OF MATHEMATICS, POLYTECHNIC OF THE SOUTH BANK, BOROUGH ROAD, LONDON, UK.

829 LAKE, J. (1974B) FUZZY SETS AND BALD MEN, DEPARTMENT OF MATHEMATICS, POLYTECHNIC OF THE SOUTH BANK, BOUROUGH ROAD, LONDON, UK.

830 LAKEOFF, G. (1972) A STUDY IN MEANING CRITERIA AND THE LOGIC OF FUZZY CONCEPTS PROC. 8TH REG. MEET. CHICAGO LINGUISTISS SOC. .

831 LAKEOFF, G. (1973) FUZZY GRAMMAR AND THE PREFORMANCE/COMPETENCE TERMINOLOGY GAME, PROCEEDINGS OF MEETING OF CHICAGO LINGUISTICS SOCIETY, 271-291.

832 LAKEOFF, G. (1973A) NOTES ON WHAT IT WOULD TAKE TO UNDERSTAND HOW ONE ADVERB WORKS, MONIST, 57, 328-343.

833 LAKEOFF, G. (1973B) PRAGMATICS IN NATURAL LOGIC, IN KEENAN, E.L. (ED.), FORMAL SEMANTICS OF NATURAL LANGUAGE, CAMBRIDGE UNIVERSITY PRESS, 253-286.

834 LAKEOFF, G. (1973C) HEDGES: A STUDY IN MEANING CRITERIA AND THE LOGIC OF FUZZY CONCEPTS, J. PHILOS. LOGIC, 2, 458-508.

835 LAKSHMIVARAHAN, S., AND RAGASETHUPATHY, K.S. (1974) CONSIDERATIONS FOR FUZZIFYING FORMAL LANGUAGES AND

SYNTHESIS OF FUZZY GRAMMARS, INDIAN INSTITUTE OF TECHNOLOGY, MADRAS, INDIA.

836 LAKSHMIVARAN, S., AND RAJASETHUPATHY, K.S. (1978) CONSIDERATIONS FOR FUZZIFYING FORMAL LANGUAGES AND SYNTHESIS FOR FUZZY GAMES, J. OF CYBERN. 8, 83-100.

837 LARSEN, J. (1976) A MULTI-STEP FORMULATION OF VARIABLE VALUED LOGIC HYPOTHESES, PROC. 6TH INT. SYMP. MULTIPLE-VALUED LOGIC, IEEE 76CH1111-4C, 157-163.

838 LARSEN, L.E., RUSPINI, E.H., MCNEW, J.J., WALTER, D.O., AND ADEY, W.R. (1972) A TEST OF SLEEP STAGING SYSTEMS IN THE UNRESTRAINED CHIMPANZEE, BRAIN RESEARCH, 40, 319-343.

839 LAW, H.Y.H., WONG, J., AND KODANI, M. (1977) LABORATORY SOURCE SELECTION UNDER A FUZZY ENVIRNMENT, PRESENT AT ORSA/TIMS MEETING, ATLANTA, GA. .

840 LAWVERE, F.W. (ED.) (1972) TOPOSES, ALGEBRAIC GEOMETRY AND LOGIC, SPRINGER- VERLAG, BERLIN.

841 LAWVERE, F.W., MAURER, C., AND WRAITH, G.C. (EDS.) (1975) MODEL THEORY AND TOPOI, LECTURE NOTES IN MATHEMATICS, 445, SPRINGER-VERLAG, BERLIN.

842 LAZAK, D. (1977) FUZZY SETS AND ARTIFICIAL INTELLIGENCE, 6TH EURO WORKING GROUP ON FUZZY SETS, ABO, FINNLAND.

843 LEAL, A. AND PEARL, J. (1976) A COMPUTER SYSTEM FOR CONVENTIONAL ELICITATION OF PROBLEM STRUCTURES, UCLA-ENG-7665, SCHOOL OF ENGINEERING AND APPLIED SCIENCE, UNIVERSITY OF CALIFORNIA, LOS ANGELES, USA, JUNE.

844 LEAL, A., PEARL, J. (1977) AN INTERACTIVE PROGRAM FOR CONVERSATIONAL ELICATION OF DECISION STRUCTURES, IEEE TRANS. SYST., MAN, CYBERNET. 7 (5).

845 LEE, E.T. (1972A) FUZZY LANGUAGES AND THEIR RELATION TO AUTOMATA, PHD THESIS, DEPARTMENT OF ELECTRICAL ENGINEERING AND COMPUTER SCIENCE, UNIVERSITY OF CALIFORNIA, BERKELEY, CALIFORNIA, USA.

846 LEE, E.T. (1972B) PROXIMITY MEASURES FOR THE CLASSIFICATION OF GEOMETRIC FIRURES, J. CYBERNETICS, 2, 43-59.

847 LEE, E.T. (1974) AN APPLICATION OF FUZZY SETS TO THE CLASSIFICATION OF GEOMETRIC FIGURES AND CHROMOSOME IMAGES, IN US-JAPAN SEMINAR ON FUZZY SETS AND THEIR APPLICATIONS, BERKELEY, CALIFORNIA, USA.

848 LEE, E.T. (1975) SHAPE-ORIENTED CHROMOSOME CLASSIFICATION, IEEE TRANS, SYST. MAN CYBERN., SMC-5, 629-632.

849 LEE, E.T. (1977) APPLICATION OF FUZZY LANGUAGES TO PATTERN RECOGNITION, KYBERNETES 6, 167-173.

850 LEE, E.T. (1978) APPL. OF FUZZY SETS TO PATTERN DESCRIPTION CLASSIFICATION, RECOGNITION, STORAGE AND RETRIEVAL, PROC. 1978 JACC, VOL. 3, 61-68.

851 LEE, E.T., AND ZADEH, L.A. (1969) NOTES ON FUZZY LANGUAGES, INFORM. SCI., 1, 421-434.

852 LEE, E.T., AND ZADEH, L.A. (1970) FUZZY LANGUAGES AND THEIR ACCEPTANCE BY AUTOMATA,, 4TH PRINCETON CONFERENCE ON INFORMATION SCIENCE AND SYSTEMS, 399 .

853 LEE, R.C.T. (1972) FUZZY LOGIC AND THE RESOLUTION PRINCIPLE, J. ASSN. COMP. MACH., 19, 109-119.

854 LEE, R.C.T., CHANG, C.L. (1971) SOME PROPERTIES OF FUZZY LOGIC, INFORM. AND CONTROL, 19, 417-431.

855 LEE, S.C. (1969) ANALYSIS AND SYNTHESIS OF SEQUENTIAL FUZZY LOGIC CIRCUITS, PROC. 7TH ANNUAL ALLECTON CONF. ON CIRCUIT AND SYSTEM THEORY, 692-701.

856 LEE, S.C., AND LEE, E.T. (1970) FUZZY NEURONS AND AUTOMATA, PROCEEDINGS OF 4TH PRINCETON CONFERENCE ON INFORMATION SCIENCE AND SYSTEMS, 381-385.

857 LEE, S.C., AND LEE, E.T. (1974) FUZZY SETS AND NEURAL NETWORKS, J. CYBERNETICS, 4, 83-103.

858 LEENDERS, J.H. (1974) VAGE VERAMELINGEN: EEN KRITISCHE BENANDERING, KWARTAALSCHRIFT WETENSCHAPPELIJK ONDERWIJS LIMBURG (BELGIUM), 4, 441-455.

859 LEFAIVRE, R. (1974) FUZZY: A PROGRAMMING LANGUAGE FOR FUZZY PROBLEM SOLVING, REPORT PB-231813/7, WISCONSIN UNIV. .

860 LEFAIVRE, R.A. (1974) FUZZY: A PROGRAMMING LANGUAGE FOR FUZZY PROBLEM SOLVING, TECHNICAL REPORT 202, DEPT. OF COMPUTER SCI., UNIV. OF WISCONSIN, MADISON.

861 LEFAIVRE, R.A. (1974A) FUZZY PROBLEM SOLVING, TECHNICAL REPORT 37, MADISON ACADEMY COMPUTER CENTER, UNIVERSITY OF WISCONSIN, USA, AUG. .

862 LEFAIVRE, R.A. (1974B) THE REPRESENTATION OF FUZZY KNOWLEDGE, J. CYBERNETICS 4, 57-66.

863 LEFAIVRE, R.A. (1976) PROCEDURAL REPRESENTATION IN FUZZY PROBLEM SOLVING SYSTEMS, PROC. NCC .

864 LEMMON, E.J. (1966A) ALGEBRAIC SEMANTICS FOR MODAL LOGICS I, J. SYMBOLIC LOGIC, 31, 46-65.

865 LEMMON, E.J. (1966B) ALGEBRAIC SEMANTICS FOR MODAL LOGICS II, J. SYMBOLIC LOGIC, 31, 191-218.

866 LEMMON, E.J., MEREDITH, C.A., MEREDITH, D., PRIOR, A.N.,

AND THOMAS, I. (1969) CALCULI OF PURE STRICT IMPLICATION, IN DAVIS, J.W., HOCKNEY, D.J., AND FREED, W.K. (EDS.), PHILOSOPHICAL LOGIC, D. REIDEL, DORDRECHT, HOLLAND, 215-250.

867 LENDARIS, G.G., AND MARTINEZ, A.J. (1976) BIBLIOGRAPHY ON FUZZY SETS AND THEIR APPLICATIONS, REPORT OF SYSTEM SCI. PROGRAM, SS-11, PORTLAND STATE UNIV. .

868 LEVI, I. (1967) GAMBLING WITH TRUTH, MIT PRESS, CAMBRIDGE, MASS., USA.

869 LEWIS, D.K. (1969) CONVENTION: A PHILOSOPHICAL STUDY, HARVARD UNIVERSITY PRESS, CAMBRIDGE, MASS., USA.

870 LEWIS, D.K.(1973) COUNTERFACTUALS, BASIL BLACKWELL, OXFORD.

871 LIENTZ, B.P. (1972) ON TIME DEPENDENT FUZZY SETS, INFORM. SCI., 4, 367-376.

872 LIENTZ, B.P. (1977) ON THE ANALYSIS OF COMPLEX SOFTLY DESIGNED SYSTEMS, REPORT GRAD. SCHOOL OF MAN., U.C.L.A.

873 LOGINOV, V.I. (1966) PROBABILITY TREATMENT OF ZADEH MEMBERSHIP FUNCTION AND THEIR USE IN PATTERN RECOGNITION, ENGINEERING CYBERNETICS, 68-69.

874 LOMBAERDE, J. (1974) MESURES D'ENTROPIE EN THEORIE DES SOUS-ENSEMBLES FLOUS, IMAGO DISCUSSION PAPER IDP-12, CENTRE INTERFACULTAIRE IMAGO, UNIVERSITE CATHOLIQUE DE LOUVAIN, HEVERLEE, BELIQUE, JAN. .

875 LONGO, G. (1975) FUZZY SET, GRAPHS AND SOURCE CODING, IN SWIRZYNSKI, J.K. (ED.), NEW DIRECTIONS IN SIGNAL PROCESSING IN COMMUNICATIONS AND CONTROL, NOORDHOFF-LEYDEN, 27-33.

876 LOO, S.G. (1977) MEASURES OF FUZZINESS, CYBERNETICA 3, 201-207.

877 LOO, S.G. (1978) FUZZY RELATIONS IN SOCIAL AND BEHAVIORAL SCIENCES, J. OF CYBERN. 8, 1-16.

878 LOS, J., AND RYLL-NARDZEWSKI, C. (1951) ON THE APPLICATION OF TYCHNOFF'S THEOREM IN MATHEMATICAL PROOFS, FUNDAMENTA MATHEMATICAE, 38, 233-237.

879 LOWE, E.A., AND TINKER, A.M. (1977) REGULATING JUMPING FUZZY SETS, INT. CONF. ON APPL. GEN. SYS. RES., BINGHAMTON, N.Y.

880 LOWEN, R. (1974A) A THEORY OF FUZZY TOPOLOGIES, PHD THESIS, FREE UNIVERSITY OF BRUSSELS, BELGIUM.

881 LOWEN, R. (1974B) TOPOLOGIES FLOUS, C.R. ACAD. DES SCIENCES, (PARIS) 278A, 925-928.

882 LOWEN, R. (1975) CONVERGENCE FLOUS, C.R. ACAD. DES SCIENCES, (PARIS) 280, 1181-1183.

883 LOWEN, R. (1976A) FUZZY TOPOLOGICAL SPACES AND FUZZY COMPACTNESS, J. MATH. ANAL. AND APPLN.

884 LOWEN, R. (1976B) INITIAL AND FINAL FUZZINESS TOPOLOGIES AND THE FUZZY TYCHNOFF THEOREM, J. MATH. ANAL. AND APPLN.

885 LOWEN, R. (1976C) A CONPARISON OF DIFFERENT COMPACTNESS NOTIONS IN FUZZY TOPOLOGY, VRIJE UNIVERTITEIT BRUSSEL, BRUSSELS, BELGIUM.

886 LOWEN, R. (1977) A COMPARISON OF DIFFERENT COMPACTNESS NOTIONS IN FUZZY TOPOLOGICAL SPACES 1. NOTICES OF THE AMS, OCT. 1976 2. TO APPEAR IN J. MATH. ANAL. APPL. [IN 1977 OR 78].

887 LOWEN, R. (1977) FUZZY TOPOLOGY, REPORT ON THE IEEE SYMP. ON FUZZY SET THEORY AND APPL., HELD AT THE 1977 IEEE CONTROL AND DECISION CONF.

888 LOWEN, R. (1977) INITIAL AND FUZZY TOPOLOGIES AND THE FUZZY TYCHONOFF THEOREM, IN J. MATH. ANAL. APPLIC.

889 LOWEN, R. (1977) LATTICE CONVERGENCE IN FUZZY TOPOLOGICAL SPACES, NOTICES OF THE AMS, NOV. 1977.

890 LOWEN, R. (1977) ON FUZZY COMPLEMENTS, PROC. 1977 IEEE CONF. ON DECISION AND CONTROL, NEW ORLEANS, 1338-1342.

891 LOWEN, R. (1977) ON FUZZY COMPLEMENTS AND CONVERGENCE IN FUZZY TOPOLOGICAL SPACES, PROCEEDINGS OF THE SYMP. ON FUZZY SET THEORY AND APPLN., IEEE CONF. ON DECISION AND CONTROL, NEW ORLEANS.

892 LOWEN, R. (1978) CONVERGENCE IN FUZZY TOPOLOGICAL SPACES, PROC. OF 4TH PRAGUE CONF. ON GENERAL TOPOLOGY.

893 LUCAS, D., EKMAN, K., AND WEETE, F. (1979) THE APPLICATION OF FUZZY POINTERS IN MULTISENSOR AND MULTITARGET INTEGATION, 17TH IEEE CONF ON DEC. AND CONTROL, 1217-1219.

894 LUDESCHER, H., AND ROVENTA, E. (1976) SUR LES TOPOLOGIES FLOUES DEFINIES A L'AIDE DES VOISINAGES, C.R. ACAD. SCI. 283, 575-577.

895 LUSCHEI, E.C. (1962) THE LOGICAL SYSTEMS OF LESNIEWSKI, NORTH-HOLLAND, AMSTERDAM.

896 LYSVAG, B. (1975) VERBS OF HEDGING, IN KIMBALL, J.P. (ED.), SYNTAX AND SEMANTICS VOL. 4, ACADEMIC PRESS, NEW YORK, 125-154.

897 MAARSCHALK, C.G.D. (1975) EXACT AND FUZZY CONCEPTS SUPERIMPOSED ON THE GST (A META THEORY), PROCEEDINGS

OF 3RD INTERNATIONAL CONGRESS OF CYBERNETICS AND SYSTEMS, BUCHAREST, RUMANIA, AUG. .

898 MAARSCHALK, C.G.D. (1976) METHOLOGY IN SYSTEMS THINKING AND SYSTEMS LANGUAGE--AN APPROACH TO FORMALIZED AND CONCEPTUAL SYSTEMS, EXACT AND FUZZY CONCEPTS AND SYSTOL (SYSTEM ORIENTED LANGUAGE), 3RD EUR. MEETING CYBERN. SYST. RES., VIENNA .

899 MACHINA, K.F. (1972) VAGUE PREDICATES, AMER. PHILOS. QUART.., 9, 225-233.

900 MACHINA, K.F. (1976) TRUTH, BELIEF AND VAGUENESS, J. PHILOS. LOGIC, 5, 47-77.

901 MACKIE, J.L. (1973) TRUTH, PROBABILITY AND PARADOX, CLARENDON PRESS, OXFORD.

902 MACLANE, S. (1971) CATEGORIES FOR THE WORKING MATHEMATICIAN, SPRINGER-VERLAG, BERLIN.

903 MACVICAR-WHELAN, P.J. (1974) FUZZY SETS, THE CONCEPT OF HEIGHT, AND THE HEDGE VERY, TECHNICAL MEMORANDUM 1, PHYSICS DEPARTMENT, GRAND VALLEY STATE COLLEGES, ALLENDALE, MICHIGAN, USA.

904 MACVICAR-WHELAN, P.J. (1975) UN MODELE DE SIGNIFICATION DE TERMES QUANTIFIANT LES DIMENSIONS: APPLICATION A LA TAILLE HUMAINE, LAAS-SMA4 75.I.49, LABORATOIRE D'AUTOMATIQUE ET D'ANALYSE DES SYSTEMES, TOULOUSE, FRANCE, DEC.

905 MACVICAR-WHELAN, P.J. (1976) FUZZY SETS FOR MAN-MACHINE INTERACTION, INT. J. MAN-MACHINE STUDIES, 8 .

906 MACVICAR-WHELAN, P.J. (1977) FUZZY AND MULTIVALUED LOGIC, PROC. 7TH INT. SYMP. MULTIPLE-VALUED LOGIC, IEEE PUB 77CH1222-9C, 98-102.

907 MACVICAR-WHELAN, P.J. (1978) FUZZY FILTERING, LABORATOIRE D'AUTOMATIQUE ET D'ANALYSE DES SYSTEMES DU CENTRE NATIONAL DE LA RECHERCHE SCIENTIFIQUE - 7, AVE. DU COLONEL ROCHE, 31400 TOULOUSE - FRANCE.

908 MACVICAR-WHELAN, P.J. (1978) FUZZY LOGIC: AN ALTERNATIVE APPROACH, FST-3-78, UCLA, 1-21.

909 MAJUMDER, D.D., AND PAL, S.K. (1976) THE CONCEPT OF FUZZY SETS AND ITS APPLICATION IN PATTERN RECOGNITION PROBLEMS, PROC. CSI, 76 CONVENTION, HYDERABAD, INDIA, NO. SD 02, JAN.

910 MAJUMDER, D.D., AND PAL, S.K. (1977) ON FUZZIFICATION, FUZZY LANGUAGE, AND MULTICATEGORY FUZZY CLASSIFIERS, PROC. 7TH INT. CONF. ON CYBERN. AND SOCIETY, WASH. D.C.

911 MAJUMDER, D.D., AND PAL, S.K. (1977) ON SOME APPLICATIONS OF FUZZY ALGORITHM IN MAN-MACHINE COMMUNICATION RESEARCH, J. INST. TELECOM. ELECTRON ENG., VOL. 23, 117-120.

912 MAJUMDER, D.D., AND PAL, S.K. FUZZY RECOGNITION SYSTEMS FOR PATTERNS OF BIOLOGICAL ORIGIN, ELECTRONICS AND COMMUNICATION SCI. UNIT, INDIAN STATISTICAL INSTITUTE, 203 BT ROAD, CALCUTTA 35, INDIA.

913 MALVACHE, N. (1975) ANALYSE ET IDENTIFICATION DES SYSTEMES VISUEL ET MANUAL EN VISION FRONTALE ET PERIPHERIQUE CHEZ L'HOMME, PHD THESIS, LILLE, FRANCE, APRIL.

914 MALVACHE, N. (1977) UTILIZATION OF FUZZY SETS FOR SYSTEM MODELLING AND CONTROL, PROCEEDINGS OF THE SYMP. ON FUZZY SET THEORY AND APPLN., IEEE CONF. ON DECISION AND CONTROL, NEW ORLEANS.

915 MALVACHE, N., AND VIDAL, P. (1974) APPLICATION DES SYSTEMES FLOUS A LA MODELISATION DES PHENOMENES DE PRISE DE DECISION ET D'APPREHENSION DES INFORMATIONS VISUELLES CHEX L'HOMME, A.T.P.-C.N.R.S 1K05, PARIS.

916 MALVACHE, N., AND WILLAYES, D. (1974) REPRENSENTATION ET MINIMISATION DE FONCTIONS FLOUS, DOC. CENTRE UNIVERSITIE DE VALENCIENNES, FRANCE.

917 MALVACHE, N., MILBRED, G., AND VIDAL, P. (1973) PERCEPTION VISUELLE: CHAMP DE VISION LATERALE, MODELE DE LA FONCTION DU REGARD: RAPPORT DE SYNTHESE, CONTRAT DRME NO. 71-251, PARIS, FRANCE.

918 MAMDANI, E.H. (1974) APPLICATIONS OF FUZZY ALGORITHMS FOR CONTROL OF SIMPLE DYNAMIC PLANT, PROC. IEEE, 121, 1585-1588.

919 MAMDANI, E.H. (1975) IFAC REPORT ON THE 2ND ROUND TABLE DISCUSSION ON FUZZY AUTOMATA AND DECISION PROCESSES, DEPT. ELECTRICAL ENGINEERING, QUEEN MARY COLLEGE, LONDON.

920 MAMDANI, E.H. (1976A) APPLICATION OF FUZZY LOGIC TO APPROXIMATE REASONING USING LINGUISTIC SYNTHESIS, PROC. 6TH INT. SYMP. MULTIPLE-VALUED LOGIC, IEEE 76CH1111-4C, MAY, 196-202.

921 MAMDANI, E.H. (1976B) ADVANCES IN THE LINGUISTIC SYNTHESIS OF FUZZY CONTROLLERS, INT. J. MAN-MACHINE STUDIES, 8, 669-678.

922 MAMDANI, E.H. (1977) APPLICATONS OF FUZZY SET THEORY TO CONTROL SYSTEMS: A SURVEY, 77-88, IN FUZZY AUTOMATA AND DECISION PROCESSES, M.M. GUPTA, G.N. SARIDIS, B.R. GAINES, EDS., NORTH-HOLLAND.

923 MAMDANI, E.H. AND BAAKLINI, N. (1975) PRESCRIPTIVE METHOD FOR DERIVING CONTROL POLICY IN A FUZZY-LOGIC CONTROLLER,

ELECTRONICS LETT., 11, 625-626.

924 MAMDANI, E.H., AND ASSILIAN, S. (1975) AN EXPERIMENT IN LINGUISTIC SYNTHESIS WITH A FUZZY LOGIC CONTROLLER, INT. J. MAN-MACHINE STUDIES, 7, 1-13.

925 MAMDANI, E.H., AND GAINES, B.R. (EDS.) (1976) DISCRETE SYSTEMS AND FUZZY REASONING, EES-MMS-DSFR-76, QUEEN MARY COLLEGE, UNIVERSITY OF LONDON, (WORKSKOP PROCEEDINGS).

926 MAMDANI, E.H., AND KING, P.J. (1975) THE APPLICATION OF FUZZY CONTROL SYSTEMS TO INDUSTRIAL PROCESSES, ROUND TABLE AT 6TH IFAC WORLD CONGRESS, BOSTON, MASS., USA.

927 MAMDANI, E.H., AND PROCYK, T.J. (1976) APPLICATION OF FUZZY LOGIC TO CONTROLLER DESIGN BASED ON LINGUISTIC PROTOCOL, 3RD EUR. MEETING CYBERN. SYST. RES., VIENNA

928 MANEK, W., AND TRACZYK, T. (1969) GENERALIZED LUKASIEWICZ ALGEBRAS, BULL. DE L'ACADEMIE POLONAISE DES SCIENCES, SER. MATH., ASTR. AND PHYS., 17, 789-792.

929 MANES, E.G. (1976) ALGEBRAIC THEORIES, SPRINGER-VERLAG.

930 MANES, E.G. (1977) TOPAS METHODS FOR QUANTAL SYSTEMS AND FUZZY SYSTEMS, INT. CONF. ON APPLIED GEN. SYS. RES., BINGHAMTON, N.Y.

931 MARCUS, RUTH BARCAN (1953) STRICT IMPLICATION, DEDUCTABILITY AND THE DEDUCTION THEOREM, J. SYMBOLIC LOGIC, 18, 234-236.

932 MARES, M. (1977) HOW TO HANDLE FUZZY QUANTITES, KYBERNETICA 13, 1, 22-40.

933 MARES, M. (1977) ON FUZZY QUANTITES WITH REAL AND INTEGER VALUES, KYBERNETICA 13, 1, 41-56.

934 MARINOS, P.N. (1966) FUZZY LOGIC, TECH, MEMO. 66-3344-1, BELL TELEPHONE LABS., HOLMDEL, NEW JERSEY, USA, AUG.

935 MARINOS, P.N. (1969) APPLICATION OF FUZZY LOGIC TO ANALOG AND HYBRID SYSTEMS, PROC. 8TH ANNUAL IEEE REGION III CONVENTION, HUNTSVILLE, ALA., 108-113.

936 MARINOS, P.N. (1969) FUZZY LOGIC AND ITS APPLICATION TO SWITCHING SYSTEMS, IEEE TRANS. COMP., C-18, 343-348.

937 MARKS, P. (1975B) FLCS: A CONTROL SYSTEM FOR FUZZY LOGIC, FUZZY LOGIC WORKING GROUP REP, 3, QUEEN MARY COLLEGE, UNIVERSITY OF LONDON, UK, NOV.

938 MARKS, P. (1976) A FUZZY LOGIC CONTROL SOFTWARE, INTERNAL REPORT, QUEEN MARY COLLEGE, LONDON.

939 MARONNA, R. (1964) A CHARACTERIZATION OF THE MORGAN LATTICES, PORTUGALIA MATHEMATICA, 23 .

940 MARTIN, J.K., AND TURKSEN, I. B. (1975) FORMATIVE EVALATION OF INFORMATION NEED ANALYSIS, DEPT. INDUSTRIAL ENGINEERING UNIVERSITY OF TORONTO, CANADA.

941 MARTIN, J.K., AND TURKSEN, I.B. (1978) A FUZZY SIMULUS RESPONSE THEORY FOR INFORMATION SYSTEM DESIGN, WORKING PAPER NO. 78-004, UNIV. OF TORONTO, DEPT. OF INDUS. ENGINEERING.

942 MARTIN, J.N. (1975) A SYNTACTIC CHARACTERISTIC OF KLEENE'S STRONG CONNECTIVES WITH TWO DESIGNATED VALUES, Z. MATH. LOGIK GRUNDLAGEN MATH., 21, 181-184.

943 MARTIN, R.L. (ED.) (1970) THE PARADOX OF THE LIAR, YALE UNIVERSITY PRESS, NEW HAVEN.

944 MARTIN, T. (1966) FUZZY ALGORITHMISCHE SCHEMATA, VORTRAGE AUS DEM PROLEMSEMINAR AUTOMATEN- UND ALGORITHMENTHEORIE, APRIL, WEISSIG, 44-51.

945 MATERNA, P. (1972) INTENSIONAL SEMANTICS OF VAGUE CONSTANTS. AN APPLICATION OF TICHY'S CONCEPT OF SEMANTICS, THEORY AND DECISION, 2, 267-273.

946 MATHAI, A.M., AND RATHIE, P.N. (1975) BASIC CONCEPTS IN INFORMATION THEORY AND STATISTICS: AXIOMATIC FOUNDATION AND APPLICATIONS, WILEY EASTERN LTD., NEW DELHI.

947 MATHESIUS, V. (1911) "O POTENCIALNOSTI JEVU JAZYKOVYCH" (ON THE POTIENTIALITY OF THE PHENOMENA OF LANGUAGE) VESTNIK KRAL., CESKE SPOLECNOSTI NAUK, TRIDA FILO-SOFICKO-HISTORICKA (PRAGUE). ENGLISH TRANSLATION IN PRAGUE SCHOOL READER IN LINGUISTICS, VACHET, J. (ED.) INDIANA UNIVERSITY PRESS, BLOOMINGTON, 1964.

948 MATROSE, E., AND MUKUNDAN, R. (1971) PLANNING OF A HEALTH CARE SYSTEM MODELED AS A TWO-PERSON GAME, INT. J. OF SYSTEM SCIENCE 3, 4, 375-383.

949 MAURER, W.D. (1974) INPUT-OUTPUT CORRECTNESS AND FUZZY CORRECTNESS, GEORGE WASHINGTON UNIVERSITY.

950 MAURO, V., BONA, B., AND INAUDI, D. (1976) A FUZZY APPROACH TO RESIDENTAL LOCATION THEORY, 3RD EUR. MEETING CYBERN. SYST. RES. VIENNA .

951 MAYDOLE, R.E. (1972) MANY-VALUED LOGIC AS A BASIS FOR SET THEORY, PHD THESIS, BOSTON UNIVERSITY, BOSTON MASS. .

952 MAYDOLE, R.E. (1975) PARADOXES AND MANY-VALUED SET THEORY, J. PHILOS. LOGIC, 4, 269-291.

953 MCCALL, S.S. (ED.) (1967) POLISH LOGIC 1920-1939, CLARENDON PRESS, OXFORD.

954 MCCAWLEY (1975) FUZZY LOGIC AND RESTRICTED QUANTIFIERS, UNIVERSITY OF CHICAGO.

955 MCKAY, A.F., AND MERRILL, D.D. (EDS.) (1976) ISSUES IN THE PHILOSOPHY OF LANGUAGE, YALE UNIVERSITY PRESS, NEW HAVEN, USA.

956 MCKINSEY, J.C.C. (1941) A SOLUTION OF THE DECISION PROBLEM FOR THE LEWIS SYSTEMS S2 AND S4, WITH AN APPLICATION TO TOPOLOGY, J. SYMBOLIC LOGIC, 6, 117-134.

957 MCKINSEY, J.C.C. (1945) ON THE SYNTACTICAL CONSTRUCTION OF SYSTEMS OF MODAL LOGIC, J. SYMBOLIC LOGIC, 10, 83-94.

958 MCKINSEY, J.C.C., AND TARSKI, A. (1944) THE ALGEBRA OF TOPOLOGY: ANNALS MATH., 45, 141-191.

959 MCKINSEY, J.C.C., AND TARSKI, A. (1948) SOME THEOREMS ABOUT THE SENTIMENTAL CALCULI OF LEWIS AND HEYTING, J. SYMBOLIC SYMBOLIC LOGIC, 13, 1-15.

960 MCNAUGHTEN, R. (1951) A THEOREM ABOUT INFINITE-VALUED SENTENTIAL LOGIC, J. SYMBOLIC, 16, 1-13.

961 MEHLBURG, H. (1958) THE REACH OF SCIENCE, UNIVERSITY OF TORONTO PRESS.

962 MENGER, K. (1951) ENSEMBLES FLOUS ET FONCTIONS ALEATOIRES, COMPTES RENDUS DE L'ACADEMIE DE SCIENCES PARIS, 232, NO. 22 (28 MAI), 2001-2003.

963 MENGES, G. (ED.) (1974) INFORMATION, INFERENCE AND DECISION, REIDEL, DORDRECHT, HOLLAND.

964 MENGES, G. (1970) ON SUBJECTIVE PROBABILITY AND RELATED PROBLEMS, THEORY AND DECISION, 1, 40-60.

965 MENGES, G., AND KOFLER, E. (1976) LINEAR PARTIAL INFORMATION AS FUZZINESS, IN BOSSEL, H., KLACZKO, S., AND MULLER, N. (EDS.), SYSTEMS THEORY IN THE SOCIAL SCIENCES, BIRKHAUSER VERLAG, BASIL, 307-322.

966 MENGES, G., AND SKALA, H.J. (1974) ON THE PROBLEM OF VAGUENESS IN THE SOCIAL SCIENCES, IN MENGES, G. (ED.), INFORMATION, INFERENCE AND DECISION, D. REIDEL, DORDRECHT, HOLLAND, 51-61.

967 MEREDITH, C.A. (1958) THE DEPENDENCE OF AN AXIOM OF LUKASIEWICZ, TRANS. AMER. MATH. SOC., 87, 54.

968 MESEGUER, J., AND SOLS, I. (1974) AUTOMATA IN SEMIMODULE CATEGORIES, PROCEEDINGS OF FIRST INTERNATIONAL SYMPOSIUM ON CATEGORY THEORY APPLIED TO COMPUTATION AND CONTROL, 196-202.

969 MESEGUER, J., AND SOLS, I. (1975) TOPOLOGY IN COMPLETE LATTICES AND CONTINUOUS FUZZY RELATIONS, UNIV. OF

SARAGOZA, SPAIN.

970 MESEGUER, J., AND SOLS, I. (1975A) FUZZY SEMANTICS IN HIGHER ORDER LOGIC AND UNIVERSAL ALGEBRA, UNIVERSITY OF ZARAGOZA, SPAIN.

971 MICHALEK, J. (1975) FUZZY TOPOLOGIES, KYBERNETIKA (PRAGUE), 11, 345-354.

972 MICHALOS, A.C. (1971) THE POPPER-CARNAP CONTROVERSY, M. NIJHOFF, THE HAJUE.

973 MICHALSKI, R.S. (1974) LEARNING BY INDUCTIVE INFERENCE, PROC. NATO ADVANCED STUDY INSTITUTE SEMINAR ON COMPUTER ORIENTED LEARNING PROCESSES, BONAS, FRANCE, AUG. .

974 MICHALSKI, R.S. (1975) VARIABLE-VALUED LOGIC AND ITS APPLICATIONS TO PATTERN RECOGNITION AND MACHINE LEARNING, IN MULTIPLE-VALUED LOGIC AND COMPUTER SCIENCE, NORTH-HOLLAND, AMSTERDAM.

975 MICHELMAN, E.H., AND HOFFMAN, L.J. (1977) SECURATE: A SECURITY EVALUATION AND ANALYSIS SYSTEM USING FUZZY METRICS, MEMO. NO. UCB/ERL M77/36, ELECTRONICS RESEARCH LAB., UNIV. OF CALIFORNIA, BERKELEY, JULY, 1977.

976 MILLER, D. (1974) POPPER'S QUALITATIVE THEORY OF VERISIMILITUDE, BRIT. J. PHILOS. SCI. 25, 166-188.

977 MIURA, S. (1972) PROBABILISTIC MODELS OF MODAL LOGICS, BULL. NAGOYA INSTITUTE OF TECHNOLOGY, 24 67-72.

978 MIZUMOTO, M. (1971) FUZZY AUTOMATA AND FUZZY GRAMMARS, PHD THESIS, FACULTY OF ENGINEERING SCIENCE, OSAKA UNIVERSITY, OSAKA, JAPAN.

979 MIZUMOTO, M. (1971) FUZZY SETS THEORY, 11TH PROFESSIONAL GROUP MEETING ON CONTROL THEORY OF SICE .

980 MIZUMOTO, M., AND TANAKA, K. (1975) ALGEBRAIC STRUCTURES OF FUZZY-FUZZY SETS, TRANS. IECE(D) 58-D, 421-428.

981 MIZUMOTO, M., AND TANAKA, K. (1976) ALGEBRAIC PROPERTIES OF FUZZY MEMBERS, INT. CONF. ON CYBERNETIC AND SOCIETY, WASH. D.C., NOV. .

982 MIZUMOTO, M., AND TANAKA, K. (1976) ALGEBRAIC PROPERTIES OF FUZZY NUMBERS, 1976 INT. CONF. ON CYBERNETICS AND SOCIETY, WASHINGTON.

983 MIZUMOTO, M., AND TANAKA, K. (1976) BOUNDED-SUM AND BOUNDED-DIFFERENCE FOR FUZZY SETS, TRANS. IECE(D) 59-D, 905-912.

984 MIZUMOTO, M., AND TANAKA, K. (1976) FOUR ARITHMETIC OPERATIONS OF FUZZY NUMBERS, TRANS. IECE(D) 59-D, 703-710.

985 MIZUMOTO, M., AND TANAKA, K. (1976) FUZZY DATA STRUCTURE AND FUZZY ARTIFICIAL INTELLIGENCE LANGUAGE, 1976 JOINT CONVENTION OF ELECTRICAL ENGINEERING.

986 MIZUMOTO, M., AND TANAKA, K. (1976) FUZZY-FUZZY AUTOMATA, KYBERNETES, 5, 107-112.

987 MIZUMOTO, M., AND TANAKA, K. (1976) SOME PROPERTIES OF FUZZY SETS OF TYPE 2, INFORM. AND CONTROL, 31, 312/340.

988 MIZUMOTO, M., AND TANAKA, K. (1976) VARIOUS KINDS OF AUTOMATA WITH WEIGHTS, J. COMP. SYST. SCI. .

989 MIZUMOTO, M., AND TANAKA, K. (1978) ALGEBRAIC PRODUCT AND ALGEBRAIC SUM OF FUZZY SETS OF TYPE 2, 4TH EUROPEAN MEETING ON CYBERNETICS AND SYSTEMS RESEARCH, LINZ.

990 MIZUMOTO, M., AND TANAKA, K. (1978) FUZZY SETS OF TYPE 2 UNDER ALGEBRAIC PRODUCT AND ALGEBRAIC SUM, PROC. INT. COLLOQ. ON FUZZY SETS, MARSEILLE.

991 MIZUMOTO, M., AND TANAKA, K. (1978) FUZZY SETS UNDER VARIOUS OPERATIONS, DEPT. OF INFORMATION AND COMPUTER SCI., FACULTY OF ENGINEERING SCI., OSAKA UNIV., TOYONAKA, OSAKA 560, JAPAN.

992 MIZUMOTO, M., AND TANAKA, K.,(TO APPEAR) ALGEBRAIC PRODUCT AND ALGEBAIC SUM OF FUZZY GRADES, TRANS. IECE(D).

993 MIZUMOTO, M., FUKAMI, S., AND TANAKA, K. (1978) FUZZY REASONING METHODS BY ZADEH AND MAMDANI AND IMPROVED METHODS, 3RD FUZZY SYMP. ON FUZZY REASONING, QUEEN MARY COLLEGE, LONDON.

994 MIZUMOTO, M., TOYODA, J., AND TANAKA, K. (1969) SOME CONSIDERATIONS ON FUZZY AUTOMATA, J. COMP. SYST. SCI., 3, 409-422.

995 MIZUMOTO, M., TOYODA, J., AND TANAKA, K. (1970) FUZZY LANGUAGES, SYSTEMS, COMPUTERS, CONTROLS, 1, 36, (ORIG. TIECE 53-C, 333-340).

996 MIZUMOTO, M., TOYODA, J., AND TANAKA, K. (1971) FUZZY ALGEBRA, RESEARCH REPORTS ON MANY-VALUED LOGIC AND ITS APPLICATIONS, KYOTO UNIV., MATHEMATICAL AND SCIENCE RESEARCH RECORDS, MARCH.

997 MIZUMOTO, M., TOYODA, J., AND TANAKA, K. (1972B) GENERAL INFORMATION OF FORMAL GRAMMARS, INFORM. SCI., 4, 87-100.

998 MIZUMOTO, M., TOYODA, J., AND TANAKA, K. (1972C) L-FUZZY LOGIC,IN RESEARCH ON MANY-VALUED LOGIC AND ITS APPLICATIONS, KYOTO UNIVERSITY, JAPAN.

999 MIZUMOTO, M., TOYODA, J., AND TANAKA, K. (1972D) FORMAL GRAMMARS WITH WEIGHTS, TRANS. INST. ELECTRON. COMM. ENG. (JAPAN), 55-D, 292-293.

1000 MIZUMOTO, M., TOYODA, J., AND TANAKA, K. (1973A) N-FOLD FUZZY GRAMMARS, INFORM. SCI., 5, 25-43.

1001 MIZUMOTO, M., TOYODA, J., AND TANAKA, K. (1973B) EXAMPLES OF FORMAL GRAMMARS WITH WEIGHT, INF. PROCESSING.LETT., 2, 74-78.

1002 MIZUMOTO, M., TOYODA, J., AND TANAKA, K. (1975) B-FUZZY GRAMMARS, COMP. MATH. 4, 343-368.

1003 MIZUMOTO, M., UMANO, M., AND TANAKA, K. (1977) IMPLEMENTATION OF A FUZZY-SET THEORETIC DATA STRUCTURE SYSTEM, 3RD INT. CONF. ON VERY LARGE DATA BASES, TOYKO.

1004 MOISEL, G.C. (1972) ESSAIS SUR LES LOGIQUES NON CRIPEPPIEUS PUL ROUMAN ACAD. OF SCI., BUCHAREST.

1005 MOISIL, G.C. (1935) RECHERCHES SUR L'ALGEBRE DE LA LOGIQUE, ANNALES SCI. DE L'UNIVERSITE DE JASSY, ROUMANIA, 22, 1-77.

1006 MOISIL, G.C. (1971) ROLE OF COMPUTERS IN THE EVOLUTION OF SCIENCE, PROCEEDINGS OF INTERNATIONAL CONFERENCE ON SCIENCE AND SOCIETY, BELGRADE, YUGOSLAVIA, 134-136.

1007 MOISIL, G.C. (1972A) LA LOGIQUE DES CONCEPTS NUANCES, IN ESSAIS SUR LES LOGIQUES NON CHRYSIPPIENNES, EDITIONS DE L'ACADEMIE DE LA REPUBLIQUE SOCIALISTE DE ROUMANIE, BUCHAREST, 157-163.

1008 MOISIL, G.C. (1972B) SUR LES ALGEBRAS DE LUKASIEWICZ O-VALENTES IN ESSAIR SUR LES LOGIQUES NON CHRYSIPIENNES, EDITIONS DE L'ACADEMIE DE LA REPUBLIQUE SOCIALISTE DE ROUMANIE, BUCHAREST, 311-324.

1009 MOISIL, G.C. (1973) ENSEMBLES FLOUS ET LOGIQUES A PLUSIEURS VALEURS, CENTRE DE RECHERCHES MATHEMATIQUES, UNIVERSITIE DE MONTREAL, CRM-286, MAI.

1010 MOISIL, G.C. (1975) LECTURES ON FUZZY LOGIC, SCIENTIFIC AND ENCYCLOPAEDIC EDITIONS, BUCHAREST, RUMANIA, (IN RUMANIAN).

1011 MOISIL, G.C. (1975) LECTURES ON THE LOGIC OF FUZZY REASONING, SCIENTIFIC EDITIONS, BUCHAREST.

1012 MOLZEN, N. (1975) FUZZY LOGIC CONTROL, PHD THESIS, TECHNICAL UNIVERSITY OF DENMARK, (IN DANISH).

1013 MONTEIRO, A.A., AND RIBEIRO, H. (1972) L'OPERATION DE FERMETURE ET SES INVARIENTS DANS LES SYSTEMES PARTIELLEMENT ORDONNES, PROTUGALIAE MATHEMATICA, 3, 171-184.

1014 MONTES, C.G., CAMACHO, E.F., AND ARACIL, J. (1976) A FUZZY ALGORITHM FOR NONLINEAR SYSTEM IDENTIFICATION, 3RD EUR. MEETING CYBERN. SYST. RES. VIENNA .

1015 MOON, R., JORDANOV, S., TURKSEN, I.B., AND PEREZ (1978) HUMAN-LIKE REASONING CAP. IN A MEDICAL DIAG. SYSTEM., THE APPL. OF FUZZY SETS TO COMP. DIAG., WORKING PAPER NO. 78-001, UNIV. OF TORONTO, DEPT. OF IND. ENG. .

1016 MORGAN, C.G. (1975) SIMILARITY AS A THEORY OF GRADED EQUALITY FOR A CLASS OF MANY-VALUED PREDICATE CALCULI, PROC. 1975 INT. SYMP. MULTIPLE-VALUED LOGIC, IEEE-75CH0959-7C, 436-449.

1017 MORGAN, C.G. (1976A) MANY-VALUED PROPOSITIONAL INTUITIONISM, PROC. 6TH INT. SYMP. MULTIPLE-VAALUED LOGIC, IEEE 76CH1111-4C, 150-156.

1018 MORGAN, C.G. (1976B) METHODS FOR AUTOMATED THEOREM PROVING IN NON-CLASSICAL LOGICS, IEEE TRANS. COMP., C-25, 852-862.

1019 MORITA, Y., AND IIDA, H. (1975) MEASUREMENT, INFORMATION AND HUMAN SUBJECTIVITY DESCRIBED BY AN ORDER RELATIONSHIP, IN SUMMARY OF PAPERS ON GENERAL FUZZY PROBLEMS, THE WORKING GROUP ON FUZZY SYSTEMS, TOKYO, JAPAN, NOV., 34-39.

1020 MORITA, Y., AND OKA, Y. (1976) ON A LOOP IN FUZZY EVALUATION AND MEASUREMENT, 87-100, SUMMARY OF PAPERS ON GENERAL FUZZY PROBLEMS, THE WORKING GROUP ON FUZZY SYSTEMS, TOYKO, JAPAN.

1021 MOROZOV, A. (1975) SOME PROBLEMS OF DECISION THEORY, EKONOMIKA I MATHEMATICESKIE METODY, 11, 252-262, (IN RUSSIAN).

1022 MORTON, A. (1975) COMPLEX INDIVIDUALS AND MULTIGRADE RELATIONS, NOUS, 9, 309-318.

1023 MOSTOWSKI, A. (1957) ON A GENERALIZATION OF QUANTIFIERS, FUNDAMENTA MATHEMATICAE, 44, 12-36.

1024 MOSTOWSKI, A. (1961) AXIOMATIZABILITY OF MANY VALUED PREDICATE CALCULI, FUNDAMENTA MATHEMATICAE, 50, 165-190.

1025 MOSTOWSKI, A. (1966) THIRTY YEARS OF FOUNDATIONAL STUDIES, BASIL BLACKWELL, OXFORD.

1026 MUKAIDONO, M. (1972) ON SOME PROPERTIES OF FUZZY LOGIC, TECHNICAL REPORT ON AUTOMATION OF IECE,

1027 MUKAIDONO, M. (1972) ON THE B-TERNARY LOGICAL FUNCTION--A TERNARY LOGIC WITH CONSIDERATION OF AMBIGUITY, TRANS. INST. ELECTRON. COMM. ENG. (JAPAN), 55-D, 355-362.

1028 MUKAIDONO, M. (1975) AN ALGEBRAIC STRUCTURE OF FUZZY LOGICAL FUNCTIONS AND ITS MINIMAL AND IRREDUNDANT FORM, TRANS. IECE(D) 58-D, 748-755.

1029 MUKAIDONO, M. (1975) AN APPLICATION OF FUZZY LOGICAL

FUNCTIONS TO PATTERN CLASSIFICATION, TECHNICAL REPORT ON PATTERN RECOGNITION AND LEARNING OF IECE, PRL75-67.

1030 MUKAIDONO, M. (1975) SOME PROPERTIES OF FUZZY LOGICS, TRANS. IECE(D) 58-D, 150-157.

1031 MUKAIDONO, M. (1976) SOME PROPERTIES ON THE RESOLVENT IN FUZZY LOGIC, TECHNICAL REPORT ON PATTERN RECOGNITION AND LEARNING OF IECE, PRL76-3.

1032 MUZYNSKI, W.,, AND JACAK, W. (1976) CONCEPTION OF DESCRIBING THE BEHAVIOUR OF THE EVENTISTIC SYSTEM BY MEANS OF THE FORMALISM OF FUZZY SETS AND RELATIONS 3RD EUR. MEETING CYBERN. SYST. RES., VIENNA .

1033 NAGAI, S. (1973) ON A SEMANICS FOR NON-CLASSICAL LOGICS, PROC. JAPAN ACAD., 49, 337-340.

1034 NAHMIAS, S. (1974) DISCRETE FUZZY RANDOM VARIABLES, UNIVERSITY OF PITTSBURGH, USA.

1035 NAHMIAS, S. (1977) FUZZY VARIABLES, FUZZY SETS AND SYST. 1, 97-110, NORTH-HOLLAND.

1036 NAHMIAS, S. (1978) FUZZY VARIABLES IN A RANDOM ENVIRONMENT, TECH. REP. NO. 39, SCHOOL OF ENGINEERING, UNIV. OF PITTS. .

1037 NAKAGAWA, Y., AND ROSENFELD, A. (1978) A NOTE ON THE USE OF LOGICAL MINIMUM AND MAXIMUM OPERATIONS IN DIGITAL PROCESSING, IEEE TRANS. ON SYST. MAN. AND CYBERN., 8, 632-635.

1038 NAKAMURA, K., AND YOSHIOKA, M. (1975) A SIMULATION MODEL OF PEDESTRIAN FLOW AND ITS INVESTIGATION, 40-45, SUMMARY OF PAPERS ON GENERAL FUZZY PROBLEMS, THE WORKING GROUP ON FUZZY SYSTEMS, TOYKO, JAPAN.

1039 NAKAMURA, K., AND YOSIOKA, M. (1976) SOME MACROSCOPIC MODELS OF COLLECTIVE HUMAN BEHAVIOUR, 79-86, SUMMARY OF PAPERS ON GENERAL FUZZY PROBLEMS, THE WORKING GROUP ON FUZZY SYSTEMS, TOYKO, JAPAN.

1040 NAKAMURA, M. (1941) CLOSURE IN GENERAL LATTICES, PROC. IMPER. ACADEMY, 17, 5-6, TOKYO.

1041 NAKATA, H., MIZUMOTO, M., TOYODA, J., AND TANAKA, K. (1972) SOME CHARACTERISTICS OF N-FOLD FUZZY OF GRAMMARS, TRANS. INST. ELECTRON. COMM. ENG. (JAPAN), 55-D, 287-288.

1042 NAPTALANOFF, N., AND LAKOV, D. (1977) DECISION MAKING IN VAGUE CONDITIONS, KYBERNITES VOL., 91-93.

1043 NASU, M., AND HONDA, N. (1968) FUZZY EVENTS REALIZED BY FINITE PROBABILISTIC AUTOMATA, INFORM. AND CONTROL, 12, 284-303.

1044 NAZAROFF, G.J. (1973) FUZZY TOPOLOGICAL POLYSYSTEMS, J. MATH. ANAL. AND APPLN., 41, 478-485.

1045 NEGOITA, C.V. (1969) INFORMATIONAL RETRIEVAL SYSTEMS, PHD THESIS, POLYTECHNIC INSTITUTE OF BUCHAREST, (IN RUMANIAN).

1046 NEGOITA, C.V. (1970) ON THE STRATEGIES IN AUTOMATIC INFORMATION SYSTEMS, 6TH INT. CONGR. CYBERNETIC SYSTEMS, NAMUR, BELGUIM.

1047 NEGOITA, C.V. (1971) INFORMATION STORAGE AND RETRIEVAL, EDITURA ACADEMIEI, BUCHAREST, (IN RUMANIAN).

1048 NEGOITA, C.V. (1972) LINEAR AND NONLINEAR INFORMATION RETRIEVAL SYSTEMS, ATLAS COMPUTER LABORATORY, DIDCOT, UK.

1049 NEGOITA, C.V. (1973) LINEAR AND NONLINEAR INFORMATION RETRIEVAL, STUDII SI CERCETARI DE DOCUMENTARE, 21-57.

1050 NEGOITA, C.V. (1973) ON THE APPLICATION OF THE FUZZY SETS SEPARATION THEOREM FOR AUTOMATIC CLASSIFICATION IN INFORMATION RETRIEVAL SYSTEMS, INFORM. SCI., 5, 279-286.

1051 NEGOITA, C.V. (1973) ON THE DECISION PROCESS IN INFORMATION RETRIEVAL, STUDII SI CERCETARI DE DOCUMENTARE, 369-381.

1052 NEGOITA, C.V. (1973) ON THE NOTION OF RELEVANCE IN INFORMATION RETRIEVAL, KYBERNETES, 2, 161-165.

1053 NEGOITA, C.V. (1975) INTRODUCTION TO FUZZY SET THEORY FOR SYSTEMS ANALYSIS, NEW YORK, JOHN WILEY AND SONS LTD.

1054 NEGOITA, C.V. (1976) FUZZINESS IN MANAGEMENT, ORSA/TIMS, MIAMI, NOV. .

1055 NEGOITA, C.V. (1976) FUZZY MODELS FOR SOCIAL PROCESSES, IN BOSSEL, H., KLACZKO, S., AND MULLER, N. (EDS.), SYSTEMS THEORY IN THE SOCIAL SCIENCES, BIRKHAUSER VERLAG, BASEL, 283-291.

1056 NEGOITA, C.V. (1976) FUZZY SYSTEMS AND MANAGEMENT SCIENCE, 3RD EUR. MEETING CYBERN. SYST. RES., VIENNA .

1057 NEGOITA, C.V. (1976) OVERLAPPING TENDENCIES IN OPERATIONS RESEARCH SYSTEM THEORY AND CYBERNETICS, PROCEEDINGS OF AN INTERNATIONAL SYMP., UNIV. OF FRIBOURG, SWITZERLAND, OCT., BIRKHAUSER-VERLAG.

1058 NEGOITA, C.V. (1977) FUZZY SYSTEMS AND SOFT SCIENCES, PAPER PRESENTED AT THE MEETING ON MATH. SYST. AND INFORMATICS IN CONTEMPORY RESEARCH, ORGANIZED BY THE UNESCO SCIENTIFIC COORPERATION BUREAU IN EUROPE, BUCHAREST, SEPT.

1059 NEGOITA, C.V. (1977) ON DYNAMICS AND FUZZINESS IN MANAGEMENT

SYSTEMS, IN ROSE, J. AND BELCIU, C. (ED), MOD. TRENDS IN CYBERN. AND SYS. SPRINGER VERLAGE, BERLIN.

1060 NEGOITA, C.V. (1977) ON THE INTERNAL MODEL PRINCIPLE, PROCEEDINGS OF THE SYMP. ON FUZZY SET THEORY AND APPLN., IEEE CONF. ON DECISION AND CONTROL, NEW ORLEANS.

1061 NEGOITA, C.V. (1977) REVIEW OF FUZZY SETS AND THEIR APPLICATION TO COGNITION AND DECISION PROCESSES, IEEE TRANS. SMC-7, NO. 2.

1062 NEGOITA, C.V. (1978) FUZZY MATHEMATICS - A NEW PARADIGM, PROCEEDINGS OF THE 22ND ANNUAL MEETING OF THE SOCIETY OF GENERAL SYSTEMS RESEARCH, WASHINGTON, FEB. .

1063 NEGOITA, C.V. (1978) ON FUZZY SYSTEMS, PRESENT AT 4TH INT. CONGRESS OF CYBERN. AND SYSTEMS, AMSTERDAM.

1064 NEGOITA, C.V. (1978) ON THE STABILITY OF FUZZY SYSTEMS, JAPAN, PROCEEDINGS OF THE 1978 CONFERENCE ON CYBERNETICS AND SOCIETY, 936-937.

1065 NEGOITA, C.V. FUZZY SETS, SYSTEM THEORY AND MANAGEMENT, IN : E. BILLETER, M. CUENOD AND S. KLAZKO, EDS. .

1066 NEGOITA, C.V., (FORTHCOMING) FUZZY SYSTEMS, LONDON: ABACUS PRESS., 1978.

1067 NEGOITA, C.V., AND FLONDOR, P. (1976) ON FUZZINESS IN INFORMATION RETRIEVAL, INT. J. MAN-MACHINE STUDIES, 8, 711-716.

1068 NEGOITA, C.V., AND RALESCU, D.A. (1974) FUZZY SYSTEMS AND ARTIFICIAL INTELLIGENCE, KYBERNETES, 3, 173-178.

1069 NEGOITA, C.V., AND RALESCU, D.A. (1974) INEXACTNESS IN DYNAMIC SYSTEMS, ECONOMIC COMPUTATION AND ECONOMIC CYBERNETICS STUDIES AND RESEARCH, 4, 69-81.

1070 NEGOITA, C.V., AND RALESCU, D.A. (1974) MULTINI VAGI APPLICABILE LOR, EDITURA TECHNICA, BUCHAREST, RUMANIA.

1071 NEGOITA, C.V., AND RALESCU, D.A. (1975) APPLICATIONS OF FUZZY SETS TO SYSTEMS ANALYSIS, NEW YORK: JOHN WILEY.

1072 NEGOITA, C.V., AND RALESCU, D.A. (1975) APPLICATIONS OF FUZZY SETS TO SYSTEMS, NEW YORK: JOHN WILEY.

1073 NEGOITA, C.V., AND RALESCU, D.A. (1975B) REPRESENTATION THEOREMS FOR FUZZY CONCEPTS, KYBERNETES, 4, 169-174.

1074 NEGOITA, C.V., AND RALESCU, D.A. (1975D) RELATIONS ON MONOIDS AND MINIMAL REALIZATION THEORY FOR DYNAMIC SYSTEMS; APPLICATIONS FOR FUZZY SYSTEMS, PROCEEDINGS OF 3RD INTERNATIONAL CONGRESS OF CYBERNETICS AND SYSTEMS, BUCHAREST, RUMANIA, AUG.

1075 NEGOITA, C.V., AND RALESCU, D.A. (1976) COMMENT ON A COMMENT ON AN ALGORITHM THAT GENERATES FUZZY PRIME IMPLICANTS BY LEE AND CHANG, INFORM. AND CONTROL, 30, 199-201.

1076 NEGOITA, C.V., AND RALESCU, D.A. (1977) ON FUZZY OPTIMIZATION, KYBERNETICS 6, 193-195.

1077 NEGOITA, C.V., AND RALESCU, D.A. (1977) SOME RESULTS IN FUZZY SYSTEMS THEORY, IN J. ROSE AND C. BILCIU (EDS.): MODERN TRENDS IN CYBERN. AND SYST., SPRINGER-VERLAG, BERLIN.

1078 NEGOITA, C.V., AND STEFANESCU, A.C. (1975) ON THE STATE EQUATION OF FUZZY SYSTEMS, KYBERNETES, 4, 231-214.

1079 NEGOITA, C.V., AND STEFANESCU, AL.,(TO BE PUBLISHED) FUZZY OBJECTS IN TOPOI: A GENERALIZATION OF FUZZY SETS.

1080 NEGOITA, C.V., AND SULARIA, M. (1978) A SELECTION METHOD OF NON-DOMINATED POINTS IN MULTI-CRITERIA DECISION PROBLEMS, ECON. COMP. AND ECON. CYBERN.: STUDIES AND RESEARCH, NO. 1, 19-23.

1081 NEGOITA, C.V., AND SULURIU, M. (1976) ON FUZZY MATHEMATICAL PROGRAMMING AND TOLERANCES IN PLANNING, ECONOMIC COMPUTATION AND ECONOMIC CYBERNETICS STUDIES AND RESEARCH, BUCHAREST, RUMANIA.

1082 NEGOITA, C.V., FLONDER, P., AND SULARIA, M. (TO APPEAR) FUZZY LINEAR PROGRAMMING .

1083 NEGOITA, C.V., FLONDOR, P., AND SULARIA, M. (1977) ON FUZZY ENVIRONMENT IN OPTIMIZATION PROBLEMS, IN J. ROSE AND C. BILCIU(EDS.): MODERN TRENDS IN CYBERNETICS AND SYSTEMS, BERLIN, SPRINGER VERLAG.

1084 NEGOITA, C.V., FLONDOR, P., AND SULARIA, M. (1977) ON FUZZY ENVIRONMENTS IN OPTIMIZATION PROBLEMS, ECON. COMPUT. ECON. CYBERNET. STUD. RES., 13-24.

1085 NEGOITA, C.V., MINOUI, S., AND STAN, E. (1976) ON CONSIDERING IMPRESSION IN DYNAMIC LINEAR PROGRAMMING, ECON. COMP. AND ECON. CYBERN., VOL. 3.

1086 NEGOITA, C.V., SULARIA, M. (1976) FUZZY LINEAR PROGRAMMING AND TOLERANCES IN PLANNING, ECON. COMPUT. ECON. CYBERNET. STUD. RES. 1, 3-15.

1087 NETTO, A.B. (1970) FUZZY CLASSES, NOTICES OF THE AMERICAN MATHEMATICAL SOCIETY, 68T-H28, 945.

1088 NEUHAUS, N.J., AND SPEVACK, M. (1975) SHAKESPEARE DICTIONARY--SOME PRELIMINARIES FOR A SEMANIC DESCRIPTION, COMPUTERS AND THE HUMANITIES, 9, 263-270.

1089 NEUSTUPNY, J.V. (1966) ON THE ANALYSIS OF LINGUISTIC VAGUENESS, TRAVAUX LINGUITIQUES DE PRAGUE, 2, 39-51.

1090 NEWPECK, F. (1977) FUZZY SETS, PROC. 1977 AIDS CONF., CHICAGO.

1091 NEWTON, L.K. (1978) FUZZY SET THEORY IN A DECISION ANAL. FORMULATION OF THE ACC. MATERIALITY DEC., PRESENTED AT JOINT NATIONAL ORSA/TIMS MEETING, LOS ANGELES.

1092 NGUYEN, C-H (1973) GENERALIZED POST ALGEBRAS AND THEIR APPLICATION TO SOME INFINITARY MANY-VALUED LOGICS, DISSERTATIONES MATHEMATICAE, 107, WARSAW.

1093 NGUYEN, H.T. (1976) A NOTE ON THE EXTENSION PRINCIPLE FOR FUZZY SETS, ELECTRONICS RESEARCH LAB. MEMO. M-611, UNIV. OF CALIFORNIA, BERKELEY.

1094 NGUYEN, H.T. (1977) ON FUZZINESS AND LINGUISTIC PROBABILITIES, J. MATH. ANAL. AND APPL. VOL. 61, NO. 3, 658-671.

1095 NGUYEN, H.T. (1977) ON RANDOM SETS AND BELIEF FUNCTIONS, MEMO ERL M77/14, UNIV. OF CALIF., BERKELEY.

1096 NGUYEN, H.T. (1977) SOME MATHEMATICAL TOOLS FOR LINGUISTIC PROBABILITIES, PROC. IEEE SYST. ON FUZZY SETS, NEW ORLEANS, 1345-1350.

1097 NGUYEN, H.T. (1978) CONDITIONING IN POSSIBILITY THEORY, UNIV. OF MASS., 1450-1453, PROC. 1978 IEEE CONF. ON DEC. AND CONTROL, INCLUDES THE 17TH SYMP. ON ADAPTIVE PROCESSES, 78CH1392-OCS, JAN., 1979, SAN DIEGO, CALIF.

1098 NGUYEN, H.T. (1978) ON CONDITIONAL POSSIBILITY DISTRIBUTIONS, FUZZY SETS AND SYST. 1, 299-309, NORTH-HOLLAND.

1099 NGUYEN, H.T. (1978) ON THE NON-INTERACTION AND CONDITIONING IN POSSIBILITY THEORY, DEPT. OF MATH. AND STATISTICS, UNIV. OF MASSACHUSETTS, PROCEEDINGS OF THE 1978 INT. CONF. ON CYBERNETICS AND SOCIETY, VOL2, 1210-1212.

1100 NGUYEN, H.T., AND OHSUGA, S. (1977) STUDY OF LINGUISTIC PROBABILITIES, PROCEEDINGS OF THE SYMP. ON FUZZY SET THEORY AND APPLN., IEEE CONF. ON DECISION AND CONTROL, NEW ORLEANS.

1101 NIEMINEN, J. (1978) FUZZY MAPPINGS AND ALGEBRAIC STRUCTURES, FUZZY SETS AND SYST. 1, 231-235, NORTH-HOLLAND.

1102 NISHIBE, T. (1977) A METHOD OF PLACING VERTICES OF A GRAPH ON PLANE IN AUTO- GRAPH-DRAWING, 118-125, SUMMARY OF PAPERS ON GENERAL FUZZY PROBLEMS, THE WORKING GROUP ON FUZZY SYSTEMS, TOYKO, JAPAN.

1103 NOGUCHI, K., UMANO, M., MIZUMOTO, M., AND TANAKA, K. (1976) IMPLEMENTATION OF FUZZY ARTIFICIAL INTELLIGENCE

LANGUAGE FLOU, TECHNICAL REPORT ON AUTOMATON AND LANGUAGE, AL76-25.

1104 NOGUCHI, Y. (1972) A PATTERN CLUSTERING METHOD ON THE BASIS OF ASSOCIATION SCHEMES, BULLETIN ELECTROTECHNICAL LABORATORY, 36, 753-767.

1105 NOVAK, J. (1968) ON PROBABILITY DEFINED ON CERTAIN CLASSES OF NON-BOOLEAN ALGEBRA, NACHRICHTEN DER OSTERREICHISCHEN MATHEMATISCHE GESELLSCHAFT, 23, 89-90.

1106 NOWAKOWSKA, M. (1976) FORMAL THEORY OF ACTIONS AND ITS APPLICATION TO SOCIAL SCIENCES, 3RD EUR. MEETING CYBERN. SYST. RES., VIENNA .

1107 NOWAKOWSKA, M. (1976) METHODOLOGICAL PROBLEMS OF MEASUREMENT OF FUZZY CONCEPTS IN SOCIAL SCIENCES, BEHAVIORAL SCIENCES

1108 NOWAKOWSKA, M. (1976) TOWARDS A FORMAL THEORY OF DIALOGUES, SEMIOTICS, 18 .

1109 NOWOKAOWSKA, M. (1978) FUZZY REASONING AND DIALOGUES, INSTIT. OF PHILOSOPHY AND SOCIOLOGY, POLISH ACADEMY OF SCI., MARZALKOWSKA 140-100, 00-061 WARZAWA, POLAND.

1110 NURMI, H. (1976) ON FUZZY GAMES, 3RD EUR. MEETING CYBERN. SYST. RES., VIENNA .

1111 NURMI, H. (1977) PROBABILITY AND FUZZINESS: SOME METODOLOGICAL CONSIDERATIONS, PRESENTED AT THE 6TH RESEARCH CONFERENCE ON SUBJECTIVE PROBABILITY, UTILITY, AND DECISION MAKING, WARZAWA, SEPT. .

1112 NURMI, H. (1978) MODELLING IMPRECISENESS IN HUMAN SYSTEMS, PRESENTED AT THE 4TH EUROPEAN MEETING ON CYBERN. AND SYSTEMS RESEARCH, LINZ, MARCH.

1113 NURMI, H. (1978) MODELLING POLITICAL VAGUENESS, IMPRECISENESS AND AMBIGUITY, PRESENTED AT THE ECPR JOINT SESSIONS OF THE WORKSHOPS, GRENOBLE, APRIL.

1114 NURMINEN, M.I. (1976) ABOUT THE FUZZINESS IN THE ANALYSIS OF INFORMATION SYSTEMS, 3RD EUR. MEETING CYBERN. SYST. RES., VIENNA .

1115 NURMINEN, M.I. (1976) STUDIES IN SYSTEMEERING ON FUZZINESS IN THE ANALYSIS OF INFORMATION SYSTEMS, DISSERTATION, INSTITUTE FOR APPLIED MATHEMATICS, UNIVERSITY OF TURKU, FINLAND.

1116 NURMINEN, M.I., AND PAASIO, A. (1976) SOME REMARKS ON THE FUZZY APPROACH TO MULTIGOAL DECISION MAKING, FINNISH JOUR. OF BUS. ECO., SPECIAL EDITION 3.

1117 ODA, M. (1977) STRUCTURE ANALYSIS AND APPLICATIONS OF FUZZY-AND-UNCERTAIN REASONING METHODS, PROCEEDINGS OF

THE SYMP. ON FUZZY SET THEORY AND APPLN., IEEE CONF. ON DECISION AND CONTROL, NEW ORLEANS.

1118 ODA, M., AND SHIMOMURA, H., CHIMURA, H. AND WOMACK, B.F. (1977) MEASUREMENT, EVALUATION, AND CONTROL OF COMMUNICATION-AND-FORMATION PROCESS OF MORALITY CONCEPT, PROC. 1977 IEEE CONF. ON DECISION AND CONTROL, NEW ORLEANS.

1119 ODEN, G. (1978) APPLICATIONS OF FUZZY SET THEORY AND FUZZY LOGIC TO PSYCHOLINGUIST PROBLEMS, PROC. 22ND MEETING OF SOCIETY FOR GEN. SYS. RES., WASH., 431-439.

1120 ODEN, G. (1978) ON THE USE OF SEMANTIC CONSTRAINTS IN GUIDING SYNTACTIC ANALYSIS WHIPP REPORT NO. 3, UNIV. OF WISCONSIN.

1121 ODEN, G., AND HOGAN, M.E. (1977) A FUZZY PROPOSITIONAL MODEL OF NEGATION ON SEMANTIC CONTINUATION, MIDWESTERN PSYCHOLOGICAL ASSOC. .

1122 ODEN, G., AND MASSARO, D.W. (1977) INTEGER OF PLACE AND VOICING INF. IN IDENTIFYING SYNTHETIC STOP- CON. SYLL., WISCON. HUMAN INF. PROC. PROGRAM REPORT, WHIP NO. 1.

1123 ODEN, G.C. (1977) FUZZINESS IN SEMANTIC MEMORY: CHOOSING EXEMPLARS OF SUBJECTIVE CATAGORIES, MEMORY AND COGNITION 5, 198-204.

1124 ODEN, G.C. (1977) INTEGRATION OF FUZZY LOGICAL INFORMATION, J. EXP. PSYCHOL. 3, 565-575.

1125 ODEN, G.C., AND ANDERSON, N.H. (1974) INTEGRATION OF SEMANTIC CONSTRAINTS, J. VERBAL LEARNING AND BEHAVIOUR, 13, 138-148.

1126 OHNISHI, M., AND MATSUMOTO, K. (1957) GENTZEN METHOD IN CALCULI, OSAKA MATH. J., 9, 113-130.

1127 OHSUGA, S. (1977) SEMANTIC INFORMATION PROCESSING IN MAN-MACHINE SYSTEMS, PROC. 1977 IEEE CONF. ON DECISION AND CONTROL, NEW ORLEANS.

1128 OKADA, N., AND TAMACHI, T. (1974) AUTOMATED EDITING OF FUZZY LINE DRAWINGS FOR PICTURE DESCRIPTION, TRANS. INST. ELECTRON. COMM. ENG. (JAPAN), 57-A, 216-223.

1129 OKADA, N., AND TAMACHI, T. (1974) THEORY OF FUZZY INTEGRALS AND ITS APPLICATIONS, THESIS, TOKYO INST. OF TECH. .

1130 OKUDA, S. (1977) STRUCTURAL ANALYSIS OF SCIENTIFIC INFORMATION USAGE, 11-18, SUMMARY OF PAPERS ON GENERAL FUZZY PROBLEMS, THE WORKING GROUP ON FUZZY SYSTEMS, TOYKO, JAPAN.

1131 OKUDA, T., TANAKA, H., AND ASAI, K. (1974) DECISION-MAKING AND INFORMATION IN FUZZY EVENTS, BULLETIN OF UNIVERSITY

OF OSAKA PREFECTURE, 23A .

1132 OKUDA, T., TANAKA, H., AND ASAI, K. (1978) A FORMULATION OF FUZZY DECISION PROBLEMS WITH FUZZY INFORMATION USING PROB. MEAS. OF FUZZY EV., INF. AND CONTROL, 38,135-147.

1133 OKUDA, T., TANAKA, H., ASAI, K. (1976) DECISION PROBLEMS AND QUANTITY OF INFORMATION IN FUZZY EVENTS, TRANS. SICE 12, 63-68.

1134 ONAGA, K., AND MAYEDA, W. (1976) BOOLEAN FLOW THEORY AND ITS APPLICATIONS TO CLUSTER ANALYSIS, FUZZY LOGICS AND PARTICLE TRANSMISSION, ALLERTON CONF. AT UNIV. OF ILLINOIS.

1135 ONICESCU, O. (1971) PRINCIPLES DE LOGIQUE ET DE PHILOSOPHIE MATHEMATIQUE, RUMANIAN ACADEMY OF SCIENCE, BUCHAREST.

1136 ORE, O. (1942) THEORY OF EQUIVALENCE RELATIONS, DUKE MATH. J., 9, 573-627.

1137 ORLOVSKY, S.A. (1977) DECISION MAKING WITH A FUZZY PREFERENCE RELATION, FUZZY SETS AND SYST. 1, 155-167, NORTH-HOLLAND.

1138 ORLOVSKY, S.A. (1977) ON PROGRAMMING WITH FUZZY CONSTRAINT SETS, KYBERNETES 6, 197-201.

1139 ORLOVSKY, S.A. (1978) DECISION MAKING WITH A FUZZY PREFERENCE RELATION IN A FUZZY SET OF ALTERATIONS, PROC. 4TH INT. CONGRESS ON CYBERN. AND SYSTEMS, AMSTERDAM, 375-376.

1140 ORLOWSKA, E. (1967) MECHANICAL PROOF PROCEDURE FOR THE N-VALUED PROPOSITIONAL CALCULUS, BULL. DE L'ACADEMIE POLONAISE DES SCIENCES, SER. DES SCIENCES MATH., ASTR. ET PHYS., 15, 537-541.

1141 ORLOWSKA, E. (1973) THEOREM-PROVING SYSTEMS, DISSERTATIONES MATHEMATICAE, 103, WARSAW.

1142 OSGOOD, C.E., SUCI, G.J., AND TANNENBAUM, P.H. (1964) THE MEASUREMENT OF MEANING, UNIVERSITY OF ILLINOIS PRESS.

1143 OSIS, J.J. (1968) FAULT DETECTION IN COMPLEX SYSTEMS USING THEORY OF FUZZY SETS, IN D.S. KRISTINKOV, J.J. OSIS, L.A. RAVTRIGIN (EDS.), KIBERNETIKA I DIAGNOSTIKA, 2, 13-18, (IN RUSSIAN) ZINATNE, RIGA, U.S.S.R. .

1144 OSTERGAARD, J.J. (1977) FUZZY LOGIC CONTROL OF A HEAT EXCHANGER PROCESS, 285-320, IN FUZZY AUTOMATA AND DECISION PROCESSES, M.M. GUPTA, G.N. SARIDIS, B.R. GAINES, EDS., NORTH-HOLLAND.

1145 OTSUKI, S. (1970) A MODEL FOR LEARNING AND RECOGIZING MACHINE, INFORMATION PROCESSING,' 11, 664-671.

1146 PAL, S.K., AND MAJUMDAR, D.D. (1977) FUZZY SETS AND DECISION-MAKING APPROACHES IN VOWEL AND SPEAKER

RECOGNITION, IEEE TRANSACTIONS ON SYSTEMS, MAN, AND CYBERNETICS, SMC-7, 625-629.

1147 PAL, S.K., AND MAJUMDER, D.D. (1978) CORRECTION TO "ON AUTOMATIC PLOSIVE IDENTIFICATION USING FUZZINESS IN PROPERTY SETS", IEEE TRANSACTIONS ON SYST., MAN, AND CYBERNETICS, VOL. SMC-8, NO. 12, 907.

1148 PAL, S.K., AND MAJUMDER, D.D. (1978) ON AUTOMATIC PLOSIVE IDENTIFICATION USING FUZZINESS IN PROPERTY SETS, VOL.8, IEEE TRANS. ON SYST., MAN, AND CYBERN., 302-307.

1149 PAL, S.K., MAJUMDER, D.D., AND CHAUDHURY, B.B. (TO APPEAR) FUZZY SETS IN HANDWRITTEN CHARACTER RECOGNITION, IN PROC. ALL INDIA INTERDISCIPLINARY SYMP. DIGITAL TECH. AND PATTERN RECOG., FEB. 15-17, I.S.I., CALCUTTA.

1150 PANDA, S.R. (1971) INVERSE PROBLEM FOR LINEAR SYSTEMS CONTAINING UNCERTAIN PARAMETERS, ASME, PAPER 71 - WA/AUT-14 FOR MEETING NOV. 28-DEC. 2, P. 12.

1151 PAPERT (-STRAUSS), D. (1968) TOPOLOGICAL LATTICES, PROC. LOND. MATH. SOC., 18, 217-230.

1152 PAPPIS, C.P., AND MAMDANI, E.H. (1976) A FUZZY LOGIC CONTROLLER FOR A TRAFFIC JUNCTION, RESEARCH REPORT, DEPT. OF ELECTRICAL ENGINEERING, QUEEN MARY COLLEGE, LONDON.

1153 PAPPIS, C.P., AND SUGENO, M. (1976) FUZZY RELATIONAL EQUATIONS AND THE INVERSE PROBLEM, INTERNAL REPORT, QUEEN MARY COLLEGE, LONDON.

1154 PARRET, H. (1974) DISCUSSING LANGUAGE, MOUTON, THE HAGUE, (SEE DIALOGUE WITH G. LAKOFF).

1155 PARRISH, E.A. (1977) ELECTROMAGNETIC INTERFERENCE SOURCE IDENTIFICATION THROUGH FUZZY CLUSTERING, PROC. 1977 IEEE CONF. ON DECISION AND CONTROL, NEW ORLEANS.

1156 PARSONS, C. (1974) THE LIAR PARADOX, J. PHILOS. LOGIC, 3, 381-412.

1157 PASK, G. (ED.) (1976) CURRENT SCIENTIFIC APPROACHES TO DECISION MAKING IN COMPLEX SYSTEMS, SYSTEM RESEARCH LTD., RICHMOND, UK, APRIL.

1158 PASK, G. (1975A) THE CYBERNETICS OF HUMAN LEARNING AND PERFORMANCE, HUTCHISON, LONDON.

1159 PASK, G. (1975B) CONVERSATION, COGNITION AND LEARNING, ELSEVIER, AMSTERDAM.

1160 PAVLIDIS, T. (1975) FUZZY REPRESENTATIONS AS MEANS OF OVERCOMING THE OVER- COMMITMENT OF SEGMENTATION, PROC. CONF. ON COMPUTER GRAPHICS, PATTERN RECOGNITION AND DATA STRUCTURES, LOS ANGELES, CALIF.

1161 PAVLIDIS, TH. (1977) APPLICATION OF FUZZY SETS IN CURVE FITTING, PROC. 1977 IEEE CONF. ON DECISION AND CONTROL, NEW ORLEANS.

1162 PAZ, A. (1967) FUZZY STAR FUNCTIONS, PROBABILISTIC AUTOMATA AND THEIR APPROXIMATION BY NONPROBABILISTIC AUTOMATA, J. COMP. SYST. SCI., 1, 371-389.

1163 PEARL, J. (1974) PROBLEM PRESENTATION RESEARCH, UCLA-ENG-7404, SCHOOL OF ENGINEERING AND APPLIED SCIENCE, UNIVERSITY OF CALIFORNIA, LOS ANGELES, USA.

1164 PEARL, J. (1975A) ON THE COMPLEXITY OF COMPUTING PROBABILITIC ASSERTIONS, UCLA- ENG-7562, SCHOOL OF ENGINEERING AND APPLIED SCIENCE, UNIVERSITY OF CALIFORNIA, LOS ANGELES, USA, JULY.

1165 PEARL, J. (1975B) ON THE COMPLEXITY OF INEXACT COMPUTATIONS, UCLA-ENG- PAPER-0775, SCHOOL OF ENGINEERING AND APPLIED SCIENCE, UNIVERSITY OF CALIFORNIA, LOS ANGELES, USA, JULY.

1166 PEARL, J. (1975C) AN ECONOMIC BASIS FOR CERTAIN METHODS OF EVALUATING PROBABILISTIC FORECASTS, UCLA-ENG-REP-7561, SCHOOL OF ENGINEERING AND APPLIED SCIENCE, UNIVERSITY OF CALIFORNIA, LOS ANGELES, USA, JULY.

1167 PEARL, J. (1975D) ON THE STORAGE ECONOMY OF INFERENTIAL QUESTION-ANSWERING SYSTEMS, IEEE TRANS. SYST. MAN CYBERN., SMC- 5, 595-602.

1168 PEARL, J. (1975E) STATE COMPLEXITY OF IMPRECISE CASUAL MODELS, UCLA-ENG- REP-7560, SCHOOL OF ENGINEERING AND APPLIED SCIENCE, UNIVERSITY OF CALIFORNIA, LOS ANGELES, USA, DEC. .

1169 PEARL, J. (1976A) A NOTE ON THE MANAGEMENT OF PROBABILITY ASSESSORS, UCLA-ENG- REP-7664, SCHOOL OF ENGINEERING AND APPLIED SCIENCE, UNIVERSITY OF CALIFORNIA, LOS ANGELES, USA, FEB. .

1170 PEARL, J. (1976B) A FRAMEWORK FOR PROCESSING VALUE JUDGEMENTS UCLA-REP-7622, SCHOOL OF ENGINEERING AND APPLIED SCIENCE, UNIVERSITY OF CALIFORNIA, LOS ANGELES, USA, MARCH.

1171 PERRY, K.E., AND WADDELL, J.J. (1972) THE ROTARY CEMENT KILN, THE CHEMICAL PUBLISHING CO., NEW YORK.

1172 PESCHEL, M. (1975) SOME REMARKS TO "FUZZY SYSTEMS" AS A COMPLEMENT TO THE TOPIC PAPER FROM L.A. ZADEH, BERLIN, FEB. .

1173 PETRESCU, I. (1971) ALGEBRES DE MORGEN INJECTIVES, IN MOISIL, G.C. (ED.), LOGIQUE, AUTOMATIQUE, INFORMATIQUE, BUCHAREST, 171-176.

1174 PETROV, E.G., AND ZATOV, V.G. (1977) A SELF THEORETICAL

METHOD FOR ESTIMATION OF THE EFFICIENCY OF COMPLEX SYSTEMS, SOVIET AUTOMATIC CONTROLS, VOL. 9, 57-65.

1175 PINKAVA, V. (1965) ON THE NATURE OF SOME LOGICAL PARADOXES, KYBERNETIKA (PRAGUE), 1, 111-121, (IN CZECH., ENG. SUMMARY).

1176 PINKAVA, V. (1975) SOME FURTHER PROPERTIES OF THE PI-LOGICS, PROC. 1975 INT. SYMP. MULTIPLE-VALUED LOGIC, IEEE 75CH0959-7C, 20-26.

1177 PINKAVA, V. (1976A) "FUZZIFICATION" OF BINARY AND FINITE MULTIVALUED LOGICAL CALCULI, INT. J. MAN-MACHINE STUDIES, 8, 717-730.

1178 PINKAVA, V. (1976B) ON THE NATURE OF SOME LOGICAL PARADOXES, INT. J. MAN-MACHINE STUDIES, (TO APPEAR).

1179 PINKAVA, V., AND KOHOUT, L.J. (1976) ENUMERABLY INFINITE-VALUED FUNCTIONALLY COMPLETE PI-LOGIC ALGEBRAS, IN MAMDANI, E.H., AND GAINES, B.R. (EDS.), DISCRETE SYSTEMS AND FUZZY REASONING, EES-MMS-DSFR-76, QUEEN MARY COLLEGE, UNIVERSITY OF LONDON, (WORKSHOP PROCEEDINGS).

1180 POLLATSCHEK, M.A. (1977) HIERARCHICAL SYSTEMS AND FUZZY-SET THEORY, PROC. 1977 IEEE CONF. ON DECISION AND CONTROL, NEW ORLEANS.

1181 PONSARD, C. (1975) CONTRIBUTION A UNE THEORIE DES ESPACES ECONOMIQUES IMPRECIS, DOCUMENT DE TRAVAIL IME, UNIVERSITY OF DIJON.

1182 PONSARD, C. (1975) L'IMPRESSION ET SON TRAITMENT EN ANALYSE ECONOMIQUE, DOCUMENT DE TRAVAIL IME, UNIVERSITY OF DIJON.

1183 POOLOCK, J.L. (1975) FOUR KINDS OF CONDITIONALS, AMER. PHILOS. QUART., 12, 51-59.

1184 POPE, S. (1978) MULTIOBJECTIVE DECISION ANALYSIS OF THE VALUE OF TEST MARKETING RESEARCH, PRES. AT ORSA/TIMS MEETING, LOS ANGELES.

1185 POPPER, K.R. (1963) CONJECTURES AND REFUTATIONS, ROUTLEDGE AND KEGAN PAUL, LONDON.

1186 POPPER, K.R. (1972A) OBJECTIVE KNOWLEDGE, CLARENDON PRESS, OXFORD.

1187 POPPER, K.R. (1972B) THE LOGIC OF SCIENTIFIC DISCOVERY, HUTCHINSON, LONDON, (1ST ED. 1959).

1188 POPPER, K.R. (1976A) A NOTE ON VERISIMILITUDE, BRIT. J. PHILOS. SCI., 27, 147-164.

1189 POPPER, K.R. (1976B) UNENDED QUEST, FONTANA, LONDON.

1190 POSELOV, D.A. (1976) SEMOITIC MODELS: SUCESSES AND PERSPECTIVES, CYBERN., VOL. 12, NO. 6, 929-937.

1191 POSPISIL, B. (1937) REMARK ON BICOMPACT SPACES, ANNALS OF MATH., 38, 845-846.

1192 POSPISIL, B. (1939A) ON BICOMPACT SPACES, PUBLICATIONS DE LA FACULTE DES SCIENCES DE L'UNIVERSITE MASARYK, NO. 270, BRNO, CZECH., 3-16.

1193 POSPISIL, B. (1939B) PRIMIDEALE IN VOLLSTANDIGEN RINGEN, FUNDAMENTA MATHEMATICAE, 33, 66-74, (WHOLE VOL. PUBLISHED IN DEC. 1945).

1194 POSPISIL, B. (1940) UBER DIE MESSBAREN FUNKTIONEN, MATHEMATISCHE ANNALEN, 117, 327-355.

1195 POSPISIL, B. (1941A) EINE BEMERKUNG UBER VOLLSTANDIGE RAUME, CASOPIS PRO PESTOVANI MATHEMATIKY A FYSIKY (PRAGUE), 70, 38-41.

1196 POSPISIL, B. (1941B) VON DEN VERTEILUNGEN AUF BOOLESCHEN RINGEN, MATHEMATISCHE ANNALEN, 118, 32-40.

1197 POSPISIL, B. (1941C) EINE BEMERKUNG UBER STETIGE VERTAILUNG, CASOPIS PRO PESTOVANI MATHEMATIKY A FYSIKY (PRAGUE), 70, 68-72.

1198 POSPISIL, B. (1941D) EINE BENERKUNG UBER FUNKTIONENFOLGEN, CASOPIS PRO PESTOVANI MATEMATIKY A FYSIKY, 70, 119-121.

1199 POST, J.F. (1973) SHADES OF THE LIAR, PHILOS. LOGIC, 2, 370-386.

1200 POSTON, T. (1971A) FUZZY GEOMETRY, PHD THESIS, UNIVERSITY OF WARWICK, UK.

1201 POSTON, T. (1971B) FUZZY GEOMETRY, MANIFOLD, 10, UNIVERSITY OF NOTTINGHAM.

1202 PRADE, H. (1978) FUZZY LOGICS: A SURVEY, PRESENT AT HIGHER ORDER SYSTEM.

1203 PRADE, H. (1978) USING FUZZY SET THEORY IN A SCHUDULING PROBLEM: A CASE STUDY SUBMITTED TO J. OF FUZZY SETS AND SYSTEMS.

1204 PRADE, H. (1978) WHY FUZZY THEORY DOESN'T SEEM VERY USEFUL FOR INDUSTRIAL ROBOTIC SYSTEMS, REPORT STANFORD ART. INTELL. LAB.

1205 PRELECKI, M. (1958) W SPRAWIE TERMINOW NIEOSTRYCH, STUDIA LOGICA, 8 .

1206 PREPARATA, F.P., AND YEH, R.T. (1970) A THEORY OF CONTINUOUSLY VALUED LOGIC, TECH. REP. 89, UNIVERSITY OF TEXAS, AUSTIN, USA, JUNE.

1207 PREPARATA, F.P., AND YEH, R.T. (1971) ON A THEORY OF CONTINUOUSLY VALUED LOGIC, CONFERENCE RECORD OF 1971 SYMPOSIUM ON THEORY AND APPLICATIONS OF MULTIPLE-VALUED LOGIC DESIGN, 124-132.

1208 PREPARATA, F.P., AND YEH, R.T. (1972) CONTINUOUSLY VALUED LOGIC, J. COMP. SYST. SCI. 6, 397-418.

1209 PRIOR, A.N. (1953) ON PROPOSITIONS NEITHER NECESSARY NOR IMPOSSIBLE, J. SYMBOLIC LOGIC, 18, 105-108.

1210 PRIOR, A.N. (1954) THE INTERPRETATION OF TWO SYSTEMS OF MODAL LOGIC, J. COMP. SYST., 4, 201-208.

1211 PRIOR, A.N. (1955A) MANY-VALUED AND MODAL SYSTEMS: AN INTUITIVE APPROACH, PHILOS. REV., 64, 626-630.

1212 PRIOR, A.N. (1955B) CURRY'S PARADOX AND 3-VALUED LOGIC, AUSTRALASIAN JOURNAL OF PHILOSOPHY, 33, 177-182.

1213 PRIOR, A.N. (1957) TIME AND MODALITY, CLARENDON PRESS, OXFORD.

1214 PRIOR, A.N. (1962) FORMAL LOGIC, CLARENDON PRESS, OXFORD, (2ND ED.).

1215 PRIOR, A.N. (1967) PAST, PRESENT AND FUTURE, CLARENDON PRESS, OXFORD.

1216 PRIOR,A.N. (1971) OBJECTS OF THOUGHT, CLARENDON PRESS, OXFORD, (EDITED BY GEACH AND KENNY).

1217 PROCTOR, C. (1977) REVIEW OF: INTRO. TO THE THEORY OF FUZZY SUBSETS, VOL. 1, KAUFMAN J.A.S.A., MARCH.

1218 PROCYK, T.J. (1974) THE CONTROL OF SYSTEMS POSSESSING DELAY USING FUZZY SET THEORY, FUZZY LOGIC WORKING GROUP ROP., QUEEN MARY COLLEGE, UNIVERSITY OF LONDON, UK, DEC.

1219 PROCYK, T.J. (1976A) LINGUISTIC REPRESENTATION OF FUZZY VARIABLES, FUZZY LOGIC WORKING GROUP REP. 3, QUEEN MARY COLLEGE, UNIVERSITY OF LONDON, UK.

1220 PROCYK, T.J. (1976B) A FUZZY LOGIC LEARNING SYSTEM FOR A SINGLE INPUT SINGLE OUTPUT PLANT, FUZZY LOGIC WORKING GROUP REP. 3, QUEEN MARY COLLEGE, UNIVERSITY OF LONDON, UK.

1221 PRUGOVECKI, E. (1973) A POSTULATIONAL FRAMEWORK FOR THEORIES OF SIMUTANEOUS MEASUREMENT OF SEVERAL OBSERVABLES, FOUNDATIONS OF PHYSICS, 3, 3-18.

1222 PRUGOVECKI, E. (1974) FUZZY SETS IN THE THEORY OF MEASUREMENT OF INCOMPATIBLE OBSERVABLES, FOUNDATIONS OF PHYSICS, 4, 9-18.

1223 PRUGOVECKI, E. (1975) MEASUREMENT IN QUANTUM MECHANICS AS A STOCHASTIC PROCESS ON SPACES OF FUZZY EVENTS, FOUNDATIONS OF PHYSICS, 5, 557-571.

1224 PRUGOVECKI, E. (1976) LOCALIZABILITY OF RELATIVISTIC PARTICLES IN FUZZY PHASE SPACE, J. PHYS. MATH., VOL. 9, NO. 11.

1225 PRUGOVECKI, E. (1976A) PROBABILITY MEASURES ON FUZZY EVENTS IN PHASE SPACE, J. MATH. PHYSICS, 17, 517-523.

1226 PRUGOVECKI, E. (1976B) QUANTUM TWO-PARTICLE SCATTERING IN FUZZY PHASE SPACE, DEPARTMENT OF MATHEMATICS, UNIVERSITY OF TORONTO, CANADA, JAN.

1227 PRUGOVECKI, E. (1977) ON FUZZY SPIN SPACES, J. PHYS. A: MATH., VOL. 10, NO. 4.

1228 PUDLAK, P. (1975A) THE OBSERVATIONAL PREDICATE CALCULUS AND COMPLEXITY OF COMPUTATIONS, COMMENTATIONES MATH. UNIVERSITATIS CAROLINAE, 16, 395-398.

1229 PUDLAK, P. (1975B) POLYNOMIALLY COMPLETE PROBLEMS IN THE LOGIC OF AUTOMATED DISCOVERY, BECVAR, J. (ED.), LECTURE NOTES IN COMPUTER SCIENCE, 32, SPRINGER-VERLAG, BERLIN, 358,361.

1230 PULTR, A. (1976) CLOSED CATEGORIES OF L-FUZZY SETS, VORTRAGE AUS DEM PROBLEMSEMINAR AUTOMATEN- UND ALGORITHMENTHEORIE, APRIL, WEISSIG.

1231 PUN, L. (1975) EXPERIENCE IN THE USE OF FUZZY FORMALISM IN PROBLEMS WITH VARIOUS DEGREES OF SUBJECTIVITY, IN SPECIAL INTEREST DISCUSSION ON FUZZY AUTOMATA AND DECISION PROCESSES, 6TH IFAC WORLD CONGRESS, BOSTON, MASS., USA., AUG.

1232 PUN, L. (1977) USE OF FUZZY FORMALISM IN PROBLEMS WITH VARIOUS DEGREES OF SUBJECTIVITY, 357-378, IN FUZZY AUTOMATA AND DECISION PROCESSES, M.M. GUPTA, G.N. SARIDIS, B.R. GAINES, EDS., NORTH-HOLLAND.

1233 PUTNAM, H. (1957) THREE-VALUED LOGIC, PHILOS. STUD., 8, 73-80.

1234 RADECKI, T. (1976) APPLICATION OF FUZZY SETS THEORY TO THE DESCRIPTION OF INFORMATION RETRIEVAL PROCESS, REPORTS OF THE MAIN LIBRARY AND SCIENTIFIC INFORMATION CENTRE, TECH. UNIV. OF WROCLAW, POLAND, SER. A, NO. 56.

1235 RADECKI, T. (1976) MATH. MODEL OF INFO. RETRIEVAL SYSTEM BASED ON THE CONCEPT OF FUZZY THESAURUS, INF. PROCESSING AND MANAGEMENT 12, 313-318.

1236 RADECKI, T. (1977) FUZZY SET THEORETICAL APPROACH TO DOCUMENT RETRIEVAL, 6TH CRANFIELD INT. CONF. ON MECHANIZED

INFO. STORAGE AND RET. SYST., CRANFIELD.

1237 RADECKI, T. (1977) LEVEL FUZZY SETS, J. OF CYBERNETICS, 7: 189-198, TECH. UNIV. OF WROCLAW, POLAND.

1238 RADECKI, T. (1978) A MODEL OF DOCUMENT RETRIEVAL SYS. BASED ON THE CONCEPT OF SEM. DISJ. NORM. FORM, TECH. UNIV. OF WROCLAW, POLAND.

1239 RAGADE, R.K. (1973) A MULTIATTRIBUTE PERCEPTION AND CLASSIFICATION OF (VISUAL) SIMILARITIES, S-001-73, SYSTEMS RESEARCH AND PLANNING, BELL-NORTHERN RESEARCH, OTTAWA, CANADA, NOV.

1240 RAGADE, R.K. (1973) ON SOME ASPECTS OF FUZZINESS IN COMMUNICATION: I FUZZY ENTROPIES, W-002-73, SYSTEMS RESEARCH AND PLANNING, BELL-NORTHERN RESEARCH, OTTAWA, CANADA, NOV.

1241 RAGADE, R.K. (1973) ON SOME ASPECTS OF FUZZINESS IN COMMUNICATION: II A NOTE ON FUZZY ENTROPIES ASSOCIATED WITH A FUZZY CHANNEL, W-006-73, SYSTEMS RESEARCH AND PLANNING, BELL-NORTHERN RESEARCH, OTTAWA, CANADA, NOV.

1242 RAGADE, R.K. (1973) ON SOME ASPECTS OF FUZZINESS IN COMMUNICATION: III FUZZY CONCEPT COMMUNICATION, W-005-73, SYSTEMS RESEARCH AND PLANNING, BELL-NORTHERN RESEARCH, OTTAWA, CANADA, DEC.

1243 RAGADE, R.K. (1974) INCERTITUDE CHARACTERIZATION OF THE RETRIEVER-SYSTEM COMMUNICATION PROCESS, PROC. 37TH ANNUAL MEETING AMERICAN SOCIETY INFORMATION SCIENCES, ATLANTA, GEORGIA, USA, OCT.

1244 RAGADE, R.K. (1974) NAIVE USERS AND ILL-FORMED PROBLEMS IN INTERACTIVE SYSTEMS, TECH. REP., BELL-NORTHERN RESEARCH, DEC. .

1245 RAGADE, R.K. (1975) BENEFIT COST ANALYSIS UNDER IMPRECSIE CONDITIONS, 1-S-040675, DEPARTMENT OF SYSTEMS DESIGN, UNIVERSITY OF WATERLOO, ONTARIO, CANADA, JUNE.

1246 RAGADE, R.K. (1976) FUZZY GAMES IN THE ANALYSIS OF OPTIONS, J. CYBERNETICS .

1247 RAGADE, R.K. (1976) FUZZY INTERPRETIVE STRUCTURAL MODELLING, J. CYBERNETICS .

1248 RAGADE, R.K. (1976) FUZZY MODELS IN MULTI-OBJECTIVE CONFLICT ANALYSIS, SYST. SCI. CENTER AND THE DEPT. OF MATH., UNIV. OF LOUIS, ALSO PRESENTED AT THE TIMS-ORSA CONF. IN MIAMI, FLA., NOV.

1249 RAGADE, R.K. (1976) FUZZY SET THEORY AND THE MATHEMATICAL PROBABLIITY THEORY OF KOLMOGOROV: SOME OBSERVATIONS, UNPUBLISHED NOTE.

1250 RAGADE, R.K. (1976) FUZZY SETS IN COMMUNICATION SYSTEMS AND CONSENSUS FORMATION SYSTEMS, TIMS-ORSA JOINT MEETING, PHILADELPIA, USA, APRIL.

1251 RAGADE, R.K. (1976) A DIFFERENTIAL GAME FORMULATION OF FUZZY CONSENSUS, PRESENTED AT THE TIMS-ORSA CONF. IN MIAMI BEACH, FLA.

1252 RAGADE, R.K. (1977) SYSTEMS ANALYSIS: DETERMINISTIC, STOCHASTIC OR FUZZY ?, PRESENTED AT THE TIMS-ORSA CONF. IN ATLANTA, GEORGIA.

1253 RAGADE, R.K. (1977) A MATH. MODEL OF APPROX. COMMUNICATION IN INFO. SYST., J. WHITE ED., THE GENERAL SYST. PARADIGM, 334-346.

1254 RAGADE, R.K. (1977) PROFILE TRANSFORMATION ALGEBRA AND GROUP CONSENSUS FORMATION THROUGH FUZZY SETS, 331-356, IN FUZZY AUTOMATA AND DECISION PROCESSES, M.M. GUPTA, G.N. SARIDIS, B.R. GAINES, EDS., NORTH-HOLLAND.

1255 RAGADE, R.K. (1978) TOWARDS MULTI-ATTRIBUTE MODELS BY FUZZY SET THEORY, PROC. OF 22ND MEETING OF SOCIETY FOR GEN. SYS. RES., WASH., 412-417.

1256 RAGADE, R.K., AND GUPTA, M.M. (1977) FUZZY SET THEORY: INTRODUCTION, 105-132, IN FUZZY AUTOMATA AND DECISION PROCESSES, M.M. GUPTA, G.N. SARIDIS, B.R. GAINES, EDS., NORTH-HOLLAND.

1257 RAGADE, R.K., AND WOMAK, B.F. (1977) FUZZY GRAPHS IN SOCIETIC MODELING, PRESENTED AT THE SYMP. ON FUZZY SET THEORY AND APPLN., IEEE CONF. ON DECISION AND CONTROL, NEW ORLEANS.

1258 RAGADE,R.K., HIPEL, AND UNNY (1975) NON-QUANTATIVE METHODS IN WATER RESOURCE MANAGEMENT, ASCE SPECIALITY CONFERENCE ON WATER RESOURCES MANAGEMENT, JULY.

1259 RAJASETHUPATHY, K.S., AND LAKSHMIVARAHAN, S. (1974) CONNECTEDNESS IN FUZZY TOPOLOGY, DEPARTMENT OF MATHEMATICS, VIVEKANAMDHA COLLEGE, MADRAS, INDIA.

1260 RALESCU, D.A., INTEGRATION ON FUZZY SETS, SUBMITTED TO THE J. FOR MATH. ANAL. AND APPLN., DEPT. OF MATH., INDIANA UNIV., BLOOMINGTON, INDIANA, 1-12.

1261 RALESCU, D.A. (1974) FUZZY SETS AND THEIR APPLICATIONS, ED. TEHNICA, BUCHAREST (IN ROMANIAN).

1262 RALESCU, D.A. (1974) ON FUZZY CHARACTERS AND SUBOBJECTS, SEMINARUL DE SISTEME FUZZY, DEPT. ECONOMIC CYBERNETICS, ACADEMY OF ECONOMIC STUDIES, BUCHAREST.

1263 RALESCU, D.A. (1975) DECOMPOSITION THEOREMS FOR FUZZY AUTOMATA, SEMINARUL DE SISTEME FUZZY, DEPT. ECONOMIC CYBERNETICS, ACADEMY OF ECONOMIC STUDIES, BUCHAREST.

1264 RALESCU, D.A. (1976) FUZZY SETS AND FLOU SETS .

1265 RALESCU, D.A. (1976) L-FUZZY SETS AND L-FLOU SETS, ELEKTRONISCHE INFORMAT. UND KYBERNETIK, 12, 599-605.

1266 RALESCU, D.A. (1976) ON FUZZY SYSTEMS, PROC. OF THE 3RD INT. CONGRESS ON CYBERNETICS.

1267 RALESCU, D.A. (1977) INEXACT SOLUTIONS FOR LARGE SCALE CONTROL PROBLEMS, PROC. OF THE 1ST INT. CONG. ON MATH. AT THE SERVICE OF MAN, BARCELONA, SPAIN.

1268 RALESCU, D.A. (1978) FUZZY SUBOBJECTS IN A CATEGORY AND THE THEORY OF C-SETS, FUZZY SETS AND SYST. 1, 193-202, NORTH-HOLLAND.

1269 RALESCU, D.A. (1978) ORDERING, PREFERENCES AND FUZZY OPTIMIZATION, PROC. 4TH INT. CONF. CYBER. AND SYSTEMS, AMSTERDAM, 377-378.

1270 RALESCU, D.A. (1978) THE INTERFACE BETWEEN ORDERINGS AND FUZZY OPTIMIZATION, PAPER PRESENTED AT THE ORSA/TIMS JOINT NAT. METTING, LOS ANGELES, CALIF., NOV. .

1271 RALESCU, D.A.,(FORTHCOMING) REPRESENTATION OF FUZZY CONCEPTS: A SURVEY OF THE APPLICATIONS .

1272 RALESCU, D.A.,(FORTHCOMING) TOWARD A GENERAL THEORY OF FUZZY VARIABLES.

1273 RASIOWA, H. (1974) AN ALGEBRAIC APPROACH TO NON-CLASSICAL LOGICS, NORTH-HOLLAND, AMSTERDAM.

1274 RASIOWA, H., AND SIKORSKI, R. (1970) THE MATHEMATICS OF METAMATHEMATICS, WARSAW, POLAND.

1275 RAUCH, J. (1975) EIN BEITRAG ZU DER GUHA METHODE IN DER DREIWERTIGEN LOGIK, KYBERNETIKA (PRAGUE), 11, 101-113.

1276 REBROVA, M.P. (1976) FUZZY SETS IN CLASSIFICATION THEORY, AUTOMATIC DOCUMENTATION AND MATHEMATICAL LINGUISTICS, 10, 4.

1277 REDDY, D. (1972) REFERENCE AND METAPHOR IN HUMAN LANGUAGE, PHD THESIS, DEPARTMENT OF ENGLISH, UNIVERSITY OF CHICAGO.

1278 REIGER, B. (1974) EINE "TOLERANTE" LEXIKONSTRUKTUR. ZUR ABBILDUNG NATURLICH-SPRACHLICHER BEDEUTUNG AUF "UNSCHARFE" MENGEN IN TOLERANZRAUMEN, ZEITSCHRIFT FUR LITERATURWISSENSCHAFT UND LINGUISTIK, 16, 31-47.

1279 REIGER, L. (1949A) A NOTE ON TOPOLOGICAL REPRESENTATION OF DISTRIBUTIVE LATTICES, CASOPIS PRO PESTOVANI MATEMATIKY A FYSIKY (PRAGUE), 74, 55-61.

1280 REISINGER, L. (1974) ON FUZZY THESAURI, PROC. COMP. STAT., VIENNA.

1281 RESCHER, N. (1963) A PROBABILISTIC APPROACH TO MODAL LOGIC, ACTA PHILOSOPHICA FENNICA, 16, 215-226.

1282 RESCHER, N. (1964) QUANTIFIERS IN MANY-VALUED LOGIC, LOGIQUE ET ANALYSE, 7, 181-184.

1283 RESCHER, N. (1967) SEMANTIC FOUNDATIONS FOR THE LOGIC OF PREFERENCE, IN RESCHER, N. (ED.), THE LOGIC OF DECISION AND ACTION, UNIVERSITY OF PITTSBURGH PRESS, 37-79.

1284 RESCHER, N. (1968) TOPICS IN PHILOSOPHICAL LOGIC, D. REIDEL, HOLLAND.

1285 RESCHER, N. (1969) MANY-VALUED LOGIC, MCGRAW-HILL, NEW YORK.

1286 RESCHER, N. (1973) THE COHERENCE THEORY OF TRUTH, CLARENDON PRESS, OXFORD.

1287 RESCHER, N., AND MANOR, R. (1970) ON INFERENCE FROM INCONSISTANT PREMISES, THEORY AND DECISION, 1, 179-217.

1288 RICKMAN, S.M., AND KANDEL, A. (1977) COLUMN TABLE APPROACH FOR THE MINIMIZATION OF FUZZY FUNCTIONS, INF. SCI. VOL. 12, 2,111-128 (N.M. INST. OF MIN. AND TECH. 1975: ALSO).

1289 RIEGER, B. (1975) ON A TOLERANCE TOPOLOGY MODEL OF NATURAL LANGUAGE MEANING, GERMANIC INSTITUTE, TECH. HOCHSCHULE, AACHEN, GERMANY.

1290 RIEGER, B. (1976) ZUM DER REPRASENTATION UND ANALYSE VAGER BEDEUTUNGEN, WORKING PAPER, INSTITUT FUR MATHEMATISCH EMPIRISCHE SYSTEMFORSCHUNG, RWTH AACHEN.

1291 RIEGER, B. (1976A) THEORIE DER UNSCHARFEN MENGEN UND EMPIRISCHE TEXTANALYZE, DEUTSCHER GERMANISTENTAG 76, DUSSELDORF, APRIL.

1292 RIEGER, B. (1976B) FUZZY STRUCTURAL SEMANTICS. ON A GENERATIVE MODEL OF VAGUE NATURAL LANGUAGE MEANING, 3RD EUR. MEETING CYBERN. SYST. RES., VIENNA .

1293 RIEGER, B. (1977) ANALYZING AND REPRESENTING VAGUE LEXICAL MEANING ON A GENERATIVE MODEL OF FUZZY-STRUCTURAL SEMANTIC, 3RD INT. CONF. ON COMPUTING IN THE HUMANTIES, UNIV. OF WATERLOO, ONTARIO, CANADA (AUG. 2-5,1977).

1294 RIEGER, B. (1977) COLING 76: CONCEPTS, FRAMES, AND SCRIPTS IN AID OF SEMANTIC NETWORKS, KNOWLEDGE SYSTEMS AND FANTASIES, IN SPRACHE UND DATENVERARBEITEN, MAX NIEMEYER VERLAG, TUBINGEN.

1295 RIEGER, B. (1977) VAGHEIT ALS PROBLEM DER LINGUISTISCHEN SEMANIK, IN: SEMANTIC UND PRAGMATIK, AKTEN DES 11, LINGUISTISCHEN KOLLOQUIUMS AACHEN 1976, VOL. 2, MAX NIEMEYER VERLAG, TUBINGEN.

1296 RIEGER, L. (1949B) ON THE LATTICE THEORIE OF BROUWERIAN PROPOSITIONAL LOGIC, ACTA FACULTATIS RERUM NATURALIUM UNIVERSITATIS CAROLINAE (PRAGUE), 189 .

1297 RIEGER, L. (1967) ALGEBRAIC METHODS OF MATHEMATICAL LOGIC, ACADEMIA, PRAGUE AND ACADEMIC PRESS, NEW YORK.

1298 RINE, D. (1978) POSSIBILITY THEORY: AS A MEANS FOR MODELING COMPUTER SECUTITY AND PROTECTION, INFORMATION SCIENCE PROGRAM, PROCEEDINGS THE 8TH INT. SYMP. ON MULTIPLE-VALUED LOGIC, 78CH1366-4C, 276-286.

1299 RINKS, D., AND STEINBERG, E. (1977) LINEAR ORDERINGS OVER FUZZY PREFERENCES IN A SOCIAL WALFARE SETTING, PROC. 1977 AIDS CONF., CHICAGO.

1300 RINKS, D., AND STEINBERG, E. (1978) APPROX. REASONING AND THE PRODUCTION SCHEDULING PROBLEM, PRESENTED AT JOINT ORSA/TIMS MEETING, LOS ANGELES.

1301 ROBERTS, F.S. (1973) TOLERANCE GEOMETRY, NOTRE DAME J. FORMAL LOGIC, 14, 68-76.

1302 ROCK, H. (1977) INTERACTIVE FUZZY DECISION MAKING IN A DISCRETE MATH. PROG. FRAMEWORK, REPORT TECH. UNIV., BERLIN.

1303 RODDER, R., AND ZIMMERMANN, H.J. (1977) ANALYSE BESCHREIBUNG UND OPTIMIERUNG VON UNSCHARF FORMULIERTEN PROBLEMEN, Z. OPERATIONS RESEARCH 21, 1-18.

1304 RODDER, W. (1975) ON "AND" AND "OR" CONNECTIVES IN FUZZY SET THROEY, EURO 1, LEHRSTUHL FUR UNTERNEHMENSFORSCHIG RWTH AACHEN, GERMANY.

1305 RODDER, W., AND ZIMMERMANN, H.J. (1977) DUALITY IN FUZZY PROGRAMMING, INT. SYMP. ON EXTERNAL METHODS AND SYST. ANALYSIS, UNIV. OF TEXAS, AUSTIN, TX, SEPT.

1306 ROSCH, E.H. (1973) ON THE INTERVAL STRUCTURE OF PERCEPTUAL AND SEMANTIC CATEGORIES, T.E. MOORE (ED) COGN. DEV. AND THE ACQUIS. OF LANGUAGE, 111-144, N.Y.

1307 ROSCH, E.H. (1974) UNIVERSALS AND CULTURAL SPECIFICS IN HUMAN CATEGORIZATION, IN R. BRISTIN, S. BOCHNOR, W. BONNOR, W. BONNOR(EDS); CROSS-CULT. PERSP. ON LEARNING.

1308 ROSE, A. (1950) COMPLETENESS OF LUKASIEWICZ-TARSKI PROPOSITIONAL CALCULUS, MATHEMATISCHE ANNALEN, 122, 296-298.

1309 ROSE, A. (1951A) THE DEGREE OF COMPLETENESS OF SOME LUKASIEWICZ-TARSKI PROPOSITIONAL CALCULI, J. LONDON MATH. SOC., 26, 47-49.

1310 ROSE, A. (1951B) AXIOM SYSTEMS FOR 3-VALUED LOGIC, J.LONDON MATH. SOC., 26, 50-58.

1311 ROSE, A. (1952) THE DEGREE OF COMPLETENESS OF THE M-VALUED LUKASIEWICZ PROPOSITIONAL CALCULUS, J. LONDON MATH. SOC., 27, 92-102.

1312 ROSE, A. (1953) THE DEGREE OF COMPLETENESS OF THE LAMDA-ZERO-VALUED LUKASIEWICZ PROPOSITIONAL CALCULUS, J. LONDON MATH. SOC., 28, 176-184.

1313 ROSE, A. (1958) MANY-VALUED LOGICAL MACHINES, PROC. CAMBRIDGE PHILOSOPHICAL SOC., 54, 307-321.

1314 ROSE, A., AND ROSSER, J.B. (1958) FRAGMENTS OF MANY-VALUED STATEMENT CALCULI, TRANS. AMER. MATH. SOC., 87, 1-53.

1315 ROSEN, R. (1974) PLANNING, MANAGEMENT POLICIES AND STRATEGIES: FOUR FUZZY CONCEPTS, INT. J. GENERAL SYST., 1, 245-252.

1316 ROSENFELD, A. (1971) FUZZY GROUPS, J. MATH. ANAL. AND APPLN., 35, 512-517.

1317 ROSENFELD, A. (1975) FUZZY GRAPHS, IN ZADEH, L.A., FU, K.S., TANAKA, K., AND SHIMURA, M. (EDS.), FUZZY SETS AND THEIR APPLICATIONS TO COGNITIVE DECISION PROCESSES, ACADEMIC PRESS, NEW YORK, 77-95.

1318 ROSENFELD, A. (1977) FUZZY DIGITAL TOPOLOGY, TR-573, DEPT. OF COMPUTER SCI. ANNUAL REPORT, UNIV. OF MARYLAND.

1319 ROSENFELD, A., HUMMEL, R.A., AND ZUCKER, S.W. (1976) SCENE LABELING BY RELAXATION OPERATIONS, IEEE TRANS. SYST., MAN CYBERN., SMC-6, 420-433.

1320 ROSSER, J.B. (1960) AXIOMATIZATION OF INFINITE VALUED LOGICS, LOGIQUE ET ANALYSE, 3, 137-153.

1321 ROSSER, J.B., AND TURQUETTE, A.R. (1945) AXIOM SCHEMES FOR M-VALUED PROPOSITIONAL CALCULI, J. SYMBOLIC LOGIC, 10, 61-82.

1322 ROSSER, J.B., AND TURQUETTE, A.R. (1952) MANY VALUED LOGICS, NORTH-HOLLAND, AMSTERDAM.

1323 ROUBENS, M. (1977) PATTERN CLASSIFICATION PROBLEMS AND FUZZY SETS, FUZZY SETS AND SYST. 1, 239-253, NORTH-HOLLAND.

1324 RUBIN, H. (1969) A NEW APPROACH TO FOUNDATIONS OF PROBABILITY, IN BULLOF, J.J., HOLYOKE, T.C., AND HAHA, S.W. (EDS.), FOUNDATIONS OF MATHEMATICS, SYMPOSIUM PAPERS COMMEMERATING THE 60TH BIRTHDAY OF K. GODEL, SPRINGER, NEW YORK.

1325 RUSPINI, E. (1969) A NEW APPROACH TO CLUSTERING, INFORM. AND CONTROL, 15, 22-32.

1326 RUSPINI, E. (1970) NUMERICAL METHODS FOR FUZZY CLUSTERING, INFORM. SCI., 2, 319-350.

1327 RUSPINI, E.H. (1972) OPTIMIZATION IN SAMPLE DESCRIPTIONS: DATA REDUCTION AND PATTERN RECOGNITION USING FUZZY CLUSTERING, IEEE TRANS. SYST. MAN CYBERN., SMC-2, 541.

1328 RUSPINI, E.H. (1973) NEW EXPERIMENTAL RESULTS IN FUZZY CLUSTERING, INFORM. SCI., 6, 273-284.

1329 RUSPINI, E.H. (1977) A THEORY OF FUZZY CLUSTERING, PROC. 1977 IEEE CONF. ON DECISION AND CONTROL, NEW ORLEANS, 1378-1383.

1330 RUSPINI, E.H. (1977) A THEORY OF NUMERICAL CLASSIFICATION, PROCEEDINGS OF THE SYMP. ON FUZZY SET THEORY AND APPLN., IEEE CONF. ON DECISION AND CONTROL, NEW ORLEANS.

1331 RUSSELL, B. (1923) VAGUENESS, AUSTRIAN JOURNAL OF PHILOSOPHY, 1, 84-92.

1332 RUTHERFORD, D.A. (1976) THE IMPLEMENTATION AND EVALUATION OF A FUZZY CONTROL ALGORITHM FOR A SINTER PLANT, IN MAMDANI, E.H., AND GAINES, B.R. (EDS.), DISCRETE SYSTEMS AND FUZZY REASONING, EES-MMS-DSFR-76, QUEEN MARY COLLEGE, UNIVERSITY OF LONDON, (WORKSHOP PROCEEDINGS).

1333 RUTHERFORD, D.A., AND BLOORE, G.C. (1975) THE IMPLEMENTATION OF FUZZY ALGORITHMS FOR CONTROL, CONTROL SYSTEMS CENTRE, UNIVERSITY OF MANCHESTER INSTITUTE OF SCIENCE AND TECHNOLOGY, MANCHESTER, UK.

1334 RUTHERFORD, D.A., AND RAO, G.P. (1979) APPROXIMATE RECONSTRUCTION OF MAPPING FUNCTIONS FROM LINGUISTIC DESCRIPTIONS IN PROBLEMS OF FUZZY LOGIC APPLIED TO SYSTEM CONTROL, CONTROL SYSTEM CENTRE, U.M.I.S.T. MANCHESTER U.K.

1335 SAATY, T.L. (1974) MEASURING THE FUZZINESS OF SETS, J. CYBERNET., 4, 53-61.

1336 SAATY, T.L. (1977) EXPLORING THE INTERFACE BETWEEN HIERARCHIES, MULTIPLE OBJECTIVES AND FUZZY SETS, FUZZY SETS AND SYST. 1, 57-68, NORTH-HOLLAND .

1337 SADOVSKII, V.N. (1974) OSNOVANIJA OBSCEI TEORII SISTEM, NAUKA, MOSKOW, (IN RUSSIAN, ON FOUNDATIONS OF GENERAL SYSTEMS THEORIES).

1338 SAGAAMA, S. (1976) SUBJECTIVE PROBABILITIES, FUZZY SETS AND DECISION MAKING, 3RD EUR. MEETING CYBERN. SYST. RES., VIENNA .

1339 SAITO, T. (1975) CHRONOLOGY ANALYSIS OF A SOCIAL CONFLICT, IN SUMMARY OF PAPERS ON GENERAL FUZZY PROBLEMS, THE WORKING GROUP ON FUZZY SYSTEMS, TOKYO, JAPAN, NOV., 46-48.

1340 SALOMAA, A. (1959) ON MANY-VALUED SYSTEMS OF LOGIC, AJATUS, 22, 115-119.

1341 SANCHEZ, E. (1974) EQUATIONS DE RELATIONS FLOUES, THESIS DE DOCTORAT EN BIOLOGIE HUMAINE, FACULTE DE MEDECINE DE MARSEILLE, FRANCE, JULY.

1342 SANCHEZ, E. (1974) FUZZY RELATIONS, FACULTY OF MEDICINE, UNIVERSITY OF MARSEILLE, FRANCE.

1343 SANCHEZ, E. (1976) EIGEN FUZZY SETS, NATIONAL COMPUTER CONFERENCE, NEW YORK, JUNE.

1344 SANCHEZ, E. (1976) RESOLUTION OF COMPOSITE FUZZY RELATION EQUATIONS, INFORM. AND CONTROL, 30, 38-47.

1345 SANCHEZ, E. (1977) EIGEN FUZZY SETS AND FUZZY RELATIONS, UNIV. OF CALIF. AT BERKELEY, ERL MEMO M77-20.

1346 SANCHEZ, E. (1977) INVERSES OF FUZZY RELATIONS AND POSSIBILITY-QUALIFICATION APPLICATION TO MEDICAL DIAGNOSIS, PROCEEDINGS OF THE SYMP. ON FUZZY SET THEORY AND APPLN., IEEE CONF. ON DECISION AND CONTROL, NEW ORLEANS.

1347 SANCHEZ, E. (1977) INVERSES OF FUZZY RELATIONS, APPLICATION TO POSSIBILITY DISTRIBUTIONS AND MEDICAL DIAGNOSIS, PROC. 1977 IEEE CONF. ON DECISION AND CONTROL, NEW ORLEANS, 1384-1389.

1348 SANCHEZ, E. (1977) ON POSSIBILISTIC-QUALIFICATION IN NATURAL LANGUAGES, MEMO M77/28. ELECTRONICS RESEARCH LAB., UNIV. OF CALIFORNIA, BERKELEY.

1349 SANCHEZ, E. (1977) RESOLUTION OF EIGEN FUZZY SETS EQUATIONS, FUZZY SETS AND SYST. 1, 69-74, NORTH-HOLLAND.

1350 SANCHEZ, E. (1977) SOLUTIONS IN COMPOSITE FUZZY RELATION EQUATIONS: APPLICATION TO MEDICAL DIAGNOSIS IN BROUWERIAN LOGIC, 221-234, IN FUZZY AUTOMATA AND DECISION PROCESSES, M.M. GUPTA, G.N. SARIDIS, B.R. GAINES, EDS., NORTH-HOLLAND.

1351 SANCHEZ, E. (1978) FUZZY RELATIONS, POSSIBILITY DISTRIBUTIONS, AND MEDICAL DIAGNOSIS, FACULTE DE MEDECINE DE MARSEILLE, PROC. 1978 IEEE CONF. ON DEC. AND CONTROL, INCLUDES THE 17TH SYMP. ON ADAPTIVE PROCESSES, 78CH1392-OCS, JAN., 1979, SAN DIEGO, CALIF.

1352 SANCHEZ, E. (1978) ON TRUTH-QUALIFICATION IN NATURAL LANGUAGES, LAB. DE BIOMATHEMATIQUES, MARSEILLE, FRANCE, PROCEEDINGS OF THE 1978 INT. CONF. ON CYBERNETICS AND SOCIETY, VOL2, 1233-1236.

1353 SANCHEZ, E., AND SAMBUC, R. (1976) RELATIONS FLOUES. FONCTIONS O-FLOUES. APPLICATION A L'AIDE AUDIAGNOSTIC EN PATHEOLOGIE THYROIDIENNE, IRIA MEDICAL DATA PROCESSING SYMPOSIUM, TAYLOR AND FRANCIS, TOULOUSE.

1354 SANFORD, D.H. (1975A) BORDERLINE LOGIC, AMER. PHILOS. QUART., 12, 29-39.

1355 SANFORD, D.H. (1975B) INFINITY AND VAGUENESS, PHILOSOPHICAL REVIEW, 84, 520-535.

1356 SANTOS, E.S. (1968A) MAXIMIN, MINIMAX AND COMPOSITE SEQUENTIAL MACHINES, J. MATH. ANAL. AND APPLN., 24, 246-259.

1357 SANTOS, E.S. (1968B) MAXIMIN AUTOMATA, INFORM. AND CONTROL, 13, 363-377.

1358 SANTOS, E.S. (1969A) MAXIMIN SEQUENTIAL CHAINS, J. MATH. ANALS. AND APPLN., 26, 28-38.

1359 SANTOS, E.S. (1969B) MAXIMIN SEQUENTIAL-LIKE MACHINES AND CHAINS, MATHEMATICAL SYSTEMS THEORY, 3, 300-309.

1360 SANTOS, E.S. (1970) FUZZY ALGORITHMS, INFORM. AND CONTROL, 17, 326-339.

1361 SANTOS, E.S. (1972A) MAX-PRODUCT MACHINES, J. MATH. ANAL. AND APPLN., 37, 677-686.

1362 SANTOS, E.S. (1972B) ON REDUCTIONS OF MAXIMIN MACHINES, J. MATH. ANAL. AND APPLN., 40, 60-78.

1363 SANTOS, E.S. (1973) FUZZY SEQUENTIAL FUNCTIONS, J. CYBERNETICS, 3, 15-31.

1364 SANTOS, E.S. (1974) CONTEXT-FREE FUZZY LANGUAGES, INFORM. AND CONTROL, 26, 1-11.

1365 SANTOS, E.S. (1975A) REALIZATION OF FUZZY LANGUAGES BY PROBABILISTIC MAX-PRODUCT AND MAXIMIN AUTOMATA, INFORM. SCI., 8, 39-53.

1366 SANTOS, E.S. (1975B) MAX-PRODUCT GRAMMARS AND LANGUAGES, INFORM. SCI., 9, 1-23.

1367 SANTOS, E.S. (1975C) FUZZY PROGRAMS, IN SPECIAL INTEREST DISCUSSION SESSION ON FUZZY AUTOMATA AND DECISION PROCESSES, 6TH IFAC WORLD CONGRESS, BOSTON, MASS., USA, AUG. .

1368 SANTOS, E.S. (1976A) FUZZY AUTOMATA AND LANGUAGES, INFORM. SCI., 10, 193-197.

1369 SANTOS, E.S. (1977) FUZZY AND PROBABILISTIC PROGRAMS, 133-148, IN FUZZY AUTOMATA AND DECISION PROCESSES, M.M. GUPTA, G.N. SARIDIS, B.R. GAINES, EDS., NORTH-HOLLAND.

1370 SANTOS, E.S. (1977) REGULAR FUZZY EXPRESSIONS, 169-176, IN FUZZY AUTOMATA AND DECISION PROCESSES, M.M. GUPTA, G.N. SARIDIS, B.R. GAINES, EDS., NORTH-HOLLAND.

1371 SANTOS, E.S., AND WEE, W.G. (1968) GENERAL FORMULATION OF SEQUENTIAL MACHINES, INFORM. AND CONTROL, 12, 5-10.

1372 SARIDIS, G.N. (1974) FUZZY NOTIONS IN NONLINEAR SYSTEM CLASSIFICATION, J. CYBERNETICS, 4, 67-82.

1373 SARIDIS, G.N. (1975) FUZZY DECISION MAKING IN PROTHETIC DEVICES AND OTHER APPLICATIONS, IN SPECIAL INTEREST DISCUSSION SESSION ON FUZZY AUTOMATA AND DECISION PROCESSES, 6TH IFAC WORLD CONGRESS, BOSTON, MASS., USA, AUG. .

1374 SARIDIS, G.N., AND STEPHANOU, H.E. (1977) FUZZY DECISION-MAKING IN PROSTHETIC DEVICES, 387-402, IN FUZZY AUTOMATA AND DECISION PROCESSES, M.M. GUPTA, G.N. SARIDIS, B.R. GAINES, EDS., NORTH-HOLLAND.

1375 SASAMA, H. (1975) FUZZY SET MODEL FOR TRAIN COMPOSITION IN MARSHALLING YARD, IN SUMMARY OF PAPERS ON GENERAL FUZZY PROBLEMS, THE WORKING GROUP ON FUZZY SYSTEMS, TOKYO, JAPAN, NOV., 49-54.

1376 SASAMA, H. (1977) A LEARNING MODEL TO DISTINGUISH THE SEX OF A HUMAN NAME, 19-24, SUMMARY OF PAPERS ON GENERAL FUZZY PROBLEMS, THE WORKING GROUP ON FUZZY SYSTEMS, TOYKO, JAPAN.

1377 SAVAGE, L.J. (1971) ELICITATION OF PERSONAL PROBABILITIES AND EXPECTATIONS, J. AMER. STATIST. ASSN., 66, 783-801.

1378 SCARPELLINI, B. (1962) DIE NICHT-AXIOMATISIERBARKEIT DES UNENDLICHWERTIGEN PRAEDIKATENKALKULS VON LUKASIEWICZ, J. SYMBOLIC LOGIC, 27, 159-170.

1379 SCHEK, H.J. (1977) TOLERATING FUZZINESS IN KEYWORDS BY SIMILIARITY SEARCHES, KYBERNETES 6, 175-184.

1380 SCHOCK, R. (1964A) ON FINITELY MANY-VALUED LOGICS, LOGIQUE ET ANALYSE, 28, 43-58.

1381 SCHOCK, R. (1964B) ON DENUMERABLY MANY-VALUED LOGICS, LOGIQUE ET ANALYSE, 28, 190-195.

1382 SCHOCK, R. (1965) SOME THEOREMS ON THE RELATIVE STRENGTHS OF MANY-VALUED LOGICS, LOGIQUE ET ANALYSE, 30, 101-104.

1383 SCHOTCH, P.K. (1975) FUZZY MODAL LOGIC, PROC. 1975 INT. SYMP. MULTIPLE-VALUED LOGIC, IEEE 75CH0959-7C, MAY, 176-182.

1384 SCHUH, E. (1973) MANY-VALUED LOGICS AND THE LEWIS PARADOXES, NOTRE DAME J. FORMAL LOGIC, 14, 250-252.

1385 SCHUTZENBERGER, M.P. (1962) ON A THEOREM OF R. JUNGEN, PROC. AMERICAN MATHEMATICAL SOCIETY, 13, 885-890.

1386 SCHWARZ, D. (1972) MENGENLEHRE UBER VORGEGEBENEN ALGEBRAISCHEN SYSTEMEN, MATH. NACHR., 53 , 365-370.

1387 SCHWEDE, G. (1976) N-VARIABLE FUZZY MAPS WITH APPLICATION

TO DISJUNCTIVE DECOMPOSITION OF FUZZY SWITCHING FUNCTIONS, PROC. 6TH INT. SYMP. MULTIPLE-VALUED LOGIC, IEEE 76CH1111-4C, MAY, 203-216.

1388 SCHWEDE, G.W., AND KANDEL, A. (1977) : FUZZY MAPS, IEEE TRANS. ON SYST., MAN, AND CYBERETICS, SMC-7, 619-674.

1389 SCOTT, D. (1974) COMPLETENESS AND AXIOMATIZABILITY IN MANY-VALUED LOGIC, IN HENKIN, L. (ED.), PROCEEDINGS OF THE TARSKI SYMPOSIUM, AMERICAN MATHEMATICAL SOCIETY, RHODE ISLAND, USA, 412-435.

1390 SCOTT, D. (1976) DOES MANY-VALUED LOGIC HAVE ANY USE?, IN KORNER (1976B), 64-88.

1391 SCOTT, D., AND KRAUSS, P. (1966) ASSIGNING PROBABILITIES TO LOGICAL FORMULAS, IN HINTIKKI, J., AND SUPPES, P. (EDS.), ASPECTS OF INDUCTIVE LOGIC, 219-264.

1392 SEGERBERG, K. (1967) SOME MODAL LOGICS BASED ON A THREE-VALUED LOGIC, THEORIA, 33, 53-71.

1393 SEKITA, Y. (1975) A CONSIDERATION OF THE FUZZY EVALUATION OF COMPLEX SOCIAL SYSTEMS, MEMOIRS OF ECONOMICS OF OSAKA UNIV. 25, 312-325.

1394 SEKITA, Y. (1976) A CONSIDERATION OF IDENTIFYING FUZZY MEASURES, MEMOIRS OF ECONOMICS OF OSAKA UNIV. 25, 133-138.

1395 SEMBI, B.S., AND MAMDANI, E.H., ON THE NATURE OF IMPLICATION IN FUZZY LOGIC, DEPT. OF ELECTRICAL AND ELECTRONIC ENGINEERING, QUEEN MARY COLLEGE, UNIV. OF LONDON, MILE END RD., LONDON E1 4NS.

1396 SERFATI, M. (1974) ALGEBRES DE BOOLE AVEC UNE INTROTUCTION A LA THEORIE DES GRAPHES ORIENTES ET AUX SOUS-ENSEMBLES FLOUS, ED. C.D.U., PARIS.

1397 SERIWAZA, M. (1973) A SEARCH TECHNIQUE OF CONTROL ROD PATTERN OF SMOOTHING CARE POWER DISTRIBUTIONS BY FUZZY AUTOMATON, JOURNAL OF NUCLEAR SCIENCE AND TECHNOLOGY, 10 .

1398 SHACKLE, G.L.S. (1949) EXPECTATION IN ECONOMICS, CAMBRIDGE UNIVERSITY PRESS, CAMBRIDGE, UK.

1399 SHACKLE, G.L.S. (1961) DECISION ORDER AND TIME IN HUMAN AFFAIRS, CAMBRIDGE UIVERSITY PRESS, CAMBRIDGE, UK, (2ND ED. 1969).

1400 SHAKET, E. (1975) FUZZY SEMANTICS FOR A NATURAL-LIKE LANGUAGE DEFINED OVER A WORLD OF BLOCKS, MSC THESIS, COMPUTER SCIENCE DEPT., UCLA, LOS ANGELES, CALIF., USA

1401 SHAKET, E. (1977) FUZZY SET SEMANTICS FOR A NATURAL-LIKE

LANGUAGE, PROCEEDINGS OF THE SYMP. ON FUZZY SET THEORY AND APPLN., IEEE CONF. ON DECISION AND CONTROL, NEW ORLEANS.

1402 SHAPIRO, D.I., AND TORGOV, J.I. (1978) FUZZY INTEGRAL GAMES, PROC. 4TH INT. CONF. CYBER. AND SYST., AMSTERDAM, 379-380.

1403 SHAW-KWEI, MOH. (1954) LOGICAL PARADOXES FOR MANY-VALUED SYSTEMS, J. SYMBOLIC LOGIC, 19, 37-39.

1404 SHEPPARD, D. (1954) THE ADEQUACY OF EVERYDAY QUANTITATIVE EXPRESSIONS AS MEASUREMENTS OF QUALITIES, BRIT. J. PSYCHOL, 45, 40-50.

1405 SHIBATA, H. (1976) A COMMENT ON FUZZY SET, 101-106, SUMMARY OF PAPERS ON GENERAL FUZZY PROBLEMS, THE WORKING GROUP ON FUZZY SYSTEMS, TOYKO, JAPAN.

1406 SHIBATA, H., AND TSUTSUMI, T. (1977) ON LOGICAL TREATMENT AND EVALUATION OF REGULATORY STATEMENTS, 172-180, SUMMARY OF PAPERS ON GENERAL FUZZY PROBLEMS, THE WORKING GROUP ON FUZZY SYSTEMS, TOYKO, JAPAN.

1407 SHIMURA, M. (1972) APPLICATION OF FUZZY FUNCTIONS TO PATTERN CLASSIFICATION, TRANS. INST. ELECTRON. COMM. ENG. (JAPAN), 55-D, 218-225.

1408 SHIMURA, M. (1973) FUZZY SETS CONCEPT IN RANK ORDERING OBJECTS, J. MATH. ANAL. AND APPLN., 43, 717-733.

1409 SHIMURA, M. (1975) AN APPROACH TO PATTERN RECOGNITION AND ASSOCIATIVE MEMORIES USING FUZZY LOGIC IN ZADEH, L.A., FU, K.S., TANAKA, K., AND SHIMURA, M. (EDS.), FUZZY SETS AND THEIR APPLICATIONS TO COGNITIVE AND DECISION PROCESSES, ACADEMIC PRESS, NEW YORK, 449-476.

1410 SHIMURA, M. (1975) APPLICATIONS OF FUZZY SETS THEORY TO PATTERN RECOGNITION, J. JAACE, 19, 243-248.

1411 SHIRAI, T. (1937) ON THE PSEUDO-SET, MEMOIRS OF THE COLLEGE OF SCIENCE KYOTO IMPERIAL UNIVERSITY, 20A, 153-156.

1412 SHORTLIFFE, E.H. (1976) COMPUTER-BASED MEDICAL CONSULTATION: MYCIN, ELSEVIER, NEW YORK.

1413 SHORTLIFFE, E.H., AND BUCHANAN, B.G. (1975) A MODEL OF INEXACT REASONING IN MEDICINE, MATH. BIOSCIENCES, 23, 351-379.

1414 SHUFORD, E.H., ALBERT, A., AND MASSENGILL, H.E. (1966) ADMISSIBLE PROBABILITY MEASUREMENT PROCEDURES, PSYCHOMETRIKA, 31, 125-145.

1415 SHUFORD, E.H., AND BROWN, T.A. (1975) ELICITATION OF PERSONAL PROBABILITIES AND THEIR ASSESSMENT, INSTRUCTIONAL SCIENCE, 4, 137-188.

1416 SIKORSKI, R. (1964) BOOLEAN ALGEBRAS, SPRINGER-VERLAG, BERLIN.

1417 SIMON, H.A. (1967) THE LOGIC OF HEURISTIC DECISION MAKING, IN RESCHER, N. (ED.), THE LOGIC OF DECISION AND ACTION, UNIVERSITY OF PITTSBURGH PRESS, 1-35.

1418 SIMONS, H.W. (1976) PERSUASION, ADDISON-WESLEY, READING, MASS., USA.

1419 SINHA, N.K., AND WRIGHT, J.D. (1977) APPLICATION OF FUZZY CONTROL TO A HEAT EXCHANGER SYSTEM, PROC. 1977 IEEE CONF. ON DECISION AND CONTROL, NEW ORLEANS.

1420 SIY, P. (1972) FUZZY LOGIC AND HARDWRITTEN CHARACTER RECOGNITION, PH.D. DISSERTATION, DEPT. OF ELECTRICAL ENGINEERING, UNIV. OF AKRON, OHIO.

1421 SIY, P. (1973) FUZZY LOGIC FOR HANDWRITTEN CHARACTER RECOGNITION, PHD THESIS, DEPARTMENT OF ELECTRICAL ENGINEERING, UNIVERSITY OF AKRON, OHIO, USA, JUNE .

1422 SIY, P., AND CHEN, C.S. (1971) FUZZY LOGIC APPROACH TO HANDWRITTEN CHARACTER RECOGNITION PROBLEM, PROC. IEEE CONF. SYST., MAN, CYBERN., ANAHEIM, CALIF., 113-117.

1423 SIY, P., AND CHEN, C.S. (1972) MINIMIZATION OF FUZZY FUNCTIONS, IEEE TRANS. COMP., C-21, 100-102.

1424 SIY, P., AND CHEN, C.S. (1974) FUZZY LOGIC FOR HANDWRITTEN NUMERICAL CHARACTER RECOGNITION, IEEE TRANS. SYST. MAN CYBERN., SMC-4, 570-575.

1425 SKALA, H.J. (1974) ON THE PROBLEM OF IMPRECISION, DORDRECHT, NETHERLANDS.

1426 SKALA, H.J. (1975) NON-ARCHIMEDEAN UTILITY THEORY, D. REIDEL, DORDRECHT.

1427 SKALA, H.J. (1976A) FUZZY CONCEPTS: LOGIC, MOTIVATION, APPLICATION, IN BOSSEL, H., KLACKO, S., AND MULLER, N. (EDS.), SYSTEMS THEORY IN THE SOCIAL SCIENCES, BIRKHAUSER VERLAG, BASIL, 292-306.

1428 SKALA, H.J. (1976B) NOT NECESSARILY ADDITIVE REALIZATIONS OF COMPARATIVE PROBABILITY RELATIONS .

1429 SKALA, H.J. (1977) ON MANY-VALUED LOGICS, FUZZY SETS, FUZZY LOGICS AND THEIR APPLICATIONS, FUZZY SETS AND SYST. 1, 129-149, NORTH-HOLLAND.

1430 SKALICKA, V. (1935) ZUR UNGARISCHEN GRAMMATIK, PRAGUE.

1431 SKOLEM, TH. (1957) BEMERKUNGEN ZUM KOMPREHENSIONSAXIOM, Z. MATH. LOGIK GRUNDLAGEN MATH., 3, 1-17.

1432 SKOLEM, TH. (1960) A SET THEORY BASED ON A CERTAIN 3-VALUED

LOGIC, MATH. SCAND., 8, 127-136.

1433 SKOLEM, TH. (1962) ABSTRACT SET THEORY, NOTRE DAME PRESS, INDIANA, USA.

1434 SKYRMS, B. (1970) RETURN OF THE LIAR: THREE:VALUED LOGIC AND THE CONCEPT OF TRUTH, AMER. PHILOS. QUART., 7, 153-161.

1435 SLACK, J.M.V. (1976A) A FUZZY SET-THEORTIC APPROACH TO SEMANTIC MEMORY: A RESOLUTION TO THE SET-THEORTIC VERSUS NETWORK MODEL CONTROVERSY, IN MAMDANI, E.H., AND GAINES, B.R. (EDS.), DISCRETE SYSTEMS AND FUZZY REASONING, EES-MMS-DSFR-76, QUEEN MARY COLLEGE, UNIVERSITY OF LONDON, (WORKSHOP PROCEEDINGS).

1436 SLACK, J.M.V. (1976B) POSSIBLE APPLICATIONS OF THE THEORY OF FUZZY SETS TO THE STUDY OF SEMANTIC MEMORY IN MAMDANI, E.H., AND GAINES, B.R. (EDS.), DISCRETE SYSTEMS AND FUZZY REASONING, EES-MMS-DSFR-76, QUEEN MARY COLLEGE, UNIVERSITY OF LONDON, (WORKSHOP PROCEEDINGS).

1437 SLUPECKI, J. (1958) TOWARDS A GENERALIZED MEREOLOGY OF LESNIEWSKI, STUDIA LOGICA, 131-154.

1438 SMITH, C.A.B. (1961) CONSISTENCY IN STATISTICAL INFERENCE AND DECISION, J. ROY. STATIST. SOC. (SER. B), 23, 1-37.

1439 SMITH, C.A.B. (1965) PERSONAL PROBABILITY AND STATISTICAL ANALYSIS, J. ROY. STATIST. SOC. (SER.A), 128, 469-499.

1440 SMITH, R.E. (1970) MEASURE THEORY ON FUZZY SETS, PHD THESIS, UNIVERSITY OF SASKATCHEWAN, SASKATOON, CANADA.

1441 SMULLYAN, R.M. (1957) LANGUAGES IN WHICH SELF-REFERENCE IS POSSIBLE, J. SYMBOLIC LOGIC, 22, 55-67.

1442 SNYDER, D.P. (1971) MODAL LOGIC, VAN NOSTRAND REINHOLD, NEW YORK.

1443 SOBER, E. (1975) SIMPLICITY, CLARENDON PRESS, OXFORD.

1444 SOBOLEWSKI, M. (1976) CLASSIFICATION SYSTEM SEMANTICS IN TERMS OF FUZZY SETS, 3RD EUR. MEETING CYBERN. SYST. RES., VIENNA .

1445 SOLS, I. (1975) TOPOLOGY IN COMPLETE LATTICES AND CONTINUOUS FUZZY RELATIONS, FACULTAD DE CIENCIAS, ZARAGOZA, ESPANA.

1446 SOLS, I. (1975A) FUZZY UNIVERSAL ALGEBRA AND APPLICATIONS, DEPARTMENT OF GEOMETRY, FACULTY OF SCIENCES, ZARAGOZA, SPAIN.

1447 SOLS, I. (1975B) APORTACIONES A LA TEORIA DE TOPOS, AL ALGEBRA UNIVERSAL Y A LAS MATHEMATICS FUZZY, PHD THESIS, ZARAGOZA, SPAIN.

1448 SOLS, I. (1975C) UNMARCO UNIFICATO PARA LA TEORIA DE AUTOMATAS, DEPARTMENT OF GEOMETRY, FACULTY OF SCIENCES, ZARAGOZA, SPAIN.

1449 SOMMER, G. (1976) A FUZZY PROGRAMMING APPROACH TO AN AIR POLLUTION REGULATION PROBLEM, 3RD MEETING CYBERN. SYST. RES., VIENNA .

1450 SOMMER, G. (1977) AN ALGORITHM FOR CHOOSING THE OPTIMAL CRISP SOLUTION OUT OF OPTIMAL FUZZY SETS, LEHRSTUHL FUR OR NO. 77/06, AACHEN.

1451 SOMMER, G. (1978) ON FUZZY INFORMATION RETRIEVAL, PROC. 4TH INT. CONF. CYBER. AND SYSTEMS, AMSTERDAM, 380-382.

1452 SPELLMAN, B., BEZDEK, J., AND SPELLMAN, R. (1977) DEVELOPMENT OF AN INSTRUMENT FOR THE DYNAMIC MEASUREMENT OF CONSENSUS, REPORT UTAH STATE UNIV. .

1453 SPILLMAN, B., BEZDEK, J., AND SPILLMAN, R. (TO APPEAR) A STUDY OF COALITION FORMATION IN DECISION MAKING GROUPS: AN APPLICATION OF FUZZY MATHEMATICS, KYBERNETES.

1454 SPILLMAN, B., SPILLMAN, R., AND BEZDEK, J., DYNAMIC MEASUREMENT OF SUBGROUP DOMINANCE AND COALITION FORMATION VIA FUZZY PREFERENCE RELATIONS, IN REVIEW, BEHAV. SCI. .

1455 STALLING, V. (1977) FUZZY SET THEORY VERSUS BAYESIAN STATISTICS, IEEE TRANS SYSTEMS, MAN, AND CYBERN., VOL. SMC-7, 216-219.

1456 STALNAKER, R. (1970) PROBABILITY AND CONDITIONALS, PHILOS. SCI., 37, 64-80.

1457 STALNAKER, R.C., AND THOMASON, R.H. (1970) A SEMANTIC ANALYSIS OF CONDITIONAL LOGIC, THEORIA, 36, 23-42.

1458 STATE, L. (1971) QUELQUES PROPRIETES DES ALGEBRES DE MORGAN, IN MOISIL, G.C. (ED.), LOGIQUE, AUTOMATIQUE, INFORMATIQUE, BUCHAREST, 195-207.

1459 STEFENESCU, A.C. (1975) CATEGORY SETF(L), SEMINARUL DE TEORIA SISTEMELOR, DEPT. ECONOMIC CYBERNETICS, ACADEMY OF ECONOMIC STUDIES, BUCHAREST.

1460 STEINBERG, E., AND RINKS, D. (1977) LINEAR ORDERING OVER FUZZY PREFERENCES IN A SOCIAL WELFARE SETTING, PROC. 9TH DEC. SCIENCES CONF., 604.

1461 STEPHANOUS, H.E., AND SARIDIS, G.N. (1976) HIERARCHICAL CONTROL IN A FUZZY ENVIRONMENT, PROC. IEEE CONF. SYS., MAN AND CYBERNETICS, WASH. .

1462 STICKEL, M.E. (1978) FUZZY FOUR-VALUED LOGIC FOR INCONSISTENCY AND UNCERTAINTY, DEPT. OF COMPUTER SCIENCE, UNIV. OF ARIZONA, TUCSON, ARIZONA 85721, 91-94.

1463 STOCIA, M., SLANCU-MINASIAN, I., AND SCARLAT, E. (1977) ON LARGE SCALE CLASSIFICATION PROBLEMS USING FUZZY SETS, ECON. COMP. AND CYBER. STUDIES AND RESEARCH, VOL. 1, 93-100.

1464 STOICA, M., AND SCARLAT, E. (1975A) FUZZY ALGORITHMS IN ECONOMIC SYSTEMS, ECONOMIC COMPUTATION AND ECONOMIC CYBERNETIC STUDIES AND RESEARCH, 3, CENTRE OF ECONOMIC COMPUTATION AND ECONOMIC CYBERNETICS, BUCHAREST, ROUMANIA, 239-247.

1465 STOICA, M., AND SCARLAT, E. (1975B) FUZZY CONCEPTS IN THE CONTROL OF PRODUCTION SYSTEMS, PROCEEDINGS OF 3RD INTERNATIONAL CONGRESS OF CYBERNETICS AND SYSTEMS, BUCHAREST, RUMANIA, AUG. .

1466 STONE, M.H. (1937-38) TOPOLOGICAL REPRESENTATIONS OF DISTRIBUTIVE LATTICES AND BROUWERIAN LOGICS, CASOPIS PRO PESTOVANI MATEMATIKY A FYSIKY, 67, 1-25.

1467 STOVE, D.C. (1973) PROBABILITY AND HUME'S INDUCTIVE SCEPTICISM, CLARENDON PRESS, OXFORD.

1468 SUGENO, M. (1971) ON FUZZY NONDETERMINISTIC PROBLEMS, ANNUAL CONFERENCE RECORD OF SICE .

1469 SUGENO, M. (1972B) EVALUATION OF SIMILARITY OF PATTERNS BY FUZZY INTEGRALS, ANNUAL CONFERENCE RECORDS OF SICE

1470 SUGENO, M. (1973) CONSTRUCTING FUZZY MEASURE AND GRADING SIMILARITY OF PATTERNS BY FUZZY INTEGRALS, TRANSACTION SICE, 9, 359-367.

1471 SUGENO, M. (1974) SUBJECTIVE EVALUATION OF FUZZY OBJECTS, PROC. OF IFAC SYMPO. ON STOCHASTIC CONTROL, BUDAPEST.

1472 SUGENO, M. (1974) THEORY OF FUZZY INTEGRALS AND ITS APPLICATIONS, PHD THESIS, TOKYO INSTITUTE OF TECHNOLOGY, TOKYO, JAPAN.

1473 SUGENO, M. (1975) FUZZY MEASURES AND FUZZY INTEGRALS, 55-60, SUMMARY OF PAPERS ON GENERAL FUZZY PROBLEMS, THE WORKING GROUP ON FUZZY SYSTEMS, TOYKO, JAPAN.

1474 SUGENO, M. (1975A) THEORETICAL DEVELOPMENTS OF FUZZY SETS, J. JAACE, 19, 229-234.

1475 SUGENO, M. (1975B) INVERSE OPERATION OF FUZZY INTEGRALS AND CONDITIONAL FUZZY MEASURES, TRANS. SICE, 11, 32-37.

1476 SUGENO, M. (1975C) FUZZY DECISION-MAKING PROBLEMS, TRANS. SICE, 11, 709-714.

1477 SUGENO, M. (1975D) FUZZY MEASURES AND FUZZY INTEGRALS, IN SUMMARY OF PAPERS ON GENERAL FUZZY PROBLEMS, THE WORKING GROUP ON FUZZY SYSTEMS, TOKYO, JAPAN, NOV., 55-60.

1478 SUGENO, M. (1977) FUZZY MEASURES AND FUZZY INTEGRALS: A SURVEY, 89-102, IN FUZZY AUTOMATA AND DECISION PROCESSES, M.M. GUPTA, G.N. SARIDIS, B.R. GAINES, EDS., NORTH-HOLLAND.

1479 SUGENO, M. (1977) FUZZY SYSTEMS WITH UNDERLYING DETERMINISTIC SYSTEMS, 25-52, SUMMARY OF PAPERS ON GENERAL FUZZY PROBLEMS, THE WORKING GROUP ON FUZZY SYSTEMS, TOYKO, JAPAN.

1480 SUGENO, M., AND TERANO, T. (1973) AN APPROACH TO THE IDENTIFICATION OF HUMAN CHARACTERISTICS BY APPLYING FUZZY INTEGRALS, PROCEEDING OF 3RD IFAC SYMPOSIUM ON IDENTIFICATION AND SYSTEM PARAMETER ESTIMATION, HAGUE.

1481 SUGENO, M., AND TERANO, T. (1977) A MODEL OF LEARNING BASED ON FUZZY INFORMATION, KYBERNETES 6, 157-166.

1482 SUGENO, M., AND TERANO, T. (1977) ANALYTICAL REPRESENTATION OF FUZZY SYSTEMS, 177-190, IN FUZZY AUTOMATA AND DECISION PROCESSES, M.M. GUPTA, G.N. SARIDIS, B.R. GAINES, EDS., NORTH-HOLLAND.

1483 SUGENO, M., AND TERNANO, T. (1977) ANALYTICAL REPRESENTATION OF FUZZY SETS, IN GUPTA ET AL., FUZZY AUTO. AND DECISION PROC., N.Y. ACAD., PRESS.

1484 SUGENO, M., TSUKAMOTO, Y., AND TERANO, T. (1974) SUBJECTIVE EVALUATION OF FUZZY OBJECTS, IFAC SYMPOSIUM ON STOCHASTIC CONTROL .

1485 SUGENO, S. (1976) FUZZY SYSTEMS AND PATTERN RECOGNITION, WORKSHOP ON DISCRETE SYSTEMS AND FUZZY REASONING, QUEEN MARY COLLEGE, UNIV. OF LONDON.

1486 SULARIA, M. (1977) ON FUZZY PROGRAMMING IN PLANNING, KYBERNETES, 6, 230.

1487 SUZUKI, T. (1976) TRIP AS TRIPLES MANIPULATING COMPUTER LANGUAGE, 48-63, SUMMARY OF PAPERS ON GENERAL FUZZY PROBLEMS, THE WORKING GROUP ON FUZZY SYSTEMS, TOYKO, JAPAN.

1488 SWINBURNE, R. (1973) AN INTRODUCTION TO CONFIRMATION THEORY, METHUEN, LONDON.

1489 TAHANI, V. (1971) FUZZY SETS IN INFORMATION RETRIEVAL, PHD THESIS, DEPARTMENT OF ELECTRICAL ENGINEERING AND COMPUTER SCIENCE, UNIVERSITY OF CALIFORNIA, BERKELEY, CALIFORNIA, USA.

1490 TAKAHARA, Y. (1977) SOME TOPOLOGICAL CONSIDERATIONS FOR GENERAL SYSTEMS THEORY, 53-69, SUMMARY OF PAPERS ON GENERAL FUZZY PROBLEMS, THE WORKING GROUP ON FUZZY SYSTEMS, TOYKO, JAPAN.

1491 TAKEUTI, G., AND ZARING, W.M. (1973) AXIOMATIC SET THEORY, SPRINGER-VERLAG, BERLIN.

1492 TAMURA, S. (1971) FUZZY PATTERN CLASSIFICATION, PROCEEDINGS OF A SYMPOSIUM ON FUZZINESS IN SYSTEMS AND ITS PROCESSING, PROFESSIONAL GROUP OF SYSTEM ENGINEERING OF SICE

1493 TAMURA, S., AND TANAKA, K. (1973) LEARNING OF FUZZY FORMAL LANGUAGE, IEEE TRANS. SYST. MAN CYBERN, SMC-3, 98-102.

1494 TAMURA, S., HIGUCHI, S., AND TANAKA, K. (1971) PATTERN CLASSIFICATION BASED ON FUZZY RELATIONS, IEEE TRANS. SYST. MAN CYBERN., SMC-1, 61-66.

1495 TANAKA, H., AND KANEKU, S. (1973) A FUZZY DECODING PROCEDURE ON ERROR-CORRECTING CODES, TECHNICAL REPORT ON INFORMATION OF IECE, MARCH.

1496 TANAKA, H., AND KANEKU, S. (1974) ON A FUZZY DECODING PROCEDURE FOR CYCLIC CODES, TRANS. IEECE(A) 57-A, 505-510.

1497 TANAKA, H., AND SOMMER, G. (1977) ON POSTERIOR PROBABILITIES CONCERNING A FUZZY INFORMATION, LEHRSTUHL FUR O.R., NO. 77/02, AACHEN.

1498 TANAKA, H., OKUDA, T., AND ASAI, K. (1974) DECISION-MAKING AND ITS GOAL IN A FUZZY ENVIRONMENT, US-JAPAN SEMINAR ON FUZZY SETS AND THEIR APPLICATIONS, BERKELEY, CALIFORNIA, USA, JULY.

1499 TANAKA, H., OKUDA, T., AND ASAI, K. (1974) ON FUZZY MATHEMATICAL PROGRAMMING, J. CYBERNET. 3, 37-46.

1500 TANAKA, H., OKUDA, T., AND ASAI, K. (1976) A FORMULATION OF FUZZY DECISION PROBLEMS AND ITS APPLICATION TO AN INVESTMENT PROBLEM. KYBERNETICS, 5, 25-30.

1501 TANAKA, K. (1972) ANALOGY AND FUZZY LOGIC, MATHEMATICAL SCIENCES .

1502 TANAKA, K. (1975) FUZZY SETS THEORY AND ITS APPLICATION, JOURNAL OF JAACE, 19, 227-228.

1503 TANAKA, K. (1976) FUZZY CONCEPT AND INTELLECTUAL INFORMATION PROCESSING, 1976 JOINT CONVENTION OF FOUR INSTITUTES OF ELECTRICAL ENGINEERS OF JAPAN.

1504 TANAKA, K. (1976) LEARNING IN FUZZY MACHINES, COMPUTER ORIENTED LEARNING PROCESSES, (ED. J.C. SIMON), 109-148, NOORDHOFF.

1505 TANAKA, K., AND ASAI, K. (1973) FUZZY MATHEMATICAL PROGRAMMING, TRANSACTIONS OF SICE, 9, 109-115.

1506 TANAKA, K., AND MIZUMOTO, M. (1975) FUZZY PROGRAMS AND THEIR EXECUTION, IN ZADEH, L.A., FU, K.S., TANAKA, K., AND SHIMURA, M. (EDS.), FUZZY SETS AND THEIR APPLICATIONS TO COGNITIVE AND DECISION PROCESSES, ACADEMIC PRESS, NEW YORK, 41-76.

1507 TANAKA, K., OKUDA, T., AND ASAI, K. (1972) ON THE FUZZY MATHEMATICAL PROGRAMMING, ANNUAL CONFERENCE RECORDS OF SICE

1508 TANAKA, K., TOYODA, J., MIZUMOTO, M., AND TSUJI, H. (1970) FUZZY AUTOMATA THEORY AND ITS APPLICATION TO AUTOMATIC CONTROLS, JOURNAL OF JAACE, 14, 541-550.

1509 TANIE, K., TACHI, S., AND ABE, M. (1977) STUDY ON THE SUBJECTIVE MAGNITUDE IN ELECTROCUTANEOUS STIMULATION, 79-86, SUMMARY OF PAPERS ON GENERAL FUZZY PROBLEMS, THE WORKING GROUP ON FUZZY SYSTEMS, TOYKO, JAPAN.

1510 TARANU, C. (1976) FUZZY ASPECTS IN COST THEORY, 3RD EUR. MEETING CYBERN, SYST. RES., VIENNA .

1511 TARSKI, A. (1956) LOGIC, SEMANTICS, METAMATHEMATICS, CLARENDON PRESS, OXFORD.

1512 TASHIRO, T. (1977) METHOD OF SOLUTION TO INVERSE PROBLEM OF FUZZY COORESPONDENCE MODEL, 70-78, SUMMARY OF PAPERS ON GENERAL FUZZY PROBLEMS, THE WORKING GROUP ON FUZZY SYSTEMS, TOYKO, JAPAN.

1513 TASHIRO, T., TERANO, T., AND TSUKAMOTO, Y. (1978) INVERSE OF FUZZY COORESPONDENCE AND EVOLUTIONARY DIAGNOSIS, TOKYO INSTITUTE OF TECHNOLOGY, IN PROCEEDINGS OF THE 1978 INT. CONF. ON CYBERNETICS AND SOCIETY, 938-941.

1514 TASNADI, A. (1978) FUZZY SYSTEMS AND LEARNING MODELS, PROC. 4TH INT. CONF. CYBER. AND SYSTEMS, AMSTERDAM, 382-383.

1515 TAZAKI, E. (1975) HEURISTIC SYNTHESIS IN A CLASS OF SYSTEMS BY USING FUZZY AUTOMATA, 61-66, SUMMARY OF PAPERS ON GENERAL FUZZY PROBLEMS, THE WORKING GROUP ON FUZZY SYSTEMS, TOYKO, JAPAN.

1516 TAZAKI, E., AMAGASA, M., AND TAKIZAWA, M. (1977) FUZZY STRUCTURAL MODELING, TRAN. THE OPERATIONS RESEARCH SOC. JAPAN, TO APPEAR.

1517 TAZAKI, E., AMAGASA, M., AND TAKIZAWA, M. (1977) STRUCTURAL MODELING IN A CLASS OF SYSTEMS BY FUZZY SETS THEORY, J. THE OPERATIONS RESEARCH SOCIETY OF JAPAN, VOL.20, 285-310.

1518 TAZAKI, E., AND AMAGASA, M. (1975) HEURISTIC STRUCTURE SYNTHESIS IN A CLASS OF SYSTEMS BY USING FUZZY AUTOMATA, IEEE SYST., MAN AND CYBERN. .

1519 TAZAKI, E., AND AMAGASA, M. (1976) HEURISTIC STRUCTURE SYNTHESIS IN A CLASS OF SYSTEMS BY FUZZY AUTOMATA, SYSTEMS AND CONTROL 20, 212-219.

1520 TAZAKI, E., AND AMAGASA, M. (1977) HEURISTIC STRUCTURE SYNTHESIS IN A CLASS OF SYSTEMS USING A FUZZY AUTOMATION,

PROC. 1977 IEEE CONF. ON DECISION AND CONTROL, NEW ORLEANS.

1521 TAZAKI, E., AND AMAGASA, M. (1977) STRUCTURAL MODDELLING IN A CLASS OF SYSTEMS USING FUZZY SET THEORY, PROC. 1977 IEEE CONF. ON DECISION AND CONTROL, NEW ORLEANS.

1522 TERANO, T. (1971) FUZZINESS AND ITS CONCEPT, PROCEEDINGS OF A SYMPOSIUM ON FUZZINESS IN SYSTEMS AND ITS PROCEEDINGS, PROFESSIONAL GROUP OF SYSTEM ENGINEERING OF SICE .

1523 TERANO, T. (1972) FUZZINESS OF SYSTEMS, NIKKA-GIREN ENGINEERS, 21-25.

1524 TERANO, T. (1975) METHOLOGY OF FUZZY SYSTEMS, J. OF IECE 58, 875-876.

1525 TERANO, T. (1976) APPLICATION OF FUZZY CONCEPT TO CONTROLS, 1976 JOINT CONVENTION OF FOUR INSTITUTES OF ELECTRICAL ENGINEERING.

1526 TERANO, T. (1976) STRUCTURAL MODEL FOR COMPLEX SOCIAL PROBLEMS, 70-78, SUMMARY OF PAPERS ON GENERAL FUZZY PROBLEMS, THE WORKING GROUP ON FUZZY SYSTEMS, TOYKO, JAPAN.

1527 TERANO, T. (1978) ET.AL : DIAGNOSIS OF ENGINE TROUBLE BY FUZZY LOGIC, IFAC 7TH WORLD CONGRESS, 1621-1628, HELINSKI.

1528 TERANO, T., AND SUGENO, M. (1977) MACROSCOPIC OPTIMIZATION USING CONDITIONAL FUZZY MEASURES, 197-208, IN FUZZY AUTOMATA AND DECISION PROCESSES, M.M. GUPTA, G.N. SARIDIS, B.R. GAINES, EDS., NORTH-HOLLAND.

1529 TERANO, T., AND TSUKAMOTO, Y. (1977) FAILURE DIAGNOSIS BY USING FUZZY LOGIC, PROCEEDINGS OF THE SYMP. ON FUZZY SET THEORY AND APPLN., IEEE CONF. ON DECISION AND CONTROL, NEW ORLEANS.

1530 TERANO, T., KUROSU, K., MURAYAMA, Y., AND INASAKA, F. (1977) PRINCIPAL COMPONENT ANALYSIS OF ENGINE TROUBLE, 103-108, SUMMARY OF PAPERS ON GENERAL FUZZY PROBLEMS, THE WORKING GROUP ON FUZZY SYSTEMS, TOYKO, JAPAN.

1531 TERASAKA, H. (1937) THEORIE DER TOPOLOGISCHEN VERBANDE, PROC. IMPERIAL ACADEMY, 13, TOKYO.

1532 TERNANI, T., AND SUGENO, M. (1975) CONDITIONAL FUZZY MEASURES AND THEIR APPLICATION, IN ZADEH, L.A., FU,K.S., TANAKA, K., AND SHIMURA, M. (EDS.), FUZZY SETS AND THEIR APPLICATIONS TO COGNITIVE AND DECISION PROCESSES, ACADEMIC PRESS, NEW YORK, 151-170.

1533 TERNANO, T., AND SUGENO, M. (1975) CONDITIONAL FUZZY MEASURES AND THEIR APPLICATIONS, IN ZADEH, FU, TANAKA, AND SHIMURA, ACAD. PRESS.

1534 TERNANO, T., AND SUGENO, M. (1975) MACROSCOPIC OPTIMIZATION BY USING FUZZY CONDITION MEASURES, IN SUMM. OF PAPERS ON GENERAL FUZZY PROBLEMS, WORKING GROUP ON FUZZY SETS, TOKYO.

1535 TERNANO, T., AND SUGENO, M. (1977) MACROSCOPIC OPTIMIZATION USING FUZZY SETS, IN GUPTA ET AL., FUZZY AUTO. AND DECISION PROC., N.Y., ACAD. PRESS.

1536 THARP, L. (1975) WHICH LOGIC IS THE RIGHT LOGIC?, SYNTHESE, 31, 1-21.

1537 THOMASON, M.G. (1973) FINITE FUZZY AUTOMATA, REGULAR FUZZY LANGUAGES AND PATTERN RECOGNITION, PATTERN RECOGNITION, 5, 383-389.

1538 THOMASON, M.G. (1974) FUZZY SYNTAX-DIRECTED TRANSLATIONS, J. CYBERNETICS, 4, 87-94.

1539 THOMASON, M.G. (1974) THE EFFECT OF LOGIC OPERATIONS ON FUZZY LOGIC DISTRIBUTIONS, IEEE TRANS. SYST. MAN CYBERN., SMC-4, 309-310.

1540 THOMASON, M.G. (1975) FINITE FUZZY AUTOMATA, REGULAR LANGUAGES AND PATTERN RECOGNITION, PATTERN RECOGNITION, 5, 383-390.

1541 THOMASON, M.G., AND MARINOS, P.N. (1972) FUZZY LOGIC RELATIONS AND THEIR UTILITY IN ROLE THEORY, PROCEEDINGS FROM 1972 IEEE INT. CONF. ON CYBERNETICS AND SOCIETY, WASH., D.C. .

1542 THOMASON, M.G., AND MARIONOS, P.N. (1974) DETERMINISTIC ACCEPTORS OF REGULAR FUZZY LANGUAGES, IEEE TRANS. SYST. MAN CYBERN., SMC-4, 228-230.

1543 TICHY, P. (1969) INTENSION IN TERMS OF TURING MACHINES, STUDIA LOGICA, 24, 7-25.

1544 TICHY, P. (1974) ON POPPER'S DEFINITION OF VERISIMILITUDE, BRIT. J. PHILOS. SCI., 25, 155-160.

1545 TICHY, P. (1976) VERISIMILITUDE REDEFINED, BRIT. J. PHILOS. SCI., 27, 25-42.

1546 TONG, R.M. (1976) AN ASSESSMENT OF A FUZZY CONTROL ALGORITHM FOR A NON-LINEAR MULTI-VARIABLE SYSTEM, PROC. WORKSHOP ON DISCRETE SYSTEMS AND FUZZY REASONING, QUEEN MARY COLLEGE, LONDON.

1547 TONG, R.M. (1976A) AN ASSESSMENT OF A FUZZY CONTROL ALGORITHM FOR A NONLINEAR MULTIVARIABLE PLANT, IN MAMDANI, E.H., AND GAINES, B.R. (EDS.), DISCRETE SYSTEMS AND FUZZY REASONING, EES-MMS-DSFR-76, QUEEN MARY COLLEGE, UNIVERSITY OF LONDON, (WORKSHOP PROCEEDINGS).

1548 TONG, R.M. (1976B) SOME PROBLEMS WITH THE DESIGN AND

IMPLEMENTATION OF FUZZY CONTROLLERS, CUED/F-CAMS/TR127(1976), CAMBRIDGE UNIVERSITY CONTROL ENGINEERING DEPT. .

1549 TONG, R.M. (1976C) ANALYSIS OF FUZZY CONTROL ALGORITHMS USING THE RELATION MATRIX, INT. J. MAN-MACHINE STUDIES, 8 .

1550 TONG, R.M. (1977) A CONTROL ENGINEERING REVIEW OF FUZZY SYSTEMS, AUTOMATICA, VOL. 13, 559-569.

1551 TONG, R.M. (1977) SYNTHESIS OF FUZZY MODELS FOR INDUSTRIAL PROCESSES: SOME RECENT RESULTS, INT. CONF. ON APPLIED GEN. SYSTEMS RESEARCH, BINGHAMTON, N.Y. .

1552 TONG, R.M. (1978) AN ANALYSIS OF FUZZY MODELS AND A DISCUSSION OF THEIR LIMITATIONS, CONTROL MANAGEMENT SYST. DIV., UNIV. ENGINEERING DEPT., MILL LANE, CAMBRIDGE CB2 1RX, ENGLAND.

1553 TONG, R.M. (1978) SYNTHESIS OF FUZZY MODELS FOR INDUSTRIAL PROCESSES - SOME RECENT RESULTS, INT. J. GENERAL SYSTEMS 4, 143-162.

1554 TRILLAS, E., AND RIERA, T. (1978) ENTROPIES IN FINITE FUZZY SETS, INFOR. SCIENCES, 15, 159-168.

1555 TSICHRITZIS, D. (1969) FUZZY COMPUTABILITY, PROC. PRINCETON CONF. INFORMATION SCIENCES AND SYSTEMS, 157-162.

1556 TSICHRITZIS, D. (1969) FUZZY PROPERTIES AND ALMOST SOLVABLE PROBLEMS, TECH. REP. 70, COMPUTER SCIENCE LABORATORY, DEPARTMENT OF ELECTRICAL ENGINEERING, PRINCETON UNIVERSITY.

1557 TSICHRITZIS, D. (1969) MEASURES ON COUNTABLE SETS, TECHNICAL REPORT 8, DEPARTMENT OF COMPUTER SCIENCE, UNIVERSITY OF TORONTO, CANADA.

1558 TSICHRITZIS, D. (1971) APPROXIMATION AND COMPLEXITY OF FUNCTIONS ON THE INTEGERS, INFORM. SCI., 70-86.

1559 TSICHRITZIS, D. (1971) PARTICIPATION MEASURES, MATH. ANAL. AND APPLN., 36, 60-72.

1560 TSICHRITZIS, D. (1973) A MODEL FOR ITERATIVE COMPUTATION, INFORM. SCI., 5, 187-197.

1561 TSICHRITZIS, D. (1973) APPROXIMATE LOGIC, PROC. SYMP. MULTIVALUED LOGIC, MAY.

1562 TSUJI, H., MIZUMOTO, M., TOYODA, J., AND TANAKA, K. (1972) INTERACTION BETWEEN RANDOM ENVIRONMENTS AND FUZZY AUTOMATA WITH VARIABLE STRUCTURES, TRANS. INST. ELECTRON. COMM. ENG.(JAPAN), 55-D, 143-144.

1563 TSUJI, H., MIZUMOTO, M., TOYODA, J., AND TANAKA, K. (1973) LINEAR FUZZY AUTOMATON, TRANS. INST. ELECTRON. COMM.

ENG. (JAPAN), 56-A, 256-257.

1564 TSUKAMOTO, T., AND TASHIRO, T., METHOD OF SOLUTION TO FUZZY INVERSE PROBLEM, TRAN. OF SICE, TO APPEAR.

1565 TSUKAMOTO, T., AND TERANO, T. (1977) FAULT DIAGNOSIS BY USING FUZZY LOGIC, 109-117, SUMMARY OF PAPERS ON GENERAL FUZZY PROBLEMS, THE WORKING GROUP ON FUZZY SYSTEMS, TOYKO, JAPAN.

1566 TSUKAMOTO, Y. (1972) IDENTIFICATION OF PREFERENCE MEASURE BY MEANS OF FUZZY INTEGRAL, PREPRINT ON JORS (IN JAPANESE).

1567 TSUKAMOTO, Y. (1975) A SUBJECTIVE EVALUATION ON ATTRACTIVITY OF SIGHTSEEING ZONES, IN SUMMARY OF PAPERS ON GENERAL FUZZY PROBLEMS, THE WORKING GROUP ON FUZZY SYSTEMS, TOKYO, JAPAN, NOV., 73-76.

1568 TSUKAMOTO, Y., AND IIDA, H. (1973) EVALUATION MODELS OF FUZZY SYSTEMS, ANNUAL CONFERENCE RECORDS OF SICE .

1569 TSUKAMOTO, Y., AND TERANO, T. (1977) FAILURE DIAGNOSIS BY USING FUZZY LOGIC, PROC. 1977 CONF. ON DECISION AND CONTROL, NEW ORLEANS.

1570 TSUKAMOTO, Y., TAKAGI, T., AND SUGENO, M. (1978) FUZZIFICATION OF L ALEPH-1 AND ITS APPLICATION TO CONTROL, TOKYO INSTITUTE OF TECHNOLOGY, PROCEEDINGS OF THE 1978 INT. CONF. ON CYBERNETICS AND SOCIETY, VOL2, 1217-1221.

1571 TSUTSUMI, T. (1976) ENGINEERING CODE PROCESSING BY COMPUTER, 64-69, SUMMARY OF PAPERS ON GENERAL FUZZY PROBLEMS, THE WORKING GROUP ON FUZZY SYSTEMS, TOYKO, JAPAN.

1572 TURKSEN, I.B., AND MARTIN, J.K. (1976) DECISION-INFORMATION SYSTEMS, A CONCEPTUAL FRAMEWORK, 76-010, DEPARTMENT OF INDUSTRIAL ENGINEERING, UNIVERSITY OF TORONTO, CANADA.

1573 TURQUETTE, A.R. (1954) MANY-VALUED LOGICS AND SYSTEMS OF STRICT IMPLICATION, PHILOS. REV., 63, 365-379.

1574 TURQUETTE, A.R. (1963) INDEPENDENT AXIOMS FOR INFINITE-VALUED LOGIC, J. SYMBOLIC LOGIC, 28, 217-221.

1575 TWAREQUE, ALI, S., AND PRUGOVECKI, E. (1977) SYSTEMS OF IMPRIM. AND REPRESENTATION OF QUANTUM MECH. ON FUZZY PHASE SPACES, J. OF MATH PHYSICS, VOL. 18, NO. 2.

1576 UHR, L. (1975) TOWARD INTEGRATED COGNITIVE SYSTEMS WHICH MUST MAKE FUZZY DECISIONS ABOUT FUZZY PROBLEMS, IN ZADEH, L.A., FU, K.S., TANAKA, K., AND SHIMURA, M. (EDS.), FUZZY SETS AND THEIR APPLICATION TO COGNITIVE AND DECISION PROCESSES, ACADEMIC PRESS, NEW YORK, 353-393.

1577 UMANO, M., MIZUMOTO, M., AND TANAKA, K. (1976) IMPLEMENTATION

OF FUZZY SETS MANIPULATION SYSTEM, TECHNICAL REPORT ON AUTOMATON AND LANGUAGE OF IECE, AL76-26.

1578 UMANO, M., MIZUMOTO, M., AND TANAKA, K. (1978) FSTDS: A FUZZY SET MANIPULATION SYSTEM, INFORMATION SCIENCE, 14, 115-159.

1579 UMANO, M., MIZUMOTO, M., TOYODA, J., AND TANAKA, K. (1975) ON INTERPRETATION AND EXECUTION OF FUZZY PROGRAMS, TECHNICAL REPORT ON AUTOMATON AND LANGUAGE OF IECE, AL75-27.

1580 UNO, K., ITAKURA, H., SANNOMIYA, N., AND NISHIKAWA, Y. (1976) LEARNING CONTROLS THAT USE A FUZZY CONTROLLER, SYSTEMS AND CONTROL 20, 262-268.

1581 URAGAMI, M., MIZUMOTO, M., AND TANAKA, K. (1976) FUZZY ROBOT CONTROLS, JOURNAL OF CYBERNETICS, 6, 39-64.

1582 URAGAMI, M., MIZUMOTO, M., TOYODA, J., AND TANAKA, K. (1975) ROBOT CONTROL BY FUZZY PROGRAMS, AL75-51, TECHNICAL REPORT ON AUTOMATON AND LANGUAGE OF IECE.

1583 URAZAMI, M., MIZUMOTO, M., AND TANAKA, K. (1976) FUZZY ROBOT CONTROLS, J. OF CYBER. 6, 39-64.

1584 URQUHART, A. (1973) AN INTERPRETATION OF MANY-VALUED LOGIC, Z. MATH. LOGIK GRUNDLAGEN MATH., 19, 111-114.

1585 VAAN FRASSEN, B.C. (1973) LAKEOFF'S FUZZY PROPOSITIONAL LOGIC, MIMO'D.

1586 VACHET, J. (1964A) PRAGUE PHONOLOGICAL STUDIES TODAY, TRAVAUX LINGUISTIQUES DE PRAGUE, 1, 7-20.

1587 VACHET, J. (1964B) ON SOME BASIC PRINCIPLES OF "CLASSICAL" PHONOLOGY, ZEITSCHR. FUR PONETIK, SPRACHWISSENSCHAFT U. KOMMUNIKATIONSFORSCHUNG (BERLIN), 17, 409-431.

1588 VACHET, J. (1966A) ON THE INTEGRATION OF THE PERIPHERAL ELEMENTS INTO THE SYSTEM OF LANGUAGE, TRAVAUX LINGUISTIQUE DE PRAGUE, 2, 23-37.

1589 VACHET, J. (1966B) THE LINGUISTIC SCHOOL OF PRAGUE, INDIANA UNIVERSITY PRESS, BLOOMINGTON.

1590 VAGIN, V.N., POSPELOV, D.A., AND PAPKE, W. (1977) APPLICATION OF FUZZY LOGIC IN CONTROL SYSTEMS, FOUNDATION OF CONTROL ENG., 2, 153-160.

1591 VAN FRAASSEN, B.C. (1968) PRESUPPOSITIONS, SUPERVALUATIONS AND SELF-REFERENCE, J. PHILOS., 65, 136-152.

1592 VAN FRAASSEN, B.C. (1974) HIDDEN VARIABLES IN CONDITIONAL LOGIC, THEORIA, 40, 176-190.

1593 VAN FRAASSEN, B.C. (1975) COMMENTS: LAKEOFF'S FUZZY

PROPOSITIONAL LOGIC, IN HOCKNEY, D., HARPER, W., AND FREED, B. (EDS.), CONTEMPORARY RESEARCH IN PHILOSOPHICAL LOGIC AND LINGUISTIC SEMANTICS, REIDEL, HOLLAND.

1594 VAN HEIJENCORT, J. (ED.) (1967) FROM FREGE TO GODEL: A SOURCE BOOK IN MATHEMATICAL LOGIC 1879-1937, HARVARD UNIVERSITY PRESS, CAMBRIDGE, MASS., USA.

1595 VAN VELTHOVEN, G.D. (1974B) ONDERZOEK NAAR TOEPASBAARHEID VAN DE THEORIE DER VAGE VERZAMELINGEN OP HET PARAMETRISCH ONDERZOEK INZAKE CRIMINALITEIT, DEC. .

1596 VAN VELTHOVEN, G.D. (1975A) APPLICATION OF FUZZY SETS THEORY TO CRIMINAL INVESTIGATION, PROC. FIRST EUROPEAN CONGRESS ON OPERATIONS RESEARCH, BRUSSELS, JAN. .

1597 VAN VELTHOVEN, G.D. (1975B) FUZZY MODELS IN PERSONNEL MANAGEMENT, PROC. THIRD INTERNATIONAL CONGRESS OF CYBERNETICS AND SYSTEMS, BUCHAREST, AUG., 15.

1598 VAN VELTHOVEN, G.D. (1975C) QUELQUES APPLICATIONS DE LA TAXONOMIE FLOUE, SEMINAIRE SUR LA CONTRIBUTION DES SYSTEMES FLOUS A L'AUTOMATIQUE: PROCESSUS HUMAIN ET INDUSTRIEL, CENTRE D'AUTOMATIQUE, UNIVERSITE DES SCIENCES ET TECHNIQUES DE LILLE, FRANCE, JUNE.

1599 VANEEDEN, D. (1976) FUZZY RANDOM VARIABLES IN DECISION PROBLEMS, REPORT TECHNISCHE HOGESCHOOL TWENTE.

1600 VARELA, F.J. (1975) A CALCULUS FOR SELF-REFERENCE, INT. J. GENERAL SYST., 2, 5-24.

1601 VARELA, F.J. (1976A) THE ARITHMETIC OF CLOSURE, 3RD EUR. MEETING CYBERN, SYST. RES., VIENNA, APRIL.

1602 VARELA, F.J. (1976B) THE EXTENDED CALCULUS OF INDICATIONS INTREPRETATED AS A THREE-VALUED LOGIC, NOTRE DAME J. FORMAL LOGIC, 17 .

1603 VERMA, R.R. (1970) VAGUENESS AND THE PRINCIPLE OF THE EXCLUDED MIDDLE, MIND, 79, 66-77.

1604 VICKERS, J.M. (1965) SOME REMARKS ON COHERENCE AND SUBJECTIVE PROBABILITY, PHILOS. SCI., 32, 32-38.

1605 VILLEGAS, C. (1964) ON QUANTITATIVE PROBABILITY SIGMA-ALGEBRAS, ANN. MATH. STATIST., 35, 1787-1796.

1606 VINCKE, P. (1973) LA THEORIE DES ENSEMBLES FLOUS, MEMORIE, FACULTE DE SCIENCE, UNIVERSITIE LIBRE DE BRUXELLES, BELGIUM.

1607 VINCKE, P. (1973) UNE APPLICATION DE LA THEORIE DES GRAPHES FLOUS, CAHIERS DU CENTRE D'ETUDES DE RECHERCHE OPERATIONELLE, 15(3), 375-395.

1608 VON WRIGHT, G.H. (1957) LOGICAL STUDIES, ROUTLEDGE AND

KEGAN PAUL, LONDON.

1609 VON WRIGHT, G.H. (1962) REMARKS ON THE EPISTEMOLOGY OF SUBJECTIVE PROBABILITY, IN NAGEL, E., SUPPES, P., AND TARSKI, A. (EDS.), LOGIC, METHODOLOGY AND PHILOSOPHY OF SCIENCE, STANFORD UNIVERSITY PRESS, CALIFORNIA, USA, 330-339.

1610 VON WRIGHT, G.H. (1963A) THE LOGIC OF PREFERENCE, EDINBURGH UNIVERSITY PRESS.

1611 VON WRIGHT, G.H. (1963B) NORM AND ACTION, ROUTLEDGE AND KEGAN PAUL, LONDON.

1612 VON WRIGHT, G.H. (1972) AN ESSAY IN DEOTIC LOGIC AND THE GENERAL THEORY OF ACTION, NORTH-HOLLAND, AMSTERDAM.

1613 VOPENKA, P., HAJAK, P. (1972) THEORY OF SEMISETS, NORTH-HOLLAND, AMSTERDAM.

1614 VOSSEN, P.H. (1974) FUZZY SET CONVOLUTION WITH RESPECT TO A GROUP OPERATION, MEMO 1974.06.20, DEPARTMENT OF PSYCHOLOGY, NIJMEGEN UNIVERSITY, HOLLAND.

1615 VOSSEN, P.H. (1974) NOTES FOR A THEORY OF FUZZINESS. THE EMERGENCE OF A BASIC CONCEPT IN MATHEMATICS, SCIENCE AND TECHNOLOGY, SSRG-74-01, DEPARTMENT OF PSYCHOLOGY, NIJMEGEN UNIVERSITY, HOLLAND.

1616 VOSSEN, P.H. (1975) VERTALING VAN VOORWOORD, VOORBERICHT EN INHOUDSOPGAVE VAN KAUFMANN 1973, MEMO 75-08, DEPARTMENT OF PSYCHOLOGY, NIJMEGEN UNIVERSITY, HOLLAND, (IN DUTCH).

1617 VOSSEN, P.H., AND KLABBERS, J.H.G. (1973) IN VOGELVLUCHT OVER ALGEMENE SYSTEMLEER EN VAGE VERZAMLINGENLEER, SSRG-73-00, DEPARTMENT OF PSYCHOLOGY, NIJMEGEN UNIVERSITY, HOLLAND, (IN DUTCH).

1618 VOSSEN, P.H., AND KLABBERS, J.H.G. (1974) A FORMAL AND EXPERIMENTAL INQUIRY INTO THE APPLICABILITY OF NONSTANDARD SET THEORY TO THE ANALYSIS OF VALUATION PROCESSES IN SOCIAL SYSTEMS, SSRG 74-11, DEPARTMENT OF PSYCHOLOGY, NIJMEGEN UNIVERSITY, HOLLAND.

1619 WAHLSTER, W. (1978) DIE SIMULATION VAGER INFERENZEN AUF UNSCHARFEM WISSEN: EINE ANWENDUNG DER MEHRWERTIGEN PROGRAMMIERSPRACHE FUZZY, IN: H. UECKERT AND D. RHENIUS, KOMPLEXE MENSCHLICHE INFORMATIONVERARBEITUNG, BEITRAGE ZUR TAGUNG KOGNITIVE PSYCHOLOGIE (HAMBURG).

1620 WAJSBERG, M. (1967) AXIOMATIZATION OF THE THREE-VALUED PROPOSITIONAL CALCULUS, IN MCCALL, S. (ED.), POLISH LOGIC 1920-1939, CLARENDEN PRESS, OXFORD, 264-284.

1621 WARREN, R.H. (1974B) BOUNDARY OF A FUZZY SET, APPLIED MATHEMATICS RESEARCH LABORATORY, WRIGHT-PATTERSON AFB, OHIO, USA, (PREV. CLOSURE OPERATOR AND BOUNDARY OPERATOR

FOR FUZZY TOPOLOGICAL SPACES).

1622 WARREN, R.H. (1974C) NEIGHBORHOODS, BASES AND CONTINUITY IN FUZZY TOPOLOGICAL SPACES, APPLIED MATHEMATICS RESEARCH LABORATORY, WRIGHT-PATTERSON AFB, OHIO, USA.

1623 WARREN, R.H. (1976) OPTIMALITY IN FUZZY TOPOLOGICAL POLYSYSTEMS, J. MATH. ANAL. APPL., VOL.54, 309-315.

1624 WARREN, R.H. (1977) BOUNDARY OF FUZZY SET, INDIANA UNIV. MATH. J. 26 (2), 191-197.

1625 WARREN, R.H.,(TO APPEAR) NEIGHBORHOODS, BASES AND CONTINUITY IN FUZZY TOPOLOGICAL SPACES, ROCKY MTN. J. MATH. .

1626 WARREN, R.N. (1974) OPTIMALARITY IN FUZZY TOPOLOGICAL POLYSYSTEMS, OPP. MATH. RES. LAB., WRIGHT PATTERSON, OHIO.

1627 WASHIOWSKI, R., AND WELK, T. (1978) APPLICATION OF FUZZY SETS TO DECISION MAKING IN ACTIVE SYSTEMS, MEMO.

1628 WATANABE, S. (1969) MODIFIED CONCEPTS OF LOGIC, PROBABILITY AND INFORMATION BASED ON GENERALIZED CONTINUOUS CHARACTERISTIC FUNCTION, INFORM. AND CONTROL, 15, 1-21.

1629 WATANABE, S. (1975) CREATIVE LEARNING AND PROPENSITY AUTOMATA, IEEE TRANS. SYST. MAN CYBERN., SMC-5, 603-609.

1630 WATANABE, S. (1978) A GENERALIZED FUZZY-SET THEORY, IEEE TRANSACTIONS ON SYST., MAN, AND CYBERNETICS, VOL. SMC-8, NO. 10, 756-759.

1631 WATANABE, S. (1978) FUZZIFICATION AND INVARIANCE, SOPHIA UNIV., TOKYO, JAPAN, PROCEEDINGS OF THE 1978 INT. CONF. ON CYBERNETICS AND SOCIETY, 947-951.

1632 WATSON, S.R., AND WEISS, J.J. (1978) APPLS. OF FUZZY SET THEORY TO NAVAL COMMAND AND CONTROL, REPORT PR 78-13-83 DECISION AND DESIGN INC. MC LEAN, VA. .

1633 WATSON, S.R., WEISS, J.J., AND DONNELL, M.L. (1976) FUZZY DECISION ANALYSIS, VOL. SMC-9, NO.1, IEEE TRANS. ON SYST., MAN, AND CYBERN., 1-9.

1634 WEBB, D.L. (1936) THE ALGEBRA OF N-VALUED LOGIC, COMPTES RENDUS DES SEANCES DE LA SOCIETE DES SCIENCES ET DES LETTRES DE VARSOVIE, 29, 153-168.

1635 WECHLER, W. (1974) ANALYSE UND SYNTHES ZEITVARIABLER R-FUZZY AUTOMATEN, ZKI INFORMATION (AKAD D. WISS. DE DDR), 1, 32-366.

1636 WECHLER, W. (1975B) R-FUZZY AUTOMATA WITH A TIME STRUCTURE, IN BLIKLE, A. (ED.), MATHEMATICAL FOUNDATIONS OF COMPUTER SCIENCE, 28, SPRINGER-VERLAG, BERLIN, GERMANY,

73-76.

1637 WECHLER, W. (1975C) ZUR VERALLGEMEINERGUNG DES THEOREMS VON KLEENE-SCHUTZEN- BERGER AUF ZEITVARIABLE AUTOMATEN, J. EIK, 11, 439-445.

1638 WECHLER, W. (1975D) AUTOMATEN UBER INPUTKATEGORIEN, J. EIK, 11, 681-685.

1639 WECHLER, W. (1975E) GESTEUERTE R-FUZZY AUTOMATEN, ZKI-INFORMATIONEN (AKAD. D. WISS. DER DDR), 1, 9-13.

1640 WECHLER, W. (1975F) THE CONCEPT OF FUZZINESS IN THE THEORY OF AUTOMATA, PROCEEDINGS OF 3RD INTERNATIONAL CONGRESS OF CYBERNETICS AND SYSTEMS, BUCHAREST, RUMANIA, AUG. .

1641 WECHLER, W. (1976) R-FUZZY GRAMMARS, TECHNISCHE UNIVERSITAT DRESDEN SEKTION MATHEMATIK, NR. 07-20-76, 1-12.

1642 WECHLER, W. (1976A) ZUM VERHALTEN GESTEUERTER R-FUZZY AUTOMATEN, TECH.UNIV. DRESDEN, GERMANY.

1643 WECHLER, W. (1976B) HIERARCHY OF N-RATIONAL LANGUAGES, TECH. UNIV. DRESDEN, GERMANY.

1644 WECHLER, W. (1977) MODELLING HUMANISTIC SYSTEMS BY FUZZY LOGIC, PROCEEDINGS OF THE SYMP. ON FUZZY SET THEORY AND APPLN., IEEE CONF. ON DECISION AND CONTROL, NEW ORLEANS.

1645 WECHLER, W., AND AGASANDYAN, G.A. (1974) AUTOMATA WITH A VARIABLE STRUCTURE AND METAREGULAR LANGUAGES, IZV. AKAD. NAUK. SSSR TEHN. KIBERNET., 1W, 146-148.

1646 WECHLER, W., AND DIMITROV, V. (1974) R-FUZZY AUTOMATA, IN INFORMATION PROCESSING 74, PROC. IFIP CONGRESS, NORTH-HOLLAND, AMSTERDAM, 657-660.

1647 WECHSLER, H. (1975) APPLICATIONS OF FUZZY LOGIC TO MEDICAL DIAGOSIS, PROC. 1975 INT. SYMP. MULTIPLE-VALUED LOGIC, IEEE 75CH0959-7C, MAY.

1648 WECHSLER, H. (1976) A FUZZY APPROACH TO MEDICAL DIAGNOSIS, INT. J. BIO-MEDICAL COMPUT. 7, 191-203.

1649 WEE, W.G. (1967) ON GENERALIZATIONS OF ADAPTIVE ALGORITHMS AND APPLICATIONS OF THE FUZZY SETS CONCEPT TO PATTERN CLASSIFICATION, PHD THESIS, PURDUE UNIVERSITY, LAFAYETTE, USA.

1650 WEE, W.G. (1968) SURVEY OF PATTERN RECOGNITION, IEEE PROC. 7TH SYMP. ON ADAPTIVE PROCESSES, UCLA, 13.

1651 WEE, W.G., AND FU, K.S. (1969) A FORMULATION OF FUZZY AUTOMATA AND ITS APPLICATION AS A MODEL OF LEARNING SYSTEMS, IEEE TRAN. SYST., MAN CYBERN., SMC-5, 215-223.

1652 WEISS, M.D. (1975) FIXED POINTS SEPARATION AND INDUCED TOPOLOGIES FOR FUZZY SETS, J. MATH. ANAL. AND APPLN., 50, 142-150.

1653 WEISS, S.E. (1973) THE SORITES ANTINOMY : A STUDY IN THE LOGIC OF VAGUENESS AND MEASUREMENT, PHD THESIS, UNIVERSITY OF NORTH CAROLINA, CHAPEL HILL.

1654 WEISS, S.E. (1976) THE SORITES FALLACY: WHAT DIFFERENCE DOES A PEANUT MAKE?, SYNTHESE .

1655 WENSTOP, F. (1975A) APPLICATION OF LINGUISTIC VARIABLES IN THE ANALYSIS OF ORGANIZATIONS, PHD THESIS, SCHOOL OF BUSINESS ADMINISTRATION, UNIVERSITY OF CALIFORNIA, BERKELEY, CALIFORNIA, USA.

1656 WENSTOP, F. (1975B) EVALUATION OF VERBAL ORGANIZAIONAL MODELS, NOAK 75, OSLO.

1657 WENSTOP, F. (1976) DEDUCTIVE VERBAL MODELS OF ORGANIZATIONS, J. MAN-MACHINE STUDIES, 8, 293-311.

1658 WENSTOP, F. (1976) FUZZY SET SIMULATION MODELS IN A SYSTEMS DYNAMIC PERSPECTIVE, KYBERNETES 6, 209-218.

1659 WENSTOP, F. (1977) FUZZY SETS AND DECISION MAKING, CALIFORNIA ENGINEER, 16, 20-24.

1660 WENSTOP, F. (1978) VERBAL FORMULATION OF FUZZY DYNAMIC SYSTEMS, OSLO INSTIT. OF BUSINESS ADMIN., FRYSJAVEIEN 33C, OSLO 8, NORWAY.

1661 WESCHLER, H. (1975) A FUZZY APPROACH TO MEDICAL DIAGOSIS, INT. J. BIO-MEDICAL COMPUT. 7, 191-203.

1662 WHALSTER, W. (1977) DIE REPRASENTATION VON VAGEM WISSEN IN NATURLICHSPRACHEN SYSTEMEN DER KUNSTLLICHEN INTELLIGENZ, WORKING PAPER 38, INSTITUT FUR INFORMATIK, UNIVERSITAT HAMBURG, JULY.

1663 WHITE, A.R. (1975) 'MODAL THINKING, BASIL BLACKWELL, OXFORD.

1664 WIERZCHON, S.T., AND ZALEWSKI, J. (1978) A SET OF SUBRTNS. FOR APPLIC. OF FUZZY CONC. TO DECISION MAKING PROBLEMS, INT. CONF. ON SYST. SCI., SYSTEM SCI. 5, WROCLAW.

1665 WIERZSHON, S.T., AND ZALEWSKI, J. (1978) ON SOME EQUIVALENCE BETWEEN CONTROL THEORY AND DECISION MAKING THEORY, 4TH INT. CONGRESS IN CYBERN. AND SYST., AMSTERDAM, AUG. .

1666 WIERZSHON, S.T., AND ZALEWSKI, J. (1978) ON SOME EQUIVALENCE BETWEEN CONTROL, TECH. UNIV. OF WARZAWA, WARZAWA, POLAND, 4TH INT. CONGRESS ON CYBERNETICS AND SYSTEMS, AMSTERDAM.

1667 WILKINSON, J. (1973) ARCHETYPES, LANGUAGE, DYNAMIC

PROGRAMMING AND FUZZY SETS, IN WILKINSON, J., BELLMAN, R., AND GARAUDY, R. (EDS.), THE DYNAMIC PROGRAMMING OF HUMAN SYSTEMS, MSS INFORMATION CORP., NEW YORK, USA, 44-53.

1668 WILKINSON, J. (1973) RETROSPECTIVE FUTUROLOGY, IN WILKINSON, J., BELLMAN, R., AND GARAUDY, R. (EDS.), THE DYNAMIC PROGRAMMING OF HUMAN SYSTEMS, MSS INFORMATION CORP., NEW YORK, USA, 19-33.

1669 WILKINSON, J., BELLMAN, R., AND GARAUDY, R. (EDS.) (1973) THE DYNAMIC PROGRAMMING OF HUMAN SYSTEMS, MSS INFORMATION CORP., NEW YORK, USA.

1670 WILKS, Y. (1975) PREFERENCE SEMANTICS, IN KEENAN, E.L. (ED.), FORMAL SEMANTICS OF NATURAL LANGUAGE, C, 320-348.

1671 WILLAEYS, D., AND MALVACHE, N. (1976) : UTILISATING REFERENTIAL OF FUZZY SETS ; APPLICATION TO FUZZY ALGORITHM, INT. CONF. ON SYST. SCI., WROCLAW, POLOGNE.

1672 WILLAEYS, D., AND MALVACHE, N. (1977) : UTILISATION OF FUZZY SETS FOR SYSTEM MODELLING AND CONTROL, CONGRESS INT. IEEE, NEW ORLEANS.

1673 WILLAEYS, D., AND MALVACHE, N. (1978) USE OF FUZZY MODEL FOR PROCESS CONTROL, JAPAN, PROCEEDINGS OF THE 1978 INT. CONF. ON CYBERNETICS AND SOCIETY, 942-946.

1674 WILLAEYS, D., MANGIN, P., AND MALVACHE, N. (1977) UTILISATION DES SOUS-ENSEMBLES FLOUS POUR LA MODELISATION ET LA COMMANDS DE SYSTEME : APPLICATION A LA REGUATION DE VITESSE D'UN MOTEUR SOUMIS A DE FORTES PERTURBATIONS, 5EME CONF. IFAC IFIP DIGITAL COMPUTER APPLICATIONS TO PROCESS CONTROL LA HAYE, 14-17, JUIN.

1675 WILLAEYS, D., MANGIN, P., MANVACHE, N. (1977) : USE OF FUZZY SETS FOR SYSTEMS MODELIZING AND CONTROL : APPLICATION TO THE SPEED REGULATION OF A STRONGLY PERTURBED MOTOR . 5EME CONF. INT. IFAC/IFIP, LA HAYE.

1676 WILSON, D. (1975) PRESUPPOSITIONS AND NON-TRUTH-CONDITIONAL SEMANTICS, ACADEMIC PRESS, LONDON.

1677 WINKLER, R.L. (1974) PROBABILISTIC PREDICTION: SOME EXACT EXPERIMENTAL RESULTS, J. AMER. STATIST. ASSN., 66, 625-688.

1678 WINKLER, R.L. AND MURPHY, A.H. (1968) GOOD PROBABILITY ASSESSORS, J. APPLIED METEOROLOGY, 7, 751-758.

1679 WINOGRAD, T. (1974) LAKOFF ON HEDGES, ARTIFICIAL INTELLIGENCE LABORATORY, COMPUTER SCIENCE DEPT., STANFORD UNIVERSITY, STANFORD, CALIFORNIA, USA, SEP.

1680 WIREDU, J.E. (1975) TRUTH AS A LOGICAL CONSTANT, WITH AN APPLICATION TO THE PRINCIPLE OF THE EXCLUDED MIDDLE,

PHILOS. QUART., 305-317.

1681 WITTEN, I.H. (1975) LEARNING TO CONTROL SEQUENTIAL AND NON-SEQUENTIAL ENVIRONMENTS, DEPT. OF ELECTRICAL ENGINEERING SCI., UNIV. OF ESSEX., EES-MMS-CON.75 .

1682 WOJCICKI, R. (1966) SEMANTICAL CRITERIA OF EMPIRICAL MEANINGFULNESS, STUDIA LOGICA, 19, 75-107.

1683 WOLF, R.G. (1975) A CRITICAL SURVEY OF MANY-VALUED LOGICS 1966-1974, PROC. 1975 INT. SYMP. MULTIPLE-VALUED LOGIC, IEEE 75CH0959-7C, 468-474.

1684 WOLNIEWICZ, B. (1970) FOUR NOTIONS OF INDEPENDENCE, THEORIA, 36, 161-164.

1685 WONG, C.K. (1973) COVERING PROPERTIES OF FUZZY TOPOLOGICAL SPACES, J. MATH. ANAL. AND APPLN., 43, 697-704.

1686 WONG, C.K. (1974A) FUZZY TOPOLOGY: PRODUCT AND QUOTIENT THEOREMS, J. MATH. ANALS. AND APPLN., 45, 512-521.

1687 WONG, C.K. (1974B) FUZZY POINTS AND LOGICAL PROPERTIES OF FUZZY TOPOLOGY, J. MATH. ANAL. AND APPLN., 46, 316-328.

1688 WONG, C.K. (1975) FUZZY TOPOLOGY, IN ZADEH, L.A., FU, K.S., TANAKA, K., AND SHIMURA, M. (EDS.), FUZZY SETS AND THEIR APPLICATIONS TO COGNITIVE AND DECISION PROCESSES, ACADEMIC PRESS, NEW YORK, 171-190.

1689 WONG, C.K. (1976) CATEGORIES OF FUZZY SETS AND FUZZY TOPOLOGICAL SPACES, J. MATH. ANAL. AND APPLN., 53, 704-714.

1690 WONG, G.A., AND SHENG, D.C. (1975) ON THE LEARNING BEHAVIOUR OF FUZZY AUTOMATA, IN ROSE, J. (ED.), ADVANCES IN CYBERNETICS AND SYSTEMS, 2, GORDON AND BREACH, LONDON, 885-896.

1691 WOODBURY, M. CLIVE, J., AND GARSON, A. (IN REVIEW) A GENERALIZED DITTO ALGORITHM FOR INITIAL FUZZY CLUSTERS, J. CYBERNET., 4, 111-121.

1692 WOODBURY, M.A., AND CLIVE, J. (1974) CLINICAL PURE TYPES AS A FUZZY PARTITION, J. CYBERNETICS, 4, NO. 3, 111-121.

1693 WOODHEAD, R.G. (1972) ON THE THEORY OF FUZZY SETS TO RESOLVE ILL-STRUCTURED MARINE DECISION PROBLEMS, DEPARTMENT OF NAVAL ARCHITECTURE AND SHIPBUILDING, UNIVERSITY OF NEWCASTLE UPON TYNE, UK.

1694 WOODRUFF, P.W. (1974) A MODAL INTREPRETATION OF THREE-VALUED LOGIC, J. PHILOS. LOGIC, 3, 433-439.

1695 WRIGHT, C. (1975) ON THE COHERENCE OF VAGUE PREDICATES, SYNTHESES, 30, 325-365.

1696 YAGER, R.R. (1976) AN EIGENVALUE METHOD OF OBTAINING SUBJECTIVE PROBABILITIES IN DECISION ANALYSIS.

1697 YAGER, R.R. (1976) COMPARING FUZZY CONSTRAINTS, PROC. 5TH NORTHEAST AIDS CONF., PHILADELPIA, USA.

1698 YAGER, R.R. (1976) FUZZY DECISION MAKING INCLUDING UNEQUAL IMPORTANCE OF GOALS AND CONSTRAINTS, PRESENTED AT ORSA/TIMS MEETING, MIAMI.

1699 YAGER, R.R. (1977) A FOUNDATION FOR A THEORY OF POSSIBILITY, IONA TECHNICAL REPORT RRY 77-25, IONA COLLEGE, NEW ROCHELLE, NEW YORK.

1700 YAGER, R.R. (1977) BUILDING FUZZY DECISION MODELS, PROC. 1ST INT. CONF. ON MATH. MODELING, ST. LOUIS, MO. .

1701 YAGER, R.R. (1977) DECISIONS UNDER UNCERTAINTY USING FUZZY SETS, T.J. HINDELANG (ED), PROC. OF THE 6TH N.E. DECISION SCI. CONF., 17-20.

1702 YAGER, R.R. (1977) FUZZY DECISION MAKING INCLUDING UNEQUAL OBJECTIVES, FUZZY SETS AND SYST. 1, 87-95, NORTH-HOLLAND.

1703 YAGER, R.R. (1977) FUZZY EQUATIONS, PROC. 7TH INT. CONF. ON CYBER. AND SOCIETY, WASH. D.C. .

1704 YAGER, R.R. (1977) MATHEMATICAL PROGRAMMING IN A FUZZY ENVIRONMENT, INT. SYMP. ON EXTR. METHODS AND SYSTEMS ANAL., AUSTIN, TEXAS.

1705 YAGER, R.R. (1977) MULTIPLE OBJECTIVE DECISION-MAKING USING FUZZY SETS, INT. J. MAN-MACHINE STUDIES 9, 375-382.

1706 YAGER, R.R. (1977) ON VALIDITY AND BUILDING FUZZY MODELS, REPORT ON THE IEEE SYMP. ON FUZZY SET THEORY AND APPL., HELD AT THE 1977 IEEE CONTROL AND DECISION CONF. .

1707 YAGER, R.R. (1978) ON THE MEASURE OF FUZZINESS AND NEGATION PART I: MEMBERSHIP IN THE UNIT INTERVAL, TECH. REPORT NO. RRY 78-BB, IONA COLLEGE, NEW ROCHELLE, N.Y. .

1708 YAGER, R.R. (1978) A NOTE ON FUZZINESS IN A STANDARD UNCERTAINTY LOGIC, IONA COLLEGE, NEW ROCHELLE, NEW YORK, 10801, TECH. REPORT NO. RRY 78-19.

1709 YAGER, R.R. (1978) FUZZY MODELING IN AN URBAN SOCIETY, PROC. 22ND MEETING OF SOCIETY GEN. SYS. RES., WASH., H451-455.

1710 YAGER, R.R. (1978) FUZZY SETS OVER THE SAME SPACE, PROC. 1978 JACC, VOL. 3, 362.

1711 YAGER, R.R. (1978) LINGUISTIC MODELS AND FUZZY TRUTHS, TO APPEAR INT. J. OF MAN-MACHINE STUDIES.

1712 YAGER, R.R. (1978) ON THE MEASURE OF FUZZINESS AND NEGATION

PART II: MEMBERSHIP IN LATICES, TECH. REPORT NO. RRY 78-CC, IONA COLLEGE, NEW ROCHELLE, N.Y. .

1713 YAGER, R.R. (1978) ON THE NEED FOR MEMBERSHIP GRADES, IN FUZZY SETS, IONA COLLEGE, NEW ROCHELLE, NY 1081, USA.

1714 YAGER, R.R. (1978) POSSIBILISTIC DECISION MAKING, TECH. REPORT NO. RRY 78-05B, IONA COLLEGE, SCHOOL OF BUSINESS ADMINISTRATION, NEW ROCHELLE, NEW YORK.

1715 YAGER, R.R. (1978) RANKING FUZZY SUBSETS OVER THE UNIT INTERVAL, 1435-1437, PROC. 1978 IEEE CONF. ON DEC. AND CONTROL, INCLUDES THE 17TH SYMP. ON ADAPTIVE PROCESSES, 78CH1392-OCS, JAN., 1979, SAN DIEGO, CALIF. .

1716 YAGER, R.R. (1978) TOWARD A THEORY OF SIGNAL RESPONSE BASED ON FUZZY SETS, IONA COLLEGE, NEW ROCHELL, N.Y., 921-925, PROCEEDINGS OF THE 1978 INT. CONF. ON CYBERNETICS AND SOCIETY.

1717 YAGER, R.R. (1978) VALIDATION OF FUZZY LINGUISTIC MODELS, J. OF CYBERNETICS, 8, 17-30.

1718 YAGER, R.R. (1979) A MEASUREMENT - INFORMATIONAL DISCUSSION OF FUZZY UNION AND INTERSECTION, INT. J. OF MAN-MACHINE STUDIES.

1719 YAGER, R.R. (1979) FUZZY SETS PROBABILITIES AND DECISION, J. OF CYBERNETICS.

1720 YAGER, R.R. (1979) MATHHEM. PROGRAMMING WITH FUZZY CONSTRAINTS AND A PREFERENCE ORDERING ON THE OBJECT, KYBERNETES.

1721 YAGER, R.R. (1979) ON CHOOSING BETWEEN FUZZY SUBSETS, KYBERNETES.

1722 YAGER, R.R. (1979) ON SOLVING FUZZY MATHEMATICAL RELATIONSHIPS, INF. AND CONTROL.

1723 YAGER, R.R. (1979) ON THE LACK OF INVERSES IN FUZZY ARITHMETIC, FUZZY SETS AND SYSTEMS.

1724 YAGER, R.R. (1979) PROPERTIES OF CONNECTIVES USEFUL IN LOCAL LOGICS, IN B.R. GAINES (ED), GEN. SYS. RES.: A SCI., A METHOD., A TECHNOLOGY, SGSR.

1725 YAGER, R.R., AND BASSON, D. (1975) DECISION MAKING WITH FUZZY SETS, DECISION SCIENCES 6, 590-600.

1726 YAGER, R.R., FUZZY SUBSETS OF TYPE II IN DECISIONS, J. OF CYBERN. .

1727 YAGER, R.R.,(TO APPEAR) AN APPROACH TO MULTIOBJECTIVE DECISIONS USING FUZZY SETS, IONA COLLEGE RESEARCH MEMO.

1728 YAMASHITA, T., AND HARA, F. (1977) CLASSIFICATION OF FACIAL

EXPRESSION, 167-171, SUMMARY OF PAPERS ON GENERAL FUZZY PROBLEMS, THE WORKING GROUP ON FUZZY SYSTEMS, TOYKO, JAPAN.

1729 YEH, R.T. (1974) TOWARD AN ALGEBRAIC THEORY OF FUZZY RELATIONAL SYSTEMS, PROCEEDINGS OF INTERNATIONAL CONGRESS ON CYBERNETICS, NAMUR.

1730 YEH, R.T., AND BANG, S.Y. (1975) FUZZY RELATIONS, FUZZY GRAPHS AND THEIR APPLICATIONS TO CLUSTERING ANALYSIS, IN ZADEH, L.A., FU, K.S., TANAKA, K., AND SHIMURA, M. (EDS.), FUZZY SETS AND THEIR APPLICATIONS TO COGNITIVE AND DECISION PROCESSES, ACADEMIC PRESS, NEW YORK, 125-149.

1731 ZADEH, L.A. (1965A) FUZZY SETS, INFORM. AND CONTROL, 338-353.

1732 ZADEH, L.A. (1965B) FUZZY SETS AND SYSTEMS, IN FOX, J. (ED.), SYSTEM THEORY, MICROWAVE RESEARCH INSTITUTE SYMPOSIA SERIES XV, POLYTECHNIC PRESS, BROOKLYN, NEW YORK, 29-37.

1733 ZADEH, L.A. (1966) SHADOWS OF FUZZY SETS, PROBLEMS IN TRANSMISSION OF INFORMATION, 2, 37-44, (IN RUSSIAN).

1734 ZADEH, L.A. (1968A) FUZZY ALGORITHMS, INFORM. AND CONTROL, 12, 94-102.

1735 ZADEH, L.A. (1968B) PROBABILITY MEASURES OF FUZZY EVENTS, J. MATH. ANAL. AND APPLN., 23, 421-427.

1736 ZADEH, L.A. (1969) BIOLOGICAL APPLICATIONS OF THE THEORY OF FUZZY SETS AND SYSTEMS, IN PROCTOR, L.D. (ED.), BIOCYBERNETICS OF THE CENTRAL NERVOUS SYSTEM, LITTLE, BROWN AND CO., BOSTON, MASS., USA, 199-212.

1737 ZADEH, L.A. (1969) THE CONCEPTS OF SYSTEM, AGGREGATE AND STATE IN SYSTEM THEORY, IN ZADEH AND POLAK SYSTEM THEORY, NEW YORK: MCGRAW-HILL.

1738 ZADEH, L.A. (1970) SIMILIARITY RELATIONS AND FUZZY ORDERING, E.R.L. MEMO M277, ELEC. RES. LAB., UNIV. OF CALIF., BERKELEY.

1739 ZADEH, L.A. (1971) FUZZY LANGUAGES AND THEIR RELATION TO HUMAN AND MACHINE INTELIGENCE, MEMO ERL-M 302, ELECTRONICS RES. LAB., UNIV. OF CALIFORNIA, BERKELEY.

1740 ZADEH, L.A. (1971A) TOWARDS A THEORY OF FUZZY SYSTEMS, IN KALMAN, R.E., AND DECLAIRIS, R.N. (EDS.), ASPECTS OF NETWORKS AND SYSTEMS THEORY, HOLT, RINEHART AND WINSTON, NEW YORK.

1741 ZADEH, L.A. (1971C) QUANTITATIVE FUZZY SEMANTICS, INFORM. SCI., 3, 159-176.

1742 ZADEH, L.A. (1971D) SIMILARITY RELATIONS AND FUZZY ORDERINGS, INFORM. SCI., 177-200.

1743 ZADEH, L.A. (1971E) HUMAN INTELLIGENCE VS. MACHINE INTELLIGENCE, PROCEEDINGS OF INTERNATIONAL CONFERENCE ON SCIENCE AND SOCIETY, BELGRADE, YUGOSLAVIA, 127-133.

1744 ZADEH, L.A. (1971F) TOWARDS FUZZINESS IN COMPUTER SYSTEMS-FUZZY ALGORITHMS AND LANGUAGES, IN BOULAYE, G. (ED.), ARCHITECTURE AND DESIGN OF DIGITAL COMPUTERS, DUNOD, PARIS, 9-18.

1745 ZADEH, L.A. (1972) A FUZZY SET THEORTIC INTERPRETATION OF HEDGES, MEMO. M-335, 1972, ELECTR. RES. LAB., UNIV. OF CALIFORNIA, BERKELEY.

1746 ZADEH, L.A. (1972) A FUZZY-SET-THEORETIC INTERPRETATION OF LINGUISTIC HEDGES, JOUR. OF CYBERNETICS 2, 4-34.

1747 ZADEH, L.A. (1972) ON FUZZY ALGORITHMS, MEMO ERL-M325, UNIV. OF CALIFORNIA, BERKELEY.

1748 ZADEH, L.A. (1972A) A RATIONALE FOR FUZZY CONTROL, JOURNAL OF DYNAMIC SYSTEMS, MEASUREMENT AND CONTROL, G94, 3-4.

1749 ZADEH, L.A. (1972B) FUZZY LANGUAGES AND THEIR RELATION TO HUMAN INTELLIGENCE, PROC. INTERNATIONAL CONFERENCE ON MAN AND COMPUTER, S. KARGER, BASEL 130-165.

1750 ZADEH, L.A. (1973A) OUTLINE OF A NEW APPROACH TO THE ANALYSIS OF COMPLEX SYSTEMS AND DECISION PROCESSES, IEEE TRANS. SYST. MAN CYBERN., 2, 28-44.

1751 ZADEH, L.A. (1973B) A SYSTEM-THEORETIC VIEW OF BEHAVIOUR MODIFICATION, IN WHEELER, H. (ED.), BEYOND THE PUNITIVE SOCIETY, W.H. FREEMAN, SAN FRANSISCO, 160-169.

1752 ZADEH, L.A. (1974A) A NEW APPROACH TO SYSTEM ANALYSIS, IN MAROIS, M. (ED.), 'MAN AND COMPUTER, NORTH-HOLLAND, AMSTERDAM, 55-94.

1753 ZADEH, L.A. (1974B) FUZZY LOGIC AND ITS APPLICATION TO APPROXIMATE REASONING, IN INFORMATION PROCESSING 74, PROC. IFIP CONGRESS 74, 3, NORTH-HOLLAND, AMSTERDAM, 591-594.

1754 ZADEH, L.A. (1974C) THE CONCEPT OF A LINGUISTIC VARIABLE AND ITS APPLICATION TO APPROXIMATE REASONING, IN FU, K.S., AND TOU, J.T. (EDS.), LEARNING SYSTEMS AND INTELLIGENT ROBOTS, PLENUM PRESS, NEW YORK, 1-10.

1755 ZADEH, L.A. (1975) A BIBLIOGRAPHY ON FUZZY SETS AND THEIR APPLICATION TO DECISION PROCESSES IN ZADEH, FU, ETC. (EDS.) FUZZY SETS AND THEIR APPLICATIONS, ACADEMIC PRESS.

1756 ZADEH, L.A. (1975) FUZZY SETS AND THEIR APPLICATIONS TO

COGNITIVE AND DECISION THEORY, ACADEMIC PRESS.

1757 ZADEH, L.A. (1975A) LINGUISTIC CYBERNETICS IN ROSE, J. (ED.), ADVANCES IN CYBERNETICS AND SYSTEMS, 3, GORDON AND BREACH, LONDON 1607-1615.

1758 ZADEH, L.A. (1975B) FUZZY LOGIC AND APPROXIMATE REASONING, SYNTHESE, 30, 407-428.

1759 ZADEH, L.A. (1975C) CALCULUS OF FUZZY RESTRICTIONS, IN ZADEH, L.A., FU, K.S., TANAKA, K., AND SHIMURA, M. (EDS.), FUZZY SETS AND THEIR APPLICATIONS TO COGNITIVE AND DECISION PROCESSES, ACADEMIC PRESS, NEW YORK, 1-39.

1760 ZADEH, L.A. (1975D) A RELATIONAL MODEL FOR APPROXIMATE REASONING, IEEE INTERNATIONAL CONFERENCE ON CYBERNETICS AND SOCIETY, SAN FRANCISCO, USA, SEP. .

1761 ZADEH, L.A. (1976) FUZZY SETS AND THEIR APPLICATION TO PATTERN CLASSIFICATION AND CLUSTER ANALYSIS, ERL MEMO. M607, ELECTRONICS RESEARCH LAB., UNIV. OF CALIFORNIA, BERKELEY, CA. .

1762 ZADEH, L.A. (1976) SEMANTIC INFERENCE FROM FUZZY DATA BY MATHEMATICAL PROGRAMMING, IEEE, MAN-SYSTEM CYBERN. CONF. .

1763 ZADEH, L.A. (1976A) THE LINGUISTIC APPROACH AND ITS APPLICATION TO DECISION ANALYSIS, IN HO, Y.C., AND MITTER, S.K. (EDS.), DIRECTIONS IN LARGE-SCALE SYSTEMS, PLENUM PRESS, NEW YORK.

1764 ZADEH, L.A. (1976B) SEMANTIC INFERENCE FROM FUZZY PREMISES, PROC. 6TH INT. SYMP. MULTIPLE-VALUED LOGIC, IEEE 76CH1111-4C, MAY, 217-218.

1765 ZADEH, L.A. (1976C) A FUZZY-ALGORITHMIC APPROACH TO THE DEFINITION OF COMPLEX OR IMPRECISE CONCEPTS, INT. J. MAN -MACHINE STUDIES, 8, 249-291.

1766 ZADEH, L.A. (1977) A THEORY OF APPROXIMATE REASONING (AR), MEMO. NO. UCB/ERL M77/58, ELECTRONICS RESEARCH LAB., COLLEGE OF ENGINEERING, UNIV. OF CALIFORNIA, BERKELEY, 1-71.

1767 ZADEH, L.A. (1977) FUZZY LOGIC AS A MODEL FOR HUMAN REASONING, INT. CONF. ON APPLIED GEN. SYSTEMS RESEARCH, BINGHAMTON, N.Y. .

1768 ZADEH, L.A. (1977) FUZZY SET THEORY: A PERSPECTIVE, 3-4, IN FUZZY AUTOMATA AND DECISION PROCESSES, M.M. GUPTA, G.N. SARIDIS, B.R. GAINES, EDS., NORTH-HOLLAND.

1769 ZADEH, L.A. (1977) FUZZY SETS AND THEIR APPLICATION TO CLASSIFICATION AND CLUSTERING, IN: J. VAN RYZIN (ED.) CLASS. AND CLUSTERING, ACAD. PRESS, N.Y., 251-299.

1770 ZADEH, L.A. (1977) LINGUISTIC CHARACTERIZATION OF PERFERENCE RELATIONS AS A BASIS FOR CHOICES IN SOCIAL SYSTEMS, MEMO. NO. UCB/ERL M77/24, ELECTRONICS RESEARCH LAB., COLLEGE OF ENGINEERING, UNIVERSITY OF CALIFORNIA, BERKELEY.

1771 ZADEH, L.A. (1977) POSSIBILITY THEORY VERSUS PROBABILITY THEORY IN DECISION ANALYSIS, PROCEEDINGS OF THE SYMP. ON FUZZY SET THEORY AND APPLN., 1977 IEEE CONF. ON DECISION AND CONTROL, NEW ORLEANS.

1772 ZADEH, L.A. (1977) PRUF AND ITS APPLICATION TO INFERENCE FROM FUZZY PROPOSITIONS, PROC. 1977 IEEE CONF. ON DECISION AND CONTROL, NEW ORLEANS, 1359-1360.

1773 ZADEH, L.A. (1977) PRUF--A MEANING REPRESENTATION LANGUAGE FOR NATURAL LANGUAGES, MEMO. NO. ERL-M77/61, ELECTRONICS RESEARCH LABORATORY, COLLEGE OF ENGINEERING, UNIV. OF CALIF., BERKELEY, OCT. .

1774 ZADEH, L.A. (1977) THEORY OF FUZZY REASONING AND PROBABILITY THEORY VS. POSSIBILITY THEORY IN DECISION MAKING, PROCEEDINGS OF THE SYMP. ON FUZZY SET THEORY AND APPLN., IEEE CONF. ON DECISION AND CONTROL, NEW ORLEANS.

1775 ZADEH, L.A. (1977) THEORY OF FUZZY SETS, MEMO. NO. UCB/ERL M77/1, UNIVERSITY OF CALIFORNIA, BERKELEY, CA. .

1776 ZADEH, L.A. (1978) FUZZY LOGIC AND ITS APPLICATIONS TO DECISION AND CONTROL ANALYSIS, PROC. 1978 IEEE CONF. ON DEC. AND CONTROL, INCLUDES THE 17TH SYMP. ON ADAPTIVE PROCESSES, 78CH1392-OCS, JAN., 1979, SAN DIEGO, CALIF. .

1777 ZADEH, L.A. (1978) FUZZY SETS AS A BASIS FOR A THEORY OF POSSIBILITY, FUZZY SETS AND SYSTEMS, 1, 3-28.

1778 ZADEH, L.A. (1978) FUZZY SYSTEM THEORY: A FRAMEWORK FOR THE ANALYSIS OF HUMANISTIC SYSTEMS, PRESENT AT IEEE CONF. ON CIRCUIT AND SYSTEMS.

1779 ZADEH, L.A., FU, K.S., TANAKA, K., AND SHIMURA, M., (EDS.) (1975) FUZZY SETS AND THEIR APPLICATIONS TO COGNITIVE AND DECISION PROCESSES, ACADEMIC PRESS, NEW YORK.

1780 ZELENY, M. (1974) A CONCEPT OF COMPROMISE SOLUTIONS AND THE METHOD OF DISPLACED IDEAL COMPUTERS AND O.R., VOL. 1., NO. 4, 479-496.

1781 ZELENY, M. (1976) THE THEORY OF DISPLACED IDEAL, IN ZELENY, M. (ED.), MULTIPLE CRITERIA DECISION MAKING KYOTO 1975, LECTURE NOTES IN ECONOMICS AND MATHEMATICAL SYSTEMS, 123, SPRINGER-VERLAG, BERLIN, 153-206.

1782 ZELENY, M. (1978) MEMBERSHIP FUNCTIONS AND THEIR ASSESMENT, COLUMBIA UNIV., NEW YORK, 4TH INT. CONGRESS ON CYBERNETICS AND SYSTEMS, AMSTERDAM.

1783 ZIMMERMANN, H.J. (BIBLIOGRAPHY), THEORY AND APPLICATIONS OF FUZZY SETS, RWTH AACHEN, NO. 75/16: AVAILABLE ON REQUEST FROM PROFESSOR DR. H.J. ZIMMERMANN, AACHEN.

1784 ZIMMERMANN, H.J. (1974) OPTIMIZATION IN FUZZY ENVIRONMENTS, TECHNICAL REPORT, INSTITUTE FOR OPERATIONS RESEARCH, TECHNICAL HOCHSCHULE, AACHEN, GERMANY.

1785 ZIMMERMANN, H.J. (1975) BIBLIOGRAPHY: THEORY AND APPLICATIONS OF FUZZY SETS, LEHRSTUHL FUR UNTERNEHMENSFORSCHUNG, RWTH, AACHEN, GERMANY, OCT. .

1786 ZIMMERMANN, H.J. (1975) DESCRIPTION AND ECOPTIMIZATION OF FUZZY SYSTEMS, INT. J. GENERAL SYST., 2, 209-215.

1787 ZIMMERMANN, H.J. (1975) DESCRIPTION AND OPTIMIZATION OF FUZZY SYSTEMS, INT. J. GENERAL SYST. 2 (4).

1788 ZIMMERMANN, H.J. (1975) FUZZY DECISIONS, FUZZY ALGORITHMS-A PROMISING APPROACH TO PROBLEM SOLVING, NOAK75, OSLO, OCT. .

1789 ZIMMERMANN, H.J. (1975) OPTIMALE ENTSCHEIDUNGEN BEI UNSCHARFEN PROBLEMBESCHREIBUNGEN, LEHRSTUHL FUR UNTERNEHMENSFORSCHUNG, RWTH, AACHEN, GERMANY.

1790 ZIMMERMANN, H.J. (1975) THE POTENTIAL OF FUZZY DECISION MAKING IN THE PRIVATE AND PUBLIC SECTOR, SOAK-75, LIDINGO, SWEDEN.

1791 ZIMMERMANN, H.J. (1976) UN SCHARFE ENTSCHEIDEN UND MULTI-CRITERIA-ANALYSE, IN: PROC. IN OPERATIONS RESEARCH 6, WURZBURG, 99-109, ISBN 3-7908-0180-1.

1792 ZIMMERMANN, H.J. (1977) DUALITY IN FUZZY PROGRAMMING, INT. SYMP. ON EXTERNAL METHODS AND SYSTEM ANALYSIS, AUSTIN, TEXAS.

1793 ZIMMERMANN, H.J. (1977) FUZZY PROGRAMMING AND LINEAR PROGRAMMING WITH SEVERAL OBJECTIVE FUNCTIONS, FUZZY SETS AND SYSTEMS 1, 45-55, NORTH-HOLLAND PUBLISHING CO. .

1794 ZIMMERMANN, H.J. (1977) RESULTS OF EMPIRICAL STUDIES IN FUZZY SET THEORY, IN G. KLU, ED., APPLIED GENERAL SYSTEMS RESEARCH, RECENT DEV., PLENUM PRESS, N.Y. .

1795 ZIMMERMANN, H.J., AND GEHRING, H. (1975) FUZZY INFORMATION PROFILE FOR INFORMATION SELECTION, 4TH INST. CONGRESS, AFCET, PARIS, FRANCE.

1796 ZIMMERMANN, H.J., AND POLLARSCHECK, M.A. (1976) APL ROUTINES FOR FUZZY 0-1 LINEAR PROGRAMS AND THEIR PERFORMANCE, AACHEN WORKING PAPER NO. 76/10, AACHEN UNIV. .

1797 ZIMMERMANN, H.J., AND RODDER, W. (1975) ANALYSE, BESCHREIBUNG UND OPTIMIERUNG VON UNSCHARF FORMULIERTEN PROBLEMEN,

LEHRSTUHUL FUR UNTERNEHMENSFORSCHUNG, RWTH, AACHEN, GERMANY.

1798 ZINOVEV, A.A. (1963) PHILOSOPHICAL PROBLEMS OF MANY-VALUED LOGIC, D. REIDEL, DORDRECHT, HOLLAND.

1799 ZWICK, M., SCHWARTZ, D.G., AND LENDARIS, G.G. (1978) FUZZINESS AND CATASTROPHE, PORTLAND STATE UNIV., PORTLAND, OREGON, USA, PROCEEDINGS OF THE 1978 INT. CONF. ON CYBERNETICS AND SOCIETY, VOL2, 1237-1241.

BIOGRAPHICAL INFORMATION ABOUT THE EDITORS AND CONTRIBUTING AUTHORS

KIYOJI ASAI
Department of Industrial Engineering, University of Osaka Prefecture, Sakai, Osaka 591, Japan.
Kiyoji Asai is a Professor of Industrial Engineering. He received the B.S. degree in electrical communication engineering and the Ph.D. degree in engineering from the Osaka University in 1951 and 1961 respectively. In 1951-1969, he was on the faculty of Department of Electrical Engineering of the Osaka City University, and, during 1967-1968, he stayed in the Department of Electrical Engineering and Computer Sciences at the University of California, Berkeley as a visiting research associate. His current fields of interest include fuzzy systems theory, systems optimization and man-machine systems.

JAMES BALDWIN
Engineering Mathematics Department, University of Bristol, England.
James Baldwin is a lecturer in Engineering Mathematics. He has degrees in Physics and Mathematics and has worked in the aircraft industry as a research mathematician. His research interests include Systems Theory, Optimization Theory and Management Science.

JAMES C. BEZDEK
Department of Mathematics, Utah State University, Logan, Utah 84322, (USA).
James C. Bezdek is an Associate Professor of Mathematics. He received the B.S. in civil engineering from the University of Nevada, Reno, and the Ph.D. in applied mathematics from Cornell University. His research interests include fuzzy sets and their application to cluster analysis, feature selection, unsupervised learning, classifier design, computerized medical diagnosis, and fluid mechanics.

HIRAM H. BROWNELL
Department of Psychology, University of Southern California, University Park, Los Angeles, California 90007, (USA).
Hiram H. Brownell is an Assistant Professor of Psychology. He received the A.B. degree from Stanford University, and the M.A. and Ph.D. degrees from The Johns Hopkins University. His current research interests are in language and perception.

DAN BUTNARIU
Str. Tepes-Vodă, Nr. 2, Bl. V1, sc. A, et. 4, ap.1, Iasi-6600, România.
Dan Butnariu is an assistant lecturer at the Department of Mathematics of the Politechnic Institute of Iassy, Romania. He received his degree in Mathematics from the University of Iassy in 1975. Now he is a candidate for a doctor's degree. His research interests include fuzzy sets, games, topologies, fixed points and their applications.

ALFONSO CARAMAZZA
Department of Psychology, The Johns Hopkins University, Baltimore, Maryland 21218, (USA).
Alfonso Caramazza is Associate Professor of Psychology and Neurology. He received the B.A. degree from McGill University, Canada, and the Ph.D. from The Johns Hopkins University. His research interests include problems in psychological semantics and the dissolution of language in brain-damaged adults.

ALDO DE LUCA
Laboratorio di Cibernetica del C.N.R., Arco Felice (Napoli), Italy.
Aldo de Luca is researcher in chief of the Laboratorio di Cibernetica del C.N.R., Arco Felice (Napoli) and Associate Professor of Automata Theory in the Faculty of Sciences of the Salerno University. He received the Laurea degree in Physics (1964) and successively a Specialization Diploma in Theoretical Physics (1967) from the Naples University. His research interests are mainly concerned with Automata and Computation Theory, Information Theory and Cybernetics. He is author of several technical papers and of a book: Introduzione Alla Cibernetica (F. Angeli, 1971, coauthored with L.M. Ricciardi).

JOHN DIFFENBACH
School of Management, Boston College, Chestnut Hill, Massachusetts 02167, (USA).
John Diffenbach is an Assistant Professor of Management. He received the B.S. in metallurgy at the Pennsylvania State University and the M.B.A. at the University of California (Berkeley). He spent seven years with the General Electric Company. His current research interests are in policy analysis, strategic planning, and public utility management.

MICHAEL L. DONNELL
Decisions and Designs, Inc., McLean, Virginia, (USA).
Michael L. Donnell is a Decision Analyst and a member of the Applied Decision Analysis Group at Decisions and Designs, Inc. He received a B.A. degree from Rice University in 1972. From the University of Michigan he received a M.A., Mathematics in 1974; a M.A., Psychology in 1975; and a Ph.D., Mathematical Psychology in 1977. Dr. Donnell has conducted basic research in human decision making, opponent-color theory, and human learning and memory. At DDI he develops adapts, and applies techniques of probability analysis and decision theory for a variety of sponsors.

DIDIER DUBOIS
Laboratoire d'Informatique et de Mathematiques Appliquees (IMAG), Universite de Grenoble, B.P. 53, 38041 Grenoble Cedex, France.
Didier Dubois is a research Engineer in the Computer Aided Design Group of IMAG. Born in 1952, he received the Engineer and the "Docteur-Ingenieur"degrees respectively in 1975 and 1977, from the "Ecole Nationale Superieure de l'Aeronautique et de l'Espace" (SUP AERO), Toulouse, France. He is the co-author of "Fuzzy Sets and Systems: theory and applications" (Academic Press 1979, forthcoming). His current interests include Operations Research, Decision Analysis, Artificial Intelligence.

JOSEPH FIKSEL
Arthur D. Little, Inc., Acorn Park, Cambridge, Massachusetts 02140, (USA).
Joseph Fiksel is a consultant in the areas of Risk and Policy Analysis. He received a B.S. in electrical engineering from MIT, and a Ph.D. in operations research from Stanford University. In addition to his practice of management science, he has published in the areas of mathematical psychology and sociology, and more recently in the field of risk management.

WILFRED A. FORDON
Electrical Engrg. Dept., Michigan Technological Univ., Houghton, MI 49931, (USA).
Wilfred A. Fordon is an Associate Professor of Electrical Engineering. He received the B.E.E. from CCNY, the M.A. in mathematics from Hofstra University, and the Ph.D. in electrical engineering from Purdue University. He has had more than twenty years of industrial experience. His research interests include electromagnetic field theory, bioengineering and pattern recognition.

SATORU FUKAMI
Yokosuka Electrical Communication Laboratory, Nippon Telegraph and Telephone Public Corporation, Yokosuka, Kanagawa 238-03, Japan.
Mr. S. Fukami is a researcher of Information Technology. He received the B.Eng. and M.Eng. degrees in Information and Computer Sciences from Osaka University, Japan, in 1977 and 1979, respectively. His research interests are data base management and artificial intelligence.

JOSEPH GOGUEN
Comp. Science Dept., Univ. of California, 3532 Boelter, Los Angeles, CA 90024,(USA).
Joseph Goguen is an Associate Professor of Computer Science at the University of California where he has been teaching since 1972. Dr. Goguen received his bachelor's degree from Harvard University in 1963 and Ph.D. from the University of California at Berkeley in 1968 both in Mathematics. He is a member of the academic staff of Naropa Institute in Boelter, Colorado. His current research interests include Algebraic Semantics of Computation.

MADAN M. GUPTA (Editor)
Cybernetics Research Laboratory, University of Saskatchewan, Saskatoon, Sask., Canada S7N OWO.
Madan M. Gupta is a Professor of Engineering. He studied at BITS, Pilani (India) obtaining B.E. (Hons.) and M.E. in 1961 and 1962 respectively. He was a lecturer at the University of Roorkee (India) during 1962-64. He was a recipient of Commonwealth Fellowship in U.K. and studied at the University of Warwick where he obtained his Ph.D. in 1967 in Adaptive Control Systems.
His interests lie in the area of Systems and Adaptive Control, Signal processing, detection and diagnosis of early cardiac abnormalities, fuzzy automata and technology and socioeconomic problems. He has published extensively internationally, and has edited a book entitled "Fuzzy Automata and Decision Processes," North-Holland 1979. He is an advisory editor of International Journal of Fuzzy Sets and Systems, and a Senior Member of the IEEE.

DONALD E. GUSTAFSON
Scientific Systems, Inc., Cambridge, Massachusetts 02138, (USA).
Donald Gustafson is Vice President of Scientific Systems,Inc. He received the Ph.D. degree in Engineering from Massachusetts Institute of Technology in 1971. From 1966 to 1976 he was employed by the Charles Stark Draper Laboratory, where he applied modern estimation and control theory to aerospace guidance, navigation and control problems. He joined Scientific Systems in 1977, where he leads efforts in biomedical signal processing, pattern recognition, estimation and data analysis.

HARRY M. HERSH
Pattern Analysis and Recognition Corporation, Rome, New York 13440, (USA).
Harry M. Hersh is a Senior Analyst in the Language Research Laboratory. He received the B.S.E.E. from Northeastern University and the M.A. and Ph.D. degrees from The Johns Hopkins University. His current research interests are in language processing, human decision making, and man-computer interaction.

KAORU HIROTA
Dept. of Information Science, Sagami Institute of Technology, Fujisawa-city, Japan
Kaoru Hirota is a full-time lecturer of Information Science. He is also a researcher of Tokyo Institute of Technology (Tokyo, Japan). He was born in Niigata Prefecture, Japan, on Jan. 6, 1950. He graduated from Nagaoka Technical College

in 1970, and he received the B.S., M.S. and Ph.D. degrees in Engineering from Tokyo Institute of Technology, in 1974, 1976 and 1979, respectively. His research interests include fuzzy theory, information theory, pattern recognition and computer architecture.

ELLEN HISDAL
Institute of Informatics, University of Oslo, Box 1080, Blindern, Oslo 3, Norway.
Ellen Hisdal is an Associate Professor of Informatics. She received her B.A. and M.A. degrees in physics from the University of California in Berkeley, and the Ph.D. degree from the University of Oslo. Her current research interests include fuzzy set theory and applications, structural and fuzzy pattern recognition, information and physical entropy.

RAMESH JAIN
Department of Computer Science, Wayne State University, Detroit, MI 48202, (USA).
Ramesh Jain is an Assistant Professor of Computer Science at Wayne State University. He received his B.E. and Ph.D. from Nagpur University and Indian Institute of Technology, Kharagpur in 1969 and 1975, respectively. His current research interests are Computer Vision, Software Engineering, and Fuzzy Sets.

ABRAHAM KANDEL
Department of Mathematics, Florida State University, Tallahassee, Florida 32306, (USA).
Abraham Kandel is an Associate Professor and Director of Computer Science. He received the B.Sc. from the Technion-Israel Institute of Technology and M.S. from the University of California, Santa Barbara, both in Electrical Engineering, and the Ph.D. from the University of New Mexico in Electrical Engineering and Computer Science. From 1970 to 1978 he was with the Computer Science Department at the New Mexico Institute of Mining and Technology. During 1976-77 he was with Ben Gurion University of the Negev and Tel-Aviv University, Israel. His current research interests include fuzzy systems, computer architecture, pattern recognition, and theoretical computer science. He is the author of more than 50 published papers in these fields and is the coauthor (with Dr. Samuel C. Lee) of the monograph "Fuzzy Switching and Automata: Theory and Applications" (Crane, Russak & Co., N.Y., 1979).

WILLIAM C. KESSEL
Scientific Systems, Inc., Cambridge, Massachusetts 02138, (USA).
William Kessel received the S.B. degree in Electrical Engineering from Massachusetts Institute of Technology in 1977, with an emphasis in Computer Engineering. He is presently working at Scientific Systems as a Programmer/Analyst and is responsible for algorithm development and evaluation on several projects.

WALTER J.M. KICKERT
Department of Industrial Engineering, Eindhoven University of Technology, P.O. Box 513, 5600 MB Eindhoven, Netherlands.
Walter Kickert is a member of the section of Organization Science. He received his 'doctoral' degree in physics from Utrecht State University in 1974. From 1974 to 1975 he was with the department of Electrical Engineering, Queen Mary College, London University. He is the author of Fuzzy Theories on Decision-Making (Nijhoff, 1978). His current research interests are in systems theory, organization science, decision making, public policy making and the application of fuzzy set theory to these areas. He is a member of the Dutch Systems Society and an advisory editor of Fuzzy Sets and Systems.

MASASUMI KOKAWA
Systems Development Laboratory, Hitachi Ltd., 1099, Ohzenji, Tama-ku, Kawasaki, 215, Japan.
M. Kokawa received his B.S., M.S. and Ph.D. degrees from Nagoya University in 1970, 1972, and 1975, respectively. His current research interests include

artificial intelligence, modeling of human behavior, fault diagnosis, and fuzzy systems.

NOEL MALVACHE
Laboratoire d'Automatique Industrielle et Humaine, Université de Valenciennes, 59326 Valenciennes, France.
Noël Malvache is a Professor of Automatic control and head of the Automatic Control Laboratory of Valenciennes University. He was granted his "Doctorat d'Etat" from the Lille University in 1973. His research interests are chiefly about the human operator study, more particularly in the field of the visio-manual loop, vestibular perturbations and man-machine systems.

MASAHARU MIZUMOTO
Department of Management Engineering, Osaka Electro-Communication University, Neyagawa, Osaka 572, Japan.
Dr. M. Mizumoto is an Associate Professor of Information Science. He received the B.Eng., M.Eng. and Dr.Eng. degrees in Electrical Engineering from Osaka University, Japan, in 1966, 1968 and 1971, respectively.
His research interests include fuzzy sets theory and its applications to artificial intelligence, robot planning and data base systems.
He is Advisory Editor of the International Journal for Fuzzy Sets and Systems, and the Journal of the Institute of Electronics and Communication Engineers of Japan.

STEVEN NAHMIAS
Department of Industrial Engineering and Operations Research, University of Pittsburgh, Pittsburgh, Pennsylvania 15261, (USA).
Steven Nahmias is an Associate Professor of Operations Research. He received a B.A. in Mathematics and Physics from Queens College, a B.S. in Industrial Engineering from Columbia University, and the M.S. and Ph.D. degrees in Operations Research from Northwestern University. During the 1978-79 academic year he visited the Departments of Operations Research and Industrial Engineering at Stanford University. He has published a number of articles in various journals primarily on problems related to inventory control.

KAHEI NAKAMURA
Nagoya University, Faculty of Engineering, Automatic Control Laboratory, Furo-Cho, Chikusa-Ku, Nagoya, Japan.
Kahei Nakamura is a Professor in the Department of Information and Control Science and Head of Automatic Control Laboratory. He received his B.D.Electrical Engineering), Eng.D. (Control Engineering) degrees from Nagoya University. His research interests include system theory, control theory (discrete-time control, computer control, etc), control strategy (optimizing control), adaptive and learning control, and artificial intelligence.

C. V. NEGOITA
Faculty of Economic Cybernetics, Bucharest, Romania.
Constantin Virgil Negoita is a Professor of System Theory. He received his Ph.D. in Information Sciences from the Polytechnic Institute of Bucharest in 1970. He is the coauthor of Applications of Fuzzy Sets to System Analysis (Halsted Press, 1975) and author of Management Applications of System Theory (Birkhauser Verlag, 1979). He is editor of the Journal of Fuzzy Sets and Systems, and a member of the editorial board of Kybernetes and Human Systems Management. He is a member of the Society of General Systems Research.

HUNG T. NGUYEN
Department of Mathematics and Statistics, University of Massachusetts, Amherst, (USA).
Hung T. Nguyen is an Assistant Professor of Mathematics and Statistics. He received the Doctorat d'Etat ès Sciences mathematiques from Université des Sciences et Techniques de Lille (France) in 1975. His research interests include

identification of stochastic dynamical systems and modeling of uncertain systems from fuzzy set theory viewpoint.

MARIA NOWAKOWSKA
Institute of Philosophy and Sociology, Polish Academy of Sciences, Warszawa, Poland.
Maria Nowakowska is Associate Professor of Psychology, Head of Laboratory of Psychology and Action Theory. She received her Ph.D. and habilitation in Mathematical Psychology from the Jagiellonian University, Krakow, Poland. She is the author of 6 books, among them "Language of Motivation and Language of Actions" (Moutin, The Hague 1973) and "Quantitative Psychology with Elements of Scientometrics" (in Polish, PWN, Warszawa 1975), as well as over 70 articles published in American and European scientific journals. Her field of interest comprises modeling and methodology of social sciences, formal action theory, with its applications to decision theory, linguistics (dialogues, discussions), semantics, scientometrics, and semiotic systems in their relation to cognitive processes.

MORIYA ODA
Nagoya University, Faculty of Engineering, Automatic Control Laboratory, Furo-Cho, Chikusa-Ku, Nagoya, Japan.
Moriya Oda is an Associate Professor in the Department of Information and Control Engineering. He received his Ph.D. degree from Nagoya University. His research interests include control theory, artificial intelligence (learning, heuristics, concept formation), and instructional engineering (instructional system, fuzzy-theoretical systems approach to concept formation, structural learning).

GREGG C. ODEN
Department of Psychology, University of Wisconsin, Madison, Wisconsin 53706, (USA).
Gregg Oden is an Assistant Professor of Psychology. He received the B.A. degree from the University of South Dakota and the Ph.D. degree from the University of California, San Diego in 1974. He is a member of the Wisconsin Human Information Processing Program (WHIPP) and is on the planning committee for a program in Cognitive Science at the University of Wisconsin. His research interests include the processing of fuzzy and continuous semantic information, the use of semantic constraints in resolving linguistic ambiguity and the integration of featural information in pattern identification as well as other topics in cognitive psychology.

TETSUJI OKUDA
Department of Industrial Management, Osaka Institute of Technology, Omiya, Asahiku, Osaka 535, Japan.
Tetsuji Okuda is a Lecturer of Industrial Engineering. He received the B.S., M.S. and Ph.D. degrees in industrial engineering from the University of Osaka Prefecture in 1972, 1974 and 1977 respectively. His research interests include theory of fuzzy decision and fuzzy information, and their applications.

HENRI M. PRADE
Laboratoire "Langages & Systèmes Informatiques" -- Université Paul Sabatier, 118 Route de Narbonne, 31077 Toulouse Cedex, France.
Henri Prade is an "Attaché de Recherche" at the National Center for Scientific Research (C.N.R.S. -France). He was born in 1953. He received the Engineer degree and the Doctor - Engineer degree from the "Ecole Nationale Supérieure de l'Aéronautique et de l'Espace" respectively in 1975 and in 1977. He is the co-author of "Fuzzy Sets and Systems: Theory and Applications" (Academic Press 1979) and over thirty technical papers. His current research interests are in Fuzzy Set Theory, Artificial Intelligence and Operations Research.

RAMMOHAN K. RAGADE (EDITOR)
University of Louisville, Systems Science Institute, Belknap Campus, Louisville, Kentucky 40208, (USA).

Rammohan K. Ragade is an Associate Professor in Systems Science Institute and in the Department of Mathematics, University of Louisville. He obtained his Ph.D. from the Indian Institute of Technology, Kanpur in 1968. His current researh interests are specifically in decision support systems, urban energy-environmental resource systems management, and health systems. In general systems, he is keenly interested in the role of fuzzy and stochastic methods in applied systems analysis, multiobjective decision evaluation modeling, and impact assessments in policy formation and communication and information theory. He is a co-editor of a special issue of Behavioral Science on Catastrophe Theory. He is an area editor for the Yearbook of the Society for General Systems Research. He is a member of TIMS, the Society for General Systems Research, and the American Association for the Advancement of Science.

DAN RALESCU
Department of Mathematics, Indiana University, Bloomington, Indiana, 47405, (USA).
Dan Ralescu is an Instructor of Mathematics. He received the B.S. and M.S. in computer science from the University of Bucharest, Romania and is completing the Ph.D. in mathematics at Indiana University in 1979. During 1976-77 he was an Assistant Professor at the Institut de Mathematiques Economiques, Universite de Dijon, France. He is a coauthor of Applications of Fuzzy Sets to Systems Analysis (Wiley, 1975) and Introduction to Fuzzy Systems Theory (Ohm-Sha, 1978). He is an advisory editor of Fuzzy Sets and Systems and a member of the American Mathematical Society and the Institute of Mathematical Statistics.

ALAN D. RENDA
School of Management, Boston College, Chestnut Hill, MA 02167, (USA).
Alan Renda is an Instructor of Administrative Sciences. He received a B.S. and M.B.A. from Boston College and is presently a Ph.D. candidate at the Columbia University Graduate School of Business where his concentration is in management sciences. His current research includes long range forecasting procedures for technological change and modeling consumer choice behavior.

ELIE SANCHEZ
Laboratoire de Biomathématiques, Statistique, Informatique Medicale, Faculté de Médecine, 27 Bd Jean Moulin, 13385 Marseille Cedex 4, France.
Elie Sanchez is currently a "Chef de Travaux de Faculté-Assistant des Hôpitaux" at Marseille. He received a Ph.D. in Mathematics (Boolean Logic) in 1972 from the "Faculté des Sciences de Marseille" and the Ph.D. in "Biologie Humaine"(Fuzzy Sets) in 1974 from the Faculté de Medecine de Marseille." During 1976-77 he was a visiting Research Associate at the University of California. His interests include problems involving fuzzy relations and their compositions, approximate reasoning in natural languages, possibility distributions, applications of fuzzy sets to medical diagnosis and pattern classification problems in a linguistic approach.

HIDEO TANAKA
Department of Industrial Engineering, University of Osaka Prefecture, Sakai, Osaka 591, Japan.
Hideo Tanaka is a Lecturer of Industrial Engineering. He received the B.S. degree from the University of Kobe in 1962 and M.S. and Ph.D. degrees from the Osaka City University in 1966 and 1969, respectively. In 1972-1973 he was Visiting Research Associate at the University of California, Berkeley, and in 1975-77 he was an Alexander Von Humboldt Foundation Fellow at the University of Aachen in West Germany. His areas of interest are fuzzy systems and decision theory.

KOKICHI TANAKA
Department of Information & Computer Sciences, Faculty of Engineering Science, Osaka University, Toyonaka, Osaka 560, Japan.
Dr. Tanaka is a Professor of Information & Computer Sciences. He received the B.Eng. and Dr.Eng. degrees from the University of Tokyo, Japan in 1944 and 1959, respectively. His area of interest includes pattern recognition and pictorial information processing, artificial intelligence, and fuzzy set theory and its applications to approximate reasoning and data base techniques.

Dr. Tanaka is the editor-in-chief, Governing Board, and the chairman of the artificial intelligence and interactive technique research group, of the Information Processing Society of Japan. He is also a senior member of IEEE and was the former chairman of the pattern recognition & machine learning theory research group of the Institute of Electronics and Communication Engineers of Japan. He is a member of the Editorial Board of the International Journal "Fuzzy Sets and Systems."
He is the General Co-chairman of the International Joint Conference on Artificial Intelligence, held in Tokyo, August, 1979.

SETTIMO TERMINI
Laboratorio di Cibernetica del C.N.R., Arco Felice (Napoli), Italy.
Settimo Termini is a researcher of the C.N.R. (Italian National Council of Researchers). He received his Laurea degree in Physics from the University of Palermo in 1968 and joined in the following year the Laboratorio di Cibernetica of the C.N.R. He also taught Cybernetics and Information Theory at the University of Palermo. His research interests are focused on the foundational aspects and epistemological problems of some theories of Computer and Information Sciences as Automata and Formal Languages, the General Theory of Codes, Information Theory and Fuzzy Sets.

RICHARD M. TONG
Dept. of Electrical Engineering and Computer Science, University of California, Berkeley, CA 94720, (USA).
Richard Tong is a visiting NATO Research Fellow. He has a B.Sc. degree from Leeds University, a M.SC. degree from Heriot-Watt University and a Ph.D. degree from Cambridge University. From 1975-1978 he was a British Steel Corporation Research Associate at Cambridge University. His current research interests are in the modelling and control of fuzzy systems.

YAHACHIRO TUSKAMOTO
Tokyo Institute of Technology, Faculty of Engineering, Tokyo 152, Japan.
Yahachiro Tuskamoto is a Research Associate in the Department of Control Engineering, at Tokyo Institute of Technology. He obtained his Ph.D. in 1979 from Tokyo Institute of Technology. His research interest is in control systems and fuzzy logics, and its application to failure diagnosis and control.

WOLFGANG WECHLER
Sektion Mathematik, Technische Universität Dresden, DDR-8027 Dresden, German Democratic Republic.
Wolfgang Wechler is an Associate Professor of Mathematics. He received the B.S. in theoretical physics, the M.S. in algebra and the Ph.D. in cybernetics from Technical University of Dresden. He is the author of "The Concept of Fuzziness in Automata and Language Theory" (Akademie-Verlag, Berlin 1978). His current research interests are fuzzy sets, fuzzy languages and grammars, formal power series and program schemata. He is Fellow of the Department of Mathematics, Technical University of Dresden.

JONATHAN J. WEISS
Decisions and Designs, Inc., McLean, Virginia, (USA).
Jonathan J. Weiss was born in 1950 in New York City. He received the A.B. degree with honors in applied mathematics, from Harvard University, Cambridge, Massachusetts in 1970, and his M.A. and Ph.D. degrees in experimental psychology at Harvard University in 1977. Since November 1977, he has been employed as a Decision/Research Analyst by Decisions and Designs, Inc., in McLean, Virginia where he has been developing and applying decision aids, conducting traditional decision analyses, and investigating fuzzy-set theory, with an emphasis on the development of a practical new decision methodology based on fuzzy sets.

FRED WENSTØP
Bedriftsøkonomisk Institutt, Oslo, Norway.
Fred Wenstøp is a Professor of Decision Analysis. He received his M.S. in cybernetics from the University of Oslo and his Ph.D. from the University of California, Berkeley, 1975 in management science. His research interests include applications of multiattribute decision theory and formulation of models in the soft sciences. He is currently chairman of the Research Committee at Bedriftsøkonomisk Institutt.

DIDIER WILLAEYS
Laboratoire d'Automatique Industrielle et Humaine, Universite de Valenciennes, 59326 Valenciennes, France.
Didier Willaeys is a professor of Automatic Control. He was granted the "Docteur Ingenieur" degree from the Lille University in 1972. His research consists in developing tools for the use of fuzzy sets theory in the fields of Automatic control and human operator study.

RONALD R. YAGER (EDITOR)
Iona College, New Rochelle, NY 10801, (USA).
Ronald R. Yager is an Associate Professor of Management Science and Information Systems at Iona College. He received the B.Elec. Eng. from the City College of New York and the Ph.D. in Systems Science from the Polytechnic Institute of New York. He previously taught at the Polytechnic Institute and at the Pennsylvania State University. His present research interests include fuzzy set theory, decision theory, operation research, computer applications, forecasting and humanistic systems.

LOTFI A. ZADEH
Computer Science and Electronics, University of California, Berkeley, California 94720, (USA).
Dr. Zadeh is a member of the Computer Science Division of the Department of Electrical Engineering and Computer Sciences. He served as chairman of the Department from 1963 to 1968. Prior to his development of the theory of fuzzy sets, Dr. Zadeh had written extensively on system theory and was one of the leading contributors to the state-space approach, the analysis of time-varying systems by Fourier techniques and various extensions of Wiener's theory.